STRUCTURAL ANALYSIS

STRUCTURAL ANALYSIS
A UNIFIED CLASSICAL AND MATRIX APPROACH

THIRD EDITION

A. Ghali
Professor of Civil Engineering,
University of Calgary

A. M. Neville
Civil Engineering Consultant
Formerly Principal and Vice-Chancellor,
University of Dundee

London   New York
CHAPMAN AND HALL

First published in 1972 by
Intext Educational Publishers
Second edition 1978 published by
Chapman and Hall Ltd
11 New Fetter Lane, London EC4P 4EE
Published in the USA by
Chapman and Hall
29 West 35th Street, New York NY 10001
Reprinted 1979, 1983
Third edition 1989

© 1978, 1989 A. Ghali and A. M. Neville

Typeset in 10/12pt Times NR by
Thomson Press (India) Ltd, New Delhi, India
Printed in Great Britain by
T.J. Press (Padstow) Ltd, Padstow, Cornwall

ISBN 0 412 29030 8 (Hb)
ISBN 0 412 29040 5 (Pb)

British Library Cataloguing in Publication Data

Ghali, A. (Amin)
    Structural analysis: unified classical
    and matrix approach—3rd ed.
    1. Structures. Analysis
    I. Title II. Neville, A. M. (Adam Matthew),
    1923–
    624.1′71

    ISBN 0-412-29030-8
    ISBN 0-412-29040-5   Pbk

Library of Congress Cataloging-in-Publication Data

Ghali, A. (Amin)
    Structural analysis: a unified classical and matrix approach/A.
        Ghali and A. M. Neville.—3rd ed.
        p. cm.
    Bibliography: p.
    Includes index.
    ISBN 0-412-29030-8.—ISBN 0-412-29040-5 (Pbk.)
    1. Structural analysis (Engineering) I. Neville, Adam M.
II. Title.
TA645.G48   1989
624.1′71—dc19

Contents

Preface to the third edition

This book, which has been translated into four languages, appears in its third edition, including new chapters, added sections and more problems with answers. The new sections review the basic mechanics of statically determinate structures, deal with structural symmetry and expand on the analysis of effects of temperature and prestressing. The chapters on the finite element method and on the use of computers in structural analysis have been totally rewritten and greatly expanded by two additional chapters. It is proposed to prepare microcomputer programs based on this book.

The book has been written so that it can be followed by anyone familiar with the application of basic mechanics to statically determinate structures— material usually covered in early courses in engineering. The material presented is both elementary and advanced, the emphasis being on a unified presentation of the whole field of structural analysis. Indeed, the book combines the classical and modern methods of analysis of statically indeterminate structures.

The text has been developed in teaching, over a number of years, undergraduate and graduate courses at the University of Calgary, Canada. The first fourteen chapters contain basic material which should be covered in the first course or courses; from the remainder of the book, a suitable choice, depending on requirements, can be made to form a more advanced course. This arrangement was chosen in order to make the book suitable not only for the student but also for the practising engineer who wants to advance his knowledge by independent study and to obtain guidance on the most convenient methods of analysis of a variety of types of structures.

The methods of analysis introduced in the various chapters are illustrated by fully solved examples in the text. At the end of each chapter carefully selected, instructive problems are set, with answers at the end of the book. Data frequently used are presented in appendices, one of which offers a full introduction to matrix algebra for the benefit of those not already familiar with it.

Matrix algebra is particularly useful in structural analysis because it makes possible a formulation of the solution as a series of matrix operations suitable for a computer. But even more important is the fact that, by using matrices, structures of all types can be analyzed through a general approach.

Matrix formulation makes it possible to represent equations in a compact form: this not only saves space but also helps the reader to concentrate on the overall operations without being distracted by algebraic or arithmetical details. For all these reasons the matrix notation is extensively used in the book.

In the first four chapters we introduce two distinct general approaches of analysis: the force method and the displacement method. In the former, the structure is made statically determinate by the removal of restraining forces; a solution satisfying statics is then obtained, and the corresponding inconsistency in geometry is corrected by additional forces. In the displacement method, artificial restraints are added and the corresponding restraining forces are calculated; the equilibrium conditions are then satisfied by the removal of the restraining forces, thus allowing the displacements to occur and returning the structure to its actual condition. Throughout the book emphasis is placed on an overall unifying view, with these two distinct approaches in mind.

Both the force and displacement methods involve the solution of linear simultaneous equations relating forces to displacements. The emphasis in the first chapters is on the basic ideas in the two methods without obscuring the procedure by the details of derivation of the coefficients needed to form the equations. Instead, use is made of Appendices B, C, and D which give displacements due to applied unit forces, forces corresponding to unit displacements, and fixed-end forces due to various loadings. The consideration of the detailed methods of displacement calculation is thus delayed to Chapters 5 to 9, by which time the need for this material in the analysis of statically indeterminate structures is clear. This sequence of presentation of material is particularly suitable when the reader is already acquainted with some of the methods of calculating the deflection of beams. If, however, it is thought preferable first to deal with methods of calculation of displacement, Chapters 5 to 9 should be read before Chapters 2 to 4; this will not disturb the continuity.

The analysis of continuous beams, frames, and trusses is treated in Chapters 10 to 12. These include the methods of column analogy, slope-deflection and moment distribution, and are presented as applications of the force or displacement methods. The numerous solved examples included vary from simple problems to the more complicated types of structures of practical occurrence. Some techniques to speed up the analysis of building frames of several storeys or several bays are discussed in Chapter 12. In essence, these are the classical methods suitable for hand computation but their value to the structural designer is undiminished for preliminary calculations, for checking computer results, and of course when no computer is available.

Influence lines and influence coefficients are a useful means of analysis of structures subjected to moving loads or to loadings which can have varying positions. Chapters 13 and 14 deal with the methods of obtaining influence lines; Müller-Breslau's principle is explained and applied to different types of structures, including grids and interconnected bridge systems. Chapter 13 and the major part of Chapter 14 are concerned with influence lines which represent the effect of a unit gravity load. In the remainder of Chapter 14, influence lines for the effect of two other types of loading are discussed; one of these influence lines is applied in the analysis of statically indeterminate prestressed structures.

The effect of axial forces in framed structures is considered in Chapter 15. Both the effect of the change in length and of the change in stiffness characteristics are discussed and applied to the problem of secondary stresses in trusses and in the determination of the critical buckling loads of continuous frames.

Chapter 16 deals with the analysis of shear walls, commonly used in modern buildings. The chapter summarizes the present knowledge, states the simplifying assumptions usually involved and presents a method of analysis that can be applied in most practical cases.

The finite-difference method and, to a larger extent, the finite-element method are powerful tools which involve a large amount of computation. Chapter 17 deals with the use of finite differences in the analysis of structures composed of beam elements and extends the procedure to axi-symmetrical shells of revolution. In Chapter 18 the finite-difference method is used in the analysis of plates and applied to obtain influence flexibility, and stiffness coefficients useful in a variety of cases.

The finite-element method for the analysis of continuum problems is based on the energy theorems and on the displacement method of analysis applied to framed structures in earlier chapters. For this reason, it is possible to represent in Chapters 19 and 20 the basic general equations with a wide range of applications. Sufficient details of the finite-element method are included for the analysis of plates, shells and three-dimensional solids. The finite-strip and finite-prism methods discussed in Chapter 20 are extensions of the finite-element approach which offer some saving in the amount of computation and in data preparation.

Chapters 24 and 25 discuss the use of computers in the analysis of structures composed of one-dimensional elements (axial members or beams) or of two- or three-dimensional finite elements. The matrices involved in the analysis are given explicitly in Chapters 24 and 25 only for framed structures composed of one-dimensional elements, such as plane and space trusses, grids, and plane and space frames with rigid joints. The material beyond Chapter 8 is not a prerequisite to Chapters 24 and 25. Thus the two chapters can be used in a course on computer methods for framed structures without the need for

Chapters 19 and 20, which are concerned with two- and three-dimensional finite elements. The four chapters, 24, 25, 19 and 20, entirely new in this edition, can be used, in this order, in a graduate course on the fundamentals of the finite element method.

Modern design of structures is based on both the elastic and plastic analyses. The plastic analysis cannot replace the elastic analysis but supplements it by giving useful information about the collapse load and the mode of collapse. Chapters 21 and 22 deal with the plastic analysis of framed structures and slabs, respectively.

An introduction to structural dynamics is presented in Chapter 23. This is a study of the response of structures to dynamic loading produced by machinery, gusts of wind, blast, or earthquakes. First, free and forced vibrations of a system with one degree of freedom are discussed. This is then extended to multidegree-of-freedom systems, taking advantage of the matrix approach.

Many structures have one or more axes or planes of symmetry, and this can be used to reduce the number of unknowns or to limit the analysis to only a part of the structure. These techniques, often neglected, have been included in Chapters 4 and 26.

The main text does not involve any specific system of units. The same applies to most of the problems set for the reader to solve. However, there are a number of problems where it was thought advantageous to use actual dimensions of members and to specify the magnitude of forces. These problems are set both in the so-called British units (still common in the United States) and in the SI units (now accepted all over the world). The former version appears at the end of the various chapters and the latter, bearing the same number, is given in a separate section at the end of the book. It is up to the reader to choose which set of problems he wants to solve—and of course the underlying theory is the same. Naturally, answers to all the problems are listed.

Some of the solutions of the examples and problems in the new edition have been checked by Messrs M. Elbady, A. Elgabry and S. Sirosh, doctoral candidates at the University of Calgary. Mrs K. Takaoka typed the manuscript. We are grateful to them as well as to those readers whose views have helped us in the choice of new material included in the Third Edition.

Calgary, Alberta, Canada   A. Ghali
London, England   A. M. Neville
January, 1989

Notation

The following is a list of symbols which are common in the various chapters of the text; other symbols are used in individual chapters. All symbols are defined in the text when they first appear.

A	Any action, which may be a reaction or a stress resultant. A stress resultant at a section of a framed structure is an internal force: bending moment, shearing force or axial force.
a	Cross-sectional area.
D_i or D_{ij}	Displacement (rotational or translational) at coordinate i. When a second subscript j is provided it indicates the coordinate at which the force causing the displacement acts.
E	Modulus of elasticity.
EI	Flexural rigidity.
F	A generalized force: a couple or a concentrated load.
FEM	Fixed-end moment.
f_{ij}	Element of flexibility matrix.
G	Modulus of elasticity in shear.
I	Moment of inertia.
i, j, k, m, n, p, r	Integers.
J	Torsion constant (length^4), equal to the polar moment of inertia for a circular cross section.
l	Length.
M	Bending moment at a section. In beams and grids, a bending moment is positive when it causes tension in bottom fibers.
M_{AB}	Moment at end A of member AB. In plane structures, an end-moment is positive when clockwise. In general, an end-moment is positive when it can be represented by a vector in the positive direction of the axes x, y, or z.
N	Axial force at a section or in a member of a truss.
P, Q	Concentrated loads.
q	Load intensity.
R	Reaction.
S_{ij}	Element of stiffness matrix.

s	Used as a subscript indicates a statically determinate action.
T	Twisting moment at a section.
u	Used as a subscript, indicates the effect of unit forces or unit displacements.
V	Shearing force at a section.
W	Work of the external applied forces.
ε	Strain.
η	Influence ordinate.
v	Poisson's ratio.
σ	Stress.
τ	Shearing stress.
$\{\ \}$	Braces indicate a vector, i.e., a matrix of one column. To save space, the elements of a vector are sometimes listed in a row between two braces.
$[\ \ ]$	Brackets indicate a rectangular or square matrix.
$[\quad]^T_{n \times m}$	Superscript T indicates matrix transpose. $n \times m$ indicates the order of the matrix which is to be transposed resulting in an $m \times n$ matrix.
$\longrightarrow\!\!\!\!\rightarrow$	Double-headed arrow indicates a couple or a rotation; its direction is that of the rotation of a right-hand screw progressing in the direction of the arrow.
$\longrightarrow$	Single-headed arrow indicates a load or a translational displacement.
$z \longrightarrow x$ $\Big\downarrow$ y	Axes: the positive direction of the z axis points away from the reader.

The SI system
of units of measurement

Length	metre	m
	millimetre $= 10^{-3}$m	mm
Area	square metre	m^2
	square millimetre $= 10^{-6}$m^2	mm^2
Volume	cubic metre	m^3
Frequency	hertz $= 1$ cycle per second	Hz
Mass	kilogram	kg
Density	kilogram per cubic metre	kg/m^3
Force	newton	N
	$=$ a force which applied to a mass of one kilogram gives it an acceleration of one metre per second per second, i.e.	
	1N $= 1$kg m/s^2	
Stress	newton per square metre	N/m^2
	newton per square millimetre	N/mm^2
Temperature interval	degree Celsius	deg C; °C

Nomenclature for multiplication factors

10^9	giga	G
10^6	mega	M
10^3	kilo	k
10^{-3}	milli	m
10^{-6}	micro	μ
10^{-9}	nano	n

Introduction to the analysis of statically indeterminate structures

1-1 INTRODUCTION

A large part of this book is devoted to the modern methods of analysis of framed structures, that is structures consisting of members which are long in comparison to their cross section. Typical framed structures are beams, grids, plane and space frames or trusses (see Fig. 1-1). Other structures, such as walls and slabs, are considered in Chapters 16, 18, 19, 20, and 22.

In all cases, we deal with structures in which displacements — translation or rotation of any section — vary linearly with the applied forces, that is, any increment in displacement is proportional to the force causing it. All deformations are assumed to be *small*, so that the resulting displacements do not significantly affect the geometry of the structure and hence do not alter the forces in the members. Under such conditions, stresses, strains, and displacements due to different actions can be added using the principle of superposition; this topic is dealt with in Sec. 1–6. The majority of actual structures are designed so as to undergo only small deformations and they deform linearly. This is the case with metal structures, the material obeying Hooke's law; concrete structures are also usually assumed to deform linearly. We are referring of course to behavior under working loads, that is to elastic analysis; plastic analysis is considered in Chapters 21 and 22.

It is, however, possible for a straight structural member made of a material obeying Hooke's law to deform nonlinearly when the member is subjected to a lateral load and to a large axial force. This topic is dealt with in Chapter 15.

Although statical indeterminacy will be dealt with extensively in the succeeding sections, it is important at this stage to recognize the fundamental difference between statically determinate and indeterminate (hyperstatic) structures, in that the forces in the latter cannot be found from the equations of static equilibrium alone: a knowledge of some geometric conditions under load is also required.

The analysis of statically indeterminate structures generally requires the solution of linear simultaneous equations, the number of which depends on the method of analysis. Some methods avoid simultaneous equations by

1

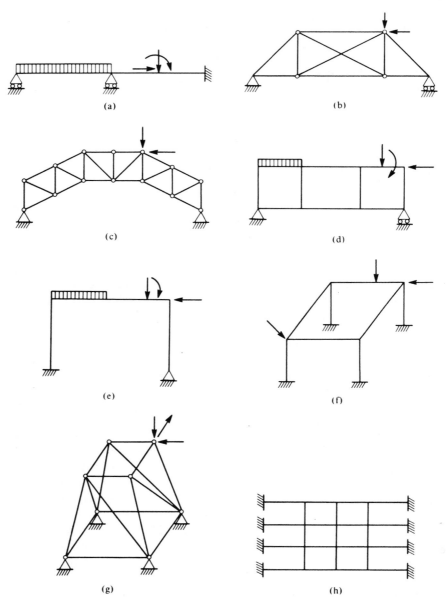

Fig. 1-1. Examples of framed structures. (a) Continuous beam. (b) and (c) Plane trusses. (d) and (e) Plane frames. (f) Space frame. (g) Space truss. (h) Horizontal grid subjected to vertical loads.

using iterative or successive correction techniques in order to reduce the amount of computation, and are suitable when the calculations are made by hand or by a hand-held or small desk calculator.

For large and complicated structures hand computation is often impracticable, and a digital computer has to be used. Its advent has shifted the emphasis from easy problem solution to efficient problem formulation: using matrices and matrix algebra, a large quantity of information can be organized and manipulated in a compact form. For this reason, in many cases, equations in this book are written in matrix form. The necessary matrix algebra is presented in Appendix A. In the text, the basic computer methods are discussed but details of programming are not given.

We should emphasize that the hand methods of solution must not be neglected. They are of value not only when a computer is not available but also for preliminary calculations and for checking of computer results.

1-2 STATICAL INDETERMINACY

Consider any free body subjected to several forces in space. The word *force* in this context means an action either of a load or of a couple. The resultant of all the forces is generally a load and a couple. For a body to be in equilibrium the components of the resultant in three orthogonal directions x, y, and z must vanish, so that the following equations of static equilibrium can be written:

$$\left. \begin{array}{ccc} \Sigma F_x = 0 & \Sigma F_y = 0 & \Sigma F_z = 0 \\ \Sigma M_x = 0 & \Sigma M_y = 0 & \Sigma M_z = 0 \end{array} \right\} \quad (1\text{-}1)$$

The summation in these equations is for all the components of the forces and of the moments about each of the three axes. Thus for a body subjected to forces in three dimensions, six equations of static equilibrium can be written. When all the forces acting on the free body are in one plane, only three of the six equations of statics are meaningful. For instance, when the forces act in the x–y plane, these equations are

$$\Sigma F_x = 0 \qquad \Sigma F_y = 0 \qquad \Sigma M_z = 0 \qquad (1\text{-}2)$$

When a structure in equilibrium is composed of several members, the equations of statics must be satisfied when applied on the structure as a whole. Each member, joint, or portion of the structure is also in equilibrium and the equations of statics must also be satisfied.

The analysis of a structure is usually carried out to determine the reactions at the supports and the internal stress resultants. As mentioned earlier, if these can be determined entirely from the equations of statics alone, then the structure is statically determinate. This book deals with statically indeterminate structures, in which there are more unknown forces than equations. The majority of structures in practice are statically indeterminate.

The indeterminacy of a structure may either be *external, internal,* or both. A structure is said to be externally indeterminate if the number of reaction components exceeds the number of equations of equilibrium. Thus a space structure is in general externally statically indeterminate when the number of reaction components is more than six. The corresponding number in a plane structure is three. The structures in Figs. 1-1a, c, e, f, g, and h are examples of external indeterminacy. Each of the beams of Figs. 1-2a and b has four reaction components. Since there are only three equations

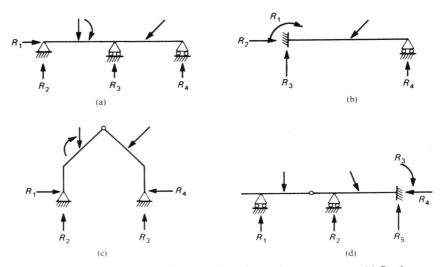

Fig. 1-2. (a), (b), (d) Externally statically indeterminate structures. (c) Statically determinate three-hinged frame.

of static equilibrium, there is one unknown force in excess of those that can be found by statics, and the beams are externally statically indeterminate. We define the degree of indeterminacy as the number of unknown forces in excess of the equations of statics. Thus the beams of Figs. 1-2a and b are indeterminate to the first degree.

Some structures are built so that the stress resultant at a certain section is known to be zero. This provides an additional equation of static equilibrium and allows the determination of an additional reaction component. For instance, the three-hinged frame of Fig. 1-2c has four reaction components, but the bending moment at the central hinge must vanish. This condition, together with the three equations of equilibrium applied to the structure as a free body, is sufficient to determine the four reaction components. Thus the frame is statically determinate. The continuous beam of Fig. 1-2d has five reaction components and one internal hinge. Four equilibrium equations can therefore be written so that the beam is externally indeterminate to the first degree.

Let us now consider structures which are externally statically determinate but internally indeterminate. For instance, in the truss[1] of Fig. 1-3a, the forces in the members cannot be determined by the equations of statics alone. If one of the two diagonal members is removed (or cut) the forces in the members can be calculated from equations of statics. Hence the truss is

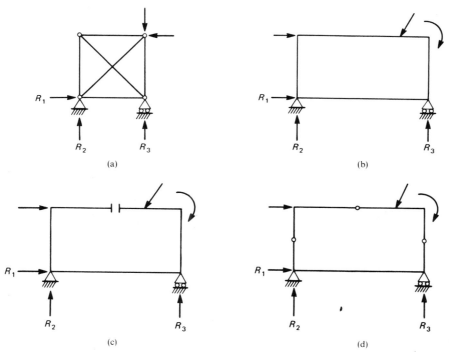

Fig. 1-3. Internally statically indeterminate structures.

internally indeterminate to the first degree, although it is externally determinate. The frame in Fig. 1-3b is internally indeterminate to the third degree: it becomes determinate if a cut is made in one of the members (Fig. 1-3c). The cut represents the removal or *release* of three stress resultants: axial force, shearing force, and bending moment. The number of releases necessary to make a structure statically determinate represents the degree of indeterminacy. The same frame becomes determinate if the releases are made by

[1]A truss is pin-jointed; a frame is rigid-jointed.

introducing three hinges as in Fig. 1-3d, thus removing the bending moment at three sections.

Structures can be statically indeterminate both internally and externally. The frame of Fig. 1-4a is externally indeterminate to the first degree, but the stress resultants cannot be determined by statics even if the reactions are assumed to have been found previously. They can, however, be determined

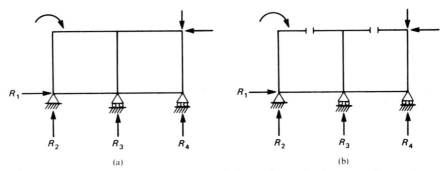

(a)   (b)

Fig. 1-4. Frame that is statically indeterminate both externally and internally.

by statics if the frame is cut at two sections, as shown in Fig. 1-4b, thus providing six releases. It follows that the frame is internally indeterminate to the sixth degree, and the total degree of indeterminacy is seven.

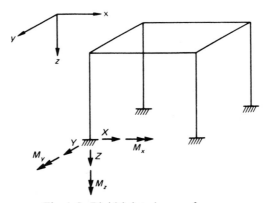

Fig. 1-5. Rigid-jointed space frame.

The space frame of Fig. 1-5 has six reaction components at each support: three components X, Y, and Z and three couples M_x, M_y, and M_z. To avoid crowding the figure the six components are shown at one of the four supports only. The moment vectors are indicated by double-headed arrows.[2] Thus the number of reaction components of the structure is 24, while the equations of equilibrium that can be written are six in number (cf. Eq. 1-1). The frame is therefore externally indeterminate to the 18th degree. If the reactions are known, the stress resultants in the four columns can be determined by statics but the beams forming a closed frame cannot be analyzed by statics alone. Cutting one of the beams at one section makes it possible to determine the stress resultants in all the beams. The number of releases in this case is six: axial force, shear in two orthogonal directions, bending moment about two axes, and twisting moment. The structure is thus internally indeterminate to the sixth degree, and the total degree of indeterminacy is 24.

The members of the horizontal grid of Fig. 1-6a are assumed to be rigidly connected (as shown in Fig. 1-6b) and to be subjected to vertical loads only. Thus both the reaction components X, Z, and M_y and the stress resultants X, Z and M_y vanish for all members of the grid. Hence, the number of equilibrium equations which can be used is three only. The reaction components at each support are Y, M_x, and M_z, so that the number of reaction components for the whole structure is $8 \times 3 = 24$. Thus it is externally statically indeterminate to the 21st degree.

If the reactions are known the stress resultants in the beams of the grid can be determined by statics alone except for the central part $ABCD$, which is internally statically indeterminate. Cutting any of the four beams of this part ($ABCD$) in one location produces three releases and makes it possible for the stress resultants to be determined by the equations of statics alone. Thus the structure is internally indeterminate to the third degree, and the total degree of indeterminacy is 24.

If the members forming the grid are not subjected to torsion — which is the case if the beams of the grid in one direction cross over the beams in the other direction with hinged connections (Fig. 1-6c) or when the torsional rigidity of the section is negligible compared with its bending stiffness — the twisting moment component (M_z in Fig. 1-6a) vanishes and the structure becomes indeterminate to the 12th degree. The grid needs at least four simple supports for stability and becomes statically determinate if the fixed supports are removed and only four hinged supports are provided. Since the number of reaction components in the original grid is 16, it is externally indeterminate to the 12th degree. There is no internal indeterminacy and, once the reactions have been determined, the internal forces in all the beams of the grid can be found by simple statics.

---

[2]All through this text, a couple (or rotation) is indicated in planar structures by an arrow in the form of an arc of a circle (see for example Figs. 1-2 and 1-3). In three-dimensional structures, a couple (or rotation) is indicated by a double-headed arrow. The direction of the couple is that of the rotation of a right-hand screw progressing in the direction of the arrow. This convention should be well understood at this stage.

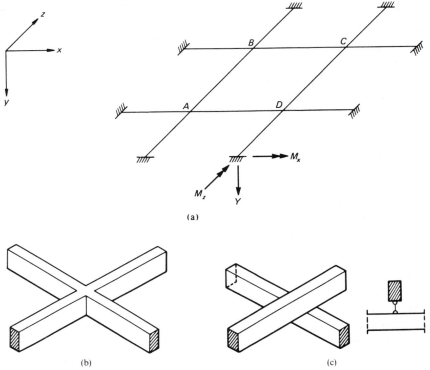

Fig. 1-6. Static indeterminacy of a grid. (a) Grid. (b) Rigid connection of beams. (c) Hinged connection of beams.

1-3   EXPRESSIONS FOR DEGREE OF INDETERMINACY

In Sec. 1-2 we found the degree of indeterminacy of various structures by inspection or from the number of releases necessary to render the structure statically determinate. For certain structures, especially those with a great many members, such an approach is difficult, and the use of a formal procedure is preferable.

Let us therefore consider a *plane* truss with three reaction components, m members and j hinged (pinned) joints (including the supports, which are also hinged). The unknown forces are the three reaction components and the force in each member – that is, $3 + m$. Now, at each joint two equations of equilibrium can be written:

$$\Sigma F_x = 0 \qquad \Sigma F_y = 0 \qquad (1\text{-}3)$$

the summation being for the components of all the external and internal forces meeting at the joint. Thus the total number of equations is $2j$.

For statical determinacy, the number of equations of statics is the same as the number of unknowns, that is,

$$2j = m + 3 \qquad (1\text{-}4)$$

Providing the structure is stable, some interchange between the number of members and the number of reaction components r is possible, so that for overall determinacy the condition

$$2j = m + r \qquad (1\text{-}5)$$

has to be satisfied. The degree of indeterminacy is then

$$i = (m + r) - 2j \qquad (1\text{-}6)$$

For the truss shown in Fig. 1-7, $r = 4$, $m = 18$, and $j = 10$. Hence $i = 2$.

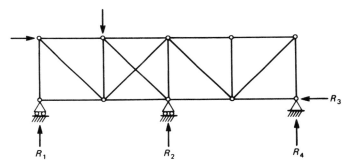

Fig. 1-7. Statically indeterminate plane truss.

In the case of a pin-jointed *space* frame, three equations of equilibrium can be written, viz.

$$\Sigma F_x = 0 \qquad \Sigma F_y = 0 \qquad \Sigma F_z = 0 \qquad (1\text{-}7)$$

the summation again being for all the internal and external forces meeting at the joint. The total number of equations is $3j$, and the condition of determinacy is

$$3j = m + r \qquad (1\text{-}8)$$

The degree of indeterminacy is

$$i = (m + r) - 3j \qquad (1\text{-}9)$$

The use of these expressions can be illustrated with reference to Fig. 1-8. For the truss of Fig. 1-8a, $j = 4$, $m = 3$, and $r = 9$, there being three reaction components at each support. Thus Eq. 1-8 is satisfied and the truss is statically determinate.

In the truss of Fig. 1-8b, $j = 10$, $m = 15$, and $r = 15$. Hence, again Eq. 1-8 is satisfied.

However, for the truss of Fig. 1-8c, $j = 8$, $m = 13$, and $r = 12$, so that from Eq. 1-9, $i = 1$. Removal of member AC would render the truss determinate.

Expressions similar to those of Eq. 1-6 and 1-9 can be established for frames with rigid joints. At a rigid joint of a *plane* frame, two resolution

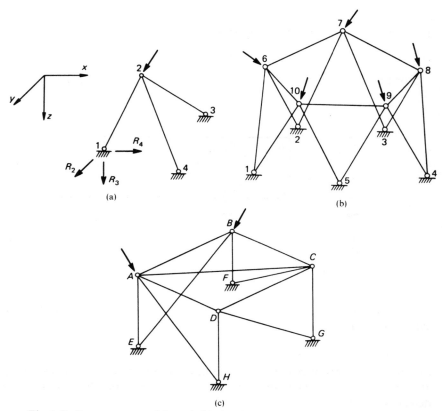

Fig. 1-8. Space trusses. (a) and (b) Statically determinate. (c) Statically indeterminate.

equations and one moment equation can be written. The stress resultants in any member of a plane frame (Fig. 1-9a) can be determined if any three of the six end-forces $F_1, F_2, ..., F_6$ are known, so that each member represents three unknown internal forces. The total number of unknowns is equal to the sum of the number of unknown reaction components r and of the unknown internal forces. Thus a rigid-jointed plane frame is statically determinate if

$$3j = 3m + r \tag{1-10}$$

and the degree of indeterminacy is   plane frame

$$i = (3m + r) - 3j \qquad (1\text{-}11)$$

In these equations j is the total number of rigid joints including the supports and m is the number of members.

If a rigid joint within the frame is replaced by a hinge, the number of equilibrium equations is reduced by one but the bending moments at the ends

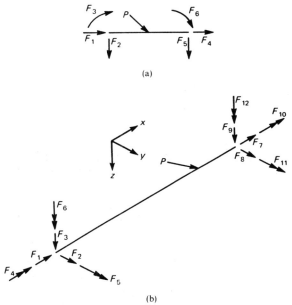

(a)

(b)

Fig. 1-9. End-forces in a member of a rigid-jointed frame. (a) Plane frame. (b) Space frame.

of the members meeting at the joint vanish, so that the number of unknowns is reduced by the number of members meeting at the hinge. This modification has to be observed when applying Eqs. 1-10 and 1-11 to plane frames with mixed type joints. We should note that at a rigid joint where more than two members meet and one of the members is connected to the joint by a hinge, the number of unknowns is reduced by one, without a reduction in the number of equilibrium equations. For example, we can verify that the frame of Prob. 1-5 is four times statically indeterminate, and the degree of indeterminacy becomes three if a hinge is inserted at the left end of member CF (just to the right of C).

As an example of rigid-jointed plane frame let us consider the frame of Fig. 1-4a: $j = 6$, $m = 7$, and $r = 4$. From Eq. 1-11 the degree of indeterminacy $i = (3 \times 7 + 4) - 3 \times 6 = 7$, which is the same result as that obtained in Sec. 1-2. For the frame of Fig. 1-2c, $j = 5$, $m = 4$ and $r = 4$.

However, one of the internal joints is a hinge, so that the number of unknowns is $(3m + r - 2) = 14$ and the number of equilibrium equations is $(3j - 1) = 14$. The frame is therefore statically determinate, as found before.

At a rigid joint of a *space* frame, three resolution and three moment equations can be written. The stress resultants in any members can be determined if any six of the twelve end-forces shown in Fig. 1-9b are known, so that each member represents six unknown forces. A space frame is statically determinate if

$$6j = 6m + r \tag{1-12}$$

and the degree of indeterminacy is ~~space frame~~.

$$i = (6m + r) - 6j \tag{1-13}$$

Applying Eq. 1-13 to the frame of Fig. 1-5, we have $m = 8$, $r = 24$ and $j = 8$. From Eq. 1-13 $i = 24$, which is of course the same as the result obtained in Sec. 1-2.

1-4 GENERAL METHODS OF ANALYSIS OF STATICALLY INDETERMINATE STRUCTURES

The objective of the analysis of structures is to determine the external forces (reaction components) and the internal forces (stress resultants). The forces must satisfy the conditions of equilibrium and produce deformations compatible with the continuity of the structure and the support conditions. As we have already seen, the equilibrium equations are not sufficient to determine the unknown forces in a statically indeterminate structure, and have to be supplemented by simple *geometrical* relations between the deformations of the structure. These relations ensure the *compatibility* of the deformations with the geometry of the structure and are called *geometry conditions* or *compatibility conditions*. An example of such conditions is that at an intermediate support of a continuous beam there can be no deflection and the rotation is the same on both sides of the support.

Two general methods of approach can be used. The first is the *force* or *flexibility* method, in which sufficient *releases* are provided to render the structure statically determinate. The released structure undergoes inconsistent deformations, and the inconsistency in geometry is then corrected by the application of additional forces.

The second approach is the *displacement* or *stiffness* method. In this method, restraints are added to prevent movement of the joints, and the forces required to produce the restraint are determined. Displacements are then allowed to take place at the joints until the fictitious restraining forces have vanished. With the joint displacements known, the forces on the structure are determined by superposition of the effects of the separate displacements.

Either the force or the displacement method can be used to analyze any structure. Since, in the force method, the solution is carried out for the forces necessary to restore consistency in geometry, the analysis generally involves the solution of a number of simultaneous equations equal to the number of unknown forces, that is the number of releases required to render the structure statically determinate. The unknowns in the displacement method are the possible joint translations and rotations. The number of the restraining forces to be added to the structure equals the number of possible joint displacements. This represents another type of indeterminacy, which may be referred to as *kinematic indeterminacy*, and is discussed in the next section. The force and displacement methods themselves are considered in more detail in Chapters 2 and 3.

1-5 KINEMATIC INDETERMINACY

When a structure composed of several members is subjected to loads, the joints undergo displacements in the form of rotation and translation. In the displacement method of analysis it is the rotation and translation of the joints that are the unknown quantities.

At a support, one or more of the displacement components are known. For instance, the continuous beam in Fig. 1-10 is fixed at C and has roller

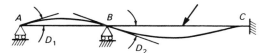

Fig. 1-10. Kinematic indeterminacy of a continuous beam.

supports at A and B. The fixity at C prevents any displacement at this end while the roller supports at A and B prevent translation in the vertical direction but allow rotation. We should note that roller supports are assumed to be capable of resisting both downward and upward forces.

If we assume that the axial stiffness of the beam is so large that the change in its length due to axial forces can be ignored, there will be no horizontal displacements at A or at B. Therefore, the only unknown displacements at the joints are the rotations D_1 and D_2 at A and B respectively (Fig. 1-10). The displacements D_1 and D_2 are independent of one another, as either can be given an arbitrary value by the introduction of appropriate forces.

A system of joint displacements is called *independent* if each displacement can be varied arbitrarily and independently of all the others. The number of the independent joint displacements in a structure is called the *degree of kinematic indeterminacy* or the *number of degrees of freedom*. This number is a sum of the degrees of freedom in rotation and in translation. The latter is sometimes called *freedom in sidesway*.

As an example of the determination of the number of degrees of freedom, let us consider the plane frame $ABCD$ of Fig. 1-11a. The joints A and D are fixed, and the joints B and C each have three components of displacement D_1, D_2, ..., D_6, as indicated in the figure. However, if the change in length of the members due to axial forces is ignored, the six displacements are not inde-

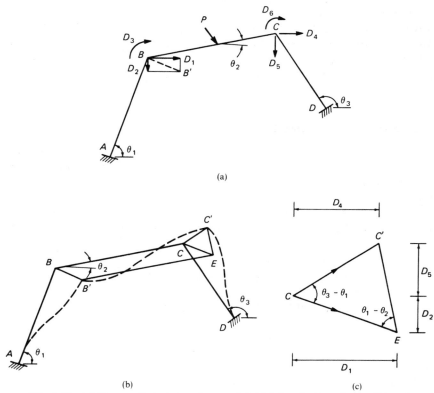

Fig. 1-11. (a) Kinematic indeterminacy of rigid-jointed plane frame. (b) and (c) Displacement diagrams.

pendent as the translation of the joints B and C is in a direction perpendicular to the original direction of the member. If an arbitrary value is assigned to any one of the translation displacements D_1, D_2, D_4, or D_5, the value of the other three is determined from geometrical relations.

For example, if D_1 is assigned a certain value, D_2 is equal to $(D_1 \cot \theta_1)$ to satisfy the condition that the resultant translation at B is perpendicular to AB. Once the position B' (joint B after displacement) is defined, the displaced position C' is defined, since it can have only one location if the lengths BC and CD are to remain unchanged. The joint displacement diagram is shown in Fig. 1-11b, in which the displacements D_2, D_4, and D_5 are determined graphically from a given value of D_1. In this figure, BB' is perpendicular to AB, $BCEB'$ is a parallelogram, and EC' and CC' are perpendicular to BC and CD respectively.

From the above discussion, it can be seen that the translation of the joints in the frame considered represents one unknown, or one degree of freedom.

We should note that the translations of the joints are very small compared with the length of the members. For this reason, the translation of the joints is assumed to be along a straight line perpendicular to the original direction of the member rather than along an arc of a circle. The triangle, $CC'E$ is drawn to a larger scale in Fig. 1-11c, from which the displacements D_2, D_4, and D_5 can be determined. The same displacements can be expressed in terms of D_1 by simple geometrical relations.

Now, the rotations of the joints at B and C are independent of one another. Thus the frame of Fig. 1-11a has one degree of freedom in sidesway and two degrees of freedom in rotation, so that the degree of kinematic indeterminacy for the frame is 3. If the axial deformations are not neglected, the four translational displacements are independent and the total degree of kinetic indeterminacy is 6.

The plane frame of Fig. 1-12 is another example of a kinematically indeterminate structure. If the axial deformation is neglected, the degree of kinetic indeterminacy is two, the unknown joint displacements being rotations at A and at B.

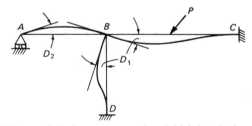

Fig. 1-12. Kinematic indeterminacy of a rigid-jointed plane frame.

We must emphasize that the kinematic indeterminacy and the statical indeterminacy must not be confused with one another. For instance, the frame of Fig. 1-12 has seven reaction components and is statically indeterminate to the fourth degree. If the fixed support at D is replaced by a hinge, the degree of statical indeterminacy will be reduced by one, but at the same time rotation at D becomes possible, thus increasing the kinematic indeterminacy by one. In general, the introduction of a release decreases the degree of statical indeterminacy and increases the degree of kinematic indeterminacy. For this reason, the higher the degree of statical indeterminacy, the more suitable the displacement method for analysis of the structure.

In a pin-jointed truss with all the forces acting at the joints, the members are subjected to an axial load only (without bending moment or shear) and therefore remain straight. The deformed shape of a plane truss is completely defined if the components of the translation in two orthogonal directions are determined for each joint, and each joint — other than a support — has two degrees of freedom.

Thus the plane structure of Fig. 1-13 is kinematically indeterminate to the second degree, as only joint A can have a displacement that can be defined by components in two orthogonal directions. From Eq. 1-6, the degree of statical indeterminacy is three. The addition to the system of an extra bar pinned at A at one end and at a support at the other would not change the degree of kinematic indeterminacy, but would increase the degree of statical indeterminacy by one.

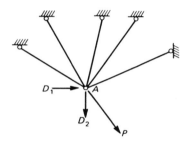

Fig. 1-13. Kinematic indeterminacy of a plane truss.

In a pin-jointed space truss loaded at the joints only, the translation of the joints can take place in any direction and can therefore be defined by components in three orthogonal directions, so that each joint — other than a support — has three degrees of freedom. It can be easily shown that the degree of kinematic indeterminacy of the truss of Fig. 1-8a is 3, that of the truss of Fig. 1-8b is 15 and that of the truss in Fig. 1-8c is 12.

Each joint of a rigid-jointed space frame can in general have six displacement components: three translations in three orthogonal directions, and three rotations, which can be represented by vectors in each of the three orthogonal directions (double-headed arrows).

Let us consider the frame of Fig. 1-14. It has eight joints, of which four are fixed in space. Each of the joints A, B, C, and D can have six displacements such as those shown at A. The degree of kinematic indeterminacy of the frame is therefore $4 \times 6 = 24$.

If the axial deformations are neglected, the lengths of the four columns remain unchanged so that the component D_3 of the translation in the vertical direction vanishes, thus reducing the unknown displacements by four. Also, since the lengths of the horizontal members do not change, the horizontal translations in the x direction of joints A and D are equal; the same applies to the joints B and C. Similarly, the translations in the y direction of joints A and B are equal; again, the same is the case for joints C and D. All this reduces the unknown displacements by four. Therefore, the degree of kinematic indeterminacy of the frame of Fig. 1-14, without axial deformation, is 16.

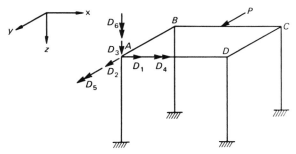

Fig. 1-14. Kinematic indeterminacy of a rigid-jointed space frame.

If a rigid-jointed grid is subjected to loads in the perpendicular direction to the plane of the grid only, each joint can have three displacement components; translation perpendicular to the plane of the grid and rotation about two orthogonal axes in the plane of the grid. Thus the grid of Fig. 1-15 is kinematically indeterminate to the sixth degree. The degree of statical indeterminacy of this grid is 15.

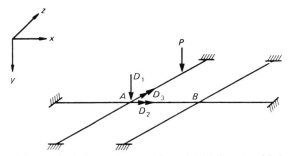

Fig. 1-15. Kinematic indeterminacy of a rigid-jointed grid loaded in a direction normal to its plane.

If the beams of a grid in one direction are hinged to the beams in the perpendicular direction in the manner shown in Fig. 1-6c, the beams will not be subjected to torsion. Hence, the degree of statical indeterminacy of the grid of Fig. 1-15 with hinged connections is eight. On the other hand, the degree of kinematic indeterminacy remains unchanged.

1-6 PRINCIPLE OF SUPERPOSITION

In Sec. 1-1, we mentioned that when deformations in a structure are proportional to the applied loads the principle of superposition holds. This principle states that the displacement due to a number of forces acting simultaneously is equal to the sum of the displacements due to each force acting separately.

In the analysis of structures it is convenient to use a notation in which a force F_j causes at a point i a displacement D_{ij}. Thus, the first subscript of a displacement describes the position and direction of the displacement, and the second subscript the position and direction of the force causing the

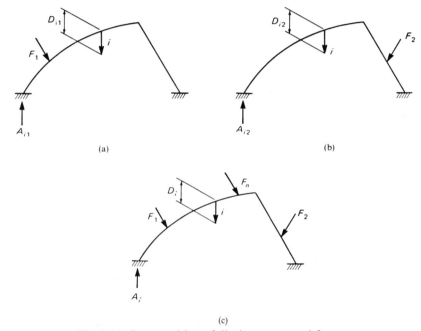

(a)   (b)

(c)

Fig. 1-16. Superposition of displacements and forces.

displacement. Each subscript refers to a *coordinate* which represents the location and direction of a force or of a displacement. The coordinates are usually indicated by arrows on the diagram of the structure.

This approach is illustrated in Fig. 1-16a. If the relation between the force applied and the resultant displacement is linear, we can write

$$D_{i1} = f_{i1} F_1 \tag{1-14}$$

where f_{i1} is the displacement at coordinate i due to a unit force at the location and direction of F_1 (coordinate 1).

If a second force F_2 is applied causing a displacement D_{i2} at i (Fig. 1-16b)

$$D_{i2} = f_{i2}F_2 \tag{1-15}$$

where f_{i2} is the displacement at i due to a unit force at coordinate 2.

If several forces $F_1, F_2, \ldots, F_n$ act simultaneously (Fig. 1-16c) the total displacement at i is

$$D_i = f_{i1}F_1 + f_{i2}F_2 + \cdots + f_{in}F_n \tag{1-16}$$

Clearly the total displacement does not depend on the order of the application of the loads. This, of course, does not hold if the stress-strain relation of the material is nonlinear.

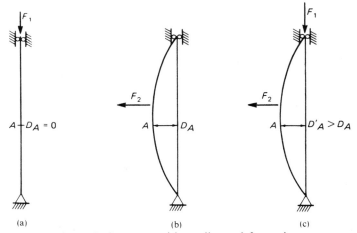

Fig. 1-17. Structure with nonlinear deformation.

A structure may behave nonlinearly even if made of a material obeying Hooke's law if changes in the geometry are caused by the applied loads. Consider the slender strut in Fig. 1-17a subjected to an axial force F_1, not large enough to cause buckling. The strut will, therefore, remain straight, and the lateral displacement at any point A is $D_A = 0$. Now if the strut is subjected to a lateral load F_2 acting alone, there will be a lateral deflection D_A at point A (Fig. 1-17b). If both F_1 and F_2 act (Fig. 1-17c), the strut will be subjected to an additional bending moment equal to F_1 multiplied by the deflection at the given section. This additional bending causes additional deflections and the deflection D'_A at A will, in this case, be greater than D_A.

No such bending moment exists of course when the loads F_1 and F_2 act separately, so that the combined effect of F_1 and F_2 is not equal to the sum of their separate effects, and the principle of superposition does not hold.

When a structure behaves linearly, the principle of superposition holds for forces as well as for displacements. Thus the internal stress resultants at any section or the reaction components of the structure in Fig. 1-16c can be determined by adding the effects of the forces F_1, F_2, ..., F_n when each acts separately.

Let the symbol A_i indicate a general *action* which may be a reaction, bending moment, shear, or thrust at any section due to the combined effect of all the forces. A general superposition equation of forces can then be written:

$$A_i = A_{ui1}F_1 + A_{ui2}F_2 + \cdots + A_{uin}F_n \qquad (1\text{-}17)$$

where A_{ui1} is the magnitude of the action A_i when a unit force is applied alone at coordinate 1. Similarly, A_{ui2}, ..., A_{uin} are the values of the action A_i when a unit force acts separately at each of the coordinates 2, ..., n.

Equation 1-17 can be written in matrix form:

$$A_i = [A_{ui}]_{1 \times n}\{F\}_{n \times 1} \qquad (1\text{-}17a)$$

We should note that the superposition of forces of Eq. 1-17 holds good for statically determinate structures regardless of the shape of the stress-strain relation of the material, provided only that the loads do not cause a distortion large enough to change appreciably the geometry of the structure. In such structures, any action can be determined by equations of statics alone without considering displacements. On the other hand, in statically indeterminate structures the superposition of forces is valid only if Hooke's law is obeyed because the internal forces depend on the deformation of the members.

1-7 INTERNAL FORCES: SIGN CONVENTION AND DIAGRAMS

As mentioned earlier, the purpose of structural analysis is to determine the reactions at the supports and the internal forces (the stress resultants) at any section. In beam and plane frames in which all the forces on the structure lie in one plane, the resultant of stresses at any section has generally three components: an axial force N, a shearing force V, and a bending moment M. The positive directions of N, V and M are shown in Fig. 1-18c, which represents an element (DE) between two closely spaced sections of the horizontal beam in Fig. 1-18a. A positive axial force N produces tension; a positive shearing force tends to push the left face of the element upwards and the right face downwards; a positive bending moment produces tensile stresses at the bottom face and bends the element in a concave shape.

In Fig. 1-18b, each of the three parts AD, DE and EC is shown as a free body subjected to a set of forces in equilibrium. To determine the internal forces at any section F (Fig. 1-18a) it is sufficient to consider only the

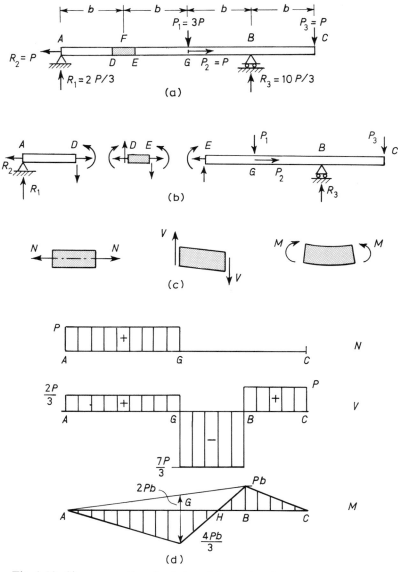

Fig. 1-18. Sign convention for internal forces in plane frames and beams.
(a) Beam. (b) Free-body diagrams. (c) Positive N, V and M (d) Axial-force,
shearing-force and bending-moment diagrams.

equilibrium of the forces on AD; thus N, V and M at F are the three forces in equilibrium with R_1 and R_2. The same internal forces are the statical equivalents of P_1, P_2, P_3 and R_3. Thus at any section F the values of N, V and M are, respectively, equal to the sums of the horizontal and vertical components and of the moments of the forces situated to the left of F. The values of N, V and M are positive when they are in the directions shown at end E of part EC in Fig. 1-18b. The internal forces at section F can also be considered as the statical equivalents of the forces situated to the right of F; in that case the positive directions of N, V and M will be as shown at end D of part AD.

The variations in N, V and M over the length of the member are presented graphically by the axial-force, shearing-force and bending-moment diagrams, respectively, in Fig. 1-18d. Positive N and V are plotted upwards, while positive M is plotted downwards. Throughout this book, the bending-moment ordinates are plotted on the tension face, that is the face where the stresses due to M are tensile. If the structure is of reinforced concrete, the reinforcement to resist bending is required near the tension face. Thus the bending-moment diagram indicates to the designer where the reinforcement is required; this is near the bottom face for part AH and near the top face for the remainder of the length (Fig. 1-18d). With this convention, it is not necessary to indicate a sign for the ordinates of the bending-moment diagram.

Calculation of the values of the internal forces at any section F in the beam of Fig. 1-18a requires knowledge of the forces situated to the left or to the right of section F. Thus, when reactions are included, they must be first determined. The values of the reactions and the internal-forces ordinates indicated in Figs 1-18a and d may now be checked.

In the above discussion we considered a horizontal beam. If the member is vertical, as for example the column of a frame, the signs of shear and bending will differ when the member is looked at from the left or the right. However, this has no effect on the sign of the axial force or on the significance of the bending-moment diagram when the ordinates are plotted on the tension side without indication of a sign. On the other hand, the signs of a shearing-force diagram will have no meaning unless we indicate in which direction the member is viewed. This will be discussed further in connection with the frame in Fig. 1-19.

Examples of shearing-force and bending-moment diagrams for a three-hinged plane frame are shown in Fig. 1-19. The three equilibrium equations (Eq. 1-2) together with the condition that the bending moment vanishes at the hinge C may be used to determine the reactions. The values indicated for the reactions and the V and M diagrams may now be checked. When determining the signs for the shearing-force diagram the nonhorizontal members are viewed with the dashed lines in Fig. 1-19 at the bottom face.

The ordinates of the shearing-force diagram for member BC (Fig. 1-19b)

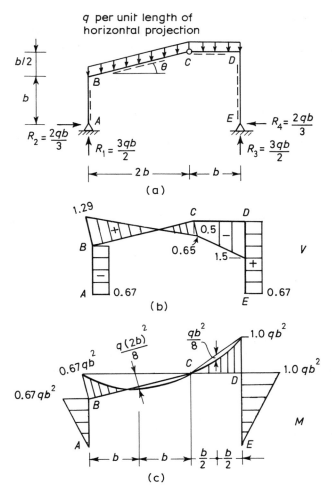

Fig. 1-19. A three-hinged plane frame. (a) Dimensions and loading. (b) Shearing-force diagram. (c) Bending-moment diagram.

may be checked as follows:

$$V_{Br} = R_1 \cos \theta - R_2 \sin \theta$$
$$V_{Cl} = [R_1 - q(2b)] \cos \theta - R_2 \sin \theta$$

where the subscripts r and l refer, respectively, to sections just to the right of B and just to the left of C; θ is the angle defined in Fig. 1-19a.

To draw diagrams of internal forces in frames with straight members, it is necessary only to plot the ordinates at member ends and at the sections where external forces are applied, and then to join these ordinates by straight lines. When a part (or the whole) of the length of a member is covered by a uniform load, the ordinates of the bending-moment diagram at the two ends of the part are to be joined by a second-degree parabola (see Fig. 1-19c). The ordinate of the parabola at the mid-point is $(qc^2/8)$, measured from the straight line joining the ordinates at the ends; q is the load intensity and c is the length of the part considered. A graphical procedure for plotting a second-degree parabola is included in Appendix F.

It is good practice to plot the ordinates perpendicular to the members and to indicate the values calculated and used to plot the diagram.

The internal forces at any section of a member of a framed structure can be easily determined if the end-forces are known. In Figs. 1-9a and b, typical members of plane and space frames are shown. The forces shown acting on each member, being the external applied force(s) and the member end-forces, represent a system in equilibrium. Thus the member may be treated as a separate structure. The internal forces at any section are the statical equivalents of the forces situated to its left or right.

In a space frame, the internal forces generally have six components: a force and a moment in the direction of x^*, y^* and z^* axes, where x^* is the centroidal axis of the member and y^* and z^* are centroidal principal axes of the cross section (Fig. 24-2).

Computer programs for the analysis of framed structures usually give the member end-forces (Figs. 1-9a and b) rather than the stress resultants at various sections. The sign convention for the end-forces usually relates to the member local axes, in the direction of the member centroidal axis and centroidal principal axes of the cross section. It is important at this stage to note that the stress resultants at the member ends may have the same magnitude as the member end-forces, but different signs, because of the difference in sign conventions. For example, at the left end of a typical member of a plane frame (Fig. 1-9a), the axial force, the shearing force and the bending moment are: $N = -F_1$; $V = -F_2$; $M = F_3$. The stress resultants at the right end are: $N = F_4$; $V = F_5$; $M = -F_6$. Here the member is viewed in a horizontal position and the positive directions of N, V and M are as indicated in Fig.1-18c.

1-8 EFFECT OF MOVING LOADS

In design of structures it is necessary to know the internal forces due to the permanent and the transient service loads. These are referred to as *dead* and *live* loads, respectively. Examples of live loads are the weight of snow on a roof, the weight of furniture and of occupants on a floor, and the wheel loads of a truck on a bridge or of a travelling crane on a crane beam. In analysis, the live load is usually represented by a uniformly distributed load or a series of concentrated forces.

Naturally, in design, we are concerned with the maximum values of the internal forces at various sections. Thus, for the maximum value of an action at any section, the live load must be placed on the structure in a position such that the maximum occurs. In many cases, the position of the load which produces a maximum is obvious. In other cases the use of *influence lines*, discussed in Sec. 1-9 and in Chapters 13 and 14, can help in determining the position of the moving load which results in the maximum value of the action considered.

We shall now consider the effects of moving loads on simple beams; the effects of moving loads on continuous beams are discussed in Sec. 2-7.

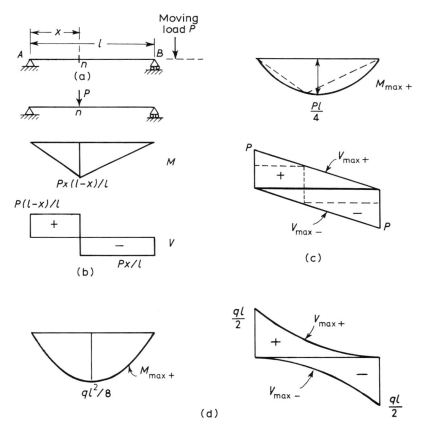

Fig. 1-20. Effect of moving loads on a simple beam. (a) Single concentrated load P. (b) Bending-moment and shearing-force diagrams when P is directly above any section n. (c) Envelopes of the M and V diagrams in (b): M_{max+}, V_{max+} and V_{max-} due to P. (d) Diagrams for maximum bending moment and shearing force due to uniform load q per unit length.

1-81   Single Load

Consider the effect of a single concentrated downward force P moving on a simple beam shown in Fig. 1-20a. At any section n, the bending moment is maximum positive when P is directly above n. The shear at n is maximum positive when P is just to the right of n, and is maximum negative when P is just to the left of n. The maximum values of bending and shear due to a single concentrated load on a simple beam may be expressed as:

$$M_{n\,\max+} = P\frac{x(l-x)}{l} \tag{1-18}$$

$$V_{n\,\max+} = P\frac{l-x}{l} \qquad V_{n\,\max-} = -P\frac{x}{l} \tag{1-19}$$

The expression *maximum negative value* is commonly used in structural design and is used here to mean a minimum value in the mathematical sense.

The bending-moment and shearing-force diagrams when P is at n are shown in Fig. 1-20b. If the load changes position, similar diagrams can be plotted and an envelope of the maximum ordinates can be constructed (Fig. 1-20c). The envelope for the maximum positive moment in the case considered is a second-degree parabola (Eq. 1-18); for shear, the envelopes for maximum positive and negative ordinates are straight lines (Eq. 1-19). The ordinates of such diagrams, which will be referred to as *maximum bending-moment* and *maximum shearing-force* diagrams, indicate in design the maximum internal forces that any section must resist.

1-82   Uniform Load

Figure 1-20d shows the maximum bending-moment and shearing-force diagrams due to a uniform load q per unit length. The load is placed over the full length of the beam or its part so as to produce the maximum effect. At any section n, the maximum bending moment occurs when q covers the full length; however, the maximum positive or negative shear occurs when q covers only the part to the right or to the left of n, respectively. The maximum values of bending and shear due to a uniform load on a simple beam are:

$$M_{n\,\max+} = q\frac{x(l-x)}{2} \tag{1-20}$$

$$V_{n\,\max+} = q\frac{(l-x)^2}{2l} \qquad V_{n\,\max-} = -q\frac{x^2}{2l} \tag{1-21}$$

It can be seen that the maximum bending-moment and shearing-force diagrams are second-degree parabolas.

1-83 Two Concentrated Loads

Consider the effects of two concentrated loads P_1 and P_2, with $P_1 \geqslant P_2$, moving on a simple beam (Fig. 1-21a). At any section n, the maximum bending moment occurs when P_1 or P_2 is directly upon the section (Fig. 1-21b or c), producing:

$$M_{n\,\text{max}+} = \frac{x(l-x)}{l}\left(P_1 + P_2\frac{l-x-s}{l-x}\right) \quad \text{with } 0 \leqslant x \leqslant (l-s) \quad (1\text{-}22)$$

or

$$M_{n\,\text{max}+} = \frac{x(l-x)}{l}\left(P_2 + P_1\frac{x-s}{x}\right) \quad \text{with } s \leqslant x \leqslant l \quad (1\text{-}23)$$

where s is the spacing between the loads and l is the span.

A third load position to be considered is shown in Fig. 1-21d, with P_2 falling outside the beam; the corresponding bending-moment at n is

$$M_{n\,\text{max}+} = P_1\frac{x(l-x)}{l} \quad \text{with } (l-s) \leqslant x \leqslant l \quad (1\text{-}24)$$

The diagram for the maximum bending moment for any part of the beam is a second-degree parabola represented by one of the above three equations, whichever has the largest ordinate. It can be shown that Eq. 1-24 governs in the central part of the beam only when $s > lP_2/(P_1 + P_2)$. The maximum bending-moment diagram in this case is composed of three parabolic parts (see Prob. 1-19). The maximum bending-moment diagram when $s \leqslant lP_2/(P_1 + P_2)$ is shown in Fig. 1-21e, which indicates the span portions to which Eqs. 1-22 and 1-23 apply. The case when $s > lP_2/(P_1 + P_2)$ is discussed in Prob. 1-19. In that problem it is indicated that the diagram for the maximum bending-moment due to specific moving load systems on a simple beam is the same as the bending-moment diagram due to virtual (unreal) stationary load systems.

The loads in the positions shown in Figs. 1-21b and c produce maximum positive and negative values of shear at section n. These are given by:

$$V_{n\,\text{max}+} = P_1\frac{l-x}{l} + P_2\frac{l-x-s}{l} \quad \text{with } 0 \leqslant x \leqslant (l-s) \quad (1\text{-}25)$$

$$V_{n\,\text{max}-} = -\left(P_1\frac{x-s}{l} + P_2\frac{x}{l}\right) \quad \text{with } s \leqslant x \leqslant l \quad (1\text{-}26)$$

The maximum shear may also be produced by P_1 or P_2 alone, placed just to the left or just to the right of n, giving (Fig. 1-21d):

$$V_{n\,\text{max}+} = P_1\frac{l-x}{l} \quad \text{with } (l-s) \leqslant x \leqslant l \quad (1\text{-}27)$$

$$V_{n\,\text{max}-} = -P_2\frac{x}{l} \quad \text{with } o \leqslant x \leqslant s \quad (1\text{-}28)$$

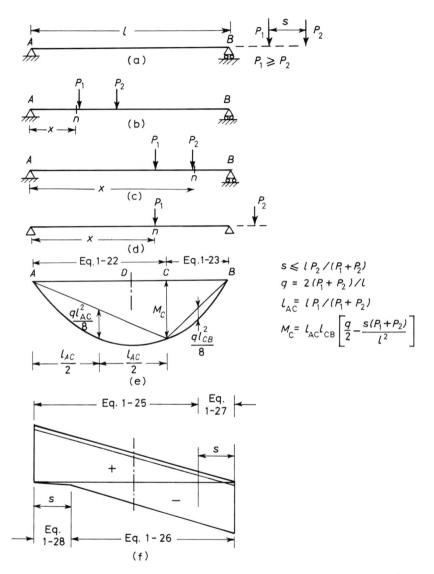

$$s \leqslant l P_2 / (P_1 + P_2)$$
$$q = 2 (P_1 + P_2) / l$$
$$l_{AC} = l P_1 / (P_1 + P_2)$$
$$M_C = l_{AC} l_{CB} \left[\frac{q}{2} - \frac{s(P_1 + P_2)}{l^2} \right]$$

Fig. 1-21. Effect of two concentrated moving loads P_1 and P_2, with $P_1 \geqslant P_2$. (a) A simple beam. (b), (c) and (d) Load positions for maximum bending moment or shear at any section n. (e) Maximum bending-moment diagram when $s \leqslant lP_2/(P_1 + P_2)$. (f) Maximum shearing-force diagram.

where x is the distance between support A and P_1 or P_2.

The maximum shearing-force diagram for any part of the beam is a straight line represented by one of the above four equations, whichever has the largest ordinate (Fig. 1-21f).

The maximum bending-moment and shearing-force diagrams shown in Fig. 1-21 are for the two forces P_1 and P_2, with the larger force P_1 to the left of P_2. If the forces can also be placed on the beam in a reversed order, with P_1 on the right-hand side of P_2, the larger of the two ordinates at any two sections symmetrically placed with respect to the center line must be considered as the maximum ordinate. Thus, assuming that the ordinates in Fig. 1-21e are larger for the left-hand half of the beam AD compared with their counterparts in the other half, the maximum bending-moment diagram should be a curve as at present shown for AD completed by its mirror image for DB.

Example 1-1 Find the maximum bending-moment diagrams for a simple beam of span l, subjected to two moving loads $P_1 = P$ and $P_2 = 0.8P$, spaced at a distance (a) $s = 0.3l$ and (b) $s = 0.6l$.

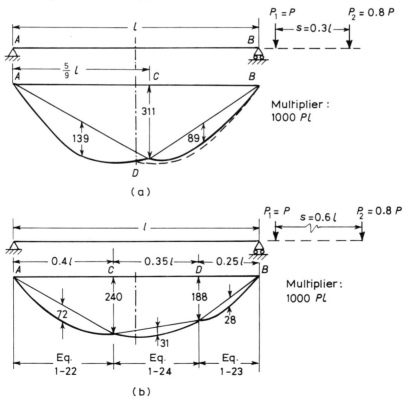

Fig. 1-22. Maximum bending-moment diagrams due to two moving concentrated loads, Example 1-1. (a) $s = 0.3l$. (b) $s = 0.6l$.

Assume that P_1 is on the left-hand side of P_2 and that this order is nonreversible. Repeat case (a) assuming that the order of loads can be reversed.

The maximum bending-moment ordinates in case (a) are calculated direct, using the equations in Fig. 1-21e because $s < lP_2/(P_1 + P_2) = l(0.8/1.8) = 0.44l$. The resulting diagram is shown in Fig. 1-22a.

For case (b), we plot three parabolas corresponding to Eqs. 1-22 to 24 and use for any part of the beam the curve with the largest ordinate (Fig. 1-22b).

With reversible loads in case (a), the maximum bending-moment diagram is the curve AD and its mirror image shown dotted in Fig. 1-22a. The ordinate at D is $0.330Pl$.

1-84   Group of Concentrated Loads

The maximum bending moment at any section n due to a system of concentrated moving loads on a simple beam (Fig. 1-23a) occurs when one of the loads is at n; by trial and error we can find which load should be at n to cause $M_{n\,\max+}$. The maximum positive shear at n occurs when P_1 is just to the right of n (Fig. 1-23b). The maximum negative shear at n occurs when the last load P_m is just to the left of n (Fig. 1-23c). It is possible, when the first (or last) load is relatively small, that the maximum shear occurs when another load is situated just to the left or to the right of n. Again, by trial and error we can find which load position produces the maximum effect at the section considered.

1-85   Absolute Maximum Effect

In the above we considered the position of moving loads to produce the maximum bending moment or shearing force at a specified section. In design, we also need to know the location and magnitude of the largest ordinate of the diagram of maximum bending moment or maximum shear. The section at which the maximum effect occurs is often identified by inspection. For example, in a simple beam, the absolute maximum shear occurs at the supports. The absolute maximum bending moment is at mid-span due to a single or uniform live load.

For a series of concentrated loads, the absolute maximum must occur directly beneath one of the loads, but it is not obvious which one. This is often determined by trial and error; usually, the absolute maximum bending moment occurs below one of the forces close to the resultant.

Assume that, in Fig. 1-23a, the absolute maximum moment occurs under load P_3. It is necessary to determine the variable x defining the position of the load system for which the bending moment M_n below P_3 is maximum. The value of M_n may be expressed as:

$$M_n = R_A(x) - P_1(s_1 + s_2) - P_2 s_2 \qquad (1\text{-}29)$$

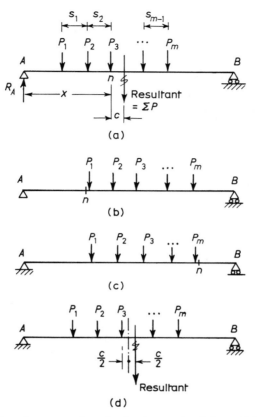

Fig. 1-23. Effect of a system of moving loads on a simple beam. (a) Position of load for $M_{n\,max\,+}$. (b) Position of load for $V_{n\,max\,+}$. (c) Position of load for $V_{n\,max\,-}$. (d) Position of load to produce largest bending moment directly beneath P_3.

and

$$R_A = \frac{l - x - c}{l} \Sigma P \qquad (1\text{-}30)$$

where c is the distance between the resultant of the system and P_3, and R_A is the reaction at A.

The maximum value of M occurs when $(dM_n/dx) = 0$. Substitution of Eq. 1-30 in Eq. 1-29 and differentiation gives:

$$\frac{dM_n}{dx} = \frac{\Sigma P}{l}(l - 2x - c) = 0$$

whence

$$x = \frac{l}{2} - \frac{c}{2} \qquad (1\text{-}31)$$

Thus, the absolute maximum bending moment due to a system of concentrated loads on a simple beam occurs under one of the loads when the center of the span is mid-way between the resultant and that particular load (Fig. 1-23d).

It should be noted that the above rule is derived on the assumption that all the forces are situated on the span. It is possible that the absolute maximum occurs when one or more of the forces are situated outside the beam. In this case the rule of the preceding paragraph may be applied using the resultant of only those forces which are located on the span. As an example of this condition, consider the beam in Fig. 1-22b: the absolute maximum is at the center line and its value is $0.25Pl$, occurring when P_1 is at mid-span and P_2 is outside the beam.

In design of cross sections of members it is often necessary to consider the interaction of two or more internal forces. For example, in the design of reinforced concrete beams for shear, we consider at any section the value of the maximum shearing-force combined with the corresponding bending moment resulting from the same loading. The value of the moment to be used in this case will obviously be smaller than the ordinate of the maximum bending-moment diagram.

In practice, with the wide use of computers, structural designers identify a number of loading cases representing the permanent loads, combined with the moving loads in the positions which are likely to produce maximum bending moment, axial force, shearing force or torsion at selected, critical sections. The designer also specifies the possible combinations of loading cases and the desired multipliers (load factors) of the individual cases. The computer can then be used to scan the internal forces due to the load combinations and give for each section the maximum positive and negative values of one of the internal forces and the values of the other internal forces occurring for the same loading. These are the values to be used in the design of sections.

Example 1-2   Find the absolute maximum bending moment in a simple beam subjected to two concentrated moving loads shown in Fig. 1-22a. What is the shearing force at the section of absolute maximum moment when the maximum occurs?

The resultant is of magnitude $1.8P$, situated at a distance $c = 0.133l$ from P_1. Thus, the absolute maximum moment occurs at a section at a distance $c/2 = 0.067l$ to the left of the center line. When P_1 is directly above this section, the reaction $R_A = 0.78P$ and the bending moment at the same section is the absolute maximum:

$$M_{\text{abs max}} = 0.78P(0.433l) = 0.338Pl$$

The corresponding shearing force to the left of P_1 is $R_A = 0.78P$.

1-9   INFLUENCE LINES FOR SIMPLE BEAMS AND TRUSSES

The subject of influence lines is treated in detail in Chapters 13 and 14. In this section we introduce only briefly the derivation of influence lines for shearing force and bending moment at a section of a simple beam and for the axial force in a member of a simply supported truss. The influence line for any of these actions is a plot of the value of the action at a given cross section as a unit load traverses the span.

Figure 1-24a shows the influence lines for the shear V_n and the bending moment M_n at any section n of a simple beam. The ordinate η at any section, at a distance x from the left-hand support, is equal to the value of V_n and M_n when

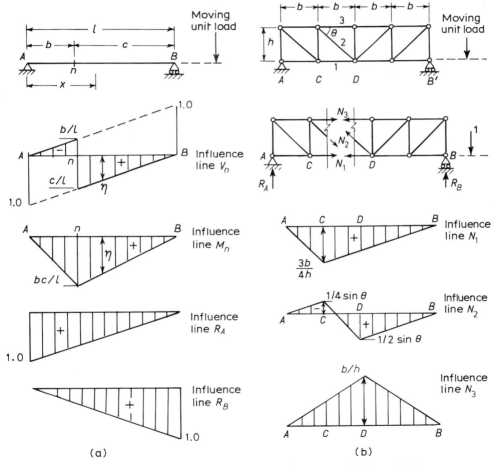

Fig. 1-24. Influence lines for simply supported structures. (a) Beam. (b) Truss.

the unit load is placed directly above the ordinate. Positive ordinates are plotted downwards.

The most effective method for obtaining an influence line uses Müller-Breslau's principle discussed in Sec. 13-3. It is based on energy theorems discussed in later chapters. We can also derive the influence lines for the structures considered here using simple statics. For this purpose, consider the influence line for the the reaction R_A. When the unit load is at a distance x from support A, the reaction is $R_A = (l - x)/l$. Thus we can express the ordinate of the influence line of R_A as:

$$\eta_{RA} = (l - x)/l$$

Similarly the ordinate of the influence line of R_B is:

$$\eta_{RB} = x/l$$

Thus the influence lines of the reactions are straight lines as shown in Fig. 1-24a. The shear at n is equal to R_A when the unit load is at any section between B and n. We can therefore use the part from B to n of the influence line for R_A to represent the influence line for V_n. Similarly, $V_n = -R_B$ when the unit load is in any position between n and A. Thus, between n and A, the influence line for V_n is the same as the influence line for R_B with reversed sign.

When the unit load is between B and n, the bending moment is $M_n = R_A b$; when the load is between n and A, $M_n = R_B c$. Thus the ordinates of the influence lines for R_A and R_B, multiplied respectively by b and c, can be used to construct the appropriate parts of the influence line for M_n.

Consider the effect of a unit load moving on the bottom chord of a simply supported truss (Fig. 1-24b). The influence lines for the forces in members 1, 2 and 3 are shown. These are constructed from the influence lines for the reactions R_A and R_B, which are the same as for the beam in Fig. 1-24a. Separating the truss into two parts by a vertical section and considering the equilibrium of each part, we can write the following equations. When the unit load is between B and D:

$$N_1 = R_A(b/h) \qquad N_2 = R_A(\sin \theta)^{-1} \qquad N_3 = R_A(-2b/h)$$

When the unit load is between C and A,

$$N_1 = R_B(3b/h) \qquad N_2 = R_B(-\sin \theta)^{-1} \qquad N_3 = R_B(-2b/h)$$

Again, the ordinates of the influence line for R_A or R_B multiplied by the quantities in brackets in the above equations give the ordinates of the influence lines for N_1, N_2 and N_3 over the length BD or CA. The influence lines between C and D are simply the straight lines joining the ordinates at C and D. This is justified because we are assuming, as usual, that the members are pin-connected and any external load is applied only at the joints. A unit moving load, when situated between two nodes such as C and D, is replaced by two statically equivalent downward forces at C and D; that is, the unit load is partitioned between C and D in inverse proportion to the distances from the load to C and D (see Sec. 13-4).

The use of influence lines to determine the effects of a series of concentrated loads is explained by the following example (see also Sec. 13-3).

Example 1-3 Find the maximum bending-moment and shearing-force values at section n for the beam in Fig. 1-25a due to the given system of three moving forces, representing the axle loads of a truck.

The influence line for M_n is shown in Fig. 1-25b. The bending-moment M_n due to the loads in any position is given by:

$$M_n = \sum_{i=1}^{3} P_i \eta_i$$

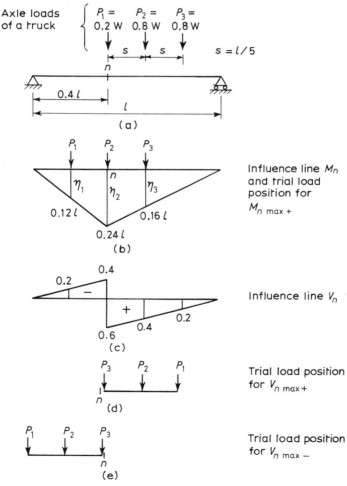

Fig. 1-25. Use of influence lines to determine the effects of a moving system of loads, Example 1-3.

where η_i is the influence-line ordinate directly beneath P_i. The maximum value of M_n is obtained by trial so that the sum of the products of the value of the individual axle load and the influence ordinate below it is as large as possible. A first trial of the truck position is shown in Fig. 1-25b; the corresponding value of M_n is:

$$(M_n)_{\text{trial 1}} = Wl[0.2(0.12) + 0.8(0.24) + 0.8(0.16)] = 0.344Wl$$

In a second trial we place the loads in a reversed order, with P_3 at n and the other two loads to the right; this gives $(M_n)_{\text{trial 2}} = 0.336Wl$. In a third trial we place P_2 at n, with P_3 at its left, giving $(M_n)_{\text{trial 3}} = 0.320Wl$. The largest value is obtained in trial 1 and hence $M_{n \, \text{max}+} = 0.344Wl$.

The influence line for V_n is shown in Figs. 1-25c, d, and e, which indicate one trial load position for the maximum positive shearing force and another trial load position for the maximum negative shearing force. These trials give:

$$V_{n \, \text{max}+} = W[0.8(0.6) + 0.8(0.4) + 0.2(0.2)] = 0.84W$$

$$V_{n \, \text{max}-} = W[0.2(0) + 0.8(-0.2) + 0.8(-0.4)] = -0.48W$$

It is clear in this case that no other trials would give larger values of V_n. Hence, we consider that the above are the maximum shear values.

We should note that influence lines give the maximum value of an action at a specified section but do not give absolute maximum values. However, if the above analysis is performed at a number of sections, a plot of the values obtained gives the maximum diagram and hence the absolute maximum ordinate can be seen.

1-10 GENERAL

The majority of modern structures are statically indeterminate, and with the flexibility method it is necessary to establish for a given structure the degree of indeterminacy, which may be external, internal, or both. In simple cases the degree of indeterminacy can be found by simple inspection, but in more complex or multispan and multibay structures it is preferable to establish the degree of indeterminacy with the aid of expressions involving the number of joints, members, and reaction components. These expressions are available for plane and space trusses (pin-jointed) and frames (rigid-jointed).

Two general methods of analysis of structures are available. One is the force (or flexibility) method, in which releases are introduced to render the structure statically determinate; the resulting displacements are computed and the inconsistencies in displacements are corrected by the application of additional forces in the direction of the releases. Hence a set of compatibility equations is obtained: its solution gives the unknown forces.

In the other method—the displacement (or stiffness) method—restraints at joints are introduced. The restraining forces required to prevent joint displacements are calculated. Displacements are then allowed to take place in the direction of the restraints until the restraints have vanished; hence a set of equilibrium equations is obtained: its solution gives the unknown displacements. The internal forces on the structure are then determined by superposition of the effects of these displacements and those of the applied loading with the displacements restrained.

The number of restraints in the stiffness method is equal to the number of possible independent joint displacements, which therefore has to be determined prior to the analysis. The number of independent displacements is the degree of kinematic indeterminacy, which has to be distinguished from the degree of statical indeterminacy. The displacements can be in the form of rotation or translation.

The analysis of structures by the force or the displacement method involves the use of the principle of superposition, which allows a simple addition of displacements (or actions) due to the individual loads (or displacements). This principle can, however, be applied only if Hooke's law is obeyed by the material of which a statically indeterminate structure is made. In all cases, the displacements must be small compared with the dimensions of the members so that no gross distortion of geometry of the structure takes place.

PROBLEMS

1-1-1-6   What is the degree of statical indeterminacy of the structures shown below? Introduce sufficient releases to render each structure statically determinate.

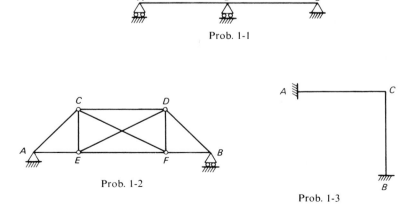

Prob. 1-1

Prob. 1-2

Prob. 1-3

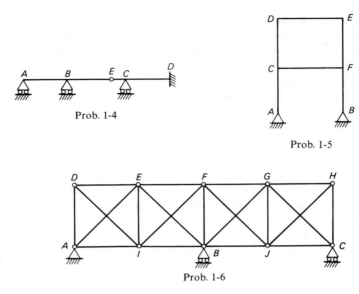

Prob. 1-4

Prob. 1-5

Prob. 1-6

1-7   The bars *AB*, *BC*, and *CD* are rigidly connected and lie in a horizontal plane. They are subjected to vertical loading. What is the degree of statical indeterminacy?

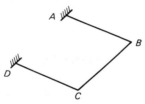

Prob. 1-7

1-8   The horizontal grid shown in the figure is subjected to vertical loads only. What is the degree of statical indeterminacy,
(a) assuming rigid connections at the joints,
(b) assuming connections of the type shown in Fig. 1-6c, i.e., a torsionless grid?
Introduce sufficient releases in each case to render the structure statically determinate.

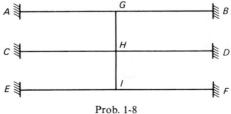

Prob. 1-8

1-9   The figure shows a pictorial view of a space truss pin-jointed to a vertical wall
       at *A*, *B*, *C*, and *D*. Determine the degree of statical indeterminacy. Introduce
       sufficient releases to make the structure statically determinate.

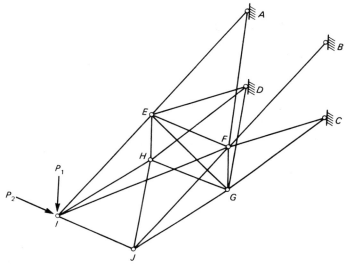

Prob. 1-9

1-10   Determine the degree of kinematic indeterminacy of the beam of Prob. 1-1
       and indicate a coordinate system of joint displacements. What is the degree
       of kinematic indeterminacy if the axial deformation is ignored?

1-11   Apply the questions of Prob. 1-10 to the frame of Prob. 1-5.

1-12   What is the degree of kinematic indeterminacy of the rigid-connected grid
       of Prob. 1-8?

1-13   Introduce sufficient releases to render the structure in the figure statically
       determinate and draw the corresponding bending-moment diagram. Draw
       the bending-moment diagram for another alternative released structure.

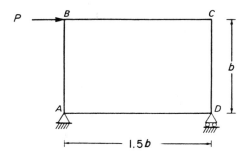

Prob. 1-13

1-14   Obtain the shearing-force and bending-moment diagrams for the statically determinate beams and frames shown.

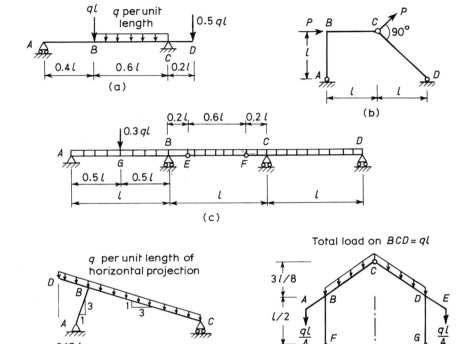

Prob. 1-14

1-15   The figure shows a space truss which has two planes of symmetry: xz and yz. Figure (a) is a typical elevation of one of the four sides of the truss. Introduce sufficient releases to make the structure statically determinate, and find the forces in the members of the released structure due to two equal forces P at the top node A.

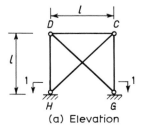

(a) Elevation

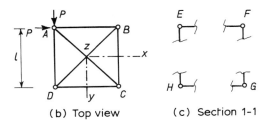

(b) Top view   (c) Section 1-1

Prob. 1-15

1-16   (a) Introduce sufficient releases to make the frame shown statically determinate. Indicate the releases by a set of coordinates.

       (b) Introduce a hinge at the middle of each member and draw the bending-moment diagram for the frame due to two horizontal forces, each equal to P, at E and C. Show by a sketch the magnitude and direction of the reaction components at A.

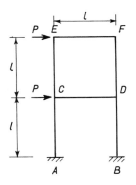

Prob. 1-16

1-17   For a simple beam of span l, determine the maximum bending moment and its location due to the following moving loads:

       (a) Two forces $P_1 = P_2 = W$, spaced at $s = 0.2l$.

       (b) Three forces $P_1 = W$, $P_2 = W$, $P_3 = W/2$, spaced at $s = 0.2l$ between P_1 and P_2 and between P_2 and P_3.

       (c) Two forces $P_1 = P_2 = W$, spaced at $s = 0.55l$.

1-18   For the simple beam and the truck axle loads specified in Example 1-3 (Fig. 1-25), determine the maximum bending moment and shearing force and their locations.

1-19   Verify that, for each of the simple beams shown, the maximum bending-moment diagram due to the moving load system in (a) is the same as the bending-moment diagram due to the virtual (unreal) stationary load system in (b).

       Hint. For any section n, place the moving load in the position which produces maximum moment and verify that M_n due to the moving load is the same as that due to the stationary load. The derivation of the virtual stationary load system is given in Wechsler, M. B., "Moment Determination for Moving Load Systems," *Journal of Structural Engineering*, American Society of Civil Engineers, III(6) (June 1985), pp. 1401–1406.

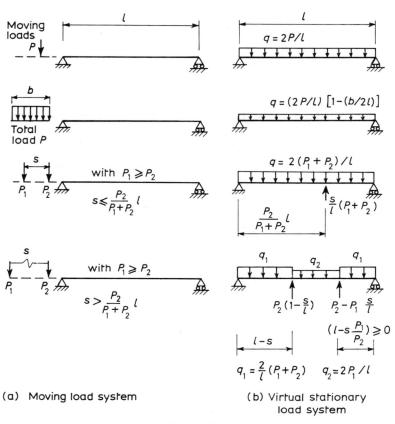

(a) Moving load system

(b) Virtual stationary load system

Prob. 1-19

1-20   Assuming that in Example 1-1 the simple beam is extended by an overhanging part of length $0.35l$ at each end, and assuming that the order of the two loads can be reversed, find the maximum bending moments.

1-21   A simple beam of span l has an overhang of length $0.35l$ at each end; the total length of the beam is thus $1.7l$. Obtain the influence lines for M and V at a section n within the span at $0.25l$ from the left-hand support. Use the influence line to find $M_{n\,max+}$, $M_{n\,max-}$ and $V_{n\,max+}$ due to: (a) two moving loads $P_1 = P$ and $P_2 = 0.8P$, separated by a distance $s = 0.3l$, assuming that the order of the loads can be reversed; and (b) a uniform moving load of q per unit length (use Eq. 13-3).

CHAPTER 2

Force method of analysis

2-1 INTRODUCTION

As mentioned in Sec. 1-4, this is one of the basic methods of analysis of structures. It is proposed to outline the procedure in this chapter, and then in Chapter 4 to compare the force and the displacement methods.

2-2 DESCRIPTION OF METHOD

1. First of all, the degree of statical indeterminacy is determined. A number of releases equal to the degree of indeterminacy is now introduced, each release being made by the removal of an external or an internal force. The releases must be chosen so that the remaining structure is stable and statically determinate. However in some cases the number of releases can be less than the degree of indeterminacy, provided the remaining statically indeterminate structure is so simple that it can be readily analyzed. In all cases, the released forces, which are also called *redundant forces*, should be carefully chosen so that the released structure is easy to analyze.

2. The releases introduce inconsistencies in displacements, and as a second step these inconsistencies or "errors" in the released structure are determined. In other words, we calculate the magnitude of the "errors" in the displacements corresponding to the redundant forces. These displacements may be due to external applied loads, settlement of supports, or temperature variation.

3. The third step consists of a determination of the displacements in the released structure due to unit values of the redundants (cf. Figs. 2-1d and e). These displacements are required at the same location and in the same direction as the error in displacements determined in step 2.

4. The values of the redundant forces necessary to eliminate the errors in the displacements are now determined. This requires the writing of superposition equations in which the effects of the separate redundants are added to the displacements of the released structure.

5. Hence, we find the forces on the original indeterminate structure: they are the sum of the correction forces (redundants) and forces on the released structure.

This brief description of the application of the force method will now be illustrated by examples.

Example 2-1   Figure 2-1a shows a beam ABC fixed at C, resting on roller supports at A and B, and carrying a uniform load of q per unit length. The beam has a constant flexural rigidity EI. Find the reactions of the beam.

The structure is statically indeterminate to the second degree, so that two redundant forces have to be removed. Several choices are possible, e.g., the moment and the vertical reaction at C, or the vertical reactions at A and B. For the purposes of this example, we shall remove the vertical reaction at B and the moment at C. The released structure is then a simple beam AC with redundant forces and displacements as shown in Fig. 2-1b. The location and direction of the various redundants and displacements is referred to as a *coordinate system*.

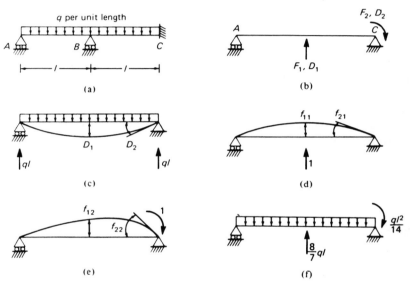

Fig. 2-1. Continuous beam considered in Example 2-1. (a) Statically indeterminate beam. (b) Coordinate system. (c) External load on released structure. (d) $F_1 = 1$. (e) $F_2 = 1$. (f) Redundants.

The positive directions of the redundants F_1 and F_2 are chosen arbitrarily but the positive directions of the displacements at the same location must always accord with those of the redundants. The arrows in Fig. 2-1b indicate the chosen positive directions in the present case and, since the arrows indicate forces as well as displacements, it is convenient in a general case to label the coordinates by numerals 1, 2, ..., n.

Following this system, Fig. 2-1c shows the displacements at B and C as D_1 and D_2 respectively. In fact, as shown in Fig. 2-1a, the actual displacements at those points are zero, so that D_1 and D_2 represent the inconsistencies in deformation.

The magnitude of D_1 and D_2 can be calculated from the behavior of the simply supported beam of Fig. 2-1c. For the present purposes, we can use the data of Appendix B. Thus

$$D_1 = -\frac{5ql^4}{24EI} \quad \text{and} \quad D_2 = -\frac{ql^3}{3EI}$$

The negative signs show that the displacements are in directions opposite to the positive directions chosen in Fig. 2-1b.

The displacements due to unit values of the redundants are shown in Figs. 2-1d and e. These displacements are as follows (again from Appendix B):

$$f_{11} = \frac{l^3}{6EI} \qquad f_{12} = \frac{l^2}{4EI}$$

$$f_{21} = \frac{l^2}{4EI} \qquad f_{22} = \frac{2l}{3EI}$$

The general coefficient f_{ij} represents the displacement at the coordinate i due to a unit redundant at the coordinate j.

The *geometry relations* express the fact that the final vertical translation at B and the rotation at C vanish. The final displacements are the result of the superposition of the effect of the external loading and of the redundants on the released structure. Thus the geometry relations can be expressed as

$$\left.\begin{array}{l} D_1 + f_{11}F_1 + f_{12}F_2 = 0 \\ D_2 + f_{21}F_1 + f_{22}F_2 = 0 \end{array}\right\} \qquad (2\text{-}1)$$

2-3 FLEXIBILITY MATRIX

The relations of Eq. 2-1 can be written in matrix form:

$$[f]\{F\} = \{-D\} \qquad (2\text{-}2)$$

where

$$\{D\} = \begin{Bmatrix} D_1 \\ D_2 \end{Bmatrix} \qquad [f] = \begin{bmatrix} f_{11} & f_{12} \\ f_{21} & f_{22} \end{bmatrix} \qquad \text{and} \qquad \{F\} = \begin{Bmatrix} F_1 \\ F_2 \end{Bmatrix}$$

(The necessary elements of the matrix algebra are given in Appendix A.)

The column vector $\{D\}$ depends on the external loading. The elements of the matrix $[f]$ are displacements due to the unit values of the redundants. Therefore, $[f]$ depends on the properties of the structure, and represents the *flexibility* of the released structure. For this reason, $[f]$ is called the flexibility matrix and its elements are called *flexibility influence coefficients*. These coefficients are used in a way similar to the ordinates of influence lines for deflections (cf. Chapter 13).

The elements of the vector $\{F\}$ are the redundants which can be obtained by solving Eq. 2-2; thus

$$\{F\} = [f]^{-1}\{-D\} \tag{2-3}$$

In the example considered, the order of the matrices $\{F\}$, $[f]$ and $\{D\}$ is 2×1, 2×2, 2×1. In general, if the number of releases is n, the order will be $n \times 1$, $n \times n$, $n \times 1$ respectively. We should note that $[f]$ is a square symmetrical matrix. The generality of this property of the flexibility matrix will be proved in Sec. 4-6.

In the example considered, the flexibility matrix and its inverse are

$$[f] = \begin{bmatrix} \dfrac{l^3}{6EI} & \dfrac{l^2}{4EI} \\[2ex] \dfrac{l^2}{4EI} & \dfrac{2l}{3EI} \end{bmatrix}$$

and

$$[f]^{-1} = \frac{12EI}{7l^3} \begin{bmatrix} 8 & -3l \\ -3l & 2l^2 \end{bmatrix}$$

The displacement vector is

$$\{D\} = \frac{ql^3}{24EI} \begin{Bmatrix} -5l \\ -8 \end{Bmatrix}$$

Substituting in Eq. 2-3, or solving Eq. 2-2, we obtain

$$\{F\} = \frac{ql}{14} \begin{Bmatrix} 16 \\ l \end{Bmatrix}$$

Therefore the redundants are

$$F_1 = \frac{8}{7}ql \qquad \text{and} \qquad F_2 = \frac{ql^2}{14}$$

The positive sign indicates that the redundants act in the positive directions chosen in Fig. 2-1b.

The final forces acting on the structure are shown in Fig. 2-1f, and any stress resultants in the structure can be determined by the ordinary methods of statics.

It is important to note that the flexibility matrix is dependent on the choice of redundants: with different redundants, the same structure would result in a different flexibility matrix.

The reactions and the internal forces can also be determined by the superposition of the effect of the external loads on the released structure and the effect of the redundants. This can be expressed by the superposition equation

$$A_i = A_{si} + (A_{ui1}F_1 + A_{ui2}F_2 + \cdots + A_{uin}F_n)$$ (2-4)

where

A_i = any action i, that is reaction at a support, shearing force, axial force, twisting moment, or bending moment at a section in the actual structure

A_{si} = same action as A_i but in the released structure subjected to the external loads

$A_{ui1}, A_{ui2}, \ldots, A_{uin}$ = corresponding action due to a unit force acting alone on the released structure at the coordinate $1, 2, \ldots, n$, respectively

$F_1, F_2, \ldots, F_n$ = redundants acting on the released structure

From Eq. 1-17, the term in parentheses in Eq. 2-4 represents the action of all the redundants applied simultaneously to the released structure.

Generally, several reactions and internal forces are required. These can be obtained by equations similar to Eq. 2-4. If the number of actions is m, the system of equations needed can be put in the matrix form

$$\{A\}_{m \times 1} = \{A_s\}_{m \times 1} + [A_u]_{m \times n}\{F\}_{n \times 1}$$ (2-5)

The order of each matrix is indicated in Eq. 2-5 but it may be helpful to write, on this occasion, the matrices in full. Thus

$$\{A\} = \begin{Bmatrix} A_1 \\ A_2 \\ \cdots \\ A_m \end{Bmatrix} \quad \{A_s\} = \begin{Bmatrix} A_{s1} \\ A_{s2} \\ \cdots \\ A_{sm} \end{Bmatrix}$$

$$[A_u] = \begin{bmatrix} A_{u11} & A_{u12} & \cdots & A_{u1n} \\ A_{u21} & A_{u22} & \cdots & A_{u2n} \\ \cdots & \cdots & \cdots & \cdots \\ A_{um1} & A_{um2} & \cdots & A_{umn} \end{bmatrix}$$

We should note that the elements of a flexibility matrix are not necessarily dimensionally homogeneous as they represent either a translation or a rotation due to a unit load or to a couple. In the above example, f_{11} is a translation, due to a unit load. If the kip-inch units are used in the analysis, then units of f_{11} are inch/kip. The coefficient f_{22} is a rotation in radians due to a unit couple; thus its units are (kip inch)$^{-1}$. Both f_{12} and f_{21} are in (kip)$^{-1}$, since f_{12} is a translation due to a unit couple and f_{21} is a rotation due to a unit load.

Since Example 2-1 represents a comparatively simple structure, the advantages of using matrix algebra are not obvious, but the approach can be seen more clearly. In more complicated structures, the use of matrices is essential, especially if different loadings are to be analyzed.

2-4 ANALYSIS FOR DIFFERENT LOADINGS

When using Eq. 2-3 to find the redundants in a given structure under a number of different loadings, the calculation of the flexibility matrix (and its inverse) need not be repeated. When the number of loadings is p the solution can be combined into one matrix equation

$$[F]_{n \times p} = [f]_{n \times n}^{-1}[-D]_{n \times p} \qquad (2\text{-}6)$$

where each column of $[F]$ and $[D]$ corresponds to one loading.

The reactions or the stress resultants in the original structure can be determined from equations similar to Eq. 2-5, viz.,

$$[A]_{m \times p} = [A_s]_{m \times p} + [A_u]_{m \times n}[F]_{n \times p} \qquad (2\text{-}7)$$

2-41 Effect of Displacement at Joints: Environmental Effects

The force method can be used to analyze a statically indeterminate structure subjected to effects other than applied loads. An example of such an effect which causes internal stresses is the movement of a support. This may be due to the settlement of foundations or to a differential temperature movement of supporting piers.

Internal forces are also developed in any structure if the free movement of a joint is prevented. For example, the temperature change of a beam with two fixed ends develops an axial force. Stresses in a structure may also be caused by a differential change in temperature.

As an example let us consider the continuous beam ABC of Fig. 2-2a when subjected to a rise in temperature varying linearly between the top and bottom faces. If support B is removed, the beam becomes statically determinate, and the rise in temperature causes it to deflect upward (Fig. 2-2b). If the beam is to remain attached to the support at B, a downward force F_B will develop so as to correct for the error in displacement, D_B. The deflected shape of the beam axis is then as shown in Fig. 2-2c.

If a member of a truss is manufactured shorter or longer than its theoretical length and then forced to fit during erection, stresses will develop in the truss. This lack of fit has a similar effect to a change in temperature of the member in question. The effect of shrinkage of concrete members on drying is also similar to the effect of a drop in temperature.

Another cause of internal forces in statically indeterminate structures is the prestrain induced in prestressed concrete members. This may be illustrated by reference to Fig. 2-3a, which shows a statically determinate concrete

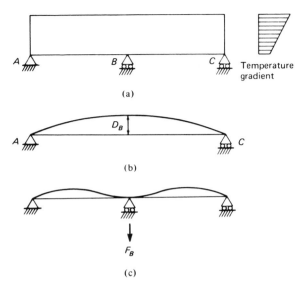

(a)

(b)

(c)

Fig. 2-2. Effect of a differential rise in temperature across a continuous beam.

beam of rectangular cross section. A cable is inserted through a duct in the lower part of the cross section. The cable is then tensioned and anchored at the ends. This produces compression in the lower part of the cross section and causes the beam to deflect upward (Fig. 2-3b). If the beam is statically

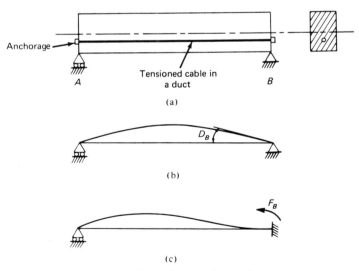

(a)

(b)

(c)

Fig. 2-3. Effect of prestrain in a beam.

indeterminate, for example if the end B is fixed, the rotation at this end cannot take place freely and a couple will develop at B so as to cause the rotation D_B to vanish (Fig. 2-3c).

In all these cases, Eq. 2-3 can be applied to calculate the redundant forces, the elements of the matrix D being the errors in the displacement of the released structure due to the given effect, or to the various effects combined. The effects of temperature, shrinkage, creep and prestressing are discussed in more detail in Secs. 4-9 to 4-11.

2-42 Effect of Displacement at Coordinates

When a displacement of a support takes place at one of the coordinates representing the redundants, a slight modification to Eq. 2-3 is required. Referring to the continuous beam of Example 2-1 (Fig. 2-1a) under the load indicated, let us assume that the supports B and C undergo displacements Δ_1 (translation) and Δ_2 (rotation) respectively in the same directions as the redundants F_1 and F_2 in Fig. 2-1b. The geometry relations (similar to Eq. 2-1) can be expressed by superposition equations which give the final displacements

$$D_1 + f_{11}F_1 + f_{12}F_2 = \Delta_1 \qquad (2\text{-}8)$$

$$D_2 + f_{21}F_1 + f_{22}F_2 = \Delta_2 \qquad (2\text{-}8a)$$

Solving for F, we obtain in matrix form

$$\{F\} = [f]^{-1}\{\Delta - D\} \qquad (2\text{-}9)$$

where Δ is a matrix of the same order as D. In the general case when the number of redundants is n,

$$\{\Delta\} = \left\{ \begin{array}{c} \Delta_1 \\ \Delta_2 \\ \cdots \\ \Delta_n \end{array} \right\}$$

— Δ is displacement specified at a coordinate.
— displacements at coordinate caused by movement elsewhere is included in $\{D\}$

We should note that the matrix $\{\Delta\}$ includes the displacement of the support if this displacement corresponds to one of the coordinates. Otherwise, the effect of the support movement on the displacement at a coordinate should be included in the calculation of the displacement $\{D\}$.

Equation 2-9 is more general than Eq. 2-3 and can in fact be used for external loading as well as for support displacements. When the analysis is to be carried out for p cases of loadings and support movement it is convenient to generalize Eq. 2-6 into

$$[F]_{n \times p} = [f]_{n \times n}^{-1}[\Delta - D]_{n \times p} \qquad (2\text{-}10)$$

2-5 FIVE STEPS OF FORCE METHOD

The analysis by the force method involves five steps which are summarized as follows:

Step 1   Introduce releases and define a system of coordinates. Also define $[A]_{m \times p}$, the required actions, and define their sign convention (if necessary).
Step 2   Due to the loadings on the released structure, determine $[D]_{n \times p}$, $[\Delta]_{n \times p}$ and $[A_s]_{m \times p}$.
Step 3   Apply unit values of the redundants one by one on the released structure and generate $[f]_{n \times n}$ and $[A_u]_{m \times n}$.
Step 4   Solve the geometry equations:

$$[f]_{n \times n}[F]_{n \times p} = [\Delta - D]_{n \times p}$$

This gives the redundants $[F]_{n \times p}$.
Step 5   Calculate the required actions by superposition:

$$[A]_{m \times p} = [A_s]_{m \times p} + [A_u]_{m \times n}[F]_{n \times p}$$

At the completion of step 3, all the matrices necessary for the analysis have been generated. The last two steps involve merely matrix algebra. Step 5 may be eliminated when no action besides the redundants is required, or when the superposition can be done by inspection after determination of the redundants (see Example 2-3). When this is the case, the matrices $[A]$, $[A_s]$ and $[A_u]$ are not required.

For quick reference the symbols used in this section are defined again as follows:

n, p, m = number of redundants, number of loading cases and number of actions required
$[A]$ = the required actions (the answers to the problem)
$[A_s]$ = values of the actions due to the loadings on the released structure
$[A_u]$ = values of the actions in the released structure due to unit forces applied separately at each coordinate
$[D]$ = displacements at the coordinates due to the loadings; these displacements represent incompatibilities to be eliminated by the redundants
$[\Delta]$ = displacements at the coordinates due to the loadings; these represent imposed displacements to be maintained
$[f]$ = flexibility matrix

Example 2-2   For the continuous beam ABC of Example 2-1 (Fig. 2-4a), find the reactions due to the following movement at the supports: (a) downward translation of $A = l/100$; (b) downward translation of $B = l/100$ together with a clockwise rotation of $C = 0.004$ radian.

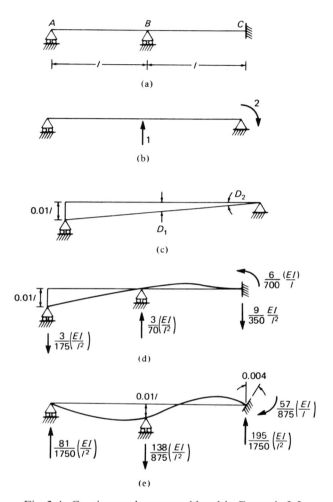

Fig. 2-4. Continuous beam considered in Example 2-2.

We can select the same redundants and positive directions of the displacements as in Example 2-1 (Fig. 2-4b) and hence the same flexibility matrix as before. Four reaction components are required, two of which are the redundants F_1 and F_2. Once these have been determined, the other two reaction components follow by inspection or by simple calculation. Thus, of the five steps of the force method, only steps 2 and 4 are needed in this example.

In case (a), the movement of A does not correspond to a redundant. The error in the displacement of the released structure can be easily seen from the geometry of Fig. 2-4c.

$$\{D\} = \begin{Bmatrix} -0.005l \\ -0.005 \end{Bmatrix} \quad \text{and} \quad \{\Delta\} = \{0\}$$

The negative signs show that the direction of the displacements is opposite to the directions chosen in Fig. 2-4b.

In case (b), the displacements correspond to the redundants. Thus

$$\{\Delta\} = \begin{Bmatrix} -0.010l \\ +0.004 \end{Bmatrix} \quad \text{and} \quad \{D\} = \{0\}$$

Substituting in Eq. 2-10, we obtain

$$[F] = \frac{12EI}{7l^3} \begin{bmatrix} 8 & -3l \\ -3l & 2l^2 \end{bmatrix} \begin{bmatrix} 0.005l & -0.010l \\ 0.005 & 0.004 \end{bmatrix} = \frac{12EI}{7l^3} \begin{bmatrix} 0.025l & -0.092l \\ -0.005l^2 & 0.038l^2 \end{bmatrix}$$

The elements in the first and second column of $[F]$ are the redundants in case (a) and (b), respectively. These redundants are applied to the released structure and the reactions at A and C are calculated from statics. The results are shown in Figs. 2-4d and e.

Example 2-3 Analyze the continuous beam of Fig. 2-5a for (a) a uniformly distributed load of intensity q on all spans; (b) a unit downward movement of support A; (c) a unit downward movement of support B. The beam has a constant flexural rigidity EI.

A statically determinate released structure can be obtained by intro-ducing a hinge over each interior support, that is, by the removal of two equal and opposite forces (moments) acting on either side of the support, so that the released structure is a series of simply supported beams (Fig. 2-5b). The released bending moments are sometimes called *connecting moments*.

The inconsistencies in displacements of the released structure are the relative rotations of the three pairs of adjacent beam ends, that is the angles between the tangent to the axis of one deflected beam and the corresponding tangent to an adjacent beam (Fig. 2-5c). The positive directions of the redun-dants and therefore of the displacements are shown in Fig. 2-5b. These are chosen so that a connecting moment is positive if it produces tension in the fibers at the bottom face of a beam.

The superposition involved in step 5 can be done by inspection or by relatively simple calculations. Thus the matrices $[A]$, $[A_s]$ and $[A_u]$ are not required. Step 1 is done in Fig. 2-5b, while steps 2, 3 and 4 are done below.

The solution of the three cases can be obtained by Eq. 2-10. In case (a), the errors in geometry of the released structure can be obtained from Appen-dix B. In cases (b) and (c), the movements of the supports do not correspond to the redundants. Therefore, the relative rotations at B, C, and D resulting from the downward movement of the supports of the released structure are included in the matrix $[D]$, and matrix $[\Delta]$ is a null matrix. These rotations are easily determined from the geometry of the released structure in Figs.

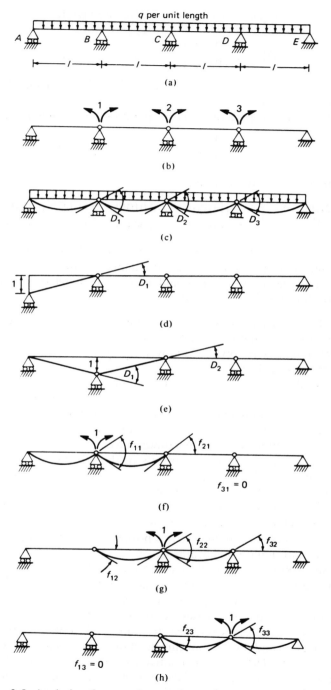

Fig. 2-5. Analysis of a continuous beam by the force method (Example 2-3).

2-5d and e. The inconsistency in the displacements for all three cases can be included in a matrix with one column per case:

$$[\Delta - D] = -[D] = - \begin{bmatrix} \dfrac{ql^3}{12EI} & \dfrac{1}{l} & -\dfrac{2}{l} \\[2ex] \dfrac{ql^3}{12EI} & 0 & \dfrac{1}{l} \\[2ex] \dfrac{ql^3}{12EI} & 0 & 0 \end{bmatrix}$$

(All the relative rotations are positive except D_1 in case (c), see Fig. 2-5e.)

Now the flexibility influence coefficients of the released structure are the relative rotations of the beam ends caused by the separate application of a pair of equal and opposite unit moments at each of the redundants at B, C, and D. The rotations at the ends of a simple beam due to a couple applied at one end can be found in Appendix B. The displacements due to the unit value of each of the redundants, shown in Figs. 2-5f, g, and h respectively, form the three columns of the flexibility matrix. It can be seen from these figures that all the relative rotations are positive. The flexibility matrix and its inverse are thus

$$[f] = \frac{l}{6EI} \begin{bmatrix} 4 & 1 & 0 \\ 1 & 4 & 1 \\ 0 & 1 & 4 \end{bmatrix} \qquad [f]^{-1} = \frac{3EI}{28l} \begin{bmatrix} 15 & -4 & 1 \\ -4 & 16 & -4 \\ 1 & -4 & 15 \end{bmatrix}$$

Substituting in Eq. 2-10,

$$[F] = -\frac{3EI}{28l} \begin{bmatrix} 15 & -4 & 1 \\ -4 & 16 & -4 \\ 1 & -4 & 15 \end{bmatrix} \begin{bmatrix} \dfrac{ql^3}{12EI} & \dfrac{1}{l} & -\dfrac{2}{l} \\[2ex] \dfrac{ql^3}{12EI} & 0 & \dfrac{1}{l} \\[2ex] \dfrac{ql^3}{12EI} & 0 & 0 \end{bmatrix}$$

whence

$$[F] = \begin{bmatrix} -\dfrac{3}{28}ql^2 & -\dfrac{45}{28}\dfrac{EI}{l^2} & \dfrac{51}{14}\dfrac{EI}{l^2} \\[2ex] -\dfrac{ql^2}{14} & \dfrac{3}{7}\dfrac{EI}{l^2} & -\dfrac{18}{7}\dfrac{EI}{l^2} \\[2ex] -\dfrac{3}{28}ql^2 & -\dfrac{3}{28}\dfrac{EI}{l^2} & \dfrac{9}{14}\dfrac{EI}{l^2} \end{bmatrix}$$

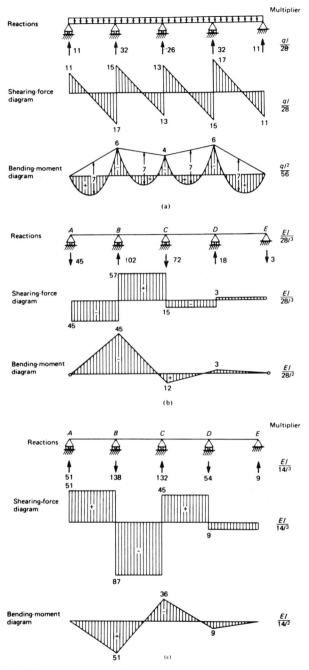

Fig. 2-6. Bending-moment and shearing-force diagrams for the continuous beam of Example 2-3. (a) Uniform load q per unit length. (b) Displacement of support A. (c) Displacement of support B.

The columns in this matrix correspond to cases (a), (b), and (c) and the three elements in each column are the bending moments at B, C, and D. Having found the values of the redundants, the bending moment and shearing force at any section can be obtained by simple statics. The reactions, and the bending-moment[1] and shearing-force diagrams for the three cases, are shown in Fig. 2-6.

2-6 EQUATION OF THREE MOMENTS

The analysis of continuous beams subjected to transverse loading and to support settlement is very common in structural design, and it is useful to simplify the general force method approach to this particular

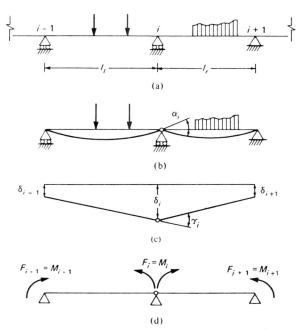

Fig. 2-7. Analysis of a continuous beam by the equation of three moments. (a) Continuous beam. (b) Deflection of released structure due to transverse loading. (c) Deflection of released structure due to support settlement. (d) Positive direction of redundants.

case. The resulting expression is known as the equation of three moments. We may note that historically this equation, developed by Clapeyron, precedes the matrix formulation of the force method.

Figure 2-7a represents two typical interior spans of a continuous beam.

---

[1]Throughout this text an ordinate representing a bending moment is plotted on the side of the beam on which it produces tension.

Let the spans to the left and to the right of an interior support i have respectively lengths l_l and l_r and flexural rigidities EI_l and EI_r, assumed to be constant within each span. The supports $i - 1$, i, and $i + 1$ are assumed to have settled in the direction of the applied loading by δ_{i-1}, δ_i, and δ_{i+1}, respectively.

A statically determinate released structure can be obtained by introducing a hinge in the beam at each support (Fig. 2-7b), so that each span deforms as a simple beam, as in Example 2-3. The same sign convention as in that example will be used, as shown in Fig. 2-7d.

As before, the inconsistencies in the displacements of the released structure are the relative rotations of adjacent beam ends, that is, with reference to Figs. 2-7b and c,

$$D_i = \alpha_i - \gamma_i \tag{2-11}$$

We should note that D_i is due both to the transverse loading and to the settlement of supports.

In an actual continuous beam, the redundants $\{F\}$ are the *connecting moments* $\{M\}$, which must be of such magnitude that the angular discontinuities vanish. A superposition equation to satisfy the continuity condition at i can be written in the form:

$$D_i + f_{i, i-1}F_{i-1} + f_{ii}F_i + f_{i, i+1}F_{i+1} = 0 \tag{2-12}$$

where the terms f represent the flexibility coefficients of the released structure.

At this stage we should consider the behavior of a beam hinged at each end and subjected to a unit moment at one end. Figure 2-8a shows such a

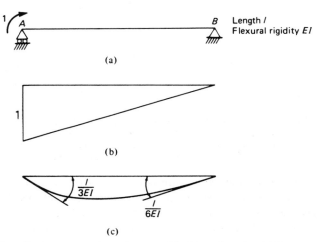

(a)

(b)

(c)

Fig. 2-8. Angular displacements in a beam with hinged ends. (a) Beam. (b) Bending-moment diagram. (c) Deflection.

beam and Fig. 2-8b the resulting bending-moment diagram. Figure 2-8c gives the deflected shape: the angular displacements are $l/3EI$ and $l/6EI$ at A and B, respectively (see Appendix B). Applying these results to the released structure of Fig. 2-7d, it can be easily seen that the flexibility coefficients are

$$f_{i,i-1} = \frac{l_l}{6EI_l} \qquad f_{ii} = \frac{l_l}{3EI_l} + \frac{l_r}{3EI_r} \qquad f_{i,i+1} = \frac{l_r}{6EI_r}$$

With these values, and using the fact that $\{F\} = \{M\}$, Eq. 2-12 yields

$$M_{i-1}\frac{l_i}{EI_l} + 2M_i\left[\frac{l_l}{EI_l} + \frac{l_r}{EI_r}\right] + M_{i+1}\frac{l_r}{EI_r} = -6D_i \qquad (2\text{-}13)$$

where l and r refer to the spans respectively to the left and right of i.

This is known as the equation of three moments. It relates the angular discontinuity at a support i to the connecting moments at this support and at a support on each side of i. The equation is valid only for continuous beams of constant flexural rigidity within each span.

For a continuous beam of constant flexural rigidity EI throughout, the equation of three moments simplifies to

$$M_{i-1}l_l + 2M_i(l_l + l_r) + M_{i+1}l_r = -6EID_i \qquad (2\text{-}14)$$

Similar equations can be written for each support at which the bending moment is not known, thus forming a system of simultaneous equations, the solution of which gives the unknown moments. These equations can, in fact, be written in the general matrix form of Eq. 2-2. We can note that the particular released structure chosen has the advantage that each row of the flexibility matrix has only three nonzero elements.

The displacement D_i in the equation of three moments can be calculated from Eq. 2-11, with the angle γ_i determined from the geometry of Fig. 2-7c

$$\gamma_i = \frac{\delta_i - \delta_{i-1}}{l_l} + \frac{\delta_i - \delta_{i+1}}{l_r} \qquad (2\text{-}15)$$

and the angle α_i calculated by the method of elastic weights (see Sec. 9-5)

$$\alpha_i = r_{il} + r_{ir} \qquad (2\text{-}16)$$

where r_{il} and r_{ir} are the reactions of the beams to the left and right of the support i loaded by the simple-beam bending moment due to lateral loading divided by the appropriate EI. For many of the practical cases the value of α_i can also be determined directly from Appendix B.

Example 2-4   Obtain the bending-moment diagrams for the beam in Fig. 2-9a due to: (a) the given vertical loads; (b) vertical settlement of $b/100$ and $b/200$ at supports B and C respectively. The beam has a constant flexural rigidity EI.

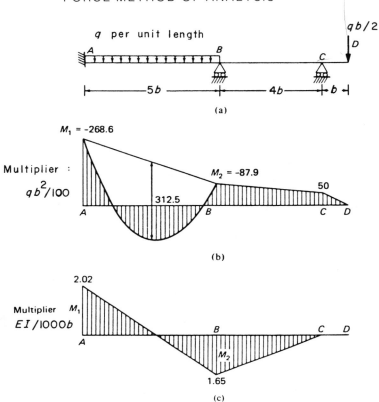

Fig. 2-9. Continuous beam with sinking supports considered in Example 2-4. (a) Continuous beam properties and loading. (b) Bending-moment diagram due to vertical loading. (c) Bending-moment diagram due to settlement of $b/100$ at support B and $b/200$ at support C.

The equation of three moments has to be applied at supports A and B to find the two unknown bending moments M_1 at A and M_2 at B, while the bending moment at C is known from simple statics. When applying the equation at the built-in end A, the beam may be considered to extend over an imaginary span to the left of A, either of infinitely small length or of infinitely large flexural rigidity.

We find $\{D\}$ from Eq. 2-11, noting that for case (a), $\{\gamma\} = \{0\}$, while for case (b), $\{\alpha\} = \{0\}$. Hence, with the aid of Appendix B, it can be readily shown that for case (a)

$$\{D\}_a = \begin{Bmatrix} 5.208 \dfrac{qb^3}{EI} \\[2ex] 5.208 \dfrac{qb^3}{EI} \end{Bmatrix}$$

For case (b), calculating γ from Eq. 2-15, we obtain

$$\{D\}_b = \begin{Bmatrix} +0.00200 \\ -0.00325 \end{Bmatrix}$$

Using Eq. 2-13, we find for case (a)

$$\frac{b}{EI}\left(10M_1 + 5M_2\right) = -6 \times 5.208 \frac{qb^3}{EI}$$

$$\frac{b}{EI}\left(5M_1 + 18M_2 - 4\frac{qb^2}{2}\right) = -6 \times 5.208 \frac{qb^3}{EI}$$

and for case (b)

$$\frac{b}{EI}\left(10M_1 + 5M_2\right) = -6 \times 0.00200$$

$$\frac{b}{EI}\left(5M_1 + 18M_2\right) = +6 \times 0.00325$$

The two systems of equations can be combined in a matrix equation:

$$\frac{b}{EI}\begin{bmatrix} 10 & 5 \\ 5 & 18 \end{bmatrix}[M] = \begin{bmatrix} -31.25\dfrac{qb^3}{EI} & -0.0120 \\ -29.25\dfrac{qb^3}{EI} & 0.0195 \end{bmatrix}$$

The same equation can be obtained following the more general procedure of the force method, with the released structure formed by introducing hinges at A and B. The square matrix on the left-hand side of the equation represents then the flexibility matrix of the released structure.

Solving for $[M]$, we find

$$[M] = \begin{bmatrix} -2.686qb^2 & -0.00202\dfrac{EI}{b} \\ -0.879qb^2 & 0.00165\dfrac{EI}{b} \end{bmatrix}$$

The bending-moment diagrams for the two cases are plotted in Figs. 2-9b and c.

2-7 MOVING LOADS ON CONTINUOUS BEAMS AND FRAMES

The live load on continuous beams and frames is often represented in design by a uniformly distributed load which may occupy any part of the structure so as to produce the maximum value of an internal force at a section or the maximum reaction at a support. We shall now discuss which parts of a continuous beam should be covered by the live load to produce these maxima.

Figures 2-10a and b show the deflected shapes, reactions and bending-moment and shearing-force diagrams for a continuous beam due to a uniform live load covering one span only. It can be seen that the deflection is largest in the loaded span and reverses sign, with much smaller values, in adjacent spans. The two reactions at either end of the loaded span are upward; the reactions on either side of the loaded span are reversed in direction and have much smaller magnitude. The values given in the figures are for the case of equal spans l and load q per unit length; EI is constant.

In a loaded span, the bending-moment is positive in the central part and negative at the supports. In the adjacent spans, the bending moment is negative over the major part of the length. The points of inflection on the deflected shapes correspond to the points O_1 and O_2 where the bending moment is zero. These points are closer to the supports C and D in case (a) and (b), respectively, than one-third of their respective spans. This is so because a couple applied at one end of a beam produces a straight-line bending-moment diagram which reverses sign at one-third of the span from the far end when that end is totally fixed. When the far end is hinged (or simply supported) the bending moment has the same sign over the whole span. The behavior of the unloaded spans BC and CD in Fig. 2-10a and of span CD in Fig. 2-10b lies between these two extremes.

The maximum absolute value of shear occurs at a section near the supports of the loaded span. Smaller values of shear of constant magnitude occur in the unloaded spans, with sign reversals as indicated.

From Figs. 2-10a and b, and perhaps two more figures corresponding to the uniform load on each of the remaining two spans, we can decide which load patterns produce the maximum values of the deflections, reactions or internal forces at any section. Figure 2-10c shows typical load patterns which produce maximum values of various actions.

It can be seen from this figure that maximum deflection and maximum positive bending moment at a section near the middle of a span occur when the load covers this span as well as alternate spans on either side. The maximum negative bending moment, the maximum positive reaction and the maximum absolute value of shear near a support occur when the load covers the two adjacent spans and alternate spans thereafter. The loading cases 3, 4 and 5 in Fig. 2-10c refer to sections just to the left or right of a support, the subscripts l and r denoting left and right respectively.

Partial loading of a span may produce maximum shear at a section as in case 6 in Fig. 2-10c. Also, partial loading may produce maximum bending moments at sections near the interior supports (closer than one-third of the span) but not at the supports. However, in practice, partial span loading is seldom considered when the maximum bending-moment values are calculated.

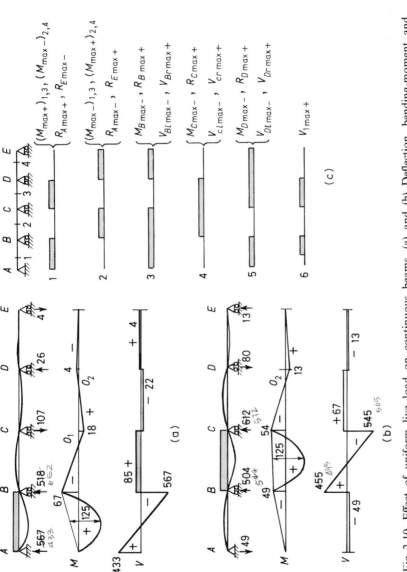

Fig. 2-10. Effect of uniform live load on continuous beams. (a) and (b) Deflection, bending-moment and shearing-force diagrams due to load p per unit length covering one span. The values given are for equal span l; the multipliers are $10^{-3}\,pl$ for reactions and shears, and $10^{-3}\,pl^2$ for moments. (c) Load patterns to produce maximum actions.

The effect of live load patterns needs to be combined with the effect of the dead load (permanent load) in order to obtain the maximum actions to be used in design. For example, if the beam in Fig. 2-10c is designed for uniform dead and live loads of intensities q and p, respectively, we can obtain the maximum bending-moment diagram by considering q on all spans combined with p according to the loading cases 1 to 5 in Fig. 2-10c. The maximum bending-moment diagram due to dead and live loads combined is shown in Fig. 2-11 for a continuous beam of four equal spans l, with $q = p$. The diagram is obtained by plotting on one graph the bending moment due to q combined with p in cases 1 to 5 in Fig. 2-10c and using for any part of the beam the curve with the highest absolute values. Additional load cases which produce a maximum positive bending moment due to p may need to be considered when p is large compared with q. For example, $M_{B\,max\,+}$ occurs when p covers CD only, and $M_{C\,max\,+}$ occurs when AB and DE are covered.

Live load p per unit length covering any part

Dead load on all spans, q per unit length ; $q = p$

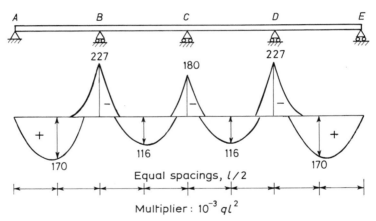

Fig. 2-11. Maximum bending-moment diagram for a continuous beam due to dead and live loads.

In practice, the live load may be ignored on spans far away from the section at which the maximum action is required. For example, the maximum negative moment and the maximum positive reaction at support B may be assumed to occur when the live load covers only the adjacent spans AB and BC, without loading on DE (see case 3, Fig. 2-10c). This may be acceptable because of the small effect of the ignored load, or on the grounds of a low probability of occurrence of the alternate load pattern with the full value of live load.

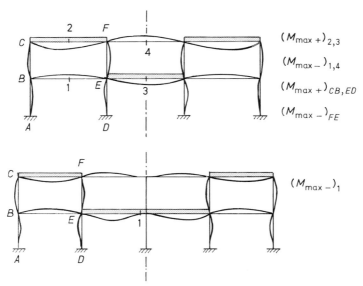

Fig. 2-12. Examples of live load patterns to produce maximum bending moments in beams or end-moments in columns of plane frames.

The alternate load patterns discussed above are typical for continuous beams and are frequently used in structural design. Similar patterns for continuous frames are shown in Fig. 2-12. The two loading cases represented produce maximum positive and negative values of the bending moments in the horizontal beams or the end-moments in the columns. As followed throughout this book, the bending moment in beams is considered positive when it produces tension at the bottom face; a clockwise member end-moment is positive.

A sketch of the deflected shape may help to determine whether or not a span should be considered loaded so as to produce a maximum effect. A span should be loaded if this results in accentuating the deflected shape in all members. However, in some cases the deflected shape is not simple to predict in all parts of the frame, particularly when sidesway occurs. Use of influence lines (see Chapters 13 and 14) helps in determining the load position for maximum effect, particularly when the live load is composed of concentrated loads.

2-8 GENERAL

The force (or flexibility) method of analysis can be applied to any structure subjected to loading or environmental effects. The solution of the compatibility equations formed directly yields the unknown forces. The number of equations involved is equal to the number of redundants. The force method is not well suited to computer use in the case of highly redundant structures, as will be discussed further in Sec. 4-3.

PROBLEMS

The following are problems on the application of the force method of analysis. At this stage, use can be made of Appendix B to determine the displacements required in the analysis, with attention directed to the procedure of the force method rather than to the methods of computation of displacements. These will be treated in subsequent chapters. Additional problems on the application of the force method can be found at the end of Chapter 6, which requires calculation of displacements by method of virtual work.

2-1   Write the flexibility matrix corresponding to coordinates 1 and 2 for the structures shown below.

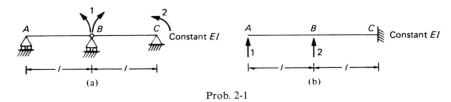

Prob. 2-1

2-2   Use the flexibility matrices derived in Prob. 2-1 to find two sets of redundant forces in two alternative solutions for the continuous beam of Example 2-1.

2-3   Use the force method to find the bending moment at the intermediate supports of the continuous beam shown in the figure.

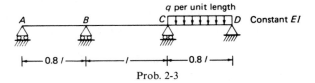

Prob. 2-3

2-4   Obtain the bending-moment diagram for the beam of Prob. 2-3 on the assumption that support B settles vertically a distance $l/1200$. (No distributed load acts in this case.)

2-5   Use the force method to find the bending moments at the supports of a continuous beam on elastic (spring) supports. The beam has a constant flexural rigidity EI, and the stiffness of the elastic supports is $K = 20EI/l^3$.

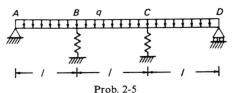

Prob. 2-5

2-6   Use the force method to find the forces in the three springs A, B, and C in the system shown. The beams DE and FG have a constant flexural rigidity EI, and the springs have the same stiffness, $100\ EI/l^3$. What are the values of the vertical reactions at D, E, F and G?

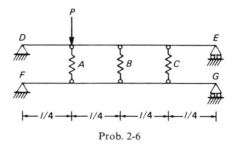

Prob. 2-6

2-7   Find the forces in the springs A and B of the system shown. Assume that the mast EF is rigid, the bars GC and HD have a constant flexural rigidity EI, and the stiffness of the springs A and B is EI/l^3. Note that in the displaced position of the mast EF, the vertical force P tends to rotate the mast about the hinge E. [*Hint*: In this problem the displacement of the mast alters the forces acting on it; therefore, the principle of superposition cannot be applied (see Sec. 1-1). However, if the vertical force P is assumed to be always acting, the superposition of the effects of two transverse loadings on the mast can be made. This idea is further discussed in Chapter 15.] For the analysis by the force method, the flexibility is determined for a released structure which has the vertical load P acting, and the inconsistency in displacement of this structure caused by the load Q is also calculated with the load P present.

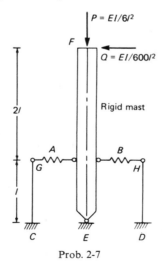

Prob. 2-7

2-8   A steel beam AB is supported by two steel cables at C and D. Using the force method, find the tension in the cables and the bending moment at D due to a load $P = 5$ k and a drop of temperature of $40°$ Fahrenheit in the two cables.

For the beam $I = 40$ in.^{4}, for the cables $a = 0.15$ in.^{2}; the modulus of elasticity for both is $E = 30 \times 10^3$ ksi, and the coefficient of thermal expansion for steel is 6.5×10^{-6} per degree Fahrenheit.

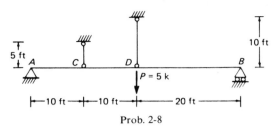

Prob. 2-8

2-9   Using the equation of three moments, find the bending-moment diagram for the beam shown.

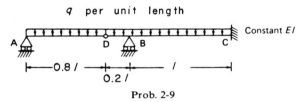

Prob. 2-9

2-10   Using the equation of three moments, obtain the bending-moment and shearing-force diagrams for the continuous beam shown.

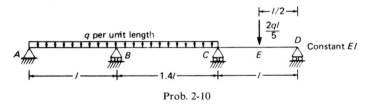

Prob. 2-10

2-11   For the beam in Fig. 2-4a, find the reactions at the supports and the bending-moment diagram due to a rise in temperature varying linearly over the beam depth h. The temperature rise in degrees at top and bottom fibers is T_t and T_b, respectively. The coefficient of thermal expansion is α per degree.

2-12   For the beam shown, obtain the bending-moment and shearing-force diagrams.

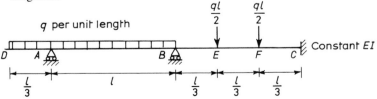

Prob. 2-12

2-13   The reinforced concrete bridge *ABC* shown in the figure is constructed in two stages. In stage 1, part *AD* is cast and its forms are removed. In stage 2, part *DC* is cast and its forms are removed; a monolithic continuous beam is obtained. Obtain the bending-moment diagrams and the reactions due to the structure self-weight, q per unit length, immediately at the end of stages 1 and 2.

Hint: At the end of stage 1, we have a simple beam with an overhang, carrying a uniform load. In stage 2, we added a load q per unit length over *DC* in a continuous beam. Superposition gives the desired answers for the end of stage 2. Creep of concrete tends gradually to make the structure behave as if it were constructed in one stage (see the references mentioned in footnote 3 of Chapter 4).

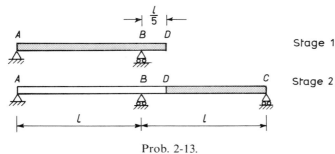

Prob. 2-13.

2-14   A continuous beam of four equal spans l and constant EI is subjected to a uniform dead load q per unit length over the whole length, combined with a uniform live load of intensity $p = q$. Determine the maximum bending-moment values over the supports and at the centers of spans. The answers to this problem are given in Fig. 2-11.

2-15   A continuous beam of three equal spans l and constant EI is subjected to a uniform dead load q per unit length over the whole length combined with a various uniform live load of intensity $p = q$. Determine:
(a) The maximum bending moments at the interior supports and mid-spans.
(b) Diagram of maximum bending moment.
(c) Maximum reaction at an interior support.
(d) Absolute maximum shearing force and its location.

2-16   For the beam of Prob. 2-12, find the reactions and the bending-moment and the shearing-force diagrams due to a unit downward settlement of support *B*. (The main answers for this problem are included in Table E-3, Appendix E. Note that the presence of the overhang *DA* has no effect.)

CHAPTER 3

Displacement method
of analysis

3-1 INTRODUCTION

The mathematical formulation of the displacement and force methods is similar, but from the point of view of economy of effort one or the other method may be preferable. This will be considered in detail in Sec. 4-3.

The displacement method can be applied to statically determinate or indeterminate structures, but it is more useful in the latter, particularly when the degree of statical indeterminacy is high.

3-2 DESCRIPTION OF METHOD

1. First of all, the degree of kinematic indeterminacy has to be found. A coordinate system is then established to identify the location and direction of the joint displacements. Restraining forces equal in number to the degree of kinematic indeterminacy are introduced at the coordinates to prevent the displacement of the joints. In some cases, the number of restraints introduced may be smaller than the degree of kinematic indeterminacy, provided that the analysis of the resulting structure is a standard one and is therefore known. (See remarks following Example 3-2).

We should note that, unlike the force method, the above procedure requires no choice to be made with respect to the restraining forces. This fact favors the use of the displacement method in general computer programs for the analysis of a structure.

2. The restraining forces are now determined as a sum of the fixed-end forces for the members meeting at a joint. For most practical cases, the fixed-end forces can be calculated with the aid of standard tables (Appendixes C and D).

We should remember that the restraining forces are those required to prevent the displacement at the coordinates due to all effects, such as external loads, temperature variation or prestrain. These effects may be considered separately or may be combined.

If the analysis is to be performed for the effect of movement of one of the joints in the structure, for example the settlement of a support, the forces at the coordinates required to hold the joint in the displaced position are included in the restraining forces.

The internal forces in the members are also determined at the required locations with the joints in the restrained position.

3. The structure is now assumed to be deformed in such a way that a displacement at one of the coordinates equals unity and all the other displacements are zero, and the forces required to hold the structure in this configuration are determined. These forces are applied at the coordinates representing the degrees of freedom. The internal forces at the required locations corresponding to this configuration are determined. The process is repeated for a unit value of displacement at each of the coordinates separately.

4. The values of the displacements necessary to eliminate the restraining forces introduced in (2) are determined. This requires superposition equations in which the effects of separate displacements on the restraining forces are added.

5. Finally, the forces on the original structure are obtained by adding the forces on the restrained structure to the forces caused by the joint displacements determined in (4).

The use of the above procedure is best explained with reference to some specific cases.

Example 3-1 The plane truss in Fig. 3-1a consists of m pin-jointed members meeting at joint A. Find the forces in the members due to the combined effect of (1) an external load P applied at A; (2) an elongation Δ_k of the kth bar caused by a rise in temperature of this bar alone.

The degree of kinematic indeterminacy of the structure is 2, because displacement can occur only at joint A, which can undergo a translation with components D_1 and D_2 in the x and y directions. The positive directions for the displacement components, as well as for the restraining forces, are arbitrarily chosen, as are indicated in Fig. 3-1b.

In case (1), the joint displacements are prevented by introducing at A a force equal and opposite to P. This force has components F_{11} and F_{21} in directions 1 and 2:

$$F_{11} = -P \cos \alpha \qquad \text{and} \qquad F_{21} = -P \sin \alpha$$

As always, the second subscript indicates the cause: here, the second subscript of the restraining force F represents case (1). The negative sign in the above equations is due to the fact that the forces act in directions opposite to the chosen coordinates in Fig. 3-1b.

Let E_i, l_i, and a_i be respectively the modulus of elasticity, length and cross-sectional area, of any bar i, and let θ_i be the angle which it makes with the x axis. Considering case (2), the elongation of the kth member, Δ_k, can be prevented by a force at A which causes a contraction of the same magnitude as the elongation. Thus, the compression in the member must be

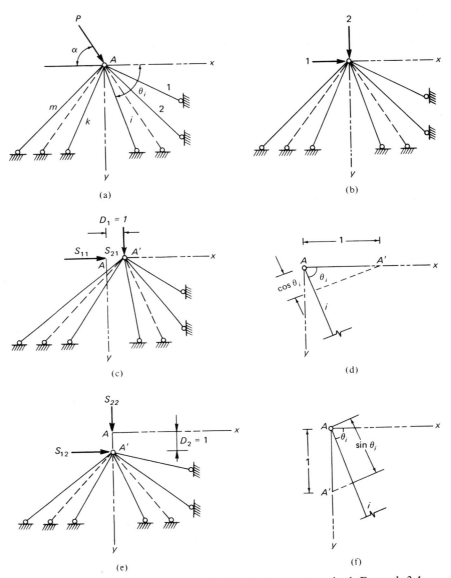

Fig. 3-1. Analysis of a plane truss by the displacement method–Example 3-1. (a) Plane truss. (b) Coordinate system. (c) $D_1 = 1$ and $D_2 = 0$. (d) Change in length in the ith member due to $D_1 = 1$. (e) $D_1 = 0$ and $D_2 = 1$. (f) Change in length in the ith member due to $D_2 = 1$.

$(a_k E_k / l_k) \Delta_k$. The components in directions 1 and 2 of the external force required to produce this compression are

$$F_{12} = \frac{a_k E_k}{l_k} \Delta_k \cos \theta_k \quad \text{and} \quad F_{22} = \frac{a_k E_k}{l_k} \Delta_k \sin \theta_k$$

where the second subscript refers to the restraint in case (2).

The total restraining forces are

$$F_1 = F_{11} + F_{12}$$

and

$$F_2 = F_{21} + F_{22}$$

We can see that when the displacements are restrained, in case (1) there are no internal forces in any of the bars. In case (2), there are no forces, except a compression of $(a_k E_k / l_k) \Delta_k$ in bar k. Denoting by $\{A_r\}$ the axial forces in the bars in the restrained condition, we have

$$A_{r1} = 0$$

$$A_{r2} = 0$$

$$\ldots$$

$$A_{rk} = -\frac{a_k E_k}{l_k} \Delta_k$$

$$\ldots$$

$$A_{rm} = 0$$

Figure 3-1c shows the forces required to hold the structure in a deformed position such that $D_1 = 1$ and $D_2 = 0$. Now from Fig. 3-1d, a unit horizontal movement of A causes a shortening of any bar i by a distance $\cos \theta_i$ and produces a compressive force of $(a_i E_i / l_i) \cos \theta_i$. Therefore, to hold joint A in the displaced position, forces $(a_i E_i / l_i) \cos^2 \theta_i$ and $(a_i E_i / l_i) \cos \theta_i \sin \theta_i$ have to be applied in directions 1 and 2, respectively. The forces required to hold all the bars in the displaced position are

$$S_{11} = \sum_{i=1}^{m} \frac{a_i E_i}{l_i} \cos^2 \theta_i$$

and

$$S_{21} = \sum_{i=1}^{m} \frac{a_i E_i}{l_i} \cos \theta_i \sin \theta_i$$

By a similar argument, the forces required to hold the joint A in the displaced position such that $D_1 = 0$ and $D_2 = 1$ (Figs. 3-1e and f) are

$$S_{12} = \sum_{i=1}^{m} \frac{a_i E_i}{l_i} \sin \theta_i \cos \theta_i$$

and

$$S_{22} = \sum_{i=1}^{m} \frac{a_i E_i}{l_i} \sin^2 \theta_i$$

The first subscript of S in the above equations indicates the coordinate of the restraining force, and the second subscript the component of the displacement which has a unit value.

In the actual structure, joint A undergoes translations D_1 and D_2 and there are no restraining forces. Therefore, the superposition of the fictitious restraints and of the effects of the actual displacements must be equal to zero. Thus, we obtain *statical relations* which express the fact that the restraining forces vanish when the displacements D_1 and D_2 take place. These statical relations can be expressed as

and
$$\left. \begin{array}{l} F_1 + S_{11}D_1 + S_{12}D_2 = 0 \\ F_2 + S_{21}D_1 + S_{22}D_2 = 0 \end{array} \right\} \tag{3-1}$$

3-3 STIFFNESS MATRIX

The statical relations of Eq. 3-1 can be written in matrix form

$$\{F\} + [S]\{D\} = \{0\}$$

or
$$[S]\{D\} = \{-F\} \tag{3-2}$$

(This equation may be compared with Eq. 2-2 for the geometry relations in the force method of analysis.)

The column vector $\{F\}$ depends on the loading on the structure. The elements of the matrix $[S]$ are forces corresponding to unit values of displacements. Therefore, $[S]$ depends on the properties of the structure, and represents its stiffness. For this reason, $[S]$ is called the *stiffness matrix* and its elements are called *stiffness coefficients*. The elements of the vector $\{D\}$ are the unknown displacements and can be determined by solving Eq. 3-2, that is,

$$\{D\} = [S]^{-1}\{-F\} \tag{3-3}$$

In a general case, if the number of restraints introduced in the structure is n, the order of the matrices $\{D\}$, $[S]$ and $\{F\}$ is $n \times 1$, $n \times n$, and $n \times 1$, respectively. The stiffness matrix $[S]$ is thus a square symmetrical matrix. This can be seen in the above example by comparing the equations for S_{21} and S_{12} but a formal proof will be given in Sec. 4-6.

The final force in any member i can be determined by superposition of the restrained condition and of the effect of the joint displacements:

$$A_i = A_{ri} + (A_{ui1}D_1 + A_{ui2}D_2 + \cdots + A_{uin}D_n) \tag{3-4}$$

The *superposition* equation for all the members in matrix form is

$$\{A\}_{m \times 1} = \{A_r\}_{m \times 1} + [A_u]_{m \times n}\{D\}_{n \times 1}$$

where the elements of $\{A\}$ are the final forces in the bars, the elements of $\{A_r\}$ are the bar forces in the restrained condition, and the elements of $[A_u]$ are the bar forces corresponding to unit displacements. Specifically the elements of column j of $[A_u]$ are the forces in the members corresponding to a displacement $D_j = 1$ while all the other displacements are zero.

Since the above equation will be used in the analysis of a variety of structures, it is useful to write it in a general form

$$\{A\} = \{A_r\} + [A_u]\{D\} \tag{3-5}$$

where the elements of $\{A\}$ are the final forces in the members, the elements of $\{A_r\}$ are the forces in members in the restrained condition, and the elements of $[A_u]$ are the forces in members corresponding to unit displacements.

In the truss of Example 3-1, with axial <u>tension in</u> a member considered positive, it can be easily seen that

$$[A_u] = \begin{bmatrix} -\dfrac{a_1 E_1}{l_1}\cos\theta_1 & -\dfrac{a_1 E_1}{l_1}\sin\theta_1 \\[2ex] -\dfrac{a_2 E_2}{l_2}\cos\theta_2 & -\dfrac{a_2 E_2}{l_2}\sin\theta_2 \\[1ex] \cdots & \cdots \\[1ex] -\dfrac{a_m E_m}{l_m}\cos\theta_m & -\dfrac{a_m E_m}{l_m}\sin\theta_m \end{bmatrix}$$

In a frame with rigid joints, we may want to find the stress resultants in any section or the reactions at the supports. For this reason, we consider the notation A in the general Eq. 3-5 to represent any action, which may be shearing force, bending moment, twisting moment, or axial force at a section or a reaction at a support.

Example 3-2 The plane frame in Fig. 3-2a consists of rigidly connected members of constant flexural rigidity EI. Obtain the bending-moment diagram for the frame due to concentrated loads P at E and F and a couple Pl at joint B. The change in length of the members can be neglected.

The degree of kinematic indeterminacy is three because there are three possible joint displacements, as shown in Fig. 3-2b, which shows also the chosen coordinate system. The restraining forces, which are equal to the sum of the end-forces at the joints, are calculated with the aid of Appendix C. As always, they are considered positive when their direction accords with that of the coordinates.

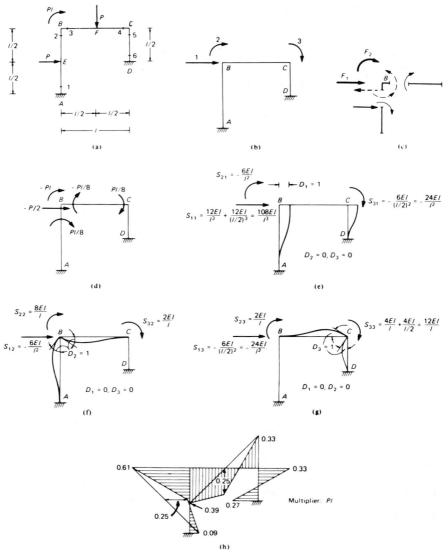

Fig. 3-2. Plane frame analyzed in Example 3-2.

To illustrate the relation between the end-forces and the restraining forces, joint B is separated from the members connected to it in Fig. 3-2c. The forces acting on the end of the members in the direction of the co-ordinate system are indicated by full-line arrows. Equal and opposite forces act on the joint, and these are shown by dotted-line arrows. For equilibrium of the joint, forces F_1 and F_2 should be applied in a direction opposite to the dotted arrows. Therefore, to obtain the restraining forces, it is sufficient to add the end-forces at each joint as indicated in Fig. 3-2d, and it is not necessary to consider the forces as in Fig. 3-2c.

The external applied couple acting at B requires an equal and opposite restraining force. Therefore

$$\{F\} = \left\{ \begin{array}{c} -\dfrac{P}{2} \\[2mm] \left(\dfrac{Pl}{8} - \dfrac{Pl}{8} - Pl\right) \\[2mm] \dfrac{Pl}{8} \end{array} \right\} = P \left\{ \begin{array}{c} -0.5 \\[2mm] -l \\[2mm] 0.125l \end{array} \right\} \tag{a}$$

To draw the bending-moment diagram the values of the moments at the ends of all the members are required, it being assumed that an end-moment is positive if it acts in a clockwise direction. The member ends 1, 2, ..., 6 are identified in Fig. 3-2a. Thus the values of the six end-moments corresponding to the restrained condition are

$$\{A_r\} = \frac{Pl}{8}\{-1, 1, -1, 1, 0, 0\}$$

Now the elements of the stiffness matrix are the forces necessary at the location in direction of the coordinates to hold the structure in the deformed shape illustrated in Figs. 3-2e, f, and g. These forces are equal to the sum of the end-forces, which are taken from Appendix D. We should note that the translation of joint B must be accompanied by an equal translation of joint C in order that the length BC remains unchanged. The stiffness matrix is

$$[S] = \frac{EI}{l} \begin{bmatrix} \dfrac{108}{l^2} & -\dfrac{6}{l} & -\dfrac{24}{l} \\[3mm] -\dfrac{6}{l} & 8 & 2 \\[3mm] -\dfrac{24}{l} & 2 & 12 \end{bmatrix} \tag{b}$$

To write the matrix of end-moments due to the unit displacements, we put the values at the beam ends 1, 2, ..., 6 in the first, second, and third column respectively for the displacements shown in Figs. 3-2e, f, and g. Thus

$$
[A_u] = \frac{EI}{l}
\begin{bmatrix}
-\dfrac{6}{l} & 2 & 0 \\[6pt]
-\dfrac{6}{l} & 4 & 0 \\[6pt]
0 & 4 & 2 \\[6pt]
0 & 2 & 4 \\[6pt]
-\dfrac{24}{l} & 0 & 8 \\[6pt]
-\dfrac{24}{l} & 0 & 4
\end{bmatrix}
\qquad\text{(c)}
$$

Append D.

Substituting Eqs. (a) and (b) into Eq. 3-2 and solving for $\{D\}$, we obtain

$$
\{D\} = \frac{Pl^2}{EI}
\begin{Bmatrix}
0.0087l \\
0.1355 \\
-0.0156
\end{Bmatrix}
\qquad\text{(d)}
$$

The final end-moments are calculated by Eq. 3-5:

$$
\{A\} = Pl
\begin{Bmatrix}
-0.125 \\
0.125 \\
-0.125 \\
0.125 \\
0 \\
0
\end{Bmatrix}
+ \frac{EI}{l}
\begin{bmatrix}
-\dfrac{6}{l} & 2 & 0 \\[6pt]
-\dfrac{6}{l} & 4 & 0 \\[6pt]
0 & 4 & 2 \\[6pt]
0 & 2 & 4 \\[6pt]
-\dfrac{24}{l} & 0 & 8 \\[6pt]
-\dfrac{24}{l} & 0 & 4
\end{bmatrix}
\frac{Pl^2}{EI}
\begin{Bmatrix}
0.0087l \\
0.1355 \\
-0.0156
\end{Bmatrix}
= Pl
\begin{Bmatrix}
0.09 \\
0.61 \\
0.39 \\
0.33 \\
-0.33 \\
-0.27
\end{Bmatrix}
\text{(e)}
$$

The bending-moment diagram is plotted in Fig. 3-2h, the ordinate appearing on the side of the tensile fiber.

The application of the preceding procedure to a frame with inclined members is illustrated by the following example.

Remarks

(1) If in the above example, the fixed end A is replaced by a hinge, the kinematic indeterminacy is increased by the rotation at A. Nevertheless, the structure can be analyzed using only the three coordinates in Fig. 3-2

because the end forces for a member hinged at one end and totally fixed at the other end are readily available (Appendices C and D).

As an exercise, we can verify the following matrices for analyzing the frame in Fig. 3-2, with support A changed to a hinge

$$\{F\} = (P/16)\{-11, -15l, 2l\}; \{A_r\} = (Pl/16)\{0, 3, -2, 2, 0, 0\}$$

$$[S] = \frac{EI}{l} \begin{bmatrix} 99/l^2 & & \text{sym.} \\ -3/l & 7 & \\ -24/l & 2 & 12 \end{bmatrix};$$

$$[A_u]^T = \frac{EI}{l} \begin{bmatrix} 0 & -3/l & 0 & 0 & -24/l & -24/l \\ 0 & 3 & 4 & 2 & 0 & 0 \\ 0 & 0 & 2 & 4 & 8 & 4 \end{bmatrix}$$

$$\{D\} = \frac{Pl^2}{EI}\{0.0058l, 0.1429, -0.0227\};$$

$$\{A\} = Pl\{0, 0.60, 0.40, 0.32, -0.32, -0.23\}.$$

(2) When a computer is used for the analysis of a plane frame, axial deformations are commonly not ignored and the unknown displacements are two translations and a rotation at a general joint. Three forces are usually determined at each member end (Fig. 24-2). These can be used to give the axial force, the shearing force and the bending moment at any section. The superposition Eq. 3-5 is applied separately to give six end forces for each member, using the six displacements at its ends. This is discussed in detail in Chapter 24 for all types of framed structures.

Example 3-3 Obtain the bending-moment diagram for the frame of Fig. 3-3a, which has a constant flexural rigidity EI. Deformations due to axial forces are to be neglected.

In Fig. 3-3b, three coordinates are defined corresponding to three independent joint displacements, that is to the degree of kinematic indeterminacy. From Appendix C, the fixed-end forces due to the applied loading are found to be as in Fig. 3-3c. The restraining couples F_1 and F_2 are obtained directly by adding the fixed-end moments. For the calculation of F_3, the shearing forces at the ends of the members meeting at joints B and C are resolved into components along the axes of the members, and F_3 is obtained by adding the components in the direction of coordinate 3 (Fig. 3-3d).

The forces at the ends of the members corresponding to a separate unit displacement at each of the coordinates are now determined from Appendix D and indicated in Figs. 3-4b, 3-5b, and 3-6c. The displacement diagram of Fig. 3-6b shows that when $D_3 = 1$, the relative translation of the ends of members AB and BC, measured in a direction normal to the member, is $D_{BA} = 1.25$ and $D_{BC} = 0.75$.

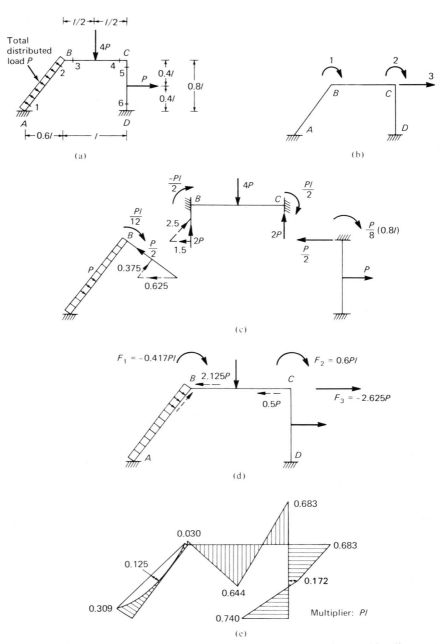

Fig. 3-3. Frame analyzed in Example 3-3. (a) Frame dimensions and loading. (b) Coordinate system. (c) Fixed-end forces. (d) Restraining forces along the coordinates. (e) Bending-moment diagram.

The forces $[S]$ forming the elements of the stiffness matrix of the structure are calculated from the end-forces and are indicated in Figs. 3-4a, 3-5a, and 3-6a.

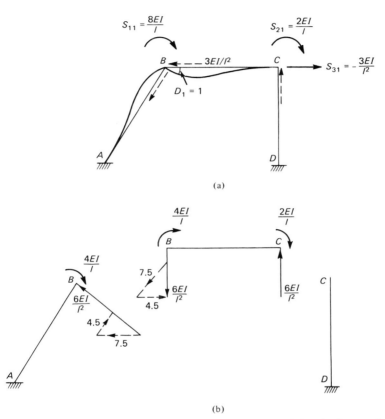

(a)

(b)

Fig. 3-4. Forces corresponding to $D_1 = 1$ and $D_2 = D_3 = 0$ for the frame of Example 3-3.

Substituting in Eq. 3-2,

$$EI \begin{bmatrix} \dfrac{8}{l} & \dfrac{2}{l} & -\dfrac{3}{l^2} \\[2mm] \dfrac{2}{l} & \dfrac{9}{l} & -\dfrac{4.875}{l^2} \\[2mm] -\dfrac{3}{l^2} & -\dfrac{4.875}{l^2} & \dfrac{48.938}{l^3} \end{bmatrix} \{D\} = P \begin{Bmatrix} 0.417l \\ -0.6l \\ 2.625 \end{Bmatrix} \qquad (a)$$

whence

$$\{D\} = \frac{P}{EI} \begin{Bmatrix} 0.0862l^2 \\ -0.0570l^2 \\ 0.0531l^3 \end{Bmatrix} \tag{b}$$

To draw the bending-moment diagram we require the end-moments $A_1, A_2, \ldots, A_6$ (see Fig. 3-3a), given by Eq. 3-5. The elements of $\{A_r\}$ and $[A_u]$ are easily determined from the end-moments in Figs. 3-3c, 3-4b, 3-5b, and 3-6c. Thus

$$\{A\} = Pl \begin{Bmatrix} -\dfrac{1}{12} \\ \dfrac{1}{12} \\ -\dfrac{1}{2} \\ \dfrac{1}{2} \\ \dfrac{1}{10} \\ -\dfrac{1}{10} \end{Bmatrix} + EI \begin{bmatrix} \dfrac{2}{l} & 0 & -\dfrac{7.5}{l^2} \\ \dfrac{4}{l} & 0 & -\dfrac{7.5}{l^2} \\ \dfrac{4}{l} & \dfrac{2}{l} & \dfrac{4.5}{l^2} \\ \dfrac{2}{l} & \dfrac{4}{l} & \dfrac{4.5}{l^2} \\ 0 & \dfrac{5}{l} & -\dfrac{9.375}{l^2} \\ 0 & \dfrac{5}{2l} & -\dfrac{9.375}{l^2} \end{bmatrix} \frac{P}{EI} \begin{Bmatrix} 0.0862l^2 \\ -0.0570l^2 \\ 0.0531l^3 \end{Bmatrix}$$

whence

$$\{A\} = Pl\{-0.309, 0.030, -0.030, 0.683, -0.683, -0.740\}$$

The bending-moment diagram is plotted in Fig. 3-3e with the ordinate on the side of the tensile fiber.

Example 3-3 is concerned with the analysis of a plane frame in which the axial deformations are ignored. The frame has three degrees of freedom (Fig. 3-3b): two rotations at B and C and one translation represented by a horizontal coordinate at C. This coordinate could have been chosen in any other direction at either B or C. Figure 3-7 shows how the restraining force F_3 can be calculated in a more general case where all three members are inclined.

With joints B and C totally fixed, determine the forces at member ends at B and C of the three members; replace the forces at B by components parallel to the members meeting at B, and do the same at C (Fig. 3-7c). Artificial restraint

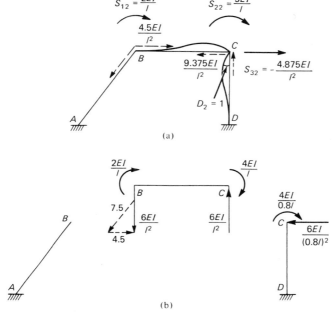

Fig. 3-5. Forces corresponding to $D_2 = 1$ and $D_1 = D_3 = 0$ for the frame of Example 3-3.

is required only for the component in the direction of BC, because in the directions of AB and CD the restraint is provided naturally by the supports A and D. Find $\sum C$, the sum of the components C_1, C_2, C_3 and C_4, and replace the force $\sum C$ (Fig. 3-7d) by two components, one parallel to CD and the other along coordinate 3 which is equal to the restraining force F_3.

The same procedure as described here can be used to calculate the force at coordinate 3 due to unit values of displacement at the three coordinates (third row of $[S]$).

Example 3-4 Find the three reaction components (vertical force, bending and twisting couples) at end A of the horizontal grid shown in Fig. 3-8a due to a uniform vertical load of intensity q acting on AC. All bars of the grid have the same cross section with the ratio of torsional and flexural rigidities GJ/EI $= 0.5$.

There are three unknown joint displacements at E, represented by the coordinates 1, 2, and 3 in Fig. 3-8b. From Appendixes C and D, it can be easily seen that the restraining forces, the stiffness matrix and the values of the

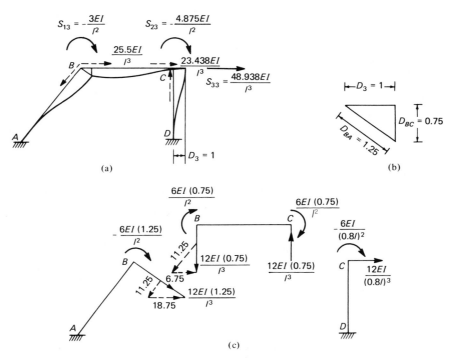

Fig. 3-6. Force corresponding to $D_3 = 1$ and $D_1 = D_2 = 0$ for the frame of example 3-3.

required actions in the restrained conditions are as follows:

$$\{F\} = \left\{ \begin{array}{cc} -\left(\dfrac{ql}{2}\right)_{AE} & -\left(\dfrac{ql}{2}\right)_{EC} \\ 0 & \\ \left(\dfrac{ql^2}{12}\right)_{AE} & -\left(\dfrac{ql^2}{12}\right)_{EC} \end{array} \right\} = q \left\{ \begin{array}{c} -\dfrac{l}{2} \\ 0 \\ \dfrac{l^2}{36} \end{array} \right\}$$

$$[S] = \left[\begin{array}{ccc} 729\dfrac{EI}{l^3} & \text{symmetrical} & \\ -40.5\dfrac{EI}{l^2} & \dfrac{20.25EI}{l} & \\ 40.5\dfrac{EI}{l^2} & 0 & \dfrac{20.25EI}{l} \end{array} \right]$$

The nonzero elements of $[S]$ are detailed as follows: $S_{11} = \sum(12EI/l^3)$, with the summation performed for the four members meeting at E; $S_{21} = -(6EI/l^2)_{BE}$

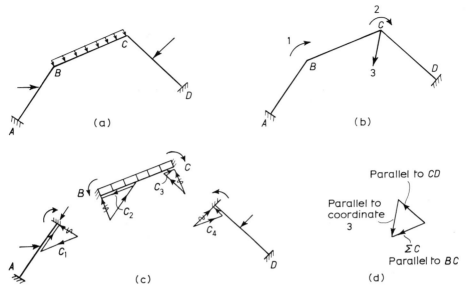

Fig. 3-7. Analysis of a plane frame with inclined members, ignoring axial deformations. (a) Frame to be analyzed. (b) Coordinate system. (c) Restraining forces at member ends at joints B and C. (d) Resolution of $\sum C$ into two components.

$+ (6EI/l^2)_{ED}$; $S_{31} = - S_{21}$ because of symmetry of the grid about a vertical plane passing through E and bisecting the angle BEC; $S_{22} = (4EI/l)_{BE}$ $+ (4EI/l)_{ED} + (GJ/l)_{AE} + (GJ/l)_{EC}$; $S_{33} = S_{22}$.

Assuming the three reaction components at A to be positive if they are represented by vectors shown in Fig. 3-8b, we write

$$\{A_r\} = \left\{ -\left(\frac{ql}{2}\right)_{AE}, 0, -\left(\frac{ql^2}{12}\right)_{AE} \right\} = q\left\{ -\frac{l}{3}, 0, \frac{-l^2}{27} \right\}$$

and

$$[A_u] = \begin{bmatrix} -\dfrac{12EI}{l^3} & 0 & \dfrac{6EI}{l^2} \\[2ex] 0 & -\dfrac{GJ}{l} & 0 \\[2ex] -\dfrac{6EI}{l^2} & 0 & \dfrac{2EI}{l} \end{bmatrix}_{AE} \begin{bmatrix} -40.5\dfrac{EI}{l^3} & 0 & 13.5\dfrac{EI}{l^2} \\[2ex] 0 & -0.75\dfrac{EI}{l} & 0 \\[2ex] -13.5\dfrac{EI}{l^2} & 0 & 3\dfrac{EI}{l} \end{bmatrix}$$

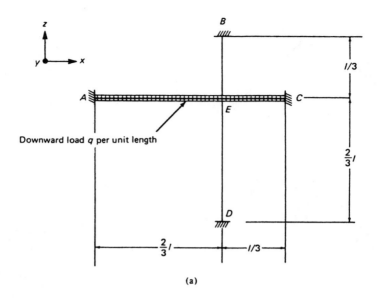

(a)

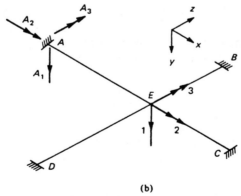

(b)

Fig. 3-8. Grid analyzed in Example 3-4. (a) Grid plan. (b) Pictorial view showing chosen coordinates 1, 2, and 3 and positive directions of the reaction components at A.

Substituting in Eq. 3-2, and solving for $\{D\}$ we obtain

$$\{D\} = q\frac{l}{EI}\left\{\begin{array}{c} 0.0010l^3 \\ 0.0020l^2 \\ -0.0034l^2 \end{array}\right\}$$

Substituting in Eq. 3-5, we find the three required actions (reaction components):

$$\{A\} = q\left\{\begin{array}{c} -\dfrac{l}{3} \\ 0 \\ -\dfrac{l^2}{27} \end{array}\right\} + \frac{EI}{l}\left[\begin{array}{ccc} -\dfrac{40.5}{l^2} & 0 & \dfrac{13.5}{l} \\ 0 & -0.75 & 0 \\ -\dfrac{13.5}{l} & 0 & 3 \end{array}\right] q\frac{l}{EI}\left\{\begin{array}{c} 0.0010l^3 \\ 0.0020l^2 \\ -0.0034l^2 \end{array}\right\}$$

whence

$$\{A\} = q\left\{\begin{array}{c} -0.4197l \\ -0.0015l^2 \\ -0.0611l^2 \end{array}\right\}$$

3-4 ANALYSIS FOR DIFFERENT LOADINGS

We have already made it clear that the stiffness matrix (and its inverse) are properties of a structure and do not depend on the system of the load applied. Therefore, if a number of different loadings are to be considered, Eq. 3-2 can be used for all of them. If the number of cases of loading is p, the solution can be combined into one matrix equation:

$$[D]_{n \times p} = [S]^{-1}_{n \times n}[-F]_{n \times p} \qquad (3-6)$$

with each column of $[D]$ and $[-F]$ corresponding to one loading.

3-5 ANALYSIS FOR ENVIRONMENTAL EFFECTS

In Secs. 2-41 and 2-42 we used the force method to analyze the separate or combined effects of temperature change, lack of fit, shrinkage, or prestrain. We shall now show how the displacement method can be used for the purpose. Equation 3-3 is directly applicable but in this case $\{F\}$ represents the forces necessary to prevent the joint displacements due to the given effects.

When the analysis is carried out for the effect of movement of support, Eq. 3-3 can also be applied, provided the movement of the support does not correspond with one of the unknown displacements forming the kinematic indeterminacy. When the movement of the support does so correspond, a modification of Eq. 3-3 is necessary. This is explained with reference to Example 3-5.

3-6 FIVE STEPS OF DISPLACEMENT METHOD

The analysis by the displacement method involves five steps which are summarized as follows:

Step 1   Define a system of coordinates representing the joint displacements to be found. Also define $[A]_{m \times p}$, the required actions as well as their sign convention (if necessary).

Step 2   With the loadings applied, calculate the restraining forces $[F]$ and $[A_r]_{m \times p}$.

Step 3   Introduce unit displacements at the coordinates one by one and generate $[S]_{n \times n}$ and $[A_u]_{m \times n}$.

Step 4   Solve the equilibrium equations:

$$[S]_{n \times n}[D]_{n \times p} = -[F]_{n \times p}$$

This gives the displacements $[D]$.

Step 5   Calculate the required actions by superposition:

$$[A]_{n \times p} = [A_r]_{m \times p} + [A_u]_{m \times n}[D]_{n \times p}$$

In a manner similar to the force method (Sec. 2-5), when step 3 is completed all the matrices necessary for the analysis have been generated. The last two steps involve merely matrix algebra.

For quick reference, the symbols used in this section are defined again as follows:

n, p, m = number of degrees of freedom, number of loading cases and number of actions required

$[A]$ = required actions (the answers to the problem)

$[A_r]$ = values of the actions due to loadings on the structure while the displacements are prevented

$[A_u]$ = values of the actions due to unit displacements introduced separately at each coordinate

$[F]$ = forces at the coordinates necessary to prevent the displacements due to the loadings

$[S]$ = stiffness matrix

Example 3-5 The continuous beam of Fig. 3-9a, of constant flexural rigidity EI, has two fixed supports A and D and two roller supports B and C. Obtain the bending-moment diagram for the beam (a) when support A settles vertically a distance $= \Delta$; (b) when the beam is rotated at B through an angle θ in a clockwise direction.

The degree of kinematic indeterminacy of the structure is two, the unknown joint displacements being the rotations D_1 and D_2 at B and C (Fig. 3-9b). The vertical translation of support A [case (a)] does not correspond to either of these. The restraining forces necessary to keep the displacements $D_1 = D_2 = 0$, obtained from Appendix D, are shown in Fig. 3-9c. Hence,

$$\{F\} = \begin{Bmatrix} F_1 \\ F_2 \end{Bmatrix} = \frac{EI}{l^2} \Delta \begin{Bmatrix} 6 \\ 0 \end{Bmatrix}$$

Considering a clockwise end-moment positive, the end-moments (identified in Fig. 3-9a) corresponding to the restrained condition of Fig. 3-9b are

$$\{A_r\} = \begin{Bmatrix} A_{r1} \\ A_{r2} \\ A_{r3} \\ A_{r4} \\ A_{r5} \\ A_{r6} \end{Bmatrix} = \frac{EI}{l^2} \Delta \begin{Bmatrix} 6 \\ 6 \\ 0 \\ 0 \\ 0 \\ 0 \end{Bmatrix}$$

The end-forces necessary to hold the members in a deformed position with D_1 or D_2 equal unity (Figs. 3-9d and e) are now determined from Appendix D, and hence the stiffness matrix of the structure is

$$[S] = \frac{EI}{l} \begin{bmatrix} 8 & 2 \\ 2 & 8 \end{bmatrix}$$

Hence

$$[S]^{-1} = \frac{l}{60EI} \begin{bmatrix} 8 & -2 \\ -2 & 8 \end{bmatrix}$$

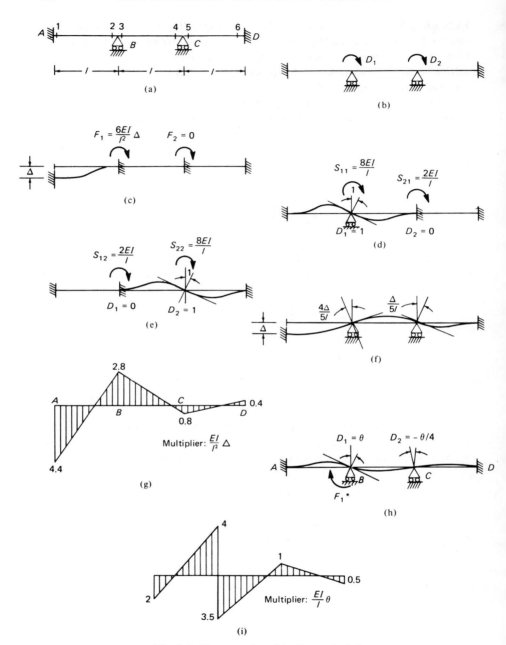

Fig. 3-9. Beam analyzed in Example 3-5.

The bending moments at the sections considered, caused by separate unit values of the displacements D_1 and D_2, are

$$[A_u] = \frac{EI}{l} \begin{bmatrix} 2 & 0 \\ 4 & 0 \\ 4 & 2 \\ 2 & 4 \\ 0 & 4 \\ 0 & 2 \end{bmatrix}$$

Applying Eq. 3-3,

$$\{D\} = \frac{l}{60EI} \begin{bmatrix} 8 & -2 \\ -2 & 8 \end{bmatrix} \frac{EI}{l^2} \Delta \begin{Bmatrix} -6 \\ 0 \end{Bmatrix} = \frac{\Delta}{5l} \begin{Bmatrix} -4 \\ 1 \end{Bmatrix}$$

that is,

$$D_1 = -\frac{4\Delta}{5l} \quad \text{and} \quad D_2 = \frac{\Delta}{5l}$$

The deflected shape of the beam caused by settlement Δ is shown in Fig. 3-9f. The corresponding bending moments at supports are obtained from Eq. 3-5.

$$\{A\} = \frac{EI}{l^2}\Delta \begin{Bmatrix} 6 \\ 6 \\ 0 \\ 0 \\ 0 \\ 0 \end{Bmatrix} + \frac{EI}{l} \begin{bmatrix} 2 & 0 \\ 4 & 0 \\ 4 & 2 \\ 2 & 4 \\ 0 & 4 \\ 0 & 2 \end{bmatrix} \frac{\Delta}{l} \begin{Bmatrix} -0.8 \\ 0.2 \end{Bmatrix} = \frac{EI}{l^2}\Delta \begin{Bmatrix} 4.4 \\ 2.8 \\ -2.8 \\ -0.8 \\ 0.8 \\ 0.4 \end{Bmatrix}$$

For equilibrium of the joints B and C, the sum of the end-moments at each joint must be zero. This fact can be used as a check of our calculations. Figure 3-9g gives the bending-moment diagram, as usual with the ordinate on the tension side of the beam.

Let us now consider case (b), in which there is a forced rotation of joint B. This would occur, for instance, if beam $ABCD$ were rigidly connected at B to a horizontal transverse beam which is twisted through an angle θ. To produce this rotation, an external couple F_1^* in the vertical plane through $ABCD$ must act at B (Fig. 3-9h). Let the external forces at the two coordinates 1 and 2 (Fig. 3-9b) corresponding to the deflected shape in Fig. 3-9h be denoted by $\{F^*\} = \{F_1^*, 0\}$. The displacements and the forces are related by

$$\begin{bmatrix} S_{11} & S_{12} \\ S_{21} & S_{22} \end{bmatrix} \begin{Bmatrix} D_1 \\ D_2 \end{Bmatrix} = \begin{Bmatrix} F_1^* \\ 0 \end{Bmatrix}$$

where the elements S_{ij} are as determined in case (a), $D_1 = \theta$, D_2 is the unknown displacement, and F_1^* the unknown external applied couple at coordinate 1.

The second of the above two equations gives

$$D_2 = -\frac{S_{21}}{S_{22}} D_1$$

Substituting, we obtain

$$D_2 = -\frac{2(EI/l)}{8(EI/l)} \theta = -\frac{\theta}{4}$$

The final end-moments are given by Eq. 3-5. However, in the present case the bending moment required is caused by the displacements $D_1 = \theta$ and $D_2 = -(\theta/4)$ only, so that the term representing the moments in the restrained condition $\{A_r\} = \{0\}$. The matrix $[A_u]$ calculated for case (a) can be used; hence the final end-moments are

$$\{A\} = \frac{EI}{l} \begin{bmatrix} 2 & 0 \\ 4 & 0 \\ 4 & 2 \\ 2 & 4 \\ 0 & 4 \\ 0 & 2 \end{bmatrix} \begin{Bmatrix} \theta \\ -\dfrac{\theta}{4} \end{Bmatrix} = \frac{EI}{l}\theta \begin{Bmatrix} 2 \\ 4 \\ 3.5 \\ 1 \\ -1 \\ -0.5 \end{Bmatrix}$$

These are shown in Fig. 3-9i.

3-7 EFFECT OF DISPLACEMENTS AT COORDINATES

The method used in case (b) of the above example will now be considered with reference to a general case of a structure with the degree of kinematic indeterminacy n, corresponding to n coordinates. Our objective is to analyze the structure for the effect of m forced displacements of magnitude Δ_1, $\Delta_2, \ldots, \Delta_m$ at m coordinates.

We write the stiffness matrix so that the coordinates corresponding to the known displacements occur in the first m rows and columns, thus

$$[S] = \left[\begin{array}{cccc:ccc} S_{11} & S_{12} & \cdots & S_{1m} & S_{1(m+1)} & \cdots & S_{1n} \\ S_{21} & S_{22} & \cdots & & \cdots & & \cdots \\ \cdots & & & & \cdots & & \cdots \\ S_{m1} & S_{m2} & \cdots & S_{mm} & S_{m(m+1)} & \cdots & S_{mn} \\ \hdashline S_{(m+1)1} & \cdots & & & \cdots & & \cdots \\ \cdots & & & & \cdots & & \cdots \\ S_{n1} & S_{n2} & \cdots & & \cdots & & S_{nn} \end{array}\right] \quad (3\text{-}7)$$

This matrix can be partitioned along the dashed lines shown above and can therefore be written in the form

$$[S] = \begin{bmatrix} [S_{11}] & \vdots & [S_{12}] \\ ---- & + & ---- \\ [S_{21}] & \vdots & [S_{22}] \end{bmatrix} \tag{3-8}$$

where $[S_{ij}]$ are the partitional matrices or submatrices. By inspection of Eq. 3-7, the order of submatrices is $[S_{11}]_{m \times m}$, $[S_{12}]_{m \times (n-m)}$, $[S_{21}]_{(n-m) \times m}$, $[S_{22}]_{(n-m) \times (n-m)}$.

To produce the displacements $\Delta_1, \Delta_2, \ldots, \Delta_m$ external forces $F_1^*, F_2^*, \ldots,$ F_m^* must be applied at the coordinates 1 to m with no forces applied at the remaining coordinates, thus allowing displacements $D_{m+1}, D_{m+2}, \ldots, D_n$ to occur. The forces and the displacements can be related by

$$\begin{bmatrix} [S_{11}] & \vdots & [S_{12}] \\ ---- & + & ---- \\ [S_{21}] & \vdots & [S_{22}] \end{bmatrix} \begin{Bmatrix} \{D_1\} \\ --- \\ \{D_2\} \end{Bmatrix} = \begin{Bmatrix} \{F_1^*\} \\ --- \\ \{0\} \end{Bmatrix} \tag{3-9}$$

where $\{D_1\}$ is a vector of the known displacements Δ, $\{D_2\}$ is a vector of the unknown displacements $D_{m+1}, D_{m+2}, \ldots, D_n$, and $\{F_1^*\}$ is a vector of the unknown forces at the coordinates $1, 2, \ldots, m$.

From the second row of the above matrix equation, we find

$$\{D_2\} = -[S_{22}]^{-1}[S_{21}]\{D_1\} \tag{3-10}$$

With the displacements at all the n coordinates known, the stress resultants at any section can be determined by the equation

$$\{A\} = [A_u]\{D\} \tag{3-11}$$

where $\{A\}$ is any action, and $[A_u]$ is the same action corresponding to a unit displacement at the given coordinate only. This equation is the same as Eq. 3-5 with $\{A_r\} = \{0\}$, since the actions required are due to the effect of the displacements $\{D\}$ only.

If the forces $\{F_1^*\}$ are required, they can be obtained from the first row of Eq. 3-9 and Eq. 3-10, as follows:

$$\{F_1^*\} = [[S_{11}] - [S_{12}][S_{22}]^{-1}[S_{21}]]\{D_1\} \tag{3-12}$$

3-8 GENERAL

The displacement (or stiffness) method of analysis can be applied to any structure but the largest economy of effort arises when the order of statical indeterminacy is high. The procedure is standardized and can be applied without difficulty to trusses, frames, grids, and other structures (see Chapter 19) subjected to external loading or to prescribed deformation, e.g., settlement of supports or temperature change. In the latter case, the effect of forced displacements at the chosen coordinates results in a slight modification in the procedure. The displacement method is extremely well suited to computer solutions.

PROBLEMS

3-1   Use the displacement method to find the forces in the members of the truss shown. Assume the value l/aE to be the same for all members.

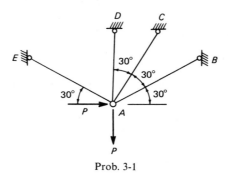

Prob. 3-1

3-2   A rigid mast AB is hinged at A, free at B and is pin-jointed to n bars of which a typical one is shown. Using the displacement method, find the force A in the ith bar if l/aE = constant for each bar. Assume that the mast and all the bars are in one plane.

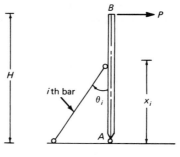

Prob. 3-2

3-3   Solve Prob. 2-7 by the displacement method.

3-4   For the truss shown in the figure, write the stiffness matrix corresponding to the four coordinates indicated. Assume a and E to be the same for all members.

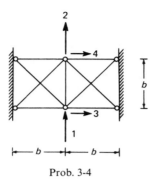

Prob. 3-4

3-5 Neglecting axial deformations, find the end-moments M_{BC} and M_{CF} for the frame shown. Consider an end-moment to be positive when it acts in a clockwise direction on the member end.

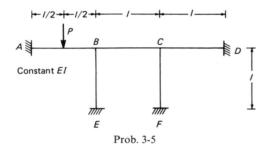

Prob. 3-5

3-6 Solve Prob. 3-5 if support E moves downward a distance Δ and the load P is not acting.

3-7 Write the stiffness matrix corresponding to the coordinates 1 and 2 of the frame in the figure. What are the values of the end-moments M_{BC} and M_{CB} due to uniform load q per unit length on BC and what are the reaction components at A? Neglect axial deformations.

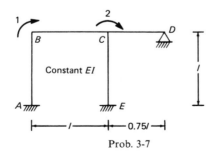

Prob. 3-7

3-8 Find the end-moments M_{BC} and M_{CB} in the frame of Prob. 3-7 due to a rise in temperature of 40° Fahrenheit in part BD only. Assume the coefficient of thermal expansion to be $\alpha = 6.5 \times 10^{-6}$ per degree Fahrenheit, and take $EI = 10^4$ k ft^2. The load q is not acting in this case. The members are assumed to have infinite axial rigidity; $l = 20$ ft.

3-9   For the frame in the figure, write the first three columns of the stiffness matrix corresponding to the six coordinates indicated. The moment of inertia and the area of the cross section are shown alongside the members.

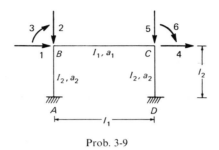

Prob. 3-9

3-10   Write the stiffness matrix for the frame of Prob. 3-9, neglecting the axial deformations. Number the coordinates in the following order: clockwise rotation at B, clockwise rotation at C, horizontal translation to the right at C.

3-11   Using the results of Prob. 3-10, obtain the bending-moment diagram for the frame due to a horizontal load P to the right at B and a downward vertical load $4P$ at the middle of BC. Take $l_1 = 2l_2$ and $I_1 = 4I_2$.

3-12   Use the displacement method to find the bending-moment and shearing-force diagrams for the frame in the figure. Neglect axial deformations.

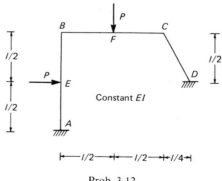

Prob. 3-12

3-13   For the grid shown in plan in the figure, calculate the displacements corresponding to the three degrees of freedom at joint B. Assume $GJ/EI = 0.8$ for AB and BC. Draw the bending-moment diagram for BC.

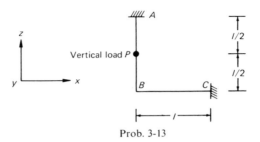

Prob. 3-13

3-14 Considering three degrees of freedom: a downward deflection, and rotations represented by vectors in the x and z directions at each of the joints I, J, K, and L of the grid in the figure, write the first three columns of the corresponding stiffness matrix. Number the twelve coordinates in the order in which they are mentioned above, that is the first three at I followed by three at J, and so on. Take $GJ/EI = 0.5$ for all members.

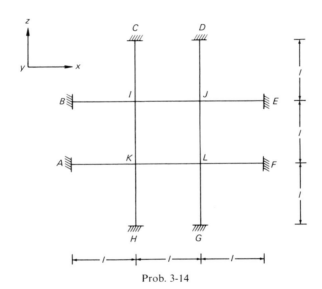

Prob. 3-14

3-15 Solve Prob. 2-5 by the displacement method. Consider the vertical deflections at B and C as the unknown displacements and use Appendix E.

3-16 Neglecting torsion, find the bending moment in girders BE and AF of the grid of Prob. 3-14 subjected to a vertical load P at joint I. Consider the vertical deflections at the joints as the unknown displacements and use Appendix E.

3-17 A horizontal grid is composed of n members meeting at one joint A of which a typical member i is shown in the figure. Write the stiffness matrix corresponding to the coordinates 1, 2, and 3 indicated, with coordinate 1 representing a downward deflection, and 2 and 3 being rotations. The length of the ith member, its torsional rigidity, and its flexural rigidity in the vertical plane through its axis are l_i, GJ_i, and EI_i respectively. (*Hint*: The stiffness matrix of the grid can be obtained by the summation of the stiffness matrices of individual members to obtain the first column of the stiffness matrix of the ith member, a unit downward displacement is introduced and the end-forces are taken from Appendix D. These forces are resolved into components at the three coordinates. For the second and third column, a unit rotation at coordinate 2 and 3 is resolved into two rotations represented by vectors along and normal to the beam axis and the corresponding end-forces are taken from Appendix D. These forces are then resolved into components at the coordinates to obtain elements of the stiffness matrix.)

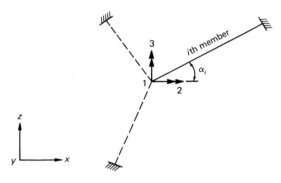

Prob. 3-17

3-18   Figure E-1, Appendix E shows a prismatic continuous beam of three equal spans l with a unit settlement at one of the interior supports. Using the displacement method, verify the bending moments and the reactions given in the figure. Two coordinates only need be used, representing the rotation at each of the interior supports.

3-19   Solve Example 3-2 with the supports at A and D changed to hinges (Fig. 3.2a).

3-20   Solve Example 3-3 with the support at A changed to a hinge (Fig. 3-3a).

CHAPTER 4

Use of force and displacement methods

4-1 INTRODUCTION

The preceding two chapters outlined two basic methods of analysis, and obviously either can be used to analyze any structure. It is worthwhile, however, to see clearly the advantages of each method so as to choose the one more economical in terms of effort in any particular case.

4-2 RELATION BETWEEN FLEXIBILITY AND STIFFNESS MATRICES

First of all, the relation between the two matrices, flexibility and stiffness, should be established. This will be done with reference to Fig. 4-1a, which shows a beam AB with a system of n coordinates representing the location and direction of displacements $D_1, D_2, \ldots, D_n$ and of forces $F_1, F_2, \ldots, F_n$. The displacements $\{D\}$ can be expressed in terms of the displacements of the forces acting separately by the superposition equations

$$D_1 = f_{11}F_1 + f_{12}F_2 + \cdots + f_{1n}F_n$$

$$D_2 = f_{21}F_1 + f_{22}F_2 + \cdots + f_{2n}F_n$$

$$\cdots \qquad \cdots$$

$$D_n = f_{n1}F_1 + f_{n2}F_2 + \cdots + f_{nn}F_n$$

The coefficients f in the above equations are the flexibility influence coefficients, f_{ij} being the displacement along the coordinate i due to a unit force applied along the coordinate j (Fig. 4-1b). The above equations can be put in matrix form

$$[f]_{n \times n}\{F\}_{n \times 1} = \{D\}_{n \times 1} \qquad (4\text{-}1)$$

This equation must not be confused with Eq. 2-2, $[f]\{F\} = \{-D\}$, used in the force method of analysis where we apply unknown redundants $\{F\}$ of such a magnitude as will produce displacements $\{-D\}$ to correct for inconsistencies $\{D\}$ of the released structure.

The forces $\{F\}$ can be expressed in terms of the displacements by solving Eq. 4-1,

$$\{F\}_{n \times 1} = [f]_{n \times n}^{-1}\{D\}_{n \times 1} \qquad (4\text{-}2)$$

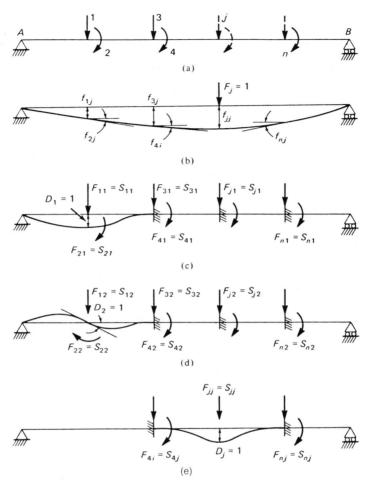

Fig. 4-1. Illustration of the relation between the flexibility and stiffness coefficients.

Equation 4-2 can be used to determine the forces forming the elements of the stiffness matrix of the same structure. If the structure is deformed by forces $F_{11}, F_{21}, \ldots, F_{n1}$ along the coordinates in a way such that the displacement $D_1 = 1$, while all the other displacements $D_2 = D_3 = \cdots = D_n = 0$ (as shown in Fig. 4-1c), then

$$\begin{Bmatrix} F_{11} \\ F_{21} \\ \cdots \\ F_{n1} \end{Bmatrix} = [f]^{-1} \begin{Bmatrix} 1 \\ 0 \\ \cdots \\ 0 \end{Bmatrix}$$

Similarly, the forces required to hold the structure in the deformed configuration with $D_2 = 1$, while all other displacements $D_1 = D_3 = \ldots = D_n = 0$ (Fig. 4-1d), are

$$
\begin{Bmatrix} F_{12} \\ F_{22} \\ \ldots \\ F_{n2} \end{Bmatrix} = [f]^{-1} \begin{Bmatrix} 0 \\ 1 \\ \ldots \\ 0 \end{Bmatrix}
$$

In the general case, if a structure is deformed so that the displacement $D_j = 1$, while all the other displacements are equal to zero (Fig. 4-1e), a set of equations similar to the above two equations can be written. All these equations can be combined into one matrix equation:

$$
\begin{bmatrix} F_{11} & F_{12} & \ldots & F_{1n} \\ F_{21} & F_{22} & & F_{2n} \\ \ldots & \ldots & & \\ F_{n1} & F_{n2} & \ldots & F_{nn} \end{bmatrix} = [f]^{-1} \begin{bmatrix} 1 & 0 & \ldots & 0 \\ 0 & 1 & \ldots & 0 \\ \ldots & \ldots & & \\ 0 & 0 & \ldots & 1 \end{bmatrix}
$$

The forces F_{ij} on the left-hand side of this equation are, in fact, the elements of the stiffness matrix required. The last matrix on the right-hand side of the above equation is a unit matrix $[I]$. Therefore, this equation can be written in the form

$$
[S] = [f]^{-1} \tag{4-3}
$$

where $[S]$ is the stiffness matrix corresponding to the given coordinate system. Inverting both sides of Eq. 4-3,

$$
[S]^{-1} = [f] \tag{4-4}
$$

Equations 4-3 and 4-4 show that the stiffness matrix is the inverse of the flexibility matrix, and vice versa, provided the same coordinate system of forces and displacement is used in the formation of the two matrices.

We may recall that in the force method of analysis, releases are introduced to render the structure statically determinate. The coordinate system represents the location and direction of these released forces. Now in the displacement method of analysis, restraining forces are added to prevent joint displacements. The coordinate system in this case represents the location and direction of the unknown displacements. It follows that the two coordinate systems cannot be the same for the same structure. Therefore, the inverse of the flexibility matrix used in the force method is a matrix, whose elements are stiffness coefficients, but not those used in the displacement method of analysis. Similarly, the inverse of the stiffness matrix used in the displacement method is a flexibility matrix, but not the matrix used in the force method.

4-3 CHOICE OF FORCE OR DISPLACEMENT METHOD

In some structures the formation of one of the matrices — stiffness or flexibility — may be easier than the formation of the other. For example, in the structure of Fig. 4-1a, the elements of the stiffness matrix can be determined by a simple calculation using Appendix D. A unit displacement at any coordinate j (Fig. 4-1e) produces forces at j and at the coordinate adjacent to j only. For example, in Fig. 4-1c, only the forces S_{11}, S_{21}, S_{31} and S_{41} need to be calculated, all the other forces being zero. On the other hand, to generate the jth column of the flexibility matrix, a unit force is applied at the coordinate j and the displacements calculated at all the n coordinates. It is obvious that, in this case, none of the flexibility coefficients vanishes and they require more calculation than the stiffness coefficients. This is, however, not always the case, and in Example 4-1 we deal with a case where the flexibility matrix is easier to generate than the stiffness matrix.

This situation arises from the following general considerations. In the force method, the choice of the released structure may affect the amount of calculation. For example, in the analysis of a continuous beam, the introduction of hinges above intermediate supports produces a released structure formed of a series of simple beams (see Example 2-3), so that the application of a unit value of the redundants has a local effect on the two adjacent spans only. In structures other than continuous beams, it may not be possible to find a released structure for which the redundants have a local effect only, and usually a redundant acting separately produces displacements at all the coordinates.

In the displacement method, generally all joint displacements are prevented regardless of the choice of the unknown displacements. Generation of the stiffness matrix is usually not difficult because of the localized effect discussed earlier. A displacement of a joint affects only the members meeting at the given joint. These two properties generally make the displacement method easy to formulate, and it is for these two reasons that the displacement method is more suitable for computer programming.

When the analysis is performed by hand, it is important to reduce the number of simultaneous equations to be solved, and thus the choice of the force or displacement method may depend on which is smaller: the degree of static or kinematic indeterminacy. No general rule can be established, and when a structure is statically as well as kinematically indeterminate to a low degree either method may involve about the same amount of computation.

When the computation is done by hand, it is possible in the displacement method to reduce the number of simultaneous equations by preventing only some of the joint displacements, provided that the resulting structure can be readily analyzed. This is illustrated in Example 4-2.

Example 4-1   Consider the beam of Fig. 4-2a, in which three coordinates are defined. To write the flexibility matrix, we obtain displacements such as those

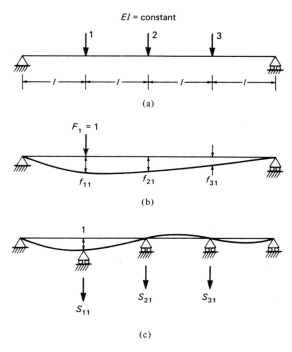

Fig. 4-2. Deflected configuration to generate the first column of $[f]$ and $[S]$ in Example 4-1.

shown in Fig. 4-2b (for the first column of the matrix), their values being taken from Appendix B. Thus

$$[f] = \frac{l^3}{12EI} \begin{bmatrix} 9 & 11 & 7 \\ 11 & 16 & 11 \\ 7 & 11 & 9 \end{bmatrix}$$

To generate the stiffness matrix requires the analysis of a statically indeterminate structure for each column of $[S]$. For example, the elements of the first column of $[S]$ are the forces necessary to hold the structure in the deformed shape illustrated in Fig. 4-2c, and this requires the solution of a structure statically indeterminate to the third degree. This exercise is left to the reader; its solution will provide a check on the stiffness matrix[1] obtained by the inversion of the flexibility matrix.

$$[S] = [f]^{-1} = \frac{EI}{l^3} \begin{bmatrix} 9.857 & -9.429 & 3.857 \\ -9.429 & 13.714 & -9.429 \\ 3.857 & -9.429 & 9.857 \end{bmatrix}$$

Example 4-2   The grid in Fig. 4-3a is formed by four simply supported main girders and one cross-girder of a bridge deck, with a flexural rigidity in the ratio $EI_m:EI_c = 3:1$. The torsional rigidity is neglected. Find the bending-moment diagram for a cross-girder due to a concentrated vertical load P acting at joint 1.

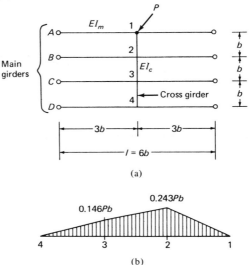

Fig. 4-3. Grid considered in Example 4-2.

There are three degrees of freedom at any joint of the grid: vertical translation, and rotation about two perpendicular axes in the plane of the grid. However, if the vertical translation of the joints is prevented, the grid becomes a system of continuous beams, and, using Appendix E, we can analyze for the effect of movement of one joint without the necessity to know the rotations.

---

[1]The elements of the stiffness matrix corresponding to the coordinates chosen in this example are often used in the analysis of continuous beams with equal spans on elastic supports or for the analysis of grids with hinged connections (torsionless grids). These elements are the reactions at the supports caused by the vertical translation of one support only. Appendix E gives the values of the reactions and bending moments at the supports for continuous beams of 2 to 5 equal spans.

Let the coordinate system be the four vertical deflections at 1, 2, 3, and 4, considered positive downward. The stiffness matrix can be generated by using the tabulated values of the reactions in Appendix E for the case of two equal spans for the main girders, and three equal spans for the cross-girder. The elements of the first row of the stiffness matrix of the grid are calculated as follows.

The displacements $D_1 = 1$ and $D_2 = D_3 = D_4 = 0$ deform the main girder A and the cross-girder, but girders B, C, and D are not deflected. The rotations at the joints are allowed to take place freely. Then, adding the vertical forces required to hold the girder A and the cross-girder in this deflected shape, we obtain

$$S_{11} = 6.0 \frac{EI_m}{(3b)^3} + 1.6 \frac{EI_c}{b^3} = 2.267 \frac{EI_c}{b^3}$$

$$S_{21} = -3.6 \frac{EI_c}{b^3} \qquad S_{31} = 2.4 \frac{EI_c}{b^3} \qquad S_{41} = -0.4 \frac{EI_c}{b^3}$$

The elements of other columns of the stiffness matrix are determined in a similar way, so that finally

$$[S] = \frac{EI_c}{b^3} \begin{bmatrix} 2.267 & -3.600 & 2.400 & -0.400 \\ -3.600 & 10.267 & -8.400 & 2.400 \\ 2.400 & -8.400 & 10.267 & -3.600 \\ -0.400 & 2.400 & -3.600 & 2.267 \end{bmatrix}$$

With the load P at coordinate 1, we need only an equal and opposite force at this coordinate to prevent the joint displacements. Thus $\{F\} = \{-P, 0, 0, 0\}$. Substituting in Eq. 3-2 and solving for the displacements, we find

$$\{D\} = \frac{Pb^3}{EI_c} \begin{Bmatrix} 1.133 \\ 0.511 \\ 0.076 \\ -0.221 \end{Bmatrix}$$

The bending moment in the cross-girder is zero at the ends 1 and 4, so that moments at 2 and 3 only have to be determined. Considering the bending moment positive if it causes tension in the bottom fiber we find the moment from Eq. 3-5. In the present case, the bending moment in the restrained structure $\{A_r\} = \{0\}$ because the load P is applied at a coordinate. If the load were applied in any other position, then the determination of $\{A_r\}$

would require an analysis of a continuous beam with equal spans over rigid supports.[2]

The elements of $[A_u]$ are obtained from Appendix E. We find

$$[A_u] = \frac{EI_c}{b^2}\begin{bmatrix} -1.6 & 3.6 & -2.4 & 0.4 \\ 0.4 & -2.4 & 3.6 & -1.6 \end{bmatrix}$$

and

$$\{A\} = \frac{EI_c}{b^2}\begin{bmatrix} -1.6 & 3.6 & -2.4 & 0.4 \\ 0.4 & -2.4 & 3.6 & -1.6 \end{bmatrix}\frac{Pb^3}{EI_c}\begin{Bmatrix} 1.133 \\ 0.511 \\ 0.076 \\ -0.221 \end{Bmatrix} = Pb\begin{Bmatrix} -0.243 \\ -0.146 \end{Bmatrix}$$

Hence the bending-moment diagram for the cross-girder is as shown in Fig. 4-3b.

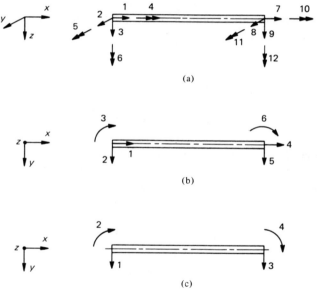

(a)

(b)

(c)

Fig. 4.4. Coordinate systems corresponding to stiffness matrices: (a) Eq. 4-5, (b) Eq. 4-6, and (c) Eq. 4-7.

---

[2]This can be carried out with aid of standard tables. See for example, *Moments, Shears and Reactions, Continuous Highway Bridge Tables*, American Institute of Steel Construction, New York, 1959, or G. Anger, *Ten Division Influence Lines for Continuous Beams*, Wilhelm Ernst and Son, Berlin, 1956.

4-4 STIFFNESS MATRIX FOR A PRISMATIC BEAM

In the examples in this chapter and in Chapter 3, we have seen that the elements of the stiffness matrix of a structure are obtained by adding the forces at the ends of the members which meet at a joint. These end-forces are elements of the stiffness matrix for individual members and are derived by the use of Appendix D. In this section, the stiffness matrix for a prismatic member is generated because it is often needed in the analysis of framed structures. We consider 12 coordinates at the ends, representing translations and rotations about three rectangular axes x, y, and z (Fig. 4-4a), with the y and z axes chosen to coincide with the principal axes of the cross section. The beam is assumed to be of length l and cross-sectional area a, and to have second moments of area I_z and I_y about the z and y axes, respectively; the modulus of elasticity of the material is E, and the torsional rigidity GJ.

If we neglect shear deformations and warping caused by twisting, all the elements of the stiffness matrix can be taken from Appendix D. The elements in any column j are equal to the forces at the coordinates produced by a displacement $D_j = 1$ at coordinate j only. The resulting stiffness matrix is

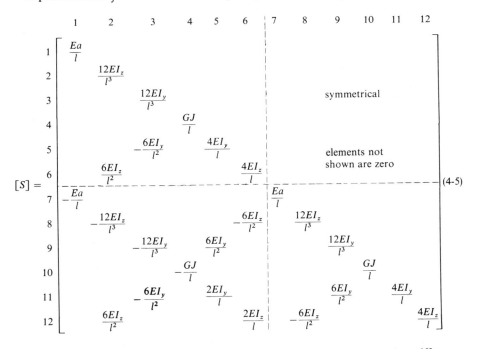

For two-dimensional problems of a frame in the x–y plane, the stiffness matrix needs to be considered for six coordinates only: 1, 2, 6, 7, 8, and 12 (Fig. 4-4a). Deletion of the columns and rows numbered 3, 4, 5, 9, 10, and 11

from the matrix in Eq. 4-5 results in the following stiffness matrix of a prismatic member corresponding to the six coordinates in Fig. 4-4b to be used in the analysis of plane frames:

$$[S] = \begin{array}{c} \\ 1 \\ 2 \\ 3 \\ 4 \\ 5 \\ 6 \end{array} \begin{array}{cccccc} 1 & 2 & 3 & 4 & 5 & 6 \\ \left[\begin{array}{cccccc} \dfrac{Ea}{l} & & & & \text{symmetrical} & \\[2mm] & \dfrac{12EI}{l^3} & & & & \\[2mm] & \dfrac{6EI}{l^2} & \dfrac{4EI}{l} & \text{elements not} & & \\ & & & \text{shown are zero} & & \\[2mm] -\dfrac{Ea}{l} & & & \dfrac{Ea}{l} & & \\[2mm] & -\dfrac{12EI}{l^3} & -\dfrac{6EI}{l^2} & & \dfrac{12EI}{l^3} & \\[2mm] & \dfrac{6EI}{l^2} & \dfrac{2EI}{l} & & -\dfrac{6EI}{l^2} & \dfrac{4EI}{l} \end{array}\right] \end{array} \quad (4\text{-}6)$$

where $I = I_z$.

If, in a plane frame, the axial deformations are ignored, the coordinates 1 and 4 in Fig. 4-4b need not be considered, and the stiffness matrix of a prismatic member corresponding to the four coordinates in Fig. 4-4c becomes

$$[S] = \begin{array}{c} \\ 1 \\ 2 \\ 3 \\ 4 \end{array} \begin{array}{cccc} 1 & 2 & 3 & 4 \\ \left[\begin{array}{cccc} \dfrac{12EI}{l^3} & & \text{symmetrical} & \\[2mm] \dfrac{6EI}{l^2} & \dfrac{4EI}{l} & & \\[2mm] -\dfrac{12EI}{l^3} & -\dfrac{6EI}{l^2} & \dfrac{12EI}{l^3} & \\[2mm] \dfrac{6EI}{l^2} & \dfrac{2EI}{l} & -\dfrac{6EI}{l^2} & \dfrac{4EI}{l} \end{array}\right] \end{array} \quad (4\text{-}7)$$

The stiffness matrix for members in which the shear deformations are not ignored is derived in Sec. 16-2. The presence of a high axial force in a deflected member causes an additional bending moment, and if this effect is to be taken into account, the above stiffness matrices must be modified. This is discussed in Chapter 15.

The stiffness matrices derived above correspond to coordinates which coincide with the beam axis or with the principal axes of its cross section. However, in the analysis of structures composed of a number of members running in arbitrary directions, the coordinates may be taken parallel to a set of *global axes* and thus the coordinates may not coincide with the principal axes of a given member. In such a case, the stiffness matrices given above (corresponding to the coordinates coinciding with the principal axes of the members) will have to be transformed to stiffness matrices corresponding to another set of coordinates by the use of transformation matrices formed by geometrical relations between the two sets of coordinates. This will be further discussed in Sec. 8-4.

4-5 CONDENSATION OF STIFFNESS MATRICES

We recall that a stiffness matrix relates displacements $\{D\}$ at a number of coordinates to the forces $\{F\}$ applied at the same coordinates by the equation

$$[S]\{D\} = \{F\} \tag{4-8}$$

If the displacement at a number of coordinates is prevented by the introduction of supports, and the matrices in the above equations are arranged in such a way that the equations corresponding to these coordinates appear at the end, we can write Eq. 4-8 in the partitioned form

$$\begin{bmatrix} [S_{11}] & [S_{12}] \\ [S_{21}] & [S_{22}] \end{bmatrix} \begin{Bmatrix} \{D_1\} \\ \{D_2\} \end{Bmatrix} = \begin{Bmatrix} \{F_1\} \\ \{F_2\} \end{Bmatrix} \tag{4-9}$$

where $\{D_2\} = \{0\}$ represents the prevented displacements. From this equation, we write

$$[S_{11}]\{D_1\} = \{F_1\} \tag{4-10}$$

and

$$[S_{21}]\{D_1\} = \{F_2\} \tag{4-11}$$

It is apparent from Eq. 4-10 that if a support is introduced at a number of the coordinates, the stiffness matrix of the resulting structure can be obtained simply by deleting the columns and the rows corresponding to these coordinates, resulting in a matrix of a lower order. If the displacements $\{D_1\}$ are known, Eq. 4-11 can be used to calculate the reactions at the supports preventing the displacements $\{D_2\}$.

As a simple example, consider the beam in Fig. 4-4c and assume that the vertical displacements at coordinates 1 and 3 are prevented as in the case of a simple beam; the stiffness matrix corresponding to the remaining two coordinates (2 and 4) is obtained by deletion of columns and rows numbered 1 and 3 in the matrix Eq. 4-7:

$$[S^*] = \begin{bmatrix} \dfrac{4EI}{l} & \text{symmetrical} \\ \dfrac{2EI}{l} & \dfrac{4EI}{l} \end{bmatrix} \tag{4-12}$$

The vertical reactions $\{F_1, F_3\}$ at coordinates 1 and 3 can be calculated from Eq. 4-11 by rearrangement of the elements in Eq. 4-7 as described for Eq. 4-9. Thus

$$\begin{bmatrix} \dfrac{6EI}{l^2} & \dfrac{6EI}{l^2} \\[2mm] -\dfrac{6EI}{l^2} & -\dfrac{6EI}{l^2} \end{bmatrix} \begin{Bmatrix} D_2 \\ D_4 \end{Bmatrix} = \begin{Bmatrix} F_1 \\ F_3 \end{Bmatrix} \tag{4-13}$$

where the subscripts of D and F refer to the coordinates in Fig. 4-4c.

If the forces are known to be zero at some of the coordinates, that is the displacements at these coordinates can take place freely, the stiffness matrix corresponding to the remaining coordinates can be derived from the partitioned matrix Eq. 4-9. In this case, we consider that the equations below the horizontal dashed line relate forces $\{F_2\}$, assumed to be zero, to the displacements $\{D_2\} \neq \{0\}$ at the corresponding coordinates. Substituting $\{F_2\} = \{0\}$ in Eq. 4-9, we write

and

$$\left.\begin{aligned} [S_{11}]\{D_1\} + [S_{12}]\{D_2\} &= \{F_1\} \\[3mm] [S_{21}]\{D_1\} + [S_{22}]\{D_2\} &= \{0\} \end{aligned}\right\} \tag{4-14}$$

Using the second equation to eliminate $\{D_2\}$ from the first, we obtain

$$\big[[S_{11}] - [S_{12}][S_{22}]^{-1}[S_{21}]\big]\{D_1\} = \{F_1\} \tag{4-15}$$

which is the same as Eq. 3-12. Equation 4-15 can be written in the form

$$[S^*]\{D_1\} = \{F_1\} \tag{4-16}$$

where $[S^*]$ is a condensed stiffness matrix relating forces $\{F_1\}$ to displacements $\{D_1\}$, and is given by

$$[S^*] = [S_{11}] - [S_{12}][S_{22}]^{-1}[S_{21}] \tag{4-17}$$

The stiffness matrix of a structure composed of a number of members is usually derived from the stiffness matrix of the individual members, and relates the forces at all the degrees of freedom to the corresponding displacements. In many cases, however, the external forces on the actual structure are limited to a small number of coordinates, and it may therefore be useful to derive a matrix of a lower order $[S^*]$ corresponding to these coordinates only, using Eq. 4-17. A simple example of the application of Eq. 4-17 is given as a problem on matrix algebra in Prob. A-7 of Appendix A.

4-6 PROPERTIES OF FLEXIBILITY AND STIFFNESS MATRICES

Consider a force F_i applied gradually to a structure, so that the kinetic energy of the mass of the structure is zero. Let the resulting displacement at the location and in the direction of F_i be D_i. If the structure is elastic the force-displacement curve follows the same path on loading and unloading, as shown in Fig. 4-5a.

Assume now that at some stage of loading, the force F_i is increased by ΔF_i, and the corresponding increase in the displacement D_i is ΔD_i. The work done by this load increment is

$$\Delta W \simeq F_i \Delta D_i$$

This is shown as the hatched rectangle in Fig. 4-5a. If the increments are sufficiently small, it can be seen that the total external work done by F_i during the displacement D_i is the area below the curve between 0 and D_i.

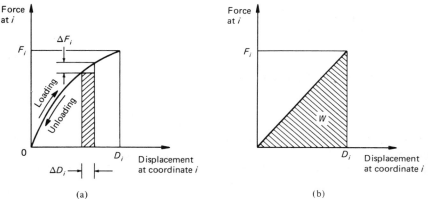

(a)   (b)

Fig. 4-5. Force-displacement relations.

When the material in the structure obeys Hooke's law, the curve in Fig. 4-5a is replaced by a straight line (Fig. 4-5b), and the work done by the force F_i becomes

$$W = \frac{1}{2} F_i D_i$$

If the structure is subjected to a system of forces $F_1, F_2, \ldots, F_n$ increased gradually from zero to their final value and causing displacements $D_1, D_2, \ldots, D_n$ at the location and in the direction of the forces, then the total external work is

$$W = \frac{1}{2}(F_1 D_1 + F_2 D_2 + \cdots + F_n D_n) = \frac{1}{2} \sum_{i=1}^{n} F_i D_i \qquad (4\text{-}18)$$

This equation can be written in the form

$$[W]_{1 \times 1} = \frac{1}{2}\{F\}_{n \times 1}^T \{D\}_{n \times 1} \qquad (4\text{-}19)$$

where $\{F\}^T$ is the transpose of the column vector $\{F\}$ representing the forces. Work done is a scalar quantity whose dimensions are (force × length).

The displacements and the forces are related by Eq. 4-1. Substituting in Eq. 4-19,

$$[W]_{1 \times 1} = \frac{1}{2}\{F\}_{n \times 1}^T [f]_{n \times n}\{F\}_{n \times 1} \qquad (4\text{-}20)$$

Taking the transpose of both sides does not change the left-hand side of the equation. The right-hand side becomes the product of the transpose of the matrices on this side but in reverse order (see Appendix A). Therefore

$$[W]_{1 \times 1} = \frac{1}{2}\{F\}_{n \times 1}^T [f]_{n \times n}^T\{F\}_{n \times 1} \qquad (4\text{-}21)$$

From Eqs. 4-20 and 4-21 it follows that the flexibility matrix and its transpose are equal, that is

$$[f]^T = [f] \qquad (4\text{-}22)$$

This means that for a general element of the flexibility matrix

$$f_{ij} = f_{ji} \qquad (4\text{-}23)$$

and is known as *Maxwell's reciprocal relation*. In other words, the flexibility matrix is a symmetrical matrix. This property is useful in forming the flexibility matrix because some of the coefficients need not be calculated or, if they are, a check is obtained. The property of symmetry can also be used to save a part of the computational effort required for matrix inversion or for a solution of equations.

Equation 4-3 tells us that the stiffness matrix is the inverse of the flexibility matrix. Since the inverse of a symmetrical matrix is also symmetrical, the stiffness matrix $[S]$ is a symmetrical matrix. Thus for a general stiffness coefficient,

$$S_{ij} = S_{ji} \qquad (4\text{-}24)$$

This property can be used in the same way as in the case of the flexibility matrix.

Another important property of the flexibility and the stiffness matrices is that the elements on the main diagonal, f_{ii} or S_{ii}, must be positive. The element f_{ii} is the deflection at coordinate i due to a unit force at i. Obviously, the force and the displacement must be in the same direction: f_{ii} is therefore positive. The element S_{ii} is the force required at coordinate i to cause a unit displacement at i. Here again, the force and the displacement must be in the same direction so that the stiffness coefficient S_{ii} is positive.

We should note, however, that in unstable structures, for example a strut subjected to an axial force reaching the buckling load, the stiffness coefficient S_{ii} can be negative. This is discussed further in Chapter 15.

Let us now revert to Eq. 4-20, which expresses the external work in terms of the force vector and the flexibility matrix. If we substitute the force vector from Eq. 4-8 in Eq. 4-19, the work can also be expressed in terms of the displacement vector and the stiffness matrix, thus

$$W = \frac{1}{2}\{F\}^T[f]\{F\} \qquad (4\text{-}25)$$

or

$$W = \frac{1}{2}\{D\}^T[S]\{D\} \qquad (4\text{-}26)$$

The quantity on the right-hand side of these equations is referred to as the *quadratic form* in variable F or D. A quadratic form is said to be *positive definite* if it assumes positive values for any nonzero vector of the variable, and moreover is zero only when the vector of the variables is zero ($\{F\}$ or $\{D\} = \{0\}$). It can also be proved that the determinant of a positive definite symmetrical matrix is greater than zero.

From the above discussion, we can see that the quadratic forms in Eqs. 4-25 and 4-26 represent the external work of a system of forces producing a system of displacements, and this quantity must be positive in a stable structure. Physically this means that work is required to produce any set of displacements $\{D\}$ by the application of a set of forces $\{F\}$. Thus the quadratic forms $(\frac{1}{2})\{F\}^T[f]\{F\}$ and $(\frac{1}{2})\{D\}^T[S]\{D\}$ are positive definite and the matrices $[f]$ and $[S]$ are said to be *positive definite matrices*. It therefore follows that, for a stable structure, the stiffness and flexibility matrices must be positive definite and the systems of linear equations

$$[S]\{D\} = \{F\}$$

and

$$[f]\{F\} = \{D\}$$

are to be positive definite. Further, since the determinants $|S|$ or $|f|$ must be greater than zero, for any nonzero vector on the right-hand side of the equations, each system has a single unique solution, i.e., there is only one set of D_i or F_i values which satisfies the first and second sets of equations, respectively.

The stiffness matrix of a free (unsupported) structure can be readily generated, as for example Eqs. 4-5 to 4-7 for the beam in Fig. 4-4. However, such a matrix is singular and cannot be inverted. Thus no flexibility matrix can be found unless sufficient restraining forces are introduced for equilibrium.

The criterion of nonsingularity of stiffness matrices for stable structures can be used for the determination of buckling loads, as will be discussed in Chapter 15. We shall see that the stiffness of members of a frame is affected by the presence of high axial forces and the frame is stable only if its stiffness matrix is positive definite and thus its determinant is greater than zero. If the determinant is put equal to zero, a condition is obtained from which the buckling load can be calculated.

If a symmetrical matrix is positive definite, the determinants of all its minors are also positive. This has a physical significance relevant to the stiffness matrix of a stable structure. If the displacement D_i is prevented at a coordinate i, for example by introduction of a support, the stiffness matrix of the resulting structure can be obtained simply by deletion of ith row and column from the stiffness matrix of the original structure. Addition of a support to a stable structure results in a structure which is also stable and thus the determinant of its stiffness matrix is positive. This determinant is that of a minor of the original stiffness matrix.

4-7 ANALYSIS OF SYMMETRICAL STRUCTURES BY FORCE METHOD

In practice, many structures have one or more axes or planes of symmetry, which divide the structure into identical parts. If, in addition, the forces applied to the structure are symmetrical, the reactions and internal forces are also symmetrical. This symmetry can be used to reduce the number of unknown redundants or displacements when the analysis is by the force or the displacement method. The analysis by the force method of several symmetrical structures is discussed below.

When the structure to be analyzed is symmetrical and symmetrically loaded, it is logical to select the releases so that the released structure is also symmetrical. In Figs. 4-6a to d, releases are suggested for a number of plane structures with a vertical axis of symmetry. Equal redundants on opposite sides of the axis of symmetry are given the same number; see for example the coordinates at A and D in each of Figs. 4-6a and b. Coordinate 1 in these figures represents the redundant force or the displacement at A (and at D). The flexibility coefficients f_{11} and f_{21} represent, respectively, displacements at coordinate 1 (at A and at D) and at coordinate 2 due to unit forces applied simultaneously at A and at D. The flexibility matrices for the two released structures in Figs. 4-6a and b are (Appendix B)

$$[f]_{\text{Fig. 4-6a}} = \begin{bmatrix} \left(\dfrac{l}{3EI}\right)_{AB} & \left(\dfrac{l}{6EI}\right)_{AB} \\ \left(\dfrac{l}{6EI}\right)_{AB} & \left(\dfrac{l}{3EI}\right)_{AB} + \left(\dfrac{l}{2EI}\right)_{BC} \end{bmatrix}$$

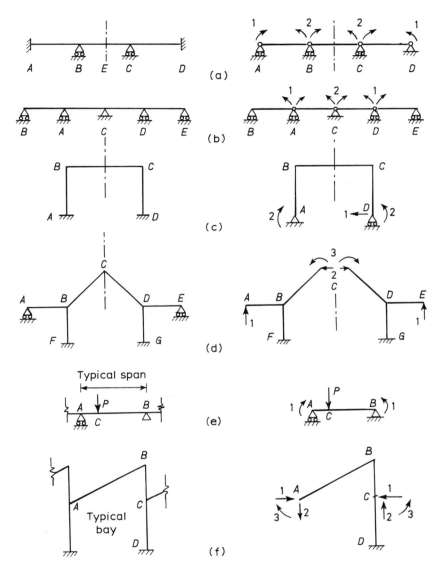

Fig. 4-6. Examples of releases for analysis by the force method of symmetrical continuous beams and frames subjected to symmetrical loading.

$$[f]_{\text{Fig. 4-6b}} = \begin{bmatrix} \left(\dfrac{l}{3EI}\right)_{BA} + \left(\dfrac{l}{3EI}\right)_{AC} & \left(\dfrac{l}{6EI}\right)_{AC} \\ 2\left(\dfrac{l}{6EI}\right)_{AC} & 2\left(\dfrac{l}{3EI}\right)_{AC} \end{bmatrix}$$

In the released structure in Fig. 4-6b, there are two coordinates with the number 1 but only one coordinate with the number 2. This causes f_{21} to be equal to $2f_{12}$ and thus the flexibility matrix is not symmetrical. However, symmetry of the geometry equations $[f]\,\{F\} = -\{D\}$ may be restored by division of the second row by 2.

The released structure in Fig. 4-6a has a roller support at A and a hinge support at D, but this does not disturb the symmetry because under symmetrical loading the horizontal reaction at D must be zero.

Because of symmetry, the slope of the deflected shape of the beam at C in Fig. 4-6b must be horizontal. Thus no rotation or deflection can occur at C; hence the analysis can be done for half the beam only, say BAC, with end C encastré. Moreover, this cannot be done for the beam in Fig. 4-6a because, at the center line, the rotation is zero but the deflection is not.

The frame in Fig. 4-6c may be released by cutting BC at its middle, thus separating the structure into two identical cantilevers. Cutting a member of a plane frame generally releases an axial force, a shearing-force and a bending moment. Moreover, because of symmetry, the shear at the middle of BC must be zero; hence, the analysis needs to determine two unknown redundants instead of three. The same frame may also be released by changing A into a hinge and D into a roller. This does not disturb the symmetry of forces because, when symmetrical loads are applied on the released structure, the horizontal reaction at A is zero and any force introduced at coordinate 1 must produce a symmetrical force at A.

The frame in Fig. 4-6d is released by removal of the roller supports at A and E and by cutting the frame at C. Because of symmetry, only two components represent the internal forces at C, with the vertical component being zero. It may be noted that the released structure will again have the flexibility coefficient f_{21} equal to $2f_{12}$.

Figure 4-6e represents a typical span of a continuous beam having many spans of the same length and the same loading. One span only needs to be analyzed, using the released structure as a simple beam with one unknown redundant representing the connecting moments which must be equal and opposite at the two ends. Here the displacement D_1 and the flexibility coefficient f_{11} represent relative rotations of the two ends of the simple beam AB.

The frame in Fig. 4-6f has an infinite number of identical bays. The analysis may be done for one bay, using the released structure and the three unknown redundants as shown. Here again, the displacements and the flexibility coefficients represent relative translations and rotations of A and C in the released structure.

4-8 ANALYSIS OF SYMMETRICAL STRUCTURES BY DISPLACEMENT METHOD

Advantage can be taken of symmetry to reduce the number of unknown displacement components when the analysis is done by the displacement method. Because of symmetry, the displacement magnitude at a coordinate is zero or is equal to the value at one (or more) coordinate(s). For a zero displacement, the coordinate may be omitted; any two coordinates where the displacement magnitudes are equal may be given the same number. Figure 4-7 shows examples of coordinate systems which may be used in the analysis by the displacement method of symmetrical structures subjected to symmetrical loading.

Figures 4-7a to e represent plane frames in which axial deformations are ignored. In Fig. 4-7a the rotations at B and C are equal and are therefore given the same coordinate number, 1. Because sidesway cannot occur under a symmetrical load, the corresponding coordinate is omitted. In Fig. 4-7b the rotations at B and F are equal, while the rotation at D and the sidesway are zero; hence this frame has only one unknown displacement component. No coordinate systems are shown for the structures in Figs. 4-7c and d, because the displacement components are zero at all nodes. Thus no analysis is needed; the member end-forces are readily available from Appendix C.

The beam over spring supports in Fig. 4-7e has only two unknown displacement components representing the vertical translation at the top of the springs. Because of symmetry, no rotation occurs at B. Also, no coordinates are shown for the rotations at A and C because the end-forces are readily available for a member with one end hinged and the other fixed (Appendices C and D). The stiffness matrix for the structure is (assuming equal spans l and EI = constant)

$$[S] = \begin{bmatrix} \dfrac{3EI}{l^3} + K & -\dfrac{3EI}{l^3} \\ -2\left(\dfrac{3EI}{l^3}\right) & 2\left(\dfrac{3EI}{l^3}\right) + K \end{bmatrix}$$

where K is the spring stiffness (assumed the same for the three springs). The element S_{11} represents the force at coordinate 1 (at A and C) when unit downward displacement is introduced simultaneously at A and at C; S_{21} is the corresponding force at 2. We should note that $S_{21} = 2S_{12}$ so that the stiffness matrix is not symmetrical. This is so because the system has two coordinates numbered 1 but only one coordinate numbered 2. The symmetry of the

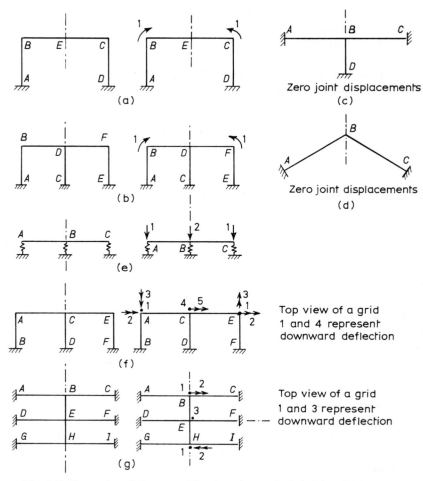

Fig. 4-7. Examples of degrees of freedom for analysis by the displacement method of symmetrical plane frames and grids subjected to symmetrical loading.

equilibrium equations, $[S]\{D\} = -\{F\}$, can be restored by division of the second row by 2.

If axial deformations are considered, additional coordinates must be used. In Fig. 4-7a, a horizontal and a vertical arrow will have to be added at B and at C. Because of symmetry, each of the two corresponding arrows takes the same number, bringing the number of unknown displacements to three. Each of the frames in Figs. 4-7c and d will have one unknown displacement: a vertical translation at B.

The horizontal grids shown in Figs. 4-7f and g have one or more vertical planes of symmetry. The coordinate systems shown may be used for the analysis of the effects of symmetrical loads.

Each of the structures in Figs. 4-7a to g may be analyzed by considering only one-half (or one-quarter) of the structure through separating it at the axis or plane of symmetry. The members situated on the axis or plane of symmetry for the part analyzed should have properties such as A, I, J or K equal to half the values in the actual structure. The same coordinate systems shown on one-half (or one-quarter) of the structure may be used. An exception is the frame in Fig. 4-7a: separation at E will result in a new node at which the vertical translation is unknown, requiring an additional coordinate.

When the analysis of small structures is done by hand or by a calculator, with the matrices generated by the analyst, consideration of one-half or one-quarter of the structure may represent no advantage. However, in large structures with many members and nodes, the analysis is usually performed entirely by computer, considering as small a part of the structure as possible and taking full advantage of symmetry. This is further discussed in Secs. 25-4 to 25-6.

The structure shown in Fig. 4-6f, representing a typical bay of a frame with an infinite number of bays, may be analyzed by the displacement method, using three degrees of freedom at each of A, B and C: a translation in the horizontal and vertical directions, and a rotation. The corresponding three displacements at A and C have the same magnitude and direction; hence there are only six unknown displacements.

4-9   EFFECT OF NONLINEAR TEMPERATURE VARIATION

Analysis of changes in stresses and internal forces in structures due to a variation in temperature or due to shrinkage or creep can be done in the same way. The distribution of temperature over the cross section of members is generally nonlinear, as shown in Fig. 4-8b for a bridge girder (Fig. 4-8a) exposed to the radiation of the sun. In a cross section composed of different materials such as concrete and steel, the components tend to contract or expand differently because of shrinkage and creep. However, contraction and expansion cannot occur freely and changes in stresses occur. In the following, we consider the effect of temperature rise varying nonlinearly over the cross section of members of a framed structure. The temperature rise is assumed constant over the length of individual members.

In a statically determinate frame, no stresses are produced when the temperature variation is linear; in this case the thermal expansion occurs freely, without restraint. This results in changes in length or in curvature of the members, but produces no changes in the reactions or in the internal forces. When the temperature variation is nonlinear, each fiber, being attached to

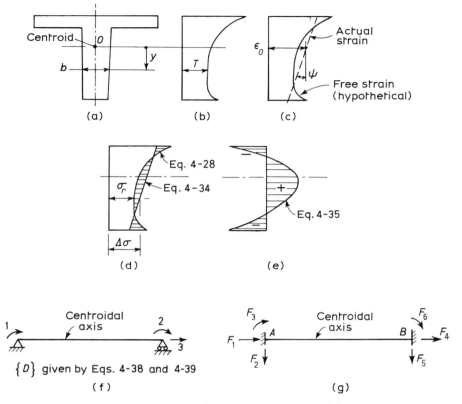

Fig. 4-8. Analysis of the effects of nonlinear temperature variation. (a) Cross section of a member. (b) Distribution of temperature rise. (c) Strain distribution. (d) Stresses σ_r and $\Delta\sigma$. (e) Self-equilibrating stresses. (f) Displacements due to a temperature rise in a simple beam. (g) Fixed-end forces due to a temperature rise.

adjacent fibers, is not free to undergo the full expansion, and this induces stresses. These stresses must be self-equilibrating in an individual cross section as long as the structure is statically determinate. The *self-equilibrating stresses* caused by nonlinear temperature (or shrinkage) variation over the cross section of a statically determinate frame are sometimes referred to as the *eigenstresses.*

If the structure is statically indeterminate, the elongations and the rotations at the member ends may be restrained or prevented. This results in changes in the reactions and in the internal forces which can be determined by an analysis using the force or the displacement method. We should note that the reactions produced by temperature must represent a set of forces in equilibrium.

Let us now analyze the self-equilibrating stresses in a statically determinate member, e.g. a simple beam of homogeneous material, subjected to a nonlinear rise in temperature (Fig. 4-8a). The hypothetical strain which would occur in each fiber if it were free is

$$\varepsilon_f = \alpha T \tag{4-27}$$

where α is the coefficient of thermal expansion and $T = T(y)$ is the temperature rise in any fiber at a distance y below the centroid O. If the expansion is artificially prevented, the stress in the restrained condition will be

$$\sigma_r = -E\varepsilon_f \tag{4-28}$$

where E is the modulus of elasticity. Tensile stress and the corresponding strain are considered positive.

The resultant of σ_r may be represented by a normal force N at O and a moment M about the horizontal axis about O, given by

$$N = \int \sigma_r \, da \tag{4-29}$$

$$M = \int \sigma_r y \, da \tag{4-30}$$

N is considered positive when tensile, and M is positive when it produces tension in the bottom fiber; the corresponding curvature ψ is positive.

To eliminate the artificial restraint, apply N and M in opposite directions, resulting in the following changes in strain at O and in curvature:

$$\varepsilon_O = -\frac{N}{Ea} \tag{4-31}$$

$$\psi = -\frac{M}{EI} \tag{4-32}$$

where a and I are the area of the cross section and its second moment about a horizontal axis through O, respectively. The corresponding strain and stress at any fiber are

$$\varepsilon = \varepsilon_O + y\psi \tag{4-33}$$
$$\Delta\sigma = E(\varepsilon_O + y\psi) \tag{4-34}$$

The addition of σ_r to $\Delta\sigma$ gives the self-equilibrating stress due to temperature:

$$\sigma_s = E(-\alpha T + \varepsilon_O + y\psi) \tag{4-35}$$

The stress σ_s must have a zero resultant because its components σ_r and $\Delta\sigma$ have equal and opposite resultants. The distribution of the self-equilibrating stress is shown in Fig. 4-8e; the ordinates of this graph are equal to the ordinates between the curve σ_r and the straight line $\Delta\sigma$ in Fig. 4-8d.

The changes in axial strain and curvature due to temperature are derived from Eqs. 4-27 to 32:

$$\varepsilon_O = \frac{\alpha}{a}\int Tb\,dy \tag{4-36}$$

$$\psi = \frac{\alpha}{I}\int Tby\,dy \tag{4-37}$$

where $b = b(y)$ is the width of the section. The actual strain distribution over the depth of the section is presented in Fig. 4-8c by a dashed line defined by the values $\Delta\varepsilon_O$ and $\Delta\psi$. The two values may be used to calculate the displacements at the coordinates in Fig. 4-8f (see Appendix B):

$$D_1 = -D_2 = \psi\frac{l}{2} \tag{4-38}$$

$$D_3 = \varepsilon_O l \tag{4-39}$$

When using Eq. 4-38 we should note that, according to the sign convention adopted, ψ in Fig. 4-8c is negative.

If the structure is statically indeterminate, the displacements $\{D\}$, such as those given above, may be used in the force method for the analysis of statically indeterminate reactions and internal forces.

When the analysis is by the displacement method, the values ε_O and ψ (Eqs. 4-36 and 4-37) can be used to determine the internal forces in a member in the restrained condition and the corresponding member end-forces (Fig. 4-8g):

$$N = -Ea\varepsilon_O \tag{4-40}$$

$$M = -EI\psi \tag{4-41}$$

$$\{F\} = E\{a\varepsilon_O, 0, -I\psi, -a\varepsilon_O, 0, I\psi\} \tag{4-42}$$

In the special case when the rise in temperature varies linearly from T_{top} to T_{bot} at top and bottom fibers in a member of constant cross section, the fixed-end forces (Fig. 4-8g) may be calculated by Eq. 4-42, with $\varepsilon_O = \alpha T_O$ and $\psi = \alpha(T_{bot} - T_{top})/h$; here T_O is the temperature at the cross-section centroid and h is the section depth.

The forces F_1 and F_3 are along the centroidal axis, and the other forces are along centroidal principal axes of the member cross section. The six forces are self-equilibrating. The restraining forces at the ends of individual members meeting at a joint should be transformed in the directions of the global axes and summed to give the external restraining forces which will artificially prevent the joint displacements of the structure. (The assemblage of end-forces is discussed further in Sec. 24-9.) In the restrained condition, the stress in any fiber may be calculated by Eq. 4-28.

When the temperature rise varies from section to section or when the member has a variable cross section, ε_O and ψ will vary over the length of the member. Equations 4-40 and 4-41 may be applied at any section to give the variables N and M in the restrained condition; the member forces are given by

$$\{F\} = E\{\{a\varepsilon_O, 0, -I\psi\}_A, \{-a\varepsilon_O, 0, I\psi\}_B\} \qquad (4\text{-}43)$$

The subscripts A and B refer to the member ends (Fig. 4-8g). For equilibrium, a distributed axial load p and a transverse load q must exist. The load intensities (force per length) are given by

$$p = E\frac{d(a\varepsilon_O)}{dx} \qquad (4\text{-}44)$$

$$q = E\frac{d^2(I\psi)}{dx^2} \qquad (4\text{-}45)$$

Positive p is in the direction A to B, and positive q is downwards; x is the distance from A to any section. Equations 4-44 and 4-45 can be derived by considering the equilibrium of a small length of the beam separated by two sections dx apart.

The restraining forces given by Eqs. 4-43 to 4-45 represent a system in equilibrium. The displacements due to temperature can be analyzed by considering the effect of these restraining forces applied in reversed directions. In the restrained condition, the displacement at *all* sections is zero and the internal forces are given by Eqs. 4-40 and 4-41. These internal forces must be superimposed on the internal forces resulting from the application of the reversed self-equilibrating restraining forces in order to give the total internal forces due to temperature.

Example 4-3   The continuous concrete beam in Fig. 4-9a is subjected to a rise of temperature which is constant over the beam length but varies over the depth as follows:

$$T = T_0 + 4.21\, T_{top}\left(\frac{7}{16} - \frac{y}{h}\right)^5 \quad \text{for} -\frac{5h}{16} \leqslant y \leqslant \frac{7h}{16}$$

$$T = T_0 \qquad\qquad\qquad \text{for} \quad \frac{7h}{16} \leqslant y \leqslant \frac{11h}{16}$$

where T is the temperature rise in degrees, T_{top} is the temperature rise in the top fiber, and $T_0 = $ constant. The beam has a cross section as shown in Fig. 4-9b,

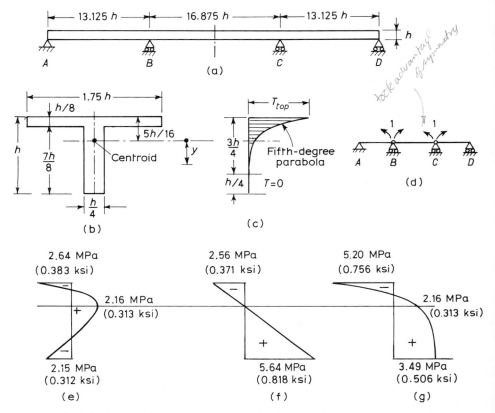

Fig. 4-9. Stresses due to a temperature rise in a continuous beam, Example 4-3. (a) Beam elevation. (b) Beam cross-section. (c) Temperature rise. (d) Released structure and coordinate system. (e) Self-equilibrating stresses. (f) Continuity stresses. (g) Total stresses.

with an area $a = 0.4375h^2$ and a second moment of area about the centroidal axis $I = 0.0416h^4$. Find the stress distribution due to the temperature rise in the section at support B.

Consider $h = 1.6\,\text{m}$ (63 in), $E = 30\,\text{GPa}$ (4350 ksi), $\alpha = 1 \times 10^{-5}$ per degree Celsius ($(5/9) \times 10^{-5}$ per degree Fahrenheit) and $T_{top} = 25°$ Celsius (45° Fahrenheit).

The above equations represent the temperature distribution which can occur in a bridge girder on a hot summer's day. When the temperature rise is constant, the length of the beam will increase freely, without inducing any stress or deflection. Hence, to solve the problem, we may put $T_0 = 0$; the temperature rise will then vary as shown in Fig. 4-9c.

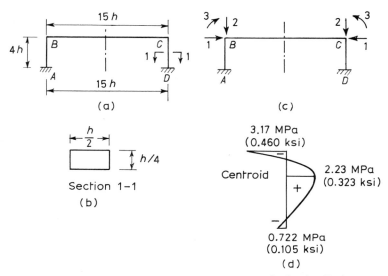

Fig. 4-10. Analysis of stresses due to a temperature rise by the displacement method, Example 4-4. (a) Plane frame with a cross section of BC and a temperature rise as in Figs. 4-9b and c. (b) Cross section of columns AB and BC. (c) Coordinate system. (d) Stress distribution at any cross section of BC.

Step 2   The values $\varepsilon_O = 0.356\alpha T_{top}$ and $\psi = -0.931\alpha T_{top}h^{-1}$ determined in Example 4-3 apply here to member BC. The first three elements of the vector in Eq. 4-42 are the restraining forces at the left-hand end of BC. Because AB and CD are not subjected to temperature change, no restraining forces need to be determined for these two members. The restraining forces at the three coordinates at B or C are

$$\{F\} = E \left\{ \begin{array}{c} 0.4375h^2(0.356\alpha T_{top}) \\ 0 \\ -0.0416h^4(-0.931\alpha T_{top}h^{-1}) \end{array} \right\} = E\alpha T_{top}h^2 \left\{ \begin{array}{c} 0.156 \\ 0 \\ 0.0387h \end{array} \right\}$$

The stress at any fiber of member BC with the joint displacement prevented is the same as determined in Example 4-3. Thus

$$A_r = \sigma_r = -4.21 E\alpha T_{top} \left(\frac{7}{16} - \frac{y}{h} \right)^5 \quad \text{for} \quad -\frac{5h}{16} \leqslant h \leqslant \frac{7h}{16}$$

$$A_r = 0 \qquad\qquad\qquad\qquad\qquad \text{for remainder of depth}$$

Step 3   The stiffness matrix of the structure (using Appendix D) is

$$[S] = \begin{bmatrix} \left(\frac{12EI}{l^3}\right)_{AB} + \left(2\frac{Ea}{l}\right)_{BC} & & \text{symmetrical} \\ 0 & \left(\frac{Ea}{l}\right)_{AB} & \\ -\left(\frac{6EI}{l^2}\right)_{AB} & 0 & \left(\frac{4EI}{l}\right)_{AB} + \left(\frac{2EI}{l}\right)_{BC} \end{bmatrix}$$

Substitute: for AB, $l = 4h$, $a = 0.125h^2$ and $I = 2.60 \times 10^{-3}h^4$; for BC, $l = 15h$, $a = 0.4375h^2$ and $I = 0.0416h^4$.
Hence

$$[S] = E \begin{bmatrix} 58.82 \times 10^{-3}h & & \text{symmetrical} \\ 0 & 31.25 \times 10^{-3}h & \\ -975.0 \times 10^{-6}h^2 & 0 & 8.147 \times 10^{-3}h^3 \end{bmatrix}$$

The stress at any fiber due to unit displacements at the coordinates is

$$[A_u] = E \left[-\left(\frac{2}{l}\right)_{BC} \quad 0 \quad \left(\frac{2EI}{l}\right)_{BC} \frac{y}{I_{BC}} \right]$$

$$[A_u] = E \left[-\frac{2}{15h} \quad 0 \quad \frac{2y}{15h} \right]$$

Step 4   Substitution for $[S]$ and $\{F\}$ and solution of the equilibrium equation $[S]\{D\} = -\{F\}$ gives

$$\{D\} = \alpha T_{top} \begin{Bmatrix} -2.736h \\ 0 \\ -5.078 \end{Bmatrix}$$

Step 5   By superposition, the stress at any fiber is

$$A = A_r + [A_u]\{D\}$$

Substitution for A_r, $[A_u]$ and $\{D\}$ gives

$$\sigma = \sigma_r + E\alpha T_{top}\left(0.365 - 0.677\frac{y}{h}\right)$$

$$\sigma = E\alpha T_{top}\left[-4.21\left(\frac{7}{16} - \frac{y}{h}\right)^5 + 0.365 - 0.677\frac{y}{h}\right] \quad \text{for } -\frac{5h}{16} \leqslant h \leqslant \frac{7h}{16}$$

$$\sigma = E\alpha T_{top}\left(0.365 - 0.677\frac{y}{h}\right) \qquad\qquad \text{for } \frac{7h}{16} \leqslant h \leqslant \frac{11h}{16}$$

Substituting for y the values $-5h/16, 0$ and $11h/16$ and using the values of E, α and T_{top} from the data for Example 4-3 gives the stress values indicated in Fig. 4-10d.

4-10 EFFECT OF SHRINKAGE AND CREEP

The phenomena of shrinkage and creep occur in different materials, but in the following discussion we shall refer mainly to concrete because it is so widely used in structures.

Shrinkage of concrete is a reduction in volume associated with drying in air. As with a temperature drop, if the change in volume is restrained by the difference in shrinkage of various parts of the structure or by the supports or by the reinforcing steel, stresses develop.

If we imagine a material which shrinks without creep, the analysis for the effect of shrinkage can be performed using the equations of Sec. 4-9, but replacing the product αT by ε_f, where ε_f is the free (unrestrained) shrinkage. The effect of swelling can be treated in the same manner as shrinkage but with a reversed sign. Swelling occurs in concrete under water.

The strain which occurs during the application of stress, or within a few seconds thereafter, may be referred to as the *instantaneous strain*. For some materials, the strain continues to increase gradually when the stress is sustained without a change in magnitude. The increase in strain with time, under a sustained stress, is referred to as *creep*. For concrete, creep is two to four times larger than the instantaneous strain, depending upon the quality of concrete, the ambient humidity and temperature, the size of the element considered, the age of concrete when the stress is applied, and the length of the period during which the stress is sustained.

If creep is assumed to be equal to the instantaneous strain multiplied by a constant coefficient, creep will have no effect on the internal forces or stresses in a structure made of a homogeneous material. Creep will cause larger displacements, which can be accounted for by the use of a reduced (effective) E, but this has no effect on the reactions even when the structure is statically indeterminate.

When a concrete structure is constructed and loaded in stages, or when member cross sections contain reinforcement, or when the section is composed of a concrete part connected to structural steel, the creep which is different in various components cannot occur freely. Similarly to temperature expansion, restrained creep induces stresses. In statically determinate structures, creep changes the distribution of stresses within a section without changing the reactions or the stress resultants. This is not so in statically indeterminate structures, where creep influences also the reactions and the internal forces.

In concrete structures, shrinkage and creep occur simultaneously. The stress changes caused by these two phenomena develop gradually over long periods, and with these changes there is associated additional creep. Hence, the analysis must account also for the creep effect of the stress which is gradually introduced. Analysis of the time-dependent stresses and deformations in reinforced and prestressed concrete structures is treated in more detail in books devoted to this subject.[3]

4-11 EFFECT OF PRESTRESSING

In Sec. 2-41, we discussed the effect of prestressing a concrete beam by a cable inserted through a duct and then anchored at the ends. This method is referred to as post-tensioning. In this section, we shall discuss the effects of a post-tensioned tendon which has a nonlinear profile. For simplicity of presentation, we ignore the friction which commonly exists between the tendon and the inner wall of the duct; thus we assume that the tensile force in the tendon is constant over its length. Let P represent the absolute value of the force in the tendon.

A straight tendon as in Fig. 2-3a produces two inward horizontal forces on the end sections, each equal to P. The two forces represent a system in equilibrium, and the reactions in the statically determinate beam are zero. The internal forces at any section are an axial force $-P$ and a bending moment $-Pe$. The sign convention used here is indicated in Fig. 1-18c. The eccentricity e is measured downward from the centroidal axis. The ordinates between the centroidal axis and the tendon profile represent the bending-moment diagram with a multiplier $-P$.

Usually, the tendon profile is selected so that the prestressing partly counteracts the effects of forces which the structure has to carry. Whenever a tendon changes direction, a transverse force is exerted by the tendon on the member. The tendon shown in Fig. 4-11a produces at the end anchorages two inward forces of magnitude P along the tangents to the tendon profile. In addition, an upward force is produced at point C. Thus the profile in Fig. 4-11a may be used in a simple beam carrying a heavy downward concentrated load.

The three forces shown in Fig. 4-11a represent a set of forces in equilibrium. In most practical cases the angle θ between the centroidal axis and the tangent to the tendon is small, so that we need to consider only the axial component of the prestressing force, P, and the perpendicular component, $P\theta$. It follows that the force at C is equal to $P\,\Delta\theta$, where $\Delta\theta$ is the absolute value of the change in slope.

A parabolic tendon produces a uniform transverse load and is thus suitable to counteract the effect of the self-weight and other distributed gravity loads. Appendix K gives the magnitude and direction of the forces produced

[3]See for example: Ghali, A and Favre, R., *Concrete Structures: Stresses and Deformations*, Chapman and Hall, London and New York, 1986; Chapters 16–20 of Neville, A. M., Dilger, W. H. and Brooks, J. J., *Creep of Plain and Structural Concrete*, Longman, London and New York, 1983, pp. 246–349.

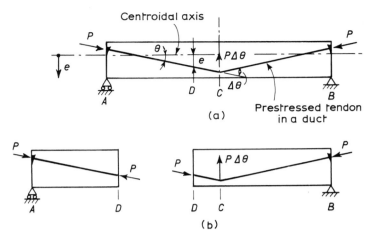

Fig. 4-11. Forces due to prestressing in a statically determine beam.
(a) Representation of prestressing by a system of forces in equilibrium.
(b) Free-body diagrams showing the stress resultant at section D.

by tendons having profiles commonly used in practice. The forces shown represent the effect of the prestressing tendon on the other components which constitute the member. The forces produced by a prestressed tendon must constitute a system in equilibrium.

Prestressing of a statically determinate structure produces no reactions. It can be shown that the resultant of the internal forces at any section is a force P along the tangent of the tendon profile. This can be seen in Fig. 4-11b, where the beam is separated into two parts in order to show the internal forces at an arbitrary section D. The ordinate e between the centroidal axis and the tendon profile represents the bending-moment ordinate with a multiplier $-P$. The axial force is $-P$ and the shear is $-P\theta$.

The internal forces, determined as outlined in the preceding paragraph, are referred to as *primary forces*. Prestressing of statically indeterminate structures produces reactions and hence induces additional internal forces referred to as *secondary forces*. The reactions also represent a system of forces in equilibrium. When the structure is statically determinate the force in the prestressed steel at any section is equal and opposite to the resultant of stresses in other components of the section, and thus the total stress resultant on the whole section is zero. This is not so in a statically indeterminate structure.

In the analysis of the effects of prestressing it is not necessary to separate the primary and secondary effects. The total effect of prestressing may be directly determined by representing the prestressing by a system of external applied self-equilibrating forces (see Appendix K). The analysis is then performed in the usual way by either the force or the displacement method.

Example 4-5 Find the reactions and the bending-moment diagram due to prestressing for the continuous beam shown in Fig. 4-12a. The prestressing tendon profile for each half of the beam is composed of two second-degree parabolas, *ACD* and *DB*, with a common tangent at *D* (Fig. 4-12b). The parabolas have horizontal tangents at *B* and *C*. Assume a constant prestressing force *P*.

The profile shown in Fig. 4-12b is often used in practice for the end span of a continuous beam. The condition that the two parabolas have a common tangent at *D* is required to avoid a sudden change in slope, which would produce an undesired concentrated transverse force at *D*. In design, the

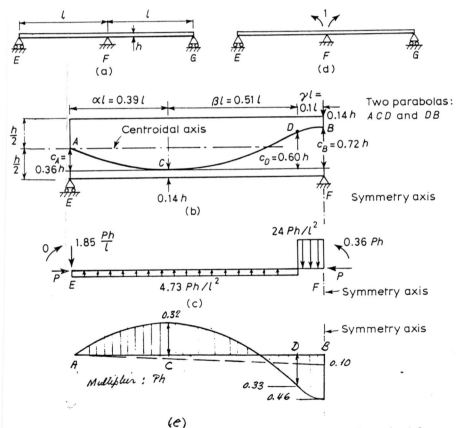

Fig. 4-12. Effect of prestressing on a continuous beam, Example 4-5. (a) Beam elevation. (b) Prestressing tendon profile in one-half of the structure. (c) Self-equilibrating forces produced by prestressing. (d) Released structure and coordinate system. (e) Bending-moment diagram.

geometry of the profile can be obtained by choosing α, c_A and c_B arbitrarily and determining β and c_D so that

$$\beta = \gamma \frac{c_D}{c_B - c_D} \qquad c_D = c_A \frac{\beta^2}{\alpha^2}$$

These two geometrical relations ensure that the slope of the tangents to the two parabolas at D is the same and that ACD is one parabola with a horizontal tangent at C. Solution of the two equations for the unknowns β and c_D may be done by trial and error, noting that $\alpha + \beta + \gamma = 1$ (Newton-Raphson method).

The forces produced by the tendon are calculated by the equations of Appendix K and are shown in Fig. 4-12c for the left-hand half of the beam.

The five steps of the force method (Sec. 2-5) are applied to determine the statically indeterminate reactions:

Step 1   The released structure and the coordinate system are shown in Fig. 4-12d. The actions required are

$$\{A\} = \{R_E, R_F, R_G\}$$

A positive reaction is upwards.

Step 2   The forces in Fig. 4-12c when applied on the released structure give the following displacement (Appendix B):

$$D_1 = 2\frac{Phl}{24EI}\{24(0.1)^2[4\text{-}4(0.1) + (0.1)^2]$$

$$- 4.73(0.9)^2[2 - (0.9)^2]\} + 2(0.36Ph)\frac{l}{3EI}$$

$$= -0.068\frac{Phl}{EI}$$

The applied forces are self-equilibrating and hence produce zero reactions in the released structure; thus

$$\{A_s\} = \{0\}$$

Step 3   The flexibility coefficient (Appendix B) is

$$f_{11} = \frac{2l}{3EI}$$

A unit redundant, $F_1 = 1$, produces the following reactions:

$$[A_u] = \frac{1}{l}\begin{bmatrix} 1 \\ -2 \\ 1 \end{bmatrix}$$

Step 4

$$f_{11}F_1 = -D_1$$

$$F_1 = -\left(\frac{2l}{3EI}\right)^{-1}\left(-0.068\frac{Phl}{EI}\right) = 0.102\,Ph$$

Step 5   Superposition gives the required reactions:

$$\{A\} = \{A_s\} + [A_u]\{F\}$$

$$\begin{Bmatrix} R_E \\ R_F \\ R_G \end{Bmatrix} = \{0\} + \frac{0.102\,Ph}{l}\begin{Bmatrix} 1 \\ -2 \\ 1 \end{Bmatrix} = \frac{Ph}{l}\begin{Bmatrix} 0.102 \\ -0.203 \\ 0.102 \end{Bmatrix}$$

The bending-moment diagram is plotted in Fig. 4-12e. Its ordinate at any section of span *EF* may be expressed as

$$M = -Pe + 0.102\frac{Phx}{l}$$

where *x* is the horizontal distance from *E* to the section and *e* is the vertical distance from the centroidal axis to the tendon profile; *e* is positive where the tendon is below the centroid. Following the convention used throughout this book, the bending-moment ordinates are plotted on the tension face of the beam. The dashed line in Fig. 4-12e is the statically indeterminate bending moment due to prestressing, called the *secondary bending moment*.

4-12   GENERAL

The stiffness and flexibility matrices are related by the fact that one is the inverse of the other, provided the same coordinate system of forces and displacements is used in their formation. However, because the coordinate systems chosen in the displacement and force methods of analysis are not the same, the relation is not valid between the stiffness and flexibility matrices involved in the analyses.

The choice of the method of analysis depends on the problem in hand and also on whether a computer is to be used. The displacement method is generally more suitable for computer programming.

Stiffness matrices for a prismatic member in a three- and two-dimensional frame, given in this chapter, are of value in standardizing the operations. Some other properties discussed are also of use in various respects, including the calculations of buckling loads, considered in Chapter 15.

PROBLEMS

4-1   Ignoring torsion, find by the use of Appendix E the stiffness matrix for the horizontal grid in Prob. 3-14 corresponding to four downward coordinates at I, J, K, and L in this order. Use this matrix to find the deflection at the coordinates due to downward equal forces P at I and J. Take advantage of the symmetry of the structure and of the loading. Draw the bending-moment diagram for beam DG.

4-2 Write the stiffness matrix corresponding to 4 downward coordinates at A, B, C and D in the horizontal grid shown in the figure. Ignore torsion and make use of Appendix E. Find the bending moment in beams AD and BF due to a pair of equal downward loads P at B and C. Take advantage of the symmetry of the structure and of the loading to reduce the number of equations to be solved.

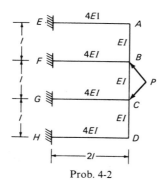

Prob. 4-2

4-3 The figure shows three coordinate systems for a beam of constant flexural rigidity EI. Write the stiffness matrix corresponding to the four coordinates in (a). Condense this stiffness matrix to obtain the stiffness matrices corresponding to the three coordinates in (b) and to the two coordinates in (c).

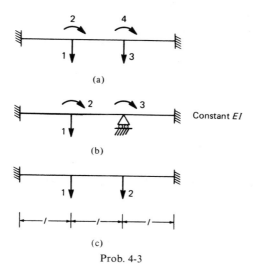

Prob. 4-3

4-4   Apply the requirements of Prob. 4-3 to the member shown in the figure.

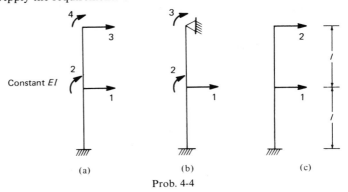

Prob. 4-4

4-5   The figure represents a beam on three spring supports of stiffness $K_1 = K_2 = K_3 = EI/l^3$. Using Appendix E, derive the stiffness matrix corresponding to the three coordinates in the figure. Use this matrix to find the deflection at the three coordinates due to loads $\{F\} = \{3P, P, 0\}$.

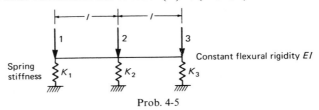

Prob. 4-5

4-6   If the stiffness of any two of the spring supports in Prob. 4-5 is made equal to zero, the stiffness matrix becomes singular. Verify this and explain why.

4-7   Neglecting axial deformations and using Appendix D, write the stiffness matrix for the frame in the figure corresponding to the four coordinates in (a). Condense this matrix to find the 2×2 stiffness matrix corresponding to the coordinates in (b).

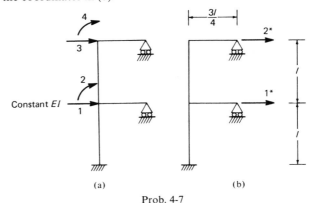

Prob. 4-7

4-8   In Prob. 3-10 we derived the stiffness matrix $[S]_{3 \times 3}$ for a portal frame. Use this stiffness matrix to derive the stiffness coefficient S^*_{11} corresponding to the one-coordinate system shown in the figure. Express the answer in terms of the elements S_{ij} of the matrix $[S]$.

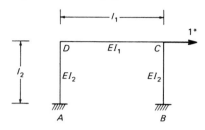

Prob. 4-8

4-9   What is the value of the force P which makes the structure in Prob. 2-7 unstable? (*Hint*: Write the stiffness matrix in terms of P; then find the value of P which makes the determinant vanish.)

4-10   Find the smallest value of P (in terms of Kl) which will make the system in the figure unstable. The bars AB, BC, and CD are rigid. See hints to Probs. 2-7 and 4-9.

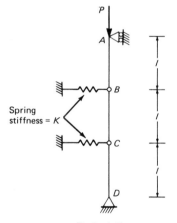

Prob. 4-10

4-11   Find the bending-moment and shearing-force diagrams for the beam in Fig. 4-6a due to a concentrated downward force P at E, the middle of BC. Assume all spans have the same length l, and $EI = $ constant.

4-12   Find the bending moments at A, C and D in the beam of Fig. 4-6b due to a uniform load q per unit length over the whole length, taking advantage of symmetry. The answers to this problem are given in Fig. 2-6.

4-13   Figure 4-6e represents a typical span of a continuous beam having an infinite number of spans. Obtain the bending-moment and shearing-force diagrams and the reactions. Assume the lengths of members are $AB = l$, $AC = l/3$, $CB = 2l/3$, and $EI = $ constant.

4-14   Obtain the bending-moment diagram and the reaction components at A for the frame of Fig. 4-7a due to a uniform downward load q per unit length on BC. Assume the lengths of members are $AB = b$ and $BC = 1.5b$, with EI constant. Consider bending deformations only.

4-15   Obtain the bending-moment diagram and the reaction components at A and C for the frame in Fig. 4-7b due to a uniform downward load q per unit length on BF. Consider bending deformations only. Assume that all members have the same length l, and $EI = $ constant.

4-16   Obtain the bending-moment and shearing-force diagrams for member AB of the frame in Fig. 4-7d, ignoring axial deformations. The frame is subjected to a uniform downward load covering ABC of total magnitude $2ql$, where $2l$ is the length AC. Assume that the inclination of AB to the horizontal is θ.

4-17   The horizontal grid in Fig. 4-7f is subjected to a downward concentrated load P at C. Write the equations of equilibrium at the five coordinates shown. Use the values of $\{D\}$ given in the Answers to check the equations. Consider that all members have the same length l and the same cross section, with $GJ/EI = 0.5$.

4-18   The horizontal grid in Fig. 4-7g is subjected to a uniform downward load q per unit length on DF only. Write the equations of equilibrium at the three coordinates shown. Use the values of $\{D\}$ given in the Answers to check the equations. Consider that all members have the same cross section, with $GJ/EI = 0.5$. Assume the lengths of members are $AB = BC = l$, $BE = EH = l/2$.

4-19   A simple beam of length l and rectangular cross section of width b and depth d is subjected to a rise of temperature which is constant over the length of the beam but varies over the depth of the section. The temperature rise at the top is T and varies linearly to zero at mid-depth; the rise of temperature is zero for the lower half of the section. Determine the stress distribution at any section, the change in length of the centroidal axis and the deflection at mid-span. Modulus of elasticity is E and coefficient of thermal expansion is α.

4-20   If the beam of Prob. 4-19 is continuous over two spans, each of length l, with one support hinged and the other two on rollers, obtain the bending-moment diagrams, the reactions and the stress distribution at the central support due to the same rise of temperature.

4-21   Member BC of the frame in Fig. 4-7a is subjected to a rise of temperature which is constant over the length of the beam but varies linearly over the depth d, from T at the top to zero at the bottom. Obtain the bending-moment diagram and the reaction components at A. Ignore deformations due to axial and shear forces. Assume the lengths of members are $AB = 10d$ and $BC = 15d$, and there is a constant rectangular cross section for all members. The second moment of area is I, and the coefficient of thermal expansion is α.
    Hint: Because of symmetry and because the deformation due to the axial force is ignored, the translation of B and of C is horizontal outward, each with a magnitude of $\alpha T(15d)/4$ (half the elongation of the centroidal axis BC). Thus the structure has only one unknown joint displacement: the rotation at B or C.

4-22   Find the bending-moment diagram, the reactions and the deflection at the middle of span AB due to prestressing of the continuous beam of Prob. 14-16. Assume the prestressing force P is constant and the tendon profile is a second-degree parabola in each span. The answers for this problem are included in the answers for Prob. 14-16.

CHAPTER 5

Strain energy and virtual work

5-1 INTRODUCTION

We have already seen that the knowledge of the magnitude of displacements in a structure is necessary in the analysis of statically indeterminate structures, and in the preceding two chapters we used for the purpose either the displacements due to forces or the forces induced by imposed displacements. Displacements are of course also of interest in design, and, in fact, in some cases the consideration of deflections under design loads may be the controlling factor in proportioning of members.

Calculation of displacements of structures made of materials obeying Hooke's law requires the knowledge of the modulus of elasticity in tension and compression (usually identical) E, and of the shear modulus G. When the stress-strain relation is nonlinear, it is necessary to develop an expression relating forces and deformations, in terms of stress and strain, axial load and extension, or moment and curvature. In this book, we deal mainly with linear structures as these are most common.

When the displacements are required solely for the solution of statically indeterminate linear structures, we need to know only the relative values of Ea, EI, and GJ at all cross sections and for all members of the structures, where a is the area of the cross section, I its second moment of area, and J a torsional constant, with a dimension (length)^4, equal to the polar moment of inertia in the case of a solid or a hollow circular bar. (For other cross sections, see Appendix G).

If the actual values of the displacements are required or if the structure is to be analyzed for the effect of settlement of supports or for a temperature variation, it becomes necessary to know the values of E or G, or both. For some materials, such as metals, these moduli have standard values, given in a handbook or guaranteed by the manufacturer to vary within fixed limits. In the case of concrete,[1] the elastic properties are a function of many variables in the material itself and in its ambient conditions; in reinforced concrete, the amount and type of reinforcement also affect the apparent overall cross-section properties. When cracking occurs, the rigidity of a reinforced concrete member is greatly reduced.[2] It is obvious that, under

[1] See, for example, A. M. Neville, *Properties of Concrete*, Longman, London, 1981.
[2] See Ghali, A. and Favre, R., *Concrete Structures: Stresses and Deformations*, Chapman and Hall, London, New York, 1986.

such circumstances, the accuracy of the structural analysis depends on the reliability of the assumed values of E, G and cross-section properties.

Several methods are available for the determination of displacements, of which the most versatile is the method of virtual work. It is particularly suitable when the displacements at only a few locations are required. However, when the complete deformed shape of a structure is required,[3] the method of virtual work becomes rather laborious and is really suitable only if a computer is used. In such a case, the other methods discussed in Chapter 9 are more advantageous to use.

5-2 GEOMETRY OF DISPLACEMENTS

In most cases, we deal with structures in which the deflections are small compared with the length of the members and the angular rotations result in small translation of the joints. This assumption is essential for much of what follows, but for clarity the displacements will be sketched to a much larger scale than the structure itself. Thus the distortion in the geometry of the deformed structure will be exaggerated.

Consider the frame ABC of Fig. 5-1a, in which the members are connected at B by a rigid joint. This means that the angle ABC between the tangents to BA and BC at B is unaffected by loading, even though the position, direction and shape of the members AB and BC change. The assumption that the displacements are small leads to a simplification in the calculation of the magnitude of translation of the joints. For example, suppose that we want to determine the displaced location of joints B and C in the frame of Fig. 5-1a caused by an angular displacement of θ radians at A in the plane of the frame. Joint B will move through an arc BB_1 of length $l_1\theta$, where l_1 is the length of AB. Since θ and BB_1 are small, the arc BB_1 can be replaced by BB_2, perpendicular to AB and of the same length as BB_1. Thus, while member AB undergoes a rotation only, member BC undergoes a translation and a rotation. The translation means that C moves to C_1, such that $CC_1 = BB_2 = l_1\theta$. The rotation—which must be equal to θ since the angle ABC is unchanged—causes a further translation $C_1C_2 = l_2\theta$, where l_2 is the length of BC.

The total displacement of joint C can also be determined directly by drawing CC_2 perpendicular to the line joining A and C, with a length $CC_2 = l_3\theta$ (see Fig. 5-1b). In other words, the rotation at A is considered to cause a rotation of the whole frame as a rigid body. This is correct if the members of the frame are not subjected to bending deformation, but, even if they are and the axes of the members assume a deflected shape, the trans-

---

[3] In Chapter 13 it will be shown that the deflected shape caused by a specific loading represents an influence line of the structure.

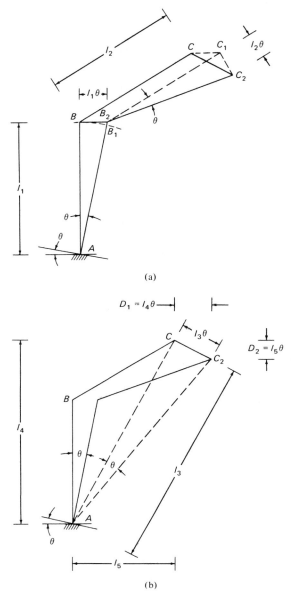

Fig. 5-1. Displacements caused by rotation through a small angle.

lation of joints B and C due to the rotation θ at A alone can still be determined in the same manner.

It is often convenient to express the displacements in terms of components in the direction of rectangular axes. From Fig. 5-1b we can see that the horizontal and vertical components of the displacement CC_2 are

$$D_1 = l_4\theta \qquad \text{and} \qquad D_2 = l_5\theta$$

where l_4 and l_5 are as defined in the figure.

As stated before, this type of calculation of displacements caused by a rotation θ at A is valid for a small θ only. In the presence of axial forces, the contribution of the rotation θ at A to the displacement of C can still be determined by the expressions given above with l_4 and l_5 equal to the original lengths.

From the above discussion we conclude that in a plane frame, a small angular rotation θ at a section causes a relative displacement at any other section equal to $l\theta$, where l is the original length of the line joining the centres of the two sections considered. This displacement is a translation in a direction perpendicular to the line joining the two points.

Because the displacements are small, they can be assumed not to cause gross distortions of the geometry of the structure so that the equilibrium equations can be based on the original directions and relative position of the external forces and of the members. This is a reasonable assumption in the majority of structures but, in some cases, the distorted structure is appreciably changed in geometry so that the equilibrium conditions based on the original geometry of the structure no longer hold. As a result, the structure behaves nonlinearly even if the stress-strain relation of the material is linear. This means that equal increments of external load will not always produce equal increments of displacement: any additional displacement depends upon the total load already acting.

5-3 STRAIN ENERGY

In Sec. 4-6 we derived an expression for the work done on a structure subjected to a system of loads $\{F\}$. This work will be completely stored in the elastic structure in the form of strain energy, provided that no work is lost in the form of kinetic energy causing vibration of the structure, or of heat energy causing a rise in its temperature. In other words, the load must be applied gradually, and the stresses must not exceed the elastic limit of the material. When the structure is gradually unloaded the internal energy is recovered, causing the structure to regain its original shape. Therefore, the external work W and the internal energy U are equal to one another:

$$W = U \qquad\qquad (5\text{-}1)$$

This relation can be used to calculate deflections or forces, but we must first consider the method of calculating the internal strain energy.

Consider a small element of a linear elastic structure in the form of a prism of cross sectional area da and length dl. The area da can be subjected to either a normal stress σ (Fig. 5-2a), or to a shear stress τ (Fig. 5-2b). Assume

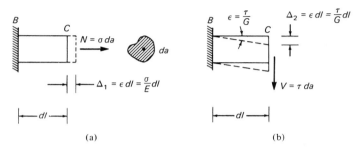

(a)   (b)

Fig. 5-2. Deformation of an element due to (a) normal stress, and (b) shearing stress.

that the left-hand end B of the element is fixed while the right-hand end C is free. The displacement of C under the two types of stress is then

$$\Delta_1 = \frac{\sigma}{E}\, dl \quad \text{and} \quad \Delta_2 = \frac{\tau}{G}\, dl$$

where E is the modulus of elasticity in tension or compression, and G is the modulus of elasticity in shear. When the forces $\sigma\, da$ and $\tau\, da$, which cause the above displacements, are applied gradually, the energy stored in the two elements is

$$dU_1 = \frac{1}{2}(\sigma\, da)\Delta_1 = \frac{1}{2}\frac{\sigma^2}{E}\, dl\, da$$

$$dU_2 = \frac{1}{2}(\tau\, da)\Delta_2 = \frac{1}{2}\frac{\tau^2}{G}\, dl\, da$$

Using ε as a general symbol for strain, the above equations can be put in the general form

$$dU = \frac{1}{2}\sigma\varepsilon\, dv \tag{5-2}$$

where $dv = dl\, da = $ volume of the element, and σ represents a generalized stress, that is either a normal or a shearing stress.

The strain ε in Eq. 5-2 is either due to a normal stress and has magnitude $\varepsilon = \sigma/E$, or due to a shearing stress, in which case $\varepsilon = \tau/G$. But G and E are related by

$$G = \frac{E}{2(1 + v)}$$

where v is Poisson's ratio, so that the strain due to the shearing stress can be written $\varepsilon = 2(\tau/E)(1 + v)$.

The increase in strain energy in any elastic element of volume dv due to a change in strain from $\varepsilon = 0$ to $\varepsilon = \varepsilon_f$ is

$$dU = dv \int_0^{\varepsilon_f} \sigma \, d\varepsilon \qquad (5\text{-}3)$$

where the integral $\int_0^{\varepsilon_f} \sigma \, d\varepsilon$ is called the *strain energy density* and is equal to the area under the stress-strain curve for the material (Fig. 5-3a). If the

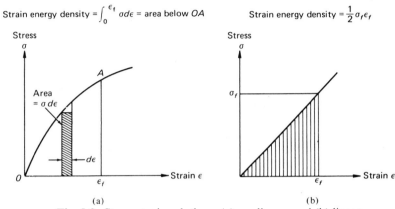

Fig. 5-3. Stress-strain relations: (a) nonlinear, and (b) linear.

material obeys Hooke's law, the stress-strain curve is a straight line (Fig. 5-3b), and the strain energy density is $(\frac{1}{2})\sigma_f \varepsilon_f$.

Any structure can be considered to consist of small elements of the type shown in Fig. 5-4 subjected to normal stresses σ_x, σ_y, σ_z and to shearing stresses τ_{xy}, τ_{xz}, and τ_{yz}, with resulting strains ε_x, ε_y, ε_z, γ_{xy}, γ_{xz}, and γ_{yz}, where

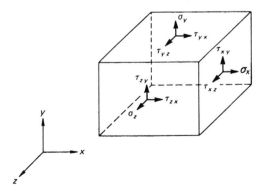

Fig. 5-4. Stress components on an element (stresses on the hidden faces act in the opposite directions).

the subscripts x, y, and z refer to rectangular cartesian coordinate axes. The total strain energy in a linear structure is then

$$U = \frac{1}{2} \sum_{m=1}^{6} \int_{v} \sigma_m \varepsilon_m \, dv \tag{5-4}$$

where m refers to the type of stress and to the corresponding strain. This means that the integration has to be carried out over the volume of the structure for each type of stress separately.

In the case of a nonlinear structure, the strain energy equation corresponding to Eq. 5-4 is obtained by integration of Eq. 5-3 for the six stress and strain components:

$$U = \sum_{m=1}^{6} \int_{v} \int_{0}^{\varepsilon_{fm}} \sigma_m \, d\varepsilon_m \, dv \tag{5-5}$$

where ε_{fm} is the final value of each strain component.

Some care is required in relating the stress and strain. The expression $\varepsilon = (\sigma/E)$ for a linear material applies when the stress is applied normal to one plane only. In the more general case, when six types of stress act (Fig. 5-4), the stress-strain relation for a homogeneous and isotropic material obeying Hooke's law can be written in matrix form

$$\{\varepsilon\} = [e]\{\sigma\} \tag{5-6}$$

where $\{\varepsilon\}$ is a column vector of the six types of strain, that is,

$$\{\varepsilon\} = \{\varepsilon_x, \varepsilon_y, \varepsilon_z, \gamma_{xy}, \gamma_{xz}, \gamma_{yz}\} \tag{5-7}$$

$\{\sigma\}$ is the stress vector, that is,

$$\{\sigma\} = \{\sigma_x, \sigma_y, \sigma_z, \tau_{xy}, \tau_{xz}, \tau_{yz}\} \tag{5-8}$$

and $[e]$ is a square symmetrical matrix representing the flexibility of the element

$$[e] = \frac{1}{E} \begin{bmatrix} 1 & -v & -v & 0 & 0 & 0 \\ -v & 1 & -v & 0 & 0 & 0 \\ -v & -v & 1 & 0 & 0 & 0 \\ 0 & 0 & 0 & 2(1+v) & 0 & 0 \\ 0 & 0 & 0 & 0 & 2(1+v) & 0 \\ 0 & 0 & 0 & 0 & 0 & 2(1+v) \end{bmatrix} \tag{5-9}$$

Equation 5-6 does no more than write in a succinct form the well-known equations of elasticity. For instance, the first equation represented by Eq. 5-6 is

$$\varepsilon_x = \frac{\sigma_x}{E} - \frac{v(\sigma_z + \sigma_y)}{E}$$

Likewise, the fourth one is

$$\gamma_{xy} = \tau_{xy} \frac{2(1 + v)}{E}$$

Inversion of Eq. 5-6 enables us to express stress in terms of strain thus:

$$\{\sigma\} = [d]\{\varepsilon\} \tag{5-10}$$

where $[d] = [e]^{-1}$ is a square symmetrical matrix representing the rigidity of the element. The matrix $[d]$ is referred to as the *elasticity matrix*; for a three-dimensional isotropic solid,

$$[d] = \frac{E}{(1 + v)(1 - 2v)} \begin{bmatrix} (1 - v) & v & v & 0 & 0 & 0 \\ v & (1 - v) & v & 0 & 0 & 0 \\ v & v & (1 - v) & 0 & 0 & 0 \\ 0 & 0 & 0 & \dfrac{(1 - 2v)}{2} & 0 & 0 \\ 0 & 0 & 0 & 0 & \dfrac{(1 - 2v)}{2} & 0 \\ 0 & 0 & 0 & 0 & 0 & \dfrac{(1 - 2v)}{2} \end{bmatrix}$$

$$\tag{5-11}$$

Using the notation of Eqs. 5-7 and 5-8, Eq. 5-4 can be written in the form

$$U = \frac{1}{2} \int_v \{\sigma\}^T \{\varepsilon\} \, dv \tag{5-12}$$

or

$$U = \frac{1}{2} \int_v \{\varepsilon\}^T \{\sigma\} \, dv \tag{5-12a}$$

Substituting Eq. 5-6 or Eq. 5-10 into Eqs. 5-12 and 5-12a, we obtain respectively

$$U = \frac{1}{2} \int_v \{\sigma\}^T [e]\{\sigma\} \, dv \tag{5-13}$$

and

$$U = \frac{1}{2} \int_v \{\varepsilon\}^T [d]\{\varepsilon\} \, dv \tag{5-14}$$

These equations are general for a linear elastic structure of any type. However, in framed structures the strain energy due to different types of stress resultants is best determined separately – in the manner discussed below.

5.31   Strain Energy Due to Axial Force

Consider a segment dl of a member of cross-sectional area a and length l subjected to an axial force N (Fig. 5-5a). The normal stress is $\sigma = N/a$, the strain is $\varepsilon = N/Ea$, and there is no shear. From Eq. 5-2, the total strain energy is

$$U = \frac{1}{2}\int_l \frac{N^2}{Ea}\, dl \tag{5-15}$$

For a prismatic member this becomes

$$U = \frac{1}{2}\frac{N^2 l}{Ea} \tag{5-16}$$

5-32   Strain Energy Due to Bending Moment

Consider a segment dl subjected to a bending moment M about the z axis — one of the principal axes of the cross section (Fig. 5-5b). The normal stress on an element da at a distance $\bar{y}$ from the z axis is $\sigma = M\bar{y}/I$, where I is the second moment of area about the x axis. The corresponding strain is $\varepsilon = \sigma/E = M\bar{y}/EI$. From Eq. 5-2, the strain energy of the element is

$$dU = \frac{1}{2}\frac{M^2\bar{y}^2}{EI^2}\, dv = \frac{1}{2}\frac{M^2\bar{y}^2}{EI^2}da\, dl$$

Integrating over the cross section of the segment dl, we find the strain energy to be

$$\Delta U = \frac{1}{2}\frac{M^2}{EI^2}dl\int_a \bar{y}^2\, da$$

The integral in the above equation is equal to I, so that

$$\Delta U = \frac{1}{2}\frac{M^2}{EI}\, dl \tag{5-17}$$

Hence for the whole structure the strain energy due to bending is

$$U = \frac{1}{2}\int \frac{M^2\, dl}{EI} \tag{5-18}$$

the integration being carried out over the entire length of each member of the structure.

Referring to Fig. 5-5b, we can see that the two sections limiting the segment dl rotate relative to one another by an angle $-d\theta = -(d^2y/dx^2)\, dl$

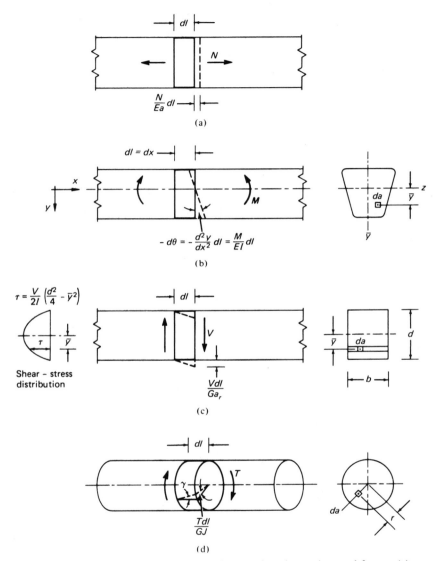

Fig. 5-5. Deformation of a segment of a member due to internal forces. (a) Axial force. (b) Bending moment. (c) Shearing force. (d) Twisting moment.

where y is the downward deflection. The external work done by a couple M moving through $-d\theta$ is

$$\Delta W = -\frac{1}{2} M \, d\theta \tag{5-19}$$

The minus sign is included in this equation because a positive bending moment M (causing tensile stress at bottom fibers) produces a decrease of the slope $\theta = dy/dx$ of the deflected axis of the beam. The difference in slope of the tangent of the deflected beam axis at the right-hand and left-hand ends of the segment dx is $-d\theta = -d^2y/dx^2$.

Since the internal and external work on the segment are equal to one another, that is $\Delta W = \Delta U$, we find from Eqs. 5-17 and 5-19

$$d\theta = -\frac{M}{EI} dl \tag{5-20}$$

Substituting in Eq. 5-18, the strain energy due to bending for the whole structure can also be given in the form

$$U = -\frac{1}{2} \int M \, d\theta \tag{5-21}$$

5-33 Strain Energy Due to Shear

Consider the segment dl of Fig. 5-5c, subjected to a shearing force V. If the shearing stress induced is τ, the shearing strain ε in an element of area da is given by τ/G. Then the strain energy for the segment dl is (from Eq. 5-2)

$$\Delta U = \frac{1}{2} \int \frac{\tau^2}{G} dl \, da \tag{5-22}$$

The integration is carried out over the whole cross section, so that the value of dU depends on the form of variation in the shearing stress over the cross section. For example, it is well known that in a rectangular cross section, the shearing stress distribution is parabolic and can be expressed by

$$\tau = \frac{V}{2I}\left(\frac{d^2}{4} - \bar{y}^2\right) \tag{5-23}$$

where b is the breadth of the section, d its depth, $\bar{y}$ is the distance of the fiber considered from the centroidal axis, and $I = (bd^3/12)$.

Substituting Eq. 5-23 in Eq. 5-22,

$$\Delta U = \frac{1}{2} \times \frac{V^2 \, dl}{4GI^2} \int_{-d/2}^{d/2} \left(\frac{d^2}{4} - \bar{y}^2\right)^2 b \, d\bar{y}$$

Recognizing that $bd = a$ is the area of the cross section, we obtain

$$\Delta U = \frac{1}{2} \left(1.2 \frac{V^2}{Ga} \right) dl \tag{5-24}$$

This applies to a rectangular section. For an arbitrary section, we can write Eq. 5-24 in the general form

$$\Delta U = \frac{1}{2} \left(\frac{V^2}{Ga_r} \right) dl \tag{5-25}$$

where a_r is the reduced area of the cross section. (Thus for a rectangular cross section, $a_r = a/1.2$.) In general, the factor relating the cross sectional area and the reduced area depends upon the shearing stress distribution, which in turn depends upon the shape of the cross section. For rolled steel I-sections, $a_r \simeq$ area of the web. The values of the coefficient for other cross sections are given in standard textbooks on strength of materials.[4]

From Eq. 5-25 we can write the strain energy due to shear for the whole structure as

$$U = \frac{1}{2} \int \frac{V^2}{Ga_r} dl \tag{5-26}$$

where the integration is carried out over the entire length of each member of the structure.

5-34 Strain Energy Due to Torsion

Figure 5-5d shows a segment dl of a circular bar subjected to a twisting moment T. The shearing stress at any point distance r from the center is $\tau = (Tr/J)$, where J is the polar moment of inertia. The corresponding strain is $\varepsilon = (\tau/G)$. From Eq. 5-2, the strain energy in the segment dl is

$$\Delta U = \frac{1}{2} \int \frac{T^2 r^2}{GJ^2} \, da \, dl = \frac{1}{2} \frac{T^2 \, dl}{GJ^2} \int r^2 \, da$$

The integral $\int r^2 \, da = J$ is the polar moment of inertia. Therefore, $\Delta U = (\frac{1}{2})(T^2 \, dl/GJ)$. Thus for the whole structure the strain energy due to torsion is

$$U = \frac{1}{2} \int \frac{T^2 \, dl}{GJ}. \tag{5-27}$$

This equation can be used for members with cross sections other than circular, but in this case J is a torsion constant [of dimensions (length)^4]

---

[4]See, for example, S. P. Timoshenko and J. M. Gere, *Mechanics of Materials*, Van Nostrand, New York, 1972; see p. 372.

which depends on the shape of the cross section.[5] Expressions for J for some structural sections are given in Appendix G.

5-35 Total Strain Energy

In a structure in which all the four types of internal forces discussed above are present, the values of energy obtained by Eqs. 5-15, 5-18, 5-26, and 5-27 are added to give the total strain energy

$$U = \frac{1}{2} \int \frac{N^2 \, dl}{Ea} + \frac{1}{2} \int \frac{M^2 \, dl}{EI} + \frac{1}{2} \int \frac{V^2 \, dl}{Ga_r} + \frac{1}{2} \int \frac{T^2 \, dl}{GJ} \tag{5-28}$$

The integration being carried out along the whole length of each member of the structure. We should note that each integral involves a product of an internal force N, M, V, or T acting on a segment dl and of the relative displacement of the cross section at the two ends of the segment. These displacements are $N\,dl/(Ea)$, $M\,dl/(EI)$, $V\,dl/(Ga_r)$, and $T\,dl/(GJ)$ (see Fig. 5-5).

5-4 COMPLEMENTARY ENERGY AND COMPLEMENTARY WORK

The concept of complementary energy is general and, like strain energy, can be applied to any type of structure. Here, complementary energy will be considered only with reference to a bar of a pin-jointed truss subjected to an axial force. Assume that the bar undergoes an extension e due to an axial tension N, and an extension e_t due to other environmental changes such as temperature, shrinkage, etc. Let $\Delta = e_t + e$. Assume further a relation between the tension N and the extension Δ as shown in Fig. 5-6. Then, for a gradually applied force reaching a final value N_f and causing a final extension e_f, so that $\Delta_f = e_t + e_f$, the strain energy is $U = \int_0^{e_f} N \, de$, which is equal to the area shaded horizontally in Fig. 5-6.

We can now define the complementary energy as

$$U^* = \int_0^{N_f} \Delta \, dN \tag{5-29}$$

or

$$U^* = N_f e_t + \int_0^{N_f} e \, dN \tag{5-29a}$$

which is the area shaded vertically in Fig. 5-6. The complementary energy has no physical meaning and the concept is used only because of its convenience in structural analysis.

[5]For the theory of torsion of noncircular sections, see S. P. Timoshenko, *Strength of Materials*, Part II, 3d ed., Van Nostrand, New York, 1956.

From Fig. 5-6 it is apparent that the sum of complementary energy and strain energy is equal to the area of the rectangle $N_f(e_t + e_f) = U + U^*$

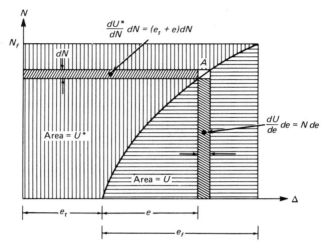

Fig. 5-6. Force-extension relation.

From the same figure, it can also be seen that if the extension is increased from e to $(e + de)$, U increases by an amount $N\,de$, so that

$$\frac{dU}{de} = N \tag{5-30}$$

The corresponding derivative of U^* with respect to N is equal to the extension Δ, that is,

$$\frac{dU^*}{dN} = e_t + e = \Delta \tag{5-31}$$

If a force-displacement diagram (similar to the force-extension diagram of Fig. 5-6) is drawn, the area to the left of the curve is defined as complementary work W^*. Thus

$$W^* = \int_0^{F_f} D\,dF \tag{5-32}$$

where F_f is the final value of the force, F, and $D = d_t + d$ is the total displacement, d is the displacement caused by the force F, and d_t is the displacement due to environmental effects such as temperature, shrinkage, etc.

If a set of forces $F_1, F_2, \ldots, F_n$ is gradually applied to a structure and the total displacements at the location and direction of these forces caused by the forces and other environmental effects are D_1, D_2, ..., D_n, then the complementary work is

$$W^* = \sum_{i=1}^{n} \int_0^{F_{if}} D_i \, dF_i \tag{5-33}$$

where F_{if} is the final value of F_i.

The complementary work W^* is equal to the external work W when the structure is linearly elastic and the displacement D is caused by the external applied loads only (that is, $d_t = 0$); thus

$$W^* = W = \frac{1}{2} \sum_{i=1}^{n} F_i D_i \tag{5-34}$$

5-5 PRINCIPLE OF VIRTUAL WORK

This principle relates a system of forces in equilibrium to a compatible system of displacements in a linear or nonlinear structure. The name of the principle is derived from the fact that a fictitious (virtual) system of forces in equilibrium or of small virtual displacements is applied to the structure and related to the actual displacements or actual forces, respectively. Any system of virtual forces or displacements can be used, but it is necessary that the condition of equilibrium of the virtual forces or of compatibility of the virtual displacements be satisfied. This means that the virtual displacements can be any geometrically possible infinitesimal displacements: they must be continuous within the structure boundary and must satisfy the boundary conditions. With an appropriate choice of virtual forces or displacements, the principle of virtual work can be used to compute displacements or forces.

Let us consider a structure deformed by the effect of external applied forces and of environmental causes such as temperature variation or shrinkage. Let the *actual* total strain at any point be ε, and the corresponding (actual) displacements at n chosen coordinates be D_1, D_2, ..., D_n. Suppose now that before these actual loads and deformations have been introduced, the structure was subjected to a system of *virtual* forces F_1, F_2, ..., F_n at the coordinates 1, 2, ..., n causing a stress σ at any point. The system of virtual forces is in equilibrium but it need not correspond to the actual displacements $\{D\}$. The principle of virtual work states that the product of the actual displacements and the corresponding virtual forces (which is the virtual complementary work) is equal to the product of the actual internal displacements and the corresponding virtual internal forces (which is the virtual complementary energy). Thus

Virtual complementary work = Virtual complementary energy

This can be expressed in a general form

$$\sum_{i=1}^{n} F_i D_i = \int_v \{\sigma\}^T \{\varepsilon\} \, dv \tag{5-35}$$

where σ is a stress corresponding to virtual forces F, and ε is a real strain compatible with the real displacements $\{D\}$. The integration is carried out over the volume of the structure and the summation is for all the virtual forces $\{F\}$. Equation 5-35 states that the values of the complementary work of the virtual external forces and of the complementary energy of the virtual internal forces while *moving* along the real displacements are equal. In other words,

$$\sum_{i=1}^{n} \binom{\text{virtual force}}{\text{at } i} \binom{\text{actual displacement}}{\text{at } i}$$

$$= \int_v \binom{\text{virtual internal}}{\text{forces}} \binom{\text{actual internal}}{\text{displacements}} dv$$

The principle of virtual work in this form is used in Sec. 5-6 to calculate the displacement at any coordinate from the strains due to known actual internal forces. The principle can also be used to determine the external force at a coodinate from the internal forces. In the latter case, the structure is assumed to acquire virtual displacements $\{D\}$ compatible with virtual strain pattern ε at any point. The product of the actual external forces $\{F\}$ and the virtual displacements $\{D\}$ is equal to the product of the actual internal forces and the virtual internal displacements compatible with $\{D\}$. This relation can be written

$$\text{Virtual work} = \text{Virtual strain energy}$$

which is also expressed by Eq. 5-35:

$$\sum_{i=1}^{n} F_i D_i = \int_v \{\sigma\}^T \{\varepsilon\} \, dv$$

but with σ being the actual stress corresponding to the actual forces $\{F\}$ and ε being the virtual strain compatible with the virtual displacements $\{D\}$. In this case, Eq. 5-35 states that the external and internal virtual work of the real forces while moving along the virtual displacements is the same. The same equation can be written in words as follows:

$$\sum_{i=1}^{n} \binom{\text{real force}}{\text{at } i} \binom{\text{virtual displacement}}{\text{at } i}$$

$$= \int_v \binom{\text{real internal}}{\text{forces}} \binom{\text{virtual internal}}{\text{displacements}} dv$$

When the principle of virtual work is used for the calculation of a displacement (or a force) the virtual loads (or the virtual displacements) are chosen in such a manner that the right-hand side of Eq. 5-35 directly gives the desired quantity. This is achieved by the so-called unit-load theorem or the unit-displacement theorem for the calculation of the displacement and force respectively.

5-6   UNIT-LOAD AND UNIT-DISPLACEMENT THEOREMS

When the principle of virtual work is used to calculate the displacement D_j at a coordinate j, the system of virtual forces $\{F\}$ is chosen so as to consist only of a unit force at the coordinate j. Equation 5-35 becomes

$$1 \times D_j = \int_v \{\sigma_{uj}\}^T \{\varepsilon\}\, dv$$

or

$$D_j = \int_v \{\sigma_{uj}\}^T \{\varepsilon\}\, dv \tag{5-36}$$

where σ_{uj} is the virtual stress corresponding to a unit virtual force at j, and ε is the real strain due to the actual loading. This equation is known as the *unit-load theorem*, and is the general form (not limited by linearity) of Eq. 6-4 developed later for linear elastic framed structures.

We should observe that the principle of virtual work as used in the unit-load theorem achieves a transformation of an actual geometrical problem into a fictitious equilibrium problem. Advantages of this will be considered with reference to examples in Sec. 5-7.

The principle of virtual work can also be used to determine the force at a coordinate j if the distribution of the real stresses or of the internal forces is known. The structure is assumed to acquire a virtual displacement D_j at the coordinate j, but the displacement at the points of application of all the other forces remains unaltered. The corresponding compatible internal displacements are now determined. The external and internal virtual work of the real forces while moving along the virtual displacements is the same, that is we can write from Eq. 5-35:

$$F_j \times D_j = \int_v \{\sigma\}^T \{\varepsilon\}\, dv \tag{5-37}$$

Thus $\{\sigma\}$ are the actual stress components due to the system of real loads, and $\{\varepsilon\}$ are the virtual strain components compatible with the configuration of the virtual displacements.

In a linear elastic structure, the strain component ε at any point is proportional to the magnitude of displacement at j, so that

$$\varepsilon = \varepsilon_{uj} D_j \tag{5-38}$$

where ε_{uj} is the strain compatible with a unit displacement at j, there being no other displacement at the points of application of other forces. Equation 5-37 becomes

$$F_j = \int_v \{\sigma\}^T \{\varepsilon_{uj}\}\, dv \tag{5-39}$$

This equation is known as the *unit-displacement theorem*.

The unit-displacement theorem is valid for linear elastic structures only, owing to the limitation of Eq. 5-38. No such limitation is imposed on the unit-load theorem because it is always possible to find a statically determinate virtual system of forces in equilibrium which results in linear relations between virtual external and internal forces, even though the material does not obey Hooke's law.

The unit-displacement theorem is the basis of calculation of the stiffness properties of structural elements used in the finite element method of analysis (see Chapter 19), where a continuous structure (e.g., a plate or a three-dimensional body) is idealized into elements with fictitious boundaries (e.g., triangles or tetrahedra). In one of the procedures used to obtain the stiffness of an element, a stress or displacement distribution within the element is assumed. The unit-displacement theorem is then used to determine the forces corresponding to a unit displacement at specified coordinates on the element.

5-7 VIRTUAL-WORK TRANSFORMATIONS

From the preceding discussion it can be seen that the principle of virtual work can be used to transform an actual geometrical problem into a fictitious equilibrium problem or an actual equilibrium problem into a fictitious geometrical problem.

The first type of transformation can be considered by reference to the plane truss in Fig. 5-7 composed of m pin-jointed members, subjected to external applied forces $\{P\}$ and to a change in temperature of t degrees in some of the members; we want to find the displacement at a coordinate j. By equations of statics the force in any member N_i can be determined and the total extension can be calculated: $\Delta_i = \alpha t_i l_i + N_i l_i/(Ea_i)$, where α is the coefficient of thermal expansion, and l_i and a_i are respectively the length and cross-sectional area of the member.

Once the elongation of all the members is determined, the displacements of the joints can be found from consideration of geometry alone, for instance by drawing a Williot-Mohr diagram (see Sec. 9-2), but this procedure is laborious and, of course, not suitable for computer work. By the unit-load theorem it is possible to transform the actual geometrical problem into a fictitious equilibrium problem, which is easier to solve. The procedure is to apply a unit virtual load at j, causing virtual internal forces N_{uij} in any member i. The virtual work Eq. 5-36 can be applied in this case in the form

$$D_j = \sum_{i=1}^{m} N_{uij} \Delta_i \qquad (5\text{-}40)$$

or in matrix form

$$D_j = \{N_u\}_j^T \{\Delta\} \qquad (5\text{-}41)$$

where $\{N_u\}_j = \{N_{u1j}, N_{u2j}, ..., N_{umj}\}$ and $\{\Delta\} = \{\Delta_1, \Delta_2, ..., \Delta_m\}$.

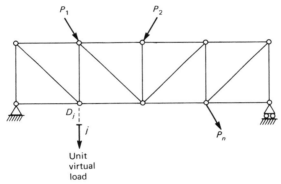

Fig. 5-7. Application of the unit-load theorem to determine displacement at coordinate j of a plane truss.

As the second type of transformation, let us consider the bridge structure shown in Fig. 5-8a, in which it is required to determine the variation in a reaction component A_j, e.g., the horizontal reaction at E when a vertical force P crosses the deck from F to G. We apply a unit virtual displacement at the coordinate j and draw the corresponding shape of the frame, as shown in Fig. 5-8b. (In constructing this figure we should note that the movement of C to C' takes place along the perpendicular to AC: a displacement diagram for joint C is shown in Fig. 5-8c.) It is obvious that, in this statically determinate structure, the members of the frame will move to their new positions as rigid bodies without strain. The vertical displacement of FG is replotted to a larger scale in Fig. 5-8d.

Consider now the unit vertical force P in any position i (Fig. 5-8a) in equilibrium with the reaction components at A and E, including A. Applying

the principle of the virtual work (Eq. 5-35) to the real system of forces in
Fig. 5-8a and the virtual displacements in Fig. 5-8b, we obtain

$$A_j \times 1 - P\eta_i = 0 \tag{5-42}$$

This equation can be easily checked by noting that only A_j and P do
work because the displacements along all the other reactions are zero. The
right-hand side of the equation is zero because no strain is induced by the
virtual displacement. Specifically, the right-hand side of Eq. 5-42 is
$\int_v \{\sigma\}^T \{\varepsilon_{uj}\} \, dv$, where $\{\varepsilon_{uj}\}$ is the strain corresponding to a unit virtual
displacement at j, and $\{\sigma\}$ is the stress in the actual structure due to real
loading, that is due to the load P acting at i. This integral is zero even if the

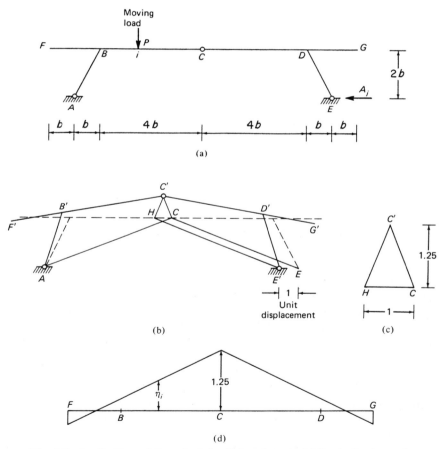

(a)

(b)

(c)

(d)

Fig. 5-8. Application of the principle of virtual work. (a) Statically determinate frame. (b) Virtual displacements. (c) Displacement diagram for joint C.
(d) Influence line for A_j.

structure is statically indeterminate, e.g., if the hinges at A and C are replaced by rigid joints.

The condition that the displacement D_j of the actual structure is zero can be written with the aid of the unit-load theorem Eq. 5-36:

$$D_j = \int_v \{\sigma_{uj}\}^T \{\varepsilon\}\, dv = 0$$

where $\{\sigma_{uj}\}$ is the stress due to a unit force at j and $\{\varepsilon\}$ is the strain in the actual structure due to the load P. But in a linear elastic structure, the stress and strain components σ_{uj} and ε_{uj} are proportional to one another at all points of the structure. Therefore, the integral $\int_v \{\sigma\}^T \{\varepsilon_{uj}\}\, dv$ vanishes also in the case of a statically indeterminate structure.

From Eq. 5-42, $A_j = P\eta_i$, and it follows that the coordinate η_i at any point i is equal to the value of the reaction component A_j due to a unit vertical load at i. The plot of η_i is called the *influence line* of the reaction component A_j. The frame dimensions and the values of A_j are given in Fig. 5-8 so that the result can be checked by considerations of statics.

Further applications of the principle of virtual work in transformation of an actual equilibrium problem to a fictitious geometrical problem will be made in the derivation of influence lines (Chapter 13) and in the analysis of the collapse mechanism of frames and slabs (Chapters 21 and 22).

5-8 GENERAL

The concept of strain energy is important in structural analysis, and it is useful to express the strain energy due to any type of stress in a general form amenable to matrix treatment. It is possible then to consider at the same time components of strain energy due to axial force, bending moment, shear, and torsion.

Complementary energy has no physical meaning, but it is of value in helping to understand some of the energy equations. The same applies to the analogous concept of complementary work.

The principle of virtual work relates a system of forces in equilibrium to a compatible system of displacements in any structure, linear or non-linear. In analysis, we apply virtual forces or virtual displacements and use the equality of the complementary work of virtual external forces and the complementary energy of the virtual internal forces moving along the real displacements. Alternatively, we utilize the equality of external and internal virtual work of the real forces moving along the virtual displacements. Unit-load and unit-displacement theorems offer a convenient formulation. It should be noted that the latter theorem is applicable only to linear structures.

We should note that the principle of virtual work makes it possible to transform an actual geometrical problem into a fictitious equilibrium problem or an actual equilibrium problem into a fictitious geometrical problem, and there exist circumstances when either transformation is desirable.

CHAPTER 6

Method of virtual work and its application to trusses

6-1 INTRODUCTION

In Chapter 5, we adopted the principle of virtual work without proof and used it (in Sec. 5-6) to derive the unit-load theorem, which is applicable to linear or nonlinear structures of any shape. In this chapter and in the succeeding one, we shall consider the method of virtual work further, first with reference to trusses and later (in Chapter 7) to beams and frames. The present chapter will also deal with evaluation of integrals for calculation of displacements by the method of virtual work, and thus serve partly as a preparation for Chapter 7.

To begin with, we shall use the unit-load theorem to calculate displacements due to external applied loading on linear framed structures and we shall also present the theory for this particular application.

6-2 CALCULATION OF DISPLACEMENT BY VIRTUAL WORK

Consider the linear elastic structure shown in Fig. 6-1 subjected to a system of forces F_1, F_2, ..., F_n, causing the stress resultants N, M, V, and T at any section. The magnitude of external and internal work is the same, so

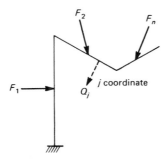

Fig. 6-1. Linear elastic structure used to illustrate the calculation of displacement by virtual work.

that from Eq. 5-1 and 5-28,

$$\frac{1}{2} \sum_{i=1}^{n} F_i D_i = \frac{1}{2} \int \frac{N^2 \, dl}{Ea} + \frac{1}{2} \int \frac{M^2 \, dl}{EI} + \frac{1}{2} \int \frac{V^2 \, dl}{Ga_r} + \frac{1}{2} \int \frac{T^2 \, dl}{GJ} \quad (6\text{-}1)$$

where D_i is the displacement at the location and in the direction of F_i, and N, M, V, and T are the stress resultants at any section due to the $\{F\}$ system.

Suppose that at the time when the forces $\{F\}$ are applied to the structure, there is already a virtual force Q_j acting at the location and in the direction of a coordinate j (Fig. 6-1). This force induces at any section internal forces N_{Qj}, M_{Qj}, V_{Qj}, and T_{Qj}. The magnitude of internal and external work during the application of the $\{F\}$ system of loads is again the same, so that

$$\frac{1}{2} \sum_{i=1}^{n} F_i D_i + Q_j D_j = \frac{1}{2} \left[\int \frac{N^2}{Ea} dl + \int \frac{M^2}{EI} dl + \int \frac{V^2}{Ga_r} dl + \int \frac{T^2}{GJ} dl \right]$$

$$+ \left[\int \frac{N_{Qj} N}{Ea} dl + \int \frac{M_{Qj} M}{EI} dl + \int \frac{V_{Qj} V}{Ga_r} dl + \int \frac{T_{Qj} T}{GJ} dl \right] \quad (6\text{-}2)$$

where D_j is the displacement at j due to the $\{F\}$ system in the direction of the virtual force Q_j. The second term on each side of Eq. 6-2 represents the work due to the force Q_j while moving along the displacement by the $\{F\}$ system. As explained in Sec. 5-5, the coefficient $\frac{1}{2}$ does not appear in these terms because the load Q_j and the corresponding internal forces act at their full value along the entire displacement by the $\{F\}$ system.

Subtracting Eq. 6-1 from Eq. 6-2, we find

$$Q_j D_j = \int \frac{N_{Qj} N}{aE} dl + \int \frac{M_{Qj} M}{EI} dl + \int \frac{V_{Qj} V}{Ga_r} dl + \int \frac{T_{Qj} T}{GJ} dl \quad (6\text{-}3)$$

To determine the deflection at any location and in any direction due to the $\{F\}$ system, we divide Eq. 6-3 by Q_j. Hence, the displacement at j is

$$D_j = \int \frac{N_{uj} N}{Ea} dl + \int \frac{M_{uj} M}{EI} dl + \int \frac{V_{uj} V}{Ga_r} dl + \int \frac{T_{uj} T}{GJ} dl \quad (6\text{-}4)$$

where

$$N_{uj} = \frac{N_{Qj}}{Q_j} \qquad M_{uj} = \frac{M_{Qj}}{Q_j} \qquad V_{uj} = \frac{V_{Qj}}{Q_j} \qquad \text{and} \qquad T_{uj} = \frac{T_{Qj}}{Q_j}$$

These are the values of the internal forces at any section due to a unit virtual force ($Q_j = 1$) applied at the coordinate j where the displacement is required.

Referring back to Eq. 5-36, we can see that Eq. 6-4 is, in fact, a particular case of the unit-load theorem applicable to framed structures.

In order to use Eq. 6-4 for the determination of the displacement at any section, the internal forces at all sections of the structure must be determined due to: (i) the actual loads, and (ii) a unit virtual force. The latter is a fictitious force or a *dummy load* introduced solely for the purpose of the analysis. Specifically, if the required displacement is a translation, the fictitious load is a concentrated unit force acting at the point and in the direction of the required deflection. If the required displacement is a rotation, the unit force is a couple acting in the same direction and at the same location as the rotation. If the relative translation of two points is to be found, two unit loads are applied in opposite directions at the given points along the line joining them. Similarly, if a relative rotation is required, two unit couples are applied in opposite directions at the two points.

The internal forces N_{uj}, M_{uj}, V_{uj}, and T_{uj} are forces per unit virtual force. If the displacement to be calculated is a translation, and the pound, inch system is used, then N_{uj}, M_{uj}, V_{uj}, and T_{uj} have respectively the dimensions: lb/lb, lb in./lb, lb/lb, and lb in./lb. When the virtual force is a couple, N_{uj}, M_{uj}, V_{uj}, and T_{uj} have respectively the dimensions: lb/lb in., lb in./lb in., lb/lb in., and lb in./lb in. A check on the units of both sides of Eq. 6-4, when used to determine translation or rotation, should easily verify the above statements.

Each of the four terms on the right-hand side of Eq. 6-4 represents the contribution of one type of internal forces to the displacement D_j. In the majority of practical cases, not all the four types of the internal forces are present, so that some of the terms in Eq. 6-4 may not be required. Furthermore, some of the terms may contribute very little compared to the others and may therefore be neglected. For example, in frames in which members are subjected to lateral loads, the effect of axial forces and shear is very small compared to bending. This is, however, not the case in members with a high depth-to-length ratio or with a certain shape of cross sections, when the displacement due to shear represents a significant percentage of the total. This will be illustrated in Example 7-2.

6-3 DISPLACEMENT OF STATICALLY INDETERMINATE STRUCTURES

As shown in Chapter 5, the principle of virtual work is applicable to any structure, whether determinate or indeterminate. However in the latter case, the internal forces induced by the real loading in all parts of the structure must be known. This requires the solution of the statically indeterminate structure by any of the methods discussed in Chapters 2 and 3.

Furthermore, we require the internal forces N_{uj}, M_{uj}, V_{uj} and T_{uj} due to a unit virtual load applied at j. These forces can be determined for any released structure satisfying the requirement of equilibrium with the unit virtual load at j. Thus it is generally sufficient to determine the internal forces due to a unit virtual load at j acting on a released stable statically determinate structure obtained by the removal of arbitrarily chosen redundants. This is so because the principle of virtual work relates a compatible system of deformations of the actual structure to a virtual system of forces in equilibrium which need not correspond to the actual system of forces (see Sec. 5-5).

As an example, we can apply the above procedure to the frame of Fig. 6-2a in order to find the horizontal displacement at C, D_4. Bending

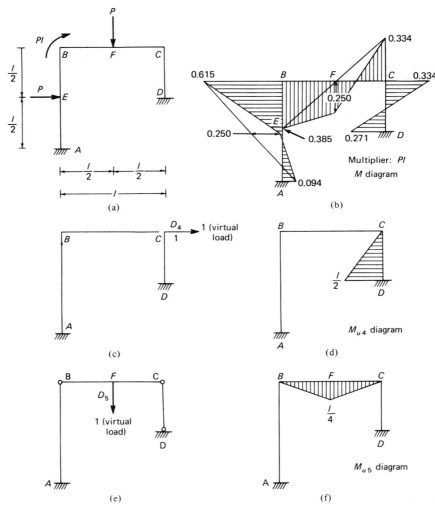

Fig. 6-2. Displacement of a statically indeterminate plane frame by virtual work.

deformations need only be considered. The frame has a constant flexural rigidity EI.

The bending-moment diagram was obtained in Example 3-2 and is shown again in Fig. 6-2b.

A unit virtual load is now applied at coordinate 4 to a statically determinate system obtained by cutting the frame just to the left of C (thus forming two cantilevers), Fig. 6-2c. The bending-moment diagram for M_{u4} is shown in Fig. 6-2d.

Applying Eq. 6-4 and considering bending only,

$$D_4 = \int \frac{M_{u4}M}{EI} dl$$

This integral needs to be evaluated for the part CD only because M_{u4} is zero in the remainder of the frame. Using Appendix H, we find

$$D_4 = \frac{1}{EI}\left[\frac{l}{2}\frac{(l/2)}{6}(2 \times 0.271Pl - 0.334Pl)\right] = 0.0087\frac{Pl^3}{EI}$$

Suppose now that we want to find the vertical deflection at F, denoted by D_5. We assume that the virtual unit load acts on the three-hinged frame of Fig. 6-2e. The corresponding bending-moment diagram M_{u5} is shown in Fig. 6-2f. By the same argument as used for D_4, we obtain

$$D_5 = \int \frac{M_{u5}M}{EI} dl$$

The integral has to be evaluated for the part BC only because M_{u5} is zero elsewhere. Using the table in Appendix H, we find

$$D_5 = 0.0240\frac{Pl^3}{EI}$$

It may be instructive to calculate D_4 and D_5 using some other choice of the virtual system of forces.

6-4 EVALUATION OF INTEGRALS FOR CALCULATION OF DISPLACEMENT BY METHOD OF VIRTUAL WORK

In the preceding section we have seen that integrals of the type

$$\int \frac{M_{uj}M}{EI} dl$$

often occur in the equations of virtual work. We shall consider here how this particular integral can be evaluated but other integrals of this type can, of course, be treated in a similar manner.

If a structure consists of m members, the integral can be replaced by a summation

$$\int \frac{M_{uj}M}{EI} \, dl = \sum_{members} \int_l M_{uj} \frac{M}{EI} \, dl \qquad (6\text{-}5)$$

so that the problem lies really in the evaluation of the integral for one member.

Let us consider, therefore, a straight member AB of length l and of variable cross section, subjected to bending (Fig. 6-3a). Figure 6-3b shows the bending-moment diagram due to an arbitrary loading. If we divide each ordinate of this diagram by the value of EI at the given section, we obtain the $M/(EI)$ diagram shown in Fig. 6-3c.

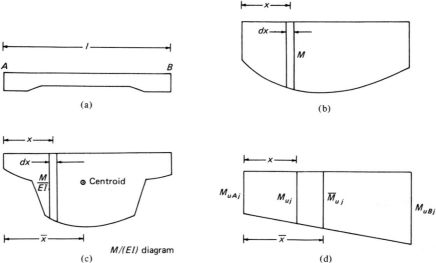

Fig. 6-3. Evaluation of integral $\int_l M_{uj} \dfrac{M}{EI} \, dl$. (a) Member AB. (b) Bending-moment diagram due to any real loading. (c) $M/(EI)$ diagram. (d) Bending-moment diagram due to a unit virtual load.

Since the member AB is straight, the plot of the bending-moment M_{uj} between A and B due to a unit virtual force applied at any coordinate j (not located between A and B) must be a straight line (Fig. 6-3d). Let the ordinates of this diagram at A and B be M_{uAj} and M_{uBj} respectively. The ordinate at any section distance x from A is then

$$M_{uj} = M_{uAj} + (M_{uBj} - M_{uAj}) \frac{x}{l}$$

Substituting this value of M_{uj} in the integral, we obtain

$$\int_l M_{uj} \frac{M}{EI} dl = M_{uAj} \int_0^l \frac{M}{EI} dx + \left(\frac{M_{uBj} - M_{uAj}}{l}\right) \int_0^l \frac{M}{EI} x\, dx \quad (6\text{-}6)$$

Let us denote the area of the $M/(EI)$ diagram by $a_M = \int_0^l (M/EI)\, dx$. If $\bar{x}$ is the distance of the centroid of the $M/(EI)$ diagram from the left-hand end A, then the first moment of the area a_M about A is $\int_0^l (M/EI) x\, dx = a_M \bar{x}$. Substituting for the integrals in Eq. 6-6,

$$\int_l \frac{M_{uj}M}{EI} dl = a_M \left[M_{uAj} + (M_{uBj} - M_{uAj})\frac{\bar{x}}{l} \right]$$

The term in square brackets in this equation is equal to the ordinate $\bar{M}_{uj}$ of the M_{uj} diagram (Fig. 6-3d) at the section through the centroid of the $M/(EI)$ diagram. Therefore

$$\int_l \frac{M_{uj}M}{EI} dl = a_M \bar{M}_{uj} \quad (6\text{-}7)$$

By parallel argument, the other integrals of similar type (see, for instance Eq. 6-4) can be evaluated thus

$$\int_0^l \frac{V_{uj}V}{Ga_r} dx = a_V \bar{V}_{uj} \quad (6\text{-}8)$$

$$\int_0^l \frac{T_{uj}T}{GJ} dx = a_T \bar{T}_{uj} \quad (6\text{-}9)$$

$$\int_0^l \frac{N_{uj}N}{Ea} dx = a_N \bar{N}_{uj} \quad (6\text{-}10)$$

The notation used on the right-hand side of Eqs. 6-7 to 6-10 can be summarized as follows:

a_M, a_V, a_T, and a_N are the areas of the $M/(EI)$, $V/(Ga_r)$, $T/(GJ)$, and $N/(Ea)$ diagrams respectively

$\bar{M}_{uj}$, $\bar{V}_{uj}$, $\bar{T}_{uj}$, and $\bar{N}_{uj}$ are the values respectively of M_{uj}, V_{uj}, T_{uj}, and N_{uj} at the centroid of each of the areas a considered above

The areas and the location of the centroid for some geometrical figures frequently needed for the application of Eqs. 6-7 to 6-10 are listed in Appendix F. For members with a constant EI, the values of the integral $\int M_u M\, dl$ for geometrical figures which commonly form bending-moment diagrams are given in Appendix H.

We should note that the value of the integral does not depend on the sign convention used for the internal forces, provided that the same convention is used for the forces due to the real and virtual loadings.

In the majority of cases, the plots of M_{uj}, V_{uj}, T_{uj} and N_{uj} are straight lines or can be divided into parts such that the plot of the ordinate is one straight line; under these circumstances, Eqs. 6-7 to 6-10 can be applied. However, in structures with curved members, these plots are not rectilinear, but it is possible to approximate a curved member by a number of straight parts over which the diagram of the internal forces due to virtual loading can be considered straight.

6-5 TRUSS DEFLECTION

In a plane or space truss composed of m pin-jointed members, with loads applied solely at joints, the only internal forces present are axial, so that Eq. 6-4 can be written

$$D_j = \sum_{i=1}^{m} \int_l \frac{N_{ui}N_i}{Ea}\,dl \qquad (6\text{-}11)$$

Generally, the cross section of any member is constant along its length. Equation 6-11 can therefore be written

$$D_j = \sum_{i=1}^{m} \frac{N_{uij}N_i}{E_i a_i}\, l_i \qquad (6\text{-}12)$$

where m is the number of members, N_{uij} is the axial force in member i due to a virtual unit load at j, and $N_i l_i/(E_i a_i)$ is the change in length of a member caused by the real loads, assuming that the material obeys Hooke's law.

Equation 6-12 can also be written in the form

$$D_j = \sum_{i=1}^{m} N_{uij}\,\Delta_i \qquad (6\text{-}13)$$

where Δ_i is the real change in length of the ith member. This form is used when the deflection due to causes other than loading is required, for example, in the case of a change in temperature of some members. A rise or drop in temperature of the ith member by t degrees causes a change in length of

$$\Delta_i = \alpha t l_i \qquad (6\text{-}14)$$

where l_i is the length of the member, and α is the coefficient of thermal expansion.

Equation 6-13 is valid for linear and nonlinear trusses.

Example 6-1 The *plane truss* of Fig. 6-4a is subjected to two equal loads P at E and D. The cross-sectional area of the members labeled 1, 2, 3, 4, and 5 is a, and that of members 6 and 7 is $1.25a$. Determine the horizontal displacement D_1 at joint C, and the relative movement, D_2, of joints B and E.

The internal forces in the members due to the actual loading are calculated by simple statics and are given in Fig. 6-4b.

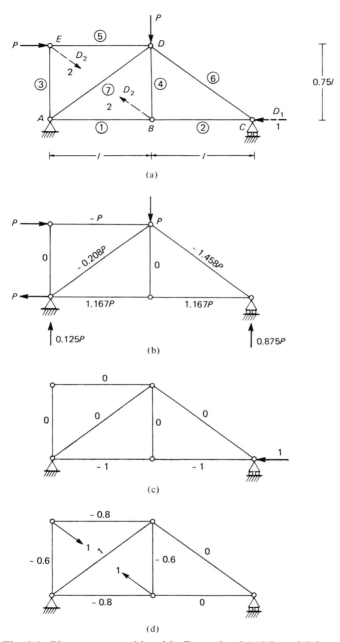

Fig. 6-4. Plane truss considered in Examples 6-1, 6-2, and 6-3.

The internal forces due to a unit virtual load at the coordinates D_1 and D_2 are then determined (Figs. 6-4c and d). The values obtained can be checked by considering the equilibrium of the joints.

We have taken axial forces as positive when tensile, which is the common practice. However, the final result is not affected by the sign convention used.

The displacements D_1 and D_2 are calculated by Eq. 6-12. It is convenient to use a tabular form, as shown in Table 6-1. The table is self-explanatory, the

TABLE 6-1. Example 6-1: Calculation of Displacements D_1 and D_2

Member	Properties of member			Actual loading	Calculation of D_1		Calculation of D_2	
	Length l	Area of cross section a	$\dfrac{l}{Ea}$	N	N_u	$\dfrac{N_u N l}{Ea}$	N_u	$\dfrac{N_u N l}{Ea}$
1	1	1	1	1.167	-1	-1.167	-0.8	-0.933
2	1	1	1	1.167	-1	-1.167	0	0
3	0.75	1	0.75	0	0	0	-0.6	0
4	0.75	1	0.75	0	0	0	-0.6	0
5	1	1	1	-1	0	0	-0.8	$+0.800$
6	1.25	1.25	1	-1.458	0	0	0	0
7	1.25	1.25	1	-0.208	0	0	1	-0.208
Multiplier	l	a	$l/(Ea)$	P	—	$Pl/(Ea)$	—	$Pl/(Ea)$
						-2.334		-0.341

displacements D_1 and D_2 being equal to the sum of the appropriate column of $N_u N l/(Ea)$. Thus, $D_1 = -2.334\ Pl/(Ea)$ and $D_2 = -0.341\ Pl/(Ea)$.

The negative sign of D_1 indicates that the displacement is in a direction opposite to the direction of the virtual load in Fig. 6-4c. This means that the horizontal translation of joint C is to the right. Likewise, the negative sign of D_2 means that the relative movement of B and E is opposite to the directions of the virtual forces assumed, that is separation.

Example 6-2 For the same truss of Fig. 6-4a, find the displacement D_2 due to a rise in temperature of members 5 and 6 by 30 degrees. (The loads P are not acting in this case.) Assume the coefficient of thermal expansion $\alpha = 0.6 \times 10^{-5}$ per degree.

A unit virtual force is applied in the same manner as in Fig. 6-4d, so that the forces in the members, indicated on the figure, can be used again.

The real change in length occurs in two members only, so that from Eq. 6-14

$$\Delta_5 = 0.6 \times 10^{-5} \times 30 \times l = 18 \times 10^{-5} l$$

and

$$\Delta_6 = 0.6 \times 10^{-5} \times 30 \times 1.25l = 22.5 \times 10^{-5} l$$

Applying Eq. 6-13 with the summation carried out for members 5 and 6 only,

$$D_2 = \sum_{i = 5, 6} N_{ui2} \, \Delta_i$$

or

$$D_2 = -0.8(18 \times 10^{-5}l) + 0 \times (22.5 \times 10^{-5}l) = -14.4 \times 10^{-5}l$$

It is clear that the same method of calculation can be used if Δ_i is due to any other cause, for example lack of fit (see Sec. 2-41).

6-6   TRUSS DEFLECTION USING MATRIX ALGEBRA

We have seen that Eq. 6-12 gives, for a truss with m members, the deflection at a joint in a direction specified by coordinate j. The same result can be obtained by the multiplication of three matrices, thus

$$D_j = \{N_u\}_{j_{m \times 1}}^T [f_M]_{m \times m} \{N\}_{m \times 1} \qquad (6\text{-}15)$$

where $\{N_u\}_j^T$ is the transpose of matrix $\{N_u\}_j$, the elements of $\{N_u\}_j$ being the forces in the members due to a unit virtual load at j. The elements of $\{N\}$ are the forces in the members caused by the real loading. Finally,

$$[f_M]_{m \times m} = \begin{bmatrix} (l/aE)_1 & & & \\ & (l/aE)_2 & & \\ \text{Elements} & & & \\ \text{not shown} & & (...) & \\ \text{are zero} & & & (l/aE)_m \end{bmatrix} \qquad (6\text{-}16)$$

where the elements on the diagonal represent the flexibility in axial deformation of the individual members, that is the change in length due to a unit axial force. The matrix $[f_M]$ is referred to as the flexibility matrix of the *unassembled structure*.

The use of matrices is particularly advantageous when the calculations are made by a computer and the deflections at several joints are required. Suppose that the deflections are required at n coordinates. A unit virtual load has to be applied at each one of these coordinates separately and the corresponding set of forces determined. These forces can be arranged in one matrix:

$$[N_u]_{m \times n} = \begin{bmatrix} N_{u11} & N_{u12} & \cdots & N_{u1n} \\ N_{u21} & N_{u22} & \cdots & N_{u2n} \\ \cdots & \cdots & \cdots & \cdots \\ N_{um1} & N_{um2} & & N_{umn} \end{bmatrix} \qquad (6\text{-}17)$$

where the elements in any column j are the forces in the members corresponding to a unit virtual load applied at j. The first subscript of N_u identifies the truss member and the second indicates the coordinate j at which the unit virtual load is applied.

Using Eq. 6-17, Eq. 6-15 can be extended to determine the deflections at all n coordinates:

$$\{D\}_{n \times 1} = [N_u]_{m \times n}^T [f_M]_{m \times m} \{N\}_{m \times 1} \tag{6-18}$$

If the deflection is required for various cases of loading, say p in number, Eq. 6-18 can be extended to

$$[D]_{n \times p} = [N_u]_{m \times n}^T [f_M]_{m \times m} [N]_{m \times p} \tag{6-19}$$

For easy reference, the notation in this equation is repeated:

D = deflection at a coordinate

N_u = force in a member due to a unit virtual load acting at the co-ordinate. The elements in each column of matrix $[N_u]$ are the forces due to a unit virtual load at the appropriate coordinate

f_M = $l/(aE)$ = flexibility of member

N = force in a member due to the real loading; elements in each column of matrix $[N]$ are the forces corresponding to one case of loading

n = number of coordinates at which the displacement is required

m = number of members

p = number of cases of loading

If the deflection is required due to the change in length of members $[\Delta]$ caused by a change in temperature, or by other deformational causes, Eq. 6-13 can be used in matrix form similar to Eq. 6-19

$$[D]_{n \times p} = [N_u]_{m \times n}^T [\Delta]_{m \times p} \tag{6-20}$$

The elements in each column of Δ are the extension of each member corresponding to one case.

Example 6-3 Solve the problem of Example 6-1 using matrices. The deflections are required at the coordinates 1 and 2 shown in Fig. 6-4a.

In this problem, we have $n = 2$, $m = 7$, and $p = 1$. The forces which form the matrices $[N]$ and $[N_u]$ are shown in Figs. 6-4b, c, and d. Thus

$$[N_u]_{7 \times 2} = \begin{bmatrix} -1 & -0.8 \\ -1 & 0 \\ 0 & -0.6 \\ 0 & -0.6 \\ 0 & -0.8 \\ 0 & 0 \\ 0 & 1 \end{bmatrix} \qquad [N]_{7 \times 1} = P \begin{bmatrix} 1.167 \\ 1.167 \\ 0 \\ 0 \\ -1 \\ -1.458 \\ -0.208 \end{bmatrix}$$

and

$$[f_M]_{7 \times 7} = \frac{l}{Ea} \begin{bmatrix} 1 & & & & & & \\ & 1 & & & & & \\ & & 0.75 & & & & \\ & & & 0.75 & & & \\ & \text{Elements} & & & 1 & & \\ & \text{not shown} & & & & 1 & \\ & \text{are zero} & & & & & 1 \end{bmatrix}$$

Substituting in Eq. 6-19,

$$[D]_{2 \times 1} = [N_u]_{7 \times 2}^T [f_M]_{7 \times 7} [N]_{7 \times 1} = \frac{Pl}{Ea} \begin{bmatrix} -2.334 \\ -0.341 \end{bmatrix}$$

6-7 GENERAL

The method of virtual work is general. It can be used for the calculation of the displacements in plane and space structures, statically determinate or indeterminate. However, in all cases the structure must first be analyzed and the stress resultants have to be determined. The basis of the method is the principle of virtual work relating a system of forces in equilibrium to a compatible system of displacements.

Calculation of displacement by virtual work involves the determination of the internal forces due to the actual loading and due to a unit load applied at each point where the displacement is required. If several displacements are required the computations can become tedious but the use of matrices systematizes the calculation procedure and adapts the problem to a computer solution.

In the general case of application of the method of virtual work, four types of stress resultants, axial force, bending-moment, shear and torsion, contribute to the displacements. In pin-jointed trusses, the axial forces are the only contributors. The virtual work expressions involve integrals of a product of two functions; the integral can be conveniently evaluated by taking advantage of the fact that one of the two functions is generally linear.

PROBLEMS

6-1   Using the method of virtual work, find D_1, the vertical deflection of joint H, and D_2, the relative translation in the direction of OB of joints O and B of the truss shown in the figure. The changes in length of the bars (inch or cm) are indicated in the figure.

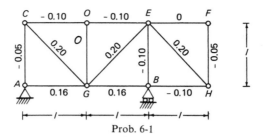

Prob. 6-1

6-2   Find the displacement at coordinate 1 which represents the relative translation of joints C and B of the truss in the figure. Assume $l/(Ea) = B$ to be
constant for all members. If a member CB is added with the same $l/(Ea)$ value
as the other members, what are the forces in the members caused by the same
loading?

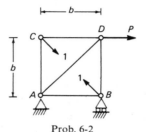

Prob. 6-2

6-3   Find the displacement along the line of action of the force P for the space
truss shown in the figure. The value of Ea is the same for all members.

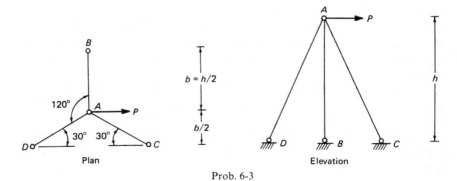

Prob. 6-3

6-4   For the plane truss shown in the figure, find: (a) the vertical deflection at E
due to the given loads, (b) the camber at E in the unloaded truss if member
EF is shortened by $b/2000$, and (c) the forces in all members if a member is
added between C and F and the truss is subjected to the given loads. Consider
Ea = constant for all members.

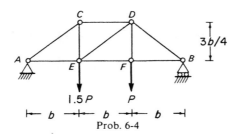

$3b/4$

$1.5P$ P

Prob. 6-4

6-5   For the plane truss shown in the figure, find: (a) the deflection at C due to the load P, (b) the vertical deflection at C in the unloaded truss if members DE and EC are each shortened by 1/8 in., and (c) the forces in all members if a member of cross-sectional area of 4 in.^{2} is added between B and D and the truss is subjected to the load P. Assume $E = 30,000$ ksi and the cross-sectional area of the members as indicated in the figure.

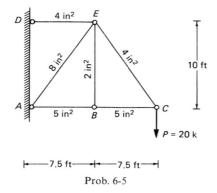

Prob. 6-5

6-6   Find the forces in all members of the truss shown in the figure. Assume l/Ea to be the same for all members.

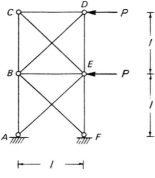

Prob. 6-6

CHAPTER 7

Further applications of method of virtual work

7-1 INTRODUCTION

We shall now extend the method introduced in the preceding chapter to other types of structures, notably beams and frames. No new principles are involved but, because actual loads may be applied at points other than joints, it is necessary to replace these loads by an equivalent loading at joints only.

7-2 EQUIVALENT JOINT LOADING

In the analysis of structures by the force method we have to know the displacements at a number of coordinates usually chosen at the joints due to several loading arrangements, and we have seen that the calculation can be conveniently carried out using matrix algebra. This requires that the loads be applied at joints only, but any loads acting between joints can be replaced by equivalent loads acting at the joints. The equivalent loads are chosen so that the resulting displacements at the joints are the same as the displacements due to the actual loading. The displacements at points other than the joints will not necessarily be equal to the displacements due to the actual loading.

Consider the beam in Fig. 7-1a for which the displacements at the coordinates 1 and 2 at B (Fig. 7-1b) are required. We consider B as a joint between members AB and BC. In Fig. 7-1c, the displacements at joints B and C are restrained and the fixed-end forces due to the actual loading on the restrained members are determined. The formulas in Appendix C may be used for this purpose. The fixed-end forces at each joint are then totalled and reversed on the actual structure, as in Fig. 7-1d. These reversed forces are statically equivalent to the actual loading on the structure and produce the same displacement at the coordinates 1 and 2 as the actual loading. The rotation at C is also equal to the rotation due to the actual loading, but this is not true for the displacements at the other joints A and D. This statement can be proved as follows.

The displacements at 1 and 2 and the rotation at C can be achieved by superposition of the displacements due to the loadings under the conditions shown in Figs. 7-1c and d. But the forces in Fig. 7-1c produce no displacements at the restrained joints B and C. Removal of the three restraining forces is equivalent to the application of the forces in Fig. 7-1d, and it follows that due to these forces the structure will undergo displacements at joints B and C equal to the displacements due to the actual loading.

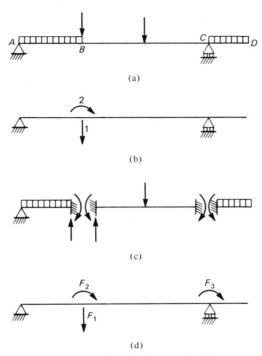

Fig. 7-1. Equivalent joint loading. (a) Beam. (b) Coordinate system. (c) Restrained condition. (d) Restraining forces reversed on actual structure.

From the above, we can see that restraints have to be introduced only at the coordinates where the deflection is required. Sometimes, additional restraints may be introduced at other convenient locations (such as C in the structure considered above) in order to facilitate the calculations of the fixed-end forces on the structure.

It is apparent that the use of the equivalent joint loading results in the same reactions at supports as the actual loading. The internal forces at the ends of the members caused by the equivalent joint loading, when added to the fixed-end forces caused by the actual loading, give the end forces in the actual condition.

The advantage of the use of equivalent loads concentrated at the joints instead of the actual loading is that the diagrams of the stress resultants become straight lines. As a result, the evaluation of the integrals of the type involved in Eq. 6-4 becomes easier and the calculations can be arranged in matrix form shown in Sec. 7-4. This is illustrated in Example 7-6.

7-3 DEFLECTION OF BEAMS AND FRAMES

The main internal forces in beams and frames are bending moments and shearing forces. Axial forces and twisting moments are either absent or they make little contribution to the lateral deflections and rotations. For this reason, in most cases, the terms in Eq. 6-4 representing the contribution of the axial force and torsion can be omitted when lateral deflections or rotations are calculated. It follows that the displacement in beams subjected to bending moment and shear is given by

$$D_j = \int \frac{M_{uj}M}{EI}\, dl + \int \frac{V_{uj}V}{Ga_r}\, dl \tag{7-1}$$

Furthermore, the cross section of beams generally used in practice is such that contribution of shear to deflection is small and can be neglected. Thus the deflection is given by

$$D_j = \int \frac{M_{uj}M}{EI}\, dl \tag{7-2}$$

From Eq. 5-20 the change in slope of the deflected axis of a beam along an element of length dl is $d\theta = -M/EI\, dl$. Substituting in Eq. 7-2,

$$D_j = -\int M_{uj}\, d\theta \tag{7-3}$$

The change in angle $d\theta$ is considered negative if in the same direction as the change in angle caused by a positive bending moment. The angle change and the bending moment indicated in Fig. 5-5b are negative and positive respectively.

Equation 7-3 can be used when we want to find displacements due to causes other than external loading, for instance a temperature differential between the top and bottom faces of a beam. The use of Eq. 7-3 will be illustrated by Example 7-4.

Example 7-1 Figure 7-2a shows a beam ABC with an overhanging end. Find the vertical deflection D_1 at C and the angular rotation D_2 at A. The beam has a constant flexural rigidity EI. Only the deformation caused by bending is required.

The bending-moment diagram is shown in Fig. 7-2c with the ordinates plotted on the tension side of the beam. Figures 7-2e and f show the bending moments due to a unit force at the coordinates 1 and 2 respectively. Because the second moment of area of the beam is constant, the calculation

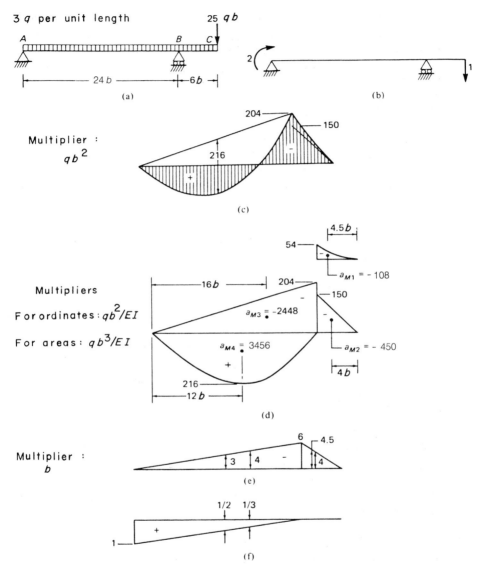

Fig. 7-2. Beam analyzed in Example 7-1. (a) Beam. (b) Coordinate system.
(c) M diagram. (d) Areas and centroids of M/EI diagram. (e) M_{u1} diagram.
(f) M_{u2} diagram.

of the areas and the location of the centroid can be carried out for the M diagram instead of the $M/(EI)$ diagram, the resulting values being divided by EI. The M diagram is divided into parts for which the areas and the centroids are easily determined, with the additional proviso that the M_{uj} diagram corresponding to each part is a straight line. These areas and their centroids are shown in Fig. 7-2d (compare Appendix F). Note that the area of a positive bending moment diagram is given a positive sign, and that only the location of the centroid along the beam need be determined.

As a next step, we require the ordinates of the M_{uj} diagrams corresponding to the centroid of each area (see Figs. 7-2e and f).

The displacements D_1 and D_2 are given by Eq. 7-2 but the integral can be evaluated with the help of Eq. 6-7 for the different parts of the M diagram. Thus

$$D_j = \sum a_M \bar{M}_{uj}$$

where $\bar{M}$ denotes the ordinate at the centroid of the bending-moment diagram. Hence

$$D_1 = \frac{qb^4}{EI}(108 \times 4.5 + 450 \times 4 + 2{,}448 \times 4 - 3{,}456 \times 3) = \frac{1{,}710}{EI}qb^4$$

and

$$D_2 = \frac{qb^3}{EI}\left(-2{,}448 \times \frac{1}{3} + 3{,}456 \times \frac{1}{2}\right) = \frac{912}{EI}qb^3$$

The positive sign of D_1 and D_2 indicates that the vertical deflection at C and the rotation at A are in the directions indicated by the coordinates in Fig. 7-2b.

Example 7-2 Find the ratio of the contribution of shear to the contribution of bending-moment in the total deflection at the center of a steel beam of I cross section, carrying a uniformly distributed load over a simple span (Fig. 7-3a). Other conditions of the problem are: the second moment of area $= I$; $a_r \simeq a_w =$ area of web; $G/E = 0.4$; span $= l$; and intensity of loading $= q$ per unit length.

Figures 7-3c and e show the bending-moment and shearing force diagrams due to the real loading. A unit virtual load is applied at coordinate 1 (Fig. 7-3b) where the deflection is required. The corresponding plots of the bending-moment M_{u1} and the shearing force V_{u1} are shown in Figs. 7-3d and f.

The M and V diagrams are now divided into two parts such that the corresponding portions of the M_{u1} and V_{u1} diagrams are straight lines. The ordinates $\bar{M}_{u1}$ and $\bar{V}_{ul}$ corresponding to the centroids of the two parts of the M and V diagrams are then determined.

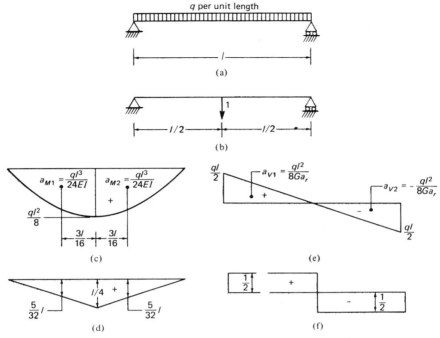

Fig. 7-3. Beam considered in Example 7-2. (a) Beam. (b) Coordinate system. (c) M diagram. (d) M_{u1} diagram. (e) V diagram. (f) V_{u1} diagram.

The total deflection at the center is

$$D_1 = \int \frac{M_{u1} M \, dl}{EI} + \int \frac{V_{u1} V \, dl}{Ga_r} \tag{7-4}$$

Of this, the deflection due to bending is

$$\int \frac{M_{u1} M \, dl}{EI} = \sum a_M \bar{M}_{u1} = 2\left(\frac{ql^3}{24EI}\right)\left(\frac{5l}{32}\right) = \frac{5}{384}\frac{ql^4}{EI}$$

and the deflection due to shear is

$$\int \frac{V_{u1} V \, dl}{Ga_r} = \sum a_V \bar{V}_{u1} = 2\left(\frac{ql^2}{8Ga_r}\right)\left(\frac{1}{2}\right) = \frac{ql^2}{8Ga_r}$$

Hence

$$\frac{\text{Deflection due to shear}}{\text{Deflection due to bending}} = 9.6\left(\frac{E}{G}\right)\left(\frac{I}{l^2 a_r}\right)$$

This equation is valid for simply supported beams of any cross section subjected to a uniform load. In our case, substituting $G = 0.4E$ and $a_r = a_w$, we find for a steel beam of I section

$$\frac{\text{Deflection due to shear}}{\text{Deflection due to bending}} = 24\frac{I}{l^2 a_w} = c\left(\frac{h}{l}\right)^2$$

where h is the height of the I section, and

$$c = \frac{24I}{a_w h^2}$$

We can see that the value of c depends on the proportions of the section. For rolled steel sections, commonly used in beams, c varies between 7 and 20. Values of c for standard rolled sections are available in various references.[1]

We may note that the depth-span ratio h/l in the majority of practical I-beams lies between 1/10 and 1/20. For uniformly loaded simple beams of rectangular cross section, with $G/E = 0.4$ and a depth-span ratio $h/l = 1/5$, 1/10, and 1/15, the magnitude of shear deflection represents respectively 9.6, 2.4, and 1.07 percent of the deflection due to bending. In plate girders the deflections due to shear can be as high as 15 to 25 percent of the deflection due to bending.

Example 7-3 Consider a uniformly loaded simple beam AB, shown in Fig. 7-4a. In Example 7-2 we found the deflection at the center using the actual loading. In this example the equivalent joint loading will be used to determine the deflection at the center and the rotation at a support. The deflection due to shear is neglected.

The coordinate system for the required displacements is shown in Fig. 7-4b, and Fig. 7-4c gives the fixed-end forces due to the actual loading on the restrained beam. The resultants of these forces are reversed and applied to the actual structure, as shown in Fig. 7-4d. The bending-moment diagram due to the reversed loading is given in Fig. 7-4e, and the bending-moment diagrams due to a unit force at coordinates 1 and 2 are plotted in Figs. 7-4f and g.

The displacements D_1 and D_2 can then be calculated from Eq. 6-7 or using the data of Appendix H. Thus

$$D_1 = \int\frac{M_{u1}M\,dl}{EI} = 2\left(\frac{ql^3}{96EI} \times \frac{l}{8} + \frac{ql^3}{32EI} \times \frac{l}{6}\right) = \frac{5ql^4}{384EI}$$

and

$$D_2 = \int\frac{M_{u2}M\,dl}{EI} = 2\left(\frac{ql^3}{96EI} + \frac{ql^3}{32EI}\right)\frac{1}{2} = \frac{ql^3}{24EI}$$

These values are, of course, the same as the values listed in Appendix B.

[1]See, for example, J. I. Parcel and B. B. Moorman, *Analysis of Statically Indeterminate Structures*, Wiley, New York, 1962, pp. 38–39.

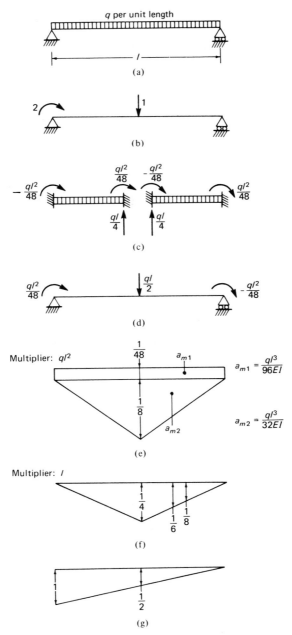

Fig. 7-4. Beam considered in Example 7-3. (a) Beam loading. (b) Coordinate system. (c) Restraint conditions. (d) Restraining forces reversed on actual structure. (e) M diagram. (f) M_{u1} diagram. (g) M_{u2} diagram.

Example 7-4 Find the deflection at the center of a simple beam of length l and depth h (Fig. 7-5a) caused by a rise in temperature which varies linearly between the top and bottom of the beam (Fig. 7-5c). The coefficient of thermal expansion is α per degree.

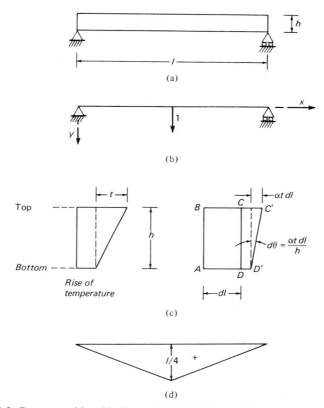

Fig. 7-5. Beam considered in Example 7-4. (a) Beam. (b) Coordinate system. (c) Deformation of an element of length dl. (d) M_{u1} diagram.

Consider an element $ABCD$ of length dl, shown in Fig. 7-5c. It is convenient to assume AB to be fixed in position. The rise of temperature will then cause a displacement of CD to $C'D'$. The angular rotation of CD with respect to AB is

$$d\theta = \frac{\alpha t \, dl}{h} \tag{7-5}$$

Figure 7-5d shows the plot of the bending moment M_{u1} due to a unit virtual load at coordinate 1 (Fig. 7-5b), corresponding to the vertical deflection at the center of the span.

Substituting Eq. 7-5 in Eq. 7-3,

$$D_1 = -\frac{\alpha t}{h} \int M_{u1} \, dl$$

The integral $\int M_{u1} \, dl$ is the area under the M_{u1} diagram — that is,

$$\int M_{u1} \, dl = \frac{l^2}{8}$$

whence

$$D_1 = -\frac{\alpha t l^2}{8h}$$

The minus sign indicates that the deflection is in a direction opposite to the coordinate 1, that is upward (as expected).

Example 7-5   The tube ABC shown in Fig. 7-6a is cantilevered with AB and BC in a horizontal plane. A vertical load is applied at the free end C. Find

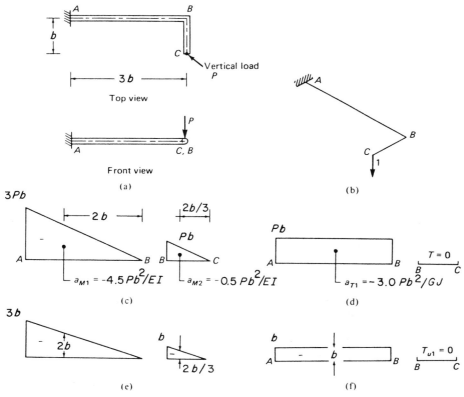

Fig. 7-6. Tube considered in Example 7-5. (a) Tube. (b) Coordinate system.
(c) M diagram. (d) T diagram. (e) M_{u1} diagram. (f) T_{u1} diagram.

the vertical deflection at C due to bending and torsion. The tube has a constant cross section. Consider $G = 0.4E$ and note that $J = 2I$.

The M and T diagrams corresponding to the real loadings are shown in Figs. 7-6c and d, in which the areas and the location of their centroids are indicated. A unit virtual load is applied at the coordinate 1, that is vertically at C, where the deflection is required. The M_{u1} and T_{u1} diagrams corresponding to the virtual loading are shown in Figs. 7-6e and f.

Applying Eq. 6-4, but including the bending- and twisting-moment terms only, we have

$$D_1 = \int \frac{M_{u1} M}{EI} dl + \int \frac{T_{u1} T}{GJ} dl$$

Evaluating the integrals for the parts AB and BC by the use of Eq. 6-7 and 6-9 and then summing, we obtain

$$D_1 = \sum a_M \bar{M}_{u1} + \sum a_T \bar{T}_{u1}$$

$$= \left(-\frac{4.5Pb^2}{EI}\right)(-2b) + \left(-\frac{0.5Pb^2}{EI}\right)\left(-\frac{2b}{3}\right) + \left(-\frac{3.0Pb^2}{GJ}\right)(-b)$$

or

$$D_1 = 9.33 \frac{Pb^3}{EI} + \frac{3.00Pb^3}{GJ}$$

Now,

$$GJ = 0.8EI$$

Substituting,

$$D_1 = 13.08 \frac{Pb^3}{EI}$$

7-4 DEFLECTION OF BEAMS AND FRAMES USING MATRIX ALGEBRA

In Sec. 7-2 we have seen that, if a given loading on a structure is replaced by an equivalent loading acting at the joints only, the resulting displacements at the joints are the same as those due to actual loading. If furthermore, the structure is composed of straight members only, then the equivalent loading results in bending moments varying linearly between the joints, and in shear and thrust constant between the joints.

Figure 7-7a shows a typical member of a plane structure composed of m members. Let A_l and A_r be the internal moments at the left- and right-hand end of the member due to any loading acting at the joints of the structure. As usual, these moments are assumed positive if clockwise. The corresponding bending-moment diagram is shown in Fig. 7-7b. The end-moments A_{ulj} and A_{urj} due to a unit virtual load at a joint and in a direction specified at an arbitrary coordinate j (not in the member lr) are indicated in Fig. 7-7c, and Fig. 7-7d shows the corresponding bending-moment diagram. If the deformation due to bending only is considered, the displacement at j can be

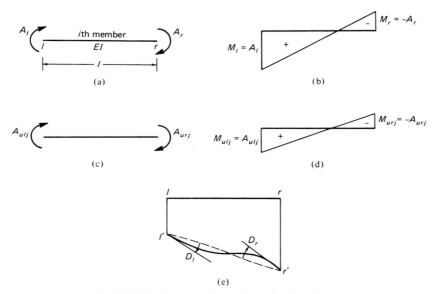

Fig. 7-7. Typical member subjected to bending.

determined by Eq. 7-2, the integration being carried out for all the members of the structure. Using Eq. 6-7 to evaluate the integral of Eq. 7-2 for the typical member in Fig. 7-7 or using Appendix H, it can be shown that the contribution of the bending of the member considered to the displacement at j is

$$\Delta D_j = \int_l \frac{M_{uj}M}{EI} \, dl = \frac{l}{6EI}(2A_lA_{ulj} - A_lA_{urj} - A_rA_{ulj} + 2A_rA_{urj})$$

The same result can be obtained by a multiplication of matrices,

$$\Delta D_j = \{A_u\}_j^T[f_M]\{A\} \tag{7-6}$$

where

$$\{A_u\}_j = \begin{Bmatrix} A_{ulj} \\ A_{urj} \end{Bmatrix} \qquad (7\text{-}7)$$

$$\{A\} = \begin{Bmatrix} A_l \\ A_r \end{Bmatrix} \qquad (7\text{-}8)$$

and

$$[f_M] = \frac{l}{6EI} \begin{bmatrix} 2 & -1 \\ -1 & 2 \end{bmatrix} \qquad (7\text{-}9)$$

The elements of $[f_M]$ are equal to the end rotations D_l and D_r relative to the chord (shown dotted in Fig. 7-7e) due to a unit couple acting at one end of the beam when simply supported (check by Appendix B). The matrix $[f_M]$ is referred to as the *bending flexibility matrix* of the member.

Equation 7-6 can also be obtained from considerations of energy. In Sec. 5-5, we saw that the displacement D_j is numerically equal to the work done by the internal forces induced by a unit virtual load at j while moving along the displacement due to the real system of loading. The deformed configuration of member lr in Fig. 7-7e can be reached by first moving the member as a rigid body to the position shown by the dotted line $l'r'$ (which does not involve bending strain energy) and then applying the end moments M_l and M_r to a simply supported bar $l'r'$. The multiplication of the last two matrices in Eq. 7-6 gives the end rotations D_l and D_r relative to the chord caused by the real loading. Thus

$$\{D\} = \begin{Bmatrix} D_l \\ D_r \end{Bmatrix} = [f_M]\{A\} \qquad (7\text{-}10)$$

The multiplication of $\{A_u\}_j^T\{D\}$ gives the work done by the end moments A_{ulj} and A_{urj} while moving along the displacement D_l and D_r. Therefore,

$$\Delta D_j = \{A_u\}_j^T\{D\} \qquad (7\text{-}11)$$

Substituting for $\{D\}$ from Eq. 7-10 into Eq. 7-11 gives Eq. 7-6.

The displacement at j can be determined by summation over all the members:

$$D_j = \sum_{i=1}^{m} \{A_u\}_{ij}^T [f_M]_i \{A\}_i \qquad (7\text{-}12)$$

Equation 7-12 can also be put in the form

$$D_j = \{A_u\}_{j\,2m\times1}^T [f_M]_{2m\times2m} \{A\}_{2m\times1} \qquad (7\text{-}13)$$

where

$$\{A_u\}_j = \left\{ \begin{array}{c} \{A_u\}_{1j} \\ \hline \{A_u\}_{2j} \\ \hline \cdots \\ \hline \{A_u\}_{mj} \end{array} \right\}$$

(7-14)

$$\{A\} = \left\{ \begin{array}{c} \{A\}_1 \\ \hline \{A\}_2 \\ \hline \cdots \\ \hline \{A\}_m \end{array} \right\}$$

(7-15)

and

$$[f_M] = \begin{bmatrix} [f_M]_1 & & \\ & [f_M]_2 & \\ \text{Submatrices} & \cdots & \\ \text{not shown} & & [f_M]_m \\ \text{are null} & & \end{bmatrix}$$

(7-16)

Matrix $[f_M]$ contains the flexibility matrices of the separate members and is referred to as the *flexibility matrix of the unassembled structure.*

The advantage of Eq. 7-12 is that the matrices require smaller storage space in a computer than when Eq. 7-13 is used.

If the displacement is required at n coordinates, a unit virtual load has to be applied at each of the coordinates separately and the corresponding set of end-moments determined. These moments can be arranged in one matrix:

$$[A_u]_{2m \times n} = \begin{bmatrix} \{A_u\}_{11} & \{A_u\}_{12} & \cdots & \{A_u\}_{1n} \\ \{A_u\}_{21} & \{A_u\}_{22} & \cdots & \{A_u\}_{2n} \\ \cdots & \cdots & \cdots & \cdots \\ \{A_u\}_{m1} & \{A_u\}_{m2} & \cdots & \{A_u\}_{mn} \end{bmatrix}$$

(7-17)

where the elements in each of the submatrices $\{A_u\}$ are the end-moments in a member. The first subscript of $\{A_u\}$ indicates the member and the second the coordinate at which a unit virtual load is applied.

If the displacement at the n coordinates is required due to p cases of loading, the two end moments $\{A\}$ have to be determined for each member in each case. These moments can be arranged in one matrix:

$$[A]_{2m \times p} = \begin{bmatrix} \{A\}_{11} & \{A\}_{12} & \cdots & \{A\}_{1p} \\ \{A\}_{21} & \{A\}_{22} & \cdots & \{A\}_{2p} \\ \cdots & \cdots & \cdots & \cdots \\ \{A\}_{m1} & \{A\}_{m2} & \cdots & \{A\}_{mp} \end{bmatrix} \tag{7-18}$$

where the first subscript of $\{A\}$ indicates the member and the second the case of loading.

The displacement of the n coordinates for the p cases of loading can be represented by one matrix equation similar to Eq. 6-19:

$$[D]_{n \times p} = [A_u]_{2m \times n}^T [f_M]_{2m \times 2m} [A]_{2m \times p} \tag{7-19}$$

For easy reference, the notation in this equation is repeated below.

D = contribution of the bending deformation to the displacement at a coordinate

$\{A_u\}$ = submatrix of order 2×1, whose elements are the two end-moments of a member due to a unit virtual load at one of the coordinates

$[f_M]_i$ = bending flexibility matrix of member i (see Eq. 7-16)

$$= \left(\frac{l}{6EI}\right)_i \begin{bmatrix} 2 & -1 \\ -1 & 2 \end{bmatrix}$$

$\{A\}$ = submatrix of the order 2×1, whose elements are the two end moments of a member due to one case of real loading

With the assumption that the members are straight and that external loading is applied at the joints only, the values of the axial force, shearing force and twist are constant along each member. If the contribution of these displacements is to be considered, one general equation can be derived in the same way as Eq. 6-19 for the deflection of trusses where the members are subjected to axial force. Thus

$$[D]_{n \times p} = [A_u]_{m \times n}^T [f_M]_{m \times m} [A]_{m \times p} \tag{7-20}$$

where D = contribution of axial force, shearing force or twisting moment to the displacement at a coordinate

A_u = axial force, shearing force or twisting moment in a member due to a unit virtual load acting at a coordinate; the elements in each column of matrix $[A_u]$ are the forces due to a unit virtual load at one of the coordinates

f_M = flexibility of member; specifically, for the effect of axial force, $f_M = l/(Ea)$, for the effect of the shearing force, $f_M = l/(Ga_r)$, for the effect of the twisting moment, $f_M = l/(GJ)$

A = axial force, shearing force or twisting moment in a member due to the real loading; elements in each column of matrix $[A]$ are the forces corresponding to one case of loading

n = number of the coordinates at which the displacement is required

m = number of members

p = number of loading cases

From a comparison of Eqs. 7-19 and 7-20, it is apparent that the contribution of the bending, axial force, shear, and twisting deformations can be represented by similar equations. The total displacement can therefore be obtained by the summation

$$[D] = \sum_{s=1}^{4} [A_u]_s^T [f_M]_s [A]_s \qquad (7\text{-}21)$$

where s represents the four contributory cases of deformation. The order of

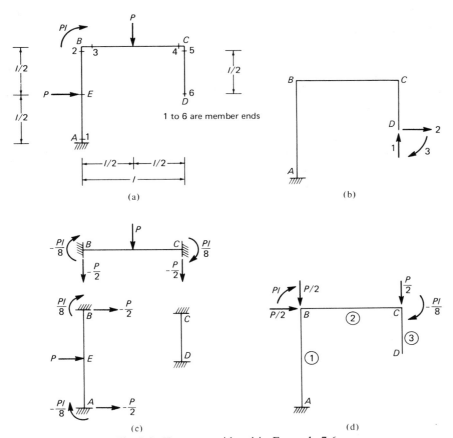

Fig. 7-8. Frame considered in Example 7-6.

the matrices in this equation is the same as that of the matrices in Eq. 7-19 for bending, and as in Eq. 7-20 for the other causes. The symbols A_u and A indicate an action or a generalized internal force, that is a force or a couple in the member considered.

Example 7-6 The rigid frame of Fig. 7-8a is made of a commonly used steel I section which has the following properties: $a = 1.34 \times 10^{-4}l^2$, $a_r = a_{web} = 0.65 \times 10^{-4}l^2$, and $I = 5.30 \times 10^{-8}l^4$. Also, $G = 0.4E$.

Determine the contribution of bending, axial force, and shear deformations to the displacement at the three coordinates in Fig. 7-8b.

First, we replace the actual forces by equivalent loads at the joints A, B, C and D. To do this, the joints are restrained (Fig. 7-8c), and the fixed-end forces are determined by the use of the formulas in Appendix C. These forces are reversed and added to the actual forces acting at the joints (only the couple Pl, in this example) to obtain the equivalent joint loading shown in Fig. 7-8d. We then proceed to find the displacements due to the equivalent loading.

The contribution of the three types of deformation is now determined by Eqs. 7-19 or 7-20, with the order of the matrices determined by $m = 3$, $n = 3$, and $p = 1$. The members AB, BC, and CD will be listed in a fixed order (indicated in Fig. 7-8d).

(a) *BENDING DEFORMATION.* We determine the end-moments in the members due to a unit virtual load acting separately at each of the coordinates and also due to the equivalent joint loading of Fig. 7-8d. These end-moments are assembled in matrices $[A_u]$ and $[A]$ (which can be checked by considerations of simple statics for the statically determinate structure $ABCD$).

$$[A_u]_{6 \times 3} = \begin{bmatrix} l & -\dfrac{l}{2} & -1 \\[2mm] -l & -\dfrac{l}{2} & 1 \\[2mm] l & \dfrac{l}{2} & -1 \\[2mm] 0 & -\dfrac{l}{2} & 1 \\[2mm] 0 & \dfrac{l}{2} & -1 \\[2mm] 0 & 0 & 1 \end{bmatrix} \qquad [A]_{6 \times 1} = Pl \begin{Bmatrix} -1.875 \\[2mm] 1.375 \\[2mm] -0.375 \\[2mm] -0.125 \\[2mm] 0 \\[2mm] 0 \end{Bmatrix}$$

It is important that the end-moments are listed in the same order and with the same sign convention in both the above matrices. Our layout is for the moment at the left-hand end to be followed by the moment at the right-hand end for each member, the members being viewed from inside of the frame. A clockwise end-moment is considered positive.

The bending flexibility matrix of the separate members can be written

$$[f_M]_{6 \times 6} = \frac{l}{6EI} \begin{bmatrix} 2 & -1 & & & & \\ -1 & 2 & & & & \\ & & 2 & -1 & & \\ & & -1 & 2 & & \\ & \text{Elements not} & & & 1 & -0.5 \\ & \text{shown are zero} & & & -0.5 & 1 \end{bmatrix}$$

Substituting in Eq. 7-19 we find the contribution of bending deformation to the displacement at the coordinates

$$[D]_{3 \times 1} = \frac{Pl^3}{6EI} \begin{Bmatrix} -10.375 \\ -0.125 \\ 10.50/l \end{Bmatrix} = \frac{P}{EI} \begin{Bmatrix} -3263 \times 10^4 \\ -39 \times 10^4 \\ 3301 \times 10^4/l \end{Bmatrix}$$

(b) *DEFORMATION DUE TO AXIAL FORCES.* The axial forces in the members due to the loading considered above are listed in matrices $[A_u]$ and $[A]$ below, in which a force is considered positive if tension

$$[A_u]_{3 \times 3} = \begin{bmatrix} 1 & 0 & 0 \\ 0 & 1 & 0 \\ -1 & 0 & 0 \end{bmatrix} \quad \text{and} \quad [A]_{3 \times 1} = P \begin{Bmatrix} -1 \\ 0 \\ 0 \end{Bmatrix}$$

The axial flexibility of the separate members can be written in matrix form as

$$[f_M]_{3 \times 3} = \frac{l}{Ea} \begin{bmatrix} 1 & 0 & 0 \\ 0 & 1 & 0 \\ 0 & 0 & 0.5 \end{bmatrix}$$

The contribution of the axial deformation to the displacement at the three coordinates, given by Eq. 7-20, is thus

$$[D]_{3 \times 1} = \frac{Pl}{Ea} \begin{Bmatrix} -1 \\ 0 \\ 0 \end{Bmatrix} = \frac{P}{EI} \begin{Bmatrix} -0.75 \times 10^4 \\ 0 \\ 0 \end{Bmatrix}$$

(c) *SHEAR DEFORMATION*. The shearing force in the members due to the same loading as above is

$$[A_u]_{3 \times 3} = \begin{bmatrix} 0 & 1 & 0 \\ -1 & 0 & 0 \\ 0 & -1 & 0 \end{bmatrix} \quad \text{and} \quad [A]_{3 \times 1} = P \begin{Bmatrix} 0.5 \\ 0.5 \\ 0 \end{Bmatrix}$$

The shear flexibility of the separate members, arranged in matrix form, is

$$[f]_{3 \times 3} = \frac{l}{Ga_r} \begin{bmatrix} 1 & 0 & 0 \\ 0 & 1 & 0 \\ 0 & 0 & 0.5 \end{bmatrix}$$

Hence the contribution of the shear deformation to the displacement at the three coordinates, calculated by Eq. 7-20, is

$$[D]_{3 \times 1} = \frac{Pl}{Ga_r} \begin{Bmatrix} -0.5 \\ 0.5 \\ 0 \end{Bmatrix} = \frac{P}{El} \begin{Bmatrix} -1.92 \times 10^4 \\ 1.92 \times 10^4 \\ 0 \end{Bmatrix}$$

A comparison of the contributions to the displacement made by the three types of deformation shows that, in the present example, the displacement caused by bending is much greater than that due to the axial forces and shear. This is true in most practical cases and, for this reason, the displacements caused by the axial and shear deformations are often neglected.

7-5 FLEXIBILITY MATRIX OF THE ASSEMBLED STRUCTURE

The flexibility matrix of a structure can be determined from the flexibilities of individual members by the use of Eq. 7-21. The elements of the flexibility matrix are the displacements at the coordinates due to a unit force acting separately at each of these coordinates. Therefore, the real and the virtual loadings are the same, that is $[A] = [A_u]$, so that Eq. 7-21 becomes

$$[f] = \sum_{s=1}^{4} [A_u]_s^T [f_M]_s [A_u]_s \qquad (7\text{-}22)$$

where $[f]$ is the flexibility matrix of the assembled structure, $[f_M]$ is the flexibility matrix of the unassembled structure (see Eq. 7-16) and the subscript s refers to the four contributory causes of deformation: bending, axial force, shear and torsion.

When only the bending deformation is considered, Eq. 7-22 becomes

$$[f]_{n \times n} = [A_u]_{2m \times n}^T [f_M]_{2m \times 2m} [A_u]_{2m \times n} \qquad (7\text{-}23)$$

For convenience, the notation is repeated:

f = element of a flexibility matrix of the assembled structure

$[f_M]_i$ = bending flexibility matrix of member i (see Eq. 7-16)

$$= \left(\frac{l}{6EI}\right)_i \begin{bmatrix} 2 & -1 \\ -1 & 2 \end{bmatrix}$$

$\{A_u\}$ = a submatrix 2×1, the elements of which are the two end-moments of a member due to a unit load at one of the coordinates

m = number of members

n = number of coordinates

Equation 7-22 can also be written in the form

$$[f] = \sum_i^m \sum_{s=1}^4 [A_u]_{is}^T [f_M]_{is} [A_u]_{is} \tag{7-24}$$

where $[A_u]_i$ and $[f_M]_i$ are matrices corresponding to member i.

The form of Eq. 7-24 is more convenient with large structures than Eq. 7-22 because recording of the zeros in $[f_M]$ is avoided.

Likewise, when only the bending deformation is considered, Eq. 7-23 can be expressed in the form

$$[f]_{n \times n} = \sum_i^m [A_u]_i^T \underset{2 \times n}{} [f_M]_i \underset{2 \times 2}{} [A_u]_i \underset{2 \times n}{} \tag{7-25}$$

Example 7-7   Determine the flexibility matrix of the frame of Example 7-6 corresponding to the three coordinates in Fig. 7-8. Consider the bending deformation only.

All the matrices required in Eq. 7-23 were determined in the calculation of the effect of bending deformation in Example 7-6. By substitution in Eq. 7-23, we find

$$[f]_{3 \times 3} = [A_u]_{6 \times 3}^T [f_M]_{6 \times 6} [A_u]_{6 \times 3}$$

or

$$[f] = \begin{bmatrix} l & -\frac{l}{2} & -1 \\ -l & -\frac{l}{2} & 1 \\ l & \frac{l}{2} & -1 \\ 0 & -\frac{l}{2} & 1 \\ 0 & -\frac{l}{2} & -1 \\ 0 & 0 & 1 \end{bmatrix}^T \frac{l}{6EI} \begin{bmatrix} 2 & -1 & & & & \\ -1 & 2 & & & & \\ & & 2 & -1 & & \\ & & -1 & 2 & & \\ \text{Elements not shown} & & 1 & -0.5 \\ \text{are zero} & & -0.5 & 1 \end{bmatrix} \begin{bmatrix} l & -\frac{l}{2} & -1 \\ -l & -\frac{l}{2} & 1 \\ l & \frac{l}{2} & -1 \\ 0 & -\frac{l}{2} & 1 \\ 0 & \frac{l}{2} & -1 \\ 0 & 0 & 1 \end{bmatrix}$$

or

$$[f] = \frac{l}{6EI} \begin{bmatrix} 8l^2 & 1.5l^2 & -9l \\ 1.5l^2 & 2.25l^2 & -3.75l \\ -9l & -3.75l & 15 \end{bmatrix}$$

7-6   GENERAL

To use the method of virtual work in frames, where loads are frequently applied at any point in the member, it is convenient to replace the actual loading by equivalent joint loading which induces the same displacements at the joints whose displacement is required as the real loading.

The procedure for the determination of deflection of beams and frames is straightforward. In frames, the displacements are mainly caused by the bending moments, and the other internal forces are often neglected. In some cases, for instance in plate girders, it may nevertheless be prudent to verify that this is justified. As in the case of trusses, matrix formulation is useful when deflection at a number of points and under a number of loadings is required and a computer is used.

The examples in this chapter are limited to structures with straight members of constant cross section. The method of virtual work can, however, be applied also to structures with curved members and a variable cross section. One way of doing this is by approximating the actual structure to a structure composed of straight segments of constant cross section.

It is worth noting that the method of virtual work can take account of all four types of internal forces while the other methods discussed in Chapter 9 account for only one or the other type of stress resultant.

PROBLEMS

7-1   Find the fixed-end moments in the beam shown.

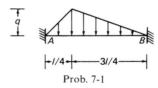

Prob. 7-1

7-2   Find the fixed-end moments in the beam shown.

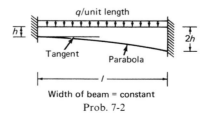

Prob. 7-2

7-3   Find the flexibility matrix corresponding to the coordinates 1 and 2 of the
beam shown in the figure: (a) by considering AB as one member of variable
EI, (b) by considering AB to be composed of two members AC and CB and
using Eq. 7-23. Neglect shear deformation.

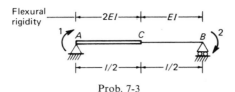

Prob. 7-3

7-4   Assuming that the displacement of the beam AB is prevented at any two of
the coordinates shown, find the flexibility matrix corresponding to the other
two. Consider both the bending and shear deformations. Use this matrix to
derive the stiffness matrix corresponding to the four coordinates. Solution
of this problem is included in Sec. 16-2.

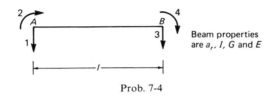

Prob. 7-4

7-5   Find the bending-moment diagram for the tied arch shown, considering only
the bending deformation in the arch and only the axial deformation in the tie.
What is the force in the tie? Ea for tie $= (43/b^2)\ EI$ for arch.

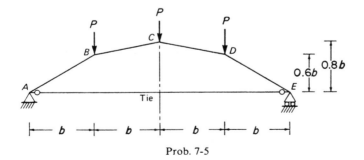

Prob. 7-5

7-6   Using the force method, find the tension in the tie and the horizontal reaction
component at A for the frame shown in the figure and draw the bending-
moment diagram. Consider only the bending deformation in the frame
$ABCDE$ and only the axial deformation in the tie BD. $E = 30{,}000$ ksi, I for
$ABCDE = 8{,}000$ in.^{4}, area of tie $= 8$ in.^{2}.

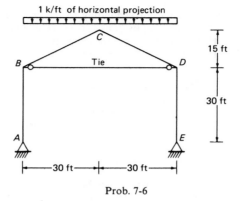

Prob. 7-6

7-7   Considering the deformations due to bending only, find the vertical deflection at *A* for the frame shown in the figure.

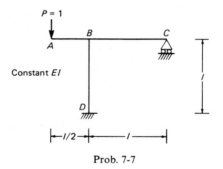

Prob. 7-7

7-8   Find the vertical deflection at *D* and the angular rotation at *A* for the beam in the figure. Consider bending deformation only.

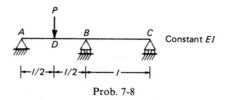

Prob. 7-8

7-9   What is the relative translation along *EC* of points *E* and *C* in the frame of Example 3-2?

7-10   Considering the tension F_1 and F_2 in the cables at C and D in Prob. 2-8 as redundants, find the flexibility matrix of the released structure using Eq. 7-23. Consider only the bending deformation in AB and only the axial deformation in the cables.

7-11   Replace the forces on the frame of Fig. 7-8 a by equivalent forces at the joints A, B, C, and D, then using Eq. 7-19 derive the displacements at the three coordinates indicated in Fig. 7-8b. Finally, use the flexibility matrix derived in Example 7-7 to find the bending moment in the frame in Fig. 3-2a.

7-12   Show that the deflection at the centre of a straight member with respect to its ends is given by

$$y_{\text{centre}} = (\psi_1 + 10\psi_2 + \psi_3) \frac{l^2}{96}$$

where ψ_1 and ψ_3 are the curvatures at the two ends; l is the distance between the two ends; and ψ_2 is the curvature at the middle. The variation of ψ is assumed to be a second-degree parabola.

This is a geometric relation which can be derived by integration of the equation $\psi = -d^2y/dx^2$; where y is the deflection and x is the distance along the member, measured from the left end. The equation can be derived more easily by the method of elastic weights, using equivalent concentrated loading (Fig. 7-11).

In practice, the expression derived can be used for continuous or simple beams having a constant or a variable cross section when the parabolic variation of ψ is acceptable.

Important energy theorems

8-1 INTRODUCTION

In Chapter 5 the concepts of strain energy and complementary energy were considered, mainly for the purpose of developing the principle of virtual work. There are several other energy theorems of interest in structural analysis, and these will now be considered.

8-2 BETTI'S AND MAXWELL'S THEOREMS

Consider any structure, such as that shown in Fig. 8-1a, with a series of coordinates $1, 2, \ldots, n, n+1, n+2, \ldots, m$ defined. The F system of forces $F_1, F_2, \ldots, F_n$ acts at coordinates 1 to n (Fig. 8-1b), and the Q system of forces $Q_{n+1}, Q_{n+2}, \ldots, Q_m$ acts at the coordinates $n+1$ to m (Fig. 8-1c). Let the displacements caused by the F system alone be $\{D_{1F}, D_{2F}, \ldots, D_{mF}\}$ and the displacements due to the Q system alone be $\{D_{1Q}, D_{2Q}, \ldots, D_{mQ}\}$.

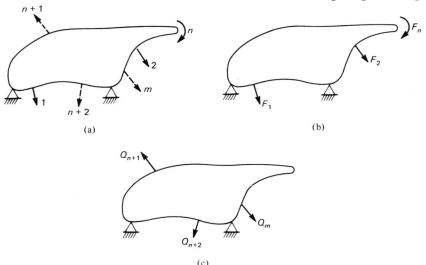

Fig. 8-1. Betti's theorem. (a) Coordinate system: $1, 2, \ldots, n, n+1, \ldots, m$. (b) F-system of forces: $F_1, F_2, \ldots, F_n$; acting at coordinate: $1, 2, \ldots, n$; causing displacements: $D_{1F}, D_{2F}, \ldots, D_{mF}$. (c) Q-system of forces: Q_{n+1}, $Q_{n+2}, \ldots, Q_m$; acting at coordinates: $n+1, n+2, \ldots, m$; causing displacements: $D_{1Q}, D_{2Q}, \ldots, D_{mQ}$.

200

Suppose that the F system alone is applied to the structure. The internal work and the external work are equal so that (see Eqs. 4-18 and 5-12)

$$\frac{1}{2}\sum_{i=1}^{n} F_i D_{iF} = \frac{1}{2}\int_v \{\sigma\}_F^T \{\varepsilon\}_F \, dv \qquad (8\text{-}1)$$

where $\{\sigma\}_F$ and $\{\varepsilon\}_F$ are the stress and strain caused by the F system.

Imagine now that when the F system is being applied to the structure, the Q system is already acting, causing stresses σ_Q at any point. The external and internal work during the application of the F system are again equal, so that

$$\frac{1}{2}\sum_{i=1}^{n} F_i D_{iF} + \sum_{i=n+1}^{m} Q_i D_{iF} = \frac{1}{2}\int_v \{\sigma\}_F^T \{\varepsilon\}_F \, dv + \int_v \{\sigma\}_Q^T \{\varepsilon\}_F \, dv \qquad (8\text{-}2)$$

The second term on each side of this equation is the work due to the Q system while moving along the displacement by the F system. From Eqs. 8-1 and 8-2,

$$\sum_{i=n+1}^{m} Q_i D_{iF} = \int_v \{\sigma\}_Q^T \{\varepsilon\}_F \, dv \qquad (8\text{-}3)$$

Now, if we assume that the F system is applied first, causing stresses $\{\sigma\}_F$, and the Q system is added subsequently, causing additional stresses $\{\sigma\}_Q$ and strains $\{\varepsilon\}_Q$, a similar equation is obtained

$$\sum_{i=1}^{n} F_i D_{iQ} = \int_v \{\sigma\}_F^T \{\varepsilon\}_Q \, dv \qquad (8\text{-}4)$$

If the material of the structure obeys Hooke's law, $\{\varepsilon\}_Q = [e]\{\sigma\}_Q$ and $\{\varepsilon\}_F = [e]\{\sigma\}_F$, where $[e]$ is constant (see Eqs. 5-6 and 5-9). Therefore, substituting for $\{\varepsilon\}$ in Eqs. 8-3 and 8-4, we find that the right-hand sides of the two equations are equal. Hence

$$\sum_{i=1}^{n} F_i D_{iQ} = \sum_{i=n+1}^{m} Q_i D_{iF} \qquad (8\text{-}5)$$

This equation is known as *Betti's theorem*, which can be expressed as follows. The sum of the products of the forces of the F system and the displacements at the corresponding coordinates caused by the Q system is equal to the sum of the products of the forces of the Q system and the displacements at the corresponding coordinates caused by the F system. We must remember that the theorem is valid only for linear elastic structures.

We shall now consider Maxwell's theorem, which is a special case of the more general Betti's theorem. Assume that there is only one force $F_i = 1$ in the F system acting at coordinate i, and one force $Q_j = 1$ in the Q system acting at j. Applying Eq. 8-5, we find

$$D_{iQ} = D_{jF} \tag{8-6}$$

Equation 8-6 can be written in the form

$$f_{ij} = f_{ji} \tag{8-7}$$

where f_{ij} is the displacement at i due to a unit force at j, and f_{ji} is the displacement at j due to a unit force at i. Equation 8-7 is called *Maxwell's reciprocal theorem*, which can be stated as follows. In a linear elastic structure, the displacement at coordinate i due to a unit force at coordinate j is equal to the displacement at j due to a unit force acting at i.

The displacements in Eq. 8-7 are the flexibility coefficients. For a structure in which m coordinates are indicated, the flexibility coefficients, when arranged in a matrix of the order $m \times m$, give the flexibility matrix of the structure. This matrix must be symmetrical, by virtue of Eq. 8-7; hence, the word "reciprocal" in the theorem. The property of symmetry of the flexibility matrix was proved in another way in Sec. 4-6.

8-3 APPLICATION OF BETTI'S THEOREM TO TRANSFORMATION OF FORCES AND DISPLACEMENTS

Betti's theorem can be used to transform the actual forces on a structure to equivalent forces at the coordinates.

Consider the structure of Fig. 8-2a in which only the bending deformation need be taken into account. The degree of kinematic indeterminacy of the frame is three, as shown in Fig. 8-2b (we recall that this represents the number of independent joint displacements). If the flexibility matrix for the frame is known, the independent displacements $\{D\}$ due to external applied forces $\{F\}$ can be determined by the equation $\{D\} = [f]\{F\}$ (Eq. 4-1), provided both the forces and the displacements are at the same three coordinates.

If the frame is subjected to an arbitrary system of forces (Fig. 8-2c), they first have to be replaced by equivalent forces at the joints, as discussed in Sec. 7-2. The fixed-end forces are shown in Fig. 8-2d; they are added at each joint and reversed to obtain the equivalent forces $\{F^*\}$ indicated in Fig. 8-2e. The forces $\{F^*\}$ act at the coordinates $\{D^*\}$ which are not independent. The displacements $\{D^*\}$ are related to the displacement $\{D\}$ by the geometry of the deformed shape of the frame. These relations can be written as

$$\{D^*\} = [C]\{D\} \tag{8-8}$$

where $[C]$ is determined from the geometry of the frame.

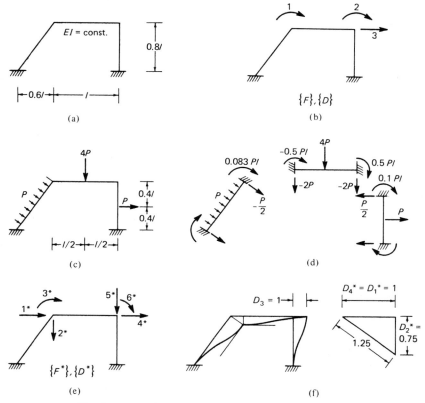

Fig. 8-2. Transformation of forces and displacements.

For the frame shown in Fig. 8-2a, we have

$$\{D^*\}_{6 \times 1} = \begin{bmatrix} 0 & 0 & 1 \\ 0 & 0 & 0.75 \\ 1 & 0 & 0 \\ 0 & 0 & 1 \\ 0 & 0 & 0 \\ 0 & 1 & 0 \end{bmatrix}_{6 \times 3} \{D\}_{3 \times 1} \qquad (8\text{-}9)$$

The elements in each column of $[C]$ are the values of the displacements at the D^* coordinates corresponding to a unit displacement at one of the D coordinates. The first two columns are obvious; the elements in the third column can be obtained with the aid of the displacement diagram in Fig. 8-2f. (The necessary principles of geometry are discussed in Secs. 1-5 and 5-2.)

Applying Betti's theorem (Eq. 8-5) to the two systems of forces in Figs. 8-2b and e, we find

$$\sum_{i=1}^{3} F_i D_i = \sum_{j=1}^{6} F_j^* D_j^*$$

The second subscript of D and D^* (which indicates the cause of the displacements) is omitted since the F and F^* forces are known to cause the same joint displacements. The matrix form of this equation is

$$\{F\}^T\{D\} = \{F^*\}^T\{D^*\} \tag{8-10}$$

Substituting for $\{D^*\}$ from Eq. 8-8, and taking the transpose of both sides of Eq. 8-10, we obtain

$$\{D\}^T\{F\} = \{D\}^T[C]^T\{F^*\}$$

whence the equation for the transformation of forces is

$$\{F\} = [C]^T\{F^*\} \tag{8-11}$$

Equation 8-11 shows that if displacements $\{D\}$ are transformed to displacements $\{D^*\}$ using a transformation matrix $[C]$ (Eq. 8-8), then $[C]^T$ can be used to transform forces $\{F^*\}$ to forces $\{F\}$. The elements in the jth column of $[C]$ in Eq. 8-8 are the displacements at the D^* coordinates corresponding to a displacement $D_j = 1$ while the displacements are zero at the other D coordinates. It is therefore apparent that $[C]$ cannot be formed unless the displacements $\{D\}$ are independent of one another (see Sec. 1-5).

As an example of the use of Eqs. 8-8 and 8-11, consider the frame in Fig. 8-3a. In a general case, when axial deformations are included, the frame

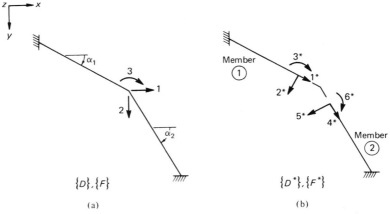

Fig. 8-3. Transformation of forces and displacements from coordinates along principal axes of member cross sections to coordinates parallel to general axes.

has three degrees of freedom as shown in the figure by the coordinates referring to displacements $\{D\}$ or forces $\{F\}$. The displacements $\{D\}$ can be transformed to displacements $\{D^*\}$ along principal axes of the member cross sections by use of Eq. 8-8. The transformation matrix $[C]$ in this case is

$$[C] = \begin{bmatrix} [t]_1 \\ [t]_2 \end{bmatrix}$$

where

$$[t]_i = \begin{bmatrix} \cos \alpha_i & \sin \alpha_i & 0 \\ -\sin \alpha_i & \cos \alpha_i & 0 \\ 0 & 0 & 1 \end{bmatrix}$$

and i refers to the number of the member. This type of transformation will be used in Chapter 24 for forces or displacements from coordinates along the principal axes of the member cross section to coordinates parallel to a common set of axes for the whole structure, referred to as global or general axes.

If we now apply Eq. 8-11, we transform forces at the ends of members $\{F^*\}$ to forces at the D coordinates, as follows:

$$\{F\} = \begin{bmatrix} \cos \alpha_1 & -\sin \alpha_1 & 0 & \cos \alpha_2 & -\sin \alpha_2 & 0 \\ \sin \alpha_1 & \cos \alpha_1 & 0 & \sin \alpha_2 & \cos \alpha_2 & 0 \\ 0 & 0 & 1 & 0 & 0 & 1 \end{bmatrix} \{F^*\}$$

The elements of the jth column of the rectangular matrix in this equation can be checked using the fact that they are the components of a force $F_j^* = 1$ at the coordinates in Fig. 8-3a.

In some cases, forces $\{F\}$ at global coordinates can be easily related to member end-forces $\{F^*\}$ at member coordinates by

$$\{F^*\} = [B]\{F\} \tag{8-12}$$

Using Betti's theorem, the corresponding displacements can be transformed by the equation:

$$\{D\} = [B]^T\{D^*\} \tag{8-13}$$

For application of Eqs. 8-12 and 8-13, consider the global and member coordinates indicated on the statically determinate frame in Figs. 8-4a and b. By simple statics, the transformation of forces in Eq. 8-12 can be made by

$$\{F^*\} = \begin{bmatrix} \cos \alpha_1 & \sin \alpha_1 & 0 \\ -\sin \alpha_1 & \cos \alpha_1 & 0 \\ -l_2 \sin \alpha_2 & l_2 \cos \alpha_2 & 1 \\ -\cos \alpha_2 & -\sin \alpha_2 & 0 \\ \sin \alpha_2 & -\cos \alpha_2 & 0 \\ l_2 \sin \alpha_2 & -l_2 \cos \alpha_2 & -1 \end{bmatrix} \{F\} \tag{8-14}$$

and using Eq. 8-13, we write

$$\{D\} = \begin{bmatrix} \cos\alpha_1 & -\sin\alpha_1 & -l_2\sin\alpha_2 & -\cos\alpha_2 & \sin\alpha_2 & l_2\sin\alpha_2 \\ \sin\alpha_1 & \cos\alpha_1 & l_2\cos\alpha_2 & -\sin\alpha_2 & -\cos\alpha_2 & -l_2\cos\alpha_2 \\ 0 & 0 & 1 & 0 & 0 & -1 \end{bmatrix} \{D^*\}$$

(8-15)

We can now check that the elements in the jth column of the rectangular matrix in this equation are the displacements $\{D\}$ corresponding to $D_j^* = 1$ with $D_i^* = 0$ when $i \neq j$.

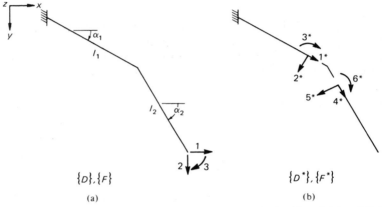

Fig. 8-4. Transformation of forces and displacements using Eqs. 8-14 and 8-15.

We note that the elements in any column of matrix $[B]$ in Eq. 8-14 are member end forces caused by a unit load at one of the D coordinates in Fig. 8-4a. Hence, by an appropriate choice of the D^* coordinates at the two ends of each member, matrix $[B]$ can become identical with matrix $[A_u]$ used in Eqs. 7-22 or 7-23 for the derivation of the flexibility of an assembled structure from flexibility of the members (see Prob. 8-2).

Example 8-1 Use Eq. 8-11 to determine the restraining forces at the three coordinates in Fig. 8-2b corresponding to the external loading in Example 3-3 (see Fig. 8-2c).

The components of the fixed-end forces (Fig. 8-2d) along the coordinates 1, 2, ..., 6 in Fig. 8-2e are

$$\{F^*\} = \begin{Bmatrix} -0.4\,P \\ -2.3\,P \\ -0.417\,Pl \\ -0.5\,P \\ -2.0\,P \\ 0.6\,Pl \end{Bmatrix}$$

Substituting for $[C]^T$ from Eq. 8-9 in Eq. 8-11, we obtain

$$\{F\} = \begin{bmatrix} 0 & 0 & 1 & 0 & 0 & 0 \\ 0 & 0 & 0 & 0 & 0 & 1 \\ 1 & 0.75 & 0 & 1 & 0 & 0 \end{bmatrix} P \begin{Bmatrix} -0.4 \\ -2.3 \\ -0.417l \\ -0.5 \\ -2.0 \\ 0.6 \end{Bmatrix} = P \begin{Bmatrix} -0.417l \\ 0.60l \\ -2.625 \end{Bmatrix}$$

which is identical with the forces calculated by resolving along the axes of the members in Example 3-3.

8-4 TRANSFORMATION OF STIFFNESS AND FLEXIBILITY MATRICES

Consider a coordinate system on a linear structure defining the location and direction of forces $\{F\}$ and displacements $\{D\}$, and let the corresponding stiffness matrix be $[S]$ and the flexibility matrix $[f]$. Another system of coordinates is defined for the same structure referring to forces $\{F*\}$ and displacements $\{D*\}$, with the stiffness and flexibility matrices $[S*]$ and $[f*]$ respectively. If the displacements or forces at the two systems of coordinates are related by

$$\{D\} = [H]\{D*\}$$

or

$$\{F*\} = [H]^T\{F\}$$
(8-16)

then the stiffness matrix $[S]$ can be transferred to $[S*]$ by the equation

$$[S*] = [H]^T[S][H]$$
(8-17)

Also, when the forces at the two coordinate systems are related:

$$\{F\} = [L]\{F*\}$$

or

$$\{D*\} = [L]^T\{D\}$$
(8-18)

the flexibility matrix $[f*]$ can be derived from $[f]$ by

$$[f*] = [L]^T[f][L]$$
(8-19)

We should note that the transformation matrices $[H]$ or $[L]^T$ are formed by geometrical relations of the displacements $\{D\}$ and $\{D*\}$, and it follows that these relations are valid regardless of the forces applied at the coordinates. **The two systems of forces $\{F\}$ and $\{F*\}$ are equivalent to each other, which** means that the forces $\{F\}$ produce displacements $\{D\}$ and $\{D^*\}$ of the same magnitude as would be caused by the forces $\{F^*\}$. Also, the systems $\{F\}$ and $\{F^*\}$ do the same work to produce the displacement $\{D\}$ or $\{D^*\}$.

For the proof of Eq. 8-17, we assume that the structure is subjected to forces $\{F\}$ and we express the work done by these forces by Eq. 4-26, viz.

$$W = \frac{1}{2}[D]^T[S]\{D\} \qquad (8\text{-}20)$$

Substituting for $\{D\}$ from Eq. 8-16, we obtain

$$W = \frac{1}{2}\{D^*\}^T[H]^T[S][H]\{D^*\}$$

Now, if we assume the structure to be subjected to forces $\{F^*\}$ and apply Eq. 4-26 again, we obtain

$$W = \frac{1}{2}\{D^*\}^T[S^*]\{D^*\}$$

A comparison of the two expressions for W gives the relation between the matrices $[S^*]$ and $[S]$, that is Eq. 8-17.

If the work is expressed in terms of the forces and flexibility (Eq. 4-25), the proof of Eq. 8-19 is established.

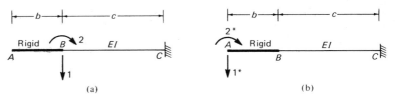

(a)   (b)

Fig. 8-5. Cantilever used to illustrate the application of Eqs. 8-17 and 8-19.
(a) Coordinate system referring to displacements $\{D\}$ and forces $\{F\}$.
(b) Coordinate system referring to displacements $\{D^*\}$ and forces $\{F^*\}$.

For an application of Eqs. 8-17 and 8-19, consider the cantilever in Fig. 8-5 with the part AB rigid and BC of flexural rigidity EI. Two coordinate systems are defined in Figs. 8-5a and b. The stiffness matrix corresponding to the coordinates in Fig. 8-5a is (see Appendix D)

$$[S] = \begin{bmatrix} \dfrac{12EI}{c^3} & \dfrac{6EI}{c^2} \\[2ex] \dfrac{6EI}{c^2} & \dfrac{4EI}{c} \end{bmatrix}$$

From geometrical relation between the displacements at the two coordinates, the transformation matrix $[H]$ in Eq. 8-16 is

$$[H] = \begin{bmatrix} 1 & b \\ 0 & 1 \end{bmatrix}$$

and the application of Eq. 8-17 gives

$$[S^*] = EI \begin{bmatrix} \dfrac{12}{c^3} & & \text{symmetrical} \\[2ex] \dfrac{12b}{c^3} + \dfrac{6}{c^2} & \dfrac{4}{c} + \dfrac{12b^2}{c^3} + \dfrac{12b}{c^2} \end{bmatrix}$$

For the application of Eq. 8-19, we first use geometrical relations to generate the transformation matrix $[L]^T$ defined in Eq. 8-18

$$[L]^T = \begin{bmatrix} 1 & -b \\ 0 & 1 \end{bmatrix}$$

and

$$[f] = \begin{bmatrix} \dfrac{c^3}{3EI} & -\dfrac{c^2}{2EI} \\[2ex] -\dfrac{c^2}{2EI} & \dfrac{c}{EI} \end{bmatrix}$$

Substituting in Eq. 8-19 gives

$$[f^*] = \dfrac{1}{EI} \begin{bmatrix} \dfrac{c^3}{3} + bc^2 + b^2c & \text{symmetrical} \\[2ex] -(c^2 + bc) & c \end{bmatrix}$$

This matrix can be derived by calculation of the displacements $\{D^*\}$ due to unit values of the forces F_1^* and F_2^*, which should, of course, give the same result.

We recall that for the validity of Eqs. 8-17 and 8-19, the geometrical relations between $\{D\}$ and $\{D^*\}$ in Eqs. 8-16 and 8-18 must be true for all values of the forces $\{F\}$ and $\{F^*\}$, and it can be easily seen that this is satisfied in the above example. If, however, the member AB is flexible, the geometrical relations do not hold when forces $\{F^*\}$ are applied and the transformation of stiffness and flexibility matrices cannot be made.

8-5   STIFFNESS MATRIX OF ASSEMBLED STRUCTURE

The stiffness matrix $[S]$ of a structure formed by the assemblage of members can be obtained from the stiffness matrices of its members.

Consider the structure shown in Fig. 8-6a and the coordinate system in Fig. 8-6b. The external work, which is also equal to the strain energy of the structure, is $W = U = \frac{1}{2}\{D\}^T[S]\{D\}$, where $[S]$ is the stiffness matrix of the structure corresponding to the coordinates in Fig. 8-6b.

The same strain energy is obtained from the sum of the values of strain energy of the individual members. This sum is equal to the work done by

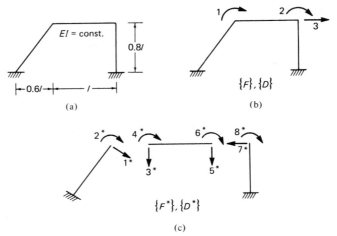

Fig. 8-6. Frame considered in Example 8-2.

the member end-forces $\{F^*\}$ undergoing the displacements $\{D^*\}$ at the coordinates in Fig. 8-6c. Thus

$$U = \frac{1}{2}\{D^*\}^T[S_M]\{D^*\} \qquad (8\text{-}21)$$

where $[S_M]$ is the stiffness matrix of the unassembled structure

$$[S_M] = \begin{bmatrix} [S_M]_1 & & \\ & [\ldots] & \\ & & [S_M]_m \end{bmatrix}$$

$[S_M]_i$ is the stiffness matrix of the ith member corresponding to the $\{D^*\}$ coordinates at its ends, and m is the number of members.

The displacements $\{D^*\}$ and $\{D\}$ are related by geometry so that $\{D^*\} = [C]\{D\}$ substituting this equation into Eq. 8-21, we obtain

$$U = \frac{1}{2}\{D\}^T[C]^T[S_M][C]\{D\} \qquad (8\text{-}22)$$

From a comparison of Eqs. 8-20 and 8-22, it can be seen that

$$[S]_{n \times n} = [C]^T_{p \times n}[S_M]_{p \times p}[C]_{p \times n} \qquad (8\text{-}23)$$

where n is the number of displacements $\{D\}$, and p is the number of displacements $\{D^*\}$.

Many of the elements of the matrices in this equation are zero. For this reason, it may be more convenient for computer programming to use Eq. 8-23 in the form

$$[S] = \sum_i^m [C]^T_i[S_M]_i[C]_i \qquad (8\text{-}24)$$

where $[C]_i$ is the matrix relating the coordinates $\{D^*\}$ at the ends of the ith member to the structure coordinates $\{D\}$.

The member coordinates D^* are usually chosen along the principal axes of the member cross section at its two ends; thus, the member stiffness $[S_M]$ is normally readily available (see for example Eqs. 4-5, 4-6, and 4-7).

Example 8-2   Find the stiffness matrix of the frame shown in Fig. 8-6a using Eq. 8-24. (The stiffness matrix of the same frame was derived by a different procedure in Example 3-3.)

From geometry of the frame, $[C]$ is determined as in Sec. 8-3:

$$[C] = \begin{bmatrix} [C]_1 \\ [C]_2 \\ [C]_3 \end{bmatrix} = \begin{bmatrix} 0 & 0 & 1.25 \\ 1 & 0 & 0 \\ \hline 0 & 0 & 0.75 \\ 1 & 0 & 0 \\ 0 & 0 & 0 \\ 0 & 1 & 0 \\ \hline 0 & 0 & -1 \\ 0 & 1 & 0 \end{bmatrix}$$

Then,

$$[S_M] = \begin{bmatrix} [S_M]_1 & & \\ & [S_M]_2 & \\ & & [S_M]_3 \end{bmatrix}$$

where

$$[S_M]_1 = \begin{bmatrix} \dfrac{12EI}{l^3} & \text{symmetrical} \\ -\dfrac{6EI}{l^2} & \dfrac{4EI}{l} \end{bmatrix}$$

$$[S_M]_2 = \begin{bmatrix} \dfrac{12EI}{l^3} & & & \text{symmetrical} \\ \dfrac{6EI}{l^2} & \dfrac{4EI}{l} & & \\ -\dfrac{12EI}{l^3} & -\dfrac{6EI}{l^2} & \dfrac{12EI}{l^3} & \\ \dfrac{6EI}{l^2} & \dfrac{2EI}{l} & -\dfrac{6EI}{l^2} & \dfrac{4EI}{l} \end{bmatrix}$$

and

$$[S_M]_3 = \begin{bmatrix} \dfrac{12EI}{(0.8l)^3} & \text{symmetrical} \\[2ex] \dfrac{6EI}{(0.8l)^2} & \dfrac{4EI}{0.8l} \end{bmatrix}$$

Applying Eq. 8-24,

$$[S] = \sum_{i=1}^{3} [C]_i{}^T[S_M]_i[C]_i = EI \begin{bmatrix} \dfrac{8}{l} & & \text{symmetrical} \\[2ex] \dfrac{2}{l} & \dfrac{9}{l} & \\[2ex] -\dfrac{3}{l^2} & -\dfrac{4.875}{l^2} & \dfrac{48.938}{l^3} \end{bmatrix}$$

8-6 ENGESSER'S THEOREM OF COMPATIBILITY

This theorem states that if the complementary energy U^* of any statically indeterminate structure is expressed in terms of redundant forces $F_1, F_2, \ldots,$ F_n, then n compatibility equations can be written in the form

$$\frac{\partial U^*}{\partial F_j} = 0 \qquad (j = 1, 2, \ldots, n) \tag{8-25}$$

The derivation of this theorem will be given in relation to a truss with m members, shown in Fig. 8-7. A statically determinate structure can be obtained if some of the reactions are removed and some of the members are

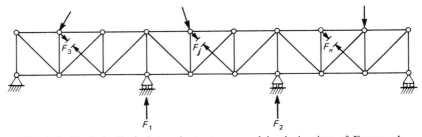

Fig. 8-7. Statistically indeterminate truss used in derivation of Engesser's theorem.

cut, as shown in the figure. The complementary energy U^* of the indeterminate truss is equal to the sum of the values of complementary energy of the members. From Eq. 5-29a,

$$U^* = \sum_{i=1}^{m} \left(N_i e_{ti} + \int_0^N e_i dN_i \right) \tag{8-26}$$

where N_i is the force in the ith member in the statically indeterminate structure, e_i is the extension due to N_i, and e_{ti} is the extension due to temperature or other environmental causes.

Differentiating with respect to the value of any redundant F_j, we can write

$$\frac{\partial U^*}{\partial F_j} = \frac{\partial U^*}{\partial N_i} \frac{\partial N_i}{\partial F_j}$$

Hence, using Eqs. 5-31 and 8-26,

$$\frac{\partial U^*}{\partial F_j} = \sum_{i=1}^{m} \frac{\partial N_i}{\partial F_j}(e_{ti} + e_i) = \sum_{i=1}^{m} \frac{\partial N_i}{\partial F_j} \Delta_i \qquad (8\text{-}27)$$

where

$$\Delta_i = e_{ti} + e_i$$

The force N_i in any member can be expressed by a superposition equation

$$N_i = N_{si} + (N_{ui1}F_1 + N_{ui2}F_2 + \ldots + N_{uin}F_n) \qquad (8\text{-}28)$$

where N_{si} is the force in member i due to the external loading on a statically determinate truss released by removal of the redundant forces, and N_{uir} is the force in the ith member corresponding to a unit value of the redundant F_r acting separately on the released structure. All the forces in Eq. 8-28 can be obtained from equations of static equilibrium alone and the superposition is valid regardless of whether the material of the structure has a linear or nonlinear stress-strain relation.

The partial derivative of the force N_i with respect to F_j is

$$\frac{\partial N_i}{\partial F_j} = N_{uij} \qquad (8\text{-}29)$$

Combining Eqs. 8-27 and 8-29,

$$\frac{\partial U^*}{\partial F_j} = \sum_{i=1}^{m} N_{uij}\Delta_i \qquad (8\text{-}30)$$

The quantity on the right-hand side of this equation is also equal to the virtual work done by the internal forces $\{N_{u1j}, N_{u2j}, \ldots, N_{umj}\}$ moving through the actual member elongations $\{\Delta\}$ of the statically indeterminate structure. This quantity is equal to the displacement D_j of the actual structure (see Eq. 5-40).

But $D_j = 0$, since no displacement occurs at an external redundant and no relative movement at a cut member. It follows that

$$\frac{\partial U^*}{\partial F_j} = 0$$

Thus the proof of Engesser's theorem is established.

We should note that the theorem was derived without stipulating a linear stress-strain relationship. Engesser's theorem can therefore be used in the analysis of nonlinear structures so long as the structure is not subjected to a gross distortion which causes an appreciable change in geometry. However, the use of the theorem presents mathematical complications in any but the simplest case.

Differentiating Eq. 8-30 with respect to the value of any other redundant F_r, we obtain

$$\frac{\partial^2 U^*}{\partial F_j \partial F_r} = \sum_{i=1}^{m} N_{uij} \frac{\partial \Delta_i}{\partial F_r} \tag{8-31}$$

The change in extension $\delta \Delta_i$ in any member due to a change δF_r in one of the redundants is

$$\delta \Delta_i = \frac{\partial \Delta_i}{\partial F_r} \delta F_r$$

In a linear structure, $\dfrac{\partial \Delta_i}{\partial F_r}$ is constant for each bar and is equal to the extension of the ith bar due to a unit value of the force F_r. Thus the right-hand side of Eq. 8-31 is equal to the displacement at j due to a unit load at r, and represents therefore the flexibility coefficient f_{jr} of the released (statically determinate) truss. We can therefore write,

$$f_{jr} = \frac{\partial^2 U^*}{\partial F_j \partial F_r} \tag{8-32}$$

If a settlement occurs at one of the supports along the line of action of the redundant F_j, or if the redundant F_j corresponds to the force in a member which is fabricated longer or shorter than its theoretical length and is forced to fit in the structure during erection, the compatibility condition becomes

$$\frac{\partial U^*}{\partial F_j} = \lambda_j \tag{8-33}$$

where λ_j is the amount of settlement or the lack of fit corresponding to the jth redundant.

From Eq. 8-33 we can see that when λ_j is zero, the complementary energy U^* has a stationary value. In a linear structure, this stationary value is clearly a minimum since $\dfrac{\partial^2 U^*}{\partial^2 F_j}$, representing the flexibility coefficient f_{jj}, is always positive.

8-7 CASTIGLIANO'S THEOREM OF COMPATIBILITY

Castigliano's theorem may be considered as a special case of Engesser's theorem of compatibility when certain conditions are fulfilled. Referring to the truss in Fig. 8-7, these conditions are

(a) The extension of the bars is caused by the bar forces only, that is, the extension e_t due to environmental causes is zero ($\Delta = e$).

(b) The force-extension relation is linear.

Castigliano's theorem states then that n compatibility equations can be written in the form

$$\frac{\partial U}{\partial F_j} = 0 \tag{8-34}$$

where U is the strain energy expressed in terms of the redundants $F_1, F_2, \ldots, F_n$.

From condition (b) above, the force-extension relation is a straight line through the origin and the complementary energy and the strain energy are equal to one another, that is,

$$U = U^* \tag{8-35}$$

Therefore, from Eq. 8-25,

$$\frac{\partial U^*}{\partial F_j} = \frac{\partial U}{\partial F_j} = 0$$

which means that the strain energy has a minimum value. This is sometimes referred to as the *principle of least work* which may be stated as follows. In a linear statically indeterminate structure the redundant caused by external applied forces are of such magnitude that the internal strain energy is minimum.

8-8 CALCULATION OF DISPLACEMENT BY COMPLEMENTARY ENERGY

The displacement D_j at a coordinate j of a linear or nonlinear structure due to the effect of external applied loads $F_1, F_2, \ldots, F_n$ and of temperature variation, shrinkage, or other environmental causes can be obtained from

$$D_j = \frac{\partial U^*}{\partial F_j} \tag{8-36}$$

where U^* is the complementary energy expressed in terms of the forces F.

This equation will be proved in relation to the truss of m members shown in Fig. 8-8, using Eq. 8-27.

The partial derivative $\dfrac{\partial N_i}{\partial F_j}$ can be considered as the change in the force in

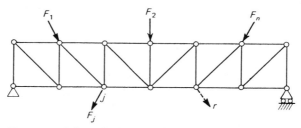

Fig. 8-8. Truss used in calculation of displacement by complementary energy.

the ith member due to a fictitious variation in the load F_j by a quantity $\partial F_j = 1$. Therefore we can write

$$\frac{\partial N_i}{\partial F_j} = N_{uij} \tag{8-37}$$

where N_{uij} is the force in the ith member due to a unit virtual load at the j coordinate. Hence, from Eq. 8-30,

$$\frac{\partial U^*}{\partial F_j} = \sum_{i=1}^{m} N_{uij} \Delta_i$$

The right-hand side of this equation is the displacement D_j, which can be obtained directly by the unit-load theorem using virtual work transformation (see Eq. 5-40):

$$D_j = \sum_{i=1}^{m} N_{uij} \Delta_i \tag{8-38}$$

From the last two equations the proof of Eq. 8-36 is established. We can note, however, that Eq. 8-36, referred to as the theorem of complementary energy is, in fact, equivalent to the unit-load theorem.

It is obvious that Eq. 8-36 can be used to determine the displacement at any coordinate r which does not necessarily represent a line of action of an external applied force.

It can be mentioned that Engesser's theorem of compatibility (Eq. 8-25) is, in fact, an application of Eq. 8-36 at a point where the displacement is known.

8-9   CASTIGLIANO'S THEOREMS

In the special case when the structure is linear elastic and the deformations are caused by external forces only ($\Delta = e$), the complementary energy U^* is equal to the strain energy U, and Eq. 8-36 becomes

$$D_j = \frac{\partial U}{\partial F_j} \tag{8-39}$$

This equation is known as Castigliano's theorem, Part II. It must be remembered that its use is limited to the calculation of displacement in linear elastic structures caused by applied loads.

The use of this theorem is equivalent to the virtual work transformation by the unit-load theorem. The application of Eq. 8-39 to express a known displacement corresponding to redundant forces in a statically indeterminate structure leads to the compatability Eq. 8-34.

Castigliano's theorem, Part I, states that if in any structure, with independent displacements $\{D_1, D_2, \ldots, D_n\}$ corresponding to external applied forces $\{F_1, F_2, \ldots, F_n\}$ along their lines of action, the strain energy U is expressed in terms of the displacements D, then n equilibrium equations can be written in the form

$$\frac{\partial U}{\partial D_j} = F_j \tag{8-40}$$

The theorem is proved by virtual work, as follows. Let the structure acquire a virtual displacement δD_j at j, while all the other displacements remain zero and environmental conditions are unchanged. The only force which does work is F_j, so that the work done during this vertical displacement is

$$\delta W = F_j \, \delta D_j$$

This is equal to the gain in the strain energy of the structure

$$\delta U = F_j \, \delta D_j \tag{8-41}$$

In the limit, when $\delta D_j \to 0$, Eq. 8-41 gives Eq. 8-40.

Castigliano's theorem, Part I is applicable to both linear and nonlinear *elastic* structures.

Referring to the truss of Fig. 8-9 which has m members and two unknown

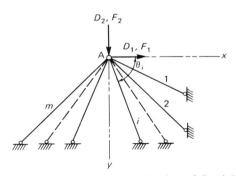

Fig. 8-9. Plane truss considered in the derivation of Castigliano's theorem, Part I.

displacements D_1 and D_2 at joint A, we can express the extension of any member i in terms of D_1 and D_2 as

$$\Delta_i = -(D_1 \cos \theta_i + D_2 \sin \theta_i) \tag{8-42}$$

where θ_i is the angle of the ith member with the horizontal. Assuming that the material obeys Hooke's law, the force in the member is

$$N_i = \frac{a_i E_i}{l_i} \Delta_i \tag{8-43}$$

where a_i is the cross-sectional area of the member and l_i its length. The strain energy of the structure is

$$U = \frac{1}{2} \sum_{i=1}^{m} N_i \Delta_i = \frac{1}{2} \sum_{i=1}^{m} \frac{a_i E_i}{l_i} \Delta_i^2 \tag{8-44}$$

Substituting Eqs. 8-42 and 8-43 into Eq. 8-44,

$$U = \frac{1}{2} \sum_{i=1}^{m} \frac{a_i E_i}{l_i} (D_1 \cos \theta_i + D_2 \sin \theta_i)^2$$

whence

$$\frac{\partial U}{\partial D_1} = \sum_{i=1}^{m} \frac{a_i E_i}{l_i} \cos \theta_i (D_1 \cos \theta_i + D_2 \sin \theta_i) \tag{8-45}$$

Applying Eq. 8-40, we obtain

$$F_1 = \left(\sum_{i=1}^{m} \frac{a_i E_i}{l_i} \cos^2 \theta_i \right) D_1 + \left(\sum_{i=1}^{m} \frac{a_i E_i}{l_i} \cos \theta_i \sin \theta_i \right) D_2 \tag{8-46}$$

and

$$F_2 = \left(\sum_{i=1}^{m} \frac{a_i E_i}{l_i} \cos \theta_i \sin \theta_i \right) D_1 + \left(\sum_{i=1}^{m} \frac{a_i E_i}{l_i} \sin^2 \theta_i \right) D_2 \tag{8-47}$$

The last two equations express the fact that the sum of the horizontal components of the bar forces at joint A is equal to the external force F_1; the same applies to the vertical components and to F_2. In Sec. 3-2, the same problem was treated by the displacement method of analysis, and it can be easily seen that Eqs. 8-46 and 8-47 are identical with the statical relations used in that method to find the unknown displacements. We should also recognize that the terms in brackets in Eqs. 8-46 and 8-47 are the stiffness coefficients of the structure.

The stiffness coefficients S_{jr} can, in general, be obtained by taking the partial derivative with respect to any displacement in Eq. 8-40. Thus

$$\frac{\partial^2 U}{\partial D_r \partial D_j} = \frac{\partial F_j}{\partial D_r}$$

In a linear structure, $\partial F_j/\partial D_r$ is the force at j due to a unit displacement at r. Therefore

$$S_{jr} = \frac{\partial^2 U}{\partial D_j \partial D_r} \tag{8-48}$$

For example, the stiffness coefficient S_{11} for the structure of Fig. 8-9 can be obtained by taking the partial derivative of Eq. 8-45

$$S_{11} = \frac{\partial^2 U}{\partial D_1^2} = \sum_{i=1}^{m} \frac{a_i E_i}{l_i} \cos^2 \theta_i$$

which is the same as the value obtained in Sec. 3-2.

8-10 POTENTIAL ENERGY

Consider an elastic structure subjected to force $\{F\}$ at n independent coordinates causing displacements $\{D\}$ at the same coordinates. If we assume the potential energy of the forces in the initial configuration to be zero, the *potential energy of the external forces* in the deformed configuration is defined as

$$V = -\sum_{i=1}^{n} F_i D_i \tag{8-49}$$

The sum of the potential energy of the external forces V and the strain energy U is called the *total potential energy*:

$$\Phi = V + U \tag{8-50}$$

Substituting Eqs. 8-49 and 5-5 in the above equation, we obtain

$$\Phi = -\sum_{i=1}^{n} F_i D_i + \sum_{m=1}^{6} \int_v \int_0^{\varepsilon_{fm}} \sigma_m d\varepsilon_m \, dv \tag{8-51}$$

where σ_m and ε_m are the values of the six stress and strain components as the forces increase from $\{0\}$ to their final values $\{F\}$, and ε_{fm} is the final value of a strain component (see Sec. 5-3).

Now let the structure acquire a configuration slightly different from the equilibrium position with the compatibility maintained at the supports, and let the corresponding change in $\{D\}$ be the small virtual displacements $\{\bar{D}\}$ and the corresponding change in strain components at any point be the virtual strain vector $\{\bar{\varepsilon}\}$. The change in the total potential energy is

$$\Delta\Phi = -\sum_{i=1}^{n} F_i \bar{D}_i + \int_v \{\sigma\}^T \{\bar{\varepsilon}\} \, dv \tag{8-52}$$

where $\{\sigma\}$ are the final values of stress components.

By the principle of virtual work (Eq. 5-35), we have

$$\sum_{i=1}^{n} F_i \bar{D}_i = \int_v \{\sigma\}^T \{\bar{\varepsilon}\}\, dv \qquad (8\text{-}53)$$

and we conclude that the right-hand side of Eq. 8-52 is zero. Hence there is no change in potential energy when the structure is given a compatible virtual small displacement from the equilibrium position.

This conclusion can serve a useful purpose when the actual deformed shape of the structure is not known. The actual displacement at any point (and hence the strain) is expressed by an assumed displacement function in terms of the unknown displacements $\{D\}$. The structure is then given a small virtual displacement ∂D_i at coordinate i, without change of the displacement at the other coordinates. Since the corresponding change in the total potential energy is zero, we can write

$$\frac{\partial \Phi}{\partial D_i} = 0 \qquad (8\text{-}54)$$

By substituting $i = 1, 2, \ldots, n$ a system of simultaneous equations can be written from which the displacements $\{D\}$ can be calculated. Equation 8-54 is, in fact, an equilibrium equation identical with Castigliano's Eq. 8-40. In other words, the use of potential energy concept leads to the same equations as Castigliano's theorem, Part I; the two approaches differ only in form.

Equation 8-54 is the *principle of stationary potential energy* and may be stated as follows. Of all deformed configurations compatible with support conditions, the one which satisfies equilibrium conditions corresponds to a stationary potential energy. It should be noted that this theorem is valid for linear and nonlinear structures.

It can be proved that the stationary value of the potential energy is minimum for stable structures and is maximum if the structure is unstable, and the principle can therefore be used for the derivation of critical loads.[1]

In the special case of a linear structure, the principle of stationary potential energy, Eq. 8-54, Castigliano's theorem, Part I, Eq. 8-40 and the unit-displacement theorem Eq. 5-39 all lead to the same equilibrium equation. We prove this by differentiating Eqs. 5-12 or 5-14 for the strain energy of a linear structure with respect to the final value D_j of the displacement at j; we obtain

$$\frac{\partial U}{\partial D_j} = \int_v \{\sigma\}^T \frac{\partial}{\partial D_j} \{\varepsilon\}\, dv \qquad (8\text{-}55)$$

[1] N. J. Hoff, *The Analysis of Structures*, Wiley, New York, 1956. See also, S. P. Timoshenko and J. M. Gere, *Theory of Elastic Stability*, McGraw-Hill, New York, 2d ed., 1961, Sec. 2-8.

where $\{\varepsilon\}$ are the final strain components. From the condition of linearity, we have

$$\frac{\partial}{\partial D_j}\{\varepsilon\} = \{\varepsilon_{uj}\} \tag{8-56}$$

where $\{\varepsilon_{uj}\}$ is the strain component at any point corresponding to a unit virtual displacement at j. Substitution of Eq. 8-56 in Eq. 8-55 and the result in Eq. 8-40 gives Eq. 5-39.

8-11 GENERAL

The energy principles are based on the law of conservation of energy which requires that the work done by external forces on an elastic structure be stored in the form of strain energy which is completely recovered when the load is removed. Betti's law derived from this law, applied to linear structures, serves a useful purpose in transformation of information given in one form into another. The equations derived in Secs. 8-3, 8-4, and 8-5 are examples of transformation equations which are useful in structural analysis.

The energy theorems presented in the remaining sections lead, in the case of linear structures, to equations which were derived in a different manner in earlier Chapters. For example, Engesser's Eqs. 8-25 and 8-33 are the same as the compatibility equations used in the force method of analysis. Similarly, Castigliano's Eq. 8-40 and the principle of stationary potential energy, Eq. 8-54, are the same as the equilibrium equations used in the displacement method. It is therefore apparent that the energy principles are not merely methods of calculation of displacements but they can form the basis for derivation of equations satisfying the requirements of equilibrium and compatibility.

Some of the energy equations derived are valid both for linear and nonlinear structures, and this was pointed out in each case. The analysis of nonlinear structures leads to some mathematical difficulties owing to the form of the stress-strain relation.

PROBLEMS

8-1   From considerations of statical equilibrium derive the matrix $[B]$ in the equation $\{F^*\} = [B]\{F\}$ relating the forces in (a) of the figure to the corresponding member end forces in (b). Then write: $\{D\} = [B]^T \{D^*\}$ and check by geometry columns 6, 9, 10 and 11 of $[B]^T$.

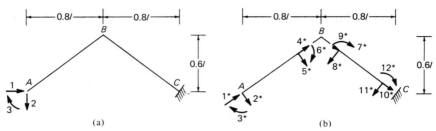

Prob. 8-1. (a) Forces $\{F\}$ and displacements $\{D\}$. (b) Member end-forces $\{F^*\}$ and displacements $\{D^*\}$.

8-2   Solve Prob. 8-1 considering the coordinates 3*, 6*, 9*, and 12* only. Then, neglecting shear and axial deformation, derive the flexibility matrix corresponding to the coordinates in Fig. (a) using the equation $[f] = [B]^T [f_M] [B]$, where $[f_M]$ is defined in Eq. 7-9 and note that this is the same as Eq. 7-25 with $[A_u] = [B]$. Assume EI constant.

8-3   Following the procedure used in Prob. 8-2, derive the flexibility matrix of the structure shown in the figure corresponding to the coordinates indicated. Ignore axial and shear deformations and assume EI constant.

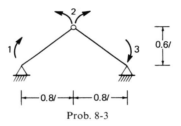

Prob. 8-3

8-4   Apply the requirements of Prob. 8-3 to the structure in the figure. Coordinates 1, 2, and 3 represent relative displacements of the two sides of a cut section at the top of the left column.

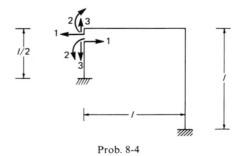

Prob. 8-4

8-5   Write the transformation matrix $[C]$ in the equation $\{D^*\} = [C]\{D\}$, then use Eq. 8-24 to derive the stiffness matrix corresponding to the D coordinates indicated in the figure. Consider only the bending deformation and assume *EI* constant.

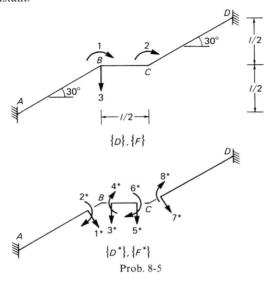

Prob. 8-5

8-6 If the structure in Prob. 8-5 is subjected to a uniform load q per unit length of horizontal projection, use Eq. 8-11 to derive the restraining forces $\{F\}$ necessary to prevent joint displacements.

8-7 Use Eq. 8-24 to solve Prob. 3-17.

8-8 The grid shown in the figure is composed of similar members of length l, flexural rigidity EI and torsional rigidity GJ. Give the matrices required to derive the stiffness matrix of the grid using Eq. 8-24. Number the members in the order shown in the figure below and take the coordinates for member 3 to be the same as the structure coordinates indicated in the figure. Number the member coordinates for other members in the same order as for member 3.

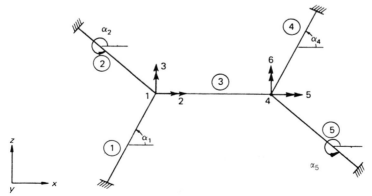

Prob. 8-8

8-9   Transform the stiffness matrix $[S]$ for a beam AB corresponding to the coordinates shown in (a) in the figure to a stiffness matrix $[S^*]$ corresponding to the coordinates shown in (b). The beam is prismatic of flexural rigidity EI and cross-sectional area a. Neglect shear deformations. In part (b) of the figure, AC and AD are two rigid arms.

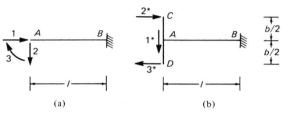

(a)   (b)

Prob. 8-9. (a) Coordinates representing $\{D\}$ and $\{F\}$. (b) Coordinates representing $\{D^*\}$ and $\{F^*\}$.

8-10   Use Eq. 8-19 to transform the flexibility matrix $[f]$ to $[f^*]$ in Prob. 8-9.

8-11   Write the stiffness matrix $[S]$ corresponding to the coordinate shown in (a), then use Eq. 8-17 to transform this matrix to a stiffness matrix $[S^*]$ corresponding to the coordinates in (b). Consider only bending deformation and take the beam flexural rigidity as EI.

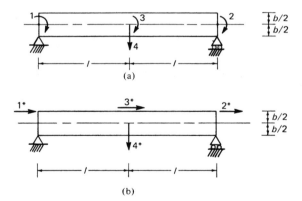

(a)

(b)

Prob. 8-11. (a) Coordinates for $\{D\}$ and $\{F\}$. (b) Coordinates for $\{D^*\}$ and $\{F^*\}$.

8-12   Use Castigliano's theorem, Part I to find the force F_1 in terms of the displacement D_1 in the beam in the figure. Assume the following approximate equation for the elastic line.

$$y = -\frac{D_1}{2} \sin \frac{\pi x}{l}$$

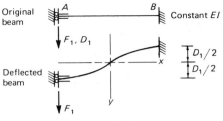

Prob. 8-12

8-13   Solve Prob. 8-12 using the principle of stationary potential energy and assuming that the beam is subjected at its two ends to two axial compressive forces $P = 4EI/l^2$ and that the horizontal movement of end A is allowed to take place freely. Assume the same approximate equation for the elastic line as in Prob. 8-12. [*Hint*: The difference between the length of an element ds of the curved beam axis and its projection dx on the x axis is

$$ds - dx = dx\sqrt{1 + \left(\frac{dy}{dx}\right)^2} - dx \simeq \frac{1}{2}\left(\frac{dy}{dx}\right)^2 dx$$

and the horizontal displacement of A is

$$D_2 = \frac{1}{2}\int_{-l/2}^{l/2}\left(\frac{dy}{dx}\right)^2 dx\,\Big]$$

8-14   Solve Prob. 3-1 using Engesser's theorem.

8-15   Solve Prob. 3-1 using Castigliano's theorem, Part I.

8-16   Solve Prob. 3-1 assuming that the stress and strain in an axial direction for any of the members are related by the equation $\sigma = C[\varepsilon - (\varepsilon^3/10)]$ where C is a constant. All members have the same length l and the same cross-sectional area a.

CHAPTER 9

Displacement of elastic structures by special methods

9-1 INTRODUCTION

The calculation of displacements in elastic structures by the method of virtual work (considered in Chapters 6 and 7) represents probably the most general approach to the problem. However, in many cases, it may be convenient to apply other, more specialized methods. Those among them which are most frequently used in practice will be developed in the present chapter.

9-2 GRAPHICAL DETERMINATION OF DEFLECTION OF A PLANE TRUSS

This method is concerned merely with the geometrical problem of determination of the displacements of joints in a plane truss in which the change in the length of all members is known. This change may be due to external applied loading, to differential variation in temperature, or to other causes, and may be elastic or plastic. The solution consists essentially of plotting the change in the length of members in a displacement diagram; two types of diagrams exist: the *Williot diagram* and the *Williot-Mohr diagram*.

As an example of the former, consider the truss of Fig. 9-1a. The table alongside gives the cross-sectional area of the members and the forces in them. The change in the length of each member, calculated by the equation: $\Delta_i = \dfrac{N_i l_i}{E a_i}$ is also tabulated and is indicated alongside each member, with the usual convention of extension being positive.

To determine the deformed shape of any triangle ABE, assume point B fixed in space and bar BE fixed in direction. In Fig. 9-1b, ABE and $A'BE'$ represent the original and the deformed shape of the triangle. The point E' is obtained direct from the elongation of member 4 (since BE is fixed in direction) but the location of A' is a function of the displacement of A relative to B and relative to E'. These displacements are shown by heavy lines. Point A' is theoretically located at the intersection of two arcs of circles with centers B and E' and radii equal respectively to the deformed lengths of AB and AE.

We should note, however, that the changes in length compared to the original length of the members are actually very much smaller than would appear from Fig. 9-1b. The angular rotations are also small, so that it is permissible to replace the arcs of circles by lines perpendicular to the original direction of the members. Hence, the relative movement of the spaces of the triangle ABE can be determined from the hatched part of Fig. 9-1b, which involves only the changes in length and does not require the original lengths of the members.

This part of the figure, shown to a larger scale in Fig. 9-1c, is called the *Williot diagram*. The vector $B'A'$ represents the displacement of joint A if the assumptions that B is a fixed point and the member BE does not change direction are correct. (The effects of this not being true are considered later.)

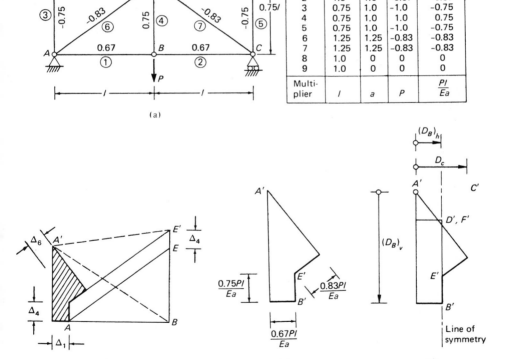

Member	Length	Area	Force	Change in Length
i	l_i	a_i	N_i	$\Delta_i = \dfrac{N_i l_i}{E a_i}$
1	1.0	1.0	0.67	0.67
2	1.0	1.0	0.67	0.67
3	0.75	1.0	-1.0	-0.75
4	0.75	1.0	1.0	0.75
5	0.75	1.0	-1.0	-0.75
6	1.25	1.25	-0.83	-0.83
7	1.25	1.25	-0.83	-0.83
8	1.0	0	0	0
9	1.0	0	0	0
Multiplier	l	a	P	$\dfrac{Pl}{Ea}$

(a)

(b)

(c)

(d)

Fig. 9-1. Displacement of joints of a truss by a graphical method (Williot diagram). (a) Loading and change in length of members. (b), (c), and (d) Williot diagrams.

A complete Williot diagram for the truss of Fig. 9-1a is shown in Fig. 9-1d. Starting from point B, assumed fixed, the length $B'E'$ equal to Δ_4 is drawn parallel to the fixed direction of BE. Since EB increases in length, point E moves upwards from B. The change in length of member EA is drawn from E', parallel to EA, to the right since the shortening of EA will cause A to move toward E. The extension of BA is plotted from B' in a similar way. The point A' is located at the intersection of the normals drawn from the ends of the two plotted changes in length. The location of point D' is determined in a similar way, using the triangle DEA, for which the joints E' and A' have already been located. Since the truss is symmetrical only one half of the Williot diagram is necessary.

The truss considered is symmetrical about the member BE, which therefore does not change direction, as assumed. It follows that all relative displacements of the joints are correct. The actual displacements depend on the fact that A is fixed in space, so that the true displacement of any joint is represented by a vector joining A' to the point representing the displaced position of the joint in question. As an example, the horizontal and vertical components of the displacement of $B[(D_B)_h, (D_B)_v]$ and the displacement of $C(D_C)$ are indicated in Fig. 9-1d. The sense of the displacements is readily apparent.

In a general case, there may be no member which remains fixed in direction. The procedure is then to assume some member to be so fixed. With this assumption we obtain the Williot diagram and then add vectorially correction displacements corresponding to rigid-body rotation of the truss about the fixed support A to obtain the true joint displacement. Since this addition to the Williot diagram was suggested by Mohr, the construction is referred to as the *Williot-Mohr diagram*.

To illustrate this procedure let us ignore the symmetry in the truss in Fig. 9-1a and assume the member DE to be fixed in direction and the joint D to be fixed in position. The Williot diagram is then constructed starting from point D' (Fig. 9-2) and points A', B', C', E', and F' are located. As a result, the displacement of joint C is represented by the vector $A'C'$ (↗). We know, however, from the physical constraints of the truss that C' has no vertical displacement relative to A'. We require therefore a rotation of the truss about A (truly fixed in space) to produce a movement of joint C perpendicular to AC, such that the total displacement of C is at the level of the support rollers. The correction displacement for point C is represented by the vector $C''A'$ (↓) and the true displacement of C is represented by the resultant vector $C''C'$ (→). The correction for any other joint is also normal to the line connecting to the joint in question to A. Hence, all the corrections are obtained by drawing the truss to scale in a 90-degree rotated position with $A'C''$ representing AC; thus points B'', D'', E'', and F'' are located. The vector joining each of these points to point A', is the correction displacement which added to the vector from A' to the corresponding point B', D', E' or F' gives the resultant true displacement. The true displacements of B, D, E, and F are therefore represented by the vectors joining B'' to B', D'' to D', E'' to E' and F'' to F, respectively. These values may now be compared with the displacements obtained by the Williot diagram in Fig. 9-1.

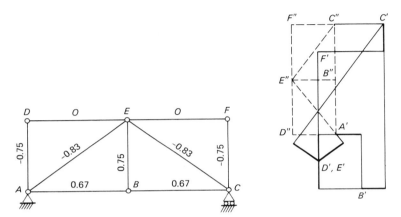

Fig. 9-2. Williot-Mohr diagram.

The Williot-Mohr diagram is thus applicable in a general case but the Williot diagram is sufficient for a symmetrical truss with a symmetrical change in length of members or in any other case when the direction of one member is known to remain unchanged.

9-3 DIFFERENTIAL EQUATION FOR DEFLECTION OF A BEAM IN BENDING

In Chapter 7 we saw that in the majority of cases the deflection of a beam is primarily due to bending. Thus it is not unreasonable to ignore the contribution of shear to deflection and to obtain the elastic deflection of a beam by solving the differential equation of the elastic line. In this section we shall derive the appropriate differential equations, and in subsequent sections deal with their solution.

Let us consider the beam of Fig. 9-3a subjected to arbitrary lateral and axial loads. Making the usual assumptions in the theory of bending, that plane transverse sections remain plane, and that the material obeys Hooke's law, we can easily show that (see Eq. 5-20)

$$M = -EI\frac{d^2y}{dx^2} \tag{9-1}$$

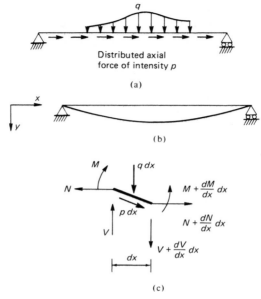

Fig. 9-3. Deflection of a beam subjected to lateral and axial load. (a) Beam and loading. (b) Positive direction of deflection. (c) Positive directions of external and internal forces.

where y is the deflection and M is the bending moment at any section x (see Fig. 9-3b); EI is the flexural rigidity, which may vary with x. The positive direction of y is indicated in Fig. 9-3b and the bending moment is considered positive if it causes tensile stress at the bottom face of the beam.

An element of the beam of length dx is in equilibrium in a deflected position under the forces shown in Fig. 9-3c. Summing the forces in the x and y directions, we obtain

and

$$\left. \begin{array}{c} \dfrac{dN}{dx} = -p \\[2em] q = -\dfrac{dV}{dx} \end{array} \right\} \qquad (9\text{-}2)$$

where N is the thrust, V is the shear, and p and q are the intensity respectively of the axial and lateral distributed load. The positive directions of q, p, y, M, V, and N are as indicated in Fig. 9-3c.

Taking moments about the right-hand edge of the element,

$$V dx - N \frac{dy}{dx} dx - \frac{dM}{dx} dx = 0 \qquad (9\text{-}3)$$

whence

$$\frac{dV}{dx} - \frac{d}{dx}\left(N\frac{dy}{dx}\right) - \frac{d^2M}{dx^2} = 0 \qquad (9\text{-}3a)$$

Substituting Eq. 9-2, we obtain the differential equation of the elastic line

$$\frac{d^2}{dx^2}\left(EI\frac{d^2y}{dx^2}\right) - \frac{d}{dx}\left(N\frac{dy}{dx}\right) = q \qquad (9\text{-}4)$$

In the absence of axial forces (i.e., when $N = 0$), Eq. 9-4 becomes,

$$\frac{d^2}{dx^2}\left(EI\frac{d^2y}{dx^2}\right) = q \qquad (9\text{-}5)$$

If, in addition, the beam has a constant flexural rigidity EI, the differential equation of the elastic line is

$$\frac{d^4y}{dx^4} = \frac{q}{EI} \qquad (9\text{-}6)$$

In a beam-column subjected to axial compressive forces P at the ends, $p = 0$ and $N = -P$, whence Eq. 9-4 becomes

$$\frac{d^2}{dx^2}\left(EI\frac{d^2y}{dx^2}\right) + P\frac{d^2y}{dx^2} = q \qquad (9\text{-}7)$$

All the equations developed in this section can be applied with a slight modification to a beam on an elastic foundation that is to a beam which, in addition to the forces already mentioned, receives transverse reaction forces proportional at every point to the deflection of the beam. Let the intensity of the distributed reaction be

$$\bar{q} = ky$$

where k is the foundation modulus with dimensions of force per (length)^2. The modulus represents the intensity of the reaction produced by the foundation on a unit length of the beam due to unit deflection. The positive direction of reaction $\bar{q}$ is upward. We can therefore use Eqs. 9-1–9-7 for a beam on an elastic foundation if the term q is replaced by the resultant lateral load intensity $q^* = (q - \bar{q}) = (q - ky)$. For example, Eq. 9-5 gives the differential equation of a beam with a variable EI on elastic foundation, subjected to lateral load q with no axial forces, as

$$\frac{d^2}{dx^2}\left(EI\frac{d^2y}{dx^2}\right) = q - ky \qquad (9\text{-}5a)$$

The differential Eqs. 9-1, 9-4, 9-5, 9-6, and 9-7 have to be solved to yield the lateral deflection y. Direct integration is possible only in a limited number of cases, considered in standard books on strength of materials. In other cases, a solution of the differential equations by other means is necessary. In the following sections, we consider the method of elastic weights, the numerical method of finite differences, and solutions by series.

9-4 MOMENT-AREA THEOREMS

The well-known Eq. 9-1 relates the deflection y to the bending moment M. Hence, we can relate the slope of the elastic line and its deflection to the area of the bending moment diagram and obtain two theorems.

(i) The difference in slope between any two points on the elastic curve is numerically equal to the area of the $M/(EI)$ diagram between these two

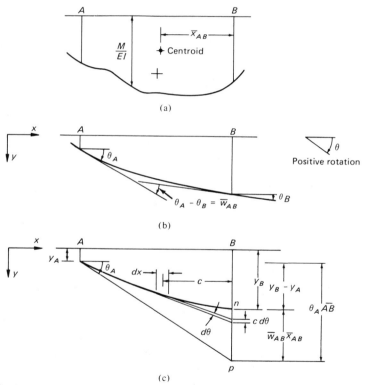

Fig. 9-4. Slope, deflection, and moment-area relations. (a) $M/(EI)$ diagram. (b) Slope/moment-area relation. (c) Deflection/moment-area relation.

points. This can be proved by integrating d^2y/dx^2 between any two points A and B in Fig. 9-4a. From Eq. 9-1,

$$\frac{d^2y}{dx^2} = \frac{d\theta}{dx} = -\frac{M}{EI}$$

Integrating,

$$\int_A^B d\theta = - \int_A^B \frac{M}{EI} dx$$

whence

$$\theta_B - \theta_A = -\bar{w}_{AB} \tag{9-8}$$

where $\theta = dy/dx$ is the slope of the elastic line, and $\bar{w}_{AB}$ is the area of the $M/(EI)$ diagram between A and B.

(ii) The deflection of B from the tangent to the elastic line at A (distance $\overline{np}$ in Fig. 9-4c) is numerically equal to the moment of $\bar{w}_{AB}$ about B. From Fig. 9-4c, we can see that

$$\overline{np} = \int_A^B c \, d\theta = \int_A^B c \left(\frac{M}{EI}\right) dx = \bar{w}_{AB}\bar{x}_{AB} \tag{9-9}$$

where c is the distance of an element dx from B, and $\bar{x}_{AB}$ is the distance of the centroid of $\bar{w}_{AB}$ from B. Note that the moment of $\bar{w}_{AB}$ is taken about the point where deflection is required.

The difference between the deflection at B and A is

$$y_B - y_A = \theta_A \overline{AB} - \bar{w}_{AB}\bar{x}_{AB} \tag{9-10}$$

According to the sign convention assumed in Eq. 9-2, the angle θ is positive if the deflection causes a rotation of the tangent to the beam in a clockwise direction. Thus the rotations θ_A and θ_B as indicated in Fig. 9-4b are positive. The area $\bar{w}_{AB}$ is positive for a positive bending moment.

Equations 9-8 and 9-10 can be used to calculate deflections in plane frames as shown by the following example.

Example 9-1 Determine the displacements D_1, D_2, and D_3 along the co-ordinates 1, 2, and 3 indicated on the frame in Fig. 9-5a. The flexural rigidity of the frame, EI is constant and only the deformations due to bending need be considered.

The bending-moment diagram is drawn in Fig. 9-5b. To conform to our sign convention when applying Eqs. 9-8 and 9-10, each member is looked at from inside of the frame as shown by the arrows in Fig. 9-5b. The deflection y represents the deflection perpendicular to the member considered.

The deflection and rotation at the fixed end A are zero. Applying Eqs. 9-8 and 9-10 to AB, we obtain

$$\theta_B = - \bar{w}_{AB} = - \frac{1.5Pb^2}{EI}$$

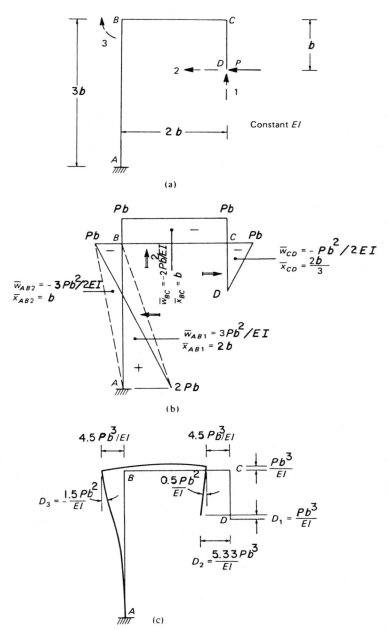

Fig. 9-5. Deflection of the plane frame of Example 9-1 by moment-area method.

and

$$y_B = - \bar{w}_{AB}\bar{x}_{AB} = - \frac{4.5Pb^3}{EI}$$

(The product $\bar{w}_{AB}\bar{x}_{AB}$ can be determined by adding algebraically the moments of the two triangles with areas $\bar{w}_{AB1}$ and $\bar{w}_{AB2}$.)

Similarly, applying the same equations to beam BC with θ_B as calculated above and the vertical deflection $y_B = 0$, we obtain

$$\theta_C = \theta_B - \bar{w}_{BC} = - \frac{1.5Pb^2}{EI} + \frac{2Pb^2}{EI} = \frac{0.5Pb^2}{EI}$$

and

$$y_C = \theta_B\overline{BC} - \bar{w}_{BC}\bar{x}_{BC} = - \frac{Pb^3}{EI}$$

Finally, we apply Eq. 9-10 to CD, with the horizontal deflection $y_C = \frac{4.5Pb^3}{EI}$, numerically equal to the horizontal translation of B, calculated earlier, hence

$$y_D = y_C + \theta_C\overline{CD} - \bar{w}_{CD}\bar{x}_{CD} = \frac{Pb^3}{EI}\left(4.5 + 0.5 \times 1.0 + 0.5 \times \frac{2}{3} \right) = \frac{5.33Pb^3}{EI}$$

The deflected shape of the frame is sketched in Fig. 9-5c, from which it can be seen that the required displacements are

$$D_1 = \frac{Pb^3}{EI} \qquad D_2 = \frac{5.33Pb^3}{EI} \qquad \text{and} \qquad D_3 = - \frac{1.5Pb^2}{EI}$$

9-5 METHOD OF ELASTIC WEIGHTS

The method of elastic weights is essentially equivalent to the moment-area method. The procedure is to calculate the rotation and the deflection respectively as the shearing force and the bending moment in a *conjugate beam* subjected to a load of intensity numerically equal to $M/(EI)$ for the actual beam. This load is referred to as elastic weight or, less aptly, as elastic load.

The conjugate beam is of the same length as the actual beam, but the conditions of support are changed, as discussed below. The $M/(EI)$ diagram of the actual beam is treated as the load on the conjugate beam, as shown in Fig. 9-6. Positive moment is taken as a positive (downward) load. From the moment-area equations (Eqs. 9-8 and 9-9) it can be shown that at any point in the beam, the shear V and moment M in the conjugate beam are equal

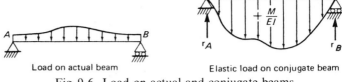

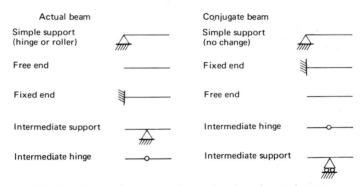

Fig. 9-6. Load on actual and conjugate beams.

respectively to the rotation θ and deflection y at the corresponding point in the actual beam.

The changes in the type of support between the actual and the conjugate beam are shown in Fig. 9-7. These changes are necessary to satisfy the known characteristics of the elastic line of the actual beam. For instance, at a fixed end of the actual beam the slope and the deflection are zero: this corresponds to no shear and no bending moment in the conjugate beam. Therefore the corresponding end of the conjugate beam must be free and unsupported.

Actual beam		Conjugate beam	
Simple support (hinge or roller)		Simple support (no change)	
Free end		Fixed end	
Fixed end		Free end	
Intermediate support		Intermediate hinge	
Intermediate hinge		Intermediate support	

Fig. 9-7. Type of support of actual and conjugate beams.

With the changes in type of support of Fig. 9-7, statically determinate beams have corresponding conjugate beams which are also statically determinate. Statically indeterminate beams appear to have unstable conjugate beams. However, such conjugate beams turn out to be in equilibrium with the particular elastic loading corresponding to the $M/(EI)$ diagram. This is illustrated in Example 9-3.

The method of elastic weights can be applied also to arches and portals. The procedure is to divide the axis of the structure into elements of length ds. The relative rotation of the sections at the two ends of the element is then $M\,ds/(EI)$. This quantity is considered as an elastic weight acting on a conjugate simply supported horizontal beam of the same span as the arch (see, for instance, Prob. 9-8). The bending moment on the conjugate beam is then equal to the vertical deflection of the arch.

Example 9-2   For the statically determinate beam of Fig. 9-8a, find the deflections at B and D and the change in slope between the left- and right-hand
sides of the hinge B. The beam has a constant value of EI.

The conjugate beam and the elastic weights on it are shown in Fig. 9-8b.
The elastic weights are obtained as areas of the different parts of the $M/(EI)$
diagram on the real beam, and they act at the respective centroids of each part.
The *elastic reactions* for these elastic loads are calculated in the usual way:

$$r_B = \frac{2.25Pb^2}{EI} \uparrow \quad \text{and} \quad r_C = \frac{7.25Pb^2}{EI} \uparrow$$

The sudden change θ_B in the slope of the elastic line at the hinge B in the
actual beam is equal to the change in shear at the support B of the conjugate
beam, which in turn is equal to the elastic reaction r_B. Therefore the change
in slope at B is $\theta_B = \dfrac{2.25Pb^2}{EI}$ (radians) in the clockwise direction because r_B
causes a sudden increase in shear.

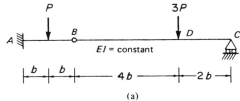

(a)

Multipliers :

Pb for moments

Pb^2/EI for elastic
weights

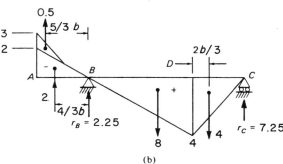

(b)

Multiplier :

Pb^3/EI

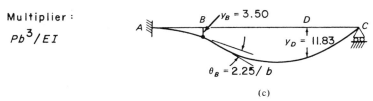

(c)

Fig. 9-8. Deflection of the beam of Example 9-2 by method of elastic weights.
(a) Actual beam. (b) Conjugate beam. (c) Elastic line.

The bending moments of the conjugate beam at B and D represent the deflections at these points:

$$y_B = \frac{Pb^2}{EI}\left(2 \times \frac{4}{3}b + 0.5 \times \frac{5}{3}b\right) = 3.5\frac{Pb^3}{EI}$$

and

$$y_D = \frac{Pb^2}{EI}\left(7.25 \times 2b - 4 \times \frac{2b}{3}\right) = 11.83\frac{Pb^3}{EI}$$

The elastic line is sketched in Fig. 9-8c.

Example 9-3 Find the deflection at B in the statically indeterminate beam with constant EI, shown in Fig. 9-9a.

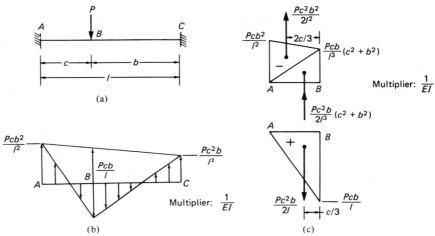

Fig. 9-9. Deflection of the statically indeterminate beam of Example 9-3 by method of elastic weights. (a) Actual beam. (b) Conjugate beam. (c) Elastic weights.

The conjugate beam is shown in Fig. 9-9b, in which the encastré ends of the actual beam are changed to free ends. The conjugate beam has therefore no supports, but it can be easily checked that the beam is in equilibrium under the elastic weights corresponding to the bending moment on the actual beam.

The elastic loading to the left of point B is equivalent to the components shown in Fig. 9-9c. The deflection y_B is equal to the moment of these components about B.

$$y_B = \frac{1}{EI}\left[\frac{Pc^2b^2}{2l^2}\frac{2c}{3} + \frac{Pc^2b}{2l^3}(c^2 + b^2)\frac{c}{3} - \frac{Pc^2b}{2l}\frac{c}{3}\right]$$

With $(c + b) = l$, this equation reduces to $y_B = \dfrac{Pc^3b^3}{3EIl^3}$.

9-51 Equivalent Concentrated Loading

The calculation of reactions and bending moments in a beam due to an irregular loading can be simplified by the use of equivalent concentrated loads. This is particularly useful in the application of the method of elastic weights.

Figure 9-10a shows any three points $i - 1$, i, and $i + 1$ in a beam subjected to irregular loading. These points will be referred to as nodes. The concentrated load Q_i equivalent to this loading is equal and opposite to the sum of the reactions at i of two simply supported beams between $i - 1$, i, and $i + 1$, carrying the same load as that on the actual beam between the nodes considered. Thus

$$Q_i = \frac{1}{\lambda_l} \int_0^{\lambda_l} x_1 q \, dx_1 + \frac{1}{\lambda_r} \int_0^{\lambda_r} x_2 q \, dx_2 \tag{9-11}$$

where λ_l and λ_r are the distances from i to the nodes $i - 1$ and $i + 1$ respectively, x_1 and x_2 are the distances indicated in Fig. 9-10b, and q is the (variable) load intensity.

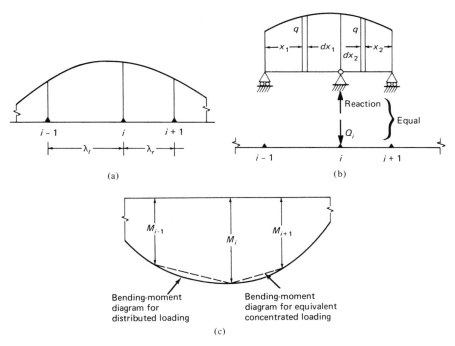

Fig. 9-10. Bending moment using equivalent concentrated loading. (a) Distributed load. (b) Equivalent concentrated load at i. (c) Loadings (a) and (b) give the same bending moment at the nodes.

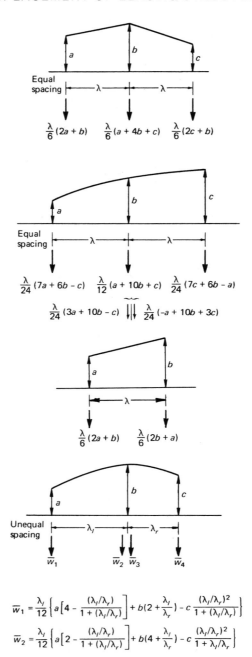

$$\frac{\lambda}{6}(2a + b) \qquad \frac{\lambda}{6}(a + 4b + c) \qquad \frac{\lambda}{6}(2c + b)$$

$$\frac{\lambda}{24}(7a + 6b - c) \qquad \frac{\lambda}{12}(a + 10b + c) \qquad \frac{\lambda}{24}(7c + 6b - a)$$

$$\frac{\lambda}{24}(3a + 10b - c) \qquad \frac{\lambda}{24}(-a + 10b + 3c)$$

$$\frac{\lambda}{6}(2a + b) \qquad \frac{\lambda}{6}(2b + a)$$

$$\overline{w}_1 = \frac{\lambda_l}{12}\left\{ a\left[4 - \frac{(\lambda_l/\lambda_r)}{1 + (\lambda_l/\lambda_r)}\right] + b(2 + \frac{\lambda_l}{\lambda_r}) - c\frac{(\lambda_l/\lambda_r)^2}{1 + (\lambda_l/\lambda_r)} \right\}$$

$$\overline{w}_2 = \frac{\lambda_l}{12}\left\{ a\left[2 - \frac{(\lambda_l/\lambda_r)}{1 + (\lambda_l/\lambda_r)}\right] + b(4 + \frac{\lambda_l}{\lambda_r}) - c\frac{(\lambda_l/\lambda_r)^2}{1 + (\lambda_l/\lambda_r)} \right\}$$

Fig. 9-11. Equivalent concentrated loading for straight-line and second-degree parabolic distribution of load.

It can be easily proved that any statically determinate structure has the same reactions and bending moments *at* the node points, regardless of whether the structure is loaded by a distributed load or by an equivalent concentrated load calculated by Eq. 9-11. However, the bending moment **between the nodes is altered, as shown in Fig. 9-10c.**

The equivalent concentrated loads (from Eq. 9-11) for straight-line and second-degree parabolic distribution are given in Fig. 9-11. The formulas for the parabolic variation can be used for other curves, because any continuous curve can be closely approximated by a series of small parabolic segments.

Figure 9-12 shows the shearing-force and the bending-moment diagrams for a beam in the vicinity of a general node i where a concentrated load Q_i

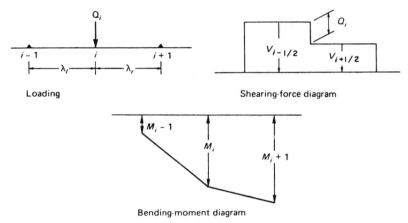

Bending-moment diagram

Fig. 9-12. Bending-moment diagram under the action of a concentrated load at nodes.

acts. The beam may be subjected to loads at other nodes, but there must be no loading between the nodes. From simple statics, the shear between the nodes is related to the bending moment at the nodes as follows:

$$V_{i - \frac{1}{2}} = (M_i - M_{i - 1})/\lambda_l$$

and

$$V_{i + \frac{1}{2}} = (M_{i + 1} - M_i)/\lambda_r$$

where $V_{i - \frac{1}{2}}$ and $V_{i + \frac{1}{2}}$ are the shear in the intervals λ_l and λ_r respectively, and

$$Q_i = V_{i - \frac{1}{2}} - V_{i + \frac{1}{2}}$$

Substituting the first two of these equations into the last one, we find

$$-\frac{M_{i-1}}{\lambda_l} + M_i\left(\frac{1}{\lambda_l} + \frac{1}{\lambda_r}\right) - \frac{M_{i+1}}{\lambda_r} = Q_i \qquad (9\text{-}12)$$

When $\lambda_l = \lambda_r = \lambda$, Eq. 9-12 becomes

$$\frac{1}{\lambda}(-M_{i-1} + 2M_i - M_{i+1}) = Q_i \tag{9-13}$$

If Eq. 9-12 or 9-13 is applied at several nodes in a beam, a system of simultaneous equations can be written, the solution of which gives the values of M.

When the deflection is calculated as the bending moment of elastic weights, we change Q_i to $\bar{w}_i$ and M to y in Eqs. 9-12 and 9-13, where $\bar{w}_i$ is the equivalent concentrated elastic weight and y is the deflection. Thus

$$-\frac{y_{i-1}}{\lambda_l} + y_i\left(\frac{1}{\lambda_l} + \frac{1}{\lambda_r}\right) - \frac{y_{i+1}}{\lambda_r} = \bar{w}_i \tag{9-14}$$

and when $\lambda_l = \lambda_r = \lambda$,

$$\bar{w}_i = \frac{1}{\lambda}(-y_{i-1} + 2y_i - y_{i+1}) \tag{9-15}$$

The equivalent concentrated elastic loading is given by Eq. 9-11:

$$\bar{w}_i = \frac{1}{\lambda_l}\int_0^{\lambda_l} x_1 \frac{M}{EI} dx_1 + \frac{1}{\lambda_r}\int_0^{\lambda_r} x_2 \frac{M}{EI} dx_2 \tag{9-16}$$

Example 9-4   Find the deflection at C and the rotations at A and B for the beam of Fig. 9-13a. The variation in the second moment of area of the beam is as shown.

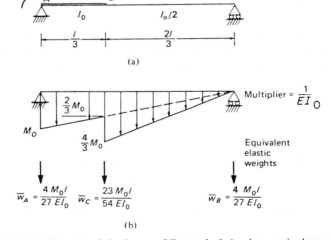

(a)

(b)

Fig. 9-13. Deflection of the beam of Example 9-4 using equivalent concentrated elastic loading. (a) Actual loading. (b) Elastic loading.

The elastic loading is shown in Fig. 9-13b with the equivalent concentrated loads, calculated by the formulas in Fig. 9-11. Applying Eq. 9-14 at C, with A, B, and C as nodes $i - 1$, i, and $i + 1$, and $y_A = 0$, we obtain

$$- 0 + y_C \left[\frac{1}{(l/3)} + \frac{1}{(2l/3)} \right] - 0 = \frac{23 M_0 l}{54 E I_0}$$

whence

$$y_C = \frac{23 M_0 l^2}{243 E I_0}$$

The values of the slope at A and B are numerically equal to the elastic reactions r_A and r_B; thus

$$\theta_A = r_A = \bar{w}_A + \frac{2}{3} \bar{w}_C = \frac{35 M_0 l}{81 E I_0}$$

$$\theta_B = -r_B = -\frac{\bar{w}_C}{3} - w_B = -\frac{47 M_0 l}{162 E I_0}$$

9-6 METHOD OF FINITE DIFFERENCES

A numerical solution of the differential equation for deflection can be obtained by finite differences. If the bending moment is known, the second derivative in Eq. 9-1 can be put in finite-difference form in terms of the unknown deflections at three consecutive points (nodes), usually equally spaced along the beam. Hence, we obtain a finite-difference equation relating the deflection at node points to the bending moment. When the finite-difference equation is applied at all the nodes where the deflection is not known, a set of simultaneous equations is obtained, the solution of which gives the deflections.

A more useful application of the method of finite differences is in cases when the bending moment is not easy to determine, a solution is then first obtained for the differential equations relating the deflection to the external loading (Eqs. 9-4–9-7), and the stress resultants and reactions are found from the deflections by differentiation. Thus a complete structural analysis can be carried out by finite differences. This method has a wide field of application, and for this reason it is treated in detail in Chapters 17 and 18.

In this section the discussion is limited to the solution of Eq. 9-1 to obtain the deflection when the bending moment is known.

Consider a simple beam of variable flexural rigidity EI shown in Fig. 9-14a. The differential equation governing the deflection is given by Eq. 9-1:

$$M = - EI \frac{d^2 y}{dx^2} \qquad\qquad [9\text{-}1]$$

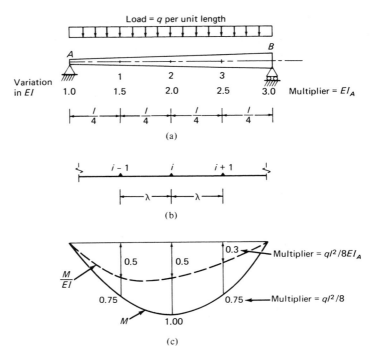

Fig. 9-14. Calculation of deflections by finite differences.

Consider three equally spaced points $i - 1$, i, $i + 1$ where the deflection is y_{i-1}, y_i, y_{i+1}, respectively (Fig. 9-14b). It can be easily shown[1] that the second derivative of the deflection at i can be approximated by

$$\frac{d^2y}{dx^2} \simeq \frac{1}{\lambda^2}(y_{i-1} - 2y_i + y_{i+1}) \tag{9-17}$$

The finite-difference form of Eq. 9-1 applied at node i is

$$\frac{1}{\lambda}(-y_{i-1} + 2y_i - y_{i+1}) \simeq \left(\frac{M}{EI}\right)_i \lambda \tag{9-18}$$

where $(M/EI)_i$ is the value of the bending moment divided by the flexural rigidity at point i. This equation is approximate, and does not take into account the manner in which $M/(EI)$ varies between the nodes.

It may be useful to compare Eq. 9-18 with Eq. 9-15, in which the elastic weight $\bar{w}_i$ replaces the right-hand side of Eq. 9-18. The latter equation is approximate and its accuracy is increased if the number of node points

[1] See Sec. 17-2.

along the span is large. However, for most practical cases, Eq. 9-18 gives sufficient accuracy even with a small number of divisions.

To compare the two expressions let us consider the beam of Fig. 9-14a. The values of $M/(EI)$ at three nodes in the beam are indicated in Fig. 9-14c. Applying Eq. 9-18 at nodes 1, 2, and 3, with the deflection at the supports $y_A = y_B = 0$, gives

$$\frac{1}{(l/4)} \begin{bmatrix} 2 & -1 & 0 \\ -1 & 2 & -1 \\ 0 & -1 & 2 \end{bmatrix} \begin{Bmatrix} y_1 \\ y_2 \\ y_3 \end{Bmatrix} \simeq \frac{ql^3}{32EI_A} \begin{Bmatrix} 0.5 \\ 0.5 \\ 0.3 \end{Bmatrix}$$

The solution of these equations is

$$y_1 = 2.10 \left(\frac{ql^4}{384EI_A} \right) \qquad y_2 = 2.70 \left(\frac{ql^4}{384EI_A} \right) \qquad \text{and} \qquad y_3 = \left(1.8 \frac{ql^4}{384EI_A} \right)$$

Now using Eq. 9-15 with the elastic weights $\bar{w}_i$ calculated by the expression:

$$\bar{w}_i = \frac{\lambda}{12} \left[\left(\frac{M}{EI} \right)_{i-1} + 10 \left(\frac{M}{EI} \right)_i + \left(\frac{M}{EI} \right)_{i+1} \right]$$

(see Fig. 9-11), we obtain

$$\frac{1}{(l/4)} \begin{bmatrix} 2 & -1 & 0 \\ -1 & 2 & -1 \\ 0 & -1 & 2 \end{bmatrix} \begin{Bmatrix} y_1 \\ y_2 \\ y_3 \end{Bmatrix} = \frac{ql^3}{384EI_A} \begin{Bmatrix} 5.5 \\ 5.8 \\ 3.5 \end{Bmatrix}$$

whence

$$y_1 = 1.975 \left(\frac{ql^4}{384EI_A} \right) \qquad y_2 = 2.575 \left(\frac{ql^4}{384EI_A} \right) \qquad \text{and} \qquad y_3 = 1.725 \left(\frac{ql^4}{384EI_A} \right)$$

9-7 REPRESENTATION OF DEFLECTIONS BY FOURIER SERIES

The equation of the deflected line of a simply supported beam under any loading can be expressed in the form of a trigonometric series

$$y = \sum_{n=1}^{\infty} a_n \sin \frac{n\pi x}{l} \tag{9-19}$$

If the loading is represented as a Fourier series,

$$q = \sum_{n=1}^{\infty} b_n \sin \frac{n\pi x}{l} \tag{9-20}$$

where

$$b_n = \frac{2}{l} \int_0^l q(x) \sin \frac{n\pi x}{l} \, dx \tag{9-21}$$

it can be easily seen that Eq. 9-19 is a solution of the differential equation $q = EI(d^4y/dx^4)$, provided that

$$a_n = \frac{l^4}{n^4\pi^4 EI} b_n \tag{9-22}$$

Equation 9-19 satisfies the end conditions of zero deflection ($y = 0$ at $x = 0, l$) and of zero moment ($-EI(d^2y/dx^2) = 0$ at $x = 0, l$).

Example 9-5   Find the deflection of the beam in Fig. 9-15.

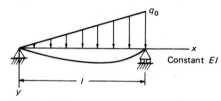

Constant EI

Fig. 9-15. Representation of elastic line of the simple beam of Example 9-5 by a Fourier series.

The intensity of load at any point distant x from the left-hand end is $q(x) = q_0 x/l$. From Eq. 9-21,

$$b_n = \frac{2q_0}{l^2} \int_0^l x \sin \frac{n\pi x}{l} \, dx = -\frac{2q_0}{n\pi} \cos n\pi$$

But $\cos n\pi = -1$ when $n = 1, 3, 5, \ldots$, and $\cos n\pi = 1$ for $n = 2, 4, 6, \ldots$. Therefore,

$$b_n = (-1)^{n+1} \frac{2q_0}{n\pi}$$

and from Eq. 9-22,

$$a_n = (-1)^{n+1} \frac{2q_0 l^4}{n^5\pi^5 EI}$$

Substituting in Eq. 9-19, we obtain the deflection

$$y = \frac{2q_0 l^4}{\pi^5 EI} \sum_{n=1}^{\infty} \frac{(-1)^{n+1}}{n^5} \sin \frac{n\pi x}{l}$$

Using three terms of the series, we find for the mid-span $y_{(x = l/2)} = 0.00652 \, q_0 l^4/(EI)$, which agrees with the exact solution to three significant places. If only one term of the series is used, we have $y_{(x = l/2)} = 0.00654$ $q_0 l^4/(EI)$. This does not differ greatly from the exact solution, and indeed in many cases sufficient accuracy is obtained by using only one term of the series.

9-8 REPRESENTATION OF DEFLECTIONS BY SERIES WITH INDETERMINATE PARAMETERS

In the preceding section, we used a series to approximate loadings, and hence obtained the deflection function. Series may also be used to approximate the deflection function directly. Each term of the series includes a parameter which can be determined by one of several numerical procedures.[2] In this section, the Rayleigh–Ritz method is used to find the deflection of beams. However, the principle of virtual work is employed instead of the minimization process of the total potential energy.

Consider the beam AB shown in Fig. 9-16a, subjected to lateral and axial forces and resting on an elastic foundation. Let the deflection be expressed in the form

$$y = a_1 g_1 + a_2 g_2 + \cdots + a_n g_n \tag{9-23}$$

where g_i is any continuous function of x between A and B which satisfies the boundary conditions (for instance, at a fixed end, $y = 0$ and $dy/dx = 0$). The equilibrium conditions at the ends (for example, the condition that shear and bending moment at a free end are zero) are not necessarily satisfied by the functions g (as shear and bending moment are derivatives of y and therefore a function of a derivative of g).

The values of the parameters $a_1, a_2, \ldots, a_n$ are determined by satisfying the energy criterion as follows.

Under the applied forces, the beam is in equilibrium in a deflected shape shown in Fig. 9-16b. The displacements $D_1, D_2, \ldots, D_m$ are the movements along the line of action of the external forces $F_1, F_2, \ldots, F_m$. The resultant lateral load is $q^* = q - ky$, where ky is the intensity of the foundation reaction. Let the beam acquire a small virtual deflection from the equilibrium position. Assume that this virtual deflection varies in magnitude along the beam according to the equation

$$dy = da_i g_i \tag{9-24}$$

where da_i represents a small increase in the parameter a_i, while the other parameters remain unchanged. Thus the virtual deflection dy varies along the beam in the same manner as the function g_i only. This is in contrast to the variation in y which depends on all the functions $g_1, g_2, \ldots, g_n$. The

[2]See, for example, S. H. Crandall, *Engineering Analysis*, McGraw-Hill, New York, 1956, pp. 230–232.

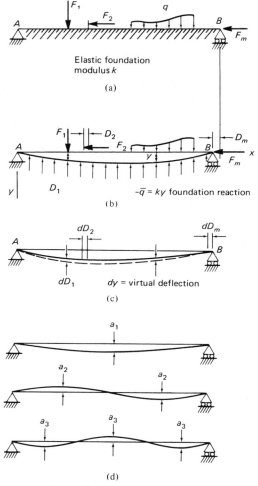

Fig. 9-16. Representation of deflections by series with indeterminate
parameters.

virtual deflection causes the forces F_1, F_2, ..., F_m to move along their lines
of action by amounts dD_1, dD_2, ..., dD_m, as shown in Fig. 9-16c.

According to the principle of virtual work (see Sec. 5-5), if an elastic
system in equilibrium undergoes a small virtual displacement, the work
done by the external forces while moving along the virtual displacement is
equal to the corresponding change in strain energy. The virtual work of
the external forces is

$$dW = \sum_{j=1}^{m} F_j dD_j + \int_{A}^{B} q^*(dy)\, dx$$

This equation may also be written in the form

$$dW = da_i \left(\sum_{j=1}^{m} F_j \frac{\partial D_j}{\partial a_i} + \int_A^B q^* g_i \, dx \right)$$

where the partial derivative $\partial D_j / \partial a_i$ represents the change in the displacement D_j corresponding to a unit change in the parameter a_i, causing a virtual deflection which varies along the beam according to Eq. 9-24.

The two sections limiting an element of length dx of the beam will rotate relative to each other during the virtual deflection by an angle

$$d\theta = da_i \frac{d^2 g_i}{dx^2} \, dx$$

The change in strain energy of this element is $-M \, d\theta$ (see Sec. 5-32), and the total change in strain energy due to bending is

$$dU = -da_i \int_A^B M \frac{d^2 g_i}{dx^2} \, dx$$

The bending moment at any section is

$$M = -EI \frac{d^2 y}{dx^2} = -EI \left(a_1 \frac{d^2 g_1}{dx^2} + a_2 \frac{d^2 g_2}{dx^2} + \cdots + a_2 \frac{d^2 g_n}{dx^2} \right)$$

Substituting for moment in the strain-energy equation and equating the external and internal work, we obtain

$$\sum_{j=1}^{m} F_j \frac{\partial D_j}{\partial a_i} + \int_A^B q^* g_i \, dx$$

$$= \int_A^B EI \left(a_1 \frac{d^2 g_1}{dx^2} + a_2 \frac{d^2 g_2}{dx^2} + \cdots + a_n \frac{d^2 g_n}{dx^2} \right) \frac{d^2 g_i}{dx^2} \, dx \qquad (9\text{-}25)$$

Equation 9-25 represents a system of equations corresponding to $i = 1$, 2, ..., n, from which the parameters $a_1, a_2, ..., a_n$ can be determined; thus by Eq. 9-23, an approximate equation to the elastic line is obtained.

The above procedure has been used extensively to treat buckling problems[3] and, to some extent, has also been applied to beams on elastic foundations.[4] For a simply supported beam, the functions g are chosen as a sinusoidal curve along the beam length l

$$g_n = \sin \frac{n\pi x}{l} \qquad \frac{d^2 g_n}{dx^2} = -\frac{n^2 \pi^2}{l^2} \sin \frac{n\pi x}{l}$$

---

[3] S. Timoshenko and J. M. Gere, *Theory of Elastic Stability*, 2d ed., McGraw-Hill, New York, 1961.

[4] M. Hetényi, *Beams on Elastic Foundation*, The University of Michigan Press, Ann Arbor, 1947.

where $n = 1, 2, 3 \ldots$. These functions satisfy the boundary conditions of a simply supported beam.[5] Thus the deflection is represented by the series

$$y = \sum_{n=1}^{\infty} a_n \sin \frac{n\pi x}{l} \qquad [9\text{-}19]$$

This means that the deflection curve may be obtained by summing the ordinates of sine curves such as those shown in Fig. 9-16d. The parameters $a_1, a_2, \ldots$ represent the maximum ordinates of these sine curves.

With this choice, and with the assumption that EI is constant the integrals of the cross-products on the right-hand side of Eq. 9-25 vanish because

$$\int_0^l \sin \frac{n\pi x}{l} \sin \frac{r\pi x}{l} \, dx = \begin{cases} 0 \text{ if } n \neq r \\ \dfrac{l}{2} \text{ if } n = r \end{cases}$$

Therefore, in the case of a prismatic beam Eq. 9-25 becomes

$$\sum_{j=1}^{m} F_j \frac{\partial D_j}{\partial a_n} + \int_0^l q^* g_n \, dx = a_n EI \int_0^l \left(\frac{d^2 g_n}{dx^2}\right)^2 dx$$

This equation has only one unknown parameter, and its solution gives the value of a_n:

$$a_n = \frac{2l^3}{n^4 \pi^4 EI} \left(\sum_{j=1}^{m} F_j \frac{\partial D_j}{\partial a_n} + \int_0^l q^* \sin \frac{n\pi x}{l} \, dx \right) \qquad (9\text{-}26)$$

Equations 9-19 and 9-26 can be used to express the deflection in a prismatic beam with hinged ends, subjected to lateral and axial loads, such as the beam in Fig. 9-16a. The same equations can be used in the analysis of beams with two hinged ends and with intermediate supports, the reactions at the intermediate supports being treated as any of the forces F. From the condition that $y = 0$ at the supports (or y has a known value), we obtain an equation from which the reaction can be determined. If the beam has intermediate spring supports (elastic supports) of stiffness k, the deflection condition at these supports is given by the equation for the reaction $F = ky$.

The above procedure is used in the following three examples for a beam loaded with a lateral load only, for a beam with lateral and axial loads, and for a beam resting on an elastic foundation.

Example 9-6 Find the deflection of a simple beam carrying one vertical load P (Fig. 9-17a). The beam is assumed to have a constant flexural rigidity EI.

[5]The functions which satisfy other end conditions are discussed in some detail in Chapter 19 in conjunction with the finite-strip method.

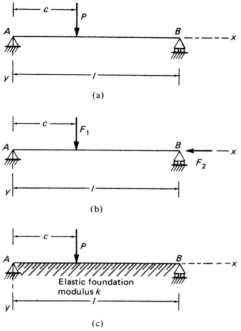

Fig. 9-17. Beams considered in Examples 9-6, 9-7, and 9-8.

Substituting in Eq. 9-26, $q^* = 0$, $F_1 = P$ and $\partial D_1/\partial a_n = \sin(n\pi c/l)$, we obtain

$$a_n = \frac{2Pl^3}{n^4\pi^4EI} \sin \frac{n\pi c}{l}$$

Substituting for a_n in Eq. 9-19, we find

$$y = \frac{2Pl^3}{\pi^4EI} \sum_{n=1}^{\infty} \frac{1}{n^4} \sin \frac{n\pi x}{l} \sin \frac{n\pi c}{l}$$

Example 9-7 Find the deflection of the beam in Fig. 9-17b, carrying a vertical force F_1 and an axial compressive force F_2.

The deflection of the beam causes the roller at B to move toward the hinge A. The movement of B is equal to the difference between the length of the curve and the length of the chord AB. The length ds corresponding to a horizontal element dx is

$$ds = \sqrt{dx^2 + dy^2}$$

Dividing both sides of this equation by dx and expanding the square root by the binomial series, and retaining the first two terms only, we obtain

$$\frac{ds}{dx} = 1 + \frac{1}{2}\left(\frac{dy}{dx}\right)^2$$

Integrating both sides,

$$\int_0^l \frac{ds}{dx}\,dx = \int_0^l dx + \frac{1}{2}\int_0^l \left(\frac{dy}{dx}\right)^2 dx$$

whence the horizontal movement of B is

$$s - l = \frac{1}{2}\int_0^l \left(\frac{dy}{dx}\right)^2 dx$$

where s is the length of the curve AB.

For the beam considered, the deflection is expressed by Eq. 9-19 as

$$y = \sum_{n=1}^{\infty} a_n \sin \frac{n\pi x}{l}$$

and

$$D_2 = s - l = \frac{1}{2}\sum_{n=1}^{\infty}\int_0^l \left(a_n \frac{n\pi}{l}\cos\frac{n\pi x}{l}\right)^2 dx = \sum_{n=1}^{\infty} \frac{a_n^2 n^2 \pi^2}{4l}$$

The partial derivative with respect to a_n is zero for all the terms in the above series except for the term containing a_n; therefore,

$$\frac{\partial D_2}{\partial a_n} = \frac{a_n n^2 \pi^2}{2l}$$

Substituting in Eq. 9-26 $\partial D_1/\partial a_n = \sin(n\pi c/l)$, $\partial D_2/\partial a_n = a_n n^2 \pi^2/(2l)$, and $q^* = 0$, and rearranging the terms, we obtain

$$a_n = \frac{2 F_1 l^3 \sin \dfrac{n\pi c}{l}}{n^2 \pi^4 EI[n^2 - F_2 l^2/(\pi^2 EI)]}$$

Therefore, from Eq. 9-19 the deflection equation is

$$y = \frac{2 F_1 l^3}{\pi^4 EI}\sum_{n=1}^{\infty}\frac{\sin\dfrac{n\pi c}{l}\sin\dfrac{n\pi x}{l}}{n^2[n^2 - F_2 l^2/(\pi^2 EI)]}$$

Example 9-8   Find the deflection of a simple beam carrying a concentrated vertical load P and resting on an elastic foundation of modulus k (Fig. 9-17c). The beam has a constant flexural rigidity EI.

The intensity of the resultant lateral load is

$$q^* = -ky = -k \sum_{n=1}^{\infty} a_n \sin \frac{n\pi x}{l}$$

Equation 9-26 gives

$$a_n = \frac{2l^3}{n^4\pi^4 EI} \left[P \sin \frac{n\pi c}{l} - k \int_0^l \left(\sum_{r=1}^{\infty} a_r \sin \frac{r\pi x}{l} \sin \frac{n\pi x}{l} \right) dx \right]$$

The integral on the right-hand side of this equation is zero when $r \neq n$ and is equal to $a_n l/2$ when $r = n$. Therefore

$$a_n = \frac{2Pl^8}{\pi^4 EI} \left(\frac{\sin \dfrac{n\pi c}{l}}{n^4 + kl^4/(\pi^4 EI)} \right)$$

Substituting in Eq. 9-19,

$$y = \frac{2Pl^3}{\pi^4 EI} \sum_{n=1}^{\infty} \frac{\sin \dfrac{n\pi c}{l} \sin \dfrac{n\pi x}{l}}{n^4 + kl^4/(\pi^4 EI)}.$$

9-9 GENERAL

When the changes in length of members of a plane truss are known, the displacements of the joints can be determined graphically by a Williot or a Williot–Mohr diagram. The Mohr addition is necessary only when there is no member of the truss known to remain fixed in direction during deformation.

Several methods can be used to find the deflection due to bending of beams. The moment-area or the elastic-weight methods can be used when the variation of the bending moment along the beam is known. Both these methods are based upon relations between the area under the bending-moment diagram and deflection. In the method of elastic weight, the calculation of angular rotation and deflection is similar to the routine calculation of shear and bending moment, except that the applied load and support conditions of the beam are generally altered.

The finite-difference method gives a numerical solution to the governing differential equations relating the deflection to bending moment or the deflection to loading. The use of finite differences is particularly advantageous when the bending moment is not known. From a numerical solution of the differential equation relating the loading to the deflection, the unknown stress resultants can be determined by differentiation. The details of this procedure are found in Chapters 17 and 18.

If the loading is represented by a Fourier series, a solution to the differential equation giving the deflection can be obtained in the form of a trigonometric series. This can be easily applied to beams with hinged ends.

The deflection can also be represented as a series including an indeterminate parameter in each term. Several numerical methods can be used to determine the parameters which give a solution that approximately satisfies the differential equation. In the method discussed in this chapter the parameters are chosen to satisfy the energy criterion. As in the case of the finite-difference method, the bending moment need not be known and the stress resultants can be determined from the deflection.

PROBLEMS

9-1   Draw a Williot diagram for the truss shown in the Prob. 9-1. Use this diagram to find the vertical displacement of joint G, and check this value by the method of virtual work. The changes in length (in.) of the members due to loading and temperature variation are indicated in the figure.

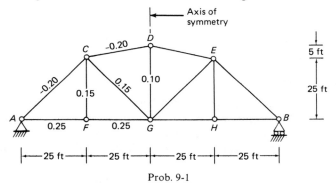

Prob. 9-1

9-2   Apply the requirements of Prob. 9-1 to the truss shown in the figure.

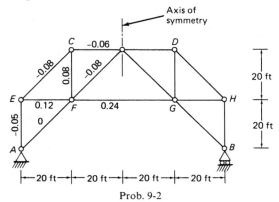

Prob. 9-2

9-3   Draw a Williot-Mohr diagram for the truss shown in the figure. The changes in lengths (in.) of the members are indicated in the figure. From the Williot-Mohr diagram find the vertical deflection of joint H.

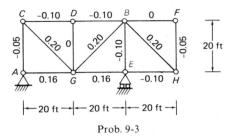

Prob. 9-3

9-4   Apply the requirements of Prob. 9-3 to the truss shown in the figure.

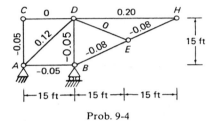

Prob. 9-4

9-5   Use the moment-area-deflection relations of Eqs. 9-8 and 9-10 to prove that for a simple beam the slope and the deflection are equal respectively to the shear and the bending-moment of the elastic weights.

9-6   Determine the deflection and the angular rotation at C for the prismatic beam shown in the figure. Assume $I = 800$ in.^{4} and $E = 30 \times 10^6$ psi.

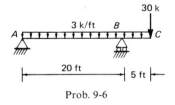

Prob. 9-6

9-7   Compute the deflection at D and C for the beam shown.

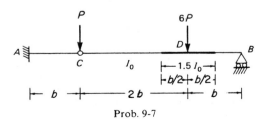

Prob. 9-7

9-8   Show that the vertical deflection due to bending of the arch AB in the figure is numerically equal to the bending-moment of the beam $A'B'$ loaded with the elastic loading $M\,ds/(EI)$, where M is the bending moment in the arch, ds is the length of a typical element along the arch axis, and EI is the flexural rigidity.

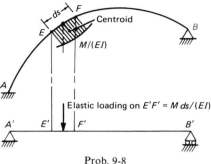

Prob. 9-8

9-9   Using the method of elastic weights, find the deflections at C, D, and E in the arch shown in the figure. Consider bending deformations only.

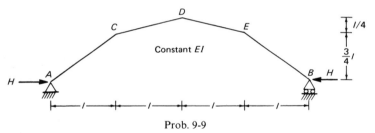

Prob. 9-9

9-10   Obtain the stiffness matrix corresponding to the coordinates 1 and 2 of the beam shown. (*Hint:* Find the angular rotations due to unit moment at each end, and form a flexibility matrix which, when inverted, gives the required stiffness matrix.)

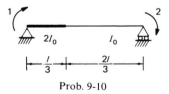

Prob. 9-10

9-11   Calculate the magnitude of two equal and opposite couples M necessary to produce an angular discontinuity of one radian at point D in the prismatic beam shown. Find the corresponding deflections at D and E. (These deflections are numerically equal to influence coefficients of the bending moment at D, see Sec. 13-3.)

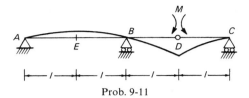

Prob. 9-11

9-12   (a) Find the deflection at D in the continuous beam on unyielding supports
shown in the figure. The beam has a constant flexural rigidity EI.
(b) The bending moments at the supports in the above case are $M_A =$
$-0.0711qb^2$ and $M_B = -0.1938qb^2$. A vertical settlement at B and a rotation
at A change these moments to $M_A = 0$ and $M_B = -0.05qb^2$. Find the amount
of the settlement and the rotation.

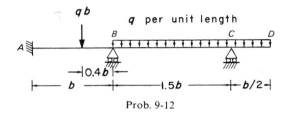

Prob. 9-12

9-13   Use the finite-difference method to calculate the deflection at nodes 1, 2, and
3 and the angular rotation at end A in the beam shown. Give the answers in
terms of EI_0. To calculate the rotation at node A, use the approximate
relation

$$\theta_i \simeq (y_{i+1} - y_i)/\lambda$$

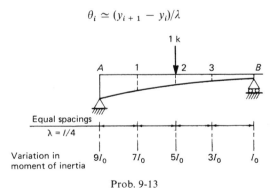

Prob. 9-13

9-14   Use the Rayleigh-Ritz procedure to determine the deflection of a simple
beam of constant EI, span l, loaded with a uniform transverse load of intensity
q and an axial compressive force P at the ends. Represent the deflection by
the series

$$y = \sum_{n=1}^{\infty} a_n \sin \frac{n\pi x}{l}$$

9-15   Find the first three parameters in the series

$$y = \sum_{n=1}^{\infty} a_n \sin \frac{n\pi x}{l}$$

which represents the deflection of the prismatic beam shown in the figure.

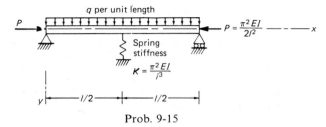

Prob. 9-15

9-16   Find the deflection of a prismatic beam on an elastic foundation symmetrically loaded by two concentrated loads as shown in the figure. Represent the deflection by the series

$$y = \sum_{n=1}^{\infty} a_n \left(1 - \cos \frac{2n\pi x}{l}\right)$$

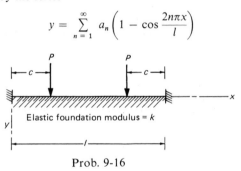

Prob. 9-16

9.17   Show that the deflection at the centre of a straight member with respect to its ends is given by

$$y_{centre} = (\psi_1 + 10\psi_2 + \psi_3)$$

where ψ_1 and ψ_3 are the curvatures at the two ends; l is the distance between the two ends; and ψ_2 is the curvature at the middle. The variation in ψ is assumed to follow a second-degree parabola.

This is a geometric relation which can be derived by integration of the equation $\psi = d^2y/dx^2$; where y is the deflection and x is the distance along the member, measured from the left end. The geometric relation can be derived more easily by the method of elastic weights, using equivalent concentrated loading (Fig. 9-11).

In practice, the expression derived can be used for continuous or simple beams having a constant or a variable cross-section when the parabolic variation of ψ is acceptable.

Application of the force method: column analogy

10-1 INTRODUCTION

In this chapter we deal with the analysis of plane frames or arches formed of one closed bent, and therefore statically indeterminate to a degree not higher than three. The analysis is by the force method with the redundants chosen at a point called the elastic center, and involves calculations similar to those for stresses in a column subjected to combined bending and direct stress. This analogy was recognized by Hardy Cross, who suggested that the calculations be arranged in the same manner as the computation of stresses in a column.

The column analogy can further be used to determine the stiffness coefficients of parts of a highly indeterminate plane frame; these stiffness coefficients are of use in the analysis of such a frame by a displacement method of analysis.

In the derivation of the column analogy method, only the bending deformations are considered, but in some cases (e.g., arches, Sec. 10-7), the effect of axial forces may be included.

10-2 ELASTIC CENTER AND THE ANALOGOUS COLUMN

We know that the fixed arch AB shown in Fig. 10-1a is statically indeterminate to the third degree. One way of making the structure statically determinate is to remove from end A three reactions: vertical and horizontal components, and a couple (Fig. 10-1b). The consistency in deformation of the actual structure can be restored by the action of three redundant forces F_1, F_2, and F_3 along the coordinates 1, 2, and 3 (Fig. 10-1c) at a point O connected to end A by a rigid (nondeformable) arm. The compatibility conditions of no translation and no rotation at A are satisfied by the equations (see Eq. 2-2)

$$[f]_{3 \times 3} \{F\}_{3 \times 1} = -\{D\}_{3 \times 1}$$

where $[f]$ is the flexibility matrix of the released structure, $\{F\}$ is the matrix of the redundant forces, and $\{D\} = \{D_1, D_2, D_3\}$ are the displacements at

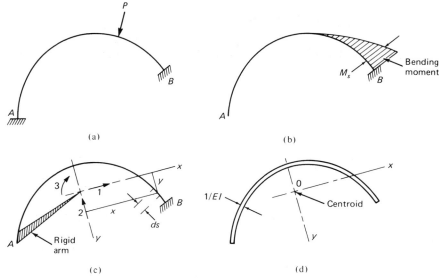

Fig. 10-1. An arch and its analogous column. (a) Actual structure. (b) Statically determinate released structure. (c) Coordinate system indicating positive direction of redundants and displacements. (d) Analogous column.

the coordinates due to the actual loading on the released structure. It will be recalled that the elements of $[f]$ are displacements due to unit values of the redundants; they can be calculated by virtual work (Eq. 7-2):

$$[f] = \begin{bmatrix} \int \dfrac{M_{u1}^2}{EI} ds & & \text{symmetrical} \\[3mm] \int \dfrac{M_{u2}M_{u1}}{EI} ds & \int \dfrac{M_{u2}^2}{EI} ds & \\[3mm] \int \dfrac{M_{u3}M_{u1}}{EI} ds & \int \dfrac{M_{u3}M_{u2}}{EI} ds & \int \dfrac{M_{u3}^2}{EI} ds \end{bmatrix} \quad (10\text{-}1)$$

where M_{uj} is the bending moment in the released structure due to a unit value of the jth redundant.

The elements of $\{D\}$ can also be calculated by virtual work:

$$\{D\} = \left\{ \begin{array}{c} \int \dfrac{M_{u1}M_s}{EI} ds \\[3mm] \int \dfrac{M_{u2}M_s}{EI} ds \\[3mm] \int \dfrac{M_{u3}M_s}{EI} ds \end{array} \right\} \quad (10\text{-}2)$$

where M_s is the statically determinate moment of the released structure due to actual loading.

If we choose the location of O at the elastic center, that is at a point such that $f_{21} = f_{31} = f_{32} = 0$, the flexibility matrix $[f]$ becomes a diagonal matrix and each of the three redundants can be determined from one equation

$$\{F\} = - \left\{ \begin{array}{c} \dfrac{D_1}{f_{11}} \\[2mm] \dfrac{D_2}{f_{22}} \\[2mm] \dfrac{D_3}{f_{33}} \end{array} \right\} \tag{10-3}$$

The location of the elastic center is determined as follows. Let O be the origin of a system of rectangular axes x–y (Fig. 10-1c). The moments at point (x, y) due to unit values of F are

$$M_{u1} = y \qquad M_{u2} = x \qquad M_{u3} = 1 \tag{10-4}$$

In order that the off-diagonal elements of $[f]$ are zero, the following conditions must be satisfied (from Eq. 10-1 and 10-4)

$$\left. \begin{array}{l} f_{21} = \displaystyle\int \dfrac{xy}{EI} ds = 0 \\[4mm] f_{31} = \displaystyle\int \dfrac{y}{EI} ds = 0 \\[4mm] f_{32} = \displaystyle\int \dfrac{x}{EI} ds = 0 \end{array} \right\} \tag{10-5}$$

We may consider the quantity $ds/(EI)$ as an element of area of a strip along the axis of the arch having a width numerically equal to $1/(EI)$ and length ds (Fig. 10-1d). The integrals

$$\int \frac{y}{EI} ds, \quad \int \frac{x}{EI} ds, \quad \int \frac{xy}{EI} ds$$

are then respectively the first moments and the product of inertia of the imaginary area about the x and y axes. These integrals are zero when O is the centroid of the "area" and x and y are the principal axes of inertia. The centroid O of the area $\int(ds/EI)$ is referred to as the elastic center.

Further, from Eq. 10-1 and 10-4, we can express the diagonal element of $[f]$ as follows:

$$
\left.\begin{aligned}
f_{11} &= \int \frac{y^2}{EI}\, ds = I_x \\[4pt]
f_{22} &= \int \frac{x^2}{EI}\, ds = I_y \\[4pt]
f_{33} &= \int \frac{ds}{EI} = a
\end{aligned}\right\} \tag{10-6}
$$

where I_x and I_y are the sum of the moments of inertia of the elemental area $ds/(EI)$ about axes x and y respectively, and a is the total area.

In a similar manner, if we think of the quantity $M_s ds/(EI)$ as an elemental load whose intensity varies as M along the axis of the arch and which acts on the elemental area $ds/(EI)$, then from Eqs. 10-2 and 10-4

$$
D_1 = \int y \frac{M_s}{EI}\, ds = M_x = \text{moment of load about the } x \text{ axis}
$$

$$
D_2 = \int x \frac{M_s}{EI}\, ds = M_y = \text{moment of load about the } y \text{ axis}
$$

$$
D_3 = \int \frac{M_s}{EI}\, ds = N = \text{total load}
$$

From the above and Eqs. 10-3 and 10-6, we obtain

$$
\left.\begin{aligned}
F_1 &= -\frac{M_x}{I_x} \\[4pt]
F_2 &= -\frac{M_y}{I_y} \\[4pt]
F_3 &= -\frac{N}{a}
\end{aligned}\right\} \tag{10-7}
$$

The bending moment at any point (x, y) on the actual structure can be obtained by superposition (see Eq. 2-4).

$$
M = M_s + (M_{u1}F_1 + M_{u2}F_2 + M_{u3}F_3)
$$

Substituting from Eq. 10-4,

$$
M = M_s + yF_1 + xF_2 + F_3 \tag{10-8}
$$

The bending moment in the actual structure can be expressed in the form

$$
M = M_s - \bar{M} \tag{10-9}
$$

where $\bar{M}$ is a statically indeterminate moment. From Eq. 10-7, 10-8, and 10-9 we can write

$$\bar{M} = \frac{N}{a} + \frac{M_x}{I_x}y + \frac{M_y}{I_y}x \qquad (10\text{-}10)$$

The form of this equation is similar to that for stress in an eccentrically loaded short column and this analogy may be exploited. Let us therefore consider a short column of area a subjected to an eccentric normal load N. The stresses due to an eccentric load are the sum of stresses due to a concentric load N and to couples $M_x = Ne_y$, and $M_y = Ne_x$, where e_x and e_y are eccentricities measured from the principal axes of inertia x, y.

The stress at any point (x, y) is then given by the well-known equation

$$\sigma = \frac{N}{a} + \frac{M_x}{I_x}y + \frac{M_y}{I_y}x \qquad (10\text{-}11)$$

Comparing Eqs. 10-10 and 10-11, we see that there is a complete correspondence — in respect to numerical values — between the statically indeterminate moment $\bar{M}$ in a frame or arch and the stresses in an imaginary analogous column loaded normal to the plane of its cross section by the bending-moment diagram M_s of the statically determinate released structure and having the area of a strip extending along the axis of the frame and having a width $1/(EI)$ (see Fig. 10-1d).

If the frame has an intermediate hinge (Fig. 10-2a), or a hinged support, the flexural stiffness at the hinge is zero, that is $1/(EI) = \infty$, and thus the analogous column has an infinite area. The centroid is therefore at the hinge and the two principal axes pass through the hinge (Fig. 10-2b), but their directions are unknown (they can be determined by the first of Eq. 10-5).

If the frame has two hinges (Fig. 10-2c), a line joining them is one of the principal axes, say the y axis (Fig. 10-2d). The statically indeterminate bending moment at any point is then given by the equation

$$\bar{M} = \frac{M_y}{I_y}x \qquad (10\text{-}12)$$

If the bent has one roller support (Fig. 10-2e), the support may be replaced by a short straight pin-jointed member perpendicular to the surface of the roller (as shown in the figure). Therefore a line perpendicular to the surface of the roller will be a principal axis (say, the y axis) of the analogous column and Eq. 10-12 can be used (see Fig. 10-2f).

If the frame is composed of a bent with extending parts, such as overhangs, in which the internal forces are statically determinate,[1] the analogous

[1]See, for example, part CE in the frame of Prob. 10-4, and parts CA and GB in Prob. 10-5.

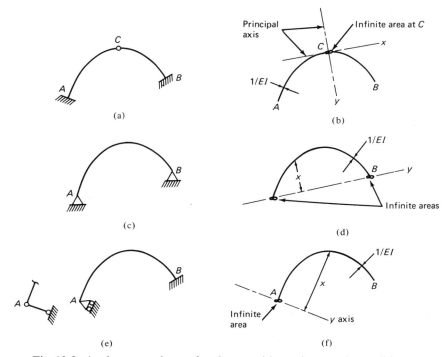

Fig. 10-2. Analogous column for bents with various end conditions.
(a) Frame with intermediate hinge. (b) Analogous column for the frame in (a).
(c) Frame with hinged supports. (d) Analogous column for the frame in (c).
(e) Frame with a roller support. (f) Analogous column for the frame in (e).

column is derived for the frame without the overhangs. The effect of the
forces on the extending parts is taken into account by applying to the frame
appropriate internal forces at the cross section where the extending part
meets the frame.

10-21 Unsymmetrical Bents

In the preceding analysis we needed to know the principal axes of the
analogous column. This presents no problem when the frame is symmetrical
but when it is not, that is in the analysis of unsymmetrical frames, the principal axes are not determined, and we take x and y as any rectangular axes
through the centroid. Equation 10-11 can of course no longer be applied
and an appropriate expression has to be derived.

Consider the area shown in Fig. 10-3a subjected to a normal tensile
force N at point (x,y) with respect to *any* rectangular axes x and y *through
the centroid* of the section. The eccentric force can be replaced by a force

N at the centroid and by two moments $M_x = Ne_y$ and $M_y = Ne_x$ (Fig. 10-3b); the normal stress at any point (x,y) is then given by[2]

$$\sigma = \frac{N}{a} + \left(\frac{M_x I_y - M_y I_{xy}}{I_x I_y - I_{xy}^2}\right) y + \left(\frac{M_y I_x - M_x I_{xy}}{I_y I_x - I_{xy}^2}\right) x \qquad (10\text{-}13)$$

where a, I_x, and I_y are defined in the previous section; and I_{xy} is the product of inertia about x and y axes.[3]

The sign convention used in the derivation of this equation is as follows: N or σ are positive if tensile; M_x, M_y are positive if in the directions shown in Fig. 10-3b.

It follows that if a tensile force is applied at a point in the first quarter, N, M_x, and M_y are all positive.

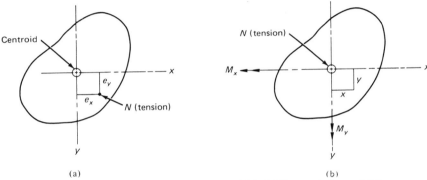

(a)   (b)

Fig. 10-3. Positive directions of N, M_x, and M_y. (a) Eccentric force. (b) Forces equivalent to the eccentric force in (a).

Referring to the arch in Fig. 10-1, the statically indeterminate moment is

$$\bar{M} = \frac{N}{a} + \left(\frac{M_x I_y - M_y I_{xy}}{I_x I_y - I_{xy}^2}\right) y + \left(\frac{M_y I_x - M_x I_{xy}}{I_y I_x - I_{xy}^2}\right) x \qquad (10\text{-}14)$$

This then is the equation to be used (instead of Eq. 10-10) when x and y are not the principal axes, but of course the origin of axes must be at the centroid of the analogous column.

The sign convention to be used in Eqs. 10-7, 10-9, 10-10, and 10-14 is as follows: positive M_s (causing tension on the inner side of the frame) corresponds to positive elastic loading N, that is tension on the analogous

---

[2]Refer to books on strength of materials–for example, F. B. Seely and J. O. Smith, *Advanced Mechanics of Materials*, Wiley, New York, 2d ed., 1952.

[3]The moments and products of inertia of plane cross sections commonly used in the column-analogy method are included in Appendix J.

column; M_x and M_y are positive if in the directions shown in Fig. 10-3b; positive (tensile) stress on the analogous column corresponds to positive $\bar{M}$.

The bending moment in the actual statically indeterminate frame, $M = M_s - \bar{M}$, when positive will cause tensile stress on the inner side of the frame.

10-3 CHOICE OF THE RELEASED STRUCTURE

It will now be shown that the bending moment M_s corresponding to *any* stable statically determinate structure may be used as the load on the analogous column to obtain $\bar{M}$ from Eqs. 10-10 or 10-14. The final moment $M = M_s - \bar{M}$ is the same for all possible choices of M_s.

The frame in Fig. 10-4a is statically equivalent to the frame in Fig. 10-4b,

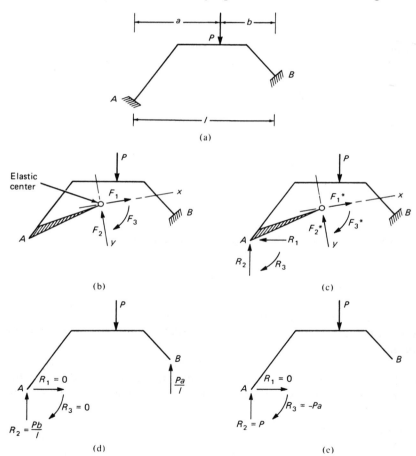

Fig. 10-4. Choice of statically determinate frame.

which is a statically determinate cantilever acted upon by the redundants F_1, F_2, and F_3 of such a magnitude that the compatibility conditions at A are satisfied. We can also select as the released structure the cantilever in Fig. 10-4c acted on by a set of arbitrarily chosen forces R_1, R_2 and R_3 at A. In this case, of course, a different set of redundants F_1^*, F_2^*, and F_3^* will be required to satisfy the compatibility condition. If both these approaches are permissible, the bending moment in the actual structure, $M = M_s - \bar{M}$, calculated from the released structure in Fig. 10-4b must be the same as the bending moment $M = M_s^* - \bar{M}^*$ calculated from the released structure in Fig. 10-4c.

By a suitable variation of the forces R_1, R_2, and R_3, a wide variety of M_s^* diagrams may be obtained. For example, if we take $R_1 = R_3 = 0$ and $R_2 = Pb/l$ (Fig. 10-4d), the M_s^* diagram will be that of a simply supported frame. If we take $R_2 = P$, $R_3 = -Pa$, and $R_1 = 0$ (Fig. 10-4e), the M_s^* diagram will be identical with that of two free cantilevers obtained by cutting the frame at any section between the load P and end B. It appears therefore that any choice of M_s^* corresponding to the actual loading may be used to calculate $\bar{M}^*$, and the difference of the two gives the moment M in the statically determinate structure.

Example 10-1 Find the bending moment in the closed frame shown in Fig. 10-5a.

The analogous column is shown in Fig. 10-5b, which indicates also the centroid of the area and the principal axes x and y. The area of the analogous column and the moments of inertia about the x and y axes are

$$a = \left(2 \times b \times 1 + 2b \times \frac{1}{2} + 2b \times \frac{1}{4}\right)\frac{1}{EI_0} = \frac{3.5b}{EI_0}$$

$$I_x = \left[\frac{2b^3}{2}(0.43)^2 + \frac{2b^3}{4}(0.57)^2 + 2\left(\frac{0.43^3}{3} + \frac{0.57^3}{3}\right)b^3\right]\frac{1}{EI_0} = \frac{0.524}{EI_0}b^3$$

$$I_y = \left[\left(\frac{2b}{2} + \frac{2b}{4}\right)\frac{(2b)^2}{12} + 2b^3\right]\frac{1}{EI_0} = \frac{2.50}{EI_0}b^3$$

A released structure can be obtained by cutting the frame at a section just to the right of C (Fig. 10-5c). The corresponding bending-moment diagram is shown in Fig. 10-5d; the loads on the analogous column and their location are indicated on the figure and in the table below it. Each of the loads N_1, N_2, N_3, and N_4 is equal to the area of the M_s diagram multiplied by the appropriate width $1/(EI)$. These loads act at points on the center line of the frame, each point of application being aligned with the centroid of the area of the appropriate M_s diagram. The moments about the x and y axes are

$$M_x = \Sigma Ny = -\frac{0.477}{EI_0}Pb^3 \quad \text{and} \quad M_y = \Sigma Nx = \frac{3.083}{EI_0}Pb^3$$

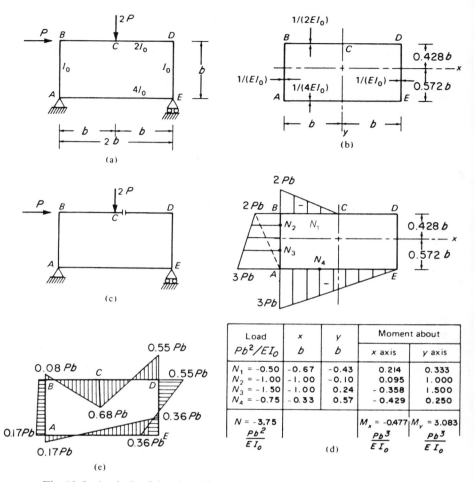

Fig. 10-5. Analysis of the closed frame in Example 10-1 by column analogy. (a) Statically indeterminate frame. (b) Analogous column. (c) Released structure. (d) M_s diagram and load on the analogous column. (e) Bending-moment diagram for the statically indeterminate structure.

The statically indeterminate moment $\bar{M}$ is given by Eq. 10-10:

$$\bar{M} = -\frac{3.75}{3.50} Pb - \frac{0.477P}{0.524} y + \frac{3.083P}{2.5} x = (-1.07b - 0.91y + 1.23x)P$$

Substituting for x and y the coordinates of A, B, C, D, and E in turn, we obtain the statistically indeterminate moments (in terms of Pb):

$$\bar{M}_A = -2.83 \qquad \bar{M}_B = -1.92 \qquad \bar{M}_C = -0.68 \qquad \bar{M}_D = 0.55 \qquad \bar{M}_E = -0.36$$

From Fig. 10-5d, the values of M_s at the same sections are (in terms of Pb):

$$M_{sA} = -3 \qquad M_{sB} = -2 \qquad M_{sC} = 0 \qquad M_{sD} = 0 \qquad M_{sE} = 0$$

Substituting in Eq. 10-9, we obtain the bending moment in the statically indeterminate frame $M = M_s - \bar{M}$. Thus (in terms of Pb)

$$M_A = -0.17 \qquad M_B = -0.08 \qquad M_C = 0.68 \qquad M_D = -0.55 \qquad M_E = +0.36$$

The bending-moment diagram is shown in Fig. 10-5e.

Example 10-2 Find the bending moment in the unsymmetrical frame shown in Fig. 10-6a. All members of the frame have a constant flexural rigidity EI.

The analogous column is shown in Fig. 10-6b. The column has an infinite area at the hinge E. The rectangular axes x and y pass through E, but they are not the principal axes of inertia. The moments and product of inertia are

$$I_x = \frac{3l^3}{8EI} \qquad I_y = \frac{4l^3}{3EI} \qquad I_{xy} = \frac{l^3}{4EI}$$

The released structure is shown in Fig. 10-6c and the corresponding bending-moment diagram in Fig. 10-6d. The loads on the analogous column, that is M_x and M_y, are calculated in the table below the diagram.
Substituting in Eq. 10-14,

$$\bar{M} = P\left(-\frac{25}{126}y + \frac{11}{84}x\right)$$

Substituting for x and y the coordinates of A, B, C, D, and E in turn, we find

$$\bar{M}_A = -0.230Pl \qquad \bar{M}_B = -0.131Pl \qquad \bar{M}_C = -0.032Pl$$

$$\bar{M}_D = 0.099Pl \qquad \bar{M}_E = 0$$

The values of M_s at the same sections are

$$M_{sA} = -0.50Pl \qquad M_{sB} = 0 \qquad M_{sC} = 0 \qquad M_{sD} = 0 \qquad M_{sE} = 0$$

Substituting in Eq. 10-9, we obtain the required bending moments,

$$M_A = -0.270Pl \qquad M_B = +0.131Pl \qquad M_C = +0.032Pl$$

$$M_D = -0.099Pl \qquad M_E = 0$$

The final bending-moment diagram is shown in Fig. 10-6e.

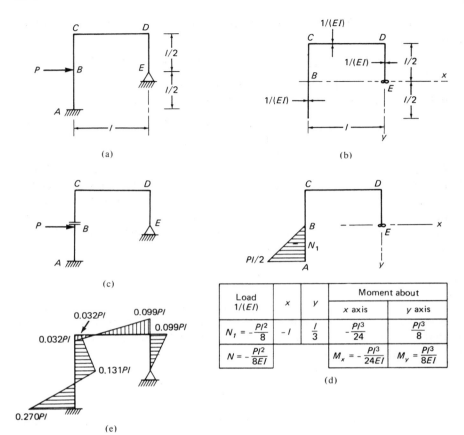

Fig. 10-6. Analysis of the unsymmetrical frame in Example 10-2 by column analogy. (a) Statically indeterminate frame. (b) Analogous column. (c) Released structure. (d) M_s diagram and loads on the analogous column. (e) Bending-moment diagram for the statically indeterminate structure.

10-4   END-STIFFNESS OF PLANE FRAMES

The column analogy can be used to calculate the elements of the stiffness matrix of parts of a larger structure, thus making its analysis possible. This approach is used for more complex structures, the stiffness matrices of the parts being used with a displacement method such as moment distribution or slope deflection (see Chapter 11).

The stiffness of a frame is given by the forces required to hold it in a deflected shape with a unit displacement at one end. Consider the bent in Fig. 10-7a with the coordinates 1, 2, ..., 6 representing the positive directions of forces or displacements. (The coordinates, 1, 2, 4, and 5 are parallel to the rectangular axes x and y, which are not necessarily parallel to the principal axes of the analogous column.) If the stiffness matrix corresponding to these coordinates is known, it can be used to analyze the multibay frame shown in

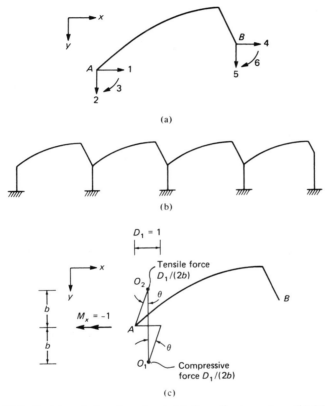

(a)

(b)

(c)

Fig. 10-7. End-translation in a bent. (a) Coordinate system. (b) Multibay frame. (c) Loading on analogous column corresponding to $D_1 = 1$ and $D_2 = D_3 = D_4 = D_5 = D_6 = 0$.

Fig. 10-7b, the part of the frame between consecutive columns being considered as one curved or polygonal member.

This approach is particularly valuable where a computer is not used, as it is then important to reduce the number of equations to be solved. The calculation of the stiffness by column analogy is also commonly used for straight or curved members of variable flexural rigidity, as will be shown by the examples in this section.

10-41 End-Rotation

As established earlier (see Eq. 5-20), the elemental load $M_s\,ds/(EI)$ on the analogous column corresponds to an angle change $d\theta$ in the actual structure. Consequently, a rotation of end A (Fig. 10-7a) through an angle $D_3 = 1$ is equivalent to a unit concentrated load on the analogous column at A. This load should be compressive, producing negative $\bar{M}$ at the end. The final moment at end A is $M = M_s - \bar{M}$, and is positive, producing tension on the inner face of the frame (M_s is zero). Similarly, it can be seen that for a rotation $D_6 = 1$ at end B, a unit concentrated tensile force is to be applied at B on the analogous column.

We can note that by this treatment we are introducing, in the first case, a displacement $D_3 = 1$ while $D_i = 0$ for $i \neq 3$.

10-42 End-Translation

A displacement $D_1 = 1$ at end A (Fig. 10-7a) can be produced by two equal and opposite rotations $\theta = \dfrac{D_1}{2b} = \dfrac{1}{2b}$ about points O_1 and O_2 at an arbitrary distance b above and below A (Fig. 10-7c). This displacement corresponds to the application of two equal and opposite loads θ on the analogous column, forming a couple $M_x = -1$ in the horizontal direction, as shown in the figure. In a similar way, each of the displacements $D_2 = 1$, $D_4 = 1$, and $D_5 = 1$ corresponds to a unit couple applied to the analogous column.

Table 10-1.   Loads on an Analogous Column
Corresponding to End-Displacements in Fig. 10-7a

Displacement	Loading on analogous column
$D_1 = 1$	$M_x = -1$
$D_2 = 1$	$M_y = 1$
$D_3 = 1$	$N\ \ = -1\,(\text{at } A)$
$D_4 = 1$	$M_x = 1$
$D_5 = 1$	$M_y = -1$
$D_6 = 1$	$N\ \ =\ \ 1\,(\text{at } B)$

Note: x and y are rectangular axes parallel to coordinates 1-4 and 2-5 respectively.

It is apparent thus that a unit displacement along one of the six coordinates in Fig. 10-7a corresponds to a unit load or a unit couple on the analogous column; Table 10-1 summarizes the quantities involved.

When the loading on the analogous column is determined, the redundant forces F_1, F_2, and F_3 at the elastic center can be calculated from Eq. 10-7. The end-forces along the coordinates 1, 2, ..., 6 can then be obtained by simple statics.

10-43 End-Forces

Consider the bent in Fig. 10-8a, for which the elastic center O and the principal axes x^* and y^* are indicated. We require the stiffness matrix corresponding to the coordinates 1, 2, ..., 6 at the ends. The coordinates 1 and 4 are parallel to the x axis and make and angle θ with the x^* axis, and are perpendicular to the coordinates 2 and 5.

The redundant forces F_1, F_2, and F_3 at the elastic center are indicated in Fig. 10-8a in their positive directions (compare with Fig. 10-1c). At coordinates, 1, 2, and 3, the forces S_1, S_2, and S_3 which are statically equivalent to F_1, F_2, and F_3 are

$$\left. \begin{aligned} S_1 &= F_1 \cos \theta - F_2 \sin \theta \\ S_2 &= -F_1 \sin \theta - F_2 \cos \theta \\ S_3 &= F_3 + F_1 y_l^* + F_2 x_l^* \end{aligned} \right\} \tag{10-15}$$

where (x_l^*, y_l^*) are the coordinates of point A, the left-hand end of the bent.

The values of x_l^* and y_l^* have to be substituted with their proper signs. Thus if the left-hand end falls in the fourth quarter with respect to the principal axes x^* and y^* (as in the structure shown in Fig. 10-8a) x_l^* has a negative value and y_l^* is positive.

The forces S_4, S_5, and S_6 are in equilibrium with the redundants F_1, F_2, and F_3. Therefore we can write

$$\left. \begin{aligned} S_4 &= -F_1 \cos \theta + F_2 \sin \theta \\ S_5 &= F_1 \sin \theta + F_2 \cos \theta \\ S_6 &= -F_3 - F_1 y_r^* - F_2 x_r^* \end{aligned} \right\} \tag{10-16}$$

where (x_r^*, y_r^*) are the coordinates of the right-hand end of the bent. When checking the third of these equations, we should note that for the structure in Fig. 10-8a the right-hand end falls in the first quarter, so that x_r^* and y_r^* are positive.

The elements of the stiffness matrix of the bent in Fig. 10-8a are derived by Eqs. 10-15 and 10-16 and are given in Eq. 10-17.

$$[S] = \begin{bmatrix}
\dfrac{c^2}{I_x^*}+\dfrac{s^2}{I_y^*} & & & & \text{symmetrical} & \\[10pt]
-\dfrac{sc}{I_x^*}+\dfrac{sc}{I_y^*} & \dfrac{s^2}{I_x^*}+\dfrac{c^2}{I_y^*} & & & & \\[10pt]
\dfrac{cy_l^*}{I_x^*}-\dfrac{sx_l^*}{I_y^*} & -\dfrac{sy_l^*}{I_x^*}-\dfrac{cx_l^*}{I_y^*} & \dfrac{1}{a}+\dfrac{y_l^{*2}}{I_x^*}+\dfrac{x_l^{*2}}{I_y^*} & & & \\[10pt]
-\dfrac{c^2}{I_x^*}-\dfrac{s^2}{I_y^*} & \dfrac{sc}{I_x^*}-\dfrac{sc}{I_y^*} & -\dfrac{cy_l^*}{I_x^*}+\dfrac{sx_l^*}{I_y^*} & \dfrac{c^2}{I_x^*}+\dfrac{s^2}{I_y^*} & & \\[10pt]
\dfrac{sc}{I_x^*}-\dfrac{sc}{I_y^*} & -\dfrac{s^2}{I_x^*}-\dfrac{c^2}{I_y^*} & \dfrac{sy_l^*}{I_x^*}+\dfrac{cx_l^*}{I_y^*} & -\dfrac{sc}{I_x^*}+\dfrac{sc}{I_y^*} & \dfrac{s^2}{I_x^*}+\dfrac{c^2}{I_y^*} & \\[10pt]
-\dfrac{cy_r^*}{I_x^*}+\dfrac{sx_r^*}{I_y^*} & \dfrac{sy_r^*}{I_x^*}+\dfrac{cx_r^*}{I_y^*} & -\dfrac{1}{a}-\dfrac{y_l^*y_r^*}{I_x^*}-\dfrac{x_l^*x_r^*}{I_y^*} & \dfrac{cy_r^*}{I_x^*}-\dfrac{sx_r^*}{I_y^*} & -\dfrac{sy_r^*}{I_x^*}-\dfrac{cx_r^*}{I_y^*} & \dfrac{1}{a}+\dfrac{y_r^{*2}}{I_x^*}+\dfrac{x_r^{*2}}{I_y^*}
\end{bmatrix}$$

$$\begin{matrix}1\\2\\3\\4\\5\\6\end{matrix}$$

(10-17)

where $s = \sin\theta$ and $c = \cos\theta$.

To explain the procedure, the elements in the third column are derived in full. These are the end forces corresponding to $D_3 = 1$ while all the other displacements are zero. The load on the analogous column is $N = -1$ at A (see Table 10-1). This load can be replaced by a load $N = -1$ at the centroid and by moments $M_{x*} = -y_I^*$ and $M_{y*} = -x_I^*$. The redundants at the elastic center, from Eq. 10-7 are

$$F_1 = \frac{y_I^*}{I_{x*}} \qquad F_2 = \frac{x_I^*}{I_{y*}} \qquad \text{and} \qquad F_3 = \frac{1}{a}$$

Substituting for F_1, F_2, and F_3 in Eqs. 10-15 and 10-16, we obtain

$$\left.\begin{aligned}
S_{13} &= \frac{y_I^*}{I_{x*}} \cos\theta - \frac{x_I^*}{I_{y*}} \sin\theta \\[2mm]
S_{23} &= -\frac{y_I^*}{I_{x*}} \sin\theta - \frac{x_I^*}{I_{y*}} \cos\theta \\[2mm]
S_{33} &= \frac{1}{a} + \frac{y_I^{*2}}{I_{x*}} + \frac{x_I^{*2}}{I_{y*}} \\[2mm]
S_{43} &= -\frac{y_I^*}{I_{x*}} \cos\theta + \frac{x_I^*}{I_{y*}} \sin\theta \\[2mm]
S_{53} &= \frac{y_I^*}{I_{x*}} \sin\theta + \frac{x_I^*}{I_{y*}} \cos\theta \\[2mm]
S_{63} &= -\frac{1}{a} - \frac{y_I^* y_r^*}{I_{x*}} - \frac{x_I^* x_r^*}{I_{y*}}
\end{aligned}\right\} \qquad (10\text{-}18)$$

10-44 Forced Displacement

The effect of change in temperature or shrinkage can be easily analyzed by the equations derived in this section. First, we compute the displacements which would occur at one end if it were free to move, then the end is forced back to its actual position by the application of end-forces. Consider, for example, the bent AB in Fig. 10-8a assumed to have two built-in ends. A uniform drop in temperature of t degrees would cause the end A when released to move to the right a distance $\alpha t(\overline{AB})_x$ and upward a distance $\alpha t(\overline{AB})_y$, where α is the coefficient of thermal expansion, and $(\overline{AB})_x$ and $(\overline{AB})_y$ are the projections of the length $\overline{AB}$ on the x and y axes. It follows that the effect of a drop in temperature is the same as the effect of displacements $D_1 = -\alpha t(\overline{AB})_x$ and $D_2 = \alpha t(\overline{AB})_y$.

10-5   STIFFNESS MATRIX FOR A BENT WITH SPECIAL END-CONDITIONS

The stiffness matrix $[S]$ in Eq. 10-17 is for a free (unsupported) bent. The elements in column j of this matrix are forces in equilibrium acting at the 6 coordinates in Fig. 10-8a which produce the displacement $D_j = 1$ with no displacement at the other 5 coordinates. From Sec. 4-6 we know that $[S]$ in Eq. 10-17 must be singular, just like the stiffness matrix for an unsupported prismatic beam given in Eq. 4-6.

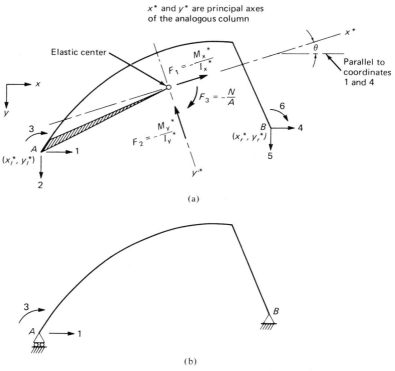

(a)

(b)

Fig. 10-8. Bent considered in Secs. 10-43, 10-45, 10-5 and 10-6. (a) Calculation of end-forces caused by end-displacement from the redundants at the elastic center. Coordinates $1, 2, \ldots, 6$ are those corresponding to the stiffness matrix in Eq. 10-17. (b) Coordinates 1 and 3 are those corresponding to the stiffness matrix $[S]$ in Eq. 10-19.

If the displacements or the forces at a number of coordinates are known to be zero, the forces and the displacement at the remaining coordinates can be related by a matrix of lower order than $[S]$, using the method considered in Sec. 4-5. If, for example, the bent in Fig. 10-8a is supported by a roller at A and by a hinge at B as shown in Fig. 10-8b, the displacements

D_2, D_4, and D_5 are prevented; thus, $D_2 = D_4 = D_5 = 0$, while the rotation D_6 is allowed to take place freely, i.e., $F_6 = 0$. The forces and displacements at the remaining coordinates 1 and 3 can be related by the condensed stiffness matrix

$$[S] = \begin{bmatrix} \left(S_{11} - \dfrac{S_{16}^2}{S_{66}} \right) & \text{symmetrical} \\[3mm] \left(S_{13} - \dfrac{S_{16}S_{36}}{S_{66}} \right) & \left(S_{33} - \dfrac{S_{36}^2}{S_{66}} \right) \end{bmatrix} \qquad (10\text{-}19)$$

where the terms S_{ij} on the right-hand side of this equation are the elements S_{ij} of the matrix $[S]$ in Eq. 10-17.

Equation 10-19 can be derived by first deleting from $[S]$ in Eq. 10-17, the columns and rows 2, 4, and 5, then applying Eq. 4-17 to the remaining matrix in a manner similar to that used in Sec. 4-5.

10-6 END-STIFFNESS OF STRAIGHT MEMBERS

The preceding derivations dealt with a bent of an arbitrary shape. Let us now consider the special case when the structure AB of Fig. 10-8a is a straight beam of variable EI. Taking into account the bending deformations only, with no change in the length of members, we have forces corresponding to the displacements along four coordinates only (Fig. 10-9). The elastic center is a point on the beam and the principal axes are along the beam (say, the x axis) and perpendicular to it. The coordinates of the ends are $A(x_l, y_l) = (-d, 0)$ and $B(x_r, y_r) = (b, 0)$, where d and b are the distances from the elastic center to the ends of the beam, as shown in Fig. 10-9. As always, the loads on the analogous column can be reduced to a force N at the elastic center and moment M_y about the y axis (there is no moment about the x axis). From the equations developed in Sec. 10-5, it can be easily seen that the stiffness matrix of a straight beam corresponding to the coordinates in Fig. 10-9 is

$$[S] = \begin{bmatrix} \dfrac{1}{I_y} & \text{symmetrical} \\[3mm] \dfrac{d}{I_y} & \dfrac{1}{a} + \dfrac{d^2}{I_y} \\[3mm] -\dfrac{1}{I_y} & -\dfrac{d}{I_y} & \dfrac{1}{I_y} \\[3mm] \dfrac{b}{I_y} & -\dfrac{1}{a} + \dfrac{bd}{I_y} & -\dfrac{b}{I_y} & \dfrac{1}{a} + \dfrac{b^2}{I_y} \end{bmatrix} \qquad (10\text{-}20)$$

where a is the area of the analogous column and I_y its moment of inertia about the y axis.

If the beam is prismatic (with a flexural rigidity EI) and of length l, the analogous column will have a constant width $1/(EI)$; thus, $b = d = l/2$, $a = l/EI$ and $I_y = l^3/(12EI)$, and Eq. 10.20 becomes the same as Eq. 4-7.

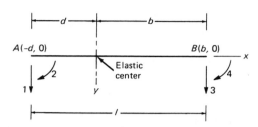

Fig. 10-9. Coordinate system corresponding to the stiffness matrix in Eq. 10-17 for a beam with a variable EI.

The element S_{22} of the above matrix is the *rotational stiffness* S_{AB} of end A, and the ratio (S_{42}/S_{22}) is referred to as the *carryover factor*, C_{AB} from A to B in the moment distribution method of structural analysis (see Chapter 11). These two quantities are given here for reference.

$$\left.\begin{aligned}
&\text{Rotational stiffness of end } A \text{ is } S_{AB} = \left(\frac{1}{a} + \frac{d^2}{I_y}\right) \\
&\text{Carryover factor from } A \text{ to } B \text{ is } C_{AB} = \left(-\frac{1}{a} + \frac{db}{I_y}\right) \Big/ \left(\frac{1}{a} + \frac{d^2}{I_y}\right) \\
&\text{Similarly,} \\
&\text{Rotational stiffness of end } B \text{ is } S_{BA} = \left(\frac{1}{a} + \frac{b^2}{I_y}\right) \\
&\text{Carryover factor from } B \text{ to } A \text{ is } C_{BA} = \left(-\frac{1}{a} + \frac{db}{I_y}\right) \Big/ \left(\frac{1}{a} + \frac{b^2}{I_y}\right)
\end{aligned}\right\} \quad (10\text{-}21)$$

Example 10-3 Find the stiffness matrix corresponding to the coordinates 1, 2, 3, and 4 of the frame in Fig. 10-10. The frame has a constant rigidity EI and the vertical movement of ends A and B is prevented. This bent represents a part of a multibay shed frame supported on columns.

The analogous column for the bent is shown in Fig. 10-10b, and has the following properties: $a = 3l/(EI)$, $I_x = 0.12l^3/(EI)$, $I_y = 1.82l^3/(EI)$. The coordinates of the ends A and B with respect to the principal axes x and y are: $A(x_l, y_l) = (-1.3l, 0.4l)$ and $B(x_r, y_r) = (1.3l, 0.4l)$.

The required stiffness matrix is of the order 4×4, and can be calculated from Eq. 10-17, reducing its order simply by omitting the 2nd and 5th columns

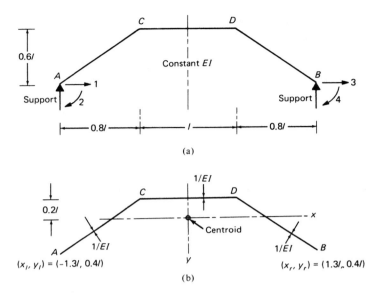

Fig. 10-10. Stiffness matrix of the bent of Example 10-3 by column analogy.
(a) Actual frame. (b) Analogous column.

and rows and putting $\theta = 0$. Therefore, the stiffness matrix corresponding
to the coordinates in Fig. 10-10a is

$$[S] = \begin{bmatrix} \dfrac{1}{I_x} & & & \text{symmetrical} \\[2ex] \dfrac{y_l}{I_x} & \dfrac{1}{a} + \dfrac{y_l^2}{I_x} + \dfrac{x_l^2}{I_y} & & \\[2ex] -\dfrac{1}{I_x} & -\dfrac{y_l}{I_x} & \dfrac{1}{I_x} & \\[2ex] -\dfrac{y_r}{I_x} & -\dfrac{1}{a} - \dfrac{y_l y_r}{I_x} - \dfrac{x_l x_r}{I_y} & \dfrac{y_r}{I_x} & \dfrac{1}{a} + \dfrac{y_r^2}{I_x} + \dfrac{x_r^2}{I_y} \end{bmatrix} \qquad (10\text{-}22)$$

By substitution, we obtain

$$[S] = \dfrac{EI}{l} \begin{bmatrix} 8.33/l^2 & & \text{symmetrical} \\ 3.33/l & 2.60 & \\ -8.33/l^2 & -3.33/l & 8.33/l^2 \\ -3.33/l & -0.74 & 3.33/l & 2.60 \end{bmatrix}$$

Example 10-4 Determine the fixed-end moments, the rotational stiffness of
the two ends, and the carryover factors for the beam of variable depth

shown in Fig. 10-11a and write the stiffness matrix corresponding to the coordinate system in Fig. 10-11d. The flexural rigidity of the beam EI is assumed to vary as the cube of its depth.

The analogous column is shown in Fig. 10-11b; its properties are calculated in Table 10-2, part (a). For this purpose the beam is divided into 4 elements as shown in Fig. 10-11b. The statically determinate bending-moment diagram M_s is taken as that of a simple beam in Fig. 10-11c. The load on the elements 1, 2, and 3 of the analogous column is approximated by the area of the element multiplied by the value of M_s at the middle of the element, and is considered to act at its middle point. For element 4 the exact load is calculated by dividing the M_s diagram as shown in Fig. 10-11c. The calculations of N and M_y are performed in Table 10-2, part (b).

Table 10-2. Example 10-4: Calculation of a, I_y, N, and M_y

	(a) Properties of Analogous Column			(b) Loading on Analogous Column		
Element	Area $l/(EI_B)$	Moment of Area about B $l^2/(EI_B)$	Moment of inertia about y Axis $10^{-6}l^3/(EI_B)$	Load $10^{-4}ql^3/(EI_B)$	x/l	Moment about y Axis $10^{-4}ql^4/(EI_B)$
1	0.0189	0.0179	54.6	$N_1 =$ 5.0	−0.536	− 2.68
2	0.0354	0.0295	64.2	$N_2 =$ 24.6	−0.424	− 10.43
3	0.0790	0.0571	78.6	$N_3 =$ 79.3	−0.313	− 24.82
4	0.6667	0.2222	285.0	$N_4 = 370.4$	−0.036	− 13.33
	0.8000	0.3267	482.4	$N_5 = 246.9$	+0.076	+ 18.76
	Centroid:			$\Sigma = 726.2$		− 32.50

$b = \dfrac{0.3267}{0.8000} l = 0.408l$

$a = 0.80l/(EI_B)$

$I_y = \dfrac{482.4l^3}{10^4 EI_B}$

$N = \dfrac{726.2ql^3}{10^4 EI_B}$

$M_y = \dfrac{-32.5ql^4}{10^4 EI_B}$

The statically indeterminate moment, given by Eq. 10-10, are

$$\bar{M} = \frac{N}{a} + \frac{M_y}{I_y} x$$

$$= \frac{726.2ql^3}{10^4 EI_B}\frac{EI_B}{0.80l} + \frac{-32.5ql^4}{10^4 EI_B}\frac{10^4 EI_B}{482.4l^3} x = ql^2\left(0.0908 - 0.0674\,\frac{x}{l}\right)$$

Substituting the value of x at the two ends, we find the statically indeterminate bending moments

$$\bar{M}_A = 0.1307ql^2 \qquad \text{and} \qquad \bar{M}_B = 0.0633ql^2$$

and with $M_{sA} = M_{sB} = 0$, Eq. 10-9 gives the bending moment at the beam

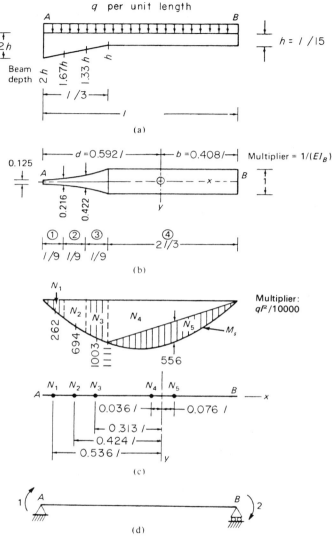

Fig. 10-11. Calculation of fixed end-moments, rotational stiffness, and carryover factors by column analogy for the beam of Example 10-4. (a) Beam and loading. (b) Analogous column. (c) Loads on analogous column. (d) Coordinate system.

ends: $M_A = -1057\,\text{k ft}$ and $M_B = -512\,\text{k ft}$. If we consider as usual the moments on the member ends to be positive when clockwise, the fixed-end moments for the beam AB are:

$$M_{AB} = -0.1307ql^2 \quad \text{and} \quad M_{BA} = 0.0633ql^2$$

The stiffness matrix corresponding to the coordinates 1 and 2 (Fig. 10-11d) can be easily derived from Eq. 10-20, thus

$$
[S] = \begin{bmatrix} \left(\dfrac{1}{a} + \dfrac{d^2}{I_y}\right) & \text{symmetrical} \\[2ex] \left(-\dfrac{1}{a} + \dfrac{bd}{I_y}\right) & \left(\dfrac{1}{a} + \dfrac{b^2}{I_v}\right) \end{bmatrix}
$$

Substituting the numerical values,

$$
[S] = \frac{EI_B}{l} \begin{bmatrix} 8.52 & 3.76 \\ 3.76 & 4.70 \end{bmatrix}
$$

The rotational end stiffnesses are

$$
S_{AB} = S_{11} = 8.52 \frac{EI_B}{l}
$$

$$
S_{BA} = S_{22} = 4.70 \frac{EI_B}{l}
$$

and the carryover factors are

$$
C_{AB} = \frac{S_{21}}{S_{11}} = \frac{3.76}{8.52} = 0.441
$$

$$
C_{BA} = \frac{S_{12}}{S_{22}} = \frac{3.76}{4.70} = 0.800
$$

10-7 CORRECTION FOR THE EFFECT OF AXIAL FORCES IN ARCHES

We recall that in deriving the equations in this chapter we considered the bending deformations only. In arches subjected to a vertical distributed load, the axial forces are high and the bending moments are usually small, and to recognize this, the effect of axial forces may be included as a correction to the analysis by column analogy.

Consider for example the parabolic arch of Fig. 10-12a subjected to uniform vertical load. If the axial deformation is ignored, there will be no bending moment in all sections of the arch. This is true no matter what is the variation in the flexural rigidity EI and can be seen by noting that if the arch is released by the introduction of three hinges (Fig. 10-12b), the shear and the statically determinate moment M_s are zero at all sections. Thus there is no loading on the analogous column ($\bar{M} = 0$) and the reactions are the same for the released and the statically indeterminate arches:

$$R_1 = R_3 = \frac{ql^2}{8h \cos \gamma} \qquad R_2 = R_4 = \frac{ql}{2} \qquad (10\text{-}23)$$

The line of action of the resultant of forces to the left or the right of any section is referred to as the *pressure line* of the arch. It can be proved by simple statics that the pressure line for the arch in Fig. 10-12b coincides

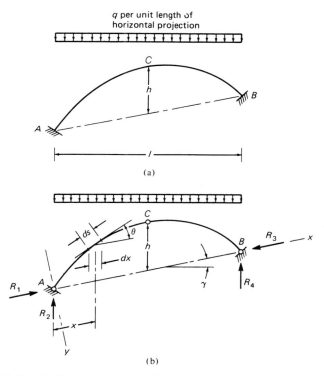

(a)

(b)

Fig. 10.12. Parabolic arch under a uniform vertical load. (a) Parabolic arch with fixed ends. (b) Released structure.

with the arch axis. Therefore the axial force N at any section at a distance x from the left-hand end is equal to the resultant of the force R_1 and an upward force $R_2 - qx \cos \gamma$. Thus the axial force at any section of the arches in Fig. 10-12a and b is

$$N = -R_1 \frac{\cos \gamma}{\cos (\theta + \gamma)} \qquad (10\text{-}24)$$

where N is (as usual) considered positive if tensile.

The change in length of an element ds as a result of the axial force is

$Nds/(aE)$, where a is the cross-sectional area and the change in length projected along the line AB is $N\,dx/(aE)$. The increase in the distance $\overline{AB}$ between the supports if they were free to move relative to each other is

$$\Delta(\overline{AB}) = \int \frac{N\,dx}{aE} \tag{10-25}$$

or

$$\Delta(\overline{AB}) = -\frac{R_1\cos\gamma}{E}\int \frac{dx}{a\cos(\theta+\gamma)} \tag{10-26}$$

Substituting for R_1 from Eq. 10-23,

$$\Delta(\overline{AB}) = -\frac{ql^2}{8Eh}\int \frac{dx}{a\cos(\theta+\gamma)} \tag{10-27}$$

The integrals in Eqs. 10-26 and 10-27 depend on the geometry of the arch and can be easily evaluated numerically.

In the actual arch in Fig. 10-12a, the relative movement of A and B is restrained, and the restraining effect is the same as that produced by the movement of A toward B by a distance $\Delta(\overline{AB})$. This movement would cause a bending moment in the arch which can be calculated by applying to the analogous column a couple of magnitude (see Table 10-1)

$$M_x = -\Delta(\overline{AB}) \tag{10-28}$$

It should be noted that the above equations for $\Delta(\overline{AB})$ involve an approximation because N is calculated by an approximate analysis in which the axial deformation is ignored. However, we are determining a correction, and an inaccuracy in the value of the correction will affect the final result only to a small extent.

The axis of the arch and the line of pressure coincide in any parabolic arch subjected to a uniform vertical load. However, if the shape of the arch is other than parabolic, or if a parabolic arch is subjected to loading other than uniform, Eqs. 10-23 and 10-27 are not applicable. Nevertheless, in most practical cases the axial force can still be approximated by Eqs. 10-24 and 10-26, and Eq. 10-28 can be used to correct for the effect of axial forces; this involves a further approximation.

10-8 GENERAL

The bending moment in a statically indeterminate frame composed of one closed bent can be expressed as the difference $M_s - \bar{M}$, where M_s is the bending moment due to the load on a released statically indeterminate structure, and $\bar{M}$ is the statically indeterminate moment. The use of the force method to determine $\bar{M}$ leads to an equation similar to the equation used to calculate the normal stresses in the cross section of an eccentrically loaded short column. This analogy is made use of to turn the computation of the statically indeterminate moments into a routine calculation of normal stresses in a section.

The moment $\bar{M}$ caused by the displacement of one end of the analogous column is numerically equal to the normal stress caused by a concentrated load or a couple on the analogous column. This equivalence is used to determine the end stiffnesses of any plane bent, curved or straight, of variable flexural rigidity. These values of stiffness can be used in the analysis of structures in which several bents are connected, such as multibay and multistorey frames.

PROBLEMS

10-1   Find the bending-moment diagrams and the reactions (if any) for the frame shown in the figure.

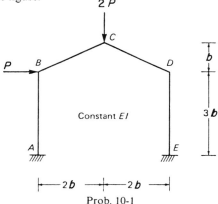

Prob. 10-1

10-2   Apply the requirements of Prob. 10-1 to the frame shown in the figure.

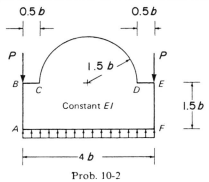

Prob. 10-2

10-3   Apply the requirements of Prob. 10-1 to the frame shown in the figure.

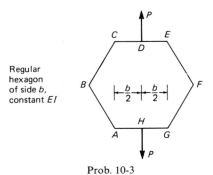

Regular
hexagon
of side b,
constant EI

Prob. 10-3

10-4   Apply the requirements of Prob. 10-1 to the frame shown in the figure.

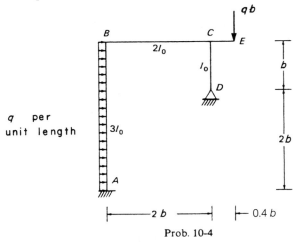

q per
unit length

Prob. 10-4

10-5   Apply the requirements of Prob. 10-1 to the frame shown in the figure.

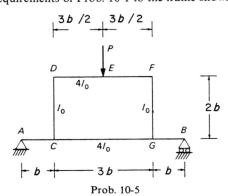

Prob. 10-5

10-6   Find the bending moment at points A, B, C, and D of the frame shown in the figure due to:

(a) a uniform shrinkage strain $= 0.0002$
(b) vertical settlement of $\frac{1}{2}$ in. at support A
(c) rotation of B in the clockwise direction by $0.20°$
Assume $I = 4$ ft^4 and $E = 2 \times 10^6$ psi.

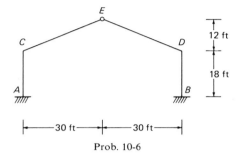

Prob. 10-6

10-7   Apply the requirements of Prob. 10-6 to the structure shown in the figure.

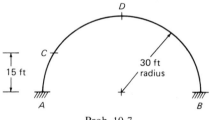

Prob. 10-7

10-8   Write the stiffness matrix corresponding to the coordinates 1, 2 and 3, 4 of the parabolic arch shown in the figure. The flexural rigidity EI is assumed to vary as the secant of the inclination of the arch axis. The properties of the analogous column are given in Appendix J.

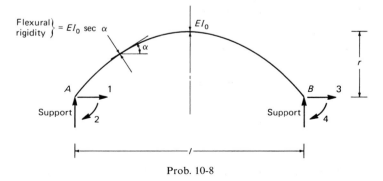

Prob. 10-8

10-9   Write the stiffness matrix corresponding to the coordinates $1, 2, \ldots, 6$ for the gable frame shown in part (a) of the figure. Use this matrix to derive the stiffness matrix corresponding to the coordinates 1 and 3 in the frame shown in part (b).

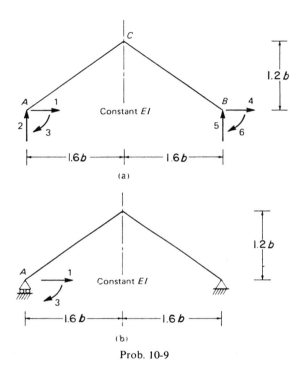

Prob. 10-9

10-10   Find the forces at the ends A and B of the gable frame in part (a) of Prob. 10-9, subjected to a uniform vertical load of intensity q per unit length of horizontal projection. The frame is assumed to be encastré at A and B.

10-11   Find the end-rotational stiffness S_A, S_B and the carryover factors C_{AB} and C_{BA} for the beam in the figure.

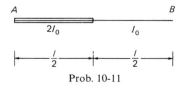

Prob. 10-11

10-12   Apply the requirements of Prob. 10-11 to the member shown in the figure.

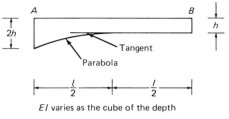

A   *B*

2*h*   *h*

Tangent

Parabola

$\dfrac{l}{2}$ ―――――――――― $\dfrac{l}{2}$

E I varies as the cube of the depth

Prob. 10-12

10-13   Find the fixed-end moments in the beam of Prob. 10-11 subjected to a uniform transverse load *q* per unit length.

10-14   Find the fixed-end moments in the beam of Prob. 10-12 loaded by a concentrated transverse force *P* at the center.

CHAPTER 11

Application of the displacement method: slope deflection and moment distribution

11-1 INTRODUCTION

The slope-deflection method has been used extensively for many years and became a standard technique long before the generalized classification of the methods of analysis into force and displacement methods. The advent of the relaxation technique of moment distribution reduced the popularity of slope deflection whose chief disadvantage is that it involves solution of a system of simultaneous equations.

The basic slope-deflection equation expresses the moment at the end of a member as the superposition of the end-moment due to external loads on the member with the ends assumed restrained and of the end-moment caused by the actual end-displacements. A set of simultaneous equations is written to express the equilibrium of the joints, the end-moments being expressed in terms of joint displacements. The solution of these equations gives the unknown joint displacements, which are then substituted in the original slope-deflection equations to give the end-moments. Hence the bending moment of the entire structure can be drawn.

The analysis by the slope-deflection method, as outlined above, is in fact an application of the more general displacement method of analysis to plane frames in which only the bending deformations are considered. There is no inherent advantage in using the slope-deflection technique but, because it is still used by engineers, a general outline of slope-deflection procedure is considered useful.

In the moment-distribution method, the joint displacements are first assumed restrained. The effect of joint displacements is then introduced by successive iterations, which can be continued to any desired precision. Thus the moment distribution is also a displacement method of analysis. There is, however, a fundamental difference in that in moment distribution generally no equations are solved to find the joint displacements; instead these displacements are allowed to take place in succession, and their effect on the end-moments is introduced as a series of successive converging corrections. This absence of the need to solve simultaneous equations has made the moment-distribution method an extremely popular one, especially when the calculations are done by a simple calculator.

The moment-distribution procedure yields bending moments, and it is the values of the moments that are generally needed for design; thus we avoid the tedious procedure of first finding the joint displacements and then calculating the moments. A further advantage of the moment-distribution procedure is that it is easily remembered and easily applied.

As in the case of the slope-deflection method, the analysis by moment distribution does not usually take into account shear deformations. Also, the effect of axial deformations is generally neglected, but may be taken into account if desired (see Chapter 15).

11-2 NOTATION AND SIGN CONVENTION

The methods of slope-deflection and moment distribution require a sign convention which must be clearly defined. Since algebraic procedures are used, correct signs are of course of paramount importance, and we shall give now the sign convention and the notation of the slope-deflection and moment-distribution methods. This is done best with reference to Fig. 11-1a, which represents the deflected shape $N'F'$ of a straight member NF of a plane frame. Clockwise moment or rotation of either end of the member is considered positive. The relative translation of the ends perpendicular to the original direction of the member, $y_F - y_N$, produces bending; the relative translation along the axis of the member x is considered in this chapter to be zero, that is it is assumed that no change in length occurs. The

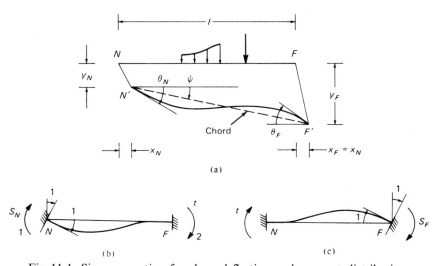

Fig. 11-1. Sign convention for slope deflection and moment distribution. (a) Positive directions of end-rotations θ_N and θ_F, and chord rotation ψ. (b) End-moments caused by a unit rotation at N. (c) End-moments caused by a unit rotation at F.

chord rotation, $\psi = (y_F - y_N)/l$, is considered positive when clockwise. Further notation is as follows.

θ_N = rotation at the end N (near end)

θ_F = rotation at the end F (far end)

S_N = rotational stiffness of end N, that is the end-moment at N corresponding to a unit rotation at N while the displacement at F is restrained (Fig. 11-1b)

S_F = rotational stiffness of end F—that is, the end-moment at F corresponding to a unit rotation at F while the displacement at end N is restrained (Fig. 11-1c)

t = carryover moment—that is, the moment at a fixed end F caused by a unit rotation of end N (Fig. 11-1b); also equal to the carryover moment at a fixed end N caused by a unit rotation of end F (Fig. 11-1c) (the equality can be proved directly by applying Betti's theorem, Sec. 8-2, to the systems of forces and displacements in Fig. 11-1b and c)

C_{NF} = carryover factor from N to F, that is

$$C_{NF} = \frac{t}{S_N} \tag{11-1}$$

C_{FN} = carryover factor from F to N, that is

$$C_{FN} = \frac{t}{S_F} \tag{11-1a}$$

The moments S_N, S_F, and t are the elements of a stiffness matrix corresponding to the coordinates 1 and 2 indicated in Fig. 11-1b.

$$[S] = \begin{bmatrix} S_N & t \\ t & S_F \end{bmatrix}$$

For a prismatic member, $S_F = S_N = 4EI/l$, $t = 2EI/l$, and the carryover factors $C_{NF} = C_{FN} = C = \frac{1}{2}$ (see Appendix D). For nonprismatic members, the stiffness and carryover factors can be calculated by Eq. 10-21.

11-3   SLOPE-DEFLECTION EQUATION FOR A STRAIGHT MEMBER

Consider the curve $N'F'$ representing the deformed shape of a member NF of any plane frame subjected to lateral loads (Fig. 11-1a). The translation part of the displacement of the member (that is its translation as a rigid body to the straight dotted line $N'F'$) produces no moments. The end-moment M_N at end N can be expressed as the sum of the moment due to the lateral load on the member with the end-displacements prevented (the fixed-end moment FEM) and of the moments induced by the rotations $(\theta_N - \psi)$ and $(\theta_F - \psi)$ at the ends N and F respectively. Thus,

$$M_N = S_N(\theta_N - \psi) + t(\theta_F - \psi) + (FEM)_N$$

where $(FEM)_N$ is the fixed-end moment at N, that is the value of the end-moment caused by the actual lateral loads on the beam with displacements at both ends prevented. This equation can be rearranged to

$$M_N = S_N\theta_N + t\theta_F - \psi(S_N + t) + (FEM)_N \qquad (11\text{-}2)$$

This is the slope-deflection equation for a prismatic or nonprismatic member. The equation can be used to express the moment at the left- or right-hand end. The symbols are arranged mnemonically so that the end for which the equation is written is referred to as the N-end (near end); the other is the F-end (far end).

The values of the stiffness S_N, the carryover factors C_{FN} and C_{NF}, and of the FEM's for many of the beam shapes and types of loading met in practice can be found in various references.[1]

When the member has a constant flexural rigidity EI,

$$S_N = \frac{4EI}{l} \qquad (11\text{-}3)$$

and

$$t = \frac{2EI}{l} \qquad (11\text{-}4)$$

The slope-deflection equation then becomes

$$M_N = \frac{EI}{l}(4\theta_N + 2\theta_F - 6\psi) + (FEM)_N \qquad (11\text{-}5)$$

11-4   EQUATIONS OF EQUILIBRIUM

The joints in a frame are subjected to forces equal and opposite to the forces acting on the ends of the members meeting at the given joint. For equilibrium of the forces (or moments) acting on any joint, the sum of their components in any direction must be zero, and equations can be written to satisfy this condition along the unknown joint displacements. The number of equations required is equal to the degree of kinematic indeterminacy. We thus obtain a system of simultaneous equations from which the joint displacements can be determined.

Once the joint displacements have been established, substitution in the slope-deflection equation (Eq. 11-2 or 11-5) gives the end-moments, and hence

[1] See, for example. *Handbook of Frame Constants*, Portland Cement Association, Skokie, Ill.

the bending moment, and other restraining forces can be determined by simple statics.

Example 11-1 The frame solved by the general displacement method in Example 3-2 is shown again in Fig. 11-2a. Let us analyze the frame by the slope-deflection procedure. The axial deformations are neglected.

There are three unknown joint displacements D_1, D_2, and D_3, indicated in Fig. 11-2b. The end-moments M_1, M_2 ..., M_6 in the actual structure can be expressed in terms of the displacements by applying the slope-deflection Eq. 11-5 at the ends of the three members AB, BC, and CD. For example, when the equation is applied at A to express M_1, $\theta_N = 0$, $\theta_F = D_2$ and $\psi = D_1/l$. When the equation is applied at B to express M_2, $\theta_N = D_2$, $\theta_F = 0$, and $\psi = D_1/l$. The chord rotation for member BC is zero because the change in the length of members is ignored, so that the level of B and C

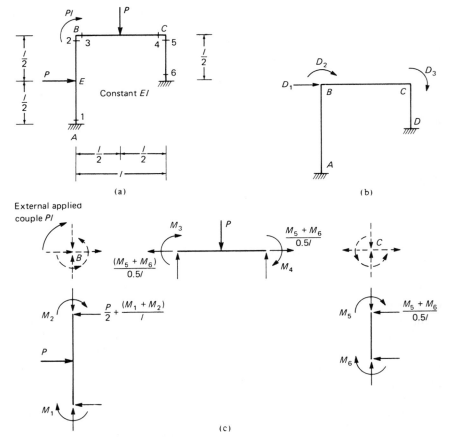

Fig. 11-2. Frame considered in Example 11-1.

is unaltered. For the same reason, the horizontal movement of B and of C is each equal to D_1; therefore the chord rotation for member CD is $D_1/(0.5l)$. The six end-moments are

$$M_1 = -\frac{Pl}{8} + \frac{EI}{l}\left(2D_2 - \frac{6D_1}{l}\right)$$

$$M_2 = +\frac{Pl}{8} + \frac{EI}{l}\left(4D_2 - \frac{6D_1}{l}\right)$$

$$M_3 = -\frac{Pl}{8} + \frac{EI}{l}(4D_2 + 2D_3)$$

$$M_4 = +\frac{Pl}{8} + \frac{EI}{l}(4D_3 + 2D_2)$$

$$M_5 = \frac{EI}{0.5l}\left(4D_3 - \frac{6D_1}{0.5l}\right)$$

$$M_6 = \frac{EI}{0.5l}\left(2D_3 - \frac{6D_1}{0.5l}\right)$$

These equations can be written in matrix form:

$$
\begin{Bmatrix} M_1 \\ M_2 \\ M_3 \\ M_4 \\ M_5 \\ M_6 \end{Bmatrix}
= Pl \begin{Bmatrix} -0.125 \\ 0.125 \\ -0.125 \\ 0.125 \\ 0 \\ 0 \end{Bmatrix}
+ \frac{EI}{l}
\begin{bmatrix}
-\dfrac{6}{l} & 2 & 0 \\
-\dfrac{6}{l} & 4 & 0 \\
0 & 4 & 2 \\
0 & 2 & 4 \\
-\dfrac{24}{l} & 0 & 8 \\
-\dfrac{24}{l} & 0 & 4
\end{bmatrix}
\begin{Bmatrix} D_1 \\ D_2 \\ D_3 \end{Bmatrix}
\qquad \text{(a)}
$$

In Fig. 11-2c the members are shown separated from the joints, and the forces acting on the ends of the members are indicated. Equal and opposite forces act on the joints, as represented by the dotted arrows on the same figure. All the end-forces in Fig. 11-2c are given in terms of the end-moments. We now consider the equilibrium of the forces at joints B and C in the directions 1, 2, and 3 of Fig. 11-2b. The equilibrium equations are

$$
\left.
\begin{aligned}
\frac{P}{2} + \frac{M_1 + M_2}{l} + \frac{M_5 + M_6}{0.5l} &= 0 \\
Pl - M_2 - M_3 &= 0 \\
- M_4 - M_5 &= 0
\end{aligned}
\right\} \qquad \text{(b)}
$$

The first of these equations expresses that the sum of the forces in the horizontal direction on joint B is zero. The second and third equations are moment-equilibrium equations at joints B and C respectively.

Substituting for the end moments in Eq. (b) from Eq. (a), we obtain

$$\frac{EI}{l} \begin{bmatrix} \dfrac{108}{l^2} & -\dfrac{6}{l} & -\dfrac{24}{l} \\[2mm] -\dfrac{6}{l} & 8 & 2 \\[2mm] -\dfrac{24}{l} & 2 & 12 \end{bmatrix} \begin{Bmatrix} D_1 \\ D_2 \\ D_3 \end{Bmatrix} = -\begin{Bmatrix} -\dfrac{P}{2} \\[2mm] -Pl \\[2mm] \dfrac{Pl}{8} \end{Bmatrix} \qquad (c)$$

The solution of this equation gives

$$\begin{Bmatrix} D_1 \\ D_2 \\ D_3 \end{Bmatrix} = \frac{Pl^2}{EI} \begin{Bmatrix} 0.00871l \\ 0.1355 \\ -0.0156 \end{Bmatrix} \qquad (d)$$

Substituting Eq. (d) into Eq. (a), we obtain the end-moments

$$\begin{Bmatrix} M_1 \\ M_2 \\ M_3 \\ M_4 \\ M_6 \\ M_6 \end{Bmatrix} = Pl \begin{Bmatrix} 0.09 \\ 0.61 \\ 0.39 \\ 0.33 \\ -0.33 \\ -0.27 \end{Bmatrix}$$

Comparing the procedure followed above with the solution of Example 3-2 we find that both lead to the same equations [compare Eqs. (a) and (c) above with Eqs. (b) and (e) in Example 3-2]. The final results are of course identical. The square matrix on the left-hand side of Eq. (c) above is the stiffness matrix of the frame corresponding to the coordinates 1, 2, and 3 of Fig. 11-2b. The elements of the vector on the right-hand side of Eq. (c) are the values of the three forces which, when applied along the three coordinates, would prevent the joint displacements.

It follows from the above that the slope-deflection method leads to the same equations as the more general procedure of the displacement method considered in Chapter 3. This is again demonstrated in the following example of a frame with inclined members.

Example 11-2 Solve the frame of Example 3-3 by the slope-deflection method.

The unknown joint displacements are indicated by the arrows 1, 2, and 3 in Fig. 11-3b. From the displacement diagram in Fig. 11-3c, we can see that the chord rotations of the members are

$$\psi_{AB} = 1.25\frac{D_3}{l} \qquad \psi_{BC} = -0.75\frac{D_3}{l} \qquad \text{and} \qquad \psi_{CD} = \frac{D_3}{0.8l}$$

Applying the slope-deflection Eq. 11-5, we obtain the end-moments

$$\begin{Bmatrix} M_1 \\ M_2 \\ M_3 \\ M_4 \\ M_5 \\ M_6 \end{Bmatrix} = Pl \begin{Bmatrix} -\dfrac{1}{12} \\ \dfrac{1}{12} \\ -\dfrac{1}{2} \\ \dfrac{1}{2} \\ \dfrac{1}{10} \\ -\dfrac{1}{10} \end{Bmatrix} + EI \begin{bmatrix} \dfrac{2}{l} & 0 & -\dfrac{7.5}{l^2} \\ \dfrac{4}{l} & 0 & -\dfrac{7.5}{l^2} \\ \dfrac{4}{l} & \dfrac{2}{l} & \dfrac{4.5}{l^2} \\ \dfrac{2}{l} & \dfrac{4}{l} & \dfrac{4.5}{l^2} \\ 0 & \dfrac{4}{0.8l} & -\dfrac{6}{(0.8l)^2} \\ 0 & \dfrac{2}{0.8l} & -\dfrac{6}{(0.8l)^2} \end{bmatrix} \begin{Bmatrix} D_1 \\ D_2 \\ D_3 \end{Bmatrix} \qquad \text{(a)}$$

The equilibrium equations express the fact that each of the following is zero: the sum of the moments on joint B; the sum of the moments on joint C; and the sum of the horizontal forces on joint C. The equilibrium equations can also be written for the forces at the ends of members meeting at a joint, since these are equal and opposite to the forces on the joint. Assuming that the positive directions of the end-forces are given by the three arrows in Fig. 11-3b, we write equations of equilibrium of the end-forces

$$\left.\begin{aligned} M_2 + M_3 &= 0 \\ M_4 + M_5 &= 0 \\ -1.25\left(\frac{P}{2} + \frac{M_1 + M_2}{l}\right) - 0.75\left(2P - \frac{M_3 + M_4}{l}\right) - \left(\frac{P}{2} + \frac{M_5 + M_6}{0.8l}\right) &= 0 \end{aligned}\right\}$$
$$\text{(b)}$$

The third equilibrium equation is written with the aid of Fig. 11-3d, in which the forces at the ends of members are resolved into components along the members. The horizontal components of the end-forces at B are transmitted to joint C as axial forces through BC. Their sum is in equilibrium with the sum of the horizontal components of the end-forces at C.

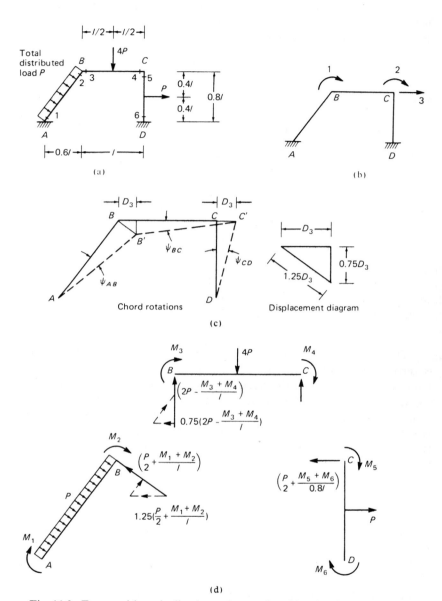

Fig. 11.3. Frame with an inclined member analyzed by the slope-deflection method in Example 11-2.

Substituting for the end-moments in Eq. (b) from Eq. (a), we obtain

$$
EI
\begin{bmatrix}
\dfrac{8}{l} & \dfrac{2}{l} & -\dfrac{3}{l^2} \\[2mm]
\dfrac{2}{l} & \dfrac{9}{l} & -\dfrac{4.875}{l^2} \\[2mm]
-\dfrac{3}{l^2} & -\dfrac{4.875}{l^2} & \dfrac{48.938}{l^3}
\end{bmatrix}
\begin{Bmatrix} D_1 \\ D_2 \\ D_3 \end{Bmatrix}
= Pl
\begin{Bmatrix} \dfrac{5}{12} \\[2mm] -\dfrac{6}{10} \\[2mm] 2.625/l \end{Bmatrix}
\qquad (c)
$$

Equation (c) is the same as Eq. (a) in Example 3-3. Solving for $\{D\}$ and then substituting in Eq. (a) gives the end-moments which will of course be the same as those obtained in Example 3-3.

11-5 PROCESS OF MOMENT DISTRIBUTION

The moment-distribution method of analysis of framed structures was introduced by Hardy Cross[2] in 1932, and was extended by others to the cases of structures subjected to high axial forces, and to axisymmetrical circular plates and shells of revolution.[3] In this chapter we deal mainly with plane frames supported in such a way that the only possible joint displacements are rotations without translation. Further developments form the content of Chapter 12.

Consider the beam ABC of Fig. 11-4a, encastré at A and C, and continuous over support B. Let us assume first that the rotation of joint B is prevented by a restraining external couple acting at B. Due to the lateral loads on the members AB and BC, with the end rotations prevented at all ends, fixed-end moments (FEM) result at the ends A and B, and B and C.

Arbitrary values are assigned to these moments in Fig. 11-4b, using the convention that a positive sign indicates a clockwise end moment. The restraining moment required to prevent the rotation of joint B is equal to the algebraic sum of the fixed-end moments of the members meeting at B, that is -50.

So far the procedure has been identical with the general displacement method considered in Sec. 3-2. However, we now recognize that the joint B is in fact not restrained, and we allow it to rotate by removing the restraining moment of -50. The same effect is achieved by applying to the joint B an external couple $M = +50$ (Fig. 11-4c), that is a moment equal and opposite to the algebraic sum of the fixed-end moments at the joint. This is known as the *balancing moment*. Its application causes the ends of the members BA

---

[2]Hardy Cross, "Analysis of Continuous Frames by Distributing Fixed-End Moments," *Trans. ASCE*, Paper 1793, Vol. 96, 1932.

[3]See List of references at pp. 247–268 of J. M. Gere, *Moment Distribution*, Van Nostrand, New York, 1963.

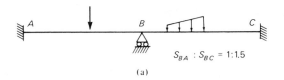

(a)

End	AB		BA	BC		CB
DF_s			0.4	0.6		
FEM's	-80		$+100$	-150		$+180$
One cycle; moment distribution and carryover	$+10$	$\longleftarrow$	$+20$	$+30$	$\longrightarrow$	$+15$
Final moments	-70		120	-120		195

(b)

Balancing moment = 50

(c)

Fig. 11.4. Analysis of a continuous beam by the moment-distribution method. (a) Beam. (b) Solution by moment distribution. (c) Beam deflection after release of joint B.

and BC meeting at B to rotate through the same angle θ_B; hence, end-moments M_{BA} and M_{BC} develop. For equilibrium of the moments acting on joint B,

$$M = M_{BA} + M_{BC} \tag{11-6}$$

The end-moments M_{BA} and M_{BC} can be expressed in terms of the rotational stiffnesses of end B of members AB and BC, viz., S_{BA} and S_{BC}, thus

$$M_{BA} = \theta_B S_{BA} \qquad \text{and} \qquad M_{BC} = \theta_B S_{BC} \tag{11-7}$$

From Eqs. 11-6 and 11-7, the end-moments can be expressed as a fraction of the balancing moment

$$
\left.
\begin{aligned}
M_{BA} &= \left(\frac{S_{BA}}{S_{BA} + S_{BC}}\right) M \\[2mm]
M_{BC} &= \left(\frac{S_{BC}}{S_{BA} + S_{BC}}\right) M
\end{aligned}
\right\} \tag{11-8}
$$

This means that the balancing moment is distributed to the ends of the members meeting at the joint, the *distributed moment* in each member being proportional to its relative rotational stiffness. The ratio of the distributed moment in a member to the balancing moment is called *the distribution factor (DF)*. It follows that the distribution factor for an end is equal to the rotational stiffness of the end divided by the sum of the rotational stiffnesses of the ends meeting at the joint, that is,

$$(DF)_i = \frac{S_i}{\sum\limits_{j=1}^{n} S_j} \tag{11-9}$$

where i refers to the near end of the member considered, and there are n members meeting at the joint.

It is clear that to determine the DF's we can use relative values of the rotational end stiffnesses rather than the actual values, so that Eq. 11-9 is valid also with S representing the *relative rotational end stiffness*. It is also evident that the sum of all the DF's of the ends meeting at a joint must be equal to unity.

Let the relative stiffnesses S_{BA} and S_{BC} for the beam in Fig. 11-4a be 1 and 1.5. The DF's are therefore: $1/(1 + 1.5) = 0.4$ for end BA, and $1.5/(1 + 1.5) = 0.6$ for end BC. The balancing moment of $+50$ will be distributed as follows: $50 \times 0.4 = 20$ to BA, and $50 \times 0.6 = 30$ to BC. The distributed moments are recorded in a table in Fig. 11-4b.

The rotation of joint B produced in the previous step induces end moments at the far fixed ends A and C. These end-moments are referred to as *carryover moments*, their values being equal to the appropriate distributed moment multiplied by the carryover factor (COF): C_{BA} from B to A, and C_{BC} from B to C. The value of a carryover factor depends upon the variation in the cross section of the member; for a prismatic member, COF $= \frac{1}{2}$. A method of calculating the COF's is described in Sec. 10-6. When the far end of a member is pinned, the COF is of course zero.

The carryover moments in the beam considered are recorded in Fig. 11-4b on the assumption that $C_{BA} = C_{BC} = \frac{1}{2}$, that is the cross section is taken to be constant within each span. The two arrows in the table pointing away from joint B indicate that the rotation at B (or moment distribution at B) causes the carryover moment of the value indicated at the head of the arrow. The use of the arrows makes it easier to follow a proper sequence of operations and also facilitates checking.

The process of moment distribution followed by carrying over is referred to as one *cycle*. In the problem considered no further cycles are required as there is no out-of-balance moment. The final end-moments are obtained by adding the end-moments in the restrained condition (FEM's) to the moments caused by the rotation of joint B in the cycle in Fig. 11-4b.

However, if rotation can occur at more than one joint, further cycles of distribution and carryover have to be performed, as shown in the following example.

Example 11-3 Use the method of moment distribution to analyze the frame in Fig. 11-5a. If the axial deformations are ignored, the only possible joint displacements are rotations at B and C. We assume the frame to have prismatic members so that the carryover factor in all cases is $\frac{1}{2}$. Then, the rotational end stiffness of any member is $S = 4EI/l$, where EI is the flexural rigidity of the cross section and l the length of the member. It follows that the relative rotational stiffnesses can be taken as $K = I/l$. The K values for all the members are shown in Fig. 11-5a. The applied load is assumed to be of such a magnitude as to produce the FEM's given in Fig. 11-5b.

The DF's are calculated by Eq. 11-9 and indicated in Fig. 11-5b. The first cycle of moment distribution and carryover is done by allowing joint

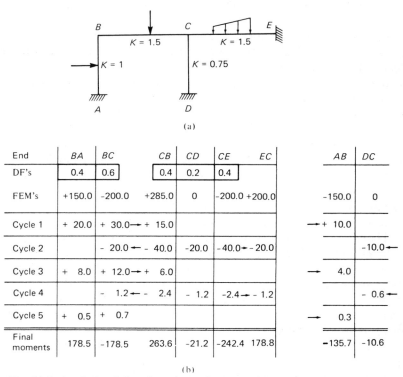

End	BA	BC		CB	CD	CE	EC		AB	DC
DF's	0.4	0.6		0.4	0.2	0.4				
FEM's	+150.0	-200.0	+285.0	0		-200.0	+200.0		-150.0	0
Cycle 1	+ 20.0	+ 30.0→	+ 15.0						→+ 10.0	
Cycle 2		- 20.0←	- 40.0	-20.0	-40.0←- 20.0					-10.0←
Cycle 3	+ 8.0	+ 12.0→	+ 6.0						→ 4.0	
Cycle 4		- 1.2←-	2.4	- 1.2	-2.4→ - 1.2					- 0.6←
Cycle 5	+ 0.5	+ 0.7							→ 0.3	
Final moments	178.5	-178.5	263.6	-21.2	-242.4	178.8			-135.7	-10.6

(b)

Fig. 11.5. Analysis of the plane frame in Example 11-3 by moment distribution without joint translation. (a) Plane frame without joint translation. (b) Moment distribution.

B to rotate; joint C continues to be restrained. This is done in the same way as in the previous example. In the second cycle, joint C is allowed to rotate, joint B now being restrained. The balancing moment for this cycle is equal to minus the algebraic sum of the FEM's at CB, CD and CE plus the carry-over moment caused by the rotation of joint B in the previous cycle – that is, $-(285 + 0 - 200 + 15) = -100$. The second cycle is terminated by the carryover of moments to the far ends of the three members meeting at C, as shown by arrows. It is evident that cycle 2 has induced an imbalance at joint B, that is, if joint B is now released, a further rotation will occur. The effect of this release is followed through in the same way as in cycle 1, but with the balancing moment equal to minus the moment carried over to BC from the previous cycle.

The carryover in the third cycle results in an unbalanced moment at joint C. In order to remove the external constraint, joint C must be balanced again. The process is repeated until the unbalanced moments at all the joints are so small as to be considered negligible. The final moments are obtained by adding the FEM's to the end-moments produced by the joint displacement allowed in *all* the cycles. Because we reach a final position of équilibrium under the applied loading, the sum of the final end-moments at any joint must be zero. This fact can be used as a check on the arithmetic in the distribution process.

It may be interesting to note that moment distribution can be carried out experimentally on a model of a structure capable of being clamped at any joint.

In the preceding example, no distribution was carried out at the fixed ends A, D, and E because the ends of the members at these points can be imagined to be attached to a body of infinite rigidity. Thus the DF for a built-in end of a frame is zero. We should also note that the moments introduced at the ends AB and DC in each cycle are equal to the COF's times the moment introduced respectively at ends BA and CD. It follows that the moments at ends AB and DC need not be recorded during the distribution process, and the final moments at these two ends can be calculated by

$$M_{AB} = (\text{FEM})_{AB} + C_{BA}[M_{BA} - (\text{FEM})_{BA}] \qquad (11\text{-}10)$$

where M_{BA} is the final moment at end BA. A similar equation can be written for the end DC.

11-6 MOMENT-DISTRIBUTION PROCEDURE FOR PLANE FRAMES WITHOUT JOINT TRANSLATION

We should recall that the process of moment distribution described in the previous section applied solely to structures in which the only possible displacement at the joints is rotation. It may be convenient to summarize the steps involved.

1. Determine the internal joints which will rotate when the external load is applied to the frame. Calculate the relative rotational stiffnesses of the ends of the members meeting at these joints, as well as the carryover factors from the joints to the far ends of these members. Determine the distribution factors by Eq. 11-9. The rotational stiffness of either end of a prismatic member is $4EI/l$ and the COF from either end to the other is $\frac{1}{2}$. If one end of a prismatic member is hinged, the rotational end stiffness of the other end is $3EI/l$, and of course no moment is carried over to the hinged end. In a frame with all members prismatic, the relative rotational end stiffness can be taken as $K = I/l$, and when one end is hinged the rotational stiffness at the other end is $(\frac{3}{4})K = (\frac{3}{4})I/l$. The rotational end stiffnesses and the carryover factors for nonprismatic members can be determined by column analogy, as shown in Sec. 10-6.

2. With all joint rotations restrained, determine the fixed-end moments due to the lateral loading on all the members.

3. Select the joints to be released in the first cycle. It may be convenient to take these as alternate internal joints. (For example, in the frame of Fig. 11-6, we can release in the first cycle either A, C, and E or D, B, and F.) Calculate the balancing moment at the selected joints; this is equal to minus the algebraic sum of the fixed-end moments. If an external clockwise couple acts at any joint, its value is simply added to the balancing moment.

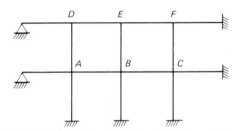

Fig. 11.6. Plane frame without joint translation.

4. Distribute the balancing moments to the ends of the members meeting at the released joints. The distributed moment is equal to the DF multiplied by the balancing moment. The distributed moments are then multiplied by the COF's to give the carryover moments at the far ends. Thus the first cycle is terminated.

5. Release the remaining internal ˙joints, while further rotation is prevented at the joints released in the first cycle. The balancing moment at any joint is equal to minus the algebraic sum of the FEM's and of the end-moments carried over in the first cycle. The balancing moments are distributed and moments are carried over to the far ends in the same way as in step 3 . This completes the second cycle.

6. The joints released in step 3 are released again, while the rotation of the other joints is prevented. The balancing moment at a joint is equal to minus the algebraic sum of the end moments carried over to the ends meeting at the joint in the previous cycle.

7. Repeat step 6 several times, for the two sets of joints in turn until the balancing moments become negligible.

8. Sum the end-moments recorded in each of the steps 2 to 7 to obtain the final-end moments. The reactions or stress resultants if required may then be calculated by simple equations of statics.

If a frame has an overhanging part, its effect is replaced by a force and a couple acting at the joint of the overhang with the rest of the structure. This is illustrated in the following example.

Example 11-4 Obtain the bending-moment diagram for the continuous beam of Fig. 11-7a.

The effect of the cantilever AB on the rest of the beam is the same as that of a downward force, qb and an anticlockwise couple of $0.6ql^2$ acting at B, as shown in Fig. 11-7b. The end B is thus free, and the distribution has to be carried out at joints C, D, and E. During the distribution, joint B (Fig. 11-7b) will be free to rotate so that B is considered to be a hinge. No moment is therefore carried over from C to B, and the relative end-rotational stiffness[4] of CB is $(\frac{3}{4})K = (\frac{3}{4})I/l$. Since I is constant throughout, it may be taken as unity. The relative rotational stiffnesses of the ends at the three joints C, D, and E are recorded in Fig. 11-7c, together with the COF's and the DF's.

The FEM's for ends BC and CB are those of a beam hinged at B, fixed at C, and subjected to a couple of $-0.6ql^2$ at B. The other FEM's are calculated in the usual way.

The moment distribution is performed at joint D in one cycle, and at joints C and E in the following cycle. The moments produced at ends CB, EF and FE are not recorded; the final moments at these ends are calculated from the values of the final moments at the other end of the respective members. First, we write $M_{CB} = -M_{CD} = -0.285ql^2$, and $M_{EF} = -M_{ED} = -0.532ql^2$. Then, by an equation similar to Eq. 11-10,

$$M_{FE} = (FEM)_{FE} + C_{EF}[M_{EF} - (FEM)_{EF}]$$

The FEM's at the two ends of member EF are zero and $C_{EF} = \frac{1}{2}$; therefore $M_{FE} = (\frac{1}{2})M_{EF} = -0.266ql^2$.

The final bending-moment diagram of the beam is shown in Fig. 11-7d.

11-7 ADJUSTED END-ROTATIONAL STIFFNESSES

The process of moment distribution can be made shorter in certain cases if adjusted end-rotational stiffnesses are used instead of the usual

---

[4]See Eq. 11-14.

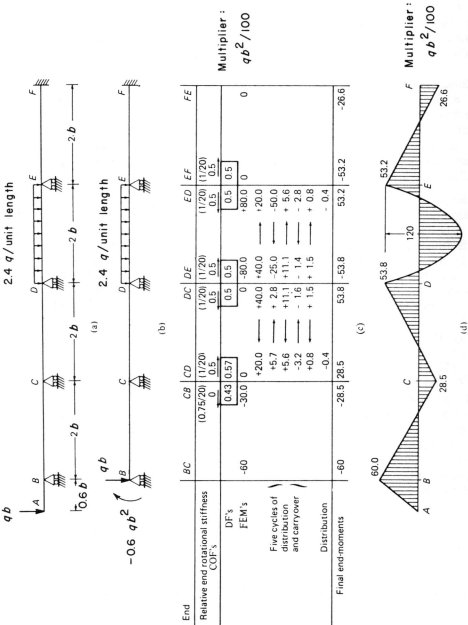

Fig. 11-7. Analysis of the continuous beam in Example 11-4 by moment distribution. (a) Continuous beam. (b) Replacement of the actual load on overhang by equivalent loading at B. (c) Moment distribution. (d) Bending moment diagram.

stiffnesses. Expressions will be derived for these adjusted end-rotational stiffnesses of nonprismatic members but, because of their frequent use, values for prismatic members will also be given.

The end-rotational stiffness S_{AB} was defined in Sec. 11-2 as the value of the moment required at A to rotate the beam end A through a unit angle while the far end B is fixed. Similarly, S_{BA} is the end-moment to produce a unit rotation at B while end A is fixed. The deflected shapes of the beam corresponding to these two conditions are shown in Fig. 11-8a. The moment t at the fixed end has the same value for the two deflected configurations. The rotated ends in Fig. 11-8b are sketched with a roller support but they can also be represented as in Figs. 11-1b and c. Both figures indicate the same conditions, that is a unit rotation without transverse translation of the end. The axial and shear deformations are ignored for the present purposes. For a prismatic[5] beam $S_{AB} = S_{BA} = 4EI/l$, and $t = 2EI/l$.

The special cases which we shall now consider are: (a) when rotation is applied at one end of a member whose far end is not fixed but hinged (Fig. 11-8b); (b) when such a member is subjected to symmetrical or antisymmetrical end-forces and rotations (Figs. 11-8c and d). The adjusted rotational end stiffnesses in these cases will be denoted by S with a subscript indicating the beam end and a superscript indicating the conditions at the far end as defined in Fig. 11-8. The adjusted stiffness $S_{AB}^{\textcircled{4}}$ is within the scope of Chapter 12 as translation of the joint is allowed to take place freely during the moment distribution.

Let us express the adjusted end rotational stiffnesses $S_{AB}^{\textcircled{1}}$, $S_{AB}^{\textcircled{2}}$, $S_{AB}^{\textcircled{3}}$, and $S_{AB}^{\textcircled{4}}$ in terms of the stiffnesses S_{AB}, S_{BA}, and t for the same beam.

The forces and displacements along the coordinates 1 and 2 in Fig. 11-8a are related by

$$[S]\{D\} = \{F\} \tag{11-11}$$

where

$$[S] = \begin{bmatrix} S_{AB} & t \\ t & S_{BA} \end{bmatrix}$$

and where $\{D\}$ are the end-rotations and $\{F\}$ the end-moments.

Putting $D_1 = 1$ and $F_2 = 0$ in Eq. 11-11 represents the conditions in Fig. 11-8b. The force F_1 in this case will be equal to the adjusted stiffness $S_{AB}^{\textcircled{1}}$. Thus

$$\begin{bmatrix} S_{AB} & t \\ t & S_{BA} \end{bmatrix} \begin{Bmatrix} 1 \\ D_2 \end{Bmatrix} = \begin{Bmatrix} S_{AB}^{\textcircled{1}} \\ 0 \end{Bmatrix}$$

---

[5]For nonprismatic beams, values of S_{AB} and S_{BA} and the COF may be taken from tables (see reference in footnote 1 of this chapter) or calculated by Eq. 10-21.

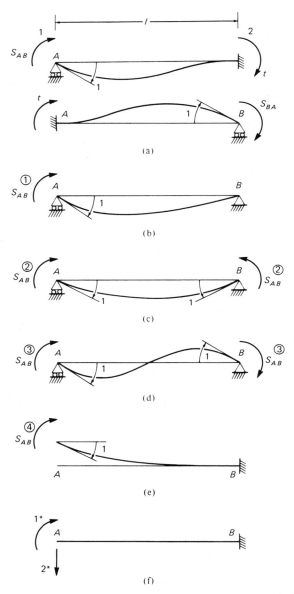

Fig. 11-8. End-rotational stiffnesses in special cases (Eqs. 11-11–11-21). (a) End-moments caused by a unit rotation at one end while the other end is fixed. (b) End-moment caused by a unit rotation at end A while end B is hinged. (c) End-moments caused by symmetrical unit rotations at ends A and B. (d) End-moments caused by antisymmetrical unit rotations at ends A and B. (e) End-moment at A caused by a unit rotation at A while B is fixed. The translation of A is not prevented. (f) Coordinates for calculation of $S_{AB}^{④}$.

Solving,

$$S_{AB}^{\textcircled{1}} = S_{AB} - \frac{t^2}{S_{BA}} \qquad (11\text{-}12)$$

The same equation can be written in terms of the COF's, $C_{AB} = t/S_{AB}$ and $C_{BA} = t/S_{BA}$, thus

$$S_{AB}^{\textcircled{1}} = S_{AB}(1 - C_{AB}C_{BA}) \qquad (11\text{-}13)$$

For a prismatic member, Eqs. 11-12 and 11-13 reduce to

$$S_{AB}^{\textcircled{1}} = \frac{3EI}{l} \qquad (11\text{-}14)$$

Referring again to Eq. 11-11, and putting $D_1 = -D_2 = 1$ represents the conditions in Fig. 11-8c. The force $F_1 = -F_2$ is equal to the end-rotational stiffness $S_{AB}^{\textcircled{2}}$. Thus

$$\begin{bmatrix} S_{AB} & t \\ t & S_{BA} \end{bmatrix} \begin{Bmatrix} 1 \\ -1 \end{Bmatrix} = \begin{Bmatrix} S_{AB}^{\textcircled{2}} \\ -S_{AB}^{\textcircled{2}} \end{Bmatrix}$$

Solving either of the above equations, we obtain the end-rotational stiffness in case of symmetry,

$$S_{AB}^{\textcircled{2}} = S_{AB} - t = S_{AB}(1 - C_{AB}) \qquad (11\text{-}15)$$

and for a prismatic member,

$$S_{AB}^{\textcircled{2}} = \frac{2EI}{l} \qquad (11\text{-}16)$$

Similarly, if we put $D_1 = D_2 = 1$, Eq. 11-11 represents the conditions in Fig. 11-8d. The end-forces are $F_1 = F_2 = S_{AB}^{\textcircled{3}}$, and the end-rotational stiffness in the antisymmetrical case is

$$S_{AB}^{\textcircled{3}} = S_{AB} + t = S_{AB}(1 + C_{AB}) \qquad (11\text{-}17)$$

For a prismatic member, this reduces to

$$S_{AB}^{\textcircled{3}} = \frac{6EI}{l} \qquad (11\text{-}18)$$

The stiffness matrix for the beam considered above corresponding to the coordinates 1 and 2 in Fig. 11-8f is

$$[S^*] = \begin{bmatrix} S_{AB} & (S_{AB} + t)/l \\ (S_{AB} + t)/l & (S_{AB} + S_{BA} + 2t)/l^2 \end{bmatrix}$$

The elements representing end-moments (S_{11}^* and S_{12}^*) of this matrix can be easily checked by applying the slope-deflection Eq. 11-2, and the elements representing the shear at $A(S_{21}^*$ and $S_{22}^*)$ are readily obtained from the end-moments by considering the equilibrium of the beam.

The forces and the displacements along the coordinates 1 and 2 in Fig. 11-8f are related by

$$[S^*]\{D^*\} = \{F^*\}$$

With $D_1^* = 1$ and $F_2^* = 0$, the above equation represents the conditions of Fig. 11-8e. Solving for $F_1^* = S_{AB}^{④}$, we obtain the end-rotational stiffness when the end-translation is not prevented while the far end is fixed:

$$S_{AB}^{④} = \frac{S_{AB}S_{BA} - t^2}{S_{AB} + S_{BA} + 2t} \tag{11-19}$$

For a prismatic member, Eq. 11-19 reduces to

$$S_{AB}^{④} = \frac{EI}{l} \tag{11-20}$$

It is obvious that for equilibrium of the beam in Fig. 11-8e, the end-moment at B must be equal and opposite to the moment at end A. Therefore, the carryover factor both for a prismatic and nonprismatic member is

$$C_{AB}^{④} = -1 \tag{11-21}$$

11-8   ADJUSTED FIXED-END MOMENTS

Figure 11-9a represents a beam with two fixed ends subjected to transverse loading. Let the end-moments for this beam be M_{AB} and M_{BA} and let the vertical reactions be F_A and F_B. The beam is assumed to be nonprismatic with end-rotational stiffnesses S_{AB} and S_{BA}, the carryover moment t, and the carryover factors C_{AB} and C_{BA}. Let us now find the end-moments for a similar beam subjected to the same transverse loading but with the support conditions of Figs. 11-9b and c. In the former, the end A is hinged; thus rotation at A is free to take place and there is only one end-moment $M_{BA}^{①}$ to be determined. In Fig. 11-9c, the translation of A is allowed to take place but rotation is prevented. Two end-moments, $M_{AB}^{②}$ and $M_{BA}^{②}$, are to be determined.

To shorten the process of moment distribution, in some cases we can use the adjusted fixed-end moments $M_{BA}^{①}$, $M_{AB}^{②}$, and $M_{BA}^{②}$, together with the adjusted end-rotational stiffnesses. Consider first the beam in Fig. 11-9a with the end-displacements restrained; the end-moments are M_{AB} and M_{BA}. Now allow the end A to rotate by an amount such that a moment $-M_{AB}$ is developed at end A. The corresponding moment developed at B is $-C_{AB}M_{AB}$. Superposing the end-moments developed by the rotation will

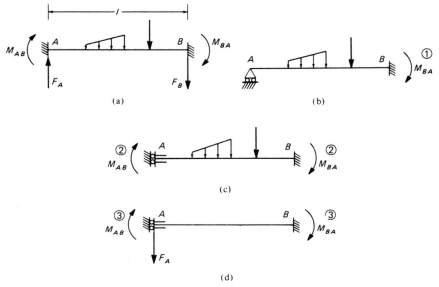

Fig. 11-9. Adjusted fixed-end moments. (a) Two ends encastré. (b) Free rotation at A. (c) Vertical movement at A is allowed without rotation. (d) Same beam as in part (c) loaded with a downward force F_A.

give a zero moment at A, and this represents the condition of the beam in Fig. 11-9b. The adjusted FEM at B when end A is hinged is therefore

$$M_{BA}^{①} = M_{BA} - C_{AB}M_{AB} \tag{11-22}$$

where M_{AB} and M_{BA} are the FEM's of the same beam with the two ends encastré, and C_{AB} is the COF from A to B. For a prismatic beam, Eq. 11-22 becomes

$$M_{BA}^{①} = M_{BA} - \frac{M_{AB}}{2} \tag{11-23}$$

Let us consider now the same beam AB subjected to a downward force F_A at end A, which is allowed to translate freely but without rotation while the other end B is encastré, as indicated in Fig. 11-9d. The end-moments in this beam, obtained by slope-deflection or by other methods, are

$$M_{AB}^{③} = F_A l \frac{(S_{AB} + t)}{(S_{AB} + S_{BA} + 2t)} \tag{11-24}$$

and

$$M_{BA}^{③} = F_A l \frac{(S_{BA} + t)}{(S_{AB} + S_{BA} + 2t)} \tag{11-25}$$

where l is the beam length and t = carryover moment = $C_{AB}S_{BA} = C_{BA}S_{BA}$. For a prismatic beam Eqs. 11-24 and 11-25 reduce to

$$M_{AB}^{③} = M_{BA}^{③} = F_A l/2 \tag{11-26}$$

The end-moments in the beam in Fig. 11-9c can be obtained by superposition of the end-moments in beams in Figs. 11-9a and 11-9d. Hence, the adjusted FEM's due to the transverse load on beam AB with end A free to translate without rotation (Fig. 11-9c) are

$$M_{AB}^{②} = M_{AB} + F_A l \frac{(S_{AB} + t)}{(S_{AB} + S_{BA} + 2t)} \tag{11-27}$$

and

$$M_{BA}^{②} = M_{BA} + F_A l \frac{(S_{BA} + t)}{(S_{AB} + S_{BA} + 2t)} \tag{11-28}$$

where M_{AB}, M_{BA}, and F_A are end forces for the same beam with the ends encastré (considered positive when in the directions indicated in Fig. 11-9a). When the beam is prismatic,

$$M_{AB}^{②} = M_{AB} + \frac{F_A l}{2} \tag{11-29}$$

and

$$M_{BA}^{②} = M_{BA} + \frac{F_A l}{2} \tag{11-30}$$

If in Fig. 11-9b the end A is encastré while the end B is on a roller, the FEM at A can be obtained by Eqs. 11-22 or 11-23 by interchanging the subscripts A and B. Similarly, if the types of support at A and B are interchanged in Fig. 11-9c, Eqs. 11-27–11-30 can be used with the subscripts A and B interchanged. They will then include F_B instead of F_A and the positive direction of F_B will be downwards.

The usefulness of the adjusted end stiffnesses and adjusted FEM's will be demonstrated by examples in this and the following chapter.

Example 11-5   Find the bending moment in the symmetrical frame shown in Fig. 11-10a by replacing the loading by equivalent symmetrical and antisymmetrical loadings.

Any load on a symmetrical structure can be replaced by the sum of a symmetrical and an antisymmetrical loading, as indicated in Figs. 11-10b and c. This procedure enables us to use the adjusted stiffnesses and FEM's, but the computational effort is comparable.

With symmetry or antisymmetry of loading, the moment distribution need be done for one-half of the frame only. Figures 11-10d and e deal with the symmetrical and antisymmetrical cases respectively. The end-rotational

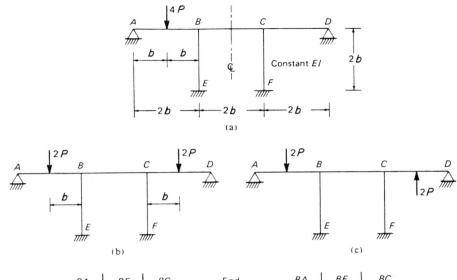

BA	BE	BC
0.33	0.45	0.22
← 0	↓ 0.5	0 →

Multiplier :

Pb/100

+75.0		
−25.0	−33.4	−16.6
+50.0	−33.4	−16.6

(d)

End	BA	BE	BC
DF's	0.23	0.31	0.46
COF's	← 0	↓ 0.5	0 →
FEM's	+75.0		
Distrib.	−17.2	−23.2	−34.6
Final end-moments	+57.8	−23.2	−34.6

(e)

End	BA	BE	BC	EB	CD	CF	CB	FC
Symmetric loading	+ 50.0	−33.4	−16.6	−16.7	−50.0	+33.4	+16.6	+16.7
Antisymmetric loading	+ 57.8	−23.2	−34.6	−11.6	+57.8	−23.2	−34.6	−11.6
Final end-moments for the frame in Fig. (a)	+107.8	−56.6	−51.2	−28.3	+7.8	+10.2	−18.0	+ 5.1

(f)

Multiplier :

Pb / 100

107.8
51.2
10.2
200
56.6
18.0
7.8
28.3
5.1

(g)

Fig. 11-10. Analysis of frame in Example 11-5. (a) Frame properties and loading. (b) Symmetrical loading. (c) Antisymmetrical loading. (d) Moment distribution for the symmetrical case. (e) Moment distribution for the antisymmetrical case. (f) Summation of end-moments calculated in parts (d) and (e). (g) Bending-moment diagram for the frame in part (a).

stiffnesses for the members meeting at B are calculated using Eqs. 11-14, 11-16, and 11-18, and we find for the symmetrical case

$$S_{BA}:S_{BE}:S_{BC} = 3K_{BA}:4K_{BE}:2K_{BC}$$

and for the antisymmetrical case

$$S_{BA}:S_{BE}:S_{BC} = 3K_{BA}:4K_{BE}:6K_{BC}$$

where $K = I/l$. In this frame, K is the same for all members. The DF's are calculated in the usual way and are given in Figs. 11-10d and e.

The FEM in end BA (for a beam with one end hinged) is obtained by the use of Appendix C and Eq. 11-22; thus $(\text{FEM})_{BA} = 0.75Pb$ for both cases.

The COF $C_{BE} = 0.5$. No moments are carried over from B to C or from B to A. Thus only one cycle of moment distribution is required at B, as shown in Figs. 11-10d and e.

It is important to note that in the symmetrical case, the end-moments in the right-hand half of the frame are equal in magnitude and opposite in sign to the end-moments in the left-hand half, while in the antisymmetrical case they are equal and of the same sign in the two halves. The summation of the end-moments in the two cases is carried out in the table in Fig. 11-10f. This gives the end-moments of the frame in Fig. 11-10a. The corresponding bending-moment diagram is shown in Fig. 11-10g.

Example 11-6 Obtain the bending-moment diagram for the symmetrical nonprismatic beam shown in Fig. 11-11a. The end-rotational stiffnesses, the COF's and the fixed end-moments are given below.

	Member AB	Member BC
End-rotational stiffness	$S_{AB} = 4.2\dfrac{EI}{l_{AB}}$ $S_{BA} = 5.0\dfrac{EI}{l_{AB}}$	$S_{BC} = S_{CB} = 5.3\dfrac{EI}{l_{BC}}$
Carryover factor	$C_{AB} = 0.57$ $C_{BA} = 0.48$	$C_{BC} = C_{CB} = 0.56$
Fixed-end moment due to a uniform load of intensity q	$M_{AB} = -0.078ql_{AB}^2$ $M_{BA} = 0.095ql_{AB}^2$	$M_{BC} = -0.089ql_{BC}^2$ $M_{CB} = 0.089ql_{BC}^2$

The beam is symmetrical and symmetrically loaded. With adjusted stiffnesses and FEM's for ends BA and BC, the moment distribution needs to be carried out at joint B only.

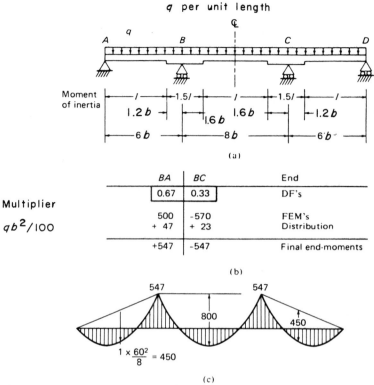

Fig. 11-11. Analysis of beam in Example 11-6. (a) Beam properties and loading. (b) Moment distribution. (c) Bending-moment diagram.

The adjusted end-rotational stiffness for BA with end A hinged is given by Eq. 11-13; in our case,

$$S_{BA}^{\text{①}} = S_{BA}(1 - C_{AB}C_{BA})$$

or

$$S_{BA}^{\text{①}} = \frac{EI}{6b}(5.0)(1 - 0.57 \times 0.48) = 0.605\frac{EI}{b}$$

The adjusted end-stiffness for the symmetrical member BC is (from Eq. 11-15)

$$S_{BC}^{\text{②}} = S_{BC}(1 - C_{BC})$$

or

$$S_{BC}^{\text{②}} = \frac{EI}{8b}(5.3)(1 - 0.56) = 0.292\frac{EI}{b}$$

The distribution factors are

$$(DF)_{BA} = \frac{0.605}{0.605 + 0.292} = 0.67$$

$$(DF)_{BC} = \frac{0.292}{0.605 + 0.292} = 0.33$$

The adjusted FEM at B for member BA with end A hinged is (from Eq. 11-22)

$$M_{BA}^{\textcircled{1}} = M_{BA} - C_{AB}M_{AB}$$

or

$$M_{BA}^{\textcircled{1}} = ql_{AB}^2(0.095 + 0.57 \times 0.078) = 0.139ql_{AB}^2 = 5.00qb^2$$

The FEM at end BC is

$$M_{BC} = -0.089ql_{BC}^2 = -5.70qb^2$$

One cycle of moment distribution is required with no carryover. This is shown in Fig. 11-11b and the bending-moment diagram is plotted in Fig. 11-11c.

11-9   GENERAL

The slope-deflection method is essentially the same as the general displacement method of analysis when applied to plane frames in which only the bending deformations are considered. Because of its generality, the displacement method outlined in Chapter 3 is considered preferable and it is likely that with time the slope-deflection method will fall into desuetude. Nevertheless, the slope-deflection equation (Eq. 11-2) can be conveniently used to find the end moments in a member whose end displacements are known.

The moment-distribution method offers great advantages in the analysis of rigid frames using a slide rule or hand computations. Solution of simultaneous equations in the case of structures not involving translation of joints is avoided. The amount of computations can be further reduced for symmetrical structures subjected to symmetrical or antisymmetrical loading (and any loading can be resolved into such components) by using adjusted end-rotational stiffnesses and FEM's.

The problems for this chapter and for the following one are given together at the end of Chapter 12.

CHAPTER 12

Moment distribution with sway: multistorey and multibay frames

12-1 INTRODUCTION

Despite the advent of computers, which can rapidly analyze large building frames, it is useful to analyze such frames by hand in preliminary design or to check the computer results. It is useful therefore to extend the method of moment distribution to frames in which translation of joints takes place, as is the case in nearly all building frames. Further, we shall consider procedures which speed up the analysis of rectilinear plane frames of several storeys and bays for which only bending deformations are considered.

12-2 GENERAL PROCEDURE FOR PLANE FRAMES WITH JOINT TRANSLATION

The procedure considered in this section is a displacement method of analysis of Chapter 3 with moment distribution used to reduce the number of simultaneous equations involved.

In the general case, the degree of kinematic indeterminacy of a structure can be expressed as

$$k = m + n$$

where m is the number of unknown joint rotations and n is the number of unknown joint translations. The number of joint translations, which is sometimes referred to as the number of *degrees of freedom in sidesway*, represents the number of equations which must be solved, and if n is much smaller than k, there is a considerable reduction in computational effort.

Figure 12-1 shows several examples of plane frames with coordinates representing the unknown joint translations indicated. In each case the degree of freedom of sidesway n and the total number of unknown joint displacements k are indicated, assuming that the length of all members remains unchanged after deformations. From a comparison of k and n, we can see that the analysis by the method to be discussed involves the solution of a much smaller number of simultaneous equations. This saving is of course not significant in a computer analysis.

317

The procedure is as follows.

1. All the possible independent joint translations are first indicated by coordinates; their number is of course equal to the number of degrees of freedom in sidesway, n. Restraining forces $\{F\}_{n \times 1}$ are then introduced along these coordinates to prevent the translation of all joints.

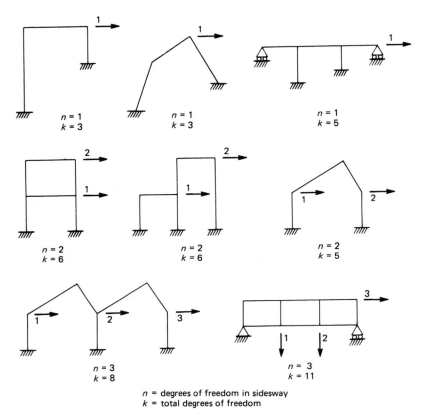

n = degrees of freedom in sidesway
k = total degrees of freedom

Fig. 12-1. Joint displacements of plane frames.

2. The frame is analyzed by moment distribution without joint translation (as in the preceding chapter), and the corresponding end moments $\{A_r\}$ are determined. These are the moments on the ends of the members with the joints allowed to rotate but with translation restrained. The n restraining forces $\{F\}$ are then calculated by considering equilibrium of the members. This usually involves no more than the simple equations of static equilibrium.

3. A unit displacement $D_1 = 1$ is introduced along the first of the n coordinates in the unloaded frame, while the translation along all the other coordinates is prevented. Assuming further that no joint rotations occur,

the FEMs in the deformed members are calculated. With no further joint translation allowed, the moment distribution is now carried out. The end-moments obtained by the distribution correspond to a unit translational displacement along the coordinate 1, all the other joint translations being prevented but joint rotations of course being allowed. From these moments, the forces required along the n coordinates in this deformed configuration are determined.

The process is repeated by allowing a unit value of translation to take place separately along each of the n coordinates. The resulting forces along the coordinates are then arranged in a matrix $[S]_{n \times n}$ representing the stiffness matrix of the frame. The end-moments are arranged in a matrix $[A_u]$ representing the end-moments caused by unit values of the translations.

4. The loaded frame in the restrained state will have sidesway $\{D_1, D_2, ..., D_n\}$, which can be determined from the equation

$$[S]\{D\} = -\{F\} \qquad (12\text{-}1)$$

5. The end-moments in the actual structure are then calculated by a superposition equation

$$\{A\} = \{A_r\} + [A_u]\{D\} \qquad (12\text{-}2)$$

It is obvious that in step (3) we could have determined forces corresponding to an arbitrary value of the translation instead of unity. This of course would lead to a matrix $[S]$ which is not the real stiffness matrix and a vector $\{D\}$ which does not represent the real displacements, but Eq. 12-2 would still give the actual end-moments in the frame.

Example 12-1   Find the end-moments in each span of the prismatic continuous beam in Fig. 12-2a. The beam is encastré at A and D and rests on elastic (spring) supports at B and C. The stiffness of the elastic supports, that is the force required to compress any of the springs by a unit distance, is $2EI/l^3$.

The beam has two degrees of freedom of joint translation: vertical displacements at B and at C. Let the positive direction of these displacements be indicated by coordinates 1 and 2 in Fig. 12-2b.

First, the beam is analyzed with the translation of B and C prevented. The end-moments on ends 1, 2, ..., 6 are

$$\{A_r\} = ql^2 \{-0.078, 0.094, -0.094, 0.044, -0.044, -0.022\}$$

These moments can be obtained by moment distribution or by other methods. The restraining forces $\{F\}$ are equal to minus the reactions at B and C if these supports are rigid (Fig. 12-2c), that is

$$F = -ql \begin{Bmatrix} 1.066 \\ 0.516 \end{Bmatrix}$$

The FEMs due to $D_1 = 1$ and $D_2 = 0$, with no joint rotation allowed at B, are given in Fig. 12-2d. The moment-distribution cycles are performed in

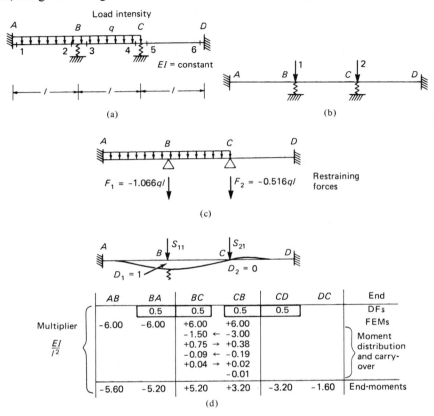

(a)

(b)

(c)

(d)

Fig. 12-2. Beam considered in Example 12-1. (a) Beam properties and loading. (b) Coordinate system. (c) Restraining forces $(D_1 = D_2 = 0)$. (d) Calculation of the end-moments corresponding to $D_1 = 1$ and $D_2 = 0$.

the usual way and the end-moments are calculated. The forces S_{11} and S_{21} required to hold the beam in this deformed configuration are calculated from the end-moments, considering the equilibrium of each span as follows:

$$S_{11} = \left(\frac{2EI}{l^3}\right) - \frac{EI}{l^3}\left[(-5.2 - 5.60) - (5.20 + 3.2)\right] = 21.2\frac{EI}{l^3}$$

and

$$S_{21} = -\frac{EI}{l^3}\left[(3.2 + 5.2) - (-3.2 - 1.6)\right] = -13.2\frac{EI}{l^3}$$

The term $2EI/l^3$ included in S_{11} is the force required at 1 to compress the spring at A by a unit amount.

Because the structure is symmetrical, we can see that the forces S_{12} and S_{22} (at 1 and 2 respectively) corresponding to the displacements $D_1 = 0$ and $D_2 = 1$ are

$$S_{12} = -13.2 \frac{EI}{l^3}$$

and

$$S_{22} = 21.2 \frac{EI}{l^3}$$

The moments on the ends, 1, 2, ..., 6 (see Fig. 12-2a) due to the two deformed configurations considered above can be arranged in a matrix:

$$[A_u] = \frac{EI}{l^2} \begin{bmatrix} -5.6 & 1.6 \\ -5.2 & 3.2 \\ 5.2 & -3.2 \\ 3.2 & -5.2 \\ -3.2 & 5.2 \\ -1.6 & 5.6 \end{bmatrix}$$

Applying Eq. 12.1,

$$\frac{EI}{l^3} \begin{bmatrix} 21.2 & -13.2 \\ -13.2 & 21.2 \end{bmatrix} \begin{Bmatrix} D_1 \\ D_2 \end{Bmatrix} = ql \begin{Bmatrix} 1.066 \\ 0.516 \end{Bmatrix}$$

Hence,

$$\{D\} = \begin{Bmatrix} D_1 \\ D_2 \end{Bmatrix} = \frac{ql^4}{EI} \begin{Bmatrix} 0.107 \\ 0.091 \end{Bmatrix}$$

The final end-moments for the beam of Fig. 12-2a are obtained from Eq. 12-2 by substituting the values calculated. Hence, the final end-moments are

$$\{A\} = ql^2 \begin{Bmatrix} -0.533 \\ -0.172 \\ 0.172 \\ -0.086 \\ 0.086 \\ 0.360 \end{Bmatrix}$$

In the present example, the end-moments due to a unit displacement at 1 or at 2 and the forces corresponding to these displacements could have been taken directly from the tables in Appendix E without the need for moment distribution. However, these tables apply only to continuous beams with equal spans.

Example 12-2   Obtain the bending-moment diagram for a two-storey frame under the vertical and horizontal loading shown in Fig. 12-3a. The relative values of $K(= I/l)$ are indicated in the figure.

The frame has two degrees of freedom in sidesway: horizontal translations along coordinates 1 and 2 (Fig. 12-3a). Without the horizontal loads, the frame is symmetrical and symmetrically loaded, and does therefore not sway. It follows that joint translations can be prevented by the application along the coordinates of restraining forces:

$$\{F\} = -\begin{Bmatrix} 4 \\ 2 \end{Bmatrix} \frac{qb}{10}$$

The end-moments for the frame without joint translations are calculated by moment distribution performed for the left-hand half of the frame in the top part of Fig. 12-3c. The relative end-rotational stiffness for beams BE and CD is taken as $2K$, which is the adjusted stiffness for a symmetrical and symmetrically loaded member (see Eq. 11-16), and for the columns as $4K$.

Forces or displacements along the coordinates 1 and 2 cause antisymmetrical bending moment in the frame. It is therefore sufficient to carry out the moment distribution for one-half of the frame (say the left-hand), with the relative end-rotational stiffness for BE and CD of $6K$. This is the adjusted end-rotational stiffness of a symmetrical member subjected to antisymmetrical end-moments (Eq. 11-18).

A displacement $D_1 = 1$, while $D_2 = 0$, causes FEMs of $6EI/l^2$ on the two ends of BC and $-(6EI/l^2)$ in AB. For any value of the displacement D_1, with $D_2 = 0$, the absolute values of the FEMs in the columns are proportional to $I/l^2 = K/l$. In Fig. 12-3c, the FEMs in the columns are taken (in terms of $qb^2/100$) as -120 in AB and $+80$ in BC, and the moment distribution is carried out. We obtain thus a set of end-moments corresponding to an arbitrary displacement along coordinate 1 while the displacement along coordinate 2 is prevented.

A similar distribution is included in Fig. 12-3c for an arbitrary displacement along the coordinate 2, while $D_1 = 0$.

The forces along the coordinates corresponding to the above two displacement patterns are calculated. It can be easily seen that the force at any floor level is equal to $\Sigma M_{ca}/h_a - \Sigma M_{cb}/h_b$, where M_c is the end-moment in a column, h its height, and the subscripts a and b refer to the columns above and below the force to be calculated; the summation is carried out for

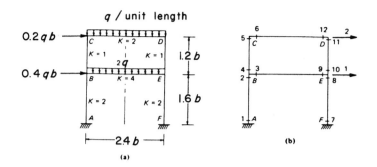

(a)

(b)

	End	1 AB	2 BA	3 BE	4 BC	5 CB	6 CD
Sideway prevented	DF's		0.4	0.4	0.2	0.5	0.5
	FEM's	0	0	-96.0	0	0	-48.0
	Distribution and carryover	16.8 ← +33.6		+33.6	+12.0 ← +24.0 +16.8 → + 8.4 - 2.1 ← - 4.2		
		0.4 ← + 0.8		+ 0.9	+ 0.4 → + 0.2 - 0.1		
	End-moments \}A,\{	+17.2	+34.4	-61.5	+27.1	+28.3	- 28.3
Displacement D_1 while $D_2 = 0$	DF's		0.22	0.67	0.11	0.25	0.75
	FEM's	- 120	- 120	0	+80	+80	0
	Distribution and carryover	6 ← +11		+33	- 10 ← - 20 + 6 → + 3 - 1		
	End-moments	- 114	- 109	+33	+76	+62	-62
Displacement D_2 while $D_1 = 0$	DF's		0.22	0.67	0.11	0.25	0.75
	FEM's	0	0	0	- 120	- 120	0
		+12 ← +23		+70 + 1	+15 ← +30 +12 → + 6 - 1 ← - 2		
	End-moments	+12	+23	+71	- 94	- 86	+86

Multiplier :

$qb^2/100$

(c)

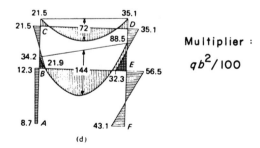

Multiplier :

$qb^2/100$

(d)

Fig. 12-3. Frame considered in Example 12-2. (a) Frame properties and loading. (b) Coordinate system. (c) Moment distribution. (d) Bending-moment diagram.

all the columns in a given floor. For example, the forces S_{11} and S_{21} corresponding to the arbitrarily chosen displacement D_1 (while $D_2 = 0$) are

$$S_{11} = 2\left[\frac{(76 + 62)}{1.2b} - \frac{(-114 - 109)}{1.6b}\right]\frac{qb^2}{100} = 5.09qb$$

and

$$S_{21} = 2\left[-\frac{(76 + 62)}{1.2b}\right]\frac{qb^2}{100} = -2.3qb$$

Applying Eq. 12-1,

$$qb\begin{bmatrix} 5.09 & -3.45 \\ -2.30 & 3.00 \end{bmatrix}\begin{Bmatrix} D_1 \\ D_2 \end{Bmatrix} = qb\begin{Bmatrix} 0.4 \\ 0.2 \end{Bmatrix}$$

whence

$$\begin{Bmatrix} D_1 \\ D_2 \end{Bmatrix} = \begin{Bmatrix} 0.259 \\ 0.265 \end{Bmatrix}$$

The final end-moments at the ends 1, 2, ..., 12 of the members indicated in Fig. 12-3b are obtained by the superposition Eq. 12-2 $\{A\} = \{A_r\} + [A_u]\{D\}$, where the elements of $\{A_r\}$ and $[A_u]$ are the end-moments calculated in Fig. 12.3c. The end-moments in the right-hand half of the frame are equal to minus the corresponding value in the left-hand half for the symmetrical case of no sidesway. In the other two cases, the bending moment is anti-symmetrical, and the corresponding end moments in the two halves are the same in sign and magnitude. Thus

$$\{A\} = \frac{qb^2}{100}\begin{Bmatrix} 17.2 \\ 34.4 \\ -61.5 \\ 27.1 \\ 28.3 \\ -28.3 \\ -17.2 \\ -34.4 \\ 61.5 \\ -27.1 \\ -28.3 \\ 28.3 \end{Bmatrix} + \frac{qb^2}{100}\begin{bmatrix} -114 & 12 \\ -109 & 23 \\ 33 & 71 \\ 76 & -94 \\ 62 & -86 \\ -62 & 86 \\ -114 & 12 \\ -109 & 23 \\ 33 & 71 \\ 76 & -94 \\ 62 & -86 \\ -62 & 86 \end{bmatrix}\begin{Bmatrix} 0.259 \\ 0.265 \end{Bmatrix} = \begin{Bmatrix} -8.7 \\ 12.3 \\ -34.2 \\ 21.9 \\ 21.6 \\ -21.6 \\ -43.1 \\ -56.5 \\ 88.5 \\ -32.3 \\ -35.0 \\ 35.0 \end{Bmatrix}\frac{qb^2}{100}$$

The bending-moment diagram is shown in Fig. 12-3d. A check on the computations can be made by verifying that the sum of the shear forces in the columns of each storey calculated from the end moments is equal to the sum of the external horizontal forces acting above the floor.

12-3 NO-SHEAR MOMENT DISTRIBUTION

Several developments[1] in the moment distribution approach have made it possible to take into account the effect of joint translation without the necessity of using simultaneous equations, as in Sec. 12-2. However, most of these methods require the application of several "rules" to obtain the adjusted stiffnesses, and these are not easy to remember. Therefore, unless the structure has to be analyzed for a number of different loading conditions, it may be easier to use the general moment-distribution procedure than to indulge in specialized techniques.

Nevertheless, it is well worthwhile to be familiar with the no-shear moment distribution, sometimes also known as the *cantilever moment distribution*. The method was originally developed for symmetrical one-bay multistorey plane frames supporting antisymmetrical loading. However, its application can be extended to multibay frames by the use of the *substitute frame* and of the *principle of multiple*, which are discussed in the succeeding sections.

In the no-shear method, the sidesway is allowed to occur freely during the moment distribution, that is no change in the forces acting at the floor level takes place when the joints are allowed to rotate; thus the shear in the columns is not changed during the distribution. The adjusted end-rotational stiffnesses and FEM's required for the no-shear moment distribution are those developed in Secs. 11-7 and 11-8.

The no-shear moment distribution procedure is best explained by examples.

Example 12-3   Obtain the bending-moment diagram for the frame of Fig. 12-4a in which the values of $K(= I/l)$ for all members are indicated.

The frame is symmetrical, and if the horizontal force at B is split into two equal forces as shown in Fig. 12-4b, the frame will be subjected to antisymmetrical loading. It is obvious that the loadings in Figs. 12-4a and b produce the same bending moment when the axial deformations are ignored. Thus the two loadings are equivalent. Under load, the joints B and C translate in the horizontal direction by the same amount, and there is a point of inflection at the center of BC. It is therefore sufficient to consider the frame in Fig. 12-4c, which will have the same bending moment as the left-hand half of the original frame.

Our task is thus to carry out the moment distribution at joint B, with the horizontal translation allowed to take place freely. The relative end-rotational stiffnesses are (from Eqs. 11-18 and 11-20): $S_{BE} = 6K$ and $S_{BA} = K$, and the COF from B to A (from Eq. 11-21) is $C_{BA} = -1$.

The FEM's are the end-moments due to the external loading with the

[1]J. M. Gere, *Moment Distribution*, Van Nostrand, New York, 1963, Chap. 4.

joint B prevented from rotation but allowed to sway. These moments are zero for BE and by Eq. 11-26 $M_{BA} = M_{AB} = -Pb/4$.

One cycle of moment distribution is required to obtain the final end-moments as shown in Fig. 12-4d. Because of antisymmetry, the end-moments in the right-hand half of the frame are the same in magnitude and sign as the corresponding moments in the left-hand half. The bending moment of the whole frame is shown in Fig. 12-4e.

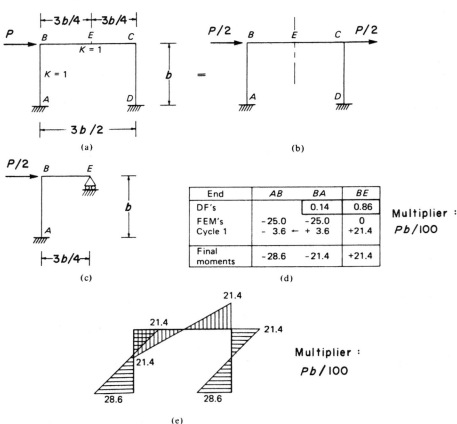

End	AB	BA	BE
DF's		0.14	0.86
FEM's	−25.0	−25.0	0
Cycle 1	− 3.6 ←	+ 3.6	+21.4
Final moments	−28.6	−21.4	+21.4

Multiplier : $Pb/100$

Fig. 12-4. Frame analyzed in Example 12-3. (a) Frame properties and loading. (b) Equivalent loading. (c) Substitute frame. (d) Moment distribution. (e) Bending-moment diagram.

Example 12-4 Consider the symmetrical two-storey frame of Example 12-2 (Fig. 12-3a). No sway occurs due to the symmetrical vertical loading; the end-moments due to it were determined by moment distribution in Fig. 12-3c and are not repeated here. The effect of the horizontal loading will now be analyzed by no-shear moment distribution.

For the same reason as discussed in the previous example, the bending moment in the frame of Fig. 12-5b is the same as for the left-hand half of the frame in Fig. 12-5a.

The moment distribution is carried out at joints B and C in Fig. 12-5c. The adjusted relative end-rotational stiffnesses are $6K$ for the beams and K for the columns, and the COF's are $C_{BA} = C_{BC} = C_{CB} = -1$.

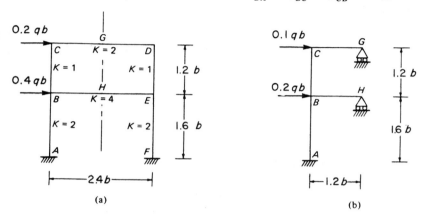

(a)

(b)

End	AB	BA	BE	BC	CB	CD
DF's		0.074	0.889	0.037	0.077	0.923
FEM's	-24.0	-24.0		- 6.0	- 6.0	
				- 0.5 ← + 0.5		
	- 2.3 ← + 2.3		+27.1	+ 1.1 → - 1.1		
					+ 0.1	
End-moments	-26.3	-21.7	+27.1	- 5.4	- 6.5	+ 6.5

Multiplier :

$qb^2/100$

(c)

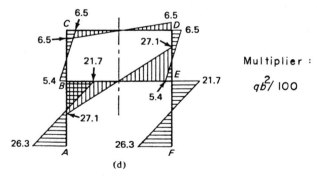

Multiplier :

$qb^2/100$

(d)

Fig. 12-5. Frame analyzed in Example 12-4. (a) Frame properties and loading. (b) The bending moment in this frame is the same as in the left-hand half of the frame in part (a). (c) Moment distribution. (d) Bending-moment diagram.

The FEMs in the columns are calculated by Eq. 11-26 which, when applied to column AB and BC, can be given in the form

$$\text{FEM} = -\frac{Fh}{2}$$

where F is the sum of the horizontal forces acting on the floor(s) above the column considered and h is the storey height. This gives the following FEMs:

$$M_{AB} = M_{BA} = -\frac{(0.1 + 0.2)1.6qb^2}{2} = -0.24qb^2$$

and

$$M_{BC} = M_{CB} = -\frac{0.1 \times 1.2qb^2}{2} = -0.06qb^2$$

These are then the moments at the ends of the columns under the horizontal load when sway is allowed to take place freely but the joint rotations at C and B are restrained.

The end-moments for the right-hand half of the frame are of the same magnitude and sign as the corresponding moments on the left-hand half, calculated in Fig. 12-5c. The bending-moment diagram due to the horizontal loading alone is shown in Fig. 12-5d.

When the end-moments calculated in Fig. 12-5c are added to the end-moments due to vertical loading, the total final end-moments are obtained, and these must of course be the same as in Fig. 12-3.

Example 12-5   Figure 12-6a represents the frame of an industrial building containing a crane gantry. The columns vary in cross section, and the second moment of area of each part of the frame is given in terms of a reference moment I_0. Obtain the bending-moment diagram due to a horizontal force of $4P$ at E.

One way of solving this problem is to consider each column as composed of two members of constant I, meeting at E and F. The frame will then have three degrees of freedom in sidesway: horizontal translations at E, F, and C (or B).

Another way of dealing with the problem is to treat each column as one member of variable I, in which case the frame has one degree of freedom of sidesway, that is the horizontal translation of C (or B). This case is solved in two stages. In the first, sidesway is prevented at the level of BC, and the resulting end moments and restraining force are calculated. In the second stage, a no-shear moment distribution is performed to find the effect of a horizontal force at C equal and opposite to the restraining force calculated in the first stage.

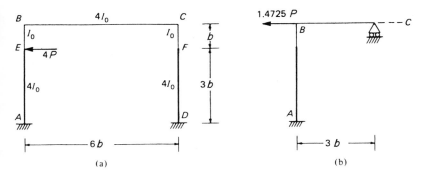

(a)   (b)

	End	AB	BA	BC	CB	CD	DC
Sidesway prevented	DF s		0.34	0.66	0.66	0.34	
	COF s	0.7 ←		0.5 ⇄		0.7 →	
	FEM s	+11.6	− 16.7	0	0	0	0
	Distribution and carryover	+ 4.5 ←	+ 6.4	+11.0 → + 5.5 − 1.8 ← − 3.6 + 1.2 → + 0.6 − 0.2 ← − 0.4 + 0.1			
	End-moments	16.1	− 10.3	+10.3	+ 2.1	− 2.1	− 1.5
Sidesway effect	DF s		0.125	0.875			
	COF s	− 1 ←		0 →			
	FEM s	+39.0 + 2.5 ←	+19.9 − 2.5				
	End-moments	+41.5	+17.4	− 17.4	− 17.4	+17.4	+41.5
	End-moments for the frame in (a)	+57.6	+ 7.1	− 7.1	− 15.3	+15.3	+40.0

Multiplier : $Pb/10$

(c)

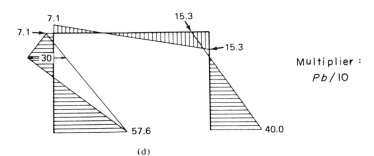

Multiplier : $Pb/10$

(d)

Fig. 12-6. Frame analyzed in Example 12-5. (a) Frame properties and loading. (b) Equivalent frame for sidesway effect. (c) Moment distribution. (d) Bending-moment diagram for the frame in part (a).

The end-rotational stiffnesses and carryover moment for the member AB are: $S_{AB} = 3.53EI_0/b$; $S_{BA} = 1.35EI_0/b$; and $t = 0.94EI_0/b$. The end-moments due to a load of $4P$ at E with ends A and B encastré are: $M_{AB} = 1.16Pb$ and $M_{BA} = -1.67Pb$.

The moment distribution with sidesway prevented is done in the top part of Fig. 12-6c. The horizontal force at the level of BC which is sufficient to prevent sidesway is:

$$F = 4P \times \frac{3}{4} - \frac{(16.1 - 10.3 - 2.1 - 1.5)Pb/10}{4b} = 2.945P \quad \text{(to the right)}$$

In the second stage, a force of $F/2$ to the left is applied to the frame of Fig. 12-6b. The adjusted end-rotational stiffness for BA when translation of B is free is calculated by Eq. 11-19

$$S_{BA}^* = \frac{S_{BA}S_{AB} - t^2}{S_{BA} + S_{AB} + 2t} = 0.574EI_0/b$$

The adjusted end-rotational stiffness S_{BC}^* for antisymmetrical loading condition is (from Eq. 11-18)

$$S_{BC}^* = \frac{6EI}{l} = \frac{6 \times 4}{6b}EI_0 = 4.0EI_0/b$$

No carryover is made from B to C, and the COF $C_{BA}^* = -1$. The moment distribution for the left-hand half of the frame is carried out in Fig. 12-6c, and equal corresponding end-moments are written for the right-hand half of the frame.

The adjusted FEM's are calculated by Eqs. 11-24 and 11-25:

$$M_{BA}^* = F_B l \frac{(S_{BA} + t)}{(S_{BA} + S_{AB} + 2t)} = 1.4725P(4b)\frac{(1.35 + 0.94)}{(1.35 + 3.53 + 2 \times 0.94)}$$

$$= 1.99Pb$$

and

$$M_{AB}^* = F_B l \frac{(S_{AB} + t)}{(S_{BA} + S_{AB} + 2t)} = 1.4725P(4b)\frac{(3.53 + 0.94)}{(1.35 + 3.53 + 2 \times 0.94)}$$

$$= 3.90Pb$$

Moment distribution for these is carried out. The final moments are those due to both cases added together as shown in Fig. 12-6d.

Example 12-6 The *Vierendeel girder* in Fig. 12-7a has top and bottom chords of equal stiffness and carries vertical loads as shown. If the axial deformations are neglected, the actual loading can be replaced by the loading in Fig. 12-7b. Furthermore, if the horizontal translation of EH is prevented,

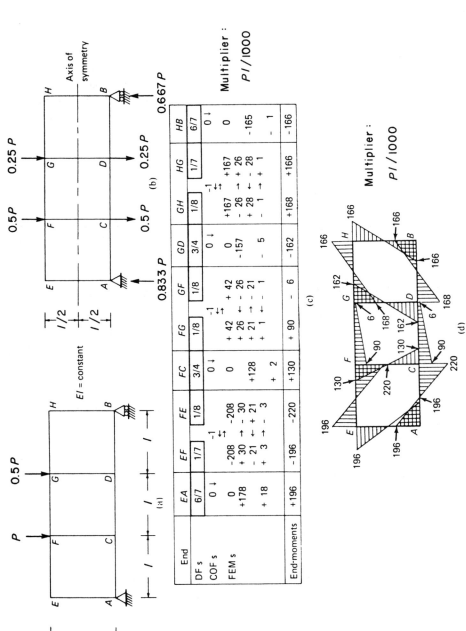

Fig. 12-7. Vierendeel girder analyzed in Example 12-6. (a) Frame properties and loading. (b) Equivalent antisymmetrical loading. (c) Moment distribution. (d) Bending-moment diagram.

the frame becomes antisymmetrical about the horizontal line joining the center point of the verticals and can be analyzed by the no-shear moment distribution procedure in the same way as the multistorey frame in Example 12-4. The effect of the actual translation of EH can be allowed for in an additional step in the calculations, but this has little effect with the type of structure and loading considered, and is ignored in this example.

The FEM's at any of the four ends of the top chord is derived from the shear:

$$FEM = -\frac{Vl}{4}$$

where l is the panel length, and V is the shear in the panel; this has the same value as the shear on a simple beam spanning between A and B and carrying the P and $P/2$ loads.

The moment distribution is carried out in Fig. 12-7c for the top half of the frame. The bending-moment diagram for the whole frame is given in Fig. 12-7d.

12-4 METHOD OF SUCCESSIVE SWAY CORRECTIONS

We recall from Sec. 12-2 in the general procedure of moment distribution for plane frames with joint translation, the sway displacements are determined from a set of simultaneous equations. Such a procedure may be convenient when there are only few degrees of freedom in sidesway but in other cases it may be preferable to deal with sway in a series of successive operations.

In the method of successive sway corrections, we allow the structure to sway under the effect of the external applied load, but the joint rotations are restrained. By moment distribution, the joints are then allowed to rotate with further sway restrained. This requires restraining forces along the sidesway coordinates. Sway is again permitted without joint rotation until the restraining forces vanish. This is referred to as sway correction. The process is repeated in cycles: moment distribution, carryover, and sway correction, until the restraining forces become small. There are no simultaneous equations to be solved, and the final stage in the general method— superposition of end-moments—is avoided.

As an illustration of the method, let us consider the structure in Fig. 12-8a which shows an intermediate storey in a multibay plane frame subjected to horizontal loads at floor levels. Assume that the frame is allowed to sway without joint rotation. Neglecting the axial deformation, the top ends of all the columns in one storey translate relative to their bottom ends by the same amount D. This sway induces end-moments and shearing forces in all the columns. For equilibrium, the sum of the shearing forces in the columns of one storey must be equal to the sum P of the horizontal forces acting on all the higher floors; thus P and the column end-moments are related by

$$P = -\frac{1}{h}\Sigma M \qquad (12\text{-}3)$$

where the summation is for the end-moments M at top and bottom of all the columns in the given storey, and h is the floor height. This equation is sometimes referred to as the *moment-sway equation*.

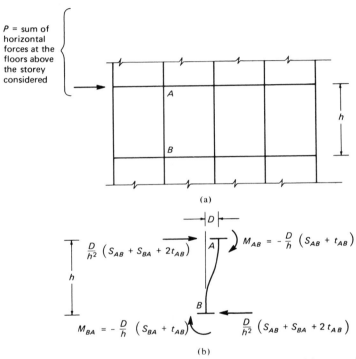

(a)

(b)

Fig. 12-8. Sway end moments. (a) Intermediate storey of a multistorey plane frame. (b) End-forces corresponding to sway without joint translation of a typical column AB.

The moments at the top and bottom ends of typical column AB can be expressed in terms of the relative displacement D (by Eq. 11-2); thus

$$M_{AB} = -\frac{D}{h}(S_{AB} + t_{AB}) \Big|$$

$$\qquad\qquad\qquad\qquad (12\text{-}4)$$

$$M_{BA} = -\frac{D}{h}(S_{BA} + t_{AB}) \Big|$$

where S_{AB} and S_{BA} are the end-rotational stiffnesses of the column AB, and t_{AB} is the carryover moment. Figure 12-8b shows the end-moments and the associated shearing forces. Substituting Eq. 12-4 in Eq. 12-3, we obtain

$$P = \frac{D}{h^2} \Sigma(S_{AB} + S_{BA} + 2t_{AB}) \tag{12-5}$$

in which the summation is for all the columns in the storey.

From Eq. 12-5 the relative translation of the top and bottom ends of any column in the storey is

$$D = P \frac{h^2}{\Sigma(S_{AB} + S_{BA} + 2t_{AB})} \tag{12-6}$$

Hence the sway end-moments at the top and bottom of a typical column AB are

$$M_{AB} = -Ph\left[\frac{S_{AB} + t_{AB}}{\Sigma(S_{AB} + S_{BA} + 2t_{AB})}\right]$$

and

$$M_{BA} = -Ph\left[\frac{S_{BA} + t_{AB}}{\Sigma(S_{AB} + S_{BA} + 2t_{AB})}\right] \tag{12-7}$$

If all the columns in a storey are prismatic, $S_{AB} = S_{BA} = 4EI_{AB}/h$, and $t_{AB} = 2EI_{AB}/h$, so that the end-moments in a typical column AB are

$$M_{AB} = M_{BA} = -Ph\left[\frac{EI_{AB}}{2\Sigma EI}\right] \tag{12-8}$$

where EI_{AB} is the flexural rigidity of column AB and ΣEI is the sum of the flexural rigidities of all the columns in the storey. The term in square brackets in Eqs. 12-7 and 12-8 is sometimes called *the sway moment-distribution factor*.

It may be convenient at this stage to summarize the steps in the calculation of moments by the method of successive sway corrections.

1. Calculate the sway end-moments with no joint rotations, using Eq. 12-7 or Eq. 12-8. The shearing forces corresponding to these end-moments in the columns of one storey are equal to the sum of the horizontal forces on higher floors.

2. Perform one cycle of moment distribution and carry over in the usual manner (without joint translation). The end-moments introduced in this cycle cause additional shear in the columns.

3. Allow new sway without joint rotation to restore the shear balance. The corresponding end-moments are obtained again by Eqs. 12-7 or 12-8, but with Ph replaced by $M(2)$ = the sum of the end-moments in the columns

of the storey introduced in the cycle of moment distribution and carry over of step 2. This sum is referred to as the *residual moment*.

4. The new zero-joint-rotation moments are again distributed as in step 2, and the routine is repeated until the residual moments become negligible.

Example 12-7 Find the end-moments in the members of the frame of Example 12-2 by the method of successive sway corrections (see Fig. 12-9a). (Since the structure is symmetrical, the method does not show the full extent of its merits.)

Because the columns in any one storey are prismatic and similar, the sway moment-distribution factors in Eq. 12-8 are all equal to 1/4. The sway initial end-moments are $-\dfrac{0.2 \times 1.2qb^2}{4} = -0.06qb^2$ for the columns of the top storey, and $-\dfrac{0.6 \times 1.6qb^2}{4} = -0.24qb^2$ for the columns of the bottom storey (Eq. 12-8). After the first moment distribution and carryover, the sum of the residual end-moments in the bottom storey (in terms of $qb^2/100$) is $(+36 + 18 - 18.9 - 9.4) = 25.7$ (see Fig. 12.9b).

The sway-correction moment at both ends of each column of this floor is $-25.7/4 = -6.4$. Similarly, the sway correction moment at the ends of the columns in the top floor is $-18.9/4 = -4.7$. The sway-correction moment after each moment distribution and carryover cycle is calculated in the same way. The calculation is terminated after the fourth moment distribution, and the end-moments are added to obtain the final end-moments in the frame. These moments (Fig. 12-9) can be compared with the results in Fig. 12-3d.

12-5 THE SUBSTITUTE-FRAME METHOD

This method, derived by Lightfoot,[2] permits the application of the no-shear moment distribution—which was originally devised for symmetrical single-bay frames subjected to antisymmetrical loading—to one-bay frames which have geometrical symmetry but are unsymmetrical in the flexural rigidities of members (e.g. the frame in Fig. 12-10a). The procedure is to replace the unsymmetrical structure by a substitute frame. This frame is composed of columns with a flexural rigidity in each storey equal to the sum of the flexural rigidities of the two columns in the original frame, and of beams of length equal to one-half the bay width and with a flexural rigidity equal to twice the beam rigidity in the original frame. The far ends of these beams are supported on rollers. Figure 12-10b illustrates

[2]E. Lightfoot, "Substitute Frames in the Analysis of Rigid-Jointed Structures," *Civil Engineering and Public Works Review*, London, Part I, Vol. 52, No. 618 (December 1957), pp. 1381–1383; Part II, Vol. 53, No. 619 (January 1958), pp. 70–72.

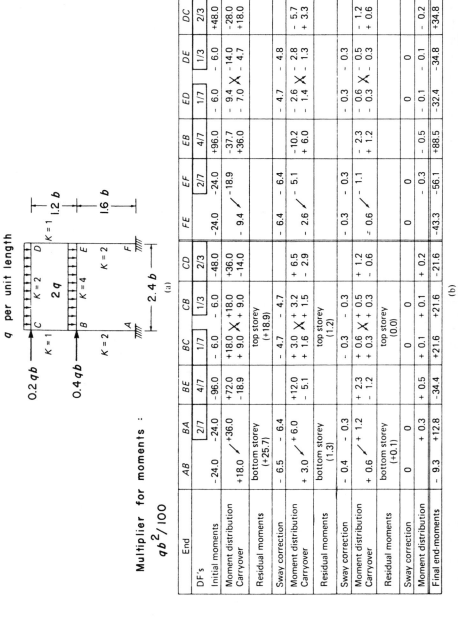

Fig. 12-9. Analysis of the frame in Example 12-7. (a) Frame properties and loading. (b) Calculation of end-moments.

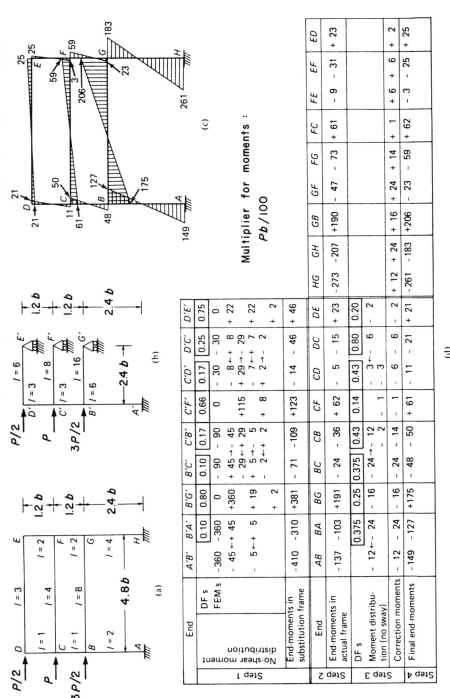

Multiplier for moments :

$Pb/100$

Step 1 — No-shear moment distribution (substitution frame)

End	A'B'	B'A'	B'G'	B'C'	C'B'	C'F'	C'D'	D'C'	D'E'
DF s		0.10	0.80	0.10	0.17	0.66	0.17	0.25	0.75
FEM s	−360	−360	0	−90	−90	0	−30	−30	0
	−45	+45	+360	+45	−45	+115	+8	−8	+22
	−5	+5	+19	−29	+29	+8	+7	−7	+22
		+2	+2	+5	−5		−2	+2	+2
				−2	+2				
End-moments in substitution frame	−410	−310	+381	−71	−109	+123	−14	−46	+46

Steps 2–4 (actual frame) — left

End	AB	BA	BG	BC	CB	CF	CD	DC	DE
Step 2 — End-moments in actual frame	−137	−103	+191	−24	−36	+62	−5	−15	+23
Step 3 — DF s	0.375	0.375	0.25	0.375	0.43	0.14	0.43	0.80	0.20
Step 3 — Moment distribution (no sway)	−12	−24	−16	−24	−12	−1	−3	−6	−2
			+2	−2	−2		−3		
Step 4 — Correction moments	−12	−24	−16	−24	−14	−1	−6	−6	−2
Step 4 — Final end-moments	−149	−127	+175	−48	−50	+61	−11	−21	+21

Steps 2–4 (actual frame) — right

End	HG	GH	GB	GF	FC	FE	FG	EF	ED
Step 2 — End-moments in actual frame	−273	−207	+190	−47	+61	−9	−73	−31	+23
Step 3 — Moment distribution (no sway)	+12	+24	+16	+24	+1	+6	+14	+6	+2
Step 4 — Correction moments	−261	−183	+206	−23	+62	−3	−59	−25	+25

(d)

Fig. 12-10. Analysis of the frame in Example 12-8. (a) Frame properties and loading. (b) Substitute frame. (c) Bending-moment diagram. (d) Moment distribution.

the derivation of a substitute frame from the original frame of Fig. 12-10a.

The no-shear moment distribution is applied to the substitute frame, and the final end-moments so obtained are apportioned between the column ends according to the actual flexural rigidities of the columns in the original frame. This procedure involves an acceptance of the sway of the substitute frame in the actual frame. The moment at the beam ends in the actual frame is taken as one-half of the end-moment in the corresponding beam of the substitute frame.

The apportionment of moments in this manner puts the moment at the joints of the actual frame out of balance and a further moment distribution is required. When this is carried out the balancing moments at the ends of the columns require a sway correction. For this, the same substitute frame may be used, with the resulting terminal moments apportioned to the ends of the actual frame; a new set of balancing moments is calculated, and the procedure is repeated until convergence has been reached.

For most practical structures, the convergence of this procedure is very rapid. The two first stages of this method — apportioning the end-moments of the substitute frame to the ends of the actual frame, and balancing the moments by moment distribution — give a sufficiently accurate result when the ratio of flexural rigidities of corresponding members does not exceed about 6; in such cases, further stages are usually unnecessary.

Example 12-8   Obtain the bending-moment diagram for the single-bay frame of Fig. 12-10a, which has unequal column flexural rigidities and is subjected to side loading at beam levels. All the members are prismatic.

The substitute frame is shown in Fig. 12-10b. Here the flexural rigidity EI of each column is equal to the sum of the flexural rigidities of the corresponding columns in the actual frame. The flexural rigidity of each beam in the substitute frame is double that of the corresponding beams in the actual frame, and the length of each beam is one-half the bay width. The adjusted relative end-rotational stiffnesses for no-shear moment distribution of the substitute frame are as follows:

$$S_{B'A'} = \frac{6}{24} = \frac{1}{4} \qquad S_{B'C'} = \frac{3}{12} = \frac{1}{4} \qquad S_{B'G'} = \frac{3 \times 16}{24} = 2$$

$$S_{C'B'} = \frac{3}{12} = \frac{1}{4} \qquad S_{C'D'} = \frac{3}{12} = \frac{1}{4} \qquad S_{C'F'} = \frac{3 \times 8}{24} = 1$$

$$S_{D'C'} = \frac{3}{12} = \frac{1}{4} \qquad S_{D'E'} = \frac{3 \times 6}{24} = \frac{3}{4}$$

The solution is presented in Fig. 12-10d and involves the following steps.

1. The initial moments (FEM's) on the substitute frame, corresponding to sway without rotation, are calculated. These moments are distributed in

proportion to the adjusted relative end-rotational stiffnesses calculated above, and the distributed moments are carried over with COF $= -1$. The moments are added together to get the end-moments in the substitute frame.

2. The end-moments in the substitute frame are apportioned to the columns according to their stiffness, and to the beam ends by halving, to obtain the first approximate solution.

3. The sum of the end-moments at each joint is now not equal to zero. Balancing moments (equal to minus the sum of the end-moments) are therefore required. They are distributed without joint translation. We should note that the balancing moments are equal and have opposite sign at the ends of each member. Since the moments to be distributed are small compared with the initial moments, it is reasonable to consider for this purpose that the frame is symmetrical with the column flexural rigidity equal to the average flexural rigidity in each storey. It is then sufficient to carry out the distribution for one-half of the frame. The DF's are calculated on the basis of the relative end-rotational stiffness of $4EI_{avge}/l$, for the columns and $2EI/l$ for the beams. (These are the adjusted end-rotational stiffnesses of a symmetrical prismatic member symmetrically loaded.)

4. The correction moments obtained in step 3 are added to the approximate moments calculated in step 2 to obtain the final end-moments.

The bending-moment diagram for the frame in Fig. 12-10a is shown in Fig. 12-10c. The end-moments in the columns can be checked by Eq. 12-3.

The above method is sufficiently accurate since the ratio of flexural rigidities of corresponding columns in the present problem is only 2. If, however, the flexural rigidities of the columns in a storey are very different, or if a more exact analysis is required, it may be necessary to carry out the moment distribution in step 3 on the actual asymmetrical structure without sway; this gives the first moment correction. The column moments will show that there is a force required at each floor level to prevent the sway during the distribution. Opposite and equal forces are then applied to the substitute frame. A repetition of the steps 1, 2, and 3 gives the second moment correction. As an alternative to this, after one cycle of moment distribution and carryover on the actual frame, we can apply the method of successive sway correction as the second and final correction stage.

12-6   THE PRINCIPLE OF MULTIPLE

Consider the frame in Fig. 12-11a with the values of $K(= I/l)$ shown. It is easy to see that the joint rotation and the column sways of the two single-bay frames in Fig. 12-11 are identical. The analysis of either of these frames will lead to the analysis of the entire structure. This division is, however, possible only because of the special relationship between the stiffnesses of the members and only for the special type of loading shown.

The general requirement is that the K pattern for the entire structure breaks down into a number of patterns $\alpha_1\{K_s\}$, $\alpha_2\{K_s\}$, ..., where $\{K_s\}$ is a pattern of K values for a single-bay frame, and α_1, α_2, ..., are constants for each subsidiary frame. The horizontal forces P are then apportioned so that any subsidiary frame i is loaded by a load $P(\alpha_i/\Sigma\alpha)$, and thus the end-moments produced in the frames are proportional to the α values.

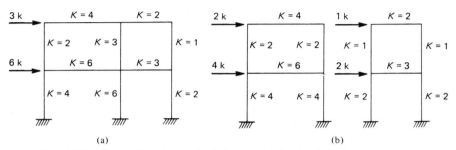

Fig. 12-11. Illustration of the principle of multiple; the three frames in parts
(a) and (b) have the same joint displacements and end-moments.

As an alternative to this, we can use a substitute frame resisting the total horizontal forces, with the end-moments obtained by the no-shear moment distribution. The adjusted relative end-rotational stiffness of the columns of the substitute frame is $\Sigma I_{col}/l_{col} = \Sigma K_{col}$, where the summation is for all the columns in a storey of the actual frame. The adjusted relative end-rotational stiffness of the beams of the substitute frame is $12\Sigma(I/l)_{beam} = 12\Sigma K_{beam}$, where the summation is for all the beams in one storey. When the end-moments in the substitute frame have been obtained, they are apportioned to the columns and beam ends according to their flexural rigidities.

Example 12-9   Obtain the bending-moment diagram for the three-bay frame of Fig. 12-12a, using the principle of multiple.

The substitute frame is shown in Fig. 12-13b and a no-shear moment distribution is carried out in Fig. 12-12c. The adjusted relative end-rotational stiffnesses are as follows:

$S_{BA} = 0$, which means that for the member AB with end A hinged and
$\qquad$ B free to sway, no moment is required to produce rotation at B

$S_{BC} = S_{CB} = \Sigma K_{col} = 2 + 4 + 4 + 2 = 12$

$S_{BE} = 12\Sigma K_{beam} = 12(4 + 4 + 4) = 144$

$S_{CD} = 12\Sigma K_{beam} = 12(3 + 3 + 3) = 108$

The moments at the ends of the columns of the substitute frame are apportioned to the ends of the columns of the actual frame in proportion to the K values. The moments at the beam ends are halved and then apportioned to the beams at any level in proportion to their K values. The final bending-moment diagram for the actual frame is shown in Fig. 12-12d.

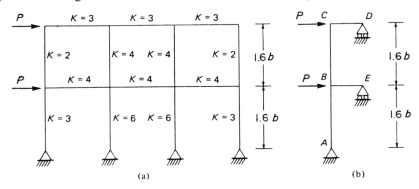

(a)   (b)

End	AB	BA	BE	BC	CB	CD
DF s		0	0.923	0.077	0.10	0.90
FEM s	0	-320		- 80	- 80	+ 72
				- 8 ← + 8		
			+377	+ 31 → - 31		+ 28
				- 3 ← + 3		
			+ 3			
End-moments		-320	+380	- 60	- 100	+100

Multiplier : $Pb/100$

(c)

Multiplier : $Pb/100$

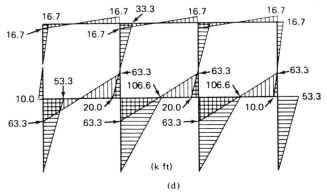

(k ft)

(d)

Fig. 12-12. Frame analyzed in Example 12-9. (a) Frame properties and loading. (b) Substitute frame. (c) No-shear moment distribution for the substitute frame. (d) Bending-moment diagram.

The procedure followed in the above example leads to end-moments which balance at each joint only in the case of frames to which the principle of multiple applies. However, the procedure can be supplemented for use in a general case, as shown in the following section.

12-7 USE OF THE SUBSTITUTE FRAME WITH ANY STIFFNESS PATTERN

In practice, structures usually have members with flexural rigidities which do not satisfy the condition necessary for the application of the principle of multiple. In such a case, the substitute frame may still be used: the end-moments are apportioned to the members according to their flexural rigidities, but the moments will not balance at the joints, even though the moment-sway Eq. 12-3 is satisified in each storey. If the unbalanced moments are calculated and summed for the joints at any floor, we find that each total equals zero. A single or double cycle of joint balance and carryover can then be performed on the actual frame, with no sidesway. The out-of-balance moments in the moment sway Eq. 12-3 are then removed by means of a second use of the substitute frame and moment apportionment. As an alternative to this last step, successive sway corrections can be used.

Example 12-10 Find the bending-moment diagram for the frame of Fig. 12-13a, to which the principle of multiple does not apply.

The relative end-rotational stiffnesses for the members of the substitute frame in Fig. 12-13a are:

$$S_{B'A'} = 6 + 6 + 6 = 18 \qquad S_{B'E'} = 12(6 + 6) = 144$$
$$S_{B'C'} = S_{C'B'} = 4 + 4 + 4 = 12 \qquad S_{C'D'} = 12(3 + 3) = 72$$

No-shear moment distribution is performed in Fig. 12-13c for the substitute frame, and in Fig. 12-13d the end-moments are apportioned to the members of the actual frame. One cycle of moment distribution and carryover is performed, followed by one sway correction. The calculations are completed by a moment distribution. The moments are added then to get the final end-moments in the actual frame. The resulting bending-moment diagram is shown in Fig. 12-13e.

12-8 GENERAL

Analysis of frames subject to sidesway due to lack of symmetry in the structure or in the loading, or in both, is seen to be more complicated than in frames considered in Chapter 11.

The no-shear moment distribution offers a quick and easy solution for one-bay symmetrical frames under antisymmetrical loading.

The method of successive sway corrections, which can be used for any rectilinear plane frame does not require the solution of simultaneous equations to determine the sway values, and avoids the last step in the general displacement method, that is the superposition of the effects of sway at different floor levels.

The no-shear moment distribution can also be extended to multibay frames by the use of a substitute frame and followed by apportionment of

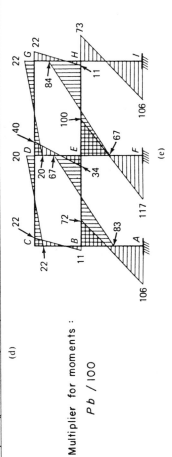

(a) Frame properties and loading

(b) Substitute frame

(c) Calculations of end-moments in substitute frame

End	A'B'	B'A'	B'E'	B'C'	C'B'	C'D'
DF's		0.10	0.83	0.07	0.14	0.86
Initial moments	-288	-288		-72	-72	
Moment distribution and carryover	-38 ← +38		+306 +4	-10 ↓ +26 ↑ -4 ↓	+10 → -26 ← +4 ←	+62 +22
End-moments in substitute frame	-326	-250	+310	-60	-84	+84

(d) Calculation of end-moments in actual frame

	AB	BA	BE	BC	CB	CD	FE	EF	EB	EH	ED	DE	DC	DG	IH	HI	HE	HG	GH	GD
DF's		3/8	3/8	1/4	4/7	3/7		3/11	3/11	3/11	2/11	2/5	3/10	3/10		3/8	3/8	1/4	4/7	3/7
Apportioned moments	-109	-83	+78	-20	-28	+21	-108	-84	+77	+77	-20	-28	+21	+21	-109	-83	+78	-20	-28	+21
Distribution Carryover	+5	+10	+9 -7	+6 +2	+4 +3	+3 -2	-7	-13	-14 +5	-14 +5	-9 -3	-6 -5	-4 +2	-4 +2	+5	+10	+10 -7	+6 +2	+4 +3	+3 -2
Sway correction	-2	-1	+3	-1	-1	0	-2	-2	-1	-1	-2	-1	+1	+1	-2	-2	+3	-1	-1	0
Distribution		+2		+2				-1								+2		+2		
Final moments	-106	-72	+83	-11	-22	+22	-117	-100	+67	+67	-34	-40	+20	+20	-106	-73	+84	-11	-22	+22

(e) Bending-moment diagram

Multiplier for moments:

$$Pb/100$$

Fig. 12-13. Frame analyzed in Example 12-10. (a) Frame properties and loading. (b) Substitute frame. (c) Calculations of end-moments in substitute frame. (d) Calculation of end-moments in actual frame. (e) Bending-moment diagram.

moments. This may be followed by one step of sway correction to improve the results. This way large frames can be analyzed rapidly, using a slide rule (particularly when they satisfy or nearly satisfy the principle of multiple). The use of a computer solution is therefore hardly worthwhile. We must remember, however, that these techniques are limited to certain types of plane frames and to certain patterns of loading.

Whenever any of the special methods considered in this chapter are used, the designer must ascertain that the structure deforms in the manner on which these techniques are based. For instance, a case to which these methods cannot be applied is a multibay frame in which the horizontal members at some floor levels are omitted.[3]

PROBLEMS

The following problems are for Chapters 11 and 12.

12-1   Find the joint displacements of the structure shown and then find the end moments in member *BC*. Use either the slope-deflection method given in Chapter 11 or the general displacement method discussed in Chapter 3. Consider bending deformations only.

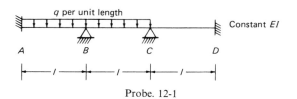

Probe. 12-1

12-2   Apply the requirements of Prob. 12-1 to the beam shown.

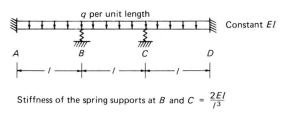

Stiffness of the spring supports at *B* and *C* = $\frac{2EI}{l^3}$

Prob. 12-2

[3]For other special methods using moment distribution in the analysis of plane frames, refer to E. Lightfoot, *Moment Distribution*, Spon Ltd., London 1961, Chaps. 5 and 6, which include lists of other references.

12-3   Apply the requirements of Prob. 12-1 to the frame shown. (*Hint:* Because of symmetry, the rotation at C is zero.)

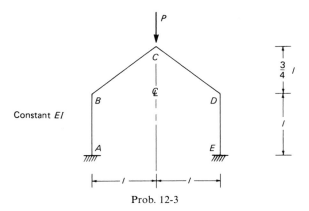

Prob. 12-3

12-4   Apply the requirements of Prob. 12-1 to the structure shown. (*Hint:* Because of symmetry no displacement occurs at D. Therefore, it is sufficient to solve the frame $ABCD$ with end D encastré.)

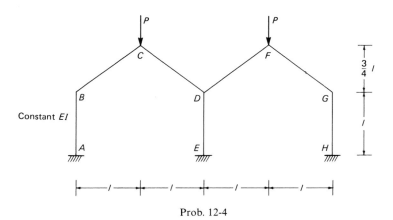

Prob. 12-4

12-5   Use the slope-deflection or the general displacement method to write the equations from which the displacement at the joints B, C, E, and F can be calculated. Use the answers given at the end of the book to check your equations.

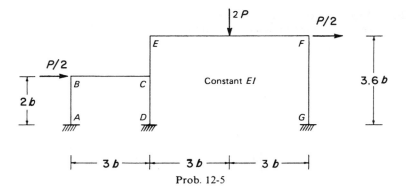

Prob. 12-5

12-6   Find the bending-moment diagram for the frame shown in the figure. Find the three reaction components at *A*.

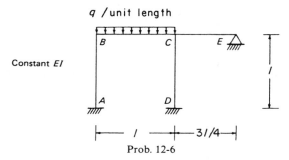

Prob. 12-6

12-7   Solve Prob. 12-6 with the end *E* supported on a roller which allows horizontal translation instead of the hinged support shown in the figure.

12-8   Draw the bending-moment diagram for the bridge frame in the figure due to a uniform load of *q* per unit length on *AD*. The end-rotational stiffnesses, the carryover moment and the FEM's for the members of variable *I* are given below the figure.

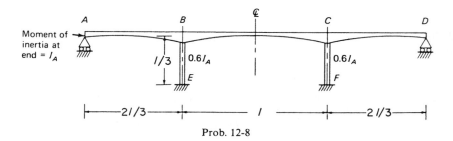

Prob. 12-8

The end-rotational stiffnesses and carryover moments are as follows:

Member AB: $S_{AB} = 5.4 \dfrac{EI_A}{l_{AB}}$, $S_{BA} = 14.6 \dfrac{EI_A}{l_{AB}}$, and $t_{AB} = 4.9 \dfrac{EI_A}{l_{AB}}$

Member BC: $S_{BC} = S_{CB} = 12.0 \dfrac{EI_A}{l_{BC}}$, and $t_{BC} = 8.3 \dfrac{EI_A}{l_{BC}}$

The FEMs due to a uniform load of intensity w:

$$M_{AB} = -0.057 \, wl_{AB}^2, \quad M_{BA} = 0.130 \, wl_{AB}^2$$
$$M_{CB} = -M_{BC} = 0.103 \, wl_{BC}^2$$

12-9 For the bridge frame of Prob. 12-8, find the bending-moment diagram due to a horizontal force H to the left, along the axis of AD.

12-10 For the bridge frame of Prob. 12.8, find the bending-moment diagram due to a unit settlement at F. Give the ordinates of the bending-moment diagram in terms of EI_A.

12-11 The beam in the figure is subjected to a dead load and a live load of intensities q and $2q$ per unit length, respectively. Draw the curve of maximum moment. *Hint:* Solve for the following 4 cases of loading:
(i)   D.L. on AD with L.L. on AB
(ii) D.L. on AD with L.L. on AC
(iii) D.L. on AD with L.L. on BC
(iv) D.L. on AD with L.L. on CD.
Draw the four bending-moment diagrams in one figure to the same scale. The curve which has the maximum ordinates at any portion of the beam is the required curve.

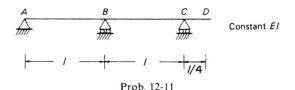

Prob. 12-11

12-12 Find the bending-moment diagram for the frame shown in the figure.

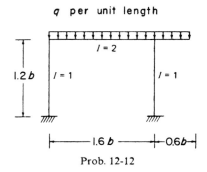

Prob. 12-12

12-13   For the frame shown in the figure:
   (a) Draw the bending-moment diagram due to a distributed vertical load
       1 k/ft of horizontal projection,
   (b) Find the reactions due to a rise of temperature of 40°F with no loading
       on the frame.

$$EI = 2 \times 10^6 \, ft^2 \, k$$

Coefficient of thermal expansion $\alpha = 0.6 \times 10^{-5}$ per °F.

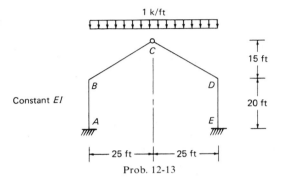

Prob. 12-13

12-14   Part (a) of the figure shows a bridge frame subjected to a horizontal force
   P at the deck level. Assuming that the deck has an infinite rigidity,
   (a) Find the bending-moment diagram in the piers.
   (b) If hinges are introduced below the deck at the top of the piers BF and DH,
       as shown in part (b), find the shearing force at the bottom of the three
       piers.

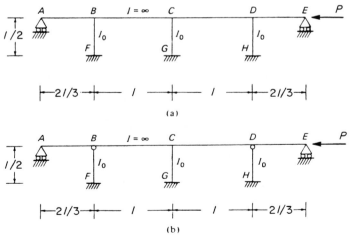

Prob. 12-14

12-15   Find the bending-moment diagram for the frame shown in the figure.

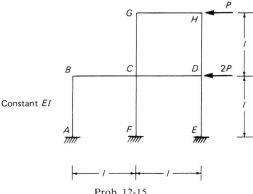

Prob. 12-15

12-16   The figure shows one of the concrete frames supporting an elevated water tank. The wind pressure causes the shown loads on the frame. Assuming that the tank elements are of infinite rigidity compared to the rigidity of the horizontal bracing beams, find the bending moment in the frame.

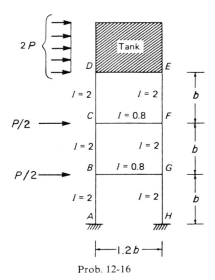

Prob. 12-16

12-17   Find the bending-moment diagram for the Vierendeel girder shown in the figure.

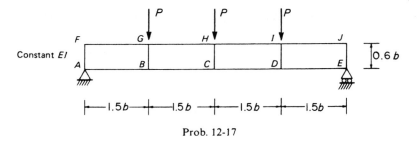

Prob. 12-17

12-18   Use the no-shear moment distribution method to find the bending-moment diagram in the frame shown.

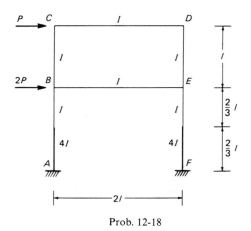

Prob. 12-18

12-19   Solve Example 12-8 by the method of successive sway correction.

12-20   Find the bending-moment diagram in the Vierendeel girder shown, which has unequal chord flexural rigidities. The relative moments of inertia are indicated alongside the members.

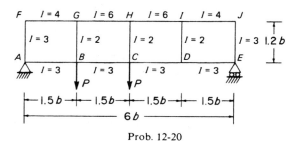

Prob. 12-20

12-21 Find the bending-moment diagram in the building frame shown. The $K = (I/l)$ value for each member is given in the figure.

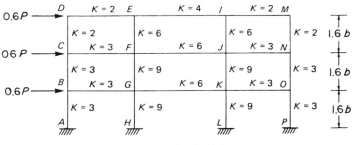

Prob. 12-21

12-22 Find the bending moment in the frame of Prob. 12-21, changing the $I/l-$ values of the two internal columns to make them the same as the external columns.

12-23 Find the bending-moment diagram for the frame shown. The $K = (I/l)$ values are indicated alongside the members.

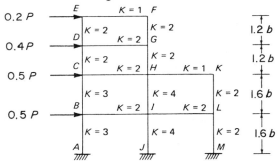

Prob. 12-23

12-24 Find the bending moment in the frame shown in the figure for which all the members are prismatic. The relative flexural rigidity is 1 for all columns and 4 for all beams.

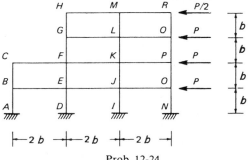

Prob. 12-24

CHAPTER 13

Influence lines for beams, frames, and grids

13-1 INTRODUCTION

The effect of live loads which can have different positions on a structure can be conveniently analyzed and succintly described in graphical form by the use of influence lines. An influence line shows the value of any action due to a unit point load moving across the structure. For example, the influence line for the bending moment at a section of a continuous beam shows the variation in the bending moment at this section as a unit transverse load traverses the beam.

In this chapter we deal with the methods of obtaining influence lines for statically indeterminate structures but, by way of introduction and review, influence lines for statically determinate structures will be first briefly discussed. The next chapter deals with further development of influence lines.

13-2 CONCEPT AND APPLICATION OF INFLUENCE LINES

A transverse concentrated load at a general position on a member of a structure causes various actions. These actions, which may be a bending moment, shearing force, thrust, or displacement at a section, or a reaction at a support, vary as the load moves across the structure.

If the values of any action A are plotted as ordinates at all the points of application of a unit transverse load, we obtain the influence line of the action A. In this chapter we use η to represent the influence ordinate — which may also be referred to as *influence coefficient* — of any action due to a unit moving concentrated load acting at right angles to the member over which the load is moving. The effects of other types of loading will be distinguished by appropriate subscripts (Sec. 14-5 onward). Our sign convention is to plot positive influence ordinates in the same direction as the applied concentrated load. Thus influence lines for gravity loads on horizontal members are drawn positive downwards.

Let us now illustrate the use of influence lines in analysis. The value of any action A due to a system of concentrated loads $P_1, P_2, ..., P_n$ (Fig. 13-1a)

352

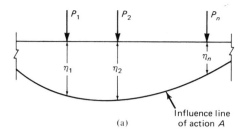

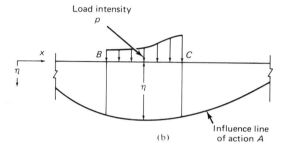

Fig. 13-1. Determination of the value of an action due to loading using the influence line.

can be obtained from the influence ordinates by

$$A = \eta_1 P_1 + \eta_2 P_2 + \ldots + \eta_n P_n \tag{13-1}$$

or

$$A = \sum_{i=1}^{n} \eta_i P_i \tag{13-1a}$$

The value of the action A due to a distributed transverse load of intensity p over a length $\overline{BC}$ (Fig. 13-1b) is

$$A = \int_B^C \eta p \, dx \tag{13-2}$$

For a uniform load of intensity q,

$$A = q \int_B^C \eta \, dx \tag{13-3}$$

The value of the integral in this equation is the area under the influence line between B and C.

The knowledge of the shape of an influence line indicates which part or parts of a structure should be loaded to obtain maximum effects. In Fig. 13-4, influence lines are sketched for a plane frame, and in Fig. 13-5 for a grid. The ordinates plotted on the column EB in Fig. 13-4 represent the value of the action considered due to a horizontal load on the column. As always, the value is positive if the load is applied in the direction of the positive ordinate. The ordinates of the influence lines for the grid are vertical to represent the effect of a unit vertical load, as shown in the pictorial view in Fig. 13-5.

We can see that, for instance, in the case of shear at section n in the frame of Fig. 13-4, a maximum negative value occurs when a distributed load covers Bn as well as the span CD, without a load on the remainder of the frame. Likewise, the bending moment at n_3 in Fig. 13-5 is maximum positive when loads cover the members CD and the central part of GI, without a load on AB, or EF.

13-3 MÜLLER-BRESLAU'S PRINCIPLE

One of the most effective methods of obtaining influence lines is by the use of Müller-Breslau's principle, which states that the ordinates of the influence line for any action in a structure are equal to those of the deflection curve obtained by releasing the restraint corresponding to this action and introducing a corresponding unit displacement in the remaining structure. The principle is applicable to any structure, statically determinate or indeterminate, and can be easily proved, using Betti's law.

Consider a loaded beam in equilibrium, as in Fig. 13-2a. Remove the support B and replace its effect by the corresponding reaction R_B, as shown in Fig. 13-2b. If the structure is now subjected to a downward load F at B such that the deflection at B equals unity, the beam will assume the deflected form in Fig. 13-2c. Because the original structure is statically determinate, the release of one restraining force turns the structure into a mechanism, and therefore the force F required to produce the displacements in Fig. 13-2c is zero. However, the release of one restraining force in a statically indeterminate structure leaves a stable structure so that the value of the force F is generally not equal to zero.

Applying Betti's law (Eq. 8-5) to the two systems of forces in Fig. 13-2b and c, we write

$$\eta_1 P_1 + \eta_2 P_2 + \ldots + \eta_n P_n - 1 \times R_B = F \times 0$$

This equation expresses the fact that the external virtual work done by the system of forces in Fig. 13-2b during the displacement by the system in Fig. 13-2c is the same as the external virtual work done by the system in Fig. 13-2c during the displacement by the system in Fig. 13-2b. This latter quantity must be zero because no deflection occurs at B in Fig. 13-2b.

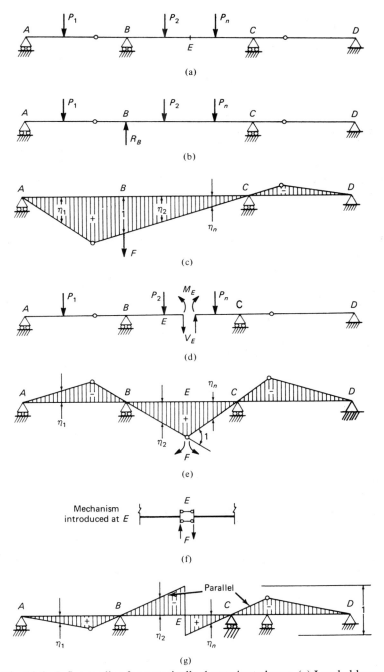

Fig. 13-2. Influence line for a statically determinate beam. (a) Loaded beam in equilibrium. (b) Support B replaced by R_B. (c) Influence line for R_B. (d) Equilibrium maintained by forces M_E. (e) Influence line for M_E. (g) Influence line for V_E.

The preceding equation can be written

$$R_B = \sum_{i=1}^{n} \eta_i P_i$$

Comparing this equation with Eq. 13-1a, we see that the deflection line in Fig. 13-2c is the influence line of the reaction R_B. This shows that the influence line of the reaction R_B can be obtained by releasing its effect, that is removing the support B, and introducing a unit displacement at B in the downward direction, that is opposite to the positive direction of the reaction.

Using simple statics, we can readily check that the deflection ordinate at any point in Fig. 13-2c is, in fact, equal to the reaction R_B if a unit load is applied at this point in the beam of Fig. 13-2a.

Let us now use Müller-Breslau's principle in the case of the influence line of the bending moment at any section E. We introduce a hinge at E, thus releasing the bending moment at this section. We then apply two equal and opposite couples F to produce a unit relative rotation of the beam ends at E (Fig. 13-2e). In order to prove that the deflection line in this case is the influence line of the bending moment at E, cut the beam in Fig. 13-2a at section E and introduce two pairs of equal and opposite forces M_E and V_E to maintain the equilibrium (Fig. 13-2d). Applying Betti's law to the systems in Figs. 13-2d and 13-2e, we can write

$$\eta_1 P_1 + \eta_2 P_2 + \ldots + \eta_n P_n - 1 \times M_E = F \times 0$$

or

$$M_E = \sum_{i=1}^{n} \eta_i P_i$$

This demonstrates that the deflection line in Fig. 13-2e is the influence line for the bending moment at E.

The influence line for shear at section E can be obtained by introducing a unit relative translation without relative rotation of the two beam ends at E (Fig. 13-2g). This is achieved by introducing at E a fictitious mechanism such as that shown in Fig. 13-2f and then applying two equal and opposite vertical forces F. With this mechanism the two ends at E remain parallel as shown in Fig. 13-2g. Applying Betti's law to the systems in Figs. 13-2d and 13-2g, we can write

$$\eta_1 P_1 + \eta_2 P_2 + \ldots + \eta_n P_n - 1 \times V_E = F \times 0$$

or

$$V_E = \sum_{i=1}^{n} \eta_i P_i$$

which shows that the deflection line in Fig. 13-2f is the influence line for the shear at E.

All the influence lines considered so far are composed of straight-line segments. This is the case for any influence line in any statically determinate structure. Thus, one computed ordinate and the known shape of the influence line are sufficient to draw it. This ordinate may be calculated from considerations of statics, or from the geometry of the influence line.

All influence lines for statically indeterminate structures are composed of curves, and therefore several ordinates must be computed. In Fig. 13-3, Müller-Breslau's principle is used to obtain the general shape of the influence lines for a reaction, bending moment, and shear at a section in a continuous beam. Sketches of influence lines for several actions in a plane frame and in a grid are deduced by Müller-Breslau's principle in Figs. 13-4 and 13-5.

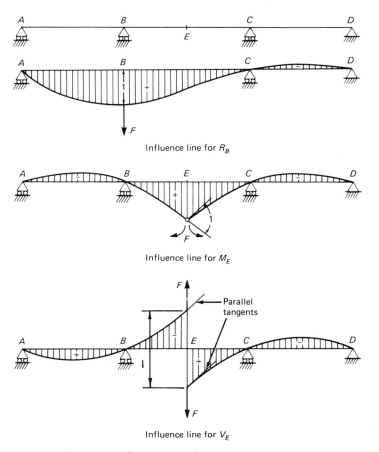

Fig. 13-3. Influence lines for a continuous beam.

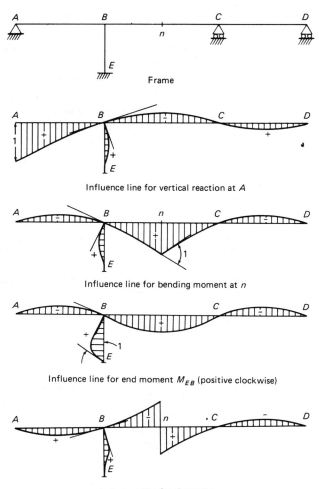

Influence line for vertical reaction at *A*

Influence line for bending moment at *n*

Influence line for end moment M_{EB} (positive clockwise)

Influence line for shear at *n*

Fig. 13-4. Shape of influence lines for a plane frame using Müller-Breslau's principle.

13-31 Procedure for Obtaining Influence Lines

The steps followed in Sec. 13-3 to obtain the influence line for any action can be summarized as follows.

1. The structure is released by removal of the restraint corresponding to the action considered. The degree of indeterminacy of the released structure compared with the original structure is reduced by one. It follows that, if the original structure is statically determinate, the released structure is a mechanism.

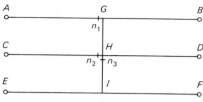

Ends A, C, E, B, D and F are simply supported

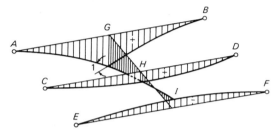

Influence line for bending moment at n_1

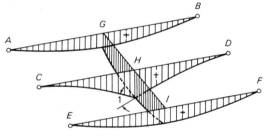

Influence line for bending moment at n_2

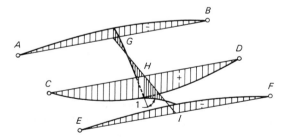

Influence line for bending moment at n_3

Fig. 13-5. Shape of influence lines for a grid using Müller-Breslau's principle.

2. Introduce a unit displacement in the released structure in a direction opposite to the positive direction of the action. This is achieved by applying a force (or a pair of equal and opposite forces) corresponding to the action.

3. The ordinates of the deflection line thus obtained are the influence ordinates of the action. The ordinates of the influence line are positive if they are in the same direction as the external applied load.

13-4 CORRECTION FOR INDIRECT LOADING

In some cases loads are not applied directly to the structure for which the influence lines are desired, but through smaller beams assumed to be simply supported on the main structure. For instance, the main girder in Fig. 13-6a supports cross-girders at nodes A, 1, 2, 3, and B, and these in turn

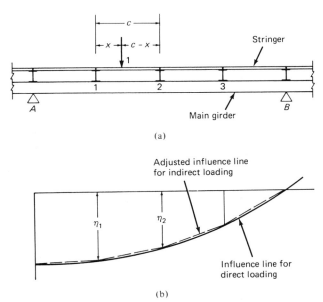

Fig. 13-6. Correction of influence lines for indirect loading. (a) Indirect loading on main girder. (b) Influence line for any action A in main girder.

carry stringers to which the live load is applied. Let the solid curve in Fig. 13-6b represent the influence line of any action A, drawn on the assumption that the unit load is applied directly to the beam. But the unit load can be transmitted to the main girder at the cross-girders only, and we have to correct the influence line accordingly.

A unit load applied at an arbitrary point between nodes 1 and 2 is transmitted to the main girder as two concentrated loads equal to $(c - x)/c$

and x/c at 1 and 2 respectively. The value of the action A due to these two loads is

$$A = \frac{x}{c}\eta_2 + \frac{c-x}{c}\eta_1 \qquad (13\text{-}4)$$

where c is the panel length and x is the distance indicated in Fig. 13-6a. This is the equation of the straight line between points 1 and 2 shown dotted in Fig. 13-6b. Thus the corrected influence line is composed of straight segments between the node points.

In pin-connected trusses, all the loads are assumed to act at the joints; thus, influence lines for trusses are composed of straight segments between the joints.

13-5   INFLUENCE LINES FOR A BEAM WITH FIXED ENDS

Let us now use Müller-Breslau's principle to find the influence lines for the end-moments of a beam with fixed ends. From these, by equations of statics, influence lines for reaction, shear, and bending moments at any section can be determined. We use — as in previous chapters — the convention that a clockwise end moment is positive.

To find the influence line for the end-moment M_{AB} in the beam in Fig. 13-7a, we introduce a hinge at A and apply there an anticlockwise moment to produce a unit angular rotation of the end A (Fig. 13-7b). This moment must be equal in magnitude to the end-rotational stiffness S_{AB}. The corrsponding end-moment at B is $t = C_{AB}S_{AB}$, where S_{AB}, C_{AB}, and t are the end-rotational stiffness, the carryover factor and the carryover moment respectively. The deflection line corresponding to the bending-moment diagram in Fig. 13-7c is the required influence line.

When the beam has a constant flexural rigidity EI and length l, the end-moments at A and B are respectively $-4EI/l$ and $-2EI/l$. These values can be substituted in the expression for the deflection[1] y in a prismatic member AB due to clockwise end-moments M_{AB} and M_{BA}

$$y = \frac{l^2}{6EI}\left[M_{AB}(2\varepsilon - 3\varepsilon^2 + \varepsilon^3) - M_{BA}(\varepsilon - \varepsilon^3)\right] \qquad (13\text{-}5)$$

where $\varepsilon = x/l$, x is the distance from the left-hand end A, and l is the length of the member. Equation 13-5 can be easily proved by the method of elastic weights (see Sec. 9-5). The values of y due to unit end moments are given in Appendix I.

Superposition of the deflections caused by an end-moment of $-4EI/l$ at A (with zero moment at B) and of the deflections caused by an end-moment

[1]When the beam ends A and B deflect, Eq. 13-5 gives the deflection measured from the straight line joining the displaced positions of A and B.

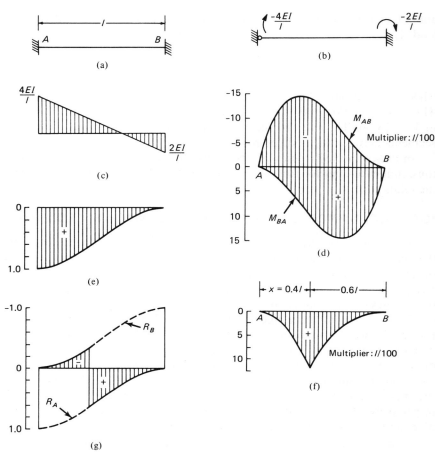

Fig. 13-7. Influence lines for a prismatic beam with fixed ends. (a) Beam. (b) End-moments corresponding to a unit angular rotation at end A. (c) Bending-moment diagram for the beam in part (b). (d) Influence lines for end-moments. (e) Influence line for R_A. (f) Influence line for $M_{x=0.4l}$. (g) Influence line for $V_{(x=0.4l)}$.

of $-2EI/l$ at B (with zero moment at A) gives the required influence line. This is conveniently done in Table 13-1.

Because the beam is symmetrical, the influence ordinates of the end-moment M_{BA} can be obtained from those for M_{AB} by reversing the sign and the order (Table 13-2).

The influence lines of the two end-moments are plotted in Fig. 13-7d. The reaction R_A may be expressed as

$$R_A = R_{As} - \frac{M_{AB} + M_{BA}}{l}$$

Table 13-1. Calculation of Ordinates of the Influence Line for the End-Moment* M_{AB}

Distance from left-hand end	0.1l	0.2l	0.3l	0.4l	0.5l	0.6l	0.7l	0.8l	0.9l	Multi-plier
Deflection due to end-moment at A of $-4EI/l$	-0.114	-0.192	-0.238	-0.256	-0.250	-0.224	-0.182	-0.128	-0.066	l
Deflection due to end-moment at B of $-2EI/l$	0.033	0.064	0.091	0.112	0.125	0.128	0.119	0.096	0.057	l
Influence ordinate for M_{AB}	-0.081	-0.128	-0.147	-0.144	-0.125	-0.096	-0.063	-0.032	-0.009	l

Table 13-2. Ordinates of the Influence Line for the End-Moment* M_{BA}

Distance from left-hand end	0.1l	0.2l	0.3l	0.4l	0.5l	0.6l	0.7l	0.8l	0.9l	Multi-plier
Influence ordinates for M_{BA}	0.009	0.032	0.063	0.096	0.125	0.144	0.147	0.128	0.081	l

*The equation to the influence lines of the two end-moments are

$$M_{AB} = -\frac{x(l-x)^2}{l^2} \quad \text{and} \quad M_{BA} = \frac{x^2(l-x)}{l^2}$$

where x is the distance from the left-hand end A.

where R_{As} is the statically determinate reaction of the beam AB if simply supported. This equation is valid for any position of a unit moving load. We can therefore write,

$$\eta_{RA} = \eta_{RAs} - \frac{1}{l}(\eta_{MAB} + \eta_{MBA}) \tag{13-6}$$

where η is the influence ordinate of the action indicated by the subscript. The influence line of R_{As} is a straight line with ordinate 1 at A and zero at

Table 13-3. Ordinates of the Influence Line for R_A

Distance from left-hand end	0	0.1l	0.2l	0.3l	0.4l	0.5l	0.6l	0.7l	0.8l	0.9l	l
η_{RAs}	1.000	0.900	0.800	0.700	0.600	0.500	0.400	0.300	0.200	0.100	0
$-\eta_{MAB}/l$	0	0.081	0.128	0.147	0.144	0.125	0.096	0.063	0.032	0.009	0
$-\eta_{MBA}/l$	0	-0.009	-0.032	-0.063	-0.096	-0.125	-0.144	-0.147	-0.128	-0.081	0
Influence ordinate for R_A	1.000	0.972	0.896	0.784	0.648	0.500	0.352	0.216	0.104	0.028	0

B. The calculation of the influence line ordinates for the reaction R_A is performed in Table 13-3, and the influence line is plotted in Fig. 13-7e.

Similarly, the influence ordinate for the bending moment at any section distance x from the left-hand end is given by

$$\eta_M = \eta_{Ms} + \frac{(l-x)}{l}\eta_{MAB} - \frac{x}{l}\eta_{MBA} \tag{13-7}$$

where η_M and η_{Ms} are the influence ordinates for the bending moment at the section for a beam with fixed ends and simply supported[2] respectively. The ordinates η_M for a section $x = 0.4l$ are calculated in Table 13-4 and Fig. 13-7f plots the relevant influence line.

Table 13-4. Ordinates of the Influence Line for $M_{(x\,=\,0.4l)}$

Distance from left-hand end	0.1l	0.2l	0.3l	0.4l	0.5l	0.6l	0.7l	0.8l	0.9l	Multiplier
η_{Ms}	0.060	0.120	0.180	0.240	0.200	0.160	0.120	0.080	0.040	l
$0.6\,\eta_{MAB}$	-0.049	-0.077	-0.088	-0.086	-0.075	-0.058	-0.038	-0.019	-0.005	l
$-0.4\,\eta_{MBA}$	-0.004	-0.013	-0.025	-0.038	-0.050	-0.058	-0.059	-0.051	-0.032	l
Influence ordinates for $M_{(x\,=\,0.4l)}$	0.007	0.030	0.067	0.116	0.075	0.044	0.023	0.010	0.003	l

The influence ordinates η_V of the shear at any section can be calculated by the equation

$$\eta_V = \eta_{Vs} - \frac{1}{l}(\eta_{MAB} + \eta_{MBA}) \tag{13-8}$$

where η_{Vs} is the influence ordinate for the shear at the same section in a simply supported beam. The influence line for shear at a section $x = 0.4l$ is shown in Fig. 13-7g. It can be seen that this influence line can be formed by parts of the influence lines for R_A and R_B.

The influence lines for continuous prismatic beams with equal spans or with unequal spans in certain ratios are given in various references[3], and in most cases they need not be calculated. On the other hand, influence lines are often calculated in the design of bridges of variable I or with irregularly varying spans forming continuous beams, also of frames and grids.

13-6 INFLUENCE LINES FOR PLANE FRAMES

In the preceding section we have seen that the influence lines for shear or bending moment at any section of a member can be determined from the

[2]The influence line for the bending moment at a section distance x from the left-hand support of a simple beam of length l is composed of two straight segments with the ordinate at x equal to $x(l-x)/l$, and zero at the two ends.

[3]See for example G. Anger, *Ten-Division Influence Lines for Continuous Beams*, W. Ernst and Son, Berlin, 1956; *Moments, Shears and Reactions, Continuous Highway Bridge Tables*, American Institute of Steel Construction.

influence lines for the end-moments by simple equations of statics. Thus influence lines for end-moments are of fundamental importance, and we shall now show how to use moment distribution to find the influence lines for the end-moments of continuous plane frames.

Let us assume that we want to find the influence line for the end-moment

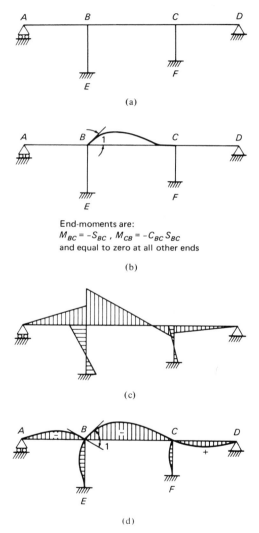

(a)

(b)

End-moments are:
$M_{BC} = -S_{BC}$, $M_{CB} = -C_{BC} S_{BC}$
and equal to zero at all other ends

(c)

(d)

Fig. 13-8. Determination of influence line for end-moment in a plane frame. (a) Plane frame. (b) Unit angular rotation of end BC without other joint displacements. (c) Bending-moment diagram corresponding to the elastic line in part (d). (d) Influence line for the end-moment M_{BC}.

M_{BC} in the frame of Fig. 13-8a. According to the Müller-Breslau principle, the influence ordinates are the ordinates of the deflected shape of the frame corresponding to a unit angular discontinuity at end BC. Assume that such a unit angular rotation is introduced at end BC without other displacements at the joints, as shown in Fig. 13-8b. The end-moments corresponding to this configuration are $-S_{BC}$ and $-t_{BC} = -C_{BC}S_{BC}$, where S_{BC} is the end rotational stiffness, t_{BC} the carryover moment, and C_{BC} the carryover factor from B to C.

We now allow joint rotations (and joint translations, if any) to take place and find the corresponding moments at the ends of the members by moment distribution in the usual way. The corresponding bending-moment diagram will be a straight line for each member (Fig 13-8c). The deflections, which are the influence line ordinates, are calculated by superposition of the deflections due to the end-moments as in the previous section.

For prismatic members, the values given in Appendix I may be used. For members of variable I, we can use the influence line ordinates of the moment at a fixed end of a member with the other end hinged[4]. To obtain the deflection due to a unit couple applied at one end, these ordinates should be divided by the adjusted end-rotational stiffness at the fixed end while the other end is hinged (see Fig. 13-9 and Eq. 11-13).

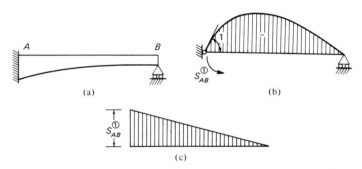

Fig. 13-9. Deflection of a nonprismatic beam due to a couple applied at one end with the other end hinged. (a) Beam. (b) Influence line for end-moment M_{AB}. (c) Bending-moment diagram corresponding to the deflection line in part (b).

The shape of the influence line for the end moment M_{BC} for the frame considered is shown in Fig. 13-8d. The ordinates plotted on the columns BE and CF can be used to find the value of M_{BC} if a unit horizontal load is applied to either of the columns. The value will be positive if the load points

---

[4]Influence-line ordinates of FEM's in beams with a variable I can be found in *Handbook of Frame Constants*, Portland Cement Association, Chicago, Ill. and R. Guldan, *Rahmentragwerke und Durchlauftrager*, Springer, Vienna, 1959.

toward the left. If however, a horizontal load on a column cannot occur, the influence ordinates on BE and CF need not be plotted.

Example 13-1 Obtain the influence line for the end-moment M_{BA} in the bridge frame in Fig. 13-10a. Use this influence line to find the influence ordinate of the bending moment M_G at the center of AB and of the shear

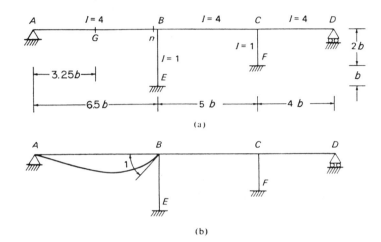

(a)

(b)

End	BA	BE	BC	CB	CF	CD
	0.29	0.21	0.50	0.39	0.24	0.37
FEM's	− 185	0	0	0	0	0
	+ 53	+ 39	+ 93 →	+ 47		
			− 9 ←	− 18	− 12	− 17
	+ 2	+ 2	+ 5 →	+ 3		
				− 1	− 1	− 1
Final end-moments	− 130	+ 41	+ 89	+ 31	− 13	− 18

Multiplier :

$I/(100b)$

(c)

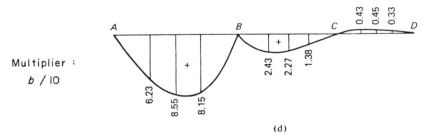

Multiplier :

$b/10$

(d)

Fig. 13-10. Influence line for an end-moment in Example 13-1. (a) Frame properties. (b) Unit angular rotation introduced at end BA. (c) Moment distribution. (d) Influence line of end-moment M_{BA}.

V_n at a point n just to the left of B. The relative values of I are shown in the figure.

A unit rotation in an anticlockwise direction is introduced at end B of BA, as shown in Fig. 13-10b. The corresponding end-moments are $M_{BA} = -3(EI/l)_{BA} = -1.85EI/b$ and zero for all the other ends. These values are the initial FEM's for which a moment distribution is carried out in Fig. 13-10c. The deflections of members AB, BC, and CD due to the final end-moments are calculated in Table 13-5 at $0.3l$, $0.5l$, and $0.7l$ of each span

Table 13-5. Ordinates of Influence Line for End-Moment M_{BA} ($b/10$)

Deflection due to end-moment at	Member *AB*			Member *BC*			Member *CD*		
	0.3*l*	0.5*l*	0.7*l*	0.3*l*	0.5*l*	0.7*l*	0.3*l*	0.5*l*	0.7*l*
Left-hand end	0	0	0	3.31	3.48	2.53	−0.43	−0.45	−0.33
Right-hand end	6.23	8.55	8.15	−0.88	−1.21	−1.15	0	0	0
Influence ordinate	6.23	8.55	8.15	2.43	2.27	1.38	−0.43	−0.45	−0.33

by the use of the tabulated values in Appendix I. These deflections, which are the influence ordinates of the end-moment M_{BA}, are plotted in Fig. 13-10d. As always, a positive sign indicates a clockwise end-moment.

Table 13-6. Ordinates of Influence Line for the Bending Moment M_G at G. ($b/10$)

Influence coefficient	Member *AB*			Member *BC*			Member *CD*		
	0.3*l*	0.5*l*	0.7*l*	0.3*l*	0.5*l*	0.3*l*	0.7*l*	0.5*l*	0.7*l*
η_{Ms}	9.75	16.25	9.75	0	0	0	0	0	0
$-\frac{1}{2}\eta_{MBA}$	−3.12	−4.28	−4.08	−1.22	−1.14	−0.69	0.22	0.23	0.17
Influence ordinate	6.63	11.97	5.67	−1.22	−1.14	−0.69	0.22	0.23	0.17

The ordinates of the influence lines for M_G and V_n are determined by superposition Eq. 13-7 and 13-8 respectively. The calculations are performed in Tables 13-6 and 13-7, and the influence lines are plotted in Figs. 13-11a and b.

Table 13-7. Ordinates of Influence Line for Shear V_n

Influence coefficient	Member *AB*				Member *BC*			Member *CD*		
	0.3*l*	0.5*l*	0.7*l*	*l*	0.3*l*	0.5*l*	0.7*l*	0.3*l*	0.5*l*	0.7*l*
η_{Vs} $-\dfrac{1}{6.5b}(\eta_{MBA})$	−0.30	−0.50	−0.70	−1.00	0	0	0	0	0	0
	−0.10	−0.13	−0.13	0	−0.04	−0.04	−0.02	0.01	0.01	0.005
Influence ordinate	−0.40	−0.63	−0.83	−1.00	−0.04	−0.04	−0.02	0.01	0.01	0.005

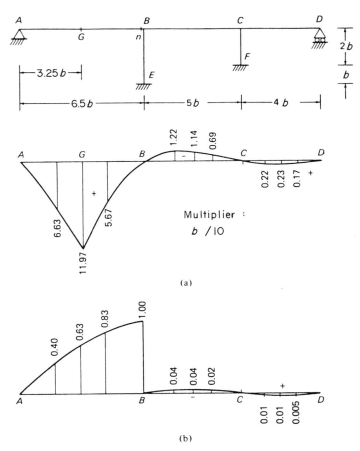

Fig. 13-11. Influence line for bending moment and shear at a section of the frame in Example 13-1. (a) Influence line for M_G. (b) Influence line for V_n.

13-7 INFLUENCE LINES FOR GRIDS

The grid in Fig. 13-12a represents the main and cross-girders of a bridge deck for which influence lines of bending moments at certain sections of the members are required. All joints are assumed to be rigid, capable of resisting bending and torsion.

For the analysis of this grid by the displacement method each of the internal joints has three unknown displacements: two rotations (vectors θ_x and θ_z) and a downward deflection δ. At each support two rotations, θ_x and θ_z are possible. One method of obtaining the influence lines is to carry out the analysis of the structure for a number of loading cases (see Sec. 3-4) with a unit vertical load at various positions. Each loading case gives one ordinate for each influence line required. This method is satisfactory when a computer is used as little additional effort in programming is required in addition to that necessary for the dead-load analysis.

Another method of obtaining the influence lines, and one which requires less computer time, is by the use of the Müller-Breslau principle, as discussed below.

We arrange the stiffness matrix of the grid in such a way that the elements corresponding to the vertical deflection occupy the first rows and the first

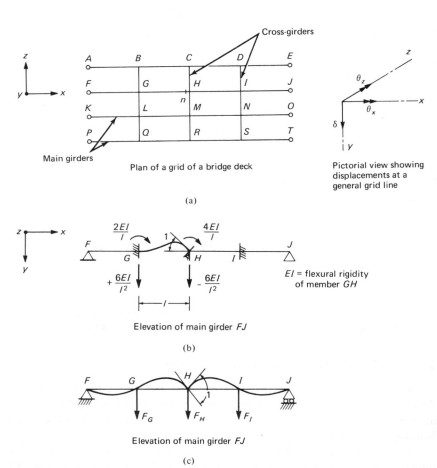

Fig. 13-12. Determination of the influence line for the bending moment at section n of a grid. (a) Grid plan. (b) Restraining forces corresponding to a unit angular discontinuity at n without joint displacements. (c) Restraining forces corresponding to a unit angular discontinuity at n with the vertical displacements prevented at the joints.

columns. Let us then partition the matrix as follows

$$[S] = \begin{bmatrix} [A_{11}] & \vdots & [A_{12}] \\ \hline [A_{21}] & \vdots & [A_{22}] \end{bmatrix}$$

(13-9)

For the grid in Fig. 13-12a, the order of $[S]$ is 52×52, and that of $[A_{11}]$ is 12×12. Assume that we require the influence line for the bending moment M_n at section n, just to the left of H. We induce a rotation at the end H of member HG in the vertical plane, as shown in the elevation of girder FJ in Fig. 13-12b. The forces required to hold the structure in this configuration are two couples and two vertical forces shown in the figure, with no forces at all the other joints. If these forces are now released, the grid will deform maintaining a unit angular discontinuity at H between the members HG and HI. The vertical deflections of the grid are therefore the ordinates of the required influence line.

The deflections can be obtained by the equation

$$\begin{bmatrix} [A_{11}] & \vdots & [A_{12}] \\ \hline [A_{21}] & \vdots & [A_{22}] \end{bmatrix} \begin{Bmatrix} \{D_1\} \\ \hline \{D_2\} \end{Bmatrix} = - \begin{Bmatrix} \{F_1\} \\ \hline \{F_2\} \end{Bmatrix}$$

(13-10)

where the elements of $\{D_1\}$ and $\{D_2\}$ are respectively the vertical deflections and rotations at the joints. The elements of $\{F_1\}$ are all zero except for the elements corresponding to the vertical forces of $6EI/l^2$ at G and $-6EI/l^2$ at H. Similarly, the elements of $\{F_2\}$ are all zero except for the elements corresponding to the couples in the vertical plane of $2EI/l$ at G and $4EI/l$ at H.

We want to solve Eq. 13-10 in order to obtain $\{D_1\}$, whose elements are the ordinates of the influence line. Using Sec. A-7 of Appendix A, we can write

$$\{D_1\} = - \left[[A_{11}] - [A_{12}][A_{22}]^{-1}[A_{21}] \right]^{-1} \left\{ \{F_1\} - [A_{12}][A_{22}]^{-1}\{F_2\} \right\}$$

(13-11)

The matrix in the large square brackets in the above equation is the stiffness matrix of the grid corresponding to a system of vertical coordinates at the joints. The vector in the large braces in the same equation represents the forces along the vertical coordinates if the joints are allowed to rotate with the vertical displacement restrained.

Equation 13-11 gives the ordinates of the influence line at the nodes. If ordinates at points between the joints are required, the end-moments[5] have to be determined first and the deflections from the straight lines joining

---

[5]These end-moments are themselves influence coefficients of another type, and are used in design of prestressed concrete structures, as discussed in Sec. 14-8.

the member ends are then determined by Eq. 13-5. The joint rotations $\{D_2\}$ required to find the end-moments can be obtained from (see Sec. A-7 of Appendix A).

$$\{D_2\} = -[A_{22}]^{-1}\left\{\{F_2\} + [A_{21}]\{D_1\}\right\} \qquad (13\text{-}12)$$

If the torsional rigidity of the members is ignored, and the beams of the grid are equally spaced in each of the x and z directions, the stiffness matrix $[S]$ corresponding to a system of vertical coordinates at the joints can be easily obtained by the use of Appendix E (see Example 4-2). For the grid in Fig. 13-12a, this matrix is of the order 12×12. To find the ordinates of the influence line M_n, a unit rotation is introduced as in Fig. 13-12b, then, the joints F, G, H, I, and J are allowed to rotate in the vertical plane without vertical displacement, and the corresponding restraining forces $\{F\}$ are determined (Fig. 13-12c). This may be conveniently done by moment distribution with the girder FJ treated as a continuous beam[6].

The deflections at the joints are determined by $[S]\{D\} = -\{F\}$. For the grid in Fig. 13-12a, the elements of $\{D\}$ are the influence ordinates at the 12 intermediate joints, and the elements of $\{F\}$ are all zero except for the three forces at joints G, H, and I. Ordinates between the joints, if required, can be calculated by Eq. 13-5. The end-moments needed for this equation can be determined from the vertical deflections, using the tabulated moments in Appendix E.

Example 13-2   Find the influence line for the bending moment at section n, just to the left of joint C, in the grid of Fig. 13-13a. The main girders are encastré. The relative moment of inertia is 4 for all main girders and 1 for the cross-girders. The ratio of the torsional rigidity GJ to the flexural rigidity EI is $1:4$ for all members.

Figure 13-13b shows the coordinate system chosen, with three coordinates at each joint. The stiffness matrix for the grid is partitioned in the following manner:

$$[S] = \begin{bmatrix} [A_{11}] & \vdots & [A_{12}] \\ (9 \times 9) & \vdots & (9 \times 18) \\ \text{-----} & + & \text{-----} \\ [A_{21}] & \vdots & [A_{22}] \\ (18 \times 9) & \vdots & (18 \times 18) \end{bmatrix}$$

---

[6]Other members of the grid are not affected during the distribution because the grid is torsionless. The same procedure can also be followed when torsion is not ignored, but the computations become more involved as the moment distribution is carried out for all the members of the grid. Refer to a paper by W. W. Ewell, S. Okubo and J. I. Abrams, "Deflections of Gridworks and Slabs," *Trans. ASCE*, Paper No. 2520, Vol. 117 (1952), pp. 859–912.

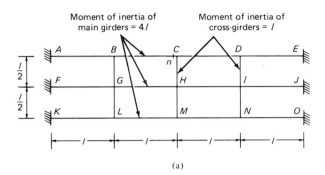

(a)

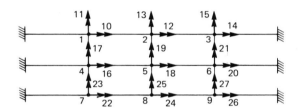

Coordinates 1 to 9 are vertical downwards
Coordinates 10 to 27 represent rotation of
a right-handed screw progressing in the
directions of the double-headed arrows

(b)

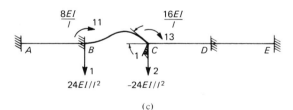

(c)

Fig. 13-13. Analysis of the grid of Example 13-2. (a) Grid plan. (b) Coordinate system. (c) Restraining forces corresponding to a unit angular discontinuity at n without joint displacements.

The order of each submatrix in this equation is indicated in brackets. The submatrices can be easily written with little calculation, and are given below. The matrices $[A_{11}]$ and $[A_{22}]$ are symmetrical, and $[A_{12}] = [A_{21}]^T$.

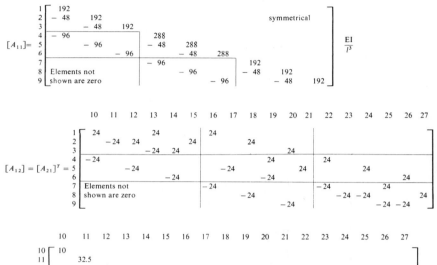

The restraining forces corresponding to a unit angular discontinuity at n without joint displacements are shown in Fig. 13-13c.

The submatrices $\{F_1\}$ and $\{F_2\}$ in Eq. 13-10 are (see the restraining forces in Fig. 13-13c):

$$\{F_1\} = \frac{EI}{l^2}\{24, -24, 0, 0, 0, 0, 0, 0, 0\}$$

$$\{F_2\} = \frac{EI}{l}\{0, 8, 0, 16, \text{ and zero for all the other elements}\}$$

Substituting in Eq. 13-11, we obtain

$$\{D_1\} = 10^{-3}\,l\{62, 350, 66, 54, 140, 51, 10, 10, 8\}$$

The elements of $\{D_1\}$ are the ordinates of the influence line for the bending moment M_n at section n. These ordinates are plotted on an elevation of the main girders in Fig. 13-14.

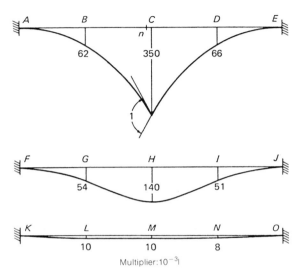

Fig. 13-14. Influence line for the bending moment M_n at section n of the grid in Example 13-2.

The rotations $\{D_2\}$ obtained by Eq. 13-12 are

$$\{D_2\} = 10^{-3}\{-35, 134, -381, -488, -49, -140, -58, 106, -331,$$
$$-10, -64, -101, -102, 10, -199, -2, -99, -9\}$$

These rotations are used when influence ordinates between the joints are required.

Example 13-3 Neglecting the torsional rigidity of the girders, find for the interconnected bridge system of Fig. 13-15a the influence lines for the following actions:

    (a) bending moment at the center of girder AB

    (b) bending moment at the center of girder CD

    (c) bending moment in the cross-girder at J

    (d) reaction R_C at support C

The main girders are simply-supported and have a moment of inertia of $4I$, where I is the moment of inertia of the cross-girder.

The stiffness matrix of the grid corresponding to vertical downward coordinates at N, J, K, and L(Fig. 13-15b) can be calculated from the values

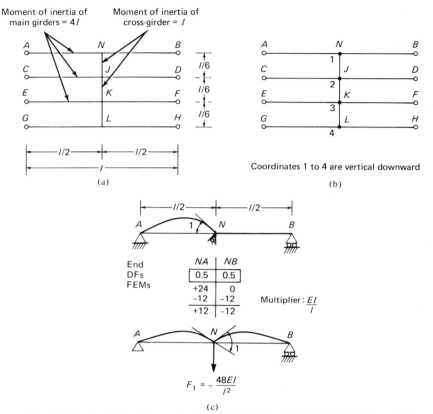

Fig. 13-15. Analysis of the torsionless grid in Example 13-3. (a) Grid plan. (b) Coordinate system. (c) Calculation of the restraining force F_1 corresponding to a unit angular discontinuity in main girder at N, with no displacements along the coordinate.

tabulated in Appendix E. We obtain

$$[S] = \frac{EI}{l^3} \begin{bmatrix} 537.6 & & & \text{symmetrical} \\ -777.6 & 2265.6 & & \\ 518.4 & -1814.4 & 2265.6 & \\ -86.4 & 518.4 & -777.6 & 537.6 \end{bmatrix} \tag{a}$$

To find the influence line of the bending moment at the center of girder AB, we introduce a unit angular discontinuity just to the left (or right) of joint N (Fig. 13-15c) with the vertical joint displacements prevented. The end-moment M_{NA} corresponding to this configuration is

$$\frac{3E(4I)}{l/2} = 24\frac{EI}{l}$$

Moment distribution for the beam ANB is carried out in Fig. 13-15c, and the restraining force F_1 required to prevent the deflection at N is calculated:

$$F_1 = -48 \frac{EI}{l^2}$$

It is obvious that no forces are required at the other three coordinates. Thus the matrix $\{F\}$ is

$$\{F\} = \frac{EI}{l^2} \{-48, 0, 0, 0\}$$

The force F_1 can also be found by the use of the reactions tabulated in Appendix E, the procedure being as follows.

In order to reach the configuration of Fig. 13-16a we can proceed in two steps. First, we introduce a unit angular discontinuity by allowing the left-hand end of the beam to lift by a distance $l/2$, as shown in Fig. 13-16b; no forces are involved. In the second step, the support A is brought back to its original level by a vertical downward force at A without a change in the angle between the ends of the members meeting at N. The value of the reaction at N due to a unit downward displacement of support A is (from Appendix E) $3E(4I)/(l/2)^3$

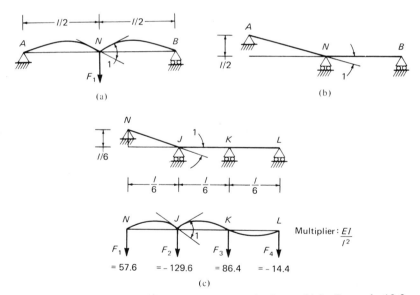

(a)   (b)

(c)

Fig. 13-16. Analysis of members of the torsionless grid in Example 13-3. (a) Unit angular discontinuity at N with the vertical deflection restrained by the force F_1. (b) Unit angular discontinuity introduced at N, without restraining forces. (c) Determination of the restraining forces corresponding to a unit angular discontinuity in the cross-girder at J by use of tabulated values in Appendix E.

upward. Therefore the value of the restraining force is $F_1 = -48(EI/l^2)$, which is the reaction at N corresponding to a downward displacement of $l/2$ at A.

Similarly, to find the influence line for the bending moment at the center of CD, we introduce a unit rotation at the end J of JC. The corresponding restraining forces are

$$\{F\} = \frac{EI}{l^2}\{0, -48, 0, 0\}$$

For the influence line for the bending moment in the cross-girder at section J, a unit angular discontinuity is introduced at this point, as shown in Fig. 13-16c, causing the lifting of end N by a distance $l/6$. The forces at N, J, K, and L required to bring joint N to its original position, determined from Appendix E, are

$$\{F\} = \frac{EI}{l^2}\{57.6, -129.6, 86.4, -14.4\}$$

For the influence line of the reaction R_C, a unit downward displacement is introduced at C. The restraining force at J is taken from Appendix E, and the other restraining forces are zero. Therefore

$$\{F\} = \frac{EI}{l^3}\{0, -96, 0, 0\}$$

The influence ordinates for each of the four effects are summarized in the equation

$$[S][D] = -\frac{EI}{l^2}\begin{bmatrix} -48 & 0 & 57.6 & 0 \\ 0 & -48 & -129.6 & -\frac{96}{l} \\ 0 & 0 & 86.4 & 0 \\ 0 & 0 & -14.4 & 0 \end{bmatrix} \qquad (b)$$

where $[S]$ is the stiffness matrix in Eq. (a). The solution of Eq. (b) gives

$$[D] = \begin{bmatrix} 0.194l & 0.082l & -0.038l & 0.164 \\ 0.082l & 0.097l & 0.055l & 0.194 \\ 0.010l & 0.062l & 0.006l & 0.124 \\ -0.034l & 0.010l & -0.024l & 0.020 \end{bmatrix}$$

The influence lines for the four actions are plotted in Fig. 13-17. They include ordinates between joints, and as an example of calculation of such ordinates, the computation for AN is given below for the influence line for

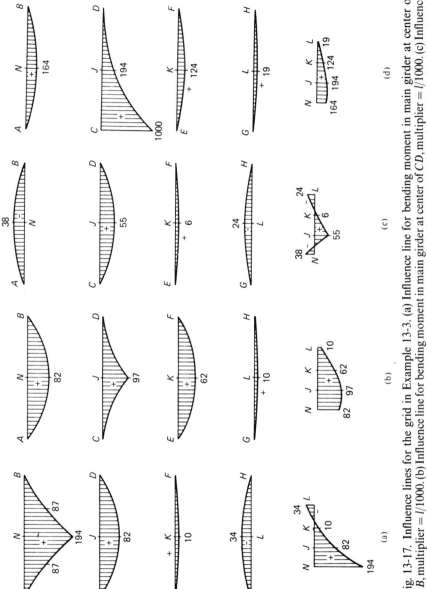

Fig. 13-17. Influence lines for the grid in Example 13-3. (a) Influence line for bending moment in main girder at center of AB, multiplier $= l/1000$. (b) Influence line for bending moment in main girder at center of CD, multiplier $= l/1000$. (c) Influence line for bending moment in cross-girder at joint J, multiplier $= l/1000$. (d) Influence line for reaction at C, multiplier $= l/1000$.

the bending moment in Fig. 13-17a. The end-moment at N of member AN corresponding to the deflected shape in Fig. 13-17a is

$$M_{NA} = 12\frac{EI}{l} - \left[\frac{3E(4I)}{(l/2)^2}\right]0.194l = 2.688\frac{EI}{l}$$

The first term in this equation is the end-moment when the vertical joint displacements are restrained, and the second term is the end-moment caused by the vertical displacement. The bending moments caused by unit values of the vertical displacement are tabulated in Appendix E. The deflection measured from the straight line between A and N can be calculated by Eq. 13-5 or by the use of Appendix I. Therefore, the equation of the influence line between A and N is

$$\eta = l[0.194\,\varepsilon - 0.028\,(\varepsilon - \varepsilon^3)]$$

where $\varepsilon = (2x/l)$, and x is the distance from A to the desired point on AN.

13-8   GENERAL

The influence line for any action in a structure can be obtained by repeating the analysis for several cases of loading with a unit load at different positions: each loading case gives one ordinate of the required influence line. This method is convenient only with a computer.

Using the Müller-Breslau principle, influence lines can be represented as deflection lines. The use of this principle makes it possible to determine without any special calculations the shape of the influence line, and this indicates the parts of the structure to be loaded to obtain the maximum effects. The principle is also used to calculate the ordinates of the influence line, and can be applied to continuous frames, grids, and trusses (see Chapter 14).

The problems for this chapter and the following one are given together at the end of Chapter 14.

CHAPTER 14

Influence lines for arches, trusses, and prestressed concrete members

14-1 INTRODUCTION

In this chapter a general superposition equation is used to derive influence lines for statically indeterminate arches and trusses. In addition, it is shown that the first and second derivatives of the influence line of any action for a unit moving point load give two additional influence lines of the same action due to other types of moving loading. The influence line corresponding to the second derivative is of particular interest in the analysis of prestressed concrete statically indeterminate structures.

14-2 GENERAL SUPERPOSITION EQUATION

The concept of adding influence coefficients for statically determinate and statically indeterminate cases in Eq. 13-7 to obtain the influence coefficient for bending moment at a section of a straight member will be extended now for any action in a statically indeterminate structure.

The influence coefficients for any action in a linearly elastic statically indeterminate structure can be obtained by adding the influence coefficients for the same action in a released structure and the influence coefficients for the redundants multiplied by the values of the action due to unit values of the redundants. Let p be the number of influence coefficients to be calculated for any action of a structure statically indeterminate to the nth degree. If a unit point load is applied at j, one of the p locations where the influence coefficients are required, the influence coefficient $\eta_j = A_j$ is the value of the action in the statically indeterminate structure determined by the superposition equation

$$A_j = A_{sj} + [F_{1j}F_{2j} \ldots F_{nj}] \{A_u\} \qquad (14\text{-}1)$$

where $A_{sj} = \eta_{sj}$ is the value of the action due to a unit load at j in a released structure, $F_{ij} = \eta_{Fij}$ is the value of the ith redundant due to a unit load at j, and the elements of $\{A_u\}$ are the values of the action considered due to unit values of the redundants on the released structure.

If the statically indeterminate structure is subjected to a unit load acting separately at each of the p locations and Eq. 14-1 is applied, we obtain the following equation of superposition of influence coefficients:

$$\{\eta\}_{p \times 1} = \{\eta_s\}_{p \times 1} + [\{\eta_{F1}\} | \{\eta_{F2}\} | \cdots | \{\eta_{Fn}\}]_{p \times n} \{A_u\}_{n \times 1} \quad (14\text{-}2)$$

in which the elements of the submatrices $\{\eta_{Fi}\}$ are the p influence coefficients of the redundants F_i.

To use Eq. 14-2 the influence lines for the redundants must first be determined. These can be obtained by an analysis for p locations of the unit load. For each position, the n redundants are determined, thus giving one of the p ordinates of the influence line for each redundant.

Influence lines for the redundants can also be determined by direct application of Müller-Breslau principle.

The use of Eq. 14-2 for arches and trusses will be considered in the following two sections.

14-3   INFLUENCE LINES FOR ARCHES

Influence lines are very useful in the analysis of arch bridges. Here the load is applied through vertical members supporting the deck (Fig. 14-1a). Let us consider the influence lines due to a unit vertical load in any position E on CD. This load is assumed to be transmitted to the arch at F vertically below E.

The fixed arch in Fig. 14-1a is statically indeterminate to the third degree. The simply supported arch in Fig. 14-1b is chosen as the released

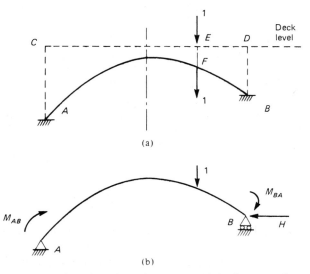

Fig. 14-1. Fixed arch and a released structure. (a) Arch supporting a bridge deck. (b) Statically determinate released structure.

structure with the clockwise end-moments M_{AB} and M_{BA}, and the inward horizontal force H as the redundants. The influence lines for the three redundants can be obtained by application of Müller-Breslau principle. For the influence line for M_{AB}, an anticlockwise unit rotation is introduced at end A: the resulting vertical displacements of the arch axis are the influence ordinates. Similarly, for the influence line for M_{BA}, we introduce an anticlockwise unit rotation at B, and for the influence line for H a unit horizontal displacement is introduced outwards at either A or B.

The bending moment M corresponding to these end displacements can be conveniently obtained by column analogy (see Sec. 10-4), and hence the corresponding vertical deflection is calculated. The method of elastic weights may be used for this purpose (see Sec. 9-5). This procedure ignores the effect of axial deformation of the arch. In extremely flat arches, however, the axial deformations may have some effect, and this can be included as a correction using the approximate method of Sec. 10-7.

The steps outlined above for finding the influence lines for the redundants are simple, but writing general algebraic expressions for the terms involved in the solution often leads to involved integrals. Simple expressions can be obtained only for a symmetrical parabolic arch in which the flexural rigidity EI is assumed to vary as the secant of the inclination of the arch axis, Fig. 14-2a. The properties of the analogous column for such an arch are given in Appendix J. The equations of the influence lines of the three redundants in such an arch (Fig. 14-2b) are

$$\eta_H = \frac{15l}{64h}(1 - \varepsilon^2)^2 \tag{14-3}$$

$$\eta_{MAB} = -\frac{l}{32}(1 - \varepsilon)^2 (5\varepsilon^2 + 6\varepsilon + 1) \tag{14-4}$$

and

$$\eta_{MBA} = \frac{l}{32}(1 + \varepsilon)^2 (5\varepsilon^2 - 6\varepsilon + 1) \tag{14-5}$$

where $\varepsilon = x/0.5l$, l is the span, and h is the rise. Equations 14-3 and 14-4 are plotted in Figs. 14-2c and d.

Tables of influence coefficients for parabolic, circular, or semielliptical prismatic and nonprismatic arches are available[1].

The influence line for the stress resultant at any section of the arch in Fig. 14-2a can be determined by Eq. 14-2. For example, to obtain the influence line of the bending moment M_C at the crown, we first find the values of this action due to unit values of the redundants on the released structure of

[1] J. Michalos, *Theory of Structural Analysis and Design*, Ronald, New York, 1958. Tables for circular arches only are also included in J. Parcel and R. B. B. Moorman, *Analysis of Statically Indeterminate Structures*, Wiley, New York, 1955.

Fig. 14-2b; $\{A_u\} = \{-h, \frac{1}{2}, -\frac{1}{2}\}$, in which the order of the redundants is H, M_{AB}, and M_{BA}. The influence line η_s for the moment M_C in the released structure is formed by two straight segments (as for a simple beam):

$$\left.\begin{array}{ll} \eta_s = \dfrac{l}{4}(1 - \varepsilon) & \text{for } 0 < \varepsilon < 1 \\[2mm] \eta_s = \dfrac{l}{4}(1 + \varepsilon) & \text{for } 0 > \varepsilon > -1 \end{array}\right\} \tag{14-6}$$

The influence ordinates η_{MC} are given by Eq. 14-2:

$$\eta_{MC} = \eta_s + (-h\eta_H + \tfrac{1}{2}\eta_{MAB} - \tfrac{1}{2}\eta_{MBA})$$

The influence ordinates on the right-hand side of this equation are given by Eq. 14-3–14-6. The shape of the influence line for M_C is shown in Fig. 14-2e.

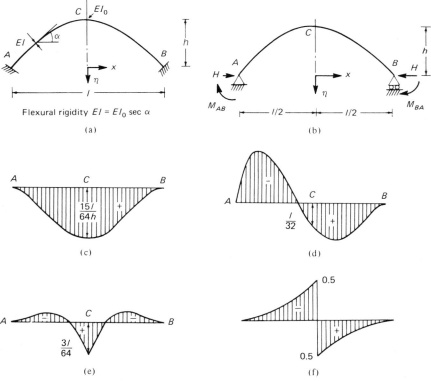

Fig. 14-2. Influence lines for a parabolic arch with secant variation of flexual rigidity. (a) Parabolic arch with secant variation in EI. (b) Positive direction of the redundants. (c) Influence line for H. (d) Influence line for M_{AB}. (e) Influence line for M_C. (f) Influence line for V_C.

The influence line for shear V_C at the crown can be obtained in a similar manner; its shape is shown in Fig. 14-2f.

It can be easily shown that the horizontal thrust H due to a uniform load w per unit length of the horizontal projection of a parabolic arch, with any variation in EI and with hinged or fixed supports, is

$$H = \frac{ql^2}{8h} \tag{14-7}$$

and that the bending moment and the shear are zero at all sections. Since the area under an influence line is equal to the value of the action considered due to a uniform load $q = 1$ (see Eq. 13-3), it follows that the area under η_H (Fig. 14-2c) is $l^2/8h$ and the area is zero under the other three influence lines in Fig. 14-2.

14-4 INFLUENCE LINES FOR TRUSSES

Influence lines for the reactions or forces in the members of pin-connected trusses can be obtained by solving for several cases with the unit load at different joints. The influence lines can also be obtained from Eq. 14-2, which applies to any linearly elastic structure. To use this equation, we need the influence line for a statically determinate released truss, the influence lines for the redundants, and the values of the action due to unit values of the redundants. The procedure is illustrated by the following example.

Example 14-1 Find the influence line for the reaction at B and the forces in the members labeled Z_1 and Z_2 in the truss of Fig. 14-3a. The unit load can act at the nodes of the lower chord only. All the members are assumed to have the same value of l/aE, l being the length, and a the cross-sectional area of the members.

A released structure is shown in Fig. 14-3b, in which the redundants F_1 and F_2 are taken as the forces in members Z_3 and Z_4, respectively. The influence ordinates η_s for the values of the required actions in the released structure are plotted in Figs. 14-3c, d, and e. These can be easily checked by simple statics.

According to Müller-Breslau principle, the influence line for F_1 can be obtained by cutting the member Z_3 and applying equal and opposite forces (causing compression in Z_3) to the remaining truss, so as to produce a relative unit displacement at the cut section. Then the deflected shape of the bottom chord of the truss gives the influence line for F_1. This is the same as finding the deformations in the actual structure due to a unit extension of member Z_3, such as that caused by a rise in temperature or a lack of fit in this member.

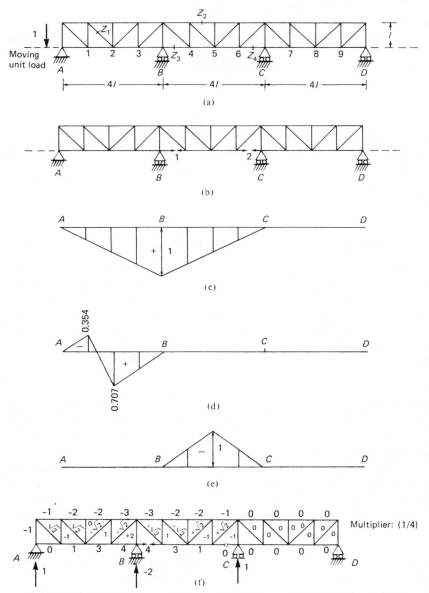

Fig. 14-3. Analysis of the continuous truss of Example 14-1. (a) Continuous truss. (b) Released structure and coordinate system. (c) Influence line for R_b in the released structure. (d) Influence line for the force in Z_1 in the released structure. (e) Influence line for the force in Z_2 in the released structure. (f) Reactions and forces in members due to $F_1 = 1$.

The flexibility matrix of the released structure is

$$[f] = \frac{l}{8Ea} \begin{bmatrix} 57 & 3 \\ 3 & 57 \end{bmatrix}$$

The elements of this matrix can be easily checked by virtual work (see Sec. 6-5). For convenience, the forces in the members due to $F_1 = 1$ are shown in Fig. 14-3f. Making use of symmetry of the structure, the forces in members due to $F_2 = 1$ can also be deduced from this figure.

A unit extension in member Z_3 produces the redundants $\{F_1, F_2\}$ given by Eq. 2-9

$$[f] \begin{Bmatrix} F_1 \\ F_2 \end{Bmatrix} = - \begin{Bmatrix} 1 \\ 0 \end{Bmatrix}$$

From this equation we can see that the redundants have the values $F_1 = -S_{11}$ and $F_2 = -S_{21}$, where S_{11} and S_{21} are the elements of the first column of the stiffness matrix $[S]$ of the released structure. Similarly, the values of the redundants corresponding to a unit extension in Z_4 are equal to minus the elements in the second column of $[S]$. In the above example,

$$[S] = [f]^{-1} = \frac{Ea}{405l} \begin{bmatrix} 57 & -3 \\ -3 & 57 \end{bmatrix}$$

With the values of the redundants $F_1 = -57Ea/405l$ and $F_2 = 3Ea/405l$, the forces in all the members can be determined. The corresponding deflections at the joints 1, 2, ..., 9 on the bottom chord give the influence ordinates of the redundant F_1. These deflections can be obtained by the methods discussed in Chapters 6 or 9. The influence line of the redundant F_1 is plotted in Fig. 14-4a. Because the structure is symmetrical, the same ordinates in reversed order are the ordinates at the nine joints of the influence line for F_2.

For the influence line of the reaction at B, we use Eq. 14-2 to determine the coordinates at the joints 1, 2, ..., 9:

$$\{\eta\}_{9 \times 1} = \{\eta_s\}_{9 \times 1} + [\{\eta_{F1}\} \,|\, \{\eta_{F2}\}]_{9 \times 2} \{A_u\}_{2 \times 1} \tag{14-8}$$

The elements of $\{A_u\}$ are the reactions at B due to $F_1 = 1$ and $F_2 = 1$ (see Fig. 14-3f)

$$\{A_u\} = \begin{Bmatrix} -0.5 \\ 0.25 \end{Bmatrix}$$

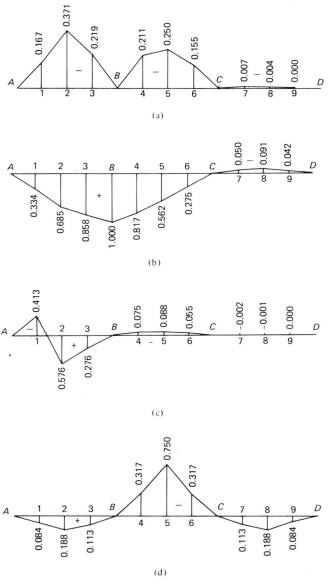

Fig. 14-4. Influence lines for the continuous truss of Fig. 14-3 (Example 14-1). (a) Influence line for redundant F_1 (force in member Z_3). (b) Influence line for vertical reaction at B. (c) Influence line for force in member Z_1. (d) Influence line for force in member Z_2.

An upward reaction is considered positive. Substituting in Eq. 14-2, we obtain the influence ordinates for R_B:

$$
\{\eta\} = \begin{Bmatrix} 0.250 \\ 0.500 \\ 0.750 \\ 0.750 \\ 0.500 \\ 0.250 \\ 0 \\ 0 \\ 0 \end{Bmatrix} + \begin{bmatrix} -0.167 & 0.000 \\ -0.371 & -0.004 \\ -0.219 & -0.007 \\ -0.211 & -0.155 \\ -0.250 & -0.250 \\ -0.155 & -0.211 \\ -0.007 & -0.219 \\ -0.004 & -0.371 \\ 0.000 & -0.167 \end{bmatrix} \begin{Bmatrix} -0.5 \\ 0.25 \end{Bmatrix} = \begin{Bmatrix} 0.334 \\ 0.685 \\ 0.858 \\ 0.817 \\ 0.562 \\ 0.275 \\ -0.050 \\ -0.091 \\ -0.042 \end{Bmatrix}
$$

The influence line for R_B is plotted in Fig. 14-4b. The influence lines for the forces in members Z_1 and Z_2 are determined by Eq. 14-2 in the same way, and the results are plotted in Figs. 14-4c and d.

14-5 PRESTRESSING MOMENT INFLUENCE COEFFICIENTS

Influence lines can be used for the analysis of the effect of prestressing in statically interminate concrete structures. The effect was discussed in Secs. 2-41 and 4-11 and will be treated here in more detail.

The beam in Fig. 14-5a is stressed by a cable in which there is a constant tension P. The cable is anchored at the beam ends, thus producing compressive forces on the concrete as shown. In most practical cases the angle θ between the direction of the cable and the axis of the beam is small, so that the axial component of the prestressing force is considered equal to the force itself, and the component perpendicular to the axis is taken as $P\theta$ (see Fig. 14-5b). Let P represent the absolute value of the prestressing force.

Where the cable changes direction by an angle $\Delta\theta$, the force exerted by the cable is $P(\Delta\theta)$ along the bisector of the angle formed by the two parts of the cable. Again, this force is considered to act at right angles to the beam axis, and its horizontal component is ignored (Fig. 14-5b). If the change in direction of the cable is made along an arc of a circle of radius R, the cable exerts on the beam a uniform radial force (Fig. 14-6) which can be approximated by a uniformly distributed load, perpendicular to the beam axis, and of intensity P/R over a length c equal to the projection of the arc on the axis. Because the curvature of the cable is usually small, it can be expressed by the equation

$$
\left| \frac{1}{R} \right| \simeq \left| \frac{d^2 e}{dx^2} \right| \tag{14-9}
$$

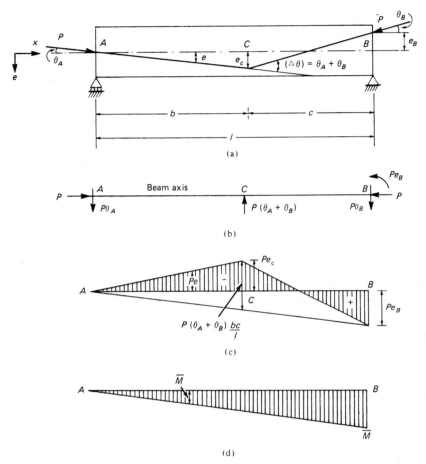

Fig. 14-5. Effects of a straight prestressing cable. (a) Prestressed beam. (b) Forces caused by prestressing the beam in part (a). (c) Statically determinate moment M_s due to prestressing. (d) Statically indeterminate moment $\bar{M}$ due to prestressing (secondary moment) occurs only when the rotation at B is restrained.

where e is the cable eccentricity, considered positive when below the centroid, and the curvature of the cable is assumed to cause a transverse load of intensity

$$q = P\frac{d^2 e}{dx^2} \tag{14-10}$$

taken as positive if q is in the positive direction of e. For the cable shown in Fig. 14-6, q is negative.

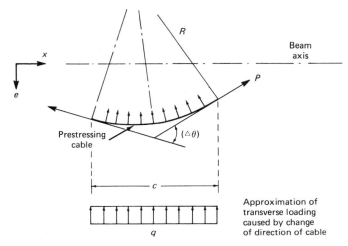

Fig. 14-6. Forces caused by a cable changing direction along an arc of a circle.

The total transverse force caused by the change in direction of the cable $\Delta\theta$ is

$$|Q| = |P(\Delta\theta)| \qquad (14\text{-}11)$$

This equation can be used also when the cable changes direction by an angle $\Delta\theta$ at one point, in which case Q is a concentrated transverse load. In the literature on prestressed concrete, the transverse loads q and Q are referred to as *balancing loads*[2] or *radial forces*.[3]

The forces in the beam of Fig. 14-5a caused by prestressing, are shown in Fig. 14-5b, and the corresponding bending moment M_s is shown in Fig. 14-5c. This bending-moment diagram can be obtained by considering the system of forces in equilibrium in Fig. 14-5b, or simply by multiplying the force P by its eccentricity.

$$M_s = -Pe \qquad (14\text{-}12)$$

As usual the bending moment is considered positive if it causes tension in the bottom fiber.

If the rotation of the beam at end B is restrained, there is a statically indeterminate end-moment at this end, and this causes an additional bending moment $\bar{M}$, as shown in Fig. 14-5d. The moment $\bar{M}$ is usually called the

[2]See for example, T. Y. Lin, "Load Balancing Method for Design and Analysis of Prestressed Concrete Structures," *Proc. ACI*, Vol. 60, No. 6 (June 1963), pp. 719–741. See also Discussion to the same paper, (December 1963), pp. 1843–1881.

[3]F. Leonhardt, *Prestressed Concrete Design and Construction*, 2d ed., W. Ernst and Son, Berlin, 1964, pp. 336–338.

secondary bending moment due to prestressing. We should note that, despite its name, the magnitude of the secondary moment may be of the same order as that of the statically determinate moment M_s. The secondary moment $\bar{M}$ is of course zero in a statically determinate structure.

Thus the bending moment at any section due to the prestressing of a statically indeterminate structure is

$$M = M_s + \bar{M} \tag{14-13}$$

A similar equation can be written for any action A. For instance, the stress resultant or reaction at a support is

$$A = A_s + \bar{A} \tag{14-14}$$

where A_s and $\bar{A}$ are respectively the statically determinate and indeterminate value of the action.

The total bending moment M (or action A) can be obtained by applying transverse loads (calculated by Eq. 14-10 or 14-11) at the locations where the cable changes direction, and also eccentric compressive forces at the sections where the cable is anchored, and then performing the analysis of the statically indeterminate structure in the usual way. In general, the force to be applied at a section where a cable is anchored can be replaced by an axial compressive force P, an end moment Pe, and an end shear $P\theta$ (see Fig. 14-5b), where e and θ are respectively the eccentricity and slope of the cable at the end of the beam.

In design practice, it is usual to assume a cable profile, and to check the corresponding moments and stresses at all sections of the structure. An improved profile is then tried, and the profile which satisfies the stress requirements is reached after several trials. If influence lines for the structure are available, they can be used for the analysis of the effect of the balancing loads. However, an additional analysis must be performed for the effect of the couple Pe at the cable anchors in cases where the cable does not end at the centroid of the section.

In dealing with this problem, we shall use a new type of influence coefficient,[4] which we shall refer to as *prestressing influence coefficient*. This can be determined for any action but is most useful for bending moments. In the latter case, we use the following definition. The prestressing moment influence coefficient at j, $\eta_{\bar{M}P}$, is defined as the secondary moment (the indeterminate moment) at a section i due to a unit prestressing force applied at unit eccentricity over an element of unit length at section j (see Fig. 14-7a).

The prestressing influence coefficients depend on the geometry of the concrete structure only, and not on the cable profile or on the value of the

[4]A. Ghali, "Bending Moments in Prestressed Concrete Structures by Prestressing Moment Influence Coefficients," *Proc. ACI*, Vol. 66, No. 6 (June 1969), pp. 494–497.

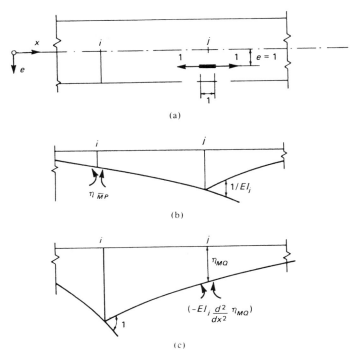

Fig. 14-7. A part of a continuous structure used to derive Eq. 14-16. (a) General sections i and j of a prestressed structure (see definition of prestressing moment influence coefficient). (b) Deflection line caused by the prestressing force in part (a). (c) Influence line for the bending moment at i due to unit transverse load.

prestressing force. Thus the use of these coefficients makes it easy to account for the variation in the prestressing force along the tendon and allows a quick estimate of the effect of prestress when the tendon profile is changed during the design. The secondary moment at any section is

$$\bar{M} = - \int P \, e \eta_{\bar{M}P} \, dx \qquad (14\text{-}15)$$

where the integral is over the length of all members of the structure. The evaluation of this integral is similar to the determination of any stress result-ant due to transverse loading of intensity Pe from its influence line (see Eq. 13-2).

In Sec. 14-7, we shall show that for frames with straight prismatic members, the influence line $\eta_{\bar{M}P}$ is composed of straight segments. Given this, the integral in Eq. 14-15 can be evaluated by the method discussed in Sec. 6-4 for the evaluation of an integral of two functions Pe and $\eta_{\bar{M}P}$, of which at least one is linear. Thus, the secondary moment due to prestressing of a prismatic beam of length l is

$$\bar{M} = - \begin{bmatrix} \text{area of the } Pe \text{ diagram} \\ \text{over the length of the beam} \end{bmatrix} \begin{bmatrix} \text{the influence ordinates} \\ \eta_{\bar{M}P} \text{ at the centroid} \\ \text{of the } Pe \text{ diagram} \end{bmatrix}$$

14-6   RELATION BETWEEN INFLUENCE LINES

We shall now show that the influence coefficient $\eta_{\bar{M}P}$ for the secondary moment at any section due to prestressing is related to the bending moment influence coefficient η_{MQ} due to a unit transverse load[5] by the equation

$$\eta_{\bar{M}P} = -\frac{d^2\eta_{MQ}}{dx^2} \qquad (14\text{-}16)$$

We recall that $\eta_{\bar{M}P}$ at j is defined as the secondary moment at section i (or the total moment, since M_s is zero at i) due to a unit prestressing force applied at a unit eccentricity over an element of unit length at j (Fig. 14-7a). The relative angular rotation of the sections at the ends of this element is $1/EI_j$, where EI_j is the flexural rigidity at section j (see Eq. 5-20). Hence, the influence coefficient $\eta_{\bar{M}P}$ represents the bending moment at i when an angular discontinuity of $1/EI_j$ is introduced at j (Fig. 14-7b).

The influence line of the bending moment at i due to a unit vertical load is sketched in Fig. 14-7c. According to Müller-Breslau principle, this is also the deflection line due to a discontinuity of one radian at section i. At j, the curvature of this deflection line is $(d^2\eta_{MQ}/dx^2)$, and the corresponding bending moment at this section is $-EI_j(d^2\eta_{MQ}/dx^2)$ (see Eq. 9-1).

Applying Betti's law (see Sec. 8-2) to the systems in Figs. 14-7b and c, and noting that the only forces that do work are the internal forces (bending moments) at i and at j in Figs. 14-7b and c respectively, we can write

$$\eta_{\bar{M}P} \times 1 = (-EI_j \frac{d^2}{dx^2}\eta_{MQ})\frac{1}{EI_j}$$

whence Eq. 14-16.

In the same manner, it can be shown that the "secondary" shear or reaction due to prestressing are related to the shear or reaction influence coefficients due to a transverse unit load by equations similar to Eq. 14-16, replacing the subscripts $\bar{M}$ and M by $\bar{V}$ and V (shear) or $\bar{R}$ and R (reaction).

---

[5]This is the usual influence coefficient treated in the earlier sections of this chapter and Chap. 13. The subscripts MQ are added to differentiate between the two types of influence lines discussed here. The first subscript M indicates the action considered, that is, the bending moment at a section; the second subscript Q indicates the cause, being a unit moving load acting at right angles to the member.

While the second derivative of the influence line of any action due to a unit transverse load gives the prestressing influence line, it can be shown that the first derivative also gives an influence line. This then is a third type of influence line. For example, if the action is a bending moment at a section of a plane structure, we can write

$$\eta_{MC} = \frac{d\eta_{MQ}}{dx} \qquad (14\text{-}17)$$

where η_{MC} is the influence line for the moment at this section due to a unit clockwise external applied couple. Equation 14-17 can be used to prove Eq. 14-16, if we consider the effect of prestressing in Fig. 14-7a to be the same as that of two equal and opposite couples at two sections a unit distance apart.

In the following sections a method of determining the prestressing moment influence coefficients is considered and the effectiveness of their use is demonstrated.

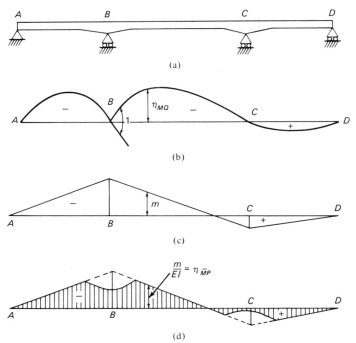

Fig. 14-8. Influence lines for the moment at a support of a continuous beam. (a) Non-prismatic continuous beam. (b) Influence line of bending moment at B due to unit vertical load. (c) Bending-moment diagram corresponding to the deflection line in part (b). (d) Influence line for the secondary bending moment at B due to prestressing.

14-7 PRESTRESSING MOMENT INFLUENCE COEFFICIENTS FOR FRAMES AND BEAMS

The influence line η_{MQ} for the bending moment at support B of the continuous beam $ABCD$ of Fig. 14-8a is sketched in Fig. 14-8b. From Müller-Breslau principle (see Sec. 13-3), this is also the deflection line corresponding to a unit angular discontinuity at B. The bending moment m corresponding to this deflection line varies linearly as indicated in Fig. 14-8c. This bending moment is related to the deflection η_{MQ} in Fig. 14-8b by the relation (see Eq. 9-1)

$$\frac{d^2\eta_{MQ}}{dx^2} = -\frac{m}{EI} \tag{14-18}$$

It follows from Eqs. 14-16 and 14-18 that the influence coefficients $\eta_{\overline{M}P}$ can be obtained simply by dividing the ordinates m of the bending-moment diagram by the EI values at each section, as shown in Fig. 14-8d that is,

$$\eta_{\overline{M}P} = \frac{m}{EI} \tag{14-19}$$

The method of obtaining the bending-moment diagram in Fig. 14-8c was discussed in Sec. 13-5 and 13-6, where this was a step in the determination of influence ordinates due to unit transverse loading, and these are the deflections due to m.

Example 14-2   Find the prestressing moment influence line $\eta_{\overline{M}P}$ for section B of the continuous beam ABC in Fig. 14-9a. The beam is assumed to have a rectangular section of constant width and a variable height, with a flexural rigidity of EI_A at end A.

The rotational stiffness of end B of the nonprismatic beam BA is $S_{BA} = 4.61\,EI_A/l$ (from tables[6]). The moment distribution is carried out in the usual way and the variation in the bending moment m is plotted in Fig. 14-9b. The ordinates of this diagram are divided by EI to obtain the influence line $\eta_{\overline{M}P}$ shown in Fig. 14-9c.

Example 14-3   Using the influence line determined in Example 14-2, find the secondary moment $\overline{M}_B$ due to a cable shown in Fig. 14-10, in which the prestressing force varies due to the friction loss. The eccentricity e of the cable profile and the value of the prestressing force P at different sections are listed in the figure.

The secondary moment at B is given by Eq. 14-15.

$$\overline{M}_B = -\int Pe\,\eta_{\overline{M}P}\,dx$$

in which the integral is over the entire length ABC. Applying Simpson's rule,

[6]*Handbook of Frame Constants*, Portland Cement Association, Chicago, Ill.

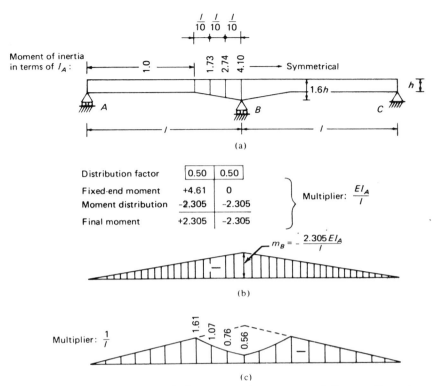

Fig. 14-9. Determination of the prestressing moment influence line at an intermediate support of the continuous beam considered in Example 14-2. (a) Beam. (b) Bending-moment diagram corresponding to a unit angular discontinuity at B. (c) Influence line $\eta_{\bar{M}P}$.

$$\int_o^n y\,dx = \frac{\lambda}{3}[(y_o + y_n) + 4(y_1 + y_3 + \ldots + y_{n-1})$$
$$+ 2(y_2 + y_4 + \ldots + y_{n-2})] \qquad (14\text{-}20)$$

where the terms y represent the values of $Pe\,\eta_{\bar{M}P}$ at $(n + 1)$ points at equal spacing λ (n being even), we obtain

$$\bar{M}_B = -2\int_A^B Pe\,\eta_{\bar{M}P}\,dx = 0.288\,P_A h$$

14-8 PRESTRESSING MOMENT INFLUENCE COEFFICIENTS FOR GRIDS

The method developed in the previous sections can be used for any continuous structure composed of beamlike members.

An example of a structure in which the use of prestressing moment influence coefficients is advantageous in the design of prestressing cables is

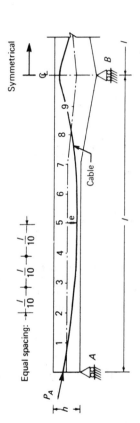

Prestressing cable profile

Section	A	1	2	3	4	5	6	7	8	9	B	Multi-plier
$\eta_{\overline{MP}}$	0	0.2304	0.4608	0.6913	0.9217	1.1521	1.3825	1.6130	1.070	0.7560	0.5630	$-\dfrac{1}{l}$
e	-0.10	+0.104	+0.256	+0.356	+0.404	+0.400	+0.344	+0.236	-0.024	-0.336	-0.568	h
P	1.00	0.992	0.983	0.974	0.965	0.956	0.947	0.938	0.929	0.920	0.891	P_A
$Pe\eta_{\overline{MP}}$	0	+0.0238	+0.1160	+0.2397	+0.3593	+0.4406	+0.4504	+0.3571	-0.2386	-0.2540	-0.2849	$-\dfrac{P_A h}{l}$

Fig. 14-10. Calculation of the secondary moment using prestressing moment influence coefficients for the beam in Example 14-3.

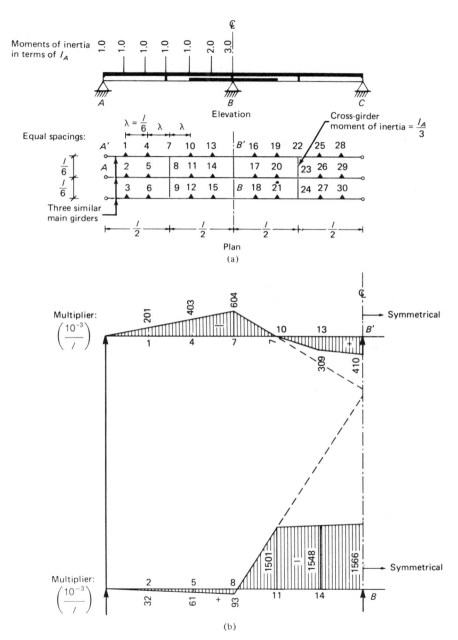

Fig. 14-11. Prestressing moment influence coefficients for a grid. (a) Bridge deck. (b) Influence line for the secondary bending moment at B.

the bridge deck shown in Fig. 14-11a; this has three main girders continuous over two spans and one cross-girder at the middle of each span. The bending rigidities of the main and cross-girders are as given in the figure and the torsional rigidity is ignored. The prestressing influence coefficients η_{MP} for the main girder at section B are plotted in Fig. 14-11b. The ordinates in this figure are the values of the bending moment causing a unit angular discontinuity in the central main girder at B divided by EI of each section. This bending-moment diagram is composed of straight lines (shown dotted) and can be determined by the method discussed in Sec. 13-7.

With the aid of the prestressing moment-influence coefficients for a structure of the type considered above, the location and the value of the prestressing force can be easily adjusted so as to satisfy the stress requirements at different sections of the structure.

14-9   GENERAL

The general superposition Eq. 14-2 is shown to facilitate the derivation of influence lines for arches and trusses.

The development of the prestressing moment-influence coefficients leads to the introduction of two new types of influence lines related to the usual type. The first derivative of the influence line of any action due to a unit transverse load gives the influence line for the same action due to a unit moving couple. This type of influence line may have little use in practice. However, the second derivative of the influence line of any action due to a transverse load gives influence coefficients of the statically indeterminate value of the action due to prestressing forces. This is useful in determining the secondary moments due to prestressing in concrete structures.

The influence coefficients of the secondary moment due to prestressing depend only on the geometry of the concrete structure, and not on the cable profile or on the magnitude of the prestressing force. Any variation in the prestressing force along the tendon or a change in the cable profile during the design can be easily allowed for by the use of these influence coefficients.

The values of the secondary moment-influence coefficients are the ordinates of a bending-moment diagram divided by the value of the flexural rigidity at all the sections. This bending-moment diagram is simple and can be obtained without much additional effort during the analysis for the influence line due to a unit transverse load.

PROBLEMS

The following problems are for Chapters 13 and 14.

14-1   Obtain the influence lines for the reaction at B, and bending moment and shear at E in the statically determinate beam shown in the figure. From the influence ordinates, determine the maximum positive values of these actions resulting from a uniform travelling live load q/unit length. The live load is to be applied on parts of the beam such that it produces maximum effect.

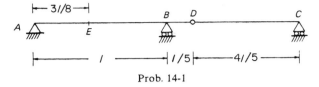

Prob. 14-1

14-2   Obtain the influence lines for the forces in the members marked Z_1 and Z_2 of the truss shown.

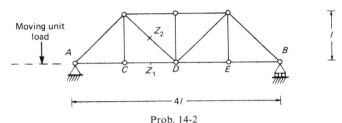

Prob. 14-2

14-3   Obtain the influence lines of the horizontal component of the reaction at A and the bending moment at D for the three-hinged parabolic arch shown in the figure. What are the maximum positive and negative values of the bending moment at D due to a uniform live load of q/unit length of horizontal projection?

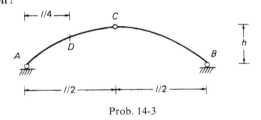

Prob. 14-3

14-4   For the bridge frame shown, calculate the influence ordinate at $0.3l$, $0.5l$, and $0.7l$ of each span and draw the influence lines for the following actions:
(a) Bending moment at n
(b) Bending moment at centre of span AB
(c) Shearing force at n

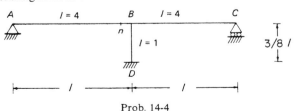

Prob. 14-4

14-5   The continuous prismatic beam shown in the figure is totally fixed at A and supported at B and C on elastic supports of stiffness EI/l^3, where EI is the flexural rigidity of the beam. Obtain the influence lines for R_A, R_B and the end moment at A. Give the ordinates at the supports and the middle of each span.

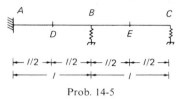

Prob. 14-5

14-6   Replace the hinged support at A of the frame of Prob. 14-4 by a roller support, then obtain the influence line of the end-moment DB. Give the ordinates at $0.3l$, $0.5l$, and $0.7l$ for each of the spans AB and BC.

14-7   Find the influence line of the bending moment at section n of the grid shown in the figure. The relative EI values are given alongside the girders. The ratio $GJ/(EI) = \frac{1}{4}$ for all members. Give ordinates at G, H, and I only. The end supports prevent the rotations about the x axis as well as the vertical displacements in the y direction.

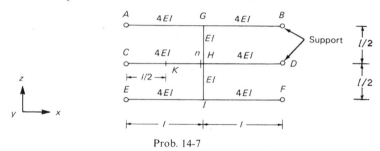

Prob. 14-7

14.8   Solve Prob. 14-7 with $GJ = 0$. Give the influence ordinates at G, H, I and the center of CH.

14-9   Prove Eqs. 13-11 and 13-12.

14-10   Prove Eq. 14-3.

14-11   Obtain the influence line for the reaction components X_A, Y_A and M_{AB} in the arch shown due to a unit vertical load. This load can be applied at the panel points A, B, C, D and E. Consider bending deformations only.

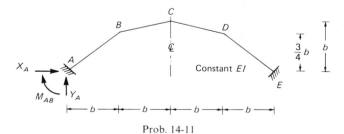

Prob. 14-11

14-12 Obtain the influence line for the component X_A of the reaction at A in the two-hinged truss in the figure. Use this influence line to obtain the influence line for the force in member DE. Assume that all the members are of the same cross section.

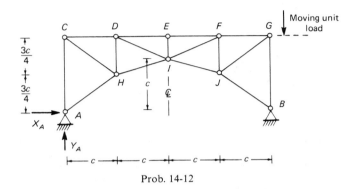

Prob. 14-12

14-13 Find the influence lines for the forces in members Z_1 and Z_2 of the truss in the figure. Assume that all the members have the same $l/(aE)$ value, l and a being the length and the cross sectional area of the member.

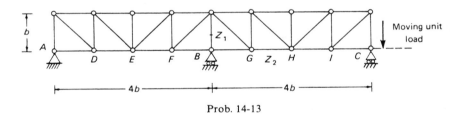

Prob. 14-13

14-14 Prove Eq. 14-17.

14-15 (a) Obtain the influence line of the secondary moment at n (due to prestressing), in the frame in Fig. 13-10a.

(b) From this influence line obtain the influence line of the statically indeterminate vertical reaction at A. [*Hint:* The bending moment corresponding to the deflection line in Fig. 13-10d when divided by $-(EI)_{BA}$ gives the influence line required in (a).]

14-16 Obtain the influence line of the secondary moment at B due to prestressing in the symmetrical beam shown in part (a) of the figure. Use this influence line to find the total moment, $M = M_s + \bar{M}$, due to the symmetrical prestressing cable in part (b). Assume the prestressing force to be constant (P) and the cable profile to be a parabola in each span.

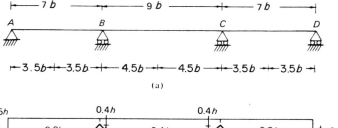

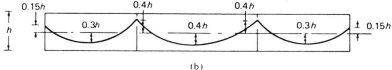

(a)

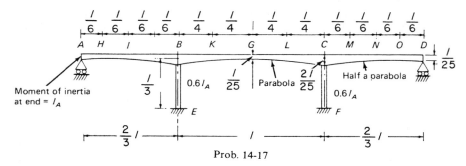

(b)

Prob. 14-16

14-17   Obtain the influence line of the secondary moment due to prestressing at a section just to the right of B in the bridge frame shown. Give the ordinates at quarter-points and mid-points of each span. The cross sections of the members are rectangles of the same width and of length indicated in the figure. The end-rotational stiffness and the carryover factors of the nonprismatic members are given in Prob. 12-8.

Prob. 14-17

14-18   Use the results obtained in Example 13-2 to obtain the influence line in girder FJ of the secondary moment at section n. Hence, calculate the secondary moment at n due to the prestressing cable shown in the figure. The prestressing force P is assumed constant. [*Hint:* Determine the bending moment in girder FJ corresponding to the displacements $\{D_1\}$ and $\{D_2\}$ calculated in Example 13-2. The bending moment divided by EI gives the required influence line.]

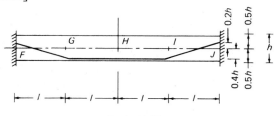

Prob. 14-18

CHAPTER 15

Effects of axial forces

15-1 INTRODUCTION

In the preceding chapters we have in many cases analyzed structures using a simplified approach which assumes that the displacements are caused by one type of internal forces only. Thus, in frames, we neglected the effect of axial forces and shear on deformation and considered displacements due to bending only. On the other hand, in trusses, we assumed the joints to be pin-connected so that the members were considered to be subject to axial forces only with no bending present. The assumptions result in a considerable simplification in the analysis but it is important to know when they are justified.

Broadly speaking, when the axial forces in members of a frame are relatively high, two effects may be of importance: a change in geometry arising from a change in length of members, and a change in stiffness arising from bending by the axial force.

The first effect means simply that the axial forces cause a change in length of the members and thus contribute to joint displacements. This effect can be included in the general force or displacement method of analysis (Chapters 2 and 3), and in most cases results in an increase in the amount of calculations making the use of a computer necessary. However, a simplified analysis using moment distribution is possible; for example, in a frame, we estimate approximately the axial forces, calculate the joint translations accordingly, and carry out the moment distribution with the starting fixed-end moments based on these displacements plus the transverse load on the members. If necessary the calculations can be repeated with more accurate values of the axial forces determined from the preceding analysis. In a truss with rigid joints, the forces in members and joint translations are first determined assuming all the joints to be pinned, then the presence of rigid joints is recognized and the member end-moments corresponding to the joint displacements obtained in the first analysis are calculated. Moment distribution is one way of calculating these end-moments.

As already mentioned, the second effect of an axial force is the change in the stiffness of a member due to bending caused by the axial force. The forces required to cause a unit rotation or translation in the transverse direction at one end of a member decrease if the member is subjected to an axial compressive force, and conversely increase if the axial force is tensile. This effect of the axial force is important in slender members only, and is referred to as *beam-column* effect.

In the following sections each of the above effects is treated separately. However, an analysis in which both effects are included can also be made.

15-2   EFFECT OF CHANGE IN LENGTH

In this section, the effect of a change in length of members of a rigid-jointed frame will be accounted for in a two-stage solution using moment distribution. The changes in length of the members are determined first and then the joints of the structure are considered to move without rotation to the corresponding displacement positions. The fixed-end moments caused by these joint displacements are added to those caused by the transverse loads on the members to give the total starting fixed-end moments. We should note that the change in length of a member depends on the magnitude of the axial force which may not be known in advance. The procedure is therefore to estimate a value of the axial force by an approximate analysis, then to determine the changes in length and carry out the moment distribution. From the results, more accurate values of axial forces are determined, and the process is repeated if necessary.

Example 15-1   The frame in Fig. 15-1a carries a uniform load of q/unit length on AB and a heavy load $P = 13qb$ directly above the column. The frame has a constant cross section with $I = 20 \times 10^{-6}b^4$ and area $a = 3.5 \times 10^{-3}b^2$. Obtain the bending-moment diagram.

First, an approximate value of the compressive force in member BD is estimated. Assume N_{BD} to be equal to P plus one-half the total uniform load, that is $N_{BD} \simeq 14qb$. The corresponding shortening of BD, that is the downward displacement of B is

$$D_{B\,\text{vert}} = \left(\frac{Nl}{aE}\right)_{BD} = 20 \times 10^3\,\frac{q}{E}$$

The axial force in AB is assumed to cause a negligible horizontal displacement at B.

The fixed-end moments due to the above joint displacement plus the transverse loading on the members are (see Appendices C and D):

$$M_{BA} = \frac{ql_1^2}{8} - \frac{3EI}{l_1^2}D_{B\,\text{vert}} = 0.2qb^2$$

$$M_{BC} = 0 + \frac{3EI}{l_1^2}D_{B\,\text{vert}} = 0.3qb^2$$

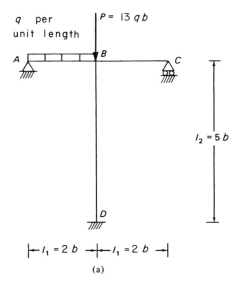

(a)

End	BA	BD	BC
DF's	0.395	0.210	0.395
FEM's	20.0	0	30.0
Distribution	-19.7	-10.5	-19.8
	0.3	-10.5	10.2

(b)

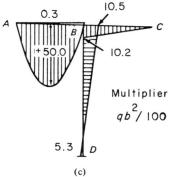

(c)

Fig. 15-1. Frame analyzed in Example 15-1. (a) Frame dimensions and loading. (b) Moment distribution. (c) Bending-moment diagram.

and the fixed-end moments at all other ends are zero. The moment distribution is carried out in the usual way in Fig. 15-1b. From the result obtained, more accurate values of the axial forces are calculated:

$$N_{BD} = 14qb + \frac{qb^2}{100} \frac{(0.3 - 10.2)}{2b} = 13.95qb \text{ (compression)}$$

and

$$N_{BA} = \frac{qb^2}{100} \frac{(10.5 + 5.3)}{5b} = 0.03qb \text{ (compression)}$$

With these new values of N the displacement of D can be calculated, and the moment distribution repeated. It is obvious that in this example the change will be negligible and repeated calculations are therefore not made. The bending moment diagram is drawn in Fig. 15-1c.

If the change in length due to axial forces is ignored, the end-moments of the members meeting at B are

$$M_{BA} = 0.303qb^2 \qquad M_{BC} = -0.198qb^2 \qquad \text{and} \qquad M_{BD} = -0.105qb^2$$

It can be seen thus that in the present example there is a large difference between the values ignoring axial forces and those obtained by the more accurate analysis because of the high axial force in BD.

15-3   SECONDARY MOMENTS IN TRUSSES

The change in length due to axial forces in the members of a truss introduces a relative translation of its ends and hence induces the so-called secondary moments. Except for special cases, the secondary moments are important only in long-span bridge trusses which have stocky members. The secondary stress due to these moments can be as high as 60 to 75 percent of the primary stress due to the axial forces. Design rules for large bridges require that the secondary stresses be considered and, when they exceed certain limits, be treated as a primary stress.

However, the secondary stresses, even when of high intensity, generally have little effect on the carrying capacity of a truss. If a high secondary stress combines with the primary stress to cause the material to yield, a plastic hinge is formed and this tends to relieve the bending stress. Nevertheless, the secondary stresses require consideration when high stresses are repeated often enough to produce fatigue and when they are likely to induce local buckling or to weaken the joint connections.

In modern practice the analysis for axial forces and moments in a truss bridge with rigid joints would most likely be made by a general displacement method using a computer. However, a solution by moment distribution is also possible; this is considered below with the following simplifying assumptions:

(a) The secondary moments do not change the primary axial forces calculated for a pin-jointed truss.

(b) The only effect of the axial forces is to change the length of the member, thereby causing bending due to the relative translation of its ends, but the stiffness characteristics of the members are unchanged. In other words, the secondary moments are significant only for heavy stocky members in which the change in stiffness due to axial force is small.

15-31 Steps in Calculation by Moment Distribution

(a) The axial forces in the members are calculated assuming the joints to be pinned. The changes in length of the members are then calculated and the joint translations determined either analytically or by a Williot diagram (see Sec. 9-2). It is not necessary to determine the absolute joint movements but only the chord rotation ψ for each member (see Fig. 15-2a). (A method of calculating ψ from the changes in member lengths is given later in this section.)

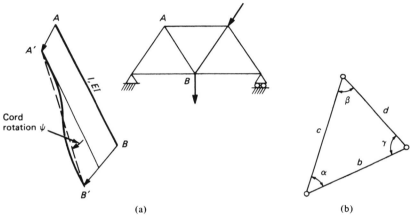

(a)   (b)

Fig. 15-2. Determination of relative chord rotation. (a) Chord rotation of a typical truss member AB. The translations AA' and BB' are determined for a pin-jointed truss. (b) Triangle used for derivation of Eqs. 15-2 and 15-3.

(b) The end-moments of the members caused by the joint translation without rotation are calculated by Eq. 11-2 (or 11-5). For any prismatic member AB (Fig. 15-2) with length l and flexural rigidity EI, the two end-moments are

$$M_{AB} = M_{BA} = -\frac{6EI}{l}\psi \tag{15-1}$$

where ψ is the chord rotation, positive if clockwise. If the member is subjected to transverse loading between the joints, the fixed-end moments due to this loading should be added to those calculated by Eq. 15-1.

(c) The joints are now allowed to rotate without further translation. This is done by balancing the end-moments calculated in (b) by moment distribution. The moments obtained in this way are sufficiently accurate for practical purposes. If, however, more accurate results are desired, we can, from the moments obtained in this step, check the truss for statical equilibrium and determine a second set of values for the axial forces. The entire analysis is then repeated.

15-32 Relative Chord Rotation

Consider any three members of a pin-jointed truss forming a triangle with sides d, b, and c and angles α, β, and γ (Fig. 15-2b). From geometry,

$$\frac{d}{\sin \alpha} = \frac{b}{\sin \beta} = \frac{c}{\sin \gamma}$$

or

$$\ln d - \ln \sin \alpha = \ln b - \ln \sin \beta = \ln c - \ln \sin \gamma$$

Differentiating,

$$\frac{\Delta d}{d} - \Delta\alpha \cot \alpha = \frac{\Delta b}{b} - \Delta\beta \cot \beta = \frac{\Delta c}{c} - \Delta\gamma \cot \gamma$$

Solving, we can express the change $\Delta\alpha$ in the angle α due to small changes in member lengths Δd, Δb, and Δc by

$$\Delta\alpha = \left[\frac{\Delta d}{d} - \frac{\Delta b}{b}\right] \cot \gamma + \left[\frac{\Delta d}{d} - \frac{\Delta c}{c}\right] \cot \beta \qquad (15\text{-}2)$$

If the axial forces in the members are N_d, N_b, and N_c (positive if tension) and a_d, a_b, and a_c are their cross-sectional areas, then

$$\Delta\alpha = \frac{1}{E}\left[\left(\frac{N_d}{a_d} - \frac{N_b}{a_b}\right) \cot \gamma + \left(\frac{N_d}{a_d} - \frac{N_c}{a_c}\right) \cot \beta\right] \qquad (15\text{-}3)$$

After the angle changes have been determined by Eq. 15-3, the chord rotations of the members can be calculated by simple arithmetic. A member is arbitrarily selected as a reference line of no rotation, and the relative chord rotations of all the other members are calculated successively from the known angle changes. These relative chord rotations can be used instead of the real rotations in Eq. 15-1 to obtain a set of end-moments which will, after moment distribution, give the real moments. If the truss is symmetrical and symmetrically loaded, a member at the center of the truss is truly fixed in direction and, if it is selected as the reference member, the calculated chord rotations are the real rotations.

Example 15-2 Analyze the rigid-jointed truss of Fig. 15-3a for axial forces and bending moments.

The axial forces N and stresses N/a in a corresponding pin-jointed truss are calculated in Fig. 15-3b. Because of symmetry, the calculations need be made only for one-half of the truss. The angle changes caused by the change in length of members are calculated in Table 15-1 by means of Eq. 15-3.

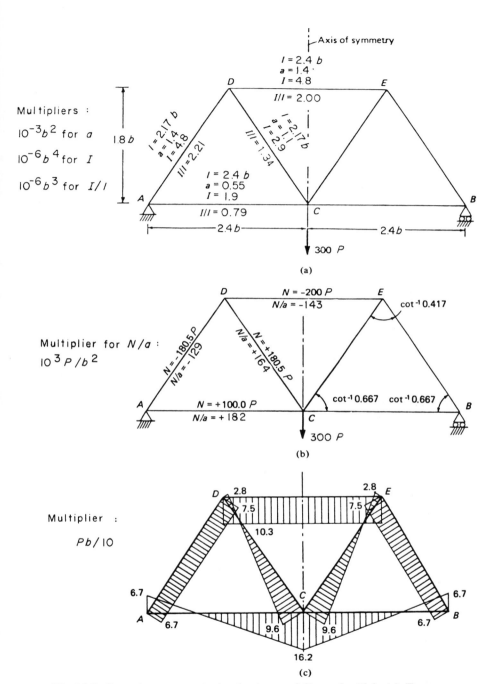

Fig. 15-3. Secondary moments in the truss of Example 15-2. (a) Truss properties and loading. (b) Stresses in members for a pin-jointed truss. (c) Bending-moment diagram.

Table 15-1. Calculation of Angle Changes by Eq. 15-3

Angle α	$\left(\dfrac{N_d}{a_d} - \dfrac{N_b}{a_b}\right)\cot\gamma$	$\left(\dfrac{N_d}{a_d} - \dfrac{N_c}{a_c}\right)\cot\beta$	$\Delta\alpha$	Check
DAC	$(164 - 182)0.667 = -\ 12$	$(164 + 129)0.417 = \quad 122$	110	
CDA	$(182 + 129)0.667 = +207$	$(182 - 164)0.667 = \quad 12$	219	} = sum zero
ACD	$(-129 - 164)0.417 = -122$	$(-129 - 182)0.667 = -207$	−329	
EDC	$(164 + 143)0.667 = \quad 205$	$(164 - 164)0.417 = \quad 0$	205	
DCE	$(-143 - 164)0.667 = -205$	$(-143 - 164)0.667 = -205$	−410	} = sum zero
CED	From symmetry		205	
Multiplier	$10^3 P/b^2$	$10^3 P/b^2$	$10^3 P/Eb^2$	

The relative chord rotations ψ are calculated in Table 15-2. The members are listed in order, starting with the member assumed to have zero rotations, proceeding to the other sides of the triangle in an anticlockwise direction and ending with a member forming a side of the adjacent triangle. If this order is followed, the summation of the angle change gives — without change in sign — the relative chord rotations. The members can be listed in a different order but the relative chord rotations will always be either the sum or the difference of the angle changes. In this truss, the member DE is known to have no rotation because of symmetry. Starting with this reference member, the chord rotations calculated are the correct absolute values. In an unsymmetrical truss, where there is no member for which the chord rotation is known to be zero, any member may be assumed to have zero rotation, and the resultant chord rotations are relative rotations. The moments based on the relative rotations do not have to be corrected, as a rotation of the truss as a rigid body would produce the true rotation of each member. An alternative to the calculations in Tables 15-1 and 15-2 is to draw a Williot diagram from which the ψ values are determined.

Table 15-2. Calculation of Relative Chord Rotations, ψ

Member	Angle between members	Change in angle $\Delta\alpha$ $10^3 P/(Eb^2)$	Relative chord rotation ψ $10^3 P/(Eb^2)$
ED			0
	EDC	205	
DC			205
	CDA	219	
DA			424
	DAC	110	
AC			534

The member end-moments are calculated by Eq. 15-1 and the moment distribution is carried out for half the truss in Table 15-3. Because of symmetry, the ends of the members meeting at C are treated as encastré and the end-rotational stiffness of DE is $S_{DE} = 2(EI/l)_{DE}$. The end-rotational stiffness of

Table 15-3.   Moment Distribution

Joint	A		D			C	
End	AC	AD	DA	DE	DC	CD	CA
DF	0.27	0.73	0.49	0.22	0.29	0	0
FEM (Pb/10)	− 25.7	− 56.5	− 56.5	0	− 16.4	− 16.4	− 25.7
Distribution	+ 22.2	+ 60.0	+ 35.7	+ 16.0	+ 21.2		
Carryover	0	+ 17.9	+ 30.0	0	0	+ 10.6	+ 11.1
Distribution	− 4.8	− 13.1	− 14.7	− 6.6	− 8.7		
Carryover	0	− 7.4	− 6.6	0	0	− 4.4	− 2.4
Distribution	+ 2.0	+ 5.4	+ 3.2	+ 1.5	+ 1.9		
Carryover	0	+ 1.6	+ 2.7	0	0	+ 1.0	+ 1.0
Distribution	− 0.4	− 1.2	− 1.3	− 0.6	− 0.8		
						− 0.4	− 0.2
Final moments, (Pb/10)	− 6.7	+ 6.7	− 7.5	+ 10.3	− 2.8	− 9.6	− 16.2

the other members is $4EI/l$. The bending-moment diagram for the truss is shown in Fig. 15-3c.

If it is desired to improve the accuracy of the results, the whole procedure can be repeated with a new set of axial forces determined by static equilibrium, taking into account the presence of the moments obtained from the first calculation.

15-4   STIFFNESS OF PRISMATIC MEMBER SUBJECTED TO AN AXIAL FORCE

We recall that the force and displacement methods of analysis considered in the previous chapters are for linear structures for which the principle of superposition holds. In such structures, the deformations are proportional to the applied loads, and the displacements or the internal forces caused by a set of effects can be obtained by superposition. For example, the slope-deflection equation (Eq. 11-5) is a superposition equation giving the end-moment of a member due to end displacements and transverse loading. Likewise, in moment distribution, the final end-moments are obtained by superposition of the effects of step-by-step joint displacements.

In Sec. 1-6 we saw that the superposition of deflections cannot be applied in the case of a strut subjected to axial compression together with a transverse load because of the additional moment caused by the change in geometry of the member by the loading. If, however, the strut is subjected to a system of loads or end displacements all acting in the presence of an unaltered axial force, the effects of these loads and the displacements can be superimposed. Thus the axial force can be looked upon as a parameter which affects the stiffness or the flexibility of the member and, once these have been determined, the methods of analysis of linear structures can be applied.

In the analysis of rigid frames with very slender members, the axial forces are generally not known at the outset of the analysis. A set of axial forces is estimated, the stiffness (or flexibility) of the members is determined accordingly, and the frame is analyzed as a linear structure. If the results show that the axial forces obtained by this analysis differ greatly from the assumed values, the calculated values are used to find new stiffness (or flexibility) values and the analysis is repeated.

Let us now consider in detail the effect of an axial compressive and tensile force on the stiffness of a prismatic member.

15-41   Effect of Axial Compression

The differential equation governing the deflection y of a prismatic member AB subjected to a compressive force P and any end restraint (Fig. 15-4a) is[1]

$$\frac{d^4y}{dx^4} + \frac{P}{EI}\frac{d^2y}{dx^2} = \frac{q}{EI} \tag{15-4}$$

where q is the intensity of transverse loading. When $q = 0$, the general solution of Eq. 15-4 is

$$y = A_1 \sin u\frac{x}{l} + A_2 \cos u\frac{x}{l} + A_3 x + A_4 \tag{15-5}$$

where

$$u = l\sqrt{\frac{P}{EI}} \tag{15-6}$$

and A_1, A_2, A_3, and A_4 are the integration constants to be determined from the boundary conditions.

Equation 15-5 will now be used to derive the stiffness matrix of an axially compressed member corresponding to the coordinates 1, 2, 3, and 4 in Fig. 15-4b. The displacements $\{D\}$ at the four coordinates

$$D_1 = (y)_{x=0} \qquad D_2 = \left(\frac{dy}{dx}\right)_{x=0} \qquad D_3 = (y)_{x=l} \qquad D_4 = \left(\frac{dy}{dx}\right)_{x=l}$$

are related to the constants $\{A\}$ by the equation

$$
\begin{Bmatrix} D_1 \\ D_2 \\ D_3 \\ D_4 \end{Bmatrix}
=
\begin{bmatrix}
0 & 1 & 0 & 1 \\
\dfrac{u}{l} & 0 & 1 & 0 \\
s & c & l & 1 \\
\dfrac{u}{l}c & -\dfrac{u}{l}s & 1 & 0
\end{bmatrix}
\begin{Bmatrix} A_1 \\ A_2 \\ A_3 \\ A_4 \end{Bmatrix}
\tag{15-7}
$$

where $s = \sin u$, and $c = \cos u$.

[1] See Eq. 9-7.

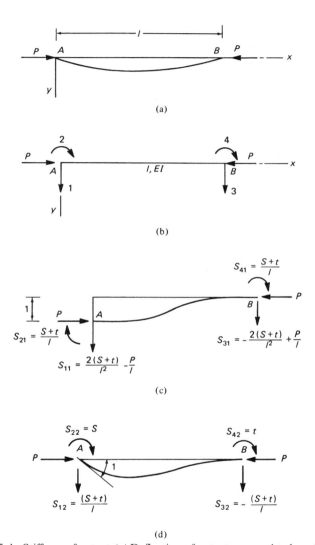

Fig. 15-4. Stiffness of a strut. (a) Deflection of a strut assumed to be subjected to end-forces and end-displacements at the coordinates indicated in part (b). (b) Coordinate system corresponding to the stiffness matrix of a strut in Eq. 15-15. (c) End-forces corresponding to $D_1 = 1$ while $D_2 = D_3 = D_4 = 0$. (d) End-forces corresponding to $D_2 = 1$ while $D_1 = D_3 = D_4 = 0$.

Equation 15-7 can be written

$$\{D\} = [B]\{A\} \tag{15-8}$$

where $[B]$ is the 4×4 matrix in Eq. 15-7.

The forces $\{F\}$ at the four coordinates are the shear and bending moment at $x = 0$ and $x = l$, thus[2]

$$
\begin{Bmatrix} F_1 \\ F_2 \\ F_3 \\ F_4 \end{Bmatrix} =
\begin{Bmatrix} (-V)_{x=0} \\ (M)_{x=0} \\ (V)_{x=l} \\ (-M)_{x=l} \end{Bmatrix} = EI
\begin{Bmatrix} \left(\dfrac{d^3y}{dx^3} + \dfrac{u^2}{l^2}\dfrac{dy}{dx}\right)_{x=0} \\[2mm] \left(-\dfrac{d^2y}{dx^2}\right)_{x=0} \\[2mm] \left(-\dfrac{d^3y}{dx^3} - \dfrac{u^2}{l^2}\dfrac{dy}{dx}\right)_{x=l} \\[2mm] \left(\dfrac{d^2y}{dx^2}\right)_{x=l} \end{Bmatrix}
$$

Differentiating Eq. 15-5 and substituting in the last equation,

$$
\begin{Bmatrix} F_1 \\ F_2 \\ F_3 \\ F_4 \end{Bmatrix} = EI
\begin{bmatrix}
0 & 0 & \dfrac{u^2}{l^2} & 0 \\[2mm]
0 & \dfrac{u^2}{l^2} & 0 & 0 \\[2mm]
0 & 0 & -\dfrac{u^2}{l^2} & 0 \\[2mm]
-\dfrac{su^2}{l^2} & -\dfrac{cu^2}{l^2} & 0 & 0
\end{bmatrix}
\begin{Bmatrix} A_1 \\ A_2 \\ A_3 \\ A_4 \end{Bmatrix}
\tag{15-9}
$$

or

$$\{F\} = [C]\{A\} \tag{15-10}$$

where $[C] = EI$ times the 4×4 matrix in Eq. 15-9.

Solving for $\{A\}$ from Eq. 15-8 and substituting into Eq. 15-10,

$$\{F\} = [C][B]^{-1}\{D\} \tag{15-11}$$

Putting

$$[S] = [C][B]^{-1} \tag{15-12}$$

[2] Equation 9-1 expresses the bending moment M in terms of the deflections. Using this equation with Eqs. 9-3 and 15-6, the shear can be expressed as

$$V = \frac{dM}{dx} - P\frac{dy}{dx} = EI\left(-\frac{d^3y}{dx^3} - \frac{u^2}{l^2}\frac{dy}{dx}\right)$$

Equation 15-11 takes the form

$$\{F\} = [S]\{D\}, \tag{15-13}$$

where $[S]$ is the required stiffness matrix. The inverse of $[B]$ is

$$[B]^{-1} = \frac{1}{2 - 2c - us}
\begin{bmatrix}
-s & \dfrac{l}{u}(1 - c - us) & s & -\dfrac{l}{u}(1 - c) \\[2mm]
(l - c) & \dfrac{l}{u}(s - uc) & -(1 - c) & \dfrac{l}{u}(u - s) \\[2mm]
\dfrac{u}{l}s & (l - c) & -\dfrac{u}{l}s & (1 - c) \\[2mm]
(1 - c - us) & -\dfrac{l}{u}(s - uc) & (l - c) & -\dfrac{l}{u}(u - s)
\end{bmatrix} \tag{15-14}$$

Substituting Eq. 15-14 into Eq. 15-12, we obtain the stiffness of a strut corresponding to the coordinates in Fig. 15-4b:

$$[S] = EI
\begin{bmatrix}
\dfrac{u^3 s}{l^3(2 - 2c - us)} & & \text{symmetrical} & \\[3mm]
\dfrac{u^2(1 - c)}{l^2(2 - 2c - us)} & \dfrac{u(s - uc)}{l(2 - 2c - us)} & & \\[3mm]
-\dfrac{u^3 s}{l^3(2 - 2c - us)} & -\dfrac{u^2(1 - c)}{l^2(2 - 2c - us)} & \dfrac{u^3 s}{l^3(2 - 2c - us)} & \\[3mm]
\dfrac{u^2(1 - c)}{l^2(2 - 2c - us)} & \dfrac{u(u - s)}{l(2 - 2c - us)} & -\dfrac{u^2(1 - c)}{l^2(2 - 2c - us)} & \dfrac{u(s - uc)}{l(2 - 2c - us)}
\end{bmatrix} \tag{15-15}$$

When the axial force P vanishes, $u \to 0$, and the above stiffness matrix becomes identical with the stiffness matrix in Eq. 4-7. Thus,

$$\lim_{u \to o} [S] =
\begin{bmatrix}
\dfrac{12EI}{l^3} & \text{symmetrical} & & \\[3mm]
\dfrac{6EI}{l^2} & \dfrac{4EI}{l} & & \\[3mm]
-\dfrac{12EI}{l^3} & -\dfrac{6EI}{l^2} & \dfrac{12EI}{l^3} & \\[3mm]
\dfrac{6EI}{l^2} & \dfrac{2EI}{l} & -\dfrac{6EI}{l^2} & \dfrac{4EI}{l}
\end{bmatrix} \tag{15-16}$$

The elements in the first and second column of $[S]$ in Eq. 15-15 are the forces necessary to hold the member in the deflected configuration shown in Figs. 15-4c and d. The two end-moments in Fig. 15-4d are the end-rotational stiffness $S(= S_{AB} = S_{BA})$ and the carryover moment t, which are needed for moment distribution or for the slope-deflection equation Eq. 11-2. Thus the end-rotational stiffness and the carryover moment for prismatic member subjected to an axial compressive force P (Fig. 15-4d) are

$$S = \frac{u(s - uc)}{(2 - 2c - us)} \frac{EI}{l} \qquad (15\text{-}17)$$

and

$$t = \frac{u(u - s)}{(2 - 2c - us)} \frac{EI}{l} \qquad (15\text{-}18)$$

where $s = \sin u$, $c = \cos u$, and $u = l\sqrt{P/(EI)}$. The carryover factor $C(= C_{AB} = C_{BA})$ is

$$C = \frac{t}{S} = \frac{u - s}{s - uc} \qquad (15\text{-}19)$$

When u tends to zero S, t, and C tend to $4EI/l$, $2EI/l$, and $1/2$, respectively. It can be seen thus that the axial compressive force reduces the value of S and increases C. When the force P reaches the critical buckling value, S becomes zero, which means that the displacement D_2 can be caused by an infinitely small value of the force F_2, and the value of C approaches infinity. From Eq. 15-17, S is zero when $s = uc$ or $\tan u = u$. The smallest value of u to satisfy this equation is $u = l\sqrt{P_{cr}/(EI)} = 4.49$. Thus the buckling load of the strut with the end conditions of Fig. 15-4d is $P_{cr} = 20.19(EI/l^2)$.

The critical buckling load corresponding to any end conditions can be easily derived from the stiffness matrix in Eq. 15-15.

The end-moments S_{21} and S_{41} in Fig. 15-4c can be expressed in terms of S and t by the slope-deflection equation (Eq. 11-2). The vertical forces at the member ends in Figs. 15-4c or d can be derived from the end-moments by static equilibrium. Thus, all the elements of $[S]$ in Eq. 15-15 can be expressed in terms of S and t, and the stiffness matrix of a strut corresponding to the coordinates in Fig. 15-4b becomes

$$[S] = \begin{bmatrix} \dfrac{2(S+t)}{l^2} - \dfrac{P}{l} & & & & \\ & & & \text{symmetrical} & \\ \dfrac{S+t}{l} & S & & & \\ -\dfrac{2(S+t)}{l^2} + \dfrac{P}{l} & -\dfrac{(S+t)}{l} & \dfrac{2(S+t)}{l^2} - \dfrac{P}{l} & \\ \dfrac{S+t}{l} & t & -\dfrac{(S+t)}{l} & S \end{bmatrix} \qquad (15\text{-}20)$$

If we substitute $P = u^2 EI/l^2$ in the above equation, it becomes apparent

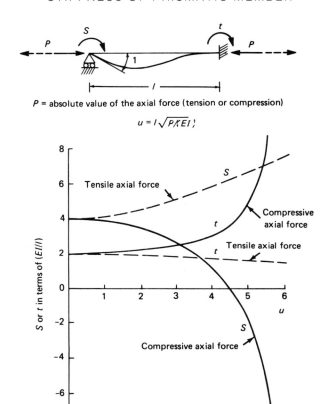

P = absolute value of the axial force (tension or compression)

$$u = l\sqrt{P/(EI)}$$

Fig. 15-5. End-rotational stiffness and carryover moment for a prismatic member subjected to an axial force.

that all the elements of the stiffness matrix are a function of the dimensionless parameter u, the member length l, and the flexural rigidity EI. Figure 15-5 shows the values of the end-rotational stiffness S and the carryover moment t in terms of the dimensionless parameter $u = l\sqrt{P/(EI)}$; Table 15-4 lists the numerical data. All these values apply to prismatic members only. The negative values of S correspond to P values higher than the critical buckling load, so that the negative sign indicates that a restraining moment is needed to hold the end rotation at a value not exceeding unity.

15-42 Effect of Axial Tension

The stiffness of a prismatic member in tension can be derived from the expressions for a member in compression by replacing P by $(-P)$, accordingly, the notation $u = l\sqrt{P/(EI)}$ is to be replaced by $l\sqrt{-P/(EI)} = iu$, where

Table 15-4. Values of End-Rotational Stiffness S and Carryover Moment t in Terms of EI/l for Prismatic Members Subjected to Axial Forces (Eqs. 15-17, 15-18, 15-21, and 15-22; see Fig. 15-5)

Axial Compressive Force P						Axial Tensile Force P					
u*	S	t	u	S	t	u	S	t	u	S	t
0.0	4.000	2.000	3.0	2.624	2.411	0.0	4.000	2.000	3.0	5.081	1.766
0.1	3.998	2.000	3.1	2.515	2.450	0.1	4.001	1.999	3.1	5.147	1.754
0.2	3.995	2.001	3.2	2.399	2.492	0.2	4.005	1.999	3.2	5.214	1.742
0.3	3.988	2.003	3.3	2.276	2.538	0.3	4.012	1.997	3.3	5.283	1.730
0.4	3.977	2.005	3.4	2.146	2.588	0.4	4.021	1.995	3.4	5.353	1.718
0.5	3.967	2.008	3.5	2.008	2.642	0.5	4.033	1.992	3.5	5.424	1.706
0.6	3.952	2.012	3.6	1.862	2.702	0.6	4.048	1.988	3.6	5.497	1.694
0.7	3.934	2.016	3.7	1.706	2.767	0.7	4.065	1.984	3.7	5.570	1.683
0.8	3.914	2.022	3.8	1.540	2.838	0.8	4.085	1.979	3.8	5.645	1.671
0.9	3.891	2.028	3.9	1.363	2.917	0.9	4.107	1.974	3.9	5.720	1.659
1.0	3.865	2.034	4.0	1.173	3.004	1.0	4.132	1.968	4.0	5.797	1.648
1.1	3.836	2.042	4.1	0.970	3.100	1.1	4.159	1.961	4.1	5.874	1.636
1.2	3.804	2.050	4.2	0.751	3.207	1.2	4.188	1.954	4.2	5.953	1.625
1.3	3.769	2.059	4.3	0.515	3.327	1.3	4.220	1.946	4.3	6.032	1.614
1.4	3.732	2.070	4.4	0.259	3.462	1.4	4.255	1.938	4.4	6.112	1.603
1.5	3.691	2.081	4.5	-0.019	3.614	1.5	4.292	1.930	4.5	6.193	1.592
1.6	3.647	2.093	4.6	-0.323	3.787	1.6	4.330	1.921	4.6	6.275	1.581
1.7	3.599	2.106	4.7	-0.658	3.984	1.7	4.372	1.912	4.7	6.357	1.571
1.8	3.548	2.120	4.8	-1.029	4.211	1.8	4.415	1.902	4.8	6.440	1.561
1.9	3.494	2.135	4.9	-1.443	4.475	1.9	4.460	1.892	4.9	6.524	1.551
2.0	3.436	2.152	5.0	-1.909	4.785	2.0	4.508	1.881	5.0	6.608	1.541
2.1	3.374	2.170	5.1	-2.439	5.151	2.1	4.557	1.871	5.1	6.693	1.531
2.2	3.309	2.189	5.2	-3.052	5.592	2.2	4.608	1.860	5.2	6.779	1.521
2.3	3.240	2.210	5.3	-3.769	6.130	2.3	4.661	1.849	5.3	6.865	1.512
2.4	3.166	2.233	5.4	-4.625	6.798	2.4	4.716	1.837	5.4	6.952	1.503
2.5	3.088	2.257	5.5	-5.673	7.647	2.5	4.773	1.826	5.5	7.039	1.494
2.6	3.005	2.283	5.6	-6.992	8.759	2.6	4.831	1.814	5.6	7.127	1.485
2.7	2.918	2.312	5.7	-8.721	10.269	2.7	4.891	1.802	5.7	7.215	1.476
2.8	2.825	2.342	5.8	-11.111	12.428	2.8	4.953	1.791	5.8	7.303	1.468
2.9	2.728	2.376	5.9	-14.671	15.745	2.9	5.016	1.779	5.9	7.392	1.460

*$u = l\sqrt{P/(EI)}$.

$i = \sqrt{-1}$. Applying this to Eqs. 15-17–15-19 and making use of the fact that

$$\sinh u = -i \sin iu$$

and

$$\cosh u = \cos iu$$

we obtain the end-rotational stiffness, carryover moment, and carryover factor for a prismatic member subjected to an axial tensile force P:

$$S = \frac{u\,(u \cosh u - \sinh u)}{(2 - 2 \cosh u + u \sinh u)} \frac{EI}{l} \tag{15-21}$$

$$t = \frac{u\,(\sinh u - u)}{(2 - 2 \cosh u + u \sinh u)} \frac{EI}{l} \tag{15-22}$$

and

$$C = \frac{t}{S} = \frac{\sinh u - u}{u \cosh u - \sinh u} \tag{15-23}$$

where $u = l\sqrt{P/(EI)}$, P being the absolute value of the axial tensile force. The values of S and t calculated by the above equations are included in Table 15-4 and are plotted against u in Fig. 15-5.

Considering equilibrium, we can readily see that the stiffness matrix corresponding to the coordinates in Fig. 15-4b for a member in tension is

$$[S] = \begin{bmatrix} \dfrac{2(S+t)}{l^2} + \dfrac{P}{l} & & \text{symmetrical} & & \\[2ex] \dfrac{S+t}{l} & S & & & \\[2ex] -\dfrac{2(S+t)}{l^2} - \dfrac{P}{l} & -\dfrac{(S+t)}{l} & \dfrac{2(S+t)}{l^2} + \dfrac{P}{l} & \\[2ex] \dfrac{S+t}{l} & t & -\dfrac{(S+t)}{l} & S \end{bmatrix} \tag{15-24}$$

15-43 General Treatment

Summarizing the above discussion, we can see that the axial force is considered as a parameter affecting the stiffness of the member and the forces corresponding to unit end-displacements are expressed in terms of S and t, which are functions of the dimensionless parameter $u = l\sqrt{P/(EI)}$. Once the stiffness of the member is known, any one of the displacement methods of analysis considered in the previous chapters can be used.

The end-forces corresponding to unit end-displacement of a prismatic member subjected to an axial force are listed in Table 15-5. All forces are

Table 15-5. End-Forces Caused by End Displacements of a Prismatic Member Subjected to Axial Compressive or Tensile Force

Bar	End-Forces	
	Axial Compressive Force P	Axial Tensile Force P
(a)	$F_1 = F_2 = \dfrac{S+t}{l}$ $F_3 = -F_4 = \dfrac{2(S+t)}{l^2} - \dfrac{P}{l}$	$F_1 = F_2 = \dfrac{S+t}{l}$ $F_3 = -F_4 = \dfrac{2(S+t)}{l^2} + \dfrac{P}{l}$
(b)	$F_1 = S$ $F_2 = t$ $F_3 = -F_4 = \dfrac{S+t}{l}$	$F_1 = S$ $F_2 = t$ $F_3 = -F_4 = \dfrac{S+t}{l}$
(c)	$F_1 = \dfrac{S-(t^2/S)}{l}$ $F_2 = -F_3 = \dfrac{S-(t^2/S)-Pl}{l^2}$	$F_1 = \dfrac{S-(t^2/S)}{l}$ $F_2 = -F_3 = \dfrac{S-(t^2/S)+Pl}{l^2}$
(d)	$F_1 = S-(t^2/S)$ $F_2 = -F_3 = \dfrac{S-(t^2/S)}{l}$	$F_1 = S-(t^2/S)$ $F_2 = -F_3 = \dfrac{S-(t^2/S)}{l}$

Note: S and t are given in terms of $u = l\sqrt{P/(EI)}$ in Table (15-4). EI is flexural rigidity and l is the length of the member. P may be expressed as $P = u^2 EI/l^2$.

given in terms of S and t, which may be defined as the end-moments F_1 and F_2 respectively corresponding to the end conditions in case (b) of the table. The values of S and t can be taken from Table 15-4 or calculated by Eqs. 15-17 and 15-18 or Eqs. 15-21 and 15-22. When the axial force is zero, S and t become $4EI/l$ and $2EI/l$, respectively, and the forces listed in Table 15-5 become the same as the forces listed in Appendix D. Table 15-5 is to be used for the same purpose and in the same way as Appendix D, but with Table 15-5 the axial force P must be known (or assumed) and the corresponding values of S and t calculated in advance.

In some cases it may be more convenient to substitute for $P = u^2(EI/l^2)$ in the equations in Table 15-5. With S and t expressed in terms of EI/l, all the forces in the table can be expressed in terms of EI and l.

15-5 ADJUSTED END-ROTATIONAL STIFFNESS FOR A PRISMATIC MEMBER SUBJECTED TO AN AXIAL FORCE

We recall that the end-rotational stiffness S of a beam is the moment required to rotate one end of the beam through unity while the far end is encastré (Fig. 15-4d). Following the same procedure as in Sec. 11-7 the adjusted end-rotational stiffness of a beam subjected to an axial force can be derived for various end conditions. In the following, the adjusted end-rotational stiffnesses which are needed for analysis by moment distribution are given for beams subjected to the conditions shown in Fig. 11-8 as well as to an axial compressive or tensile force of value P (not shown in the figure).

For the end conditions in Fig. 11-8b:

$$S_{AB}^{\textcircled{1}} = S(1 - C^2) = S - (t^2/S) \tag{15-25}$$

For the symmetrical conditions in Fig. 11-8c:

$$S_{AB}^{\textcircled{2}} = S(1 - C) = S - t \tag{15-26}$$

For the antisymmetrical case in Fig. 11-8d:

$$S_{AB}^{\textcircled{3}} = S(1 + C) = S + t \tag{15-27}$$

For the cantilever in Fig. 11-8e:

$$S_{AB}^{\textcircled{4}} = S - \frac{(S + t)}{2 \pm Pl/(S + t)} \tag{15-28}$$

The carryover factor in the last case is

$$C_{AB}^{\textcircled{4}} = \frac{t^2 - S^2 \pm tPl}{S^2 - t^2 \pm SPl} \tag{15-29}$$

The sign of the terms containing Pl $(= u^2(EI/l))$ in Eqs. 15-28 and 15-29 is plus when the axial force is tensile and minus when compressive. The values of S, t, and C to be used in Eqs. 15-25–15-29 should be calculated by Eqs. 15-17–15-19 for compression members and by Eqs. 15-21–15-23 for tension members, or by Table 15-4 for both cases.

15-6 FIXED-END MOMENTS FOR A PRISMATIC MEMBER SUBJECTED TO AN AXIAL FORCE

Consider a straight member subjected to transverse loading and an axial force (Fig. 15-6a). The end-displacement along the beam axis can take place freely but the end-rotation is prevented, so that fixed-end moments are

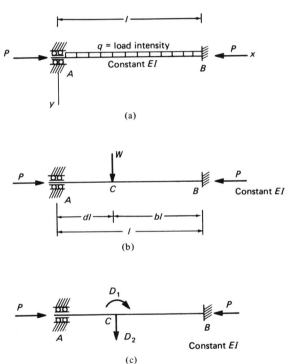

Fig. 15-6. Straight prismatic member subjected to an axial force and transverse loading. (a) Loading corresponding to the FEM Eqs. 15-32 or 15-34. (b) Loading corresponding to the FEM Eq. 15-35 and Table 15-6. (c) Degrees of freedom (D_1 and D_2) considered for the analysis of the strut in part (b).

induced. The presence of an axial compressive force causes an increase in the fixed-end moments while a tensile force results in a decrease. Two cases of loading will be considered; a uniform load and a concentrated load.

15-61   Uniform Load

The deflection of the prismatic strut AB in Fig. 15-6a carrying a transverse load of constant intensity q is governed by the differential Eq. 15-4, for which the solution is

$$y = G_1 \sin u \frac{x}{l} + G_2 \cos u \frac{x}{l} + G_3 x + G_4 + \frac{qx^2}{2P} \qquad (15\text{-}30)$$

The integration constants $\{G\}$ are determined from the boundary conditions $y = 0$ and $(dy/dx) = 0$ at $x = 0$ and $x = l$. It can be easily checked that these four conditions can be put in the form

$$[B]\begin{Bmatrix} G_1 \\ G_2 \\ G_3 \\ G_4 \end{Bmatrix} + \frac{q}{P}\begin{Bmatrix} 0 \\ 0 \\ l^2/2 \\ l \end{Bmatrix} = \{0\}$$

where $[B]$ has the same meaning as in Eq. 15-8. Solving for the integration constants, we obtain

$$\begin{Bmatrix} G_1 \\ G_2 \\ G_3 \\ G_4 \end{Bmatrix} = -\frac{ql^2}{u^2 EI}[B]^{-1}\begin{Bmatrix} 0 \\ 0 \\ l^2/2 \\ l \end{Bmatrix} \qquad (15\text{-}31)$$

where $[B]^{-1}$ is that given by Eq. 15-14.

Considering the end moments to be positive if clockwise and making use of symmetry of the member, the two fixed-end moments are related to the deflection by

$$M_{AB} = -M_{BA} = -EI\left(\frac{d^2 y}{dx^2}\right)_{x=0}$$

Differentiating twice Eq. 15-30 and substituting in the above equation, we obtain

$$M_{AB} = -M_{BA} = EI\left(\frac{u^2}{l^2}G_2 - \frac{q}{P}\right)$$

With G_2 from Eq. 15-31 substituted in the above equation, the fixed-end moments due to a uniform load on a member subjected to an axial compression are

$$M_{AB} = -M_{BA} = -ql^2\frac{1}{u^2}\left(1 - \frac{u}{2}\cot\frac{u}{2}\right) \qquad (15\text{-}32)$$

If the axial force is tensile, the expression for fixed-end moments becomes

$$M_{AB} = -M_{BA} = -ql^2 \frac{1}{u^2} \left(\frac{u}{2} \coth \frac{u}{2} - 1 \right) \qquad (15\text{-}33)$$

The fixed-end moments due to a uniform transverse load given by Eqs. 15-32 and 15-33 can be expressed in the form

$$M_{AB} = -M_{BA} = -\frac{EI}{2l(S + t)} ql^2 \qquad (15\text{-}34)$$

where the values of S and t are determined as a function of $u = l\sqrt{P/(EI)}$, P being the absolute value of the axial compressive or tensile force, by Eqs. 15-17 and 15-18, or by Eqs. 15-21 or 15-22, or by Table 15-4. When the axial force P vanishes ($u \rightarrow 0$), and Eqs. 15-32, 15-33, or 15-34 give $M_{AB} = -M_{BA} = -ql^2/12$.

15-62 Concentrated Load

Consider the prismatic strut AB in Fig. 15-6b carrying a transverse load W at a distance dl and bl from the left- and right-hand ends respectively. The strut can be treated as an assemblage of two members AC and CB with the two degrees of freedom indicated in Fig. 15-6c, for which the stiffness matrix can be easily derived from the stiffness of the individual members (Eq. 15-20). Thus

$$[S] = \begin{bmatrix} (S_d + S_b) & \text{symmetrical} \\ \left[\dfrac{(S_b + t_b)}{bl} - \dfrac{(S_d + t_d)}{dl} \right] & \left[\dfrac{2(S_d + t_d)}{d^2 l^2} + \dfrac{2(S_b + t_b)}{b^2 l^2} - \dfrac{P}{dbl} \right] \end{bmatrix}$$

where the subscripts d and b refer to members AC and BC respectively. The values of S and t for the two members are given by Eqs. 15-17 and 15-18, with du or bu in place of u.

The displacements at the two coordinates are given by

$$\begin{Bmatrix} D_1 \\ D_2 \end{Bmatrix} = [S]^{-1} \begin{Bmatrix} 0 \\ W \end{Bmatrix}$$

and the fixed-end moment at A (considered positive if clockwise) is given by

$$M_{AB} = t_d D_1 - (S_d + t_d) \frac{D_2}{dl}$$

The above procedure can also be followed for members subjected to axial tension. Thus the fixed-end moment in member AB subjected to an

Table 15-6. Influence Coefficient of the Fixed-End Moment* M_{AB} Due to a Unit Transverse Load on a Prismatic Beam Subjected to an Axial Force

$$M_{AB} = -\text{coefficient} \times l$$

(Refer to Fig. 15-6b, calculations by Eqs. 15-35 and 15-36)

Force P	$u = l\sqrt{\dfrac{P}{EI}}$	Value of d										
		0	0.1	0.2	0.3	0.4	0.5	0.6	0.7	0.8	0.9	1.0
	0	0	0.0810	0.1280	0.1470	0.1440	0.1250	0.0960	0.0630	0.0320	0.0090	0
Compressive	1.0	0	0.0815	0.1294	0.1493	0.1467	0.1276	0.0981	0.0644	0.0327	0.0091	0
	2.0	0	0.0831	0.1342	0.1569	0.1558	0.1365	0.1054	0.0692	0.0350	0.0097	0
	3.0	0	0.0863	0.1438	0.1725	0.1747	0.1552	0.1208	0.0796	0.0402	0.0111	0
	4.0	0	0.0922	0.1624	0.2037	0.2136	0.1946	0.1540	0.1022	0.0515	0.0141	0
	5.0	0	0.1057	0.2070	0.2820	0.3150	0.3009	0.2459	0.1662	0.0843	0.0229	0
	6.0	0	0.1943	0.5218	0.8689	1.1150	1.1750	1.0288	0.7282	0.3789	0.1039	0
Tensile	1.0	0	0.0805	0.1265	0.1447	0.1413	0.1224	0.0939	0.0616	0.0313	0.0088	0
	2.0	0	0.0791	0.1226	0.1386	0.1341	0.1155	0.0883	0.0579	0.0295	0.0083	0
	3.0	0	0.0770	0.1168	0.1297	0.1239	0.1058	0.0806	0.0529	0.0271	0.0077	0
	4.0	0	0.0745	0.1100	0.1196	0.1125	0.0951	0.0722	0.0474	0.0244	0.0070	0
	5.0	0	0.0718	0.1029	0.1093	0.1012	0.0848	0.0641	0.0423	0.0220	0.0064	0
	6.0	0	0.0690	0.0959	0.0996	0.0908	0.0754	0.0569	0.0377	0.0198	0.0059	0

*A fixed-end moment is positive when clockwise. To find the fixed-end moment at the right-hand end enter the value of b in lieu of d and change the sign of the moment. P is the absolute value of the axial compressive or tensile force.

axial compression (Fig. 15-6b) is

$$M_{AB} = -Wl \frac{(bu \cos u - \sin u + \sin du + \sin bu - u \cos bu + du)}{u(2 - 2 \cos u - u \sin u)} \quad (15\text{-}35)$$

and when the axial force P is tension

$$M_{AB} = -Wl \frac{(bu \cosh u - \sinh u + \sinh du + \sinh bu - u \cosh bu + du)}{u(2 - 2 \cosh u + u \sinh u)}$$

$$(15\text{-}36)$$

Equations 15-35 and 15-36 are used to calculate influence coefficients of the fixed-end moment M_{AB} in Table 15-6, by fixing a value for u and varying d (and $b = 1 - d$). The same table can be used for the fixed-end moments at the right-hand end M_{BA} by considering d in the table to represent b and changing the sign of the moment given.

15-7 ADJUSTED FIXED-END MOMENTS FOR A PRISMATIC MEMBER SUBJECTED TO AN AXIAL FORCE

With the end-moments M_{AB} and M_{BA} for a beam with two fixed ends subjected to an axial force (Fig. 15-7a), we can derive the adjusted end-moments due to the same loading but with different end conditions. The parameters involved, in addition to M_{AB} and M_{BA}, are the end-rotational stiffness S and the carryover moment t. Following the same procedure as in Sec. 11-8, we obtain for the beam in Fig. 15-7b, with a hinged end A, the end-moment at B.

$$M_{BA}^{①} = M_{BA} - CM_{AB} \quad (15\text{-}37)$$

where $C = t/S$. This equation is valid regardless of whether the axial force is compressive or tensile.

For the beam in Fig. 15-7c, with translation of end A in the transverse direction allowed but rotation prevented, the end-moments are

$$M_{AB}^{②} = M_{AB} + F_A l \frac{(S + t)}{2(S + t) \pm Pl} \quad (15\text{-}38)$$

$$M_{BA}^{②} = M_{BA} + F_A l \frac{(S + t)}{2(S + t) \pm Pl} \quad (15\text{-}39)$$

where M_{AB}, M_{BA}, and F_A are the end-moments and reaction for the same beam but with the end condition of Fig. 15-7a; the positive directions of these forces are indicated in the figure. The sign of the term $Pl[= u^2(EI/l)]$ in the denominator in the last two equations is plus when the axial force is tensile and minus when compressive.

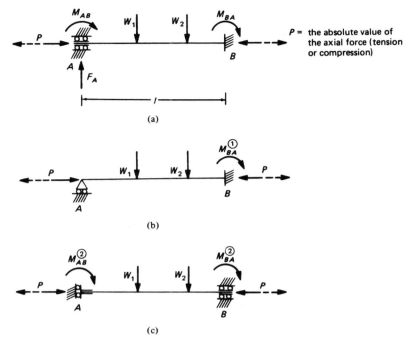

Fig. 15-7. Adjusted fixed-end moments. (a) Rotation and translation in the transverse direction prevented at both ends. (b) Free rotation at A. (c) Vertical translation at A is allowed with no rotation.

Example 15-3 Find the end-moments for the members in the frame of Fig. 15-8a, taking into account the beam-column effect.

We assume approximate values of the axial forces in the members: $P_{AB} = P_{BC} = 5qb$ tensile, and $P_{BD} = 12.5qb$ compressive. The corresponding values of $u = l\sqrt{P/(EI)}$ are: $u_{AB} = u_{BC} = 1.41$, and $u_{BD} = 2.53$.

From Eqs. 15-21 and 15-22, or from Table 15-4, we find for members AB and BC: $S = 4.26(EI/l)$; $t = 1.94(EI/l)$; and $C = t/S = 0.46$.

Similarly, from Eqs. 15-17 and 15-18, or from Table 15-4, the values of S and t for member BD are: $S = 3.06(EI/l)$, $t = 2.26(EI/l)$, and $C = t/S = 0.74$.

The adjusted rotational stiffness of end B of members BA and BC, to account for the hinged ends at A and C, is (from Eq. 15-25) $S_{BA} = S_{BC} = S(1 - C^2) = 3.36\ EI/l$.

We now substitute the relative values of EI/l for the individual members in the above expressions, then calculate the distribution factors in the usual way (Fig. 15-8b).

From Eq. 15-37, the fixed-end moment is $M_{BC}^{\textcircled{1}} = M_{BC} - CM_{CB}$, where M_{BC} and M_{CB} are the end-moments if both B and C are encastré.

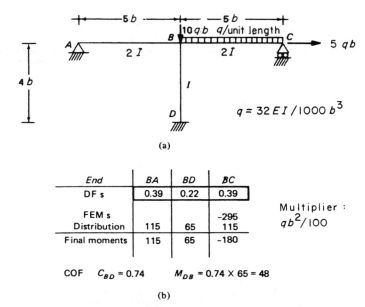

(a)

End	BA	BD	BC
DF s	0.39	0.22	0.39
FEM s			-295
Distribution	115	65	115
Final moments	115	65	-180

Multiplier :
$qb^2/100$

COF $C_{BD} = 0.74$ $M_{DB} = 0.74 \times 65 = 48$

(b)

Fig. 15-8. Analysis of the frame of Example 15-3 by moment distribution taking into account the beam-column effect. (a) Frame properties and loading. (b) Moment distribution.

Using Eq. 15-34,

$$M_{BC} = -M_{CB} = -\frac{EI}{2l(S + t)} ql^2 = -\frac{ql^2}{2(4.26 + 1.94)}$$
$$= -0.081ql^2 = -2.02qb^2$$

Thus the adjusted fixed-end moment is

$$M_{BC}^{①} = -(2.02 + 0.46 \times 2.02)qb^2 = -2.95qb^2$$

The moment distribution is carried out in the usual way in Fig. 15-8b, and the end-moments thus obtained are (in terms of $qb^2/100$): $M_{BA} = 115$, $M_{BC} = -180$, $M_{BD} = 65$, and $M_{DB} = 48$.

More accurate values of the axial forces can now be determined and the above calculations may be repeated. It is clear, however, that in this example no appreciable change in values will result.

Example 15-4   Solve Example 15-3 with the support at A being a roller instead of a hinge but for the loading in Fig. 15-9a. Note also the change in relative I values.

A solution can be obtained by moment distribution similar to the procedure followed in Example 15-3, but the adjusted end-rotational stiffness and the carryover factor have to be calculated for BD by Eqs. 15-28 and

15-29 and the fixed-end moments for ends BD and DB have to be determined by Eqs. 15-38 and 15-39. This solution is not given here but instead we analyze the structure by the general displacement method.

The approximate values of the axial forces are $P_{AB} = 0$, $P_{BC} = 0.05qb$ tensile, and $P_{BD} = 7qb$ compressive; the corresponding u values are: $u_{AB} = 0$, $u_{BC} \simeq 0$, and $u_{BD} = 1.34$. For member BD, $S = 3.75EI/l$ and $t = 2.06EI/l$. With these values, the end-forces corresponding to unit end displacements can be calculated using Table 15-5.

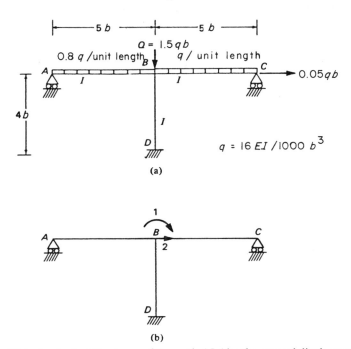

(a)

(b)

Fig. 15-9. Analysis of the frame of Example 15-4 by the general displacement method taking into account the beam-column effect. (a) Frame properties and loading. (b) Coordinate system.

The stiffness matrix of the structure corresponding to the coordinates 1 and 2 in Fig. 15-9b is

$$[S] = \begin{bmatrix} 3\left(\dfrac{EI}{l}\right)_{BA} + 3\left(\dfrac{EI}{l}\right)_{BC} + 3.75\left(\dfrac{EI}{l}\right)_{BD} & \text{symmetrical} \\[2ex] -(3.75 + 2.06)\left(\dfrac{EI}{l^2}\right)_{BD} & 2(3.75 + 2.06)\left(\dfrac{EI}{l^3}\right)_{BD} - \left(\dfrac{P}{l}\right)_{BD} \end{bmatrix}$$

The fixed-end moments at end B of members BA, BC, BD, and DB are

$2.5qb^2$, $-3.125qb^2$, 0 and 0, respectively. The restraining forces at coordinates 1 and 2 to prevent displacements at B are

$$\{F\} = \begin{Bmatrix} -0.625qb^2 \\ -0.05qb \end{Bmatrix}$$

The displacements at the coordinates are given by Eq. 3-3:

$$\{D\} = [S]^{-1}\{-F\}$$

Substituting the values of I, l, and $P(=u^2[EI/l^2])$ into $[S]$ and inverting it, then substituting in Eq. 3-3, we obtain

$$\{D\} = \frac{qb^3}{EI} \begin{Bmatrix} 0.58 \\ 1.71b \end{Bmatrix}$$

The end-moments M_{BA}, M_{BC}, M_{BD}, and M_{DB} are then determined by Eq. 3-5:

$$\{A\} = \{A_r\} + [A_u]\{D\}$$

$$\begin{Bmatrix} M_{BA} \\ M_{BC} \\ M_{BD} \\ M_{DB} \end{Bmatrix} = qb^2 \begin{Bmatrix} 2.50 \\ -3.125 \\ 0 \\ 0 \end{Bmatrix} + \begin{bmatrix} 3.00\left(\dfrac{EI}{l}\right)_{BA} & 0 \\ 3.00\left(\dfrac{EI}{l}\right)_{BC} & 0 \\ 3.75\left(\dfrac{EI}{l}\right)_{BD} & -(3.75+2.06)\left(\dfrac{EI}{l^2}\right)_{BD} \\ 2.06\left(\dfrac{EI}{l}\right)_{DB} & -(3.75+2.06)\left(\dfrac{EI}{l^2}\right)_{DB} \end{bmatrix} \{D\}$$

Substituting for $\{D\}$, the final end-moments are obtained

$$\begin{Bmatrix} M_{BA} \\ M_{BC} \\ M_{BD} \\ M_{DB} \end{Bmatrix} = \frac{qb^2}{100} \begin{Bmatrix} 285 \\ -277 \\ -8 \\ -32 \end{Bmatrix}$$

15-8   ELASTIC STABILITY OF FRAMES

In design based on a definite load factor, failure loads must be known. Broadly speaking, failure can occur by yielding of the material at a sufficient number of locations to form a mechanism (see Chapters 21 and 22) or by buckling due to axial compression without the stresses exceeding the elastic limit.

Buckling of individual members was considered in Sec. 15-4. In the present section, we shall deal with buckling of rigidly jointed plane frames in which the members are subjected to axial forces only.

Consider a plane frame subjected to a system of forces $\{Q\}$ (Fig. 15-10a) causing axial compression in some of the members. The buckling, or critical, loading $\alpha\{Q\}$ is defined by the value of the scalar α at which the structure can be given small displacements without application of disturbing forces. In other words, when the buckling loading $\alpha\{Q\}$ acts, it is possible to maintain the structure in a displaced configuration without additional loading, as shown in Fig. 15-10b.

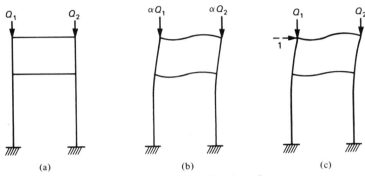

Fig. 15-10. Buckling of a plane frame.

If, for the frame considered, an estimate is made of the value of α and the corresponding values of the axial forces $\{P(\alpha)\}$ are computed, it becomes possible to find the stiffness matrix $[S(\alpha)]$ of the structure corresponding to any chosen coordinate system. The elements of this matrix are functions of the value α. Any system of forces $\{F\}$ and displacements $\{D\}$ at the coordinates are related by the equation

$$[S(\alpha)]\{D\} = \{F\}$$

If α corresponds to the critical buckling value, then it is possible to let the structure acquire some small displacements $\{\delta D\}$ without application of forces, and the last equation thus becomes

$$[S(\alpha)]\{\delta D\} = \{0\}$$

For a nontrivial solution to exist, the stiffness matrix $[S(\alpha)]$ must be singular — that is the determinant

$$|S(\alpha)| = 0 \qquad\qquad (15\text{-}40)$$

The collapse loading is that which has the smallest value of α which satisfies Eq. 15-40. In general, there is more than one value of α which satisfies this equation, and each value has associated with it values of $\{\delta D\}$

of arbitrary magnitude but in definite proportions defining an associated critical mode.

To solve Eq. 15-40 for α, the value of the determinant is calculated for a set of values of α, and the value corresponding to a zero value of the determinant is obtained by interpolation. Once α has been calculated, the associated displacement vector, if required, can be obtained in a way similar to that used in the calculation of eigenvectors in Sec. A-10 of Appendix A.

The above procedure may involve a large amount of numerical work so that access to a computer is virtually necessary. However, in most structures, it is possible to guess the form of the critical mode associated with the lowest critical load. For example, it can be assumed that when the buckling load is reached in the frame of Fig. 15-10a, a small disturbing force at coordinate 1 will cause buckling in sidesway (Fig. 15-10c). At this stage, the force at 1 required to produce a displacement at this coordinate is zero, that is $S_{11} = 0$. This condition can be used to determine the critical load. A value of α lower than the critical value is estimated and the stiffness coefficient S_{11} (α) corresponding to coordinate 1 is determined. As α is increased the value of S_{11} decreases and it vanishes when the critical value of α is attained. If a set of values of α is assumed and plotted against S_{11}, the value of α corresponding to $S_{11} = 0$ can be readily determined.

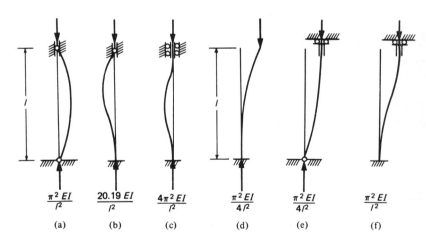

Fig. 15-11. Critical buckling load for a prismatic member.

The critical buckling loads for a straight prismatic member with various end conditions are given in Fig. 15-11. These may be used to establish lower and upper bounds to the buckling load of a frame and hence are of help in estimating the approximate values of α.

Example 15-5 Find the value of the force Q at B which causes buckling of the frame of Example 15-4 (Fig. 15-9a). The frame is not subjected to loads other than the force Q.

Let $Q = \alpha$. The axial forces in the members are $P_{BA} = P_{BC} = 0$ and $P_{BD} = \alpha$, compressive. The stiffness matrix of the frame corresponding to the coordinates in Fig. 15-9b, expressed as function of α, is

$$[S(\alpha)] = \begin{bmatrix} 3\left(\dfrac{EI}{l}\right)_{BA} + 3\left(\dfrac{EI}{l}\right)_{BC} + S_{BD} & \text{symmetrical} \\[2ex] -(S_{BD} + t_{BD})/l_{BD} & \dfrac{2(S_{BD} + t_{BD})}{l_{BD}^2} - \dfrac{\alpha}{l_{BD}} \end{bmatrix} \tag{a}$$

where S_{BD} and t_{BD} are functions of α to be calculated by Eqs. 15-17 and 15-18 or from Table 15-4.

When α corresponds to the buckling load, the determinant $|S(\alpha)| = 0$, so that

$$\text{Determinant} = S_{11}S_{22} - S_{21}^2 \tag{b}$$

where S_{ij} are elements of the stiffness matrix in Eq. (a).

Upper and lower bounds of the critical load are established if we consider that the rotation at end B of BD is partially restrained. Thus, the buckling load for BD is some value between $\pi^2 EI/(4l^2) = 0.154EI/b^2$ and $\pi^2 EI/l^2 = 0.617EI/b^2$, corresponding to the conditions in Figs. 15-11d and f respectively. As a first trial, let us take the average value between these two bounds, that is $\alpha = 0.386EI/b^2$. The corresponding value of the determinant in Eq. (b) is $0.0279(EI)^2/b^4$. In the second trial, we take $\alpha = 0.5EI/b^2$, giving a value of $-0.0354(EI)^2/b^4$ for the determinant. By linear interpolation, we choose for the third trial

$$\alpha = \left[0.386 + \frac{(0.5 - 0.386)0.0279}{0.0279 + 0.0354}\right]\frac{EI}{b^2} = 0.436\frac{EI}{b^2}$$

for which the determinant $= -0.0004(EI)^2/b^4$. This value is negligible compared with the preceding values; thus, $\alpha = 0.436EI/b^2$ can be considered as the buckling value. Therefore, the critical load is

$$Q_{cr} = 0.436\frac{EI}{b^2}$$

Example 15-6 Find the value of Q which causes buckling of the frame in Fig. 15-12a.

The stiffness matrix for the three coordinates in Fig. 15-12b is determined from Table 15-5:

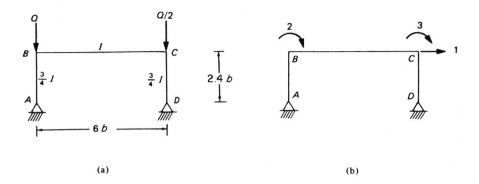

(a)   (b)

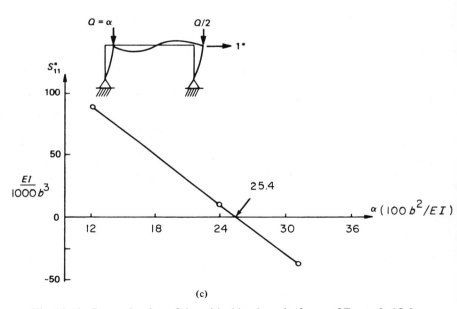

(c)

Fig. 15-12. Determination of the critical load on the frame of Example 15-6 by consideration of stiffness reduction. (a) Frame dimensions and loading. (b) Coordinate system; the corresponding stiffness matrix in presence of the Q and $Q/2$ forces is given in Eq. (a). (c) Variation of stiffness S_{11}^* with α.

$$[S] = \begin{bmatrix} \left(\dfrac{S - t^2/S - u^2EI/l}{l^2}\right)_{BA} + \\ \left(\dfrac{S - t^2/S - u^2EI/l}{l^2}\right)_{CD} & & \text{symmetrical} \\ -\left(\dfrac{S - t^2/S}{l}\right)_{BA} & (S - t^2/S)_{BA} + 4\left(\dfrac{EI}{l}\right)_{BC} \\ -\left(\dfrac{S - t^2/S}{l}\right)_{CD} & 2\left(\dfrac{EI}{l}\right)_{BC} & (S - t^2/S)_{CD} + 4\left(\dfrac{EI}{l}\right)_{BC} \end{bmatrix} \quad \text{(a)}$$

The lowest buckling load for this frame corresponds to a sidesway by a small disturbing force at coordinate 1. It can be easily shown that the stiffness S_{11}^* corresponding to coordinate 1^* (Fig. 15-12c) is given by (see Sec. 4-5, Eq. 4-17)

$$S_{11}^* = S_{11} - \frac{(S_{21}^2 S_{33} - 2S_{21}S_{31}S_{32} + S_{31}^2 S_{22})}{(S_{22}S_{33} - S_{32}^2)} \quad \text{(b)}$$

where S_{ij} are elements of the matrix in Eq. (a).

Let the buckling load occur when $Q = \alpha$ (kip). Considering that the rotation of end C of member CD is partially restrained, it can be concluded that the buckling load is lower than the value $(\pi^2 EI/4l^2)_{BA}$ which corresponds to the conditions in Fig. 15-11e. Thus an upper bound for α is

$$\alpha < \left(\frac{\pi^2 EI}{4l^2}\right)_{BA} = 0.64 \frac{EI}{b^2} \qquad Q < 0.64 \frac{EI}{b^2}$$

The values of α are assumed, the corresponding u values for BA and CD are determined, and the corresponding S and t values are taken from Table 15-4:

$\alpha(b^2/EI)$	u_{BA}	$\dfrac{S_{BA}}{(EI/l)_{BA}}$	$\dfrac{t_{BA}}{(EI/l)_{BA}}$	u_{CD}	$\dfrac{S_{CD}}{(EI/l)_{CD}}$	$\dfrac{t_{CD}}{(EI/l)_{CD}}$
0.12	0.960	3.876	2.031	0.679	3.938	2.015
0.24	1.357	3.747	2.066	0.960	3.876	2.031
0.31	1.543	3.672	2.086	1.091	3.839	2.041

Substituting into Eqs. (a) and (b), we obtain the following value of S_{11}^*

$\alpha(b^2/EI)$	$S_{11}^*(b^3/EI)$
0.12	0.0888
0.24	0.0092
0.31	−0.0376

The above values are plotted in Fig. 15-12c from which it is seen that $S_{11}^* = 0$ when $\alpha = 0.254EI/b^2$. Thus, the buckling load $Q_{cr} = 0.254EI/b^2$.

15-9 CALCULATION OF BUCKLING LOAD FOR FRAMES BY MOMENT DISTRIBUTION

We shall first consider frames without sidesway, and then adapt the method to the case when sidesway is possible. The critical buckling load of frames with no joint translation may be determined by examining the convergence of moment distribution. Consider the frame in Fig. 15-13a, for which the critical value of the force Q is required. We assume a value for Q, and determine the end-rotational stiffness of the members and the carryover factors. We now assume an arbitrary disturbing moment at one (or more) joint and carry out the moment distribution. The significance of the operation lies in the fact that the behavior under moment distribution depends on the magnitude of the load relative to its critical value. Specifically, as the load is increased to approach the critical value, the final moments approach positive or negative infinity, while if the load is less than the critical load the results converge, that is the moments become smaller after each cycle of moment distribution and finite values of the end-moments can be reached.

When the assumed load is only slightly below the critical load, the results converge slowly, and, as already stated, above the critical load the results diverge. Therefore, to find the critical load, it is not necessary to continue the moment distribution for a great many cycles: if after one cycle it is seen that the moments converge, then the load chosen is smaller than the critical load.

Example 15-7   Using moment distribution, find the value of Q which will cause buckling of the frame in Fig. 15-13a.

First, the upper and lower bounds of the critical load $Q_{cr} = \alpha$ have to be established. Member AB can rotate freely at its bottom end while the rotation at the top end is partially restrained. Therefore the buckling load for this member has a value lying between the critical loads for the conditions shown in Figs. 15-11a and b. Thus,

$$\left(\frac{\pi^2 EI}{l^2}\right)_{AB} < Q_{cr} < 20.19\left(\frac{EI}{l^2}\right)_{AB}$$

or

$$2.47\,\frac{EI}{b^2} < Q_{cr} < 5.05\,\frac{EI}{b^2}$$

Consider now the member BC. The rotation at both ends is restrained so that the buckling load for BC lies between the critical loads for the conditions in Figs. 15-11a and c. Therefore

$$\left(\frac{\pi^2 EI}{l^2}\right)_{BC} < Q_{cr} < \left(\frac{4\pi^2 EI}{l^2}\right)_{BC}$$

$$1.85\,\frac{EI}{b^2} < Q_{cr} < 7.40\,\frac{EI}{b^2}$$

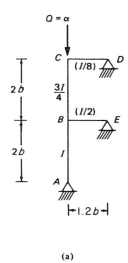

(a)

α	Joint	B				C		Remarks
	End	BA	BE	BC		CB	CD	
$2.7\dfrac{EI}{b^2}$	DF s	-0.16	0.79	0.37	1.83	0.65	0.35	$\dfrac{80.6}{100} < 1.00$
	COF s				⟵			
	Apply a clockwise moment = 100 at B							
	Distribution -	16.0	79.0	37.0	⟶	67.7		
	CO			-80.6	⟵	-44.0		$\therefore \; \alpha_{cr} > 2.7\dfrac{EI}{b^2}$
$2.9\dfrac{EI}{b^2}$	DF s	-0.41	1.01	0.40	2.27	0.61	0.39	$\dfrac{125.6}{100} > 1.00$
	COF s							
	Apply a clockwise moment = 100 at B							
		- 41.0	101.0	40.0	⟶	90.8		
				-125.6	⟵	-55.3		$\therefore \; \alpha_{cr} < 2.9\dfrac{EI}{b^2}$
$2.8\dfrac{EI}{b^2}$	DF s	-0.27	0.89	0.38	2.02	0.63	0.37	$\dfrac{97.7}{100} \simeq 1.00$
	COF s							
	Apply a clockwise moment = 100 k ft at B							
		27.0	89.0	38.0	⟶	76.8		
				- 97.7	⟵	-48.4		$\therefore \; \alpha_{cr} = 2.8\dfrac{EI}{b^2}$

(b)

Fig. 15-13. Calculation of the critical buckling load for Example 15-7.
(a) Frame dimensions and loading. (b) Moment distribution.

Hence, α lies between $1.85EI/b^2$ and $5.05EI/b^2$. For the first trial, take $\alpha = 3.6EI/b^2$; the values of u, S, and t are respectively 3.79, 1.56 $(EI/l)_{AB}$, and 2.83 $(EI/l)_{AB}$ for AB, and 4.38, 0.32 $(EI/l)_{BC}$, and 3.43 $(EI/l)_{BC}$ for BC. The rotational stiffnesses of the member ends meeting at joint B are

$$S_{BA} = \frac{EI}{2b}\left(1.56 - \frac{2.83^2}{1.56}\right) = -1.79EI/b \qquad \text{(by Eq. 15-25)}$$

$$S_{BC} = \frac{E(0.75)I}{2b} \times 0.32 \quad = \quad 0.12EI/b$$

$$S_{BE} = \frac{3.0E(0.5)I}{1.2b} \qquad\qquad = \quad \underline{1.25EI/b}$$

$$\text{Sum} = -0.42EI/b$$

The sum of the rotational stiffnesses of the ends meeting at B is negative, indicating that even with the top end of member BC totally fixed, the assumed value of α is greater than the critical value.

For a second trial, take $\alpha = 2.7EI/b^2$; the corresponding end-rotational stiffness and carryover moments are calculated as before. The corresponding distributing and carryover factors are given in Fig. 15-13b. A disturbing clockwise moment of 100 is applied at joint B, an appropriate moment is carried over to joint C, joint C is balanced, and a moment of -80.6 is carried back to joint C, thus ending the first cycle. If a second cycle of distribution were to be performed, it would begin by the distribution of a balancing moment of $+80.6$, which is less than the starting moment of 100. It is clear, therefore, that the calculation converges, indicating that the critical load has not been reached.

Two additional trials are shown in Fig. 15-13b from which we conclude that the critical load is $Q_{cr} = 2.8EI/b^2$.

Let us now consider frames with sidesway. The value of Q which causes buckling of the frame in Fig. 15-12a can be obtained by moment distribution. The approach is similar to that followed in Example 15-6 in that we seek the value of Q such that sidesway can be produced without a horizontal force at the level of BC ($F_1^* = 0$ in Fig. 15-12c).

The procedure is to assume a value of Q, then introduce a unit sidesway at the level of BC; the member end-moments are determined by moment distribution and the force F_1^* is found by static equilibrium. If F_1^* is positive, the frame is stable and the value of Q is less than the critical load. The calculation is repeated with new values of Q until F_1^* reaches a zero value.

15-10   GENERAL

In a framed structure, axial forces change the lengths of members as well as their stiffnesses. These two effects may be neglected in some structures but when the axial forces are large they have to be considered.

In most structures, the axial forces can be determined with reasonable accuracy by a simple analysis in which the above two effects are ignored. Using the joint translations, obtained from this analysis, secondary moments and shears can be calculated; this approach is used in frames and rigid-jointed trusses. The importance of secondary moments depends on the proportions of the members in the structure.

By inspection of the numerical values of S and t in Table 15-4, we can see that the stiffness of a member is appreciably affected by an axial force P only when $u = l\sqrt{P/(EI)}$ is relatively large, say 1.5. However, a lower value of u appreciably changes the shear required to produce a unit translation. For example, if the axial force is compressive, the element S_{11} of the stiffness matrix in Eq. 15-20 (see also Table 15-5, case a)

$$S_{11} = \frac{2(S + t)}{l^2} - \frac{P}{l} = \frac{2(S + t)}{l^2} - \frac{u^2 EI}{l^3}$$

is equal to 10.8 EI/l^3 when $u = 1$. The value of S_{11} is $12EI/l^3$ when the axial force vanishes.

If the effect of the change in length of a member is to be taken into consideration together with the beam-column effect, the procedure of calculation discussed in Secs. 15-2 and 15-3 may be used except that the appropriate expressions for the member stiffnesses and fixed-end moment have to be used in the analysis.

In the plane structures considered in this chapter, the deflections and bending moments are assumed to be in the plane of the structure, while displacements normal to this plane are assumed to be prevented. For space structures, similar methods of analysis can be developed but they become complicated by the possibility of torsional-flexural buckling.

When the axial force is compressive and the value of P is the critical buckling load, the force required to produce a unit displacement at a coordinate becomes zero. Thus when $S_{11} = 0$ in the stiffness matrix of Eq. 15-15, the value of the axial load is the critical buckling value for a strut with the displacement at coordinate one (Fig. 15-4b) allowed to take place freely, while the displacements at the coordinates 2, 3, and 4 are restrained. Similarly, $S_{22} = 0$ corresponds to a critical buckling load of a strut when D_2 is free to occur while $D_1 = D_3 = D_4 = 0$. At higher values of the axial loads, the elements on the diagonal of $[S]$ become negative which means that a restraining force is necessary to limit the displacement to unity at one of the coordinates.

The force or displacement methods of analysis for linear structures can be used for structures in which the axial forces in the members affect their stiffness, provided that a set of axial forces is assumed to be present in the members in all the steps of the calculations. These forces are regarded as known parameters of the members, just like E, I, and l. Tables 15-4 and 15-5 based on a dimensionless parameter can be used for the calculation of the stiffness of prismatic members subjected to axial forces.

If after carrying out the analysis the axial forces determined by considering the static equilibrium of the members differ from the assumed values, revised values of the member stiffnesses or flexibilities based on the new axial forces have to be used and the analysis is repeated. However, the repetition is usually unnecessary, as in most practical problems a reasonably accurate estimate of the axial forces in the members can be made. If a computer is used, the calculation may be repeated until the assumed and calculated axial forces agree to any desired degree.

Buckling of frames subjected to forces acting at joints only, causing axial compressive forces in the members, can be treated by a trial-and-error approach seeking the lowest load that leads to divergent moment distribution or to a singular stiffness matrix. A transverse load within the length of a member can be replaced by statically equivalent concentrated loads at the joints. The critical load thus obtained is slightly higher than that obtained from a more accurate analysis in which account is taken of the bending moment produced by the load acting at intermediate points[3].

PROBLEMS

15-1   Determine the end-moments in the frame shown in the figure, taking into account the change in length of members.

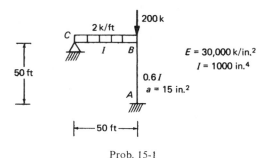

Prob. 15-1

[3]An extensive list of references on the stability analysis of frames is included in a paper by L. W. Lee, "A Survey of Literature on the Stability of Frames," Welding Research Council Bulletin 81, New York, 1962.

15-2 Apply the requirements of Prob. 15-1 to the frame shown.

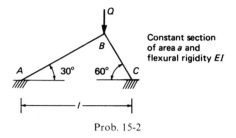

Constant section
of area a and
flexural rigidity EI

Prob. 15-2

15-3 Apply the requirements of Prob. 15-1 to the frame shown.

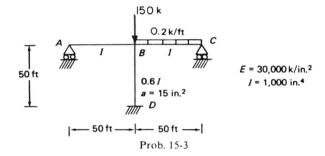

$E = 30,000 \, k/in.^2$
$I = 1,000 \, in.^4$

Prob. 15-3

15-4 Find the member end-moments (in terms of $\sigma I/l$) for the rigid-jointed truss shown in the figure, assuming the primary stress $|\sigma|$ caused by axial forces to be constant for all members. Find the ratios of the maximum bending stress at the member end to the primary stress in terms of the ratio d/l, where d is the depth of the member assuming bending to take place about an axis of symmetry. Ignore the beam-column effect.

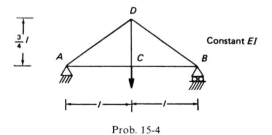

Constant EI

Prob. 15-4

15-5 Find the member end-moments for the rigid-jointed truss shown in the figure. Assume for all members $I = 2.0 \, in.^4$ and take the primary stress caused by the axial tensile or compressive forces as $15 \, k/in.^2$.

444 EFFECTS OF AXIAL FORCES

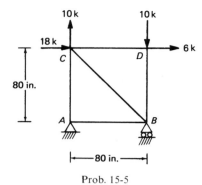

Prob. 15-5

15-6 Determine the end-moments in the members in the left-hand half of the rigid-jointed truss shown in the figure. $E = 30,000$ k/in.^{2}.

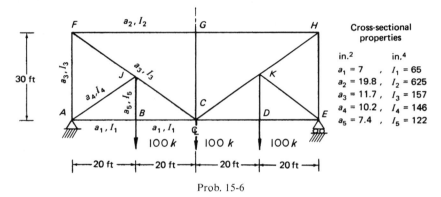

Prob. 15-6

15-7 Determine the member end-moments in the rigid-jointed truss shown in the figure. Consider the effects of change in length and change in stiffness of the members due to axial forces. For all members: $a = 1440I/b^2$.

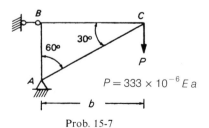

Prob. 15-7

15-8 Solve Prob. 15-1 taking into account the beam-column effect.

15-9 Solve Prob. 15-3 taking into account both the change in length of members as well as the beam-column effect.

15-10 Prove Eqs. 15-28 and 15-29.

15-11   Obtain the bending-moment diagram for beam ABC shown in the figure. The
beam is fixed at A, on a roller at C and on a spring support of stiffness =
$3EI/l^3$ at B.

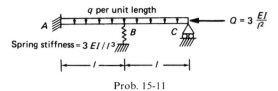

Prob. 15-11

15-12   Solve Example 15-5, with the roller support at C changed to a hinged support.

15-13   Find the value of the force Q which causes buckling of the frame of Prob. 15-2.
Assume $I_{AB} = 1.73I_{BC}$.

15-14   Find the value of Q which causes buckling of the beam in Prob. 15-11. The
beam is not subjected to distributed loads in this case.

15-15   Show that the critical load Q associated with sidesway mode for the frame in
the figure is given by the equation

$$(S + t)^2 = \left(S + 6\frac{EI_1}{l}\right)(2S + 2t - Qh)$$

where S and t are respectively the end rotational stiffness and carryover
moment for BA (S and t are defined in Fig. 15-5d).

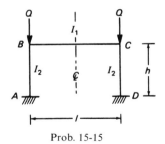

Prob. 15-15

15-16   Solve Example 15-6 by moment distribution.

15-17   Determine the value of Q which causes instability of the structure shown in
the figure. Assume E = constant.

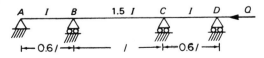

Prob. 15-17

15-18   Apply the requirements of Prob. 15-17 to the structure shown.

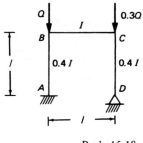

Prob. 15-18

CHAPTER 16

Analysis of shear-wall structures

16-1   INTRODUCTION

In high-rise buildings, it is important to ensure adequate stiffness to resist lateral forces induced by wind, or seismic or blast effects. These forces can develop high stresses, and produce sway movement or vibration, thereby causing discomfort to the occupants. Concrete walls, which have high in-plane stiffness, placed at convenient locations are often economically used to provide the necessary resistance to horizontal forces. This type of wall is called a shear wall. The walls may be placed in the form of assemblies surrounding lift shafts or stair wells; this box-type structure is efficient in resisting horizontal forces. Columns, of course, also resist horizontal forces, their contribution depending on their stiffness relative to the shear walls. The object of the analysis for horizontal forces is to determine in what proportion are the external loads at each floor level distributed among shear walls and the columns.

The horizontal forces are usually assumed to act at floor levels. The stiffness of the floors in the horizontal direction is very large compared with the stiffness of shear walls or columns. For this reason, it is common to assume that each floor diaphragm is displaced in its horizontal plane as a rigid body. This rigid-body movement can be defined by translations along horizontal perpendicular axes and a rotation about a vertical axis at an arbitrary point in the floor (Fig. 16-1c).

The assumption of rigid-body in-plane behavior is important in that it reduces considerably the degree of kinematic indeterminacy. However, even so, the analysis of the general case as a three-dimensional structure represents a complex problem, and further assumptions are usually made in order to produce an analysis at a reasonable cost and in moderate time. These assumptions differ with the chosen method of analysis or the type of structure, but the assumption that the floor diaphragms are rigid in their own plane is generally accepted[1].

---

[1]A collection of papers on the analysis of shear walls is available in *Tall Buildings; Proceedings of a Symposium on Tall Buildings, with Particular Reference to Shear Wall Structures*, Pergamon Press, New York, 1967. A list of earlier papers on analysis of shear wall structures is included in a paper by A. Coull and B. S. Smith. See pp. 151–155 of the above reference.

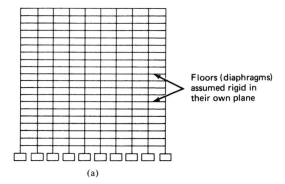

(a)

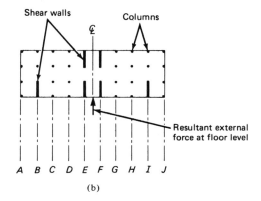

(b)

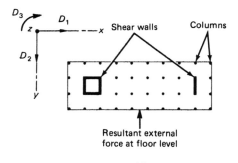

(c)

Fig. 16-1. Illustration of shear walls and some assumptions involved in their analysis. (a) Elevation of a multistorey structure. (b) Plan for a regular symmetric building. (c) Plan of unsymmetrical building (rigid-body translation of the floor is defined by displacements at the coordinates D_1, D_2, and D_3).

A major simplification of the problem is achieved if the analysis can be limited to a plane structure composed of shear walls and frames subjected to horizontal forces in their plane. This is possible when the building is laid out in a symmetrical rectangular grid pattern, so that the structure can be assumed to be made up of two sets of parallel frames acting in perpendicular directions (Fig. 16-1b). Very often, even in an irregular building, an idealized plane structure is used to obtain an approximate solution. The majority of published papers on the analysis of shear-wall structures deal with this type of two-dimensional problem.

The unsymmetrical arrangement of shear walls in the building in Fig. 16-1c causes the diaphragms to rotate and translate under the action of symmetrical horizontal forces. The shear walls in this case are subjected to twisting moments which cannot be calculated if the analysis is limited to an idealized plane structure.

In the following sections, the analysis of an idealized plane shear-wall structure is treated on the basis of certain simplifying assumptions. The three-dimensional problem is also considered, including further simplifying assumptions.

16-2   STIFFNESS OF A SHEAR-WALL ELEMENT

In the analysis to follow we shall treat shear walls as vertical deep beams transmitting loads to the foundations. The effect of shear deformations in these walls is of greater importance than in conventional beams, where the span-depth ratio is much larger. The stiffness matrix of an element of a shear wall between two adjacent floors (Fig. 16-2a) will now be derived, shear deformation being taken into consideration.

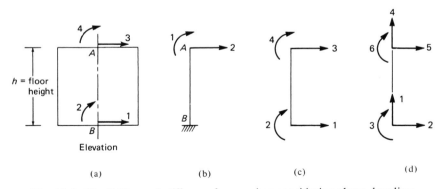

Fig. 16-2. Flexibility and stiffness of a member considering shear, bending and axial deformations. (a) Shear-wall element. (b), (c), and (d) Coordinates corresponding to the flexibility or stiffness matrices in: (b) Eqs. 16-1 and 16-2, (c) Eq. 16-4, and (d) Eq. 16-5.

Consider the cantilever AB in Fig. 16-2b. The flexibility matrix corresponding to the two coordinates indicated at A is

$$[f] = \begin{bmatrix} \dfrac{h}{EI} & \text{symmetrical} \\[2mm] \dfrac{h^2}{2EI} & \left(\dfrac{h^3}{3EI} + \dfrac{h}{Ga_r}\right) \end{bmatrix} \qquad (16\text{-}1)$$

The term h/Ga_r is the shear deflection at A due to a unit transverse load at A, with the notation: G is the shear modulus of elasticity, a_r is the effective shear area (see Sec. 5-33), and h is the floor height.

The stiffness matrix corresponding to the coordinates in Fig. 16-2b is obtained by inverting $[f]$. Thus

$$[f]^{-1} = \frac{1}{(1+\alpha)} \begin{bmatrix} (4+\alpha)\dfrac{EI}{h} & \text{symmetrical} \\[2mm] -\dfrac{6EI}{h^2} & \dfrac{12EI}{h^3} \end{bmatrix} \qquad (16\text{-}2)$$

where

$$\alpha = \frac{12EI}{h^2 Ga_r} \qquad (16\text{-}3)$$

From the elements of the stiffness matrix of Eq. 16-2 and by considering equilibrium, the stiffness matrix corresponding to the coordinates in Fig. 16-2c can be derived. Thus the stiffness matrix for a prismatic bar (Fig. 16-2c), or the shear-wall element in Fig. 16-2a, with the shear deformation considered is

$$[S] = \frac{1}{1+\alpha} \begin{bmatrix} \dfrac{12EI}{h^3} & & \text{symmetrical} \\[2mm] \dfrac{6EI}{h^2} & (4+\alpha)\dfrac{EI}{h} \\[2mm] -\dfrac{12EI}{h^3} & -\dfrac{6EI}{h^2} & \dfrac{12EI}{h^3} \\[2mm] \dfrac{6EI}{h^2} & (2-\alpha)\dfrac{EI}{h} & -\dfrac{6EI}{h^2} & (4+\alpha)\dfrac{EI}{h} \end{bmatrix} \qquad (16\text{-}4)$$

Putting $\alpha = 0$, the stiffness matrix becomes identical with the matrix in Eq. 4-7 in which the shear deformations are ignored.

In some cases it is necessary to consider the axial deformations so that the stiffness matrix has to be written for six coordinates as shown in Fig.

16-2d. The stiffness matrix corresponding to these coordinates taking into account bending, shear and axial deformation is

$$
[S] =
\begin{array}{c}
1 \\
2 \\
3 \\
4 \\
5 \\
6
\end{array}
\left[
\begin{array}{cccccc}
\dfrac{Ea}{h} & & & \text{symmetrical} & & \\
 & & & \text{Elements not shown} & & \\
 & \dfrac{12EI}{(1+\alpha)h^3} & & \text{are zero} & & \\
 & \dfrac{6EI}{(1+\alpha)h^2} & \dfrac{(4+\alpha)EI}{(1+\alpha)h} & & & \\
-\dfrac{Ea}{h} & & & \dfrac{Ea}{h} & & \\
 & -\dfrac{12EI}{(1+\alpha)h^3} & -\dfrac{6EI}{(1+\alpha)h^2} & & \dfrac{12EI}{(1+\alpha)h^3} & \\
 & \dfrac{6EI}{(1+\alpha)h^2} & \dfrac{(2-\alpha)EI}{(1+\alpha)h} & & -\dfrac{6EI}{(1+\alpha)h^2} & \dfrac{(4+\alpha)EI}{(1+\alpha)h}
\end{array}
\right]
$$

$$(16\text{-}5)$$

where a is the area of a cross section perpendicular to the axis and α is as before (Eq. 16-3). If we put $\alpha = 0$, Eq. 16-5 becomes the same as Eq. 4-6 with the shear deformations ignored.

16-3 STIFFNESS MATRIX OF A BEAM WITH RIGID END PARTS

Shear walls are usually connected by beams and for purposes of analysis we have to find the stiffness of such a beam corresponding to coordinates at the wall axis. Consider the beam AB of Fig. 16-3a. We assume that the beam has two rigid parts AA' and $B'B$ (Fig. 16-3b). The displacements $\{D^*\}$ at A and B are related to the displacements $\{D\}$ at A' and B' by geometry as follows:

$$\{D\} = [H]\{D^*\} \qquad (16\text{-}6)$$

where

$$[H] =
\begin{bmatrix}
1 & dl & 0 & 0 \\
0 & 1 & 0 & 0 \\
0 & 0 & 1 & -bl \\
0 & 0 & 0 & 1
\end{bmatrix}
\qquad (16\text{-}7)$$

The elements in the first and second column of $[H]$, which are the $\{D\}$ displacements due to $D_1^* = 1$ and $D_2^* = 1$ respectively, can be checked by examining Fig. 16-3c.

If shear deformations are to be considered, the stiffness matrix $[S]$ of

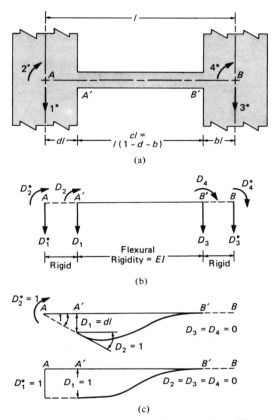

Fig. 16-3. Coordinate system corresponding to the stiffness matrices in Eqs. 16-8 and 16-10 of a beam between shear walls. (a) Evaluation of a beam between shear walls. (b) Coordinate systems. (c) Deflected configuration corresponding to $D_1^* = 1$ and $D_2^* = 1$.

the beam corresponding to the $\{D\}$ coordinates is the same as in Eq. 16-4 with cl substituted for h. Thus

$$
[S] = \frac{1}{1+\alpha}
\begin{bmatrix}
\dfrac{12EI}{c^3 l^3} & & \text{symmetrical} & \\[2ex]
\dfrac{6EI}{c^2 l^2} & (4+\alpha)\dfrac{EI}{cl} & & \\[2ex]
-\dfrac{12EI}{c^3 l^3} & -\dfrac{6EI}{c^2 l^2} & \dfrac{12EI}{c^3 l^3} & \\[2ex]
\dfrac{6EI}{c^2 l^2} & (2-\alpha)\dfrac{EI}{cl} & -\dfrac{6EI}{c^2 l^2} & (4+\alpha)\dfrac{EI}{cl}
\end{bmatrix}
\tag{16-8}
$$

where $c = 1 - d - b$, and d and b are the ratios of the lengths of the rigid parts of the beam to the total length l.

From Eq. 8-17, the stiffness matrix corresponding to the $\{D^*\}$ coordinates is given by

$$[S^*] = [H]^T [S] [H] \tag{16-9}$$

Substituting Eqs. 16-7 and 16-8 into Eq. 16-9 gives

$$[S^*] = \frac{EI}{1+\alpha} \begin{bmatrix} \dfrac{12}{c^3 l^3} & & & & & \text{symmetrical} \\[2mm] \dfrac{6}{c^2 l^2} + \dfrac{12d}{c^3 l^2} & 4 + \alpha + \dfrac{12d}{c^2 l} + \dfrac{12d^2}{c^3 l} & & & & \\[2mm] -\dfrac{12}{c^3 l^3} & -\dfrac{6}{c^2 l^2} - \dfrac{12d}{c^3 l^2} & \dfrac{12}{c^3 l^3} & & & \\[2mm] \dfrac{6}{c^2 l^2} + \dfrac{12b}{c^3 l^2} & 2 - \alpha + \dfrac{6d + 6b}{c^2 l} + \dfrac{12db}{c^3 l} & -\dfrac{6}{c^2 l^2} - \dfrac{12b}{c^3 l^2} & 4 + \alpha + \dfrac{12b}{c^2 l} + \dfrac{12b^2}{c^3 l} \end{bmatrix} \tag{16-10}$$

Equation 16-10 gives thus the stiffness matrix corresponding to $\{D^*\}$ coordinates (Fig. 16-3a) for a prismatic bar of length l with parts dl and bl at the ends of infinite flexural rigidity (Fig. 16-3b). The term $\alpha = (12EI)/(c^2 l^2 Ga_r)$ accounts for the shear deformations; if these are to be ignored α is put equal to zero.

16-4   ANALYSIS OF A PLANE FRAME WITH SHEAR WALLS

Consider the structure shown in Fig. 16-1b, composed of frames parallel to the axis of symmetry. Some of these frames include shear walls. Because of symmetry in structure and in loading, the diaphragms translate without rotation. With the diaphragms assumed rigid in their own planes, all the frames sway by the same amount D^* at a given floor level, as shown in Fig. 16-4a.

The stiffness matrix $[S^*]_i$ (of the order $n \times n$, where n is the number of floors), corresponding to the $\{D^*\}$ coordinates is calculated for each plane frame. The matrices are then added to obtain the stiffness $[S^*]$ of the entire structure

$$[S^*] = \sum_{i=1}^{m} [S^*]_i \tag{16-11}$$

where m is the number of the frames.

The sway at the floor levels is calculated by

$$[S^*]_{n \times n} \{D^*\}_{n \times 1} = \{F^*\}_{n \times 1} \tag{16-12}$$

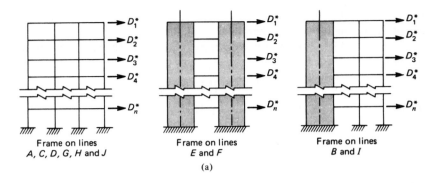

Frame on lines
A, C, D, G, H and J

Frame on lines
E and F

Frame on lines
B and I

(a)

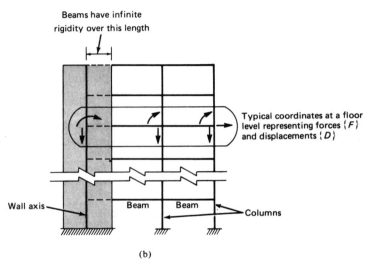

(b)

Fig. 16-4. Plane frames considered in the analysis of the symmetrical three-dimensional structure of Fig. 16-1b. (a) Frames parallel to the line of symmetry in the building of Fig. 16-1b. (d) Coordinate system corresponding to the stiffness matrix $[S]_i$ for the frame on lines B and I.

where $\{F^*\}$ are the resultant horizontal forces at floor levels, and n is the number of the floors.

In order to determine $[S^*]_i$ for any frame, say the frame on line B or I (Fig. 16-4a), coordinates are taken at the frame joints as shown in Fig. 16-4b. These represent rotation and vertical displacement at each joint and sway of the floor as a whole. The corresponding stiffness matrix $[S]_i$ (of the order $7n \times 7n$, in this case) is first derived using the stiffness of the shear wall and beam attached to it (as obtained in Secs. 16-2 and 16-3). The stiffness matrix $[S]_i$ is then condensed into matrix $[S^*]_i$ corresponding to coordinates for the sidesway at floor level (see Sec. 4-5). The elements of $[S^*]_i$ are forces at floor levels corresponding to unit horizontal displacements at alternate floors with the rotations and vertical joint displacements allowed to take place.

After solving Eq. 16-12 for $\{D^*\}$, the horizontal forces at floor levels for each plane frame are determined by

$$[S^*]_i\{D^*\} = \{F^*\}_i \qquad (16\text{-}13)$$

When the horizontal forces $\{F^*\}_i$ are applied at the floors of the ith frame, without forces at the other coordinates in Fig. 16-4b, the displacements at all the other coordinates in this figure can be calculated. From these, the stress resultants in any element can be determined.

Clough[2] et al. give details of an analysis involving the same assumptions as in the above approach and describe an appropriate computer program. It may be interesting to give here the results of their analysis of a 20-storey structure with the plan arrangement shown in Fig. 16-5a, subjected to wind loading in the direction of the x axis. Storey heights are 10 ft (3.05 m), except for the ground floor which is 15 ft (4.58 m) high. All columns and shear walls are fixed at the base. The properties of the structural members are listed in Table 16-1.

Table 16-1.   Member Properties for the Example of Fig. 16-5

	Columns		Shear Walls		Girders	
	I, in.^{4}	Area in.^{2}	I in.^{4}	Area, in.^{2}	I, in.^{4}	Area, in.^{2}
Stories 11–20	3437	221.5	1793×10^4	3155	6875	—
Stories 1–10	9604	313.5	2150×10^4	3396	11295	—

1 in.^{2} = 645 mm^2 1 in.^{4} = 416000 mm^4

For this structure, two types of frames need to be considered: frame A with five bays, with all columns considered to be of zero width in the x direction; and frame B of three bays: 26, 36, and 26 ft (7.93, 10.98 and 7.93 m) with two shear wall columns 20 ft (6.10 m) wide.

Figure 16-5b, taken from Clough's paper, shows the distribution of shear force between the columns and shear walls; we can see that the major part of the lateral resistance is provided by the shear walls.

Figure 16-5c, also reproduced from Clough et al., shows the bending moment in the shear wall and in an inner column of frame A. These results clearly demonstrate the different behavior of columns and shear walls: in Clough's words, "the shear wall is basically a cantilever column, with frame action modifying its moment diagram only slightly, whereas the single column shows essentially pure frame action." The effect of discontinuity in

[2]R. W. Clough, I. P. King and E. L. Wilson, "Structural Analysis of Multi-Storey Buildings," Proc. ASCE, Vol. 90, No. ST3, Part 1 (1964), pp. 19–34.

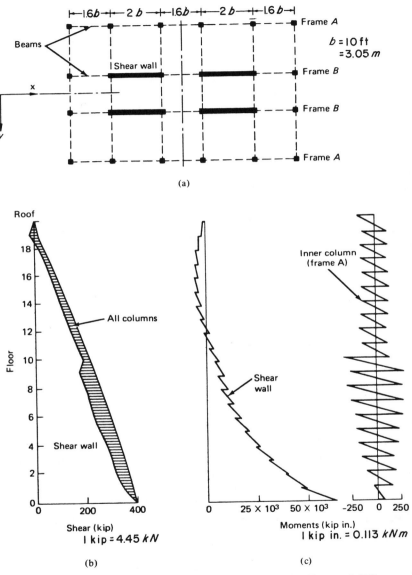

Fig. 16-5. Example of a building analyzed by Clough, King, and Wilson. (a) Plan. (b) Shearing-force distribution. (c) Moments in vertical members.

column stiffness between the 10th and 11th floor is apparent in Figs. 16-5b and c.

16-5 SIMPLIFIED APPROXIMATE ANALYSIS OF A BUILDING AS A PLANE STRUCTURE

The preceding example (see Fig. 16-5c) shows that the columns have a point of inflection within the height of each storey, while the deflection of the walls is similar to that of a cantilever. This is so because the rotation of the column ends is elastically restrained by the beams. When the walls have a very high I value compared to that of the beams, which is the case in practice, the beams cannot significantly prevent the rotation at the floor levels associated with the deflection in the form of a cantilever.

This behavior leads to the suggestion that structures of the type shown in Figs. 16-1b or 16-5a under the action of horizontal forces, can be idealized into a structure composed of the two systems indicated in Fig. 16-6a. One of these is a shear wall which has an I value in any storey equal to the sum of the I values of all the walls; the second system is an equivalent column rigidly jointed to the beams. The I_c value for the equivalent column is the sum of the I values for all the columns in a storey. The $(I/l)_b$ value for any of the beams is equal to four times the sum of (I/l) values for all the beams running in the x direction (see Secs. 12-5 and 12-7). The two systems, connected by inextensible link members, are assumed to resist the full external horizontal forces at floor levels. Further, the axial deformations of all the members are ignored. The shear deformations of the wall or the columns may or may not be included in the analysis. If they are, the reduced (effective) area is the sum of the reduced areas of the walls or columns in a storey.

The idealized structure is assumed to have n degrees of freedom representing the sidesway of the floors. The stiffness matrix $[S^*]_{n \times n}$ of this structure is obtained by summation of the stiffness matrices of the two systems. Thus,

$$[S^*] = [S^*]_w + [S^*]_r \qquad (16\text{-}14)$$

where $[S^*]_w$ and $[S^*]_r$ are the stiffness matrices respectively of the shear wall and the substitute frame, corresponding to n horizontal coordinates at the floor levels. For the determination of $[S^*]_w$ or $[S^*]_r$, two degrees of freedom (a rotation and sidesway) are considered at each floor level for the wall and at each beam-column connection in the substitute frame. A stiffness matrix $[S]_w$ or $[S]_r$ of order $2n \times 2n$ corresponding to the coordinates in Fig. 16-6b is written, then these matrices are condensed to $[S^*]_w$ and $[S^*]_r$, which relate horizontal forces to sidesway with the rotations unrestrained (see Sec. 4-5).

The sidesway at floor levels of the actual structure is then calculated by solving

$$[S^*]_{n \times n} \{D^*\}_{n \times 1} = \{F^*\}_{n \times 1} \qquad (16\text{-}15)$$

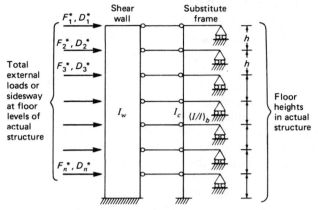

For the shear wall:
$$I_w = \Sigma I_{wi}$$

For the substitute frame:
$$I_c = \Sigma I_{ci} \text{ and } (I/l)_b = 4\Sigma (I/l)_{bi}$$

i is for all walls, columns or beams in a floor. Subscripts
w, c, and *b* refer to wall, column, and beam, respectively.

(a)

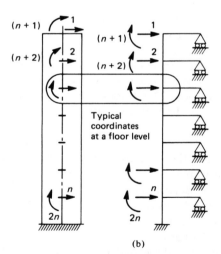

(b)

Fig. 16-6. Simplified analysis of a building frame of the type shown in Figs.
16-1b or 16-5a. (a) Idealized structure. (b) Coordinates corresponding to
stiffness matrices $[S_w]$ and $[S_r]$.

The displacements $\{D^*\}$ represent the horizontal translation at floor levels of all columns or shear walls in the building. The external forces $\{F^*\} = \{F^*\}_w + \{F^*\}_r$, where $\{F^*\}_w$ and $\{F^*\}_r$ are respectively the forces resisted by walls and by the substitute frame, can be calculated from

$$\{F^*\}_w = [S^*]_w\{D^*\} \tag{16-16}$$

and

$$\{F^*\}_r = [S^*]_r\{D^*\} \tag{16-17}$$

These forces are then applied to the shear wall and to the substitute frame, and the member end-moments are determined in each system. If these end-moments are apportioned to the walls, columns, and beams of the actual structure according to their (EI/h) or (EI/l) values, approximate values of the actual member end-moments can be obtained. The apportionment in this manner may result in unbalanced moments at some of the joints. Improved values can be rapidly reached by performing one or two cycles of moment distribution. This procedure bears, in parts, a similarity to the substitute frame method of Sec. 12-7.

An iteration procedure for calculating the sidesway of the idealized structure considered here has been suggested by Khan and Sbarounis,[3] who give also some charts to assist in practical design.

It is important to note that if the shear walls differ considerably from one wall to another or if there are variations in cross section at different levels, the above method of calculation can lead to erroneous results. In such a case, it may be necessary to consider an idealized structure composed of more than one wall attached by links to the substitute frame and to derive the stiffness of each separately; the stiffness of the idealized structure is then obtained by summation.

16-51   Special Case of Similar Columns and Beams

When the column cross section and height are the same in all storeys and $(I/l)_b$ is the same at all floors, the stiffness matrix of the substitute frame in the right-hand side of Fig. 16-6b is

$$[S]_{r_{2n \times 2n}} = \begin{bmatrix} [S_{11}]_r & [S_{12}]_r \\ \hline [S_{21}]_r & [S_{22}]_r \end{bmatrix} \tag{16-18}$$

[3]F. R. Khan and J. A. Sbarounis, "Interaction of Shear Walls and Frames," *Proc. ASCE*, Vol. 90, No. ST3, Part 1 (1964), pp. 285–335.

The submatrices are

$$[S_{11}]_r = \frac{2(S + t)}{h^2} \begin{bmatrix} 1 & -1 & & & & & \\ -1 & 2 & -1 & & & & \\ & -1 & 2 & -1 & & & \\ & & \cdots & \cdots & \cdots & & \\ \text{Elements not} & & \cdots & \cdots & \cdots & & \\ \text{shown are} & & & -1 & 2 & -1 \\ \text{zero} & & & & -1 & 2 \end{bmatrix}_{n \times n} \quad (16\text{-}19)$$

$$[S_{21}]_r = [S_{12}]_r^T = \frac{(S + t)}{h} \begin{bmatrix} -1 & 1 & & & & & \\ -1 & 0 & 1 & & & & \\ & -1 & 0 & 1 & & & \\ & & \cdots & \cdots & \cdots & & \\ \text{Elements not} & & \cdots & \cdots & \cdots & & \\ \text{shown are zero} & & & -1 & 0 & 1 \\ & & & & -1 & 0 \end{bmatrix}_{n \times n} \quad (16\text{-}20)$$

and

$$[S_{22}]_r = S \begin{bmatrix} (1 + \beta) & C & & & & & \\ C & (2 + \beta) & C & & & & \\ & C & (2 + \beta) & C & & & \\ & & \cdots & \cdots & \cdots & & \\ \text{Elements not} & & \cdots & \cdots & \cdots & & \\ \text{shown are zero} & & & C & (2 + \beta) & C \\ & & & & C & (2 + \beta) \end{bmatrix}_{n \times n}$$

$$(16\text{-}21)$$

where

$$\beta = \frac{3E}{S}(I/l)_b \quad (16\text{-}22)$$

$$S = \frac{(4 + \alpha)}{(1 + \alpha)} \frac{EI_c}{h} \quad (16\text{-}23)$$

$$t = \frac{(2 - \alpha)}{(1 + \alpha)} \frac{EI_c}{h} \quad (16\text{-}24)$$

and

$$C = (t/S) \quad (16\text{-}25)$$

The term S is the column rotational stiffness at one end with the far end fixed; t is the carryover moment (refer to Eq. 16-4); and C the carryover factor. The shear deformation of the vertical members is accounted for by the term

$$\alpha = \frac{12EI_c}{h^2 Ga_{rc}} \tag{16-26}$$

while the shear deformation of the beams is ignored. I_c and $(I/l)_b$ are the properties respectively of the column and of the beam in the substitute frame (see Fig. 16-6a). The cross-sectional area $a_{rc} = \Sigma a_{rci} =$ sum of the reduced (effective) cross-sectional areas of all the columns in the frame.

The general case in which the column cross section and height varies from storey to storey and the beams do not have the same $(I/l)_b$ – values at all floor levels is considered in Prob. 16-2.

The above equations can be used to obtain the stiffness matrix $[S]_w$ of the wall corresponding to the coordinates in Fig. 16-6b. For this purpose, we put $\beta = 0$ and substitute the subscript w for c.

Example 16-1 Find the approximate values of the end-moments in a column and a shear wall in a structure which has the same plan as in Fig. 16-5a, and has four stories of equal height $h = b$. The frame is subjected to a horizontal force in the x direction of magnitude $P/2$ at top floor and P at each of the other floor levels. The properties of members are as follows: for any column $I = 17 \times 10^{-6}b^4$, for any beam $I = 34 \times 10^{-6}b^4$, and for any wall $I = 87 \times 10^{-3}b^4$. Take $E = 2.3G$. The area of wall cross section = $222 \times 10^{-3}b^2$. Consider shear deformation in the walls only.

The wall and the substitute frame of the idealized structure are shown in Fig. 16-7a. In the actual structure there are 16 columns, four walls, 12 beams of length $1.6b$, and 4 beams of length $2b$. The properties of members of the idealized structure are:

$$I_c = \Sigma I_{ci} = 16 \times 17 \times 10^{-6}b^4 = 272 \times 10^{-6}b^4$$

$$(I/l)_b = 4\Sigma(I/l)_{bi} = 4 \times 34 \times 10^{-6}b^4\left(\frac{12}{1.6b} + \frac{4}{2.0b}\right) = 1292 \times 10^{-6}b^3$$

$$I_w = \Sigma I_{wi} = 4 \times 87 \times 10^{-3}b^4 = 348 \times 10^{-3}b^4$$

$$a_{rw} = \Sigma a_{rwi} = 4 \times \frac{5}{6} \times 222 \times 10^{-3}b^2 = 740 \times 10^{-3}b^2$$

For the derivation of the stiffness matrix of the wall and the substitute frame, the following quantities are calculated by Eqs. 16-22–16-26.

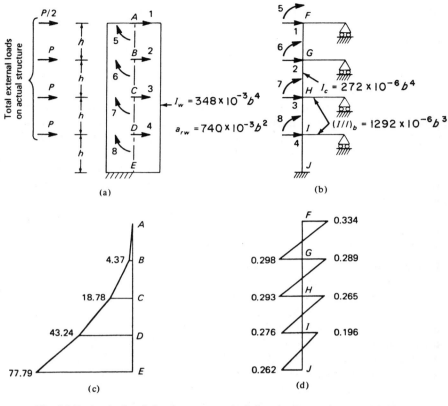

Fig. 16-7. Analysis of the four-storey building in Example 16-1. (a) Shear wall. (b) Substitute frame. (c) Sum of moments in walls in terms of $Ph/10$. (d) Sum of moments in columns in terms of $Ph/10$.

	Wall	Substitute frame
α:	12.9	0
S:	$0.423Eh^3$	$1.09 \times 10^{-3}Eh^3$
t:	$-0.273Eh^3$	$0.54 \times 10^{-3}Eh^3$
C:	-0.65	0.5
β:	—	3.56

To obtain the stiffness matrix of the substitute frame, we substitute in Eqs. 16-19–16-21 to calculate the submatrices of $[S_r]$:

$$[S_{11}]_r = 3.26 \times 10^{-3}Eh \begin{bmatrix} 1 & -1 & & \\ -1 & 2 & -1 & \\ & -1 & 2 & -1 \\ & & -1 & 2 \end{bmatrix}$$

$$[S_{21}] = [S_{12}]_r^T = 1.63 \times 10^{-3}Eh^2 \begin{bmatrix} -1 & 1 & & \\ -1 & 0 & 1 & \\ & -1 & 0 & 1 \\ & & -1 & 0 \end{bmatrix}$$

$$[S_{22}]_r = 1.09 \times 10^{-3}Eh^3 \begin{bmatrix} 4.56 & 0.5 & & \\ 0.5 & 5.56 & 0.5 & \\ & 0.5 & 5.56 & 0.5 \\ & & 0.5 & 5.56 \end{bmatrix}$$

The condensed stiffness matrix corresponding to the four sidesway coordinates is (by Eq. 4-17):

$$[S^*]_r = [S_{11}]_r - [S_{12}]_r [S_{22}]_r^{-1} [S_{21}]_r \qquad (16\text{-}27)$$

Substituting, we obtain

$$[S^*]_r = 10^{-4}Eh \begin{bmatrix} 23.55 & & \text{symmetrical} & \\ -27.16 & 55.08 & & \\ 3.93 & -31.93 & 56.03 & \\ -0.36 & 4.40 & -32.40 & 60.37 \end{bmatrix}$$

In a similar way, the stiffness matrix of the wall corresponding to horizontal coordinates at the floor level is

$$[S^*]_w = 10^{-2}Eh \begin{bmatrix} 13.98 & & \text{symmetrical} & \\ -22.76 & 51.22 & & \\ 5.70 & -31.84 & 53.77 & \\ 2.33 & 2.55 & -30.19 & 56.33 \end{bmatrix}$$

The stiffness matrix of the idealized structure (wall and substitute frame connected) is obtained by Eq. 16-14:

$$\begin{aligned}[S^*] = \\ [S^*]_w + [S^*]_r = 10^{-4}Eh\end{aligned} \begin{bmatrix} 1421.40 & & \text{symmetrical} & \\ -2303.05 & 5177.15 & & \\ 574.22 & -3215.76 & 5433.34 & \\ 232.28 & 259.64 & -3051.41 & 5692.92 \end{bmatrix}$$

Substituting $[S^*]$ from the above equation and $\{F^*\} = P\{0.5, 1.0, 1.0, 1.0\}$ in Eq. 16-15 and solving for $\{D^*\}$, we find the sidesway of the actual structure at floor levels

$$\{D^*\} = \frac{10P}{Eh} \begin{Bmatrix} 11.47 \\ 8.32 \\ 5.03 \\ 2.03 \end{Bmatrix}$$

Multiplying $[S^*]_w$ or $[S^*]_r$ by $\{D^*\}$, we obtain the forces resisted by the wall and by the substitute frame (Eqs. 16-16 and 16-17)

$$\{F^*\}_w = P \begin{Bmatrix} 0.4368 \\ 1.0050 \\ 1.0041 \\ 1.0083 \end{Bmatrix} \qquad \{F^*\}_r = P \begin{Bmatrix} 0.0632 \\ -0.0050 \\ -0.0041 \\ -0.0083 \end{Bmatrix}$$

The joint rotations of the substitute frame $\{D_2\}_r$ can be determined by

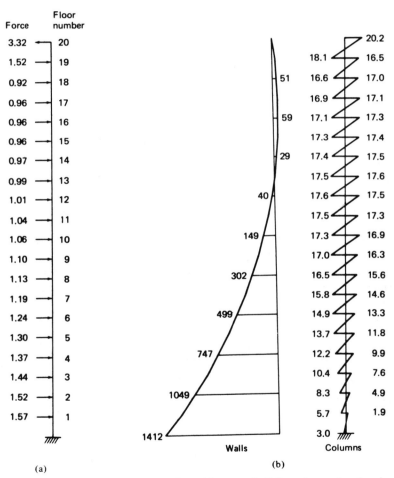

(a)   (b)

Fig. 16-8. Forces and moments in a 20-storey building shown in plan in Fig. 16-5a, using analysis in Example 16-1. (a) Forces resisted by shear walls in terms of P. (b) Moments in vertical members in terms of $Ph/10$.

$$\{D_2\}_r = -[S_{22}]_r^{-1}[S_{21}]_r\{D_1\}_r \qquad (16\text{-}28)$$

where $\{D_1\}_r = \{D^*\}$ are the displacements at the first n coordinates in Fig. 16-7b. A part of the calculations required for this equation has already been done in the operations for Eq. 16-27. From the rotation and translation of the joints in the column, the end-moments can be easily calculated. The bending-moment diagram of the substitute frame is shown in Fig. 16-7d. The moment in the shear walls is calculated by applying the forces $\{F^*\}_w$ on a cantilever and is plotted in Fig. 16-7c.

Because all the walls are of the same cross section, and likewise the columns, the moment in each wall is $\frac{1}{4}$ of the value in Fig. 16-7c, and in each column $\frac{1}{16}$ of the value in Fig. 16-7d.

The solution of the same problem with the number of stories $n = 20$, instead of 4, gives the results shown in Fig. 16-8. The external applied forces are $P/2$ on the top diaphragm and P on each of the others. All stories have the same height h.

16-6   SHEAR WALLS WITH OPENINGS

Figures 16-9a and b show the types of walls which are often used in dwelling blocks. The two types shown differ in the size of openings and in their locations. An exact treatment of this problem would require the solution of the governing plane-stress elasticity equations, but this is difficult and cannot be used in practice. A reasonable solution can be obtained by the finite element method (see Chapter 19) or by idealizing the wall into different types of latticed frames composed of small elements; however, the calculation generally requires the solution of a large number of equations. These methods are used in the analysis of walls with openings arranged in any pattern, and they give a better picture of the stress distribution than the much more simplified analysis described below.

In the simplified analysis, walls with a row of openings of the type shown in Figs. 16-9a and b are idealized to a frame composed of two wide columns connected by beams with end parts infinitely rigid. The stiffness of the elements forming such a structure and the method of analysis are given in Secs. 16-2 and 16-3. Macleod[4] showed by model testing that the idealization of a wall of this type by a frame gives a good estimate of stiffness (corresponding to sidesway) for most practical cases. It seems, therefore, that the finite element idealization offers little advantage in this respect.

The symmetrical wall in Fig. 16-9a can be easily analyzed using a suitable frame composed of one column rigidly connected to beams (see

[4]I. A. MacLeod, "Lateral Stiffness of Shear Walls with Openings, in Tall Buildings," *Proceedings of a Symposium on Tall Buildings, Southampton*, Pergamon Press, New York, 1967, pp. 223–244.

Fig. 16-7b and Sec. 12-3). Each beam has a rigid part near its connection with the column. The stiffness matrix of such a beam with one end hinged, corresponding to the coordinates in Fig. 16-9d can be easily derived from Eq. 16-10 and is

$$[\bar{S}] = \begin{bmatrix} S_{11}^* - \dfrac{S_{14}^{*2}}{S_{44}^*} & \text{symmetrical} \\[3ex] S_{21}^* - \dfrac{S_{24}^* S_{41}^*}{S_{44}^*} & S_{22}^* - \dfrac{S_{24}^{*2}}{S_{44}^*} \end{bmatrix} \tag{16-29}$$

where S_{ij}^* are elements of the stiffness matrix in Eq. 16-10 with $b = 0$. If the axial deformation is ignored in the special case of a symmetrical wall having the same height and cross section in all storeys and the beams between

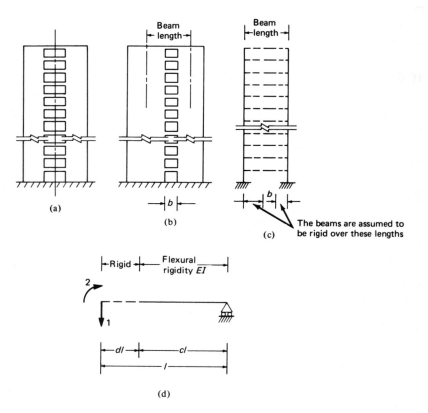

Fig. 16-9. Idealized structure for the analysis of a wall with row of openings. (a) Symmetrical wall. (b) Wall with a row of openings. (c) Idealized structure for the analysis of the wall in part (b). (d) Coordinates corresponding to the stiffness matrix $\bar{S}$ given by Eq. 16-29.

the windows are the same in all floors, Eqs. 16-18–16-26 can be used to derive the stiffness matrix of the substitute frame, but β of Eq. 16-22 must be replaced by

$$\beta' = \frac{\bar{S}_{22}}{S} \qquad (16\text{-}30)$$

where $\bar{S}_{22}$ is an element of the stiffness matrix in Eq. 16-29 and S is defined by Eq. 16-23. The reason for this is that in Eq. 16-22, the quantity $3E(I/l)_h$ is the adjusted end-rotational stiffness of a prismatic horizontal beam in the substitute frame. The corresponding quantity for the beam in Fig. 16-9d is $\bar{S}_{22}$.

16-7 THREE-DIMENSIONAL ANALYSIS

A joint in a three-dimensional framed structure has in general six degrees of freedom: three rotations and three translations in the x, y, and z directions (Fig. 16-10a). The assumption that the diaphragms in multistorey building are rigid constrains three of the displacements (D_1, D_2, and D_3) to be the same at all joints in one floor. Even with this important simplification, the analysis is involved in the case of three-dimensional structures incorporating members having an arbitrary orientation in space. Weaver and Nelson[5] have suggested an analysis using the stiffness method for a three-dimensional building in which the framing is arranged in a rectangular pattern.

This section deals with the case of a structure formed of shear walls in a random arrangement. Any two or more walls which are monolithic will be referred to as *wall assembly*. A typical wall assembly is shown in plan in Fig. 16-10b. The structure is analyzed to determine the forces resisted by different shear walls when horizontal forces in any direction are applied at floor levels. In addition to the rigid-diaphragm assumption used in the previous sections, we assume here that the floors do not restrain the joint rotations about the x and y axes (D_4 and D_5 in Fig. 16-10a). This assumption is equivalent to considering that the diaphragm has a small flexural rigidity compared with the walls and can therefore be ignored. With this additional assumption, horizontal forces result in no axial forces in the walls; thus the vertical displacements (D_6 in Fig. 16-10a) are zero.

Given all these assumptions, the analysis by the stiffness method will now be performed for a single-storey structure and then extended to a multistorey building.

16-71 One-Storey Structure

Imagine that the building shown in plan in Fig. 16-10b has one storey of height h and the walls are totally fixed at the base. The displacement of the

[5]W. Weaver, Jr., and M. F. Nelson, "Three-Dimensional Analysis of Tier Buildings," *Proc. ASCE*, Vol. 92, No. ST6 (1966), pp. 385–404.

walls at the floor level is completely defined if the displacements $\{D\}$ at the coordinates 1, 2, and 3 are known at any arbitrary point O in the floor level. A horizontal force anywhere in the plane of the floor can be analyzed into three components $\{F\}$ along the three coordinates.

The forces and displacements at the shear center at the top of any wall assembly are related by

$$[\bar{S}]_i\{q\}_i = \{Q\}_i \tag{16-31}$$

where $[\bar{S}]_i$ is the stiffness matrix of the ith wall assembly, and $\{q\}_i$ and $\{Q\}_i$ are respectively displacements and forces at three local coordinates

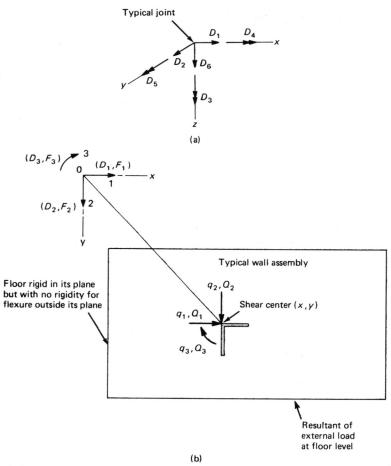

Fig. 16-10. Coordinate system for the analysis of a single-storey shear-wall structure. (a) Degrees of freedom of a typical joint in a building frame. (b) Coordinate system.

in that assembly; they represent translations (or forces) at the shear center of this wall at the floor level parallel to the x and y axes and a rotation (or a couple) about the z-axis (Fig. 16-10b).

To derive the stiffness matrix $[\bar{S}]_i$, consider any wall AB in Fig. 16-11a fixed at the base and free at the top. The flexibility matrix corresponding

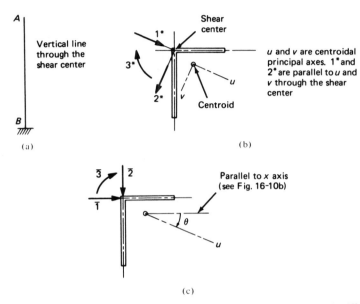

Fig. 16-11. Coordinate systems corresponding to the flexibility and stiffness matrices of a wall assembly in Eqs. 16-32, 16-34, and 16-37. (a) Elevation. (b) Plan showing coordinates corresponding to the flexibility matrix in Eq. 16-32. (c) Coordinates corresponding to matrix in Eqs. 16-34 or 16-37.

to the coordinates in Fig. 16-11b, of which 1* and 2* are parallel to the principal axes of inertia of the cross section, is

$$
[f^*]_i = \begin{bmatrix} \left(\dfrac{h^3}{3EI_v} + \dfrac{h}{Ga_{rv}}\right) \\ & \left(\dfrac{h^3}{3EI_u} + \dfrac{h}{Ga_{ru}}\right) \\ \text{Elements not shown} & & \dfrac{h - (\tanh \gamma h)/\gamma}{GJ} \\ \text{are zero} \end{bmatrix}_i
\tag{16-32}
$$

where I_v and I_u are the second moments of the cross-sectional area about the principal axes v and u respectively; and a_{rv} and a_{ru} are reduced (effective) areas of the cross section corresponding to loading in the vertical plane through axes u and v respectively. The second term in the expressions for

f_{11}^* and f_{22}^* in the above matrix accounts for shear deformation, while the second term in f_{33}^* is included because the warping of the cross section at the bottom of the wall is prevented.[6] The parameter γ is given by

$$\gamma = \sqrt{\frac{GJ}{EK}} \tag{16-33}$$

where J is the torsion constant (length[4]) and K is the warping constant (length[6]) of the cross section. If shear deformation is ignored and the warping at the two ends is not prevented, or the warping effect is ignored, the three diagonal terms in Eq. 16-32 become: $h^3/(3EI_v)$, $h^3/(3EI_u)$, and $h/(GJ)$ (which are the displacements given in Appendix B).

The nondiagonal elements in $[f^*]_i$ are all zero because the three co-ordinates are chosen through the shear center and 1* and 2* are parallel to the principal axes of the section: a force applied through the shear center produces no twisting of the cross section; further, if this force is parallel to one of the principal axes, then the deflection takes place in a plane parallel to this axis.

The stiffness matrix $[\overline{S}]_i$ corresponding to the coordinates in Fig. 16-11c can be derived by inversion of $[f^*]_i$ and transformation (see Eq. 8-17).

$$[\overline{S}]_i = \begin{bmatrix} \overline{S}_{11} & \overline{S}_{12} & 0 \\ \overline{S}_{21} & \overline{S}_{22} & 0 \\ 0 & 0 & \overline{S}_{33} \end{bmatrix}_i \tag{16-34}$$

where

$$\left. \begin{aligned} \overline{S}_{11} &= \frac{E}{h^3}\left[\frac{12\cos^2\theta}{(4+\alpha_v)} I_v + \frac{12\sin^2\theta}{(4+\alpha_u)} I_u \right] \\[4pt] \overline{S}_{22} &= \frac{E}{h^3}\left[\frac{12\sin^2\theta}{(4+\alpha_v)} I_v + \frac{12\cos^2\theta}{(4+\alpha_u)} I_u \right] \\[4pt] \overline{S}_{12} &= \overline{S}_{21} = \frac{E}{h^3}\left[\sin\theta\cos\theta\left(\frac{12I_v}{4+\alpha_v} - \frac{12I_u}{4+\alpha_u} \right) \right] \\[4pt] \overline{S}_{33} &= \frac{GJ}{h-(\tanh\gamma h)/\gamma} \end{aligned} \right\} \tag{16-35}$$

$$\left. \begin{aligned} \alpha_u &= \frac{12EI_u}{h^2 Ga_{r_u}} \\[4pt] \alpha_v &= \frac{12EI_v}{h^2 Ga_{r_v}} \end{aligned} \right\} \tag{16-36}$$

and θ = the angle between the x and u axes.

[6]See S. P. Timoshenko, *Strength of Materials*, Part II, 3d ed., Van Nostrand, New York, 1956, p. 260.

If the shear deformation and the warping effect are ignored, the stiffness matrix in Eq. 16-34 corresponding to the three coordinates in Fig. 16-11c becomes

$$[\bar{S}]_i = \begin{bmatrix} \dfrac{3EI_y}{h^3} & \text{symmetrical} & \\[2mm] \dfrac{3EI_{xy}}{h^3} & \dfrac{3EI_x}{h^3} & \\[2mm] 0 & 0 & \dfrac{GJ}{h} \end{bmatrix}_i \qquad (16\text{-}37)$$

where I_y and I_x are the moments of inertia about axes parallel to the y and x axes through the centroid, and I_{xy} is the product of inertia about the same axes.

The displacements $\{q\}_i$ of the ith wall are related to the floor displacement $\{D\}$ by geometry as follows

$$\{q\}_i = [C]_i\{D\} \qquad (16\text{-}38)$$

where

$$[C]_i = \begin{bmatrix} 1 & 0 & -y \\ 0 & 1 & x \\ 0 & 0 & 1 \end{bmatrix}_i \qquad (16\text{-}39)$$

Here $[C]_i$ is a transformation matrix for the ith wall, and x and y are the Cartesian coordinates of the shear center of this wall (Fig. 16.10b). The transpose of this matrix relates the forces $\{Q\}_i$ to equivalent forces $\{F\}_i$ at the $\{D\}$ coordinates, (see Sec. 8-3)

$$\{F\}_i = [C]_i^T\{Q\}_i$$

Applying Eq. 8-17, the stiffness matrix of the ith wall corresponding to the $\{D\}$ coordinates is

$$[S]_i = [C]_i^T[\bar{S}]_i[C]_i \qquad (16\text{-}40)$$

This equation transforms the stiffness matrix $[\bar{S}]_i$ corresponding to the $\{q\}$ coordinates to a stiffness matrix $[S]_i$ corresponding to the $\{D\}$ coordinates. Performing the multiplication in Eq. 16-40 gives

$$[S]_i = \begin{bmatrix} \bar{S}_{11} & & \text{symmetrical} \\ \bar{S}_{21} & \bar{S}_{22} & \\ (-\bar{S}_{11}y + \bar{S}_{21}x) & (-\bar{S}_{12}y + \bar{S}_{22}x) & \begin{matrix}(\bar{S}_{11}y^2 - 2\bar{S}_{21}xy \\ + \bar{S}_{22}x^2 + \bar{S}_{33})\end{matrix} \end{bmatrix}$$

$$(16\text{-}40a)$$

The stiffness matrix of the structure can now be obtained by summation, thus

$$[S] = \sum_{i=1}^{m} [S]_i \qquad (16\text{-}41)$$

where m is the number of wall assemblies.

The floor displacement can be determined by the equation

$$\{F\}_{3 \times 1} = [S]_{3 \times 3}\{D\}_{3 \times 1} \qquad (16\text{-}42)$$

and the forces $\{Q\}_i$ on each wall assembly are calculated by Eqs. 16-38 and 16-31.

The above analysis for a single-storey structure is given separately not only because of its intrinsic importance but also because it serves as an introduction to the multistorey case. Some designers use the single-storey solution to each storey of a multistorey structure in turn in order to obtain an approximate solution. However, for a rational solution, all the floors must be treated simultaneously.

The above solution can also be used to calculate the displacement of bridge decks due to horizontal forces at the deck level. This is particularly useful for skew or curved bridges. The stiffness $[S]_i$ of the elements in this case would be that of a supporting pier or elastomeric bearing pad or a combination of these (see Prob. 16-8).

Example 16-2 Find the horizontal force resisted by each of the shear walls 1, 2, and 3 in a single-storey building whose plan is shown in Fig. 16-12a. All the walls are fixed to the base. Take shear deformation into consideration but ignore warping. Assume $E = 2.3G$.

The axes x and y are chosen to pass through the center of the shaft as shown in Fig. 16-12b, in which the coordinate system is indicated. In the present case, the shear center of each wall coincides with its centroid, for which the x and y coordinates are given in brackets. (The principal axes for each wall are parallel to the x and y axes.) The values of I_u, I_v, J, a_{ru}, and a_{rv} for each wall are given **below**:

	I_u	I_v	J	a_{ru}	a_{rv}
Wall 1:	$0.03413b^4$	$0.03413b^4$	$0.0512b^4$	$0.16b^2$	$0.16b^2$
Wall 2 or 3:	$0.0342b^4$	$0.00013b^4$	$0.00053b^4$	$0.133b^2$	$0.133b^2$

Substituting in Eq. 16-34 with $\theta = 0$ and neglecting the warping term, we obtain

$$[\bar{S}]_1 = 10^{-4}Eh \begin{bmatrix} 414.2 & 0 & 0 \\ 0 & 414.2 & 0 \\ 0 & 0 & 223.0h^2 \end{bmatrix}$$

No transformation is needed for this matrix, because of the choice of the

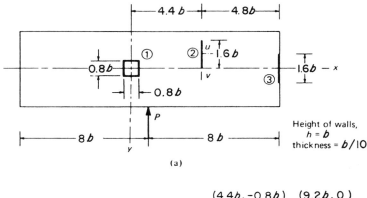

(a)

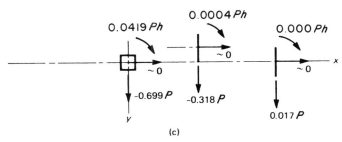

(b)

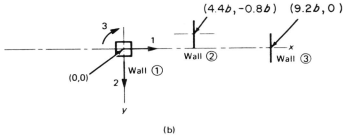

(c)

Fig. 16-12. A one-storey structure considered in Example 16-2. (a) Plan of shear walls. (b) Coordinate system. (c) Forces resisted by the various walls.

origin at the shear center of the shaft. Thus

$$[S]_1 = [\bar{S}]_1$$

Similarly, Eq. 16-34 applied to the wall 2 or 3 gives

$$[\bar{S}]_{2\ or\ 3} = 10^{-4}Eh \begin{bmatrix} 3.87 & 0 & 0 \\ 0 & 370 & 0 \\ 0 & 0 & 2.3h^2 \end{bmatrix}$$

Substituting in Eq. 16-40a, we obtain

$$[S]_2 = Eh \begin{bmatrix} 0.387 \times 10^{-3} & & \text{symmetrical} \\ 0 & 37 \times 10^{-3} & \\ 0.3099 \times 10^{-3}h & 0.1627h & 716.46 \times 10^{-3}h^2 \end{bmatrix}$$

and

$$[S]_3 = Eh \begin{bmatrix} 0.387 \times 10^{-3} & & \text{symmetrical} \\ 0 & 37 \times 10^{-3} & \\ 0 & 0.3402h & 3130.4 \times 10^{-3}h^2 \end{bmatrix}$$

The stiffness matrix of the structure is given by Eq. 16-41 with $m = 3$:

$$[S] = Eh \begin{bmatrix} 42.2 \times 10^{-3} & & \text{symmetrical} \\ 0 & 115.4 \times 10^{-3} & \\ 0.31 \times 10^{-3}h & 0.503h & 3869.1 \times 10^{-3}h^2 \end{bmatrix}$$

The forces at the coordinates equivalent to the external applied load are

$$\{F\} = P \begin{Bmatrix} 0 \\ -1 \\ -1.2h \end{Bmatrix}$$

Substituting in Eq. 16-42 and solving for $\{D\}$, we obtain

$$\{D\} = \frac{P}{Eh} \begin{Bmatrix} -0.0138 \\ -16.8782 \\ 1.884/h \end{Bmatrix}$$

Combining Eqs. 16-38 and 16-31, we find the forces resisted by any one of the wall assemblies

$$\{Q\}_i = [\bar{S}]_i [C]_i \{D\}$$

For example, the forces on wall 2 are:

$$\{Q\}_2 = [\bar{S}]_2 \begin{bmatrix} 1 & 0 & 0.8h \\ 0 & 1 & 4.4h \\ 0 & 0 & 1 \end{bmatrix} \{D\} = P \begin{bmatrix} 0.0000 \\ -0.3176 \\ 0.00044h \end{bmatrix}$$

The forces on all three walls are shown in Fig. 16-12c.

16-72 Multistorey Structure

A typical wall assembly in a building with n floors is shown in Fig. 16-13a. Figure 16-13b is an isometric view of a vertical axis through the shear center

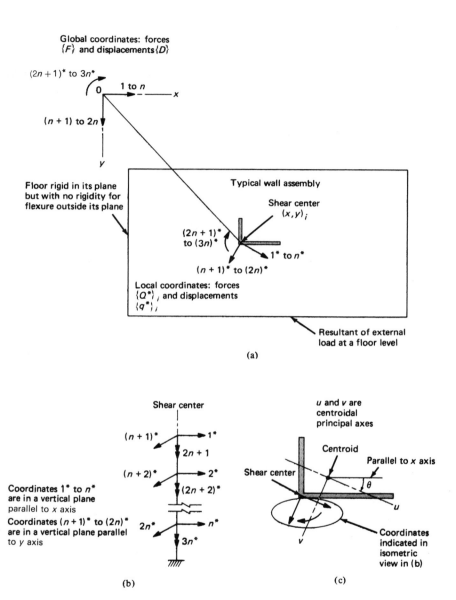

Fig. 16-13. Coordinates for the analysis of a multistorey shear wall structure. (a) Coordinate system. (b) Local coordinates for a typical wall. (c) Plan of a typical wall assembly.

of a shear wall assembly i. The local coordinates 1^* to n^* and $(n + 1)^*$ to $(2n)^*$ lie in vertical planes parallel to the principal axes of inertia of the cross section u and v, respectively (Fig. 16-13b). The directions of u and v are assumed to be the same at all floors for a given wall. The third set of coordinates, $(2n + 1)^*$ to $(3n)^*$, represent angles of twist or twisting couples. The stiffness matrix for the ith wall corresponding to these coordinates is of the form

$$[S^*]_i = \begin{bmatrix} [S_u^*] & & \\ & [S_v^*] & \\ \text{Submatrices not shown are zero} & & [S_\theta^*] \end{bmatrix}_i \qquad (16\text{-}43)$$

Each of the submatrices $[S_u^*]_i$ and $[S_v^*]_i$ can be determined separately in a way similar to that used for a plane structure in Secs. 16-4 and 16-5 (using Eq. 4-17). If warping is ignored and the wall is encastré at the base, the submatrix $[S_\theta^*]_i$ will be

$$[S_\theta^*]_i = \begin{bmatrix} \left(\dfrac{GJ}{h}\right)_1 & & & & & \\ -\left(\dfrac{GJ}{h}\right)_1 & \left(\dfrac{GJ}{h}\right)_1 + \left(\dfrac{GJ}{h}\right)_2 & & & \text{symmetrical} & \\ & -\left(\dfrac{GJ}{h}\right)_2 & \left(\dfrac{GJ}{h}\right)_2 + \left(\dfrac{GJ}{h}\right)_3 & & & \\ & & \cdots & \cdots & & \\ & & & & -\left(\dfrac{GJ}{h}\right)_{n-1} & \left(\dfrac{GJ}{h}\right)_{n-1} + \left(\dfrac{GJ}{h}\right)_n \end{bmatrix}_i$$

$$(16\text{-}43\text{a})$$

where the subscripts refer to the storey number reckoned from the top of the structure.

In order to consider the effect of warping in a rational way the angles of twist can be calculated at floor levels due to a unit twisting couple acting separately at each floor,[7] thus forming a flexibility matrix which, when inverted, gives $[S_\theta^*]$. At a fixed base, the warping is restrained while it is free to occur in the upper storeys in a high building. Thus, neglecting warping underestimates the torsional stiffness at the floors near the base.

The displacements $\{q\}_i$ and the forces $\{Q^*\}_i$ at the coordinates in Fig. 16-13b are related by

$$\{Q^*\}_i = [S^*]_i\{q^*\}_i \qquad (16\text{-}44)$$

[7]See reference in footnote 6 of this chapter: use Eq. (p), p. 260 of that reference.

A global coordinate system is defined in Fig. 16-13a, in which the coordinates 1 to n, and $(n + 1)$ to $2n$ represent translation or horizontal forces at the floor levels at an arbitrary point 0, and $(2n + 1)$ to $3n$ are floor rotations or twisting couples. The displacements $\{q^*\}_i$ and $\{D\}$ are related by geometry so that

$$\{q^*\}_{i3n \times 1} = [B]_{i3n \times 3n}\{D\}_{3n \times 1} \qquad (16\text{-}45)$$

where

$$[B]_i = \begin{bmatrix} \lceil \cos \theta, \ldots \rfloor & \lceil \sin \theta, \ldots \rfloor & \lceil (x \sin \theta - y \cos \theta), \ldots \rfloor \\ \lceil -\sin \theta, \ldots \rfloor & \lceil \cos \theta, \ldots \rfloor & \lceil (x \cos \theta + y \sin \theta), \ldots \rfloor \\ [0] & [0] & [I] \end{bmatrix}_i \qquad (16\text{-}46)$$

The special brackets used indicate diagonal submatrices each of order $n \times n$. The diagonal term for each of these submatrices is given above. The principal axes of any wall are assumed to have the same directions in all floors. If, in addition, the x and y coordinates of the shear center are the same in all floors, the submatrices have a diagonal element repeated in each row. For example,

$$[B_{11}]_i = \lceil \cos \theta, \cos \theta, \ldots, \cos \theta \rfloor_i = \cos \theta_i[I]$$

If the coordinates x and y change from floor to floor, the submatrix $[B_{13}]_i$ is

$$[B_{13}]_i = \lceil (x_1 \sin \theta - y_1 \cos \theta), \ldots, (x_j \sin \theta - y_j \cos \theta), \ldots \rfloor$$

where the subscripts of x and y indicate the floor (1 means the top floor and n the lowest floor). A similar equation can be written for $[B_{23}]_i$.

Applying Eq. 8-17, the stiffness matrix of the ith wall corresponding to the global coordinates can be obtained from

$$[S]_i = [B]_i^T[S^*]_i[B]_i \qquad (16\text{-}47)$$

If $[B]_i$ is divided into submatrices $[B_1]_i$, $[B_2]_i$, and $[B_3]_i$ (each of order $n \times 3n$) along the horizontal dotted lines in Eq. 16-46, then Eq. 16-47 can be written in a more convenient form

$$[S]_i = \sum_{r=1}^{3} [B_r]_i^T[S_r^*]_i[B_r]_i \qquad (16\text{-}47a)$$

where $[S_r^*]_i$, with $r = 1, 2, 3$ are the three submatrices $[S_u^*]$, $[S_v^*]$, and $[S_\theta^*]$ of $[S^*]_i$ defined in Eq. 16-43.

When x_i and y_i coordinates are the same at all floors, Eq. 16-47a yields

$$[S]_i = \begin{bmatrix} c^2[S_u^*] + s^2[S_v^*] & & \text{symmetrical} \\ cs([S_u^*] - [S_v^*]) & s^2[S_u^*] + c^2[S_v^*] & \\ \begin{array}{l} c(xs - yc)[S_u^*] - \\ s(xc + ys)[S_v^*] \end{array} & \begin{array}{l} s(xs - yc)[S_u^*] + \\ c(xc + ys)[S_v^*] \end{array} & \begin{array}{l} (xs - yc)^2[S_u^*] + \\ (xc + ys)^2[S_v^*] + [S_\theta^*] \end{array} \end{bmatrix}_i$$

(16-47b)

where $c = \cos \theta$ and $s = \sin \theta$.

The stiffness matrix of the structure corresponding to the global coordinates is obtained by summation

$$[S] = \sum_{i=1}^{m} [S]_i$$

(16-48)

where m is the number of wall assemblies.

The displacements at the global coordinates can now be determined by solving the equation

$$\{F\}_{3n \times 1} = [S]_{3n \times 3n} \{D\}_{3n \times 1}$$

(16-49)

where $\{F\}$ are forces at the global coordinates equivalent to the external loading. The forces on each wall assembly can be calculated by Eqs. 16-44 and 16-45.

Example 16-3   Analyze the structure of Example 16-2 but with three storeys instead of one.

Let the center of the global coordinates be chosen at the center of the shaft (wall 1 in Fig. 16-12b) with the coordinates 1, 2, and 3 representing translation in the x direction, 4, 5, and 6 translation in the y direction, and 7, 8, and 9 rotation in the clockwise direction. The stiffness matrix $[S]_2$ of wall 2 with respect to the global coordinates is derived below in some detail.

First, a stiffness matrix of order 6×6 is generated corresponding to a translation in the u direction and a rotation about the v axis at each floor level.[8] This matrix is then condensed by Eq. 4-17 to obtain

$$[S_u^*]_2 = 10^{-3} Eh \begin{bmatrix} 0.21 & \text{symmetrical} & \\ -0.47 & 1.30 & \\ 0.35 & -1.35 & 2.35 \end{bmatrix}$$

By a similar procedure, we obtain

$$[S_v^*]_2 = \frac{Eh}{10} \begin{bmatrix} 0.1965 & \text{symmetrical} & \\ -0.3425 & 0.8039 & \\ 0.1188 & -0.5255 & 0.9323 \end{bmatrix}$$

[8]See answer to Prob. 16-2 for a general form of the matrix required.

Because the cross section and the height of wall 2 are the same in all floors and warping is ignored, Eq. 16-43a gives

$$[S_\theta^*]_2 = 0.2304 \times 10^{-3} Eh^3 \begin{bmatrix} 1 & \text{symmetrical} \\ -1 & 2 \\ 0 & -1 & 2 \end{bmatrix}$$

For the transformation of the stiffness matrix $[S^*]_2$ corresponding to local coordinates into the matrix $[S]_2$ referring to the global coordinates (using Eq. 16-47) we need the transformation matrix $[B]_2$ given by Eq. 16-46. Substituting in this equation $i = 2$, $\theta = 0$, $x = 44$, and $y = -8$, we obtain

$$[B]_2 = \begin{bmatrix} [I] & [0] & h\lfloor 0.8, 0.8, 0.8 \rfloor \\ [0] & [I] & h\lfloor 4.4, 4.4, 4.4 \rfloor \\ [0] & [0] & [I] \end{bmatrix}$$

Equations 16-47 or 16-47b give

$$[S]_2 = \frac{Eh}{10^3} \begin{bmatrix} 2 \\ -5 & 13 \\ 4 & -13 & 23 \\ 0 & 0 & 0 & 196 \\ 0 & 0 & 0 & -343 & 804 \\ 0 & 0 & 0 & 119 & -526 & 932 \\ \frac{h}{10}\begin{bmatrix} 17 & -38 & 28 \\ -38 & 104 & -108 \\ 26 & -108 & 188 \end{bmatrix} & \frac{h}{10}\begin{bmatrix} 8646 & -15072 & 5226 \\ -15072 & 35370 & -23124 \\ 5226 & -23124 & 41021 \end{bmatrix} & \frac{h^2}{100}\begin{bmatrix} 380785 \\ -663706 & 155759 \\ 230167 & -1018553 & 1806900 \end{bmatrix} \end{bmatrix}$$

By a similar procedure, $[S]_1$ and $[S]_3$ for walls 1 and 3 are derived and added to $[S]_2$ to obtain the stiffness matrix of the structure (Eq. 16-48).

The forces at the global coordinates equivalent to the external applied loads are:

$$\{F\} = P\{0, 0, 0, -1, -1, -1, -1.2h, -1.2h, -1.2h\}$$

Solving for $\{D\}$ in Eq. 16-49, we obtain

$$\{D\} = \frac{P}{E}\{-0.0154, -0.0105, -0.0060, -35.4258, -21.0885,$$

$$-7.9686, 0.3909, 0.2331, 0.0882\}$$

$$\frac{P}{Eh}\{-0.154, -0.105, -0.060, -354.258, -210.885,$$

$$-79.686, 39.09/h, 23.31/h, 8.82/h\}$$

Substitution in Eqs. 16-44 and 16-45 gives the forces in the walls at each of the three storeys; these are given in Fig. 16-14.

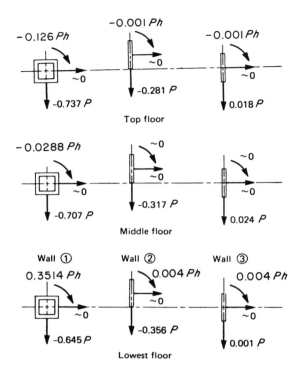

Fig. 16-14. Forces resisted by various walls in Example 16-3.

16-8 GENERAL

The analysis of the effect of horizontal forces on building frames with shear walls is simplified by the assumption that each floor is infinitely rigid in its own plane, so that the degree of kinematic indeterminacy of the frame is considerably reduced.

Some regular building frames can be analyzed as plane structures, two procedures being available (Secs. 16-4 and 16-5). In the simplified approximate method of Sec. 16-5 the problem is reduced to the analysis of one wall and one substitute frame connected by inextensible links.

A relatively simple analysis of a three-dimensional structure is possible when it is composed of frames — with or without walls — arranged in plan in a regular rectangular pattern. When the shear walls are arranged in a random manner, the analysis is rather complex, but procedures have been developed both for one-storey and multistorey frames, on the assumption that floors have a negligible flexural rigidity compared with the walls.

PROBLEMS

16-1   Assuming that the displacements at coordinates 3* and 4* for the beam in
       Fig. 16-3a are restrained, find the flexibility matrix corresponding to the
       coordinates 1* and 2*. Use the principle of virtual work to calculate the
       displacements and take shear deformation into account. Invert the derived
       matrix and compare the elements of the resulting stiffness matrix with the
       appropriate elements in Eq. 16-10.

16-2   Derive the submatrices in Eq. 16-18 in the case when the column cross
       section and height vary from storey to storey and the beams at floor levels
       do not have the same $(I/l)_b$ values.

16-3   Solve Example 16-1 with the properties of members in the top two storeys
       as follows: For any column $I = 17 \times 10^{-6}b^4$; for any beam $I =
       24.3 \times 10^{-6}b^4$; for any wall $I = 58 \times 10^{-3}b^4$, and area of wall cross section $=
       146 \times 10^{-3}b^2$. All other data unchanged.

16-4   Find the bending moment at section A–A and the end-moment in the beam
       at the lower floor in the symmetrical wall with openings shown in the figure.
       Neglect shear deformation in the beams and axial deformations in all ele-
       ments. Take $E = 2.3G$. The wall has a constant thickness. To idealize the shear
       wall, use a substitute frame similar to that shown in Fig. 16-7b. The substitute
       frame has one vertical column connected to horizontal beams. The top three
       beams are of length $3b/4$ and the lower three are of length b.

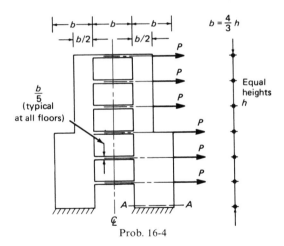

Prob. 16-4

16-5   Solve Prob. 16-4 by moment distribution. Use a substitute frame similar to
       the one in Fig. 16-7b. A solution can then be obtained by the no-shear moment
       distribution procedure of Sec. 12-3. The end-rotational stiffness for the column
       is EI_c/h and for any of the beams is given in Eq. 16-29 (element $\bar{S}_{22}$ of the
       matrix in this equation). Because of the small stiffness of the beams in the
       substitute frame compared to the stiffness of the column, the moment distri-
       bution in the above solution will converge slowly.

16-6   Find the forces on each of the shear walls in a single-storey building shown in the figure. All walls have thickness $= b/12$ and height $= b$ and are totally fixed to a rigid foundation. Neglect the warping effect of the shear walls. Take $E = 2.3G$.

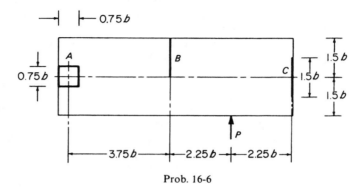

Prob. 16-6

16-7   Apply the requirements of Prob. 16-6 to the single-storey building shown in the figure.

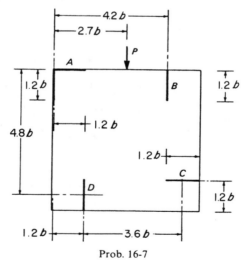

Prob. 16-7

16-8   The figure shows a plan view of a curved slab bridge deck, supported on piers at B and C and on bearing pads above rigid abutments at A and D. The piers are assumed to be pin-connected to an infinitely rigid bridge deck. Each pier has a cross section 2 ft × 8 ft, and is 40 ft high and encastré at the base. The bearing pads at A and D have an area of 4 ft², thickness 1.25 in., and shear modulus of elasticity $= 300$ lb/in.^{2}. Find the three displacement components of the deck and the forces on each of the supporting elements A, B, C, and D due to a force P at the deck level, as shown. Assume that the pier material $E = 2.3G = 4,000$ kip/in.^{2}, and the bearing pads at A and D are strips of length 8 ft, width 6 in., 1.25 in. high.

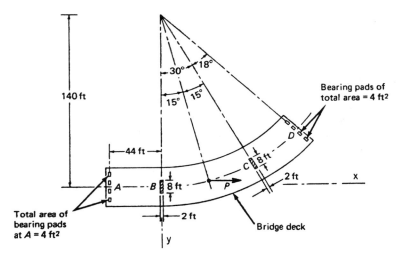

Prob. 16-8

CHAPTER 17

Method of finite differences

17-1 INTRODUCTION

We should bear in mind that all methods of structural analysis are essentially concerned with solving the basic differential equations of equilibrium and compatibility, although in some of the methods this fact may be obscured. Analytical solutions are limited to the cases when the load distribution, section properties and boundary conditions can be described by mathematical expressions, but for complex structures numerical methods are in general a more practical means of analysis.

One of these is the finite-difference method, in which a numerical solution of the differential equation for displacement or stress resultant is obtained for chosen points on the structure, referred to as *nodes* or *pivotal points*, or simply as points of division. The numerical solution is thus obtained from differential equations which are applicable to the actual continuous structure. This is different from the *finite-element* method, in which the actual continuous structure is idealized into an assembly of discrete elements, for which force-displacement relations and stress distributions are determined (or assumed), and the complete solution is obtained by combining the individual elements into an idealized structure for which the conditions of equilibrium and compatability are satisfied at the junctions of these elements.

The numerical solution by finite differences generally requires replacing the derivatives of a function by difference expressions of the function at the nodes. The differential equation governing the displacement (or stress) is applied in a difference form at each node, relating the displacement at the given node and nodes in its vicinity to the external applied load. This usually provides a sufficient number of simultaneous equations for the displacements (or stresses) to be determined. The finite-difference coefficients of the equations applied at nodes on, or close to, the boundary have to be modified, compared with the coefficients used at interior points, in order to satisfy the boundary conditions of the problem. Therein lies one of the difficulties of the method of finite differences and a disadvantage in its use compared with the finite element method. Nevertheless, the finite-difference method can be conveniently used for a variety of problems, and when it is used the number of simultaneous equations required (for a comparable degree of accuracy) is generally only about a half or a third of the number of the equations needed in the finite-element method.

17-2 REPRESENTATION OF DERIVATIVES BY FINITE DIFFERENCES

Figure 17-1 represents a function $y = f(x)$, which for our purposes can, for example, be the deflection of a beam. Consider equally spaced abscissae

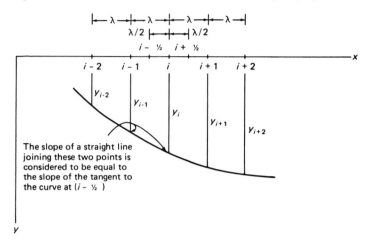

Fig. 17-1. Graph of function $y = f(x)$.

x_{i-1}, x_i, and x_{i+1} and the corresponding ordinates y_{i-1}, y_i, and y_{i+1}. The derivative (or the slope) of the curve at point $x_{i-\frac{1}{2}}$, midway between i and $i - 1$, can be approximated by

$$\left(\frac{dy}{dx}\right)_{i-\frac{1}{2}} \cong \frac{1}{\lambda}(y_i - y_{i-1}) \tag{17-1}$$

where λ is the spacing of the abscissae. Similarly, the slope of the curve midway between i and $i + 1$ is

$$\left(\frac{dy}{dx}\right)_{i+\frac{1}{2}} \cong \frac{1}{\lambda}(y_{i+1} - y_i) \tag{17-1a}$$

The second derivative at i (which is the rate of change of slope) is approximately equal to the difference between the slope at $i + \frac{1}{2}$ and at $i - \frac{1}{2}$ divided by λ; thus

$$\left(\frac{d^2y}{dx^2}\right)_i \cong \frac{1}{\lambda}\left[\left(\frac{dy}{dx}\right)_{i+\frac{1}{2}} - \left(\frac{dy}{dx}\right)_{i-\frac{1}{2}}\right]$$

Substituting from Eqs. 17-1 and 17-1a,

$$\left(\frac{d^2y}{dx^2}\right)_i \cong \frac{1}{\lambda^2}(y_{i+1} - 2y_i + y_{i-1}) \tag{17-2}$$

In the above expressions we have used *central differences* because the derivative of the function in each case was expressed in terms of the values of the function at points located symmetrically with respect to the point considered.

The process can be repeated to calculate higher derivatives, in which case the values of y at a greater number of equally spaced points are required. This is done in Table 17-1, and the finite-difference pattern of coefficients is shown in Fig. 17-2.

The first derivative at i can also be expressed in terms of y_{i-1} and y_{i+1} with interval 2λ. Similarly, the third derivative at i can be expressed from the difference of the second derivatives at $i+1$ and $i-1$ with interval 2λ. The resulting coefficients are given in the last two rows of Fig. 17-2.

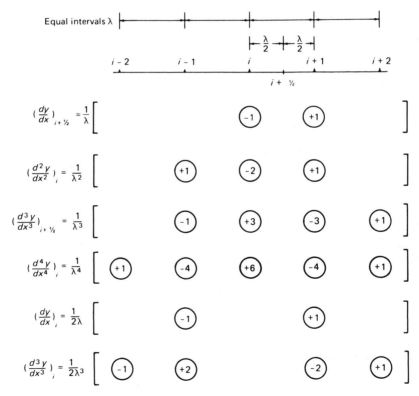

Fig. 17-2. Finite-difference pattern of coefficients, using central differences.

Table 17-1. Finite-Difference Expressions for the Derivatives $\dfrac{dy}{dx}, \dfrac{d^2y}{dx^2}, \dfrac{d^3y}{dx^3}, \dfrac{d^4y}{dx^4}$, Using Central Differences

Point	y	$\dfrac{dy}{dx}$	$\dfrac{d^2y}{dx^2}$	$\dfrac{d^3y}{dx^3}$	$\dfrac{d^4y}{dx^4}$
$i-2$	y_{i-2}				
$i-\tfrac{3}{2}$		$\dfrac{1}{\lambda}(-y_{i-2}+y_{i-1})$			
$i-1$	y_{i-1}		$\dfrac{1}{\lambda^2}(y_{i-2}-2y_{i-1}+y_i)$		
$i-\tfrac{1}{2}$		$\dfrac{1}{\lambda}(-y_{i-1}+y_i)$		$\dfrac{1}{\lambda^3}(-y_{i-2}+3y_{i-1}-3y_i+y_{i+1})$	
i	y_i		$\dfrac{1}{\lambda^2}(y_{i-1}-2y_i+y_{i+1})$		$\dfrac{1}{\lambda^4}(y_{i-2}-4y_{i-1}+6y_i-4y_{i+1}+y_{i+2})$
$i+\tfrac{1}{2}$		$\dfrac{1}{\lambda}(-y_i+y_{i+1})$		$\dfrac{1}{\lambda^3}(-y_{i-1}+3y_i-3y_{i+1}+y_{i+2})$	
$i+1$	y_{i+1}		$\dfrac{1}{\lambda^2}(y_i-2y_{i+1}+y_{i+2})$		
$i+\tfrac{3}{2}$		$\dfrac{1}{\lambda}(-y_{i+1}+y_{i+2})$			
$i+2$	y_{i+2}				

Other finite-difference expressions[1] can be obtained by considering *forward or backward differences,* in which the derivative at any point is expressed in terms of the value of the function at points in ascending or descending order with respect to the point under consideration. The central differences are more accurate than either forward or backward differences and they will be used in this chapter.

17-21 Errors in Finite-Difference Equations

It is apparent from the foregoing that the finite-difference approach involves errors compared with the continuous function, and it is useful to be aware of the magnitude of the error involved. The value of a function $y(x)$ at a point x_{i+1} can be expressed in terms of $y(x_i)$ and its derivatives by Taylor's expansion

$$y(x_{i+1}) = y(x_i) + \frac{\lambda}{1!} y'(x_i) + \frac{\lambda^2}{2!} y''(x_i) + \frac{\lambda^3}{3!} y'''(x_i) + \cdots \qquad (17\text{-}3)$$

Similarly,

$$y(x_{i-1}) = y(x_i) - \frac{\lambda}{1!} y'(x_i) + \frac{\lambda^2}{2!} y''(x_i) - \frac{\lambda^3}{3!} y'''(x_i) + \cdots \qquad (17\text{-}4)$$

Subtracting Eq. 17-4 from Eq. 17-3, we obtain

$$y'(x_i) = \frac{1}{2\lambda} \left[y(x_{i+1}) - y(x_{i-1}) \right] - \frac{\lambda^2}{3!} y'''(x_i) - \frac{\lambda^4}{5!} y^{(5)}(x_i) - \cdots \qquad (17\text{-}5)$$

or, in the notation of Fig. 17-1,

$$\left(\frac{dy}{dx}\right)_i = \frac{1}{2\lambda} (y_{i+1} - y_{i-1}) - \frac{\lambda^2}{3!} y_i''' - \frac{\lambda^4}{5!} y_i^{(5)} - \cdots \qquad (17\text{-}6)$$

Comparing Eq. 17-6 with the finite-difference expression given in the penultimate line of Fig. 17-2, we can see that the error in the finite-difference expression for the first derivative is

$$\varepsilon_1 = -\frac{\lambda^2}{6} y_i''' - \frac{\lambda^4}{120} y_i^{(5)} - \cdots \qquad (17\text{-}7)$$

If we add Eqs. 17-3 and 17-4, it can be shown that

$$\left(\frac{d^2 y}{dx^2}\right)_i = \frac{1}{\lambda^2} (y_{i-1} - 2y_i + y_{i+1}) - \frac{2\lambda^2}{4!} y_i^{(4)} - \frac{2\lambda^4}{6!} y_i^{(6)} \qquad (17\text{-}8)$$

---

[1] For a list of finite-difference expressions see E. L. Stiefel, *An Introduction to Numerical Mathematics,* Academic Press, New York, 1963.

The error in the finite-difference expression for the second derivative (Eq. 17-2) is therefore

$$\varepsilon_2 = -\frac{\lambda^2}{12} y_i^{(4)} - \frac{\lambda^4}{360} y_i^{(6)} - \cdots \tag{17-9}$$

Similarly, it can be shown that the error in the third derivative given in the last equation in Fig. 17-2 is[2]

$$\varepsilon_3 = -\frac{\lambda^2}{4} y_i^{(5)} - \cdots \tag{17-10}$$

and the error in the fourth derivative is

$$\varepsilon_4 = -\frac{\lambda^2}{6} y_i^{(6)}(x_i) - \cdots \tag{17-11}$$

From the above equations it is evident that the first error term is of the order of λ^2. The accuracy can be improved by reducing the errors to a higher order of λ.

17-3   BENDING MOMENTS AND DEFLECTIONS IN A STATICALLY DETERMINATE BEAM

Consider a simple beam AB subjected to vertical loading of varying intensity, as shown in Fig. 17-3. The bending moment M and the load intensity q are related by the differential equation (see Sec. 9-3)

$$\frac{d^2 M}{dx^2} = -q \tag{17-12}$$

which, when applied at a general point i, can be put in the finite-difference form (see Fig. 17-2)

$$[1 \quad -2 \quad 1] \begin{Bmatrix} M_{i-1} \\ M_i \\ M_{i+1} \end{Bmatrix} \cong -q_i \lambda^2 \tag{17-13}$$

The load q is positive when downward and M is positive when the bottom fiber is in tension; x is measured from the left-hand end of the beam.

The bending moment at the ends of the beam of Fig. 17-3a is $M_0 = M_4 = 0$. Writing Eq. 17-13 at each of the three interior points, we obtain

$$\begin{bmatrix} 2 & -1 & 0 \\ -1 & 2 & -1 \\ 0 & -1 & 2 \end{bmatrix} \begin{Bmatrix} M_1 \\ M_2 \\ M_3 \end{Bmatrix} \cong \lambda^2 \begin{Bmatrix} q_1 \\ q_2 \\ q_3 \end{Bmatrix} \tag{17-14}$$

[2]It can be easily shown that the third equation in Fig. 17-2 is more accurate than the last one (see Prob. 17-1).

With the load q known, the solution of these three simultaneous equations gives the moments M_1, M_2, and M_3.

In the beam considered we have found it possible to carry out the analysis by using only the finite-difference equations relating the bending moment to the external loading. This is so because the bending moment at the ends of the beam is known to be zero, so that the finite-difference equations at the internal points are sufficient in number to determine the unknown moments. If, however, the beam is statically indeterminate, e.g., when the ends A and B are encastré, the end-moments are unknown. However, the deflection and slope at the ends are known to be zero, and we therefore use finite-difference equations relating the deflection to the applied loading, discussed in Sec. 17-4.

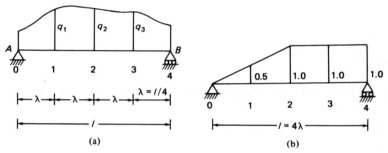

Fig. 17-3. Simple beam subjected to nonuniform loading. (a) Beam divisions used in the finite-difference Eq. 17-14. (b) Beam considered in Example 17-1.

Example 17-1   Determine the bending moments in a simple beam with the loading shown in Fig. 17-3b:

Applying Eq. 17-14 for the loading: $q_1 = 0.50$, $q_2 = 1.00$, and $q_3 = 1.00$, and solving, we obtain $\{M\} = l^2\{0.0703, 0.1094, 0.0859\}$. The exact answer is $\{M\} = l^2\{0.0677, 0.1042, 0.0833\}$.

We should note that the solution of Eq. 17-14 depends upon the intensity of loading at the nodes but does not take into account the manner in which the loading varies between these points. More accurate values would be obtained if the beam were divided into smaller intervals, thereby requiring a larger number of equations.

Exact values can be obtained if the actual loading is replaced by equivalent concentrated loads; the finite-difference Eq. 17-13 at a general point is then replaced by the exact equation

$$[1 \quad -2 \quad 1] \begin{Bmatrix} M_{i-1} \\ M_i \\ M_{i+1} \end{Bmatrix} = -Q_i\lambda \qquad (17\text{-}15)$$

where Q_i is an equivalent concentrated load replacing $(q_i \lambda)$ in Eq. 17-13. We should note that we no longer use $\cong$ but, $=$, as Eq. 17-15 is exact. The method of calculating the equivalent concentrated load Q_i and the proof of Eq. 17-15 are given in Sec. 9-51 (refer to Eq. 9-13).

The bending moment M and the deflection y are related by the differential equation (see Sec. 9-3)

$$EI \frac{d^2 y}{dx^2} = -M \qquad (17\text{-}16)$$

where EI is the flexural-rigidity of the beam, and x is the distance from its left-hand end. Comparing this equation with Eq. 17-12, we can write the approximate finite differences form of Eq. 17-16 by replacing M and q in Eq. 17-13 by y and $M/(EI)$ respectively. Similarly, an exact form can be obtained by replacing M and Q in Eq. 17-15 by y and $\bar{w}_i$, where $\bar{w}_i$ is an equivalent elastic loading (refer to Eqs. 9-15 and 9-16). An example of the use of the approximate and exact equations in the calculation of deflection is given in Sec. 9-6.

17-4 FINITE-DIFFERENCE RELATION BETWEEN BEAM DEFLECTION AND APPLIED LOADING

The equations relating deflection to the bending moment can be used to calculate the deflections when the moments are known, as is the case in statically determinate beams. In statically indeterminate structures, it is more convenient to relate the deflections to the applied loads, and solve for deflections which can then be used to determine the unknown stress resultants.

From Eqs. 17-12 and 17-16 the deflection y and the intensity of loading q can be related by the differential equation

$$\frac{d^2}{dx^2}\left(EI \frac{d^2 y}{dx^2}\right) = q \qquad (17\text{-}17)$$

This equation can be put in finite-difference form in two steps as follows. First, the term in brackets (which is equal to minus the bending moment) is replaced by finite differences, using Eq. 17-2:

$$M_i \cong -\frac{EI_i}{\lambda^2}(y_{i-1} - 2y_i + y_{i+1}) \qquad (17\text{-}18)$$

Then the second derivative of the moment is put in finite-difference form, again using Eq. 17-2:

$$\frac{d^2}{dx^2}\left(EI \frac{d^2 y}{dx^2}\right) = -\left(\frac{d^2 M}{dx^2}\right) \cong -\frac{1}{\lambda^2}(M_{i-1} - 2M_i + M_{i+1}) \qquad (17\text{-}19)$$

Position of node i	Coefficient of the deflection in terms of E/λ^3					Right-hand side	Equation number
	y_{i-2}	y_{i-1}	y_i^*	y_{i+1}	y_{i+2}		
(a)	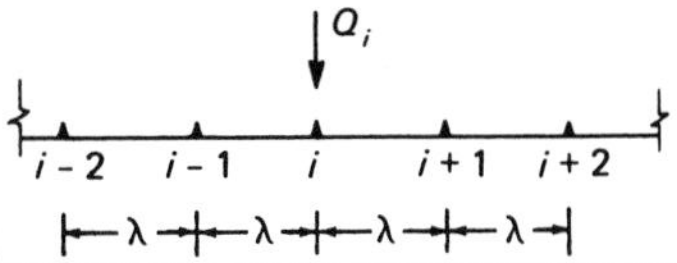 I_{i-1}	$-2(I_{i-1}+I_i)$	$(I_{i-1}+4I_i+I_{i+1})$	$-2(I_i+I_{i+1})$	I_{i+1}	$=Q_i$	17-21
(b) Hinged support 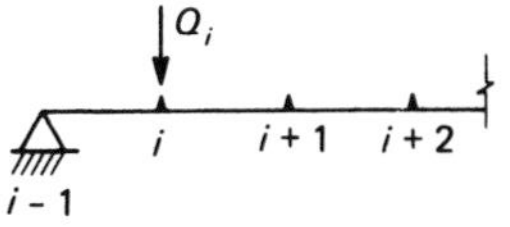	–	–	$(4I_i+I_{i+1})$	$-2(I_i+I_{i+1})$	I_{i+1}	$=Q_i$	17-27
(c) Fixed support 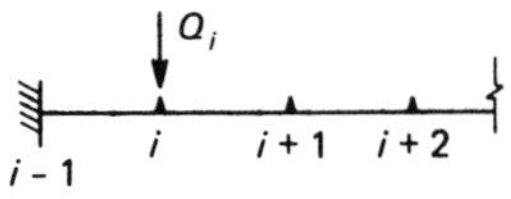	–	–	$(2I_{i-1}+4I_i+I_{i+1})$	$-2(I_i+I_{i+1})$	I_{i+1}	$=Q_i$	17-28
(d) Free end 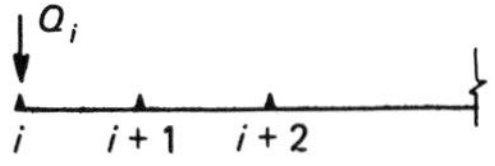	–	–	I_{i+1}	$-2I_{i+1}$	I_{i+1}	$=Q_i$	17-29
(e) Free end 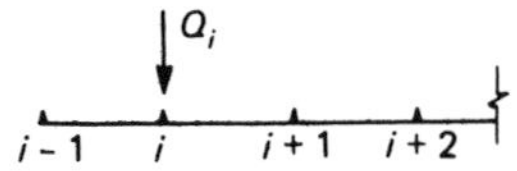	–	$-2I_i$	$(4I_i+I_{i+1})$	$-2(I_i+I_{i+1})$	I_{i+1}	$=Q_i$	17-30

*For a beam on elastic foundation of modulus k_i (force/length^2), add $k_i\lambda/2$ to the coefficient of y_i in Eq. 17-29 and $k_i\lambda$ in other equations. If the beam is on elastic spring at i of stiffness K_i (force/length), add K_i to the coefficient of y_i in all equations.

Fig. 17-4. Finite-difference equations relating beam deflection to applied load.

Position of node i	Coefficient of the deflection in terms of EI/λ^3					Right-hand side	Equation number
	y_{i-2}	y_{i-1}	y_i^*	y_{i+1}	y_{i+2}		
(a)	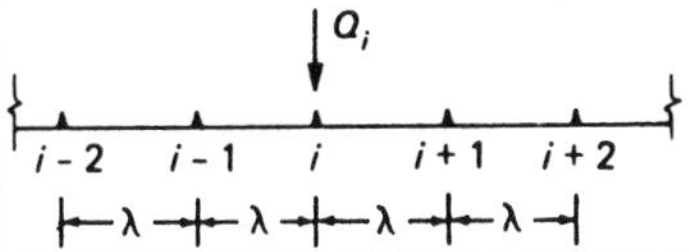 1	-4	6	-4	1	$= Q_i$	17-21a
(b) Hinged support 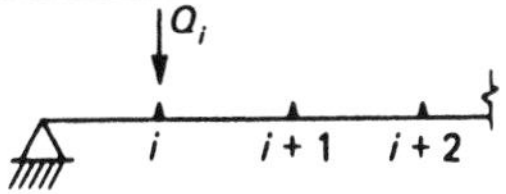	-	-	5	-4	1	$= Q_i$	17-27a
(c) Fixed support 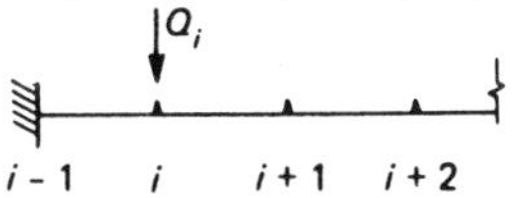	-	-	7	-4	1	$= Q_i$	17-28a
(d) Free end 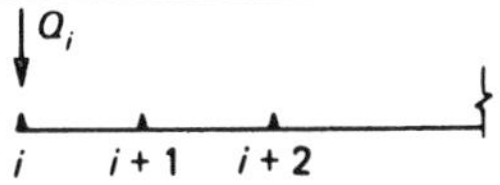	-	-	1	-2	1	$= Q_i$	17-29a
(e) Free end 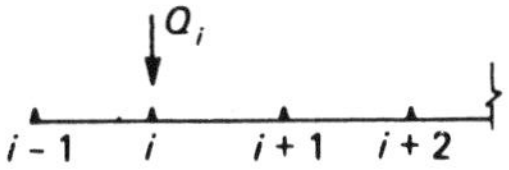	-	-2	5	-4	1	$= Q_i$	17-30a

*For a beam on elastic foundation of modulus k_i (force/length^2), add the term $k_i\lambda/2$ to the coefficient of y_i in Eq. 17-29a and $k_i\lambda$ in other equations. If the beam is on elastic spring at i of stiffness K_i force/length), add K_i to the coefficient of y_i in all equations.

Fig. 17-5. Finite-difference equations relating beam deflection to applied load, when l is constant.

493

The values of M_{i-1} and M_{i+1} can be expressed in terms of deflections by using Eq. 17-18 with $i-1$ and $i+1$ in place of i. Substituting these values in Eq. 17-19 and combining it with Eq. 17-17, we obtain the finite-difference equation applied at a general point i

$$\frac{E}{\lambda^4}\left[I_{i-1} \left| -2(I_{i-1}+I_i) \right| (I_{i-1}+4I_i+I_{i+1}) \left| -2(I_i+I_{i+1}) \right| I_{i+1} \right]$$

$$\times \left\{ \begin{array}{c} y_{i-2} \\ y_{i-1} \\ y_i \\ y_{i+1} \\ y_{i+2} \end{array} \right\} \cong q_i \qquad (17\text{-}20)$$

The accuracy of this equation is improved if the equivalent concentrated load Q_i is used instead of $q_i\lambda$. Then Eq. 17-20 becomes

$$\frac{E}{\lambda^3}\left[I_{i-1} \left| -2(I_{i-1}+I_i) \right| (I_{i-1}+4I_i+I_{i+1}) \left| -2(I_i+I_{i+1}) \right| I_{i+1} \right]$$

$$\times \left\{ \begin{array}{c} y_{i-2} \\ y_{i-1} \\ y_i \\ y_{i+1} \\ y_{i+2} \end{array} \right\} \cong Q_i \qquad (17\text{-}21)$$

This equation is, in fact, a combination of the exact Eq. 17-15, and the approximate Eq. 17-18, and is therefore an approximate relation for beams of variable I. The pattern of coefficients of Eq. 17-21 is shown in Fig. 17-4a.

When the beam is prismatic, i.e., I is constant, Eq. 17-21 simplifies to Eq. 17-21a, given in Fig. 17-5a.

17-41 Beam with a Sudden Change in Section

Consider the beam shown in Fig. 17-6, which has a sudden change in the flexural rigidity at a node i. We propose to show that, if the flexural rigidities are EI_{il} and EI_{ir} at sections just to the left and just to the right of i respectively, Eqs. 17-20 and 17-21 are applicable with an effective flexural rigidity EI_i at i, where

$$EI_i = \left(\frac{2}{1+\alpha}\right)EI_{ir} \qquad (17\text{-}22)$$

and

$$\alpha = \frac{EI_{ir}}{EI_{il}}$$

Let the deflection at three equally spaced points on the beam be y_{i-1}, y_i, and y_{i+1}. Extend the two parts AB and CB of the deflection line respectively to fictitious points C' and A' (Fig. 17-6b) whose ordinates are y'_{i+1} and

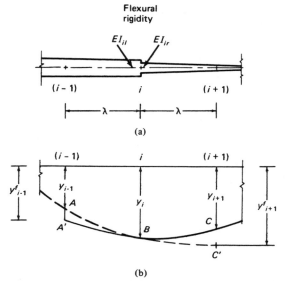

Fig. 17-6. Deflection of a beam with a sudden change in section. (a) Beam properties. (b) Deflection line.

y_{i-1}^f. For compatibility, the slope of the two curves at i must be the same. Thus

$$\frac{1}{2\lambda}(-y_{i-1}^f + y_{i+1}) = \frac{1}{2\lambda}(y_{i+1}^f - y_{i-1}) \qquad (17\text{-}23)$$

Also, for equilibrium, the bending moment just to the right and just to the left of node i must be the same—that is,

$$M_{il} = M_{ir} = M_i$$

Applying Eq. 17-18,

$$M_i = M_{il} = -\left(EI\frac{d^2y}{dx^2}\right)_{il} \cong -\frac{EI_{il}}{\lambda^2}(y_{i-1} - 2y_i + y_{i+1}^f) \qquad (17\text{-}24)$$

and

$$M_i = M_{ir} = -\left(EI\frac{d^2y}{dx^2}\right)_{ir} \cong -\frac{EI_{ir}}{\lambda^2}(y_{i-1}^f - 2y_i + y_{i+1}) \qquad (17\text{-}25)$$

From Eqs. 17-23–17-25, by eliminating the fictitious deflections, we obtain

$$M_i \cong -\left(\frac{2}{1+\alpha}\right)\frac{EI_{ir}}{\lambda^2}[1 \quad -2 \quad 1]\left\{\begin{array}{c} y_{i-1} \\ y_i \\ y_{i+1} \end{array}\right\} \qquad (17\text{-}26)$$

If the quantity $2EI_{ir}/(1+\alpha)$ in this equation is substituted by an effective flexural rigidity EI_i (as defined in Eq. 17-22), Eq. 17-26 becomes identical with Eq. 17-18.

If the derivation of Eqs. 17-20 and 17-21 is reviewed, it can be seen that the sudden variation in the flexural rigidity can be accounted for simply by using an effective flexural rigidity (calculated by Eq. 17-22) at the node where the sudden change occurs. The same applies to all the other finite-difference equations in Figs. 17-4, 17-5, and 17-7.

17-42   Boundary Conditions

When the finite-difference Eq. 17-21 relating the deflection to external loading is applied at or near a discontinuous end, the deflections at fictitious nodes outside the beam are included. These deflections are then expressed in terms of the deflections at other nodes on the beam, the procedure for different end conditions being as follows.

(a) *SIMPLE SUPPORT*. The deflection and the bending moment are zero. Referring to Fig. 17-4b, $y_{i-1} = 0$, and $M_{i-1} = 0$.

These two conditions are satisfied if the beam is considered to be continuous with another similar beam with similar loading acting in the opposite direction. Therefore, at a fictitious point $i - 2$ (not shown in the figure), the deflection $y_{i-2} = -y_i$, and the finite-difference Eq. 17-21, when applied at a point i adjacent to a simple support, takes the form of Eq. 17-27 in Fig. 17-4b.

(b) *FIXED END*. The deflection and the slope are zero. Referring to Fig. 17-4c,

$$y_{i-1} = 0 \quad \text{and} \quad \left(\frac{dy}{dx}\right)_{i-1} = 0$$

These two conditions are satisfied if the beam is considered continuous with a similar beam loaded in the same manner as the actual beam. Therefore, $y_{i-2} = y_i$, and the finite-difference Eq. 17-21, when applied at a point i adjacent to a fixed support, takes the form of Eq. 17-28 in Fig. 17-4c.

(c) *FREE END*. The bending moment is zero. Referring to Fig. 17-4d, and considering simple statics, we can write

$$M_{i+1} = -Q_i \lambda$$

Substituting for M_{i+1} in terms of the deflections from Eq. 17-18, we obtain the finite-difference Eq. 17-29 applied at point i at the free end, given in Fig. 17-4d.

Referring to Fig. 17-4e, and applying Eq. 17-15, with $M_{i-1} = 0$, we obtain

$$-\frac{1}{\lambda}(-2M_i + M_{i+1}) = Q_i$$

Substituting for M_i and M_{i+1} in terms of deflection from Eq. 17-18, we obtain the finite-difference Eq. 17-30 applied at point i adjacent to a free end, given in Fig. 17-4e.

When the beam is of constant moment of inertia, the equations listed in Fig. 17-4 simplify to the form given in Fig. 17-5.

17-5   FINITE-DIFFERENCE RELATION BETWEEN BEAM DEFLECTION AND STRESS RESULTANT OR REACTION

The finite-difference method can be used in the analysis of structures also for the purpose of determining the internal forces. First, the deflection is related to the loading by a finite-difference equation applied at each node where the deflection is unknown. The appropriate equation to be used at each node is selected from Figs. 17-4 or 17-5. Thus we obtain a system of simultaneous linear equations which can be put in the form

$$[K]\{y\} = \{Q\} \tag{17-31}$$

Examination of the finite-difference coefficients of y in the equations listed in Figs. 17-4 or 17-5 will show that the matrix $[K]$ is symmetrical. For example, the coefficient of y_{i+1} when the finite-difference equation is applied at node i in Fig. 17-4d is the same as the coefficient of y_i when the equation is applied at node $i + 1$ (see Fig. 17-4e). In Chapter 19 we shall use the matrix $[K]$ as an equivalent stiffness matrix.

A solution of the equation $[K]\{y\} = \{Q\}$ makes it possible to find the nodal deflections $\{y\}$. With the deflections known, the bending moment, shear, and reactions can be calculated by the finite-difference equations derived below.

The shear midway between node i and $i + 1$ is

$$V_{i + \frac{1}{2}} \cong \frac{1}{\lambda}(M_{i + 1} - M_i)$$

Substituting for M_i from Eq. 17-18,

$$V_{i + \frac{1}{2}} \cong \frac{E}{\lambda^3}[I_i y_{i-1} - (2I_i + I_{i+1})y_i + (I_i + 2I_{i+1})y_{i+1} - I_{i+1}y_{i+2}]$$

$$\tag{17-32}$$

The same value represents also the shear at any point between i and $i + 1$ when no load acts between the nodes i and $i + 1$.

The reaction at an intermediate support i in a continuous beam is given in Fig. 17-7a as

$$R_i = Q_i - \frac{E}{\lambda^3}[I_{i-1}y_{i-2} - 2(I_{i-1} + I_i)y_{i-1} + (I_{i-1} + 4I_i + I_{i+1})y_i$$

$$- 2(I_i + I_{i+1})y_{i+1} + I_{i+1}y_{i+2}] \tag{17-33}$$

Type of support		y_{i-2}	y_{i-1}	y_i^*	y_{i+1}	y_{i+2}	Right-hand side	Equation number
		Coefficient of the deflection in terms of EI/λ^3 when I is variable, or in terms of EI/λ^3 when I is constant						
(a) Intermediate support	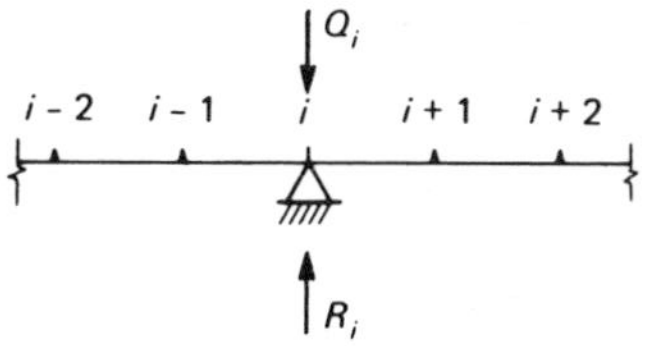 Variable I	$-I_{i-1}$	$2(I_{i-1}+I_i)$	$-(I_{i-1}+4I_i+I_{i+1})$	$2(I_i+I_{i+1})$	$-I_{i+1}$	$=(R_i-Q_i)$	17-33
	Constant I	-1	4	-6	4	-1	$=(R_i-Q_i)$	17-33a
(b) Hinged end	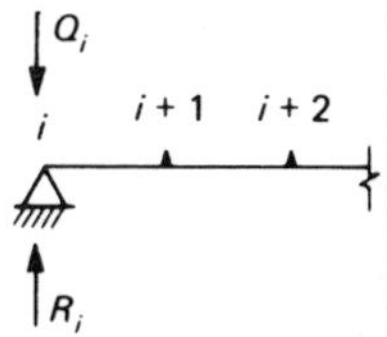 Variable I	–	–	$-I_{j+1}$	$2I_{j+1}$	$-I_{j+1}$	$=(R_i-Q_i)$	17-34
	Constant I	–	–	-1	2	-1	$=(R_i-Q_i)$	17-34a
(c) Rotation at i is prevented. Fixed-end moment M_i (given by Eq. 17-36) 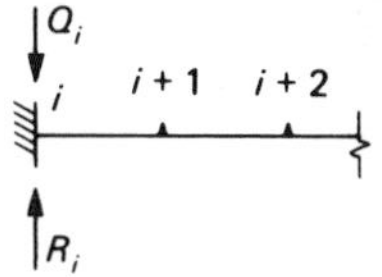	Variable I	–	–	$-(2I_i+I_{i+1})$	$2(I_i+I_{i+1})$	$-I_{i+1}$	$=(R_i-Q_i)$	17-37
	Constant I	–	–	-3	4	-1	$=(R_i-Q_i)$	17-37a

Fig. 17-7. Finite-difference equations relating beam deflection to reaction.

* For a beam on an elastic foundation of modulus k_i (force/length²), add to the coefficient of y_i the value $(-k_i\lambda)$ in the first two equations and

where Q_i is the equivalent concentrated load acting directly above the support.

The reaction at a hinged end i (Fig. 17-7b) is

$$R_i = Q_i + \frac{M_{i+1}}{\lambda}$$

Substituting for M_{i+1} from Eq. 17-18,

$$R_i \cong Q_i - \frac{EI_{i+1}}{\lambda^3}(y_i - 2y_{i+1} + y_{i+2}) \tag{17-34}$$

The reaction at a totally fixed end is

$$R_i = Q_i + \frac{1}{\lambda}(M_{i+1} - M_i) \tag{17-35}$$

The moment M_i at a fixed end i, where rotation is prevented but transverse displacement y_i can occur (Fig. 17-7c) is

$$M_i \cong \frac{EI_i}{\lambda^2}(2y_i - 2y_{i+1}) \tag{17-36}$$

The bending moment M_i is, of course, numerically equal to the fixed-end moment. A positive M_i indicates a clockwise moment for a left-hand end of the beam.

Substituting for M_{i+1} and M_i in Eq. 17-35 in terms of deflection, we obtain the reaction at a totally fixed end i

$$R_i = Q_i + \frac{E}{\lambda^3}[-(2I_i + I_{i+1})y_i + 2(I_i + I_{i+1})y_{i+1} - I_{i+1}y_{i+2}] \tag{17-37}$$

The various equations are collected in Fig. 17-7, together with the corresponding equations for the case when I is constant. When there is no settlement at the support, $y_i = 0$ in the appropriate equation.

17-6 BEAM ON AN ELASTIC FOUNDATION

The basic differential equation for a beam resting on an elastic foundation is (see Eq. 9-5a)

$$\frac{d^2}{dx^2}\left(EI\frac{d^2y}{dx^2}\right) = q - ky \tag{17-38}$$

where k is the foundation modulus, that is the foundation reaction per unit length of the beam per unit deflection (force/length^2), and the remaining notation is the same as used in Eq. 17-17.

The finite-difference form of Eq. 17-38 when applied at a general point i is

$$\frac{E}{\lambda^4}[I_{i-1}| - 2(I_{i-1} + I_i)|(I_{i-1} + 4I_i + I_{i+1})| - 2(I_i + I_{i+1})|I_{i+1}]$$

$$\times \left\{ \begin{array}{c} y_{i-2} \\ y_{i-1} \\ y_i \\ y_{i+1} \\ y_{i+2} \end{array} \right\} \cong q_i - k_i y_i \qquad (17\text{-}39)$$

Equation 17-39 can be derived in the same way as Eq. 17-20. Using the equivalent concentrated load Q_i to replace $q_i \lambda$ and rearranging terms, Eq. 17-39 becomes

$$\frac{E}{\lambda^3}[I_{i-1}| - 2(I_{i-1} + I_i)|(I_{i-1} + 4I_i + I_{i+1} + \frac{k_i \lambda^4}{E})| - 2(I_i + I_{i+1})|I_{i+1}]$$

$$\times \left\{ \begin{array}{c} y_{i-2} \\ y_{i-1} \\ y_i \\ y_{i+1} \\ y_{i+2} \end{array} \right\} \cong Q_i \qquad (17\text{-}40)$$

This equation can be applied to a beam of variable section resting on a foundation of variable modulus. Comparing Eqs. 17-21 and 17-40, we see that the presence of the elastic foundation can be accounted for simply by adding the term $k_i \lambda$ to the coefficient of y_i. If, instead of an elastic foundation a spring of stiffness K_i (force/length) is placed at point i, the term to be added to the coefficient of y_i is equal to K_i.

With a similar modification, all the other equations in Figs. 17-4 and 17-5 can be used for a beam on an elastic foundation. The modification is indicated at the bottom of these two figures.

Example 17-2   Determine the deflection, bending moment, and end support reactions for the beam of Fig. 17-8a, which has a constant EI. The beam is resting on an elastic foundation between the supports with a modulus $k = 0.1024EI/\lambda^4$, where $\lambda = l/4$.

The dimensionless term is

$$\frac{k\lambda^4}{EI} = 0.1024$$

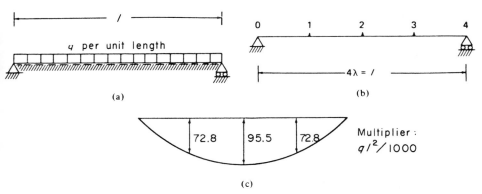

Fig. 17-8. Beam on an elastic foundation considered in Example 17-2. (a) Beam on an elastic foundation. (b) Nodes. (c) Bending-moment diagram.

Applying the appropriate finite-difference equations from Fig. 17-5 at each point of division, we can write

$$\frac{EI}{\lambda^3}\begin{bmatrix}(5+0.1024) & -4 & 1 \\ -4 & (6+0.1024) & -4 \\ 1 & -4 & (5+0.1024)\end{bmatrix}\begin{Bmatrix}y_1 \\ y_2 \\ y_3\end{Bmatrix}=q\lambda\begin{Bmatrix}1 \\ 1 \\ 1\end{Bmatrix}$$

for which the solution is

$$y_1 = y_3 = 1.928q\frac{\lambda^4}{EI} \quad \text{and} \quad y_2 = 2.692q\frac{\lambda^4}{EI}$$

or

$$y_1 = y_3 = 0.00753ql^4/EI \quad \text{and} \quad y_2 = 0.0105ql^4/EI$$

We may note in passing that we could have made use of symmetry of the structure by putting $y_1 = y_3$; hence, only two simultaneous equations would have required solving.

Applying Eq. 17-18 at points 1 and 2, we obtain

$$M_1 = -\frac{EI}{\lambda^2} \times q\frac{\lambda^4}{EI}(0 - 2 \times 1.928 + 2.692) = 1.164q\lambda^2 = 0.0728ql^2$$

and

$$M_2 = -\frac{EI}{\lambda^2} \times q\frac{\lambda^4}{EI}(1.928 - 2 \times 2.692 + 1.928) = 1.528q\lambda^2 = 0.0955ql^2$$

The bending-moment diagram for the beam is plotted in Fig. 17-8c. The reaction at 0 is given by Eq. 17-34a (Fig. 17-7b) as

$$\frac{EI}{\lambda^3}(2y_1 - y_2) = (R_0 - Q_0)$$

Substituting for y the values calculated previously and $Q_0 = q\lambda/2 = ql/8$, we find

$$R_0 = R_4 = 0.416ql$$

The exact values using an analytical solution[3] are: $y_1 = y_3 = 0.00732ql^4/EI$, $y_2 = 0.0102ql^4/EI$, $M_1 = M_3 = 0.0745ql^2$, $M_2 = 0.0978ql^2$, and $R_0 = R_4 = 0.414ql$.

17-7 AXISYMMETRICAL CIRCULAR CYLINDRICAL SHELL

Consider a thin-walled elastic cylinder subjected to any axisymmetrical radial loading. Because of symmetry, any section of the shell perpendicular to the cylinder axis will remain circular, while the radius r will undergo a change $\Delta r = y$. We need therefore to consider the deformation of only one strip parallel to the generatrix of the cylinder (Fig. 17-9a). Let the width of the strip be unity.

The radial displacement y must be accompanied by a circumferential (or hoop) force (Fig. 17-9b) whose magnitude per unit length of the generatrix is

$$N = \frac{Eh}{r} y \tag{17-41}$$

where h is the thickness of the cylinder and E is the modulus of elasticity. The hoop forces are considered positive when tensile. The radial deflection and loading are positive when outward.

The resultant of the hoop forces N on the two edges of the strip acts in the radial direction opposing the deflection, and its value per unit length of the strip is

$$-\frac{N}{r} = -\frac{Eh}{r^2} y$$

Hence the strip may be regarded as a beam on an elastic foundation whose modulus is

$$k = \frac{Eh}{r^2} \tag{17-42}$$

Because of the axial symmetry of the deformation of the wall, the edges of any strip must remain in radial planes, and lateral extension or contraction (caused by bending of the strip in a radial plane) is prevented. This restraining influence is equivalent to a bending moment in a circumferential direction.

$$M_\phi = \nu M$$

---

[3]See M. Hetényi, *Beams on Elastic Foundation*, University of Michigan Press, Ann Arbor, 1952, p. 60.

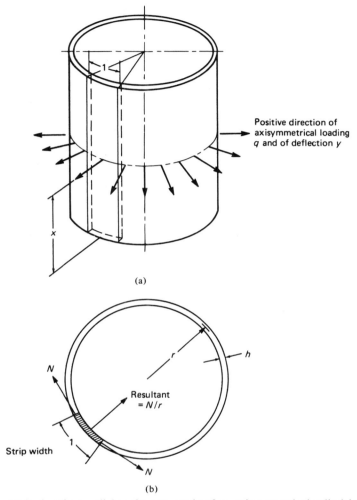

(a)

Positive direction of
axisymmetrical loading
q and of deflection y

(b)

Fig. 17-9. A strip parallel to the generatrix of an axisymmetrical cylindrical
wall considered analogous to a beam on an elastic foundation. (a) Pictorial
view. (b) Cross section.

where M is the bending moment parallel to a generatrix and v is Poisson's
ratio of the material. It can be easily shown that the stiffening effect of
M_ϕ on the bending deformation of the beam strip can be taken into account
by increasing the moment of inertia of the strip in the ratio $1/(1 - v^2)$; hence
the flexural rigidity of the strip of unit width is

$$EI = \frac{Eh^3}{12(1 - v^2)} \qquad (17\text{-}43)$$

The differential equation for the deflection of a beam resting on elastic foundation (Eq. 17-38) can thus be used for the cylinder with y indicating the deflection in the radial direction, and x the distance from the lower edge of the wall, (Fig. 17-9a); EI and k are as defined earlier, and q is the intensity of radial pressure on the wall.

When the thickness of the wall is variable both EI and k vary. However, this does not cause an appreciable difficulty in the finite differences solution.

The finite-difference equations derived for beams and listed in Figs. 17-4, 17-5, and 17-7 can be used for circular-cylindrical shells subjected to axisymmetrical loading, the equivalent concentrated load Q_i being taken for a strip of unit width.

Example 17-3   Determine the lateral radial deflection caused by hydrostatic pressure on the wall of a cylindrical water tank shown in Fig. 17-10a. Calculate also the hoop force, the bending moment along the wall, and the reaction at its top and bottom. The wall is assumed pinned at both edges. Neglect the effect of Poisson's ratio.

The nodes are chosen as shown in Fig. 17-10b, with $\lambda = H/5$. The terms $k_i \lambda^4 / E$ and Q_i required in the calculation of the coefficients of the finite-difference equations are given in Fig. 17-10b. Substituting for k_i from Eq. 17-42,

$$\frac{k_i \lambda^4}{E} = \frac{E h_i}{r^2} \frac{\lambda^4}{E} = \frac{h_i \lambda^4}{r^2}$$

The value of Q_i is calculated using the expression for the equivalent concentrated loading from Fig. 9-11.

(a) *DEFLECTIONS.* Applying the appropriate finite-difference equation from Fig. 17-4 at each of the nodes 1, 2, 3, and 4, we write

$$\left(\frac{E}{\lambda^3}\right)\left(\frac{H}{40}\right)^3 \times$$

$$\begin{bmatrix} (4 \times 0.033 + 0.043 \\ + 1.876) & -2(0.033 + 0.043) & 0.043 & 0 \\ -2(0.033 + 0.043) & (0.033 + 4 \times \\ 0.043 + 0.054 \\ + 2.048) & -2(0.043 + 0.054) & 0.054 \\ 0.043 & -2(0.043 + 0.054) & (0.043 + 4 \times \\ 0.054 + 0.068 \\ + 2.220) & -2(0.054 + 0.068) \\ 0 & 0.054 & -2(0.054 + 0.068) & (0.054 + 4 \times 0.068 \\ + 2.388) \end{bmatrix} \begin{Bmatrix} y_1 \\ y_2 \\ y_3 \\ y_4 \end{Bmatrix}$$

$$= \gamma H^2 \begin{Bmatrix} 0.040 \\ 0.080 \\ 0.120 \\ 0.160 \end{Bmatrix}$$

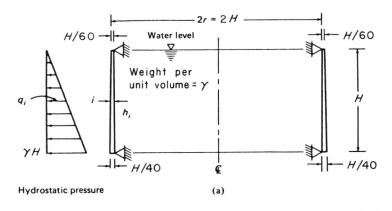

i	h_i $(H/40)$	$I_i = \dfrac{h_i^3}{12}$ $(H/40)^3$	$\dfrac{k_i\lambda^4}{E} = \dfrac{h_i\lambda^4}{r^2}$ $(H/40)^3$	Q_i (γH^2)
0	0.667	0.025		0.0067
1	0.733	0.033	1.876	0.0400
2	0.800	0.043	2.048	0.0800
3	0.867	0.054	2.220	0.1200
4	0.933	0.068	2.388	0.1600
5	1.00	0.083		0.0933

(b)

Fig. 17-10. Wall of the water tank considered in Example 17-3. (a) Wall.
(b) Calculation of $k_i\lambda^4/E$ and Q_i.

The solution of these equations is:

$$\{y_1, y_2, y_3, y_4\} = \frac{\gamma H^2}{E}\{10.88, 20.12, 28.57, 32.35\}$$

(b) *HOOP FORCES*. Substituting in Eq. 17-41 for y_i, we find:

$$\{N_1, N_2, N_3, N_4\} = \gamma H^2\{0.199, 0.402, 0.619, 0.755\}$$

(c) *BENDING MOMENT*. Applying Eq. 17-18 at points 1, 2, 3, and 4, with $y_0 = y_5 = 0$, we obtain:

$$\{M_1, M_2, M_3, M_4\} = \frac{\gamma H^3}{1000}\{0.021, 0.013, 0.099, 0.960\}$$

(d) *REACTIONS.* Applying Eq. 17-34 (see Fig. 17-7b), we find the reaction at the top support $R_0 = 0.0068\gamma H^2$, and the reaction at the bottom support $R_5 = 0.0981\gamma H^2$.

17-8 CONICAL AND SPHERICAL SHELLS

The analysis of conical shells of revolution subjected to axisymmetrical loading can also be reduced to that of a beam on an elastic foundation. A strip of the shell between two radial planes (Fig. 17-11) deforms in a similar way to a beam on an elastic foundation produced by rings whose diameter increases as we move away from the apex of the cone. This analogy ignores the stresses due to circumferential bending of the hoop rings. Because the elemental beam is tapered, the circumferential bending has a component in the radial plane through the beam axis. This component is, however, significant only in extremely flat shells, which can be calculated as circular flat

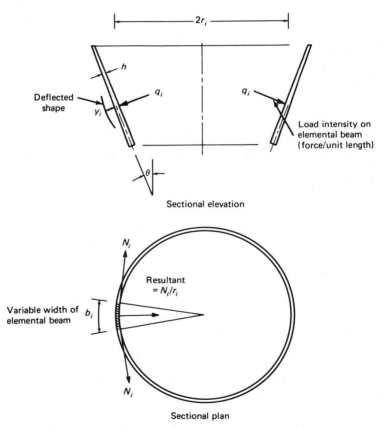

Fig. 17-11. Conical shell subjected to axisymmetrical loading.

plates. The degree of accuracy and the range of applicability of ignoring the component of the circumferential bending are discussed by Hetényi[4] and by Timoshenko and Wojnowsky-Krieger.[5]

In most practical cases, the differential Eq. 17-38 of a beam on elastic foundation can be used also for axisymmetrical conical shells with q and y being respectively the load and deflection in the direction perpendicular to a generatrix. The foundation modulus at any point i is

$$k_i = \frac{b_i E h_i \cos^2 \theta}{r_i^2} \tag{17-44}$$

and the flexural rigidity is

$$EI_i = \frac{E b_i h_i^3}{12(1 - v^2)} \tag{17-45}$$

where b_i is the breadth of the strip, h_i its thickness at point i, r_i the radius perpendicular to the cone axis at i, and θ is the angle which the elemental beam makes with the cone axis (Fig. 17-11).

The finite-difference equations of Fig. 17-4 can be used in a similar way to that employed in the previous section for cylindrical shells. The same sign convention is used, that is the deflection and loading are positive when outward. The distance x is measured from the edge of the shell closer to the cone apex. The bending moment, shear, and reaction of the beam strip can be calculated by the expressions given in Sec. 17-5. The sign convention for bending is that a positive moment causes tension at the outer face. The hoop force is positive when tensile, and is given by

$$N_i = \frac{E h_i}{r_i} y_i \cos \theta \tag{17-46}$$

In a manner similar to that used in the preceding two sections, the bending analysis of an axisymmetrical spherical dome can be reduced to the problem of a curved beam on an elastic foundation. The curved beam in this case is a meridional element of the shell (which may be of variable thickness). The element is of variable width, and its elastic foundation is produced by ring action. The foundation modulus is also variable because of the variation in the radius of the hoop circles and possibly also in the dome thickness.

If the shell is subjected to horizontal forces, uniformly distributed along a hoop circle of the shell, the horizontal radial deflection u and the applied load can be related by a finite-difference equation similar to Eq. 17-21:

[4]M. Hetényi, *Beams on Elastic Foundation*, University of Michigan Press, Ann Arbor, 1952.
[5]S. Timoshenko and S. Wojnowsky-Krieger, *Theory of Plates and Shells*, 2d ed., McGraw-Hill, New York, 1959.

$$[C_{i,i-2} \quad C_{i,i-1} \quad C_{i,i} \quad C_{i,i+1} \quad C_{i,i+2}] \begin{Bmatrix} u_{i-2} \\ u_{i-1} \\ u_i \\ u_{i+1} \\ u_{i+2} \end{Bmatrix} = Q_i \qquad (17\text{-}47)$$

where C are variable coefficients, the first subscript indicating the point i where the equation is applied, and the second representing the five consecutive equally spaced points along the meridional strip. The load Q_i is the magnitude of equivalent radial line load at point i. Since the strip width is variable, it is convenient to choose the unit width at the lower edge. The derivation of Eq. 17-47 and the expressions defining the coefficients C can be found in a paper by Glockner and Ghali.[6]

17-9   BUCKLING LOAD OF A COLUMN WITH HINGED ENDS

The differential equation for a member with pinned ends subjected to an axial load equal to the critical buckling load P_{cr} (see Fig. 17-12) is given

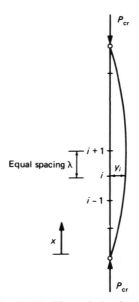

Fig. 17-12. Pin-ended column.

[6]P. G. Glockner and A. Ghali, "Numerical Analysis of Spherical Domes of Variable Thickness," IASS Symposium on Pipes and Tanks, Weimar, May 1968.

by Eq. 17-16 and by the equation:

$$M = P_{cr}y \qquad (17\text{-}48)$$

When EI is constant, Eq. 17-16 is satisfied when y varies as a sine curve, and the solution for P_{cr} is the Euler buckling value $\pi^2 EI/l^2$, where l is the length of the member.

In more complicated cases such as columns with variable section or with intermediate axial loads, the analytical solution of the above equations can become difficult. In such cases the following method is believed to give good results. The finite-difference form of Eq. 17-16, using the equivalent concentrated elastic load, is (see Sec. 9-6)

$$\frac{1}{\lambda} \begin{bmatrix} -1 & 2 & -1 \end{bmatrix} \begin{Bmatrix} y_{i-1} \\ y_i \\ y_{i+1} \end{Bmatrix} = \bar{w}_i \qquad (17\text{-}49)$$

The elastic load $\bar{w}_i$ can be expressed in terms of the deflection at three consecutive points by using the concentration formula given in Fig. 9-11. Using the parabolic distribution formula

$$\bar{w}_i = \frac{P_{cr}\lambda}{12} \begin{bmatrix} \dfrac{1}{EI_{i-1}} & \dfrac{10}{EI_i} & \dfrac{1}{EI_{i+1}} \end{bmatrix} \begin{Bmatrix} y_{i-1} \\ y_i \\ y_{i+1} \end{Bmatrix} \qquad (17\text{-}50)$$

and substituting for $\bar{w}_i$ in Eq. 17-49, we obtain

$$\frac{1}{\lambda} \begin{bmatrix} -1 & 2 & -1 \end{bmatrix} \begin{Bmatrix} y_{i-1} \\ y_1 \\ y_{i+1} \end{Bmatrix} = \frac{P_{cr}\lambda}{12E} \begin{bmatrix} \dfrac{1}{I_{i-1}} & \dfrac{10}{I_i} & \dfrac{1}{I_{i+1}} \end{bmatrix} \begin{Bmatrix} y_{i-1} \\ y_i \\ y_{i+1} \end{Bmatrix}$$

If the member is divided into equal intervals, and the above equation is applied at each of the n internal nodes, simultaneous equations can be written. These take the form

$$[A]_{n \times n} \{y\}_{n \times 1} = \frac{P_{cr}\lambda}{12} [B]_{n \times n} [C]_{n \times n} \{y\}_{n \times 1} \qquad (17\text{-}51)$$

where $[A]$ and $[B]$ are tridiagonal matrices

$$[A] = \frac{1}{\lambda} \begin{bmatrix} 2 & -1 & & & \\ -1 & 2 & -1 & & \\ & \cdots & \cdots & \cdots & \\ \text{Elements not} & -1 & 2 & -1 \\ \text{shown are} & & -1 & 2 \\ \text{zero} & & & \end{bmatrix} \qquad (17\text{-}52)$$

$$[B] = \begin{bmatrix} 10 & 1 & & & & \\ 1 & 10 & 1 & & & \\ & \cdots & \cdots & \cdots & & \\ \text{Elements not} & & 1 & 10 & 1 \\ \text{shown are} & & & 1 & 10 \\ \text{zero} & & & & \end{bmatrix} \qquad (17\text{-}53)$$

and $[C]$ is a diagonal matrix:

$$[C] = \begin{bmatrix} (1/EI_1) & & & \\ & (1/EI_2) & & \\ \text{Elements not} & & \cdots & \\ \text{shown are zero} & & & (1/EI_n) \end{bmatrix} \qquad (17\text{-}54)$$

Multiplying both sides by $[A]^{-1}$, we obtain

$$[H]\{y\} = \gamma\{y\} \qquad (17\text{-}55)$$

where

$$[H] = [A]^{-1}[B][C] \qquad (17\text{-}56)$$

and

$$\gamma = \frac{12}{P_{cr}\lambda} \qquad (17\text{-}57)$$

The solution of Eq. 17-55 is an eigenvalue problem. The buckling load can be calculated from the largest eigenvalue γ by the equation

$$P_{cr} = \frac{12}{\gamma\lambda} \qquad (17\text{-}58)$$

An iteration procedure can be used to determine the eigenvalue as follows. A value of the eigenvector $\{y\} = \{y^0\}$ is assumed, and substituted in the left-hand side of Eq. 17-55. Thus $[H]\{y^0\} = \{y^1\}$. An approximate value of γ is obtained from

$$\gamma_0 = \frac{\Sigma\, y^1}{\Sigma\, y^0}$$

If each element of $\{y^1\}$ bears the ratio γ to the corresponding element in $\{y^0\}$, then Eq. 17-55 is satisfied. If this condition is not fulfilled, the process is repeated with the eigenvector $\{y\} = (1/\gamma)\{y^1\}$ to get a vector $\{y^{II}\}$, from which a better value of γ is obtained

$$\gamma_1 = \gamma \frac{\Sigma \, y^{II}}{\Sigma \, y^{I}}$$

The procedure is repeated until Eq. 17-55 is satisfied. The convergence of the iterations may be slow unless a good choice is made for the deflection $\{y\}$ assumed to begin with.

Example 17-4 Estimate the value of the buckling load for the tapered strut[7] of Fig. 17-13. The moment of inertia of the strut is assumed to vary according to the equation $I_x = I_0[1 - 2.00(x/l)^2]^2$.

Divide the strut into five intervals so that $\lambda = l/5$. The values of I at the nodes are given in the figure as multiples of I_0, which is the moment of

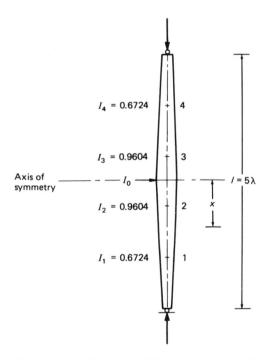

Fig. 17-13. Tapered strut considered in Example 17-4.

[7]The exact solution to this problem is given by W. L. Cowley and H. Levy, *Critical Loading of Struts and Structures*, Part IV: "On the Strength of a Strut of Variable Flexural Rigidity," Reports and Memoranda, No. 484, Advisory Committee for Aeronautics, HM. Stationery Office, London, 1918.

inertia at the center. The matrices required for Eq. 17-55 are

$$[A] = \frac{5}{l} \begin{bmatrix} 2 & -1 & 0 & 0 \\ -1 & 2 & -1 & 0 \\ 0 & -1 & 2 & -1 \\ 0 & 0 & -1 & 2 \end{bmatrix} \qquad [A]^{-1} = \frac{l}{25} \begin{bmatrix} 4 & 3 & 2 & 1 \\ 3 & 6 & 4 & 2 \\ 2 & 4 & 6 & 3 \\ 1 & 2 & 3 & 4 \end{bmatrix}$$

$$[B] = \begin{bmatrix} 10 & 1 & 0 & 0 \\ 1 & 10 & 1 & 0 \\ 0 & 1 & 10 & 1 \\ 0 & 0 & 1 & 10 \end{bmatrix} \quad \text{and} \quad [C] = \frac{1}{EI_0} \begin{bmatrix} \dfrac{1}{0.6724} & & & \\ & \dfrac{1}{0.9604} & & \\ & & \dfrac{1}{0.9604} & \\ \text{Elements not} & & \dfrac{1}{0.9604} & \\ \text{shown are zero} & & & \dfrac{1}{0.6724} \end{bmatrix}$$

We note that the elements in each column of matrix $[A]^{-1}$ are influence coefficients of the bending moment at the nodes due to a unit lateral load.

Substituting in Eq. 17-56,

$$[H] = \frac{l}{25EI_0} \begin{bmatrix} 63.9 & 37.5 & 25.0 & 17.8 \\ 53.5 & 69.8 & 50.0 & 35.7 \\ 35.7 & 50.0 & 69.8 & 53.5 \\ 17.8 & 25.0 & 37.5 & 63.9 \end{bmatrix}$$

Assume that the deflection line is parabolic with the central deflection equal to unity; therefore,

$$\{y^0\} = \begin{Bmatrix} 0.64 \\ 0.96 \\ 0.96 \\ 0.64 \end{Bmatrix} \qquad \text{and} \qquad \Sigma y^0 = 3.20$$

The left-hand side of Eq. 17-55 is

$$[H]\{y\} = \frac{l}{25EI_0} \begin{Bmatrix} 112.3 \\ 172.1 \\ 172.1 \\ 112.3 \end{Bmatrix} \qquad \text{and} \qquad \Sigma y^l = 568.8 \, \frac{l}{25EI_0}$$

and

$$\gamma_0 = \frac{568.8}{3.20 \times 25EI_0} = 7.11 \frac{l}{EI_0}$$

From Eq. 17-58,

$$P_{cr} = \frac{12 \times 5 \, EI_0}{7.11 \quad l^2} = 8.44 \frac{EI_0}{l^2}$$

The answer given in the reference mentioned in footnote 7 of this chapter is

$$P_{cr} = 8.35 \frac{EI_0}{l^2}$$

which shows good agreement.

In the above solution, no advantage was taken of symmetry in order to show the procedure in a general form. However, if in Eq. 17-55, we put $y_1 = y_4$ and $y_2 = y_3$, we can write

$$\frac{l}{25EI_0} \begin{bmatrix} 81.7 & 62.5 \\ 89.2 & 119.4 \end{bmatrix} \begin{Bmatrix} y_1 \\ y_2 \end{Bmatrix} = \gamma \begin{Bmatrix} y_1 \\ y_2 \end{Bmatrix}$$

The largest eigenvalue γ can then be obtained from the solution of the determinantal equation (see Sec. A-10 of Appendix A)

$$\frac{l}{25EI_0} \begin{vmatrix} (81.7 - \gamma) & 62.5 \\ 89.2 & (119.8 - \gamma) \end{vmatrix} = 0$$

which gives $\gamma = 7.11 \, (l/EI_0)$. Then, from Eq. 17-58, $P_{cr} = 8.44(EI_0/l^2)$.

17-10 BUCKLING LOAD OF COLUMNS WITH END RESTRAINTS

When the rotation at the ends of a loaded strut is restrained, Eq. 17-48 cannot be used in the finite-difference solution because the expression for bending moment has to include the end-moments. However, by double differentiation of the moment expression the end-moments can be eliminated so that we obtain

$$\frac{d^2M}{dx^2} = P_{cr} \frac{d^2y}{dx^2}$$

which is valid for a strut with hinged or restrained ends. Substituting for M from Eq. 17-16,

$$\frac{d^2}{dx^2}\left(EI\frac{d^2y}{dx^2}\right) = -P_{cr}\frac{d^2y}{dx^2} \tag{17-59}$$

The finite-difference form of this equation applied at a general node i is

$$\frac{E}{\lambda^3}[I_{i-1}| - 2(I_{i-1} + I_i)|(I_{i-1} + 4I_i + I_{i+1})| -$$

$$2(I_i + I_{i+1})|I_{i+1}] \left\{\begin{array}{c} y_{i-2} \\ y_{i-1} \\ y_i \\ y_{i+1} \\ y_{i+2} \end{array}\right\} = \frac{P_{cr}}{\lambda}[-1 \quad 2 \quad -1]\left\{\begin{array}{c} y_{i-1} \\ y_i \\ y_{i+1} \end{array}\right\} \tag{17-60}$$

If the strut is divided into equal intervals, and Eq. 17-60 is applied at each of the n nodes where deflection is not prevented, the resulting equations take the form

$$[K]_{n \times n}\{y\}_{n \times 1} = \frac{P_{cr}}{\lambda}[A]_{n \times n}\{y\}_{n \times 1} \tag{17-61}$$

where $[K]_{n \times n}$ is a matrix formed by the finite differences coefficients of y on the left-hand side of Eq. 17-60 with the top and lower rows modified to suit the end conditions (see Fig. 17-4 for a list of coefficients for various end conditions). The matrix $[A]$ is defined by the coefficients of y on the right-hand side of Eq. 17-60 and is of the form given in Eq. 17-52. It will be shown in Section 18-11 that $[K]$ is equivalent to the stiffness matrix of the member in the absence of an axial force.

Multiplying both sides of Eq. 17-61 by the flexibility matrix $[f] = [K]^{-1}$, we obtain

$$[G]\{y\} = \alpha\{y\} \tag{17-62}$$

where

$$[G] = [K]^{-1}[A] = [f][A] \tag{17-63}$$

and

$$\alpha = \frac{\lambda}{P_{cr}} \tag{17-64}$$

The solution of Eq. 17-62 for the largest eigenvalue α makes possible the determination of the buckling load by the equation

$$P_{cr} = \frac{\lambda}{\alpha} \tag{17-65}$$

The eigenvalue can be determined by an iteration procedure as shown in the previous section.

Equation 17-62 can also be used for struts with hinged ends, but Eq. 17-55 gives a more accurate result in that case.

An alternative to Eq. 17-62 can be obtained if both sides of Eq. 17-61 are multiplied by $[A]^{-1}$ instead of $[K]^{-1}$. In this case, we obtain

$$[L]\{y\} = \frac{1}{\alpha}\{y\} \tag{17-66}$$

where

$$[L] = [A]^{-1}[K] \tag{17-67}$$

and α is given by Eq. 17-64, as before.

Equation 17-66 is to be solved for the smallest eigenvalue $1/\alpha$.

Example 17-5 Estimate the value of the buckling load for the prismatic strut of Fig. 17-14. This simple problem is chosen in order that the answer obtained by finite differences can be compared with the known exact solution (see Fig. 15-11b).

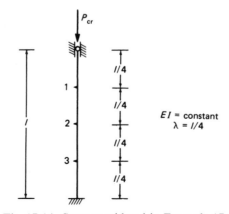

Fig. 17-14. Strut considered in Example 17-5.

Using the finite-difference coefficients of Fig. 17-5, we obtain the stiffness matrix

$$[K] = \frac{EI}{\lambda^3} \begin{bmatrix} 5 & -4 & 1 \\ -4 & 6 & -4 \\ 1 & -4 & 7 \end{bmatrix}$$

From Eq. 17-52,

$$[A] = \begin{bmatrix} 2 & -1 & 0 \\ -1 & 2 & -1 \\ 0 & -1 & 2 \end{bmatrix} \quad \text{and} \quad [A]^{-1} = \frac{1}{4} \begin{bmatrix} 3 & 2 & 1 \\ 2 & 4 & 2 \\ 1 & 2 & 3 \end{bmatrix}$$

Substituting in Eq. 17-67,

$$[L] = [A]^{-1}[K] = \frac{EI}{\lambda^3} \begin{bmatrix} 2 & -1 & 0.5 \\ -1 & 2 & 0 \\ 0 & -1 & 3.5 \end{bmatrix}$$

Solving Eq. 17-66 for the smallest eigenvalue $1/\alpha$, we find

$$\frac{1}{\alpha} = 1.111 \frac{EI}{\lambda^3}$$

From Eq. 17-65,

$$P_{cr} = 17.78 \frac{EI}{l^2}$$

The exact answer (see Fig. 15-11b) is

$$P_{cr} = 20.19 \frac{EI}{l^2}$$

17-11   GENERAL

The problems treated in this chapter, ranging from deflection and bending moment in a beam to stresses in a shell and critical load in a strut, involve the solution of ordinary differential equations using the method of finite differences. Thus, derivatives of a function are represented by finite differences and the problem reduces to the solution of a system of simultaneous algebraic equations. The number of these equations depends on the number of nodes chosen to approximate the given function. The choice of this number is governed by the fact that the accuracy of the solution is improved with a decrease in the node spacing and therefore with an increase in the number of nodes. However, in many problems, adequate accuracy can be achieved with a small number of equations.

The finite-difference method is simple to use, and coefficients are tabulated for the deflection of beams with constant or variable I on rigid supports or on an elastic foundation. The latter case can be adapted to the treatment of elemental strips of axisymmetrical shells of various shapes.

The finite-difference form of the differential equation governing the transverse deflection of a beam can be adapted to treat buckling problems of struts, both of constant cross section and tapered. This approach leads to an eigenvalue problem which can be easily formulated and solved by hand or by existing computer routines.

The accuracy of finite-difference solutions can be greatly improved by the use of Richardson's extrapolation method.[8] Two or more solutions are obtained with different numbers of node points, then extrapolation equations are used to improve the accuracy of the answers.

The use of finite differences in the solution of problems involving partial differential equations is treated in the following chapter.

PROBLEMS

17-1   Using Taylor's expansion, find the first error term in the finite-difference equation (see Fig. 17-2):

$$\left(\frac{d^3 y}{dx^3}\right)_{i+1/2} = \frac{1}{\lambda^3}(-y_{i-1} + 3y_i - 3y_{i+1} + y_{i+2}) + \text{error terms}$$

17-2   Using Taylor's expansion, find the first error term in the following expression for the first derivative at A (see figure):

$$y'_A = \frac{1}{6\lambda}(-11y_A + 18y_B - 9y_C + 2y_D) + \text{error terms}$$

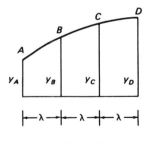

Prob. 17-2

17-3   Using Taylor's expansion, express the second derivative at point B in terms of y_A, y_B, and y_C (see figure) and find the first term of the error series.

[8] S. H. Crandall, *Engineering Analysis*, McGraw-Hill, New York, 1956.

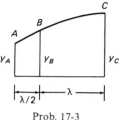

Prob. 17-3

17-4   Find the deflections at points 1, 2, and 3 in the beam in Fig. 9-14a using the approximate equations listed in Fig. 17-4. Compare the answers with the values obtained in Sec. 9-6.

17-5   Determine the deflection, the bending moment at the nodes, and the support reactions for the beam in the figure, with a constant EI. The beam is on an elastic foundation between the supports A, B, and C with a modulus $k = 1184EI/l^4$.

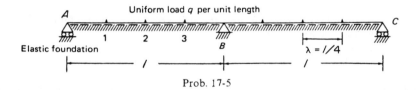

Prob. 17-5

17-6   Using finite differences, find the rotation at end A due to a unit couple applied at A for the beam in the figure and compare the result with the exact answer $(l/4EI)$. (*Hint*: Replace the couple by two equal and opposite forces at A and at node 1 and find the deflections using the equations listed in Fig. 17-5. The slope is then calculated from the deflection using the expression given in Prob. 17-2.)

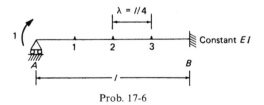

Prob. 17-6

17-7   Using finite differences, find the flexibility matrix $[f]$ corresponding to coordinates 1 and 2 of the cantilever in the figure. Compare the answer with accurate values from Appendix B. (*Hint*: A couple at coordinate 2 can be replaced by two equal and opposite forces at nodes 1 and 2. To calculate f_{21} — the rotation at coordinate 2 due to a unit force at 1 — we use the fact that the slope of the elastic line at 1 is approximately equal to the slope of the line joining the deflected position of nodes 1 and 2. This is acceptable because for this loading the bending moment, and hence the rate of change of slope at node 1, are zero.) To calculate f_{22}, use the expression given in Prob. 17-2.

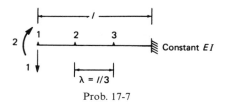

Prob. 17-7

17-8   Using finite differences, find the end-rotational stiffness at A and the carryover moment at B for the beam of Prob. 17-6. The beam is subjected to an axial compressive force $P = 9\,EI/l^2$ (not shown in the figure). Compare your results with the exact answers from Table 15-4. See the hint given for Prob. 17-6.

17-9   Using the finite differences, find the end-moment at B in the beam-column shown in the figure.

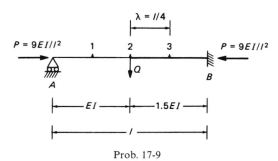

Prob. 17-9

17-10   Using finite differences, find the deflection at the nodes 1, 2, and 3 on the base of the long rectangular water tank shown in the figure. Consider a strip of unit width of the wall and base as a frame $ABCD$ of constant flexural rigidity EI. Neglect the self-weight of the frame. (*Hint:* Walls AB and DC are subjected to a linearly varying outward horizontal load of $\gamma l/2$ at bottom and zero at top, and the base BC is subjected to a uniform vertical load $\gamma l/2$.) Analyze the symmetrical beam BC by finite differences. One of the end conditions at B is

$$M_B = -\,(\text{FEM})_{BA} - S_{BA}\,y'_B = -\,EI\,y''_B$$

where $(\text{FEM})_{BA}$ is the absolute value of the fixed-end moment of BA with end A hinged, and S_{BA} is the end-rotational stiffness. To express the derivatives of y in terms of the deflections, extend BC by one interval to a fictitious node F on each side and use central differences expressions in Fig. 17-2.

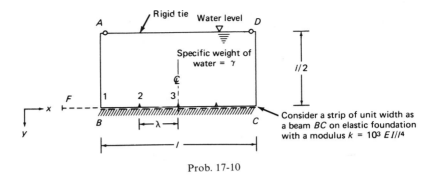

Prob. 17-10

17-11   Find the radial deflection caused by hydrostatic pressure on the wall of a circular cylindrical water tank shown in the figure. Calculate also the hoop force, the bending moment in the vertical direction along the wall height, and the reaction at the bottom. Assume Poisson's ratio $v = 1/6$. The answers to this problem are given at the end of the book, using two solutions: one with $\lambda = H/5$ as shown in the figure, and the other with $\lambda = H/20$. A comparison of the two results gives an indication of the accuracy gained by reducing λ. With $\lambda = H/5$, five simultaneous equations have to be solved. The reader may either solve the equations or use the given answers to verify his equations.

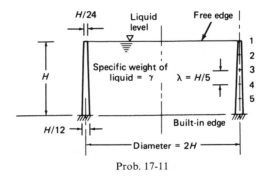

Prob. 17-11

17-12   Find the axial critical buckling load for the beam of Prob. 17-9, with both ends hinged and no transverse loading.

17-13   Find the axial critical buckling load for the beam of Prob. 17-9. No transverse loading is acting.

CHAPTER 18

Analysis of plates by
finite differences

18-1 INTRODUCTION

In this chapter the method of finite differences will be used in the analysis of thin plates. Two different problems arise depending on the type of loading.

When all the forces are applied in the plane of the plate, say the $x-y$ plane, the stresses produced are σ_x, σ_y and $\tau_{xy}(= \tau_{yx})$, while σ_z, τ_{zx}, τ_{xz}, τ_{zy} and τ_{yz} are zero (see Figs. 5-4 and 18-3). This is referred to as *plane-stress* distribution and the forces involved are called *in-plane forces*. The displacement at any point in the plane of the plate can be completely defined by two components in the x and y directions.

Thin plates subjected to transverse (normal) loading are subjected to bending and are hence referred to as *plates in bending*. They undergo transverse deflections which are small compared with the dimensions of the plate. As a result, the stretching of the middle plane of the plate is negligible, and the in-plane displacement of points on the middle plane is assumed to be zero. Thus the displacement at any point on the middle plane (or surface) of the plate, say the $x-y$ plane, can be defined by a translation in the z direction and two rotations about the x and y axes.

From the above assumptions, it follows that the two types of loading cause displacements which are *uncoupled* so that each type can be treated separately. Each of the plane-stress and the plate-bending problems involves the solution of a partial differential equation, for which the finite-difference method will be used.

The above argument applies only when the assumptions made are satisfied. When the transverse deflections are large compared with the plate thickness, or when the in-plane forces are large, the displacements caused by transverse loading and by in-plane forces are coupled and cannot be treated separately. In such a case the analysis requires the solution of two differential equations, each including the transverse deflections and a stress function ϕ which is related to the stresses σ_x, σ_y, and τ_{xy}. The treatment of this problem in its general form is outside the scope of this book.[1]

---

[1] See S. P. Timoshenko and J. M. Gere, *Theory of Elastic Stability*, 2d ed., McGraw-Hill, New York, 1961. For a numerical procedure, which is an alternative to the method of finite differences used to solve this problem as well as other problems in this chapter, see reference in footnote 4 of this chapter.

18-2 REPRESENTATION OF PARTIAL DERIVATIVES BY FINITE DIFFERENCES

Figure 18-1 shows a mesh drawn on a surface representing a function $w = f(x, y)$, which can, for example, be the deflected surface of a plate in bending. The derivative of w with respect to x or y can be expressed as a difference of the values of w at the nodes of the mesh in the same manner as the ordinary finite differences considered in Chapter 17. In the following expressions, central differences are used to express values at point O.

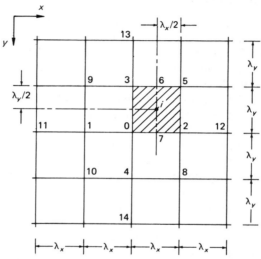

Fig. 18-1. Mesh used in a finite difference representation of the derivatives of a function $w = f(x, y)$.

The slope of the surface in the x direction is

$$\left(\frac{\partial w}{\partial x}\right)_0 \cong \frac{1}{2\lambda_x}[-1 \quad 1]\begin{Bmatrix} w_1 \\ w_2 \end{Bmatrix} \tag{18-1}$$

and the curvature in the x direction is

$$\left(\frac{\partial^2 w}{\partial x^2}\right)_0 \cong \frac{1}{\lambda_x^2}[1 \quad -2 \quad 1]\begin{Bmatrix} w_1 \\ w_0 \\ w_2 \end{Bmatrix} \tag{18-2}$$

Similar expressions can be written for derivatives with respect to y.
The Laplacian operator in the x and y variables

$$\nabla^2 = \frac{\partial^2}{\partial x^2} + \frac{\partial^2}{\partial y^2}$$

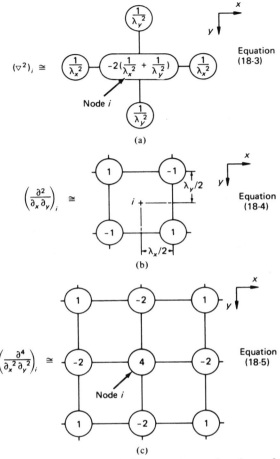

Fig. 18-2. Coefficients for central-difference approximations of partial derivatives (λ_x and λ_y are mesh widths in x and y directions). (a) Equation 18-3. Laplacian operator in two variables. (b) Equation 18-4. Multiplier for all coefficients: $1/\lambda_x\lambda_y$. (c) Equation 18-5. Multiplier for all coefficients: $1/(\lambda_x^2\lambda_y^2)$.

applied at a general point O can be put in the finite-difference form

$$(\nabla^2 w)_0 \cong \frac{1}{\lambda_x^2}\begin{bmatrix}1 & -2 & 1\end{bmatrix}\begin{Bmatrix}w_1 \\ w_0 \\ w_2\end{Bmatrix} + \frac{1}{\lambda_y^2}\begin{bmatrix}1 & -2 & 1\end{bmatrix}\begin{Bmatrix}w_3 \\ w_0 \\ w_4\end{Bmatrix} \qquad (18\text{-}3)$$

The mixed derivative at point i, the center of the hatched rectangle in Fig. 18-1, is

$$\left(\frac{\partial^2 w}{\partial x \partial y}\right)_i \cong \frac{1}{\lambda_y}\left[\left(\frac{\partial w}{\partial x}\right)_7 - \left(\frac{\partial w}{\partial x}\right)_6\right]$$

whence

$$\left(\frac{\partial^2 w}{\partial x \partial y}\right)_i \cong \frac{1}{\lambda_y \lambda_x}[1 \quad -1 \quad \quad 1 \quad -1] \begin{Bmatrix} w_2 \\ w_0 \\ w_3 \\ w_5 \end{Bmatrix} \tag{18-4}$$

The mixed derivative at node O can be expressed in a way similar to Eq. 18-4 in terms of w at the four corners of the rectangle of size $2\lambda_x \times 2\lambda_y$ whose center is the node O.

The coefficients of w at the nodes in Eqs. 18-3 and 18-4 applied at a general node i are given in Figs. 18-2a and b. The coefficients of w for the derivative $(\partial^4 w/\partial x^2 \partial y^2)_i$ are given in Eq. 18-5 in Fig. 18-2c.

18-3 GOVERNING DIFFERENTIAL EQUATIONS FOR PLATES SUBJECTED TO IN-PLANE FORCES

A solution of the plane-stress problem must satisfy the equation of equilibrium and compatibility, and the boundary conditions; these are given below.[2]

(a) *EQUILIBRIUM.* Considering the forces in the x and y directions acting on a small rectangular block A with sides dx, dy, and h, where h is the

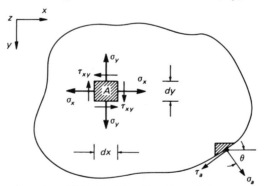

Fig. 18-3. Positive direction of stresses.

plate thickness (Fig. 18-3), we can write the following differential equations of equilibrium

$$\frac{\partial \sigma_x}{\partial x} + \frac{\partial \tau_{xy}}{\partial y} + X = 0 \tag{18-6}$$

[2]For further details of the derivation of the equations, refer to S. Timoshenko and J. N. Goodier, *Theory of Elasticity*, 2d ed., McGraw-Hill, New York, 1951.

and

$$\frac{\partial \sigma_y}{\partial y} + \frac{\partial \tau_{xy}}{\partial x} + Y = 0 \qquad (18\text{-}7)$$

where X and Y are components of body force per unit volume in the x and y directions. The positive direction of stresses is shown in Fig. 18-3.

(b) *COMPATIBILITY*. If u and v are the in-plane displacements of a general point in the x and y directions, the strain components ε_x, ε_y, and γ_{xy} can be expressed as

$$\varepsilon_x = \frac{\partial u}{\partial x} \qquad \varepsilon_y = \frac{\partial v}{\partial y} \qquad \gamma_{xy} = \frac{\partial u}{\partial y} + \frac{\partial v}{\partial x} \qquad (18\text{-}8)$$

The three strain components ε_x, ε_y, and γ_{xy} are not independent, but must satisfy the compatibility equation

$$\frac{\partial^2 \varepsilon_x}{\partial y^2} + \frac{\partial^2 \varepsilon_y}{\partial x^2} = \frac{\partial^2 \gamma_{xy}}{\partial x \partial y} \qquad (18\text{-}9)$$

which can be easily derived from Eq. 18-8.

The stress-strain relations (Hooke's law, Eq. 5-6) for an isotropic thin plate with a plane-stress distribution can be written in the form

$$\begin{Bmatrix} \varepsilon_x \\ \varepsilon_y \\ \gamma_{xy} \end{Bmatrix} = \frac{1}{E} \begin{bmatrix} 1 & -v & 0 \\ -v & 1 & 0 \\ 0 & 0 & 2(1+v) \end{bmatrix} \begin{Bmatrix} \sigma_x \\ \sigma_y \\ \tau_{xy} \end{Bmatrix} \qquad (18\text{-}10)$$

where E is the modulus of elasticity, and v the Poisson's ratio. Combining Eqs. 18-6–18-10, the compatibility condition in terms of stress components becomes

$$\nabla^2 (\sigma_x + \sigma_y) = -(1+v) \left(\frac{\partial X}{\partial x} + \frac{\partial Y}{\partial y} \right) \qquad (18\text{-}11)$$

(c) *BOUNDARY CONDITIONS*. Considering the equilibrium of forces acting on a triangular block B at the plate boundary (Fig. 18-3), the external applied stresses normal (σ_a) and tangential (τ_a) to the boundary can be related to the stress components σ_x, σ_y, and τ_{xy} near the boundary by

$$\begin{Bmatrix} \sigma_a \\ \tau_a \end{Bmatrix} = \begin{bmatrix} \cos^2 \theta & \sin^2 \theta & \sin 2\theta \\ -\dfrac{\sin 2\theta}{2} & \dfrac{\sin 2\theta}{2} & \cos 2\theta \end{bmatrix} \begin{Bmatrix} \sigma_x \\ \sigma_y \\ \tau_{xy} \end{Bmatrix} \qquad (18\text{-}12)$$

where θ is the angle between the normal to the boundary and the x axis (see Fig. 18-3).

18-4 AIRY STRESS FUNCTION

If the body forces are expressed in terms of a potential function V such that

$$X = -\frac{\partial V}{\partial x} \quad \text{and} \quad Y = -\frac{\partial V}{\partial y} \tag{18-13}$$

the equilibrium Eqs. 18-6 and 18-7 will be satisfied by

$$\sigma_x - V = \frac{\partial^2 \phi}{\partial y^2} \quad \sigma_y - V = \frac{\partial^2 \phi}{\partial x^2} \quad \text{and} \quad \tau_{xy} = -\frac{\partial^2 \phi}{\partial x \partial y} \tag{18-14}$$

where ϕ is a function called the Airy stress function. Substituting Eqs. 18-13 and 18-14 into Eq. 18-11, we obtain

$$\nabla^2 (\nabla^2 \phi) = -(1 - v)(\nabla^2 V) \tag{18-15}$$

Consider the case, common in practice, where the only body force is the self weight of the plate acting in the y direction. Then, $Y = -\partial V/\partial y = \omega$, where ω is the specific weight of the plate material, and $X = -\partial V/\partial x = \nabla^2 V = 0$. Choosing $V = 0$ at $y = 0$, we have $V = -\omega y$ and the right-hand side of Eq. 18-15 vanishes. Taking the solution $\phi = 0$, from Eq. 18-14 the stresses are

$$\sigma_x = \sigma_y = -\omega y \quad \text{and} \quad \tau_{xy} = 0 \tag{18-16}$$

This corresponds to a state of hydrostatic pressure ωy in two dimensions. This stress distribution can exist if the boundary stresses are $\sigma_a = -\omega y$ and $\tau_a = 0$ (see Eq. 18-12). The actual boundary stresses have to be superposed on a tensile boundary stress ωy, and a solution to the plane-stress problem is sought without body forces, the requisite equations being

$$\frac{\partial^4 \phi}{\partial x^4} + 2 \frac{\partial^4 \phi}{\partial x^2 \partial y^2} + \frac{\partial^4 \phi}{\partial y^4} = 0 \tag{18-17}$$

or

$$\nabla^2 (\nabla^2 \phi) = 0 \tag{18-17a}$$

and

$$\sigma_x = \frac{\partial^2 \phi}{\partial y^2} \quad \sigma_y = \frac{\partial^2 \phi}{\partial x^2} \quad \text{and} \quad \tau_{xy} = -\frac{\partial^2 \phi}{\partial x \partial y} \tag{18-18}$$

Equation 18-17 (or 18-17a) is often referred to as the *biharmonic equation*. We shall now show how it can be solved by finite differences to satisfy the boundary conditions of Eq. 18-12.

18-5   FINITE-DIFFERENCE EQUATIONS FOR PLATES SUBJECTED TO IN-PLANE FORCES

Equation 18-3 gives the finite-difference form of the Laplacian operator ∇^2 for a function of x and y, and the finite-difference coefficients are schematically represented in Fig. 18-2a. Applying Laplace's operator ∇^2 to the function $\nabla^2(\phi)$ and using the pattern of coefficients of Fig. 18-2a, we obtain the finite-difference form of the derivative $\nabla^2(\nabla^2\phi)$. The pattern of coefficients

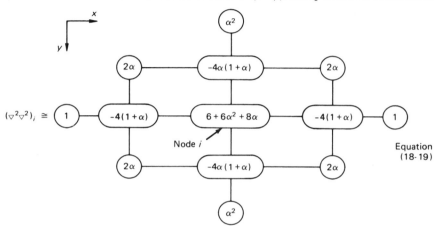

Multiplier for all coefficients: $(1/\lambda_x{}^4)$

(a)

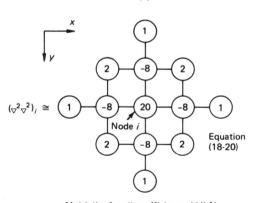

Multiplier for all coefficients: $(1/\lambda^4)$

(b)

Fig. 18-4. Central-difference approximation of the partial derivative $(\nabla^2\nabla^2)_i$. The above coefficients are to be used in the finite-difference form of Eqs. 18-17 and 18-35 as follows: [coefficients] $\{\phi\} \cong 0$ and [coefficients] $\{w\} \cong q_i/N$. (a) Rectangular mesh; $\alpha = (\lambda_x/\lambda_y)^2$; λ_x and λ_y are mesh widths in the x and y directions. (b) Square mesh; the mesh width $= \lambda$ in the x and y directions.

for the operator $\nabla^2(\nabla^2)$ applied at a node i for a function which varies with x and y is shown in Fig. 18-4a. When the mesh widths are such that $\lambda_x = \lambda_y = \lambda$, the pattern of coefficients is as shown in Fig. 18-4b. The coefficients in Eqs. 18-19 and 18-20 (given in Fig. 18-4) can be derived also from Eqs. 18-5 and 18-17 together with the finite-difference approximation for the fourth derivative given in Fig. 17-2.

Referring to the mesh in Fig. 18-1, the stresses σ_x and σ_y at any node 0, expressed in terms of the stress function ϕ, are given by the finite-difference form of Eq. 18-18 as

$$
\left.
\begin{aligned}
(\sigma_x)_0 &\cong \frac{1}{\lambda_y^2} \begin{bmatrix} 1 & -2 & 1 \end{bmatrix} \begin{Bmatrix} \phi_4 \\ \phi_0 \\ \phi_3 \end{Bmatrix} \\[2em]
(\sigma_y)_0 &\cong \frac{1}{\lambda_x^2} \begin{bmatrix} 1 & -2 & 1 \end{bmatrix} \begin{Bmatrix} \phi_2 \\ \phi_0 \\ \phi_1 \end{Bmatrix}
\end{aligned}
\right\} \tag{18-21}
$$

and the stress τ_{xy} at point 7, midway between nodes 0 and 2, is

$$
(\tau_{xy})_7 \cong -\frac{1}{2\lambda_x\lambda_y} \begin{bmatrix} 1 & -1 & 1 & -1 \end{bmatrix} \begin{Bmatrix} \phi_8 \\ \phi_4 \\ \phi_3 \\ \phi_5 \end{Bmatrix} \tag{18-22}
$$

Consider the rectangular plate of Fig. 18-5a, subjected to stresses σ_a, τ_a applied at the boundaries and distributed in an arbitrary manner, provided the equilibrium of the plate as a whole is maintained. In order to determine the state of stress in this plate, we draw a mesh and extend it to include fictitious points along a line one mesh width beyond the boundary (Fig. 18-5b).

At points on the boundary, the stress component τ_{xy} and σ_x or σ_y are equal to the known applied stress. (For example, at point A in Fig. 18-5a, $\sigma_y = \sigma_a$ and $\tau_{xy} = 0$.) These conditions can be put in finite-difference form, using Eqs. 18-21 and 18-22; hence, we obtain a system of equations relating the ϕ values at the mesh points to the applied stress. For the mesh in Fig. 18-5b, the first of Eqs. 18-21 can be applied at the 4 nodes on each of the two edges parallel to the y axis, and the second equation can be applied at the 5 nodes on each of the edges parallel to the x axis. For the shearing stress, Eq. 18-22 can be applied at points midway between each pair of adjacent boundary nodes. The total number of boundary equations in this case is 32.

At each of the interior nodes, the finite-difference form of Eq. 18-17 (see Fig. 18-4) can be applied. This brings the total number of equations to 38 for the plate in Fig. 18-5b. The equations can be written in matrix form

$$[A] \begin{Bmatrix} \{\phi_1\} \\ \text{---} \\ \{\phi_2\} \end{Bmatrix} \cong \begin{Bmatrix} \{\sigma_a\} \\ \text{---} \\ 0 \end{Bmatrix} \qquad (18\text{-}23)$$

where $\{\phi_1\}$ represents the values of the stress function at the boundary and at the fictitious nodes, $\{\phi_2\}$ is the vector of the stress functions at the interior nodes, $\{\sigma_a\}$ represents the known values of the applied stress normal or tangential to the boundary, and $[A]$ is a matrix formed by the finite-difference coefficients.

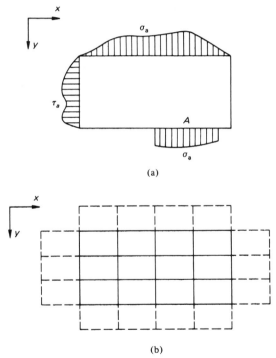

(a)

(b)

Fig. 18-5. A plate subjected to in-plane forces. (a) Plate in equilibrium under applied boundary stresses σ_a, τ_a. (b) Finite-difference mesh.

Since the stress components depend on the second derivatives of ϕ (Eq. 18-18), a function $\phi^* = \phi + Ax + By + C$, where A, B, and C are arbitrary constants, gives the same stresses as ϕ. If we imagine the ϕ function to represent a surface, a change of the constants A, B, and C corresponds to a rigid-body displacement of the surface. It follows that the value of ϕ can be arbitrarily chosen (say zero) at three points not on a straight line.

This reduces the number of unknown stress functions by three. Thus, for the plate in Fig. 18-5b, the number of unknown ϕ values is $n = 35$.

The stress values $\{\sigma_a\}$ in Eq. 18-23 are not independent because they must maintain the equilibrium of the plate as a whole. Three values of nonparallel boundary stress components can be derived from considerations of equilibrium of the plate if the other stress values are defined. Thus the number of boundary equations which need to be written for the plate in Fig. 18-5b is 29 instead of 32. These 29 boundary equations, together with the 6 equations applied at the interior points, are sufficient for the determination of the 35 ϕ-values from which the stress components can be calculated using the finite-difference forms of Eq. 18-14.

The above procedure is easy to apply if a computer is available. The number of simultaneous equations to be solved is large but it can be reduced to the number of the interior points (six for the plate in Fig. 18-5b) as follows.

18-51   Value of the Stress Function at Plate Boundary

Instead of assuming three arbitrary values of ϕ (as we did before) we can put three values:

$$\phi = 0 \qquad \frac{\partial \phi}{\partial x} = 0 \qquad \text{and} \qquad \frac{\partial \phi}{\partial y} = 0$$

all three at one boundary node, or distributed among two or three different boundary nodes.

At an edge parallel to the y axis, the relations

$$\frac{\partial^2 \phi}{\partial y^2} = \sigma_a \qquad \text{and} \qquad \frac{\partial}{\partial y}\left(\frac{\partial \phi}{\partial x}\right) = -\tau_a$$

suggest an analogy to the relation between moment and load and between shear and load in a beam. These well-known relations are: the second derivative of the bending moment is numerically equal to the intensity of load per unit area

$$\frac{d^2 M}{dx^2} = -q$$

and the first derivative of the shear is numerically equal to the intensity of load

$$\frac{dV}{dx} = -q$$

It follows from this analogy that the value of the stress function ϕ along a straight edge parallel to the y axis is equal to the bending moment due to a load of intensity $-\sigma_a$ normal to the edge. Similarly, the slope of the surface representing the stress function $\partial \phi / \partial x$ is equal to the shearing force induced by a load of intensity τ_a normal to the edge. This method has been used by Dubas[3].

The above analogy is valid also for a curved boundary. Three values of ϕ or its derivative parallel or normal to the tangent at the boundary node(s) are arbitrarily chosen (say, 0), and the values of ϕ and of the slope $\partial\phi/\partial n$ parallel to the normal at other nodes are calculated by equations of statics.

Along a straight boundary, the variation in ϕ and in $\partial\phi/\partial n$ can take one of the following forms. When the edge is free, ϕ is linear and $\partial\phi/\partial n$ = constant. At an edge subjected to normal stress only, $\partial\phi/\partial n$ = constant, while ϕ varies as the moment due to the applied stress. If the edge is subjected to a shearing stress only (without normal stresses), ϕ is linear and $\partial\phi/\partial n$ varies as the shear caused by transverse loading of magnitude equal to the shearing stress applied. Concentrated forces applied at the boundary provide no difficulty as the calculation of the bending moment and shearing force due to such forces does not require a special treatment.

The shape of the surfaces representing the stress function for plates subjected to concentrated or distributed loads is shown in Fig. 18-6. The state of stress in the plate in Fig. 18-6a is determined in the following example.

Example 18-1 Find the distribution of σ_x and τ_{xy} along lines EF and GH of the plate in Fig. 18-6a. The plate has a constant thickness h.

If ϕ is chosen equal to zero at points A, B, and C, the bending-moment diagram due to forces P/h on a rectilinear frame $ABCD$ with three hinges at the three points A, B, and C gives the variation in ϕ along the boundary (Fig. 18-6b).

Since $\tau_a = 0$ at all boundary nodes, $\partial\phi/\partial n$ = constant. At corner A,

$$\frac{\partial\phi}{\partial x} = \frac{P}{2h} \qquad \text{and} \qquad \frac{\partial\phi}{\partial y} = 0$$

Thus, at all nodes on AD

$$\frac{\partial\phi}{\partial x} = \frac{P}{2h}$$

and at nodes on AB, $\partial\phi/\partial y = 0$. Similarly, $\partial\phi/\partial x = -P/(2h)$ at nodes on BC, and $\partial\phi/\partial y = 0$ at nodes on DC.

We choose a square mesh with $\lambda = l/4$ (Fig. 18-6a). Because of symmetry, there are only four values of ϕ to be determined at nodes 1, 2, 3, and 4. In order to apply the finite-difference form of Eq. 18-17 (see Eq. 18-20 in Fig. 18-4b) at nodes 1, 2, or 3, the ϕ value at one fictitious node outside the

---

[3]See reference in footnote 4 of this chapter.

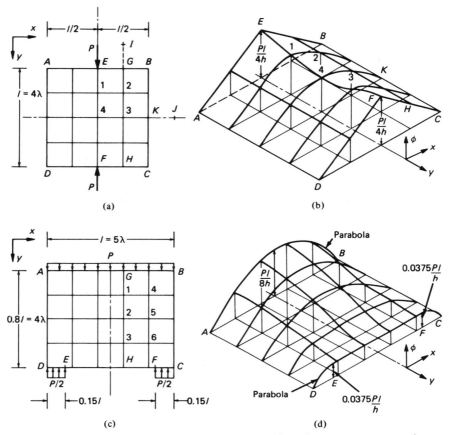

Fig. 18-6. Stress functions. (a) Square plate subjected to two concentrated forces P. Plate thickness $= h$. (b) Pictorial view of a surface representing stress function ϕ for the plate in part (a). (c) Rectangular plate subjected to distributed loads. Plate thickness $= h$. (d) Pictorial view of a surface representing stress function ϕ for the plate in part (c).

boundary will be included. This value is eliminated by using the known slope at the boundary of the surface representing the stress function. For example, at G

$$\left(\frac{\partial \phi}{\partial y}\right)_G = 0$$

thus

$$\frac{(\phi_2 - \phi_I)}{2\lambda} \cong 0 \quad \text{or} \quad \phi_I = \phi_2$$

at K,

$$\left(\frac{\partial \phi}{\partial x}\right)_K = -\frac{P}{2h}$$

thus

$$\frac{(\phi_J - \phi_3)}{2\lambda} \cong -\frac{P}{2h} \qquad \text{or} \qquad \phi_J \cong \phi_3 - \frac{P\lambda}{h}$$

Applying Eq. 18-17 in finite-difference form at each of the nodes 1 to 4, and eliminating ϕ at the fictitious points and substituting the ϕ values at the boundary nodes (see Fig. 18-6b), we obtain the following simultaneous equations:

$$\begin{bmatrix} 22 & -16 & 4 & -8 \\ -8 & 24 & -8 & 2 \\ 4 & -16 & 22 & -8 \\ -16 & 8 & -16 & 20 \end{bmatrix} \begin{Bmatrix} \phi_1 \\ \phi_2 \\ \phi_3 \\ \phi_4 \end{Bmatrix} \cong \frac{P\lambda}{h} \begin{Bmatrix} 6 \\ 3 \\ 0 \\ -2 \end{Bmatrix}$$

We should note that the coefficients of ϕ generally form (or can be made to form) a square symmetrical matrix. For instance, in the above equations, symmetry can be achieved by multiplying the first to the fourth equations by 2, 4, 2 and 1, respectively. These values represent respectively the number of nodes in the mesh which have the same ϕ_1, ϕ_2, ϕ_3, and ϕ_4. This property of symmetry is of value in detecting any mistakes made in forming the equations.

The solution of the above equations gives

$$\{\phi\} \cong \frac{P\lambda}{h} \{0.8127, \quad 0.4944, \quad 0.4794, \quad 0.7359\}$$

Application of Eqs. 18-21 and 18-22 gives the values of stresses at the nodes; these are plotted in Fig. 18-7. More accurate values can be obtained if a smaller mesh is used but this requires a greater number of equations.

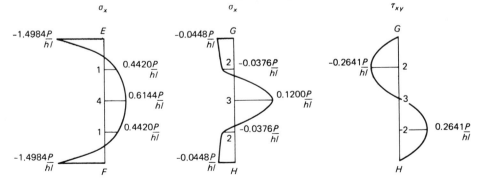

$(\tau_{xy} = 0$ for all points on $EF)$

Fig. 18-7. Variation in σ_x and τ_{xy} in the plate of Fig. 18-6a.

Accuracy can also be improved by the use of a numerical procedure[4] which involves more elaborate computations in the preparation of the coefficients in the simultaneous equations for ϕ.

18-6   GOVERNING DIFFERENTIAL EQUATION FOR PLATES IN BENDING

In this section we deal with thin plates subjected to transverse loading causing deflections which are small compared with the plate thickness (Fig. 18-8a). If the orthogonal axes x and y are chosen in the middle surface

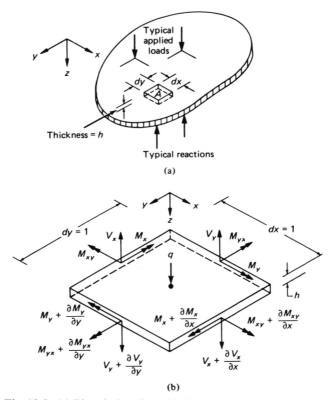

Fig. 18-8. (a) Plate in bending. (b) Forces acting on the block A in part (a). This figure also indicates the positive directions of the stress resultants.

[4]The plate in Fig. 18-6a is analyzed by this method in Pierre Dubas, *Numerical Analysis of Plates and Thin Walls*, Federal Institute of Technology, Zurich, Publication No. 27, Institute of Applied Statics, Leeman, Zurich, 1955 (in French).

of the plate, we can neglect at all points on this surface the displacements u and v parallel to the x and y axes, the stresses σ_x and σ_y, and the deformations caused by the stress σ_z. Further, we assume that a normal to the middle surface remains straight across the plate thickness and is normal to the deflected middle surface. This is analogous to the assumption in the bending theory that plane cross sections remain plane after bending, which is normally made for beams long in comparison with their cross-sectional dimensions.

Figure 18-8b shows the forces acting on a small rectangular block A of sides dx, dy, and h. If dx and dy are taken equal to unity, the internal forces M_x, M_y, M_{xy}, V_x, and V_y on the block edges are equal to the resultants of the appropriate stresses on the element sides. The positive directions of these stress resultants are indicated in Fig. 18-8b, and the positive directions of stresses are shown in Fig. 5-4. The stress resultants are related to the stresses σ_x, σ_y, and τ_{xy} as follows:

$$M_x = \int_{-h/2}^{h/2} \sigma_x z \, dz \qquad M_y = \int_{-h/2}^{h/2} \sigma_y z \, dz \qquad (18\text{-}24)$$

$$M_{yx} = -M_{xy} = \int_{-h/2}^{h/2} \tau_{xy} z \, dz \qquad (18\text{-}25)$$

and

$$V_x = \int_{-h/2}^{h/2} \tau_{xz} \, dz \qquad \text{and} \qquad V_y = \int_{-h/2}^{h/2} \tau_{yz} \, dz \qquad (18\text{-}26)$$

To solve for the stress resultants we consider the equilibrium and the deformation of the block.

(a) *EQUILIBRIUM*. The sum of the vertical forces is zero and the sum of the moments about edges dx and dy equals zero. Thus

$$\left. \begin{aligned} & \frac{\partial V_x}{\partial x} + \frac{\partial V_y}{\partial y} + q = 0 \\[2mm] & \frac{\partial M_x}{\partial x} + \frac{\partial M_{yx}}{\partial y} - V_x = 0 \\[2mm] & -\frac{\partial M_y}{\partial y} + \frac{\partial M_{xy}}{\partial x} + V_y = 0 \end{aligned} \right\} \qquad (18\text{-}27)$$

and

where q is the intensity of the applied load per unit area. These three equations can be combined in one equation

$$\frac{\partial^2 M_x}{\partial x^2} - 2\frac{\partial^2 M_{xy}}{\partial x \partial y} + \frac{\partial^2 M_y}{\partial y^2} = -q \qquad (18\text{-}28)$$

(b) *DEFORMATION*. The displacements u and v (in the direction of the x and y axes) at any point at a distance z from the middle surface can be expressed in terms of the slope of the deflected surface by

$$u = -z\frac{\partial w}{\partial x} \quad \text{and} \quad v = -z\frac{\partial w}{\partial y} \qquad (18\text{-}29)$$

where w is the deflection in the direction of the z axis. The partial derivatives represent the slope of the deflected surface which is also equal to the rotation of the normal to the middle surface.

The strain at the same point is expressed in terms of the displacements by Eq. 18-8, from which we obtain

$$\varepsilon_x = -z\frac{\partial^2 w}{\partial x^2} \qquad \varepsilon_y = -z\frac{\partial^2 w}{\partial y^2} \quad \text{and} \quad \gamma_{xy} = -2z\frac{\partial^2 w}{\partial x \partial y} \quad (18\text{-}30)$$

The stress is related to strain by Hooke's law, which for a homogeneous, isotropic plate is expressed by Eq. 18-10. Using this equation and Eq. 18-30, the stress can be expressed in terms of the deflection of the middle surface:

$$\begin{Bmatrix} \sigma_x \\ \sigma_y \\ \tau_{xy} \end{Bmatrix} = -\frac{Ez}{(1-v^2)} \begin{bmatrix} 1 & v & 0 \\ v & 1 & 0 \\ 0 & 0 & (1-v) \end{bmatrix} \begin{Bmatrix} \dfrac{\partial^2 w}{\partial x^2} \\ \dfrac{\partial^2 w}{\partial y^2} \\ \dfrac{\partial^2 w}{\partial x \partial y} \end{Bmatrix} \qquad (18\text{-}31)$$

Substituting for the stress from Eq. 18-31 into Eqs. 18-24 and 18-25, we obtain

$$\begin{Bmatrix} M_x \\ M_y \\ M_{xy} \end{Bmatrix} = -N \begin{bmatrix} 1 & v & 0 \\ v & 1 & 0 \\ 0 & 0 & -(1-v) \end{bmatrix} \begin{Bmatrix} \dfrac{\partial^2 w}{\partial x^2} \\ \dfrac{\partial^2 w}{\partial y^2} \\ \dfrac{\partial^2 w}{\partial x \partial y} \end{Bmatrix} \qquad (18\text{-}32)$$

where

$$N = \frac{Eh^3}{12(1-v^2)} \qquad (18\text{-}33)$$

represents the flexural rigidity of a strip of the plate having a unit width.

From last two of Eq. 18-27 and Eq. 18-32, the shear can also be expressed in terms of the deflection

$$V_x = -N\frac{\partial}{\partial x}(\nabla^2 w) \qquad \text{and} \qquad V_y = -N\frac{\partial}{\partial y}(\nabla^2 w) \qquad (18\text{-}34)$$

where ∇^2 is the Laplacian operator.

(c) *DERIVATION*. Substitution of Eq. 18-32 into Eq. 18-28 gives

$$\frac{\partial^4 w}{\partial x^4} + 2\frac{\partial^4 w}{\partial x^2 \partial y^2} + \frac{\partial^4 w}{\partial y^4} = \frac{q}{N} \qquad (18\text{-}35)$$

which can be written in the form

$$\nabla^2(\nabla^2 w) = q/N \qquad (18\text{-}35a)$$

In the analysis of plates in bending, deflection is obtained by solving Eq. 18-35 satisfying the conditions at the plate boundaries; the stress resultants are then determined by Eqs. 18-32 and 18-34. In the following, the solution for the deflection will be obtained by the method of finite differences.

18-7 FINITE-DIFFERENCE EQUATIONS AT AN INTERIOR NODE OF A PLATE IN BENDING

The partial derivatives of w in Eq. 18-35 are the same as the derivatives of ϕ in Eq. 18-17. It follows that the finite-difference coefficients in Eqs. 18-19 and 18-20 (given in Fig. 18-4) can be also used for Eq. 18-35. Hence, the finite-difference form of Eq. 18-35, applied at a general interior node i, is

$$[\text{coefficients}]\{w\} \cong q_i/N \qquad (18\text{-}36)$$

where the coefficients are as indicated in Fig. 18-4. This equation relates the deflection w_i at i and at 12 other nodes in its vicinity to the load intensity q_i.

The stress resultants at an interior node can be expressed in terms of the node deflections w by the finite-difference form of Eqs. 18-32 and 18-34. The pattern of coefficients for these approximations is given in Fig. 18-9.

18-8 BOUNDARY CONDITIONS OF A PLATE IN BENDING

In Fig. 18-10 the orthogonal coordinates n and t are taken normal and tangential to plate boundary. Considering equilibrium of the small block B (see Fig. 18-10b) the bending and twisting moment at the boundary can be related to M_x, M_y, and M_{xy} by

$$\begin{Bmatrix} M_n \\ M_{nt} \end{Bmatrix} = \begin{bmatrix} \cos^2\theta & \sin^2\theta & -\sin 2\theta \\ \dfrac{\sin 2\theta}{2} & -\dfrac{\sin 2\theta}{2} & \cos 2\theta \end{bmatrix} \begin{Bmatrix} M_x \\ M_y \\ M_{xy} \end{Bmatrix} \qquad (18\text{-}37)$$

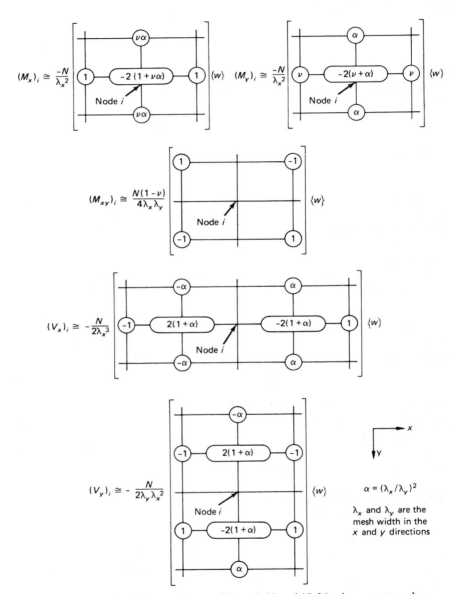

Fig. 18-9. Finite-difference forms of Eqs. 18-32 and 18-34 using a rectangular mesh.

The boundary shear V_n is related to V_x and V_y by

$$V_n = V_x \cos \theta + V_y \sin \theta \qquad (18\text{-}38)$$

At a plate edge, one or more of the stress resultants or the deflection or slope of the deflected surface can be known. This information is taken into account when writing the finite differences equations relating the deflection to the applied loads. The different edge conditions are considered below.

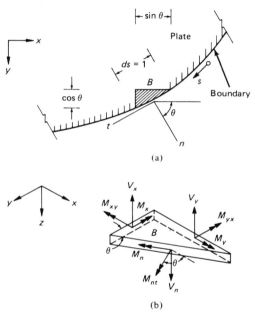

(a)

(b)

Fig. 18-10. A block near the boundary of a plate in bending considered in the derivation of Eqs. 18-37 and 18-38. (a) Top view of a plate. (b) Forces acting on block B in part (a). This figure also indicates the positive directions of V_n, M_n, and M_{nt}.

(a) *BUILT-IN EDGE*. The deflection at all points on a built-in edge and the slope of the deflected surface normal to the edge are zero, viz.,

$$w = 0 \quad \text{and} \quad \frac{\partial w}{\partial n} = 0 \qquad (18\text{-}39)$$

Consider the case when the plate edge is straight. The conditions of Eq. 18-39 will be satisfied if the plate is considered continuous with a symmetrical fictitious plate subjected to symmetrical loading so that its deflected middle surface is a mirror image of the actual deflected middle surface.

If the built-in edge is parallel to the x axis, as in Fig. 18-11a, the conditions of Eq. 18-39 give $w_D = w_B = w_E = 0$ and $w_C = w_A$. When the finite-differ-

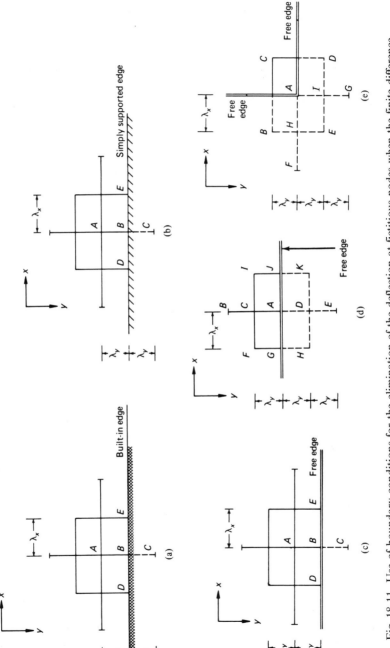

Fig. 18-11. Use of boundary conditions for the elimination of the deflection at fictitious nodes when the finite difference Eq. 18-36 is applied at a node A on or near a plate edge. (a) Node A near a fixed edge. (b) Node A near a simply-supported edge. (c) Node A on a free edge. (d) Node A near a free edge. (e) Node A at a corner.

ence Eq. 18-36 is applied at node A, the deflection at the fictitious node C is put equal to w_A and the deflection of the edge nodes is taken as zero. Hence, we obtain an equation relating the deflection at the interior nodes to the intensity of the applied load at A. The finite-difference Eq. 18-36 applied at each interior node gives a sufficient number of simultaneous equations for the nodal deflections to be calculated.

(b) *SIMPLY SUPPORTED EDGE.* The deflection w and the moment M_n vanish at all points of a simply-supported edge, viz.,

$$w = 0 \quad \text{and} \quad M_n = 0 \tag{18-40}$$

Combining Eqs. 18-37 and 18-32, the second boundary condition can be represented in terms of the deflection by

$$(1 - v)\left(\frac{\partial^2 w}{\partial x^2}\cos^2\theta + 2\frac{\partial^2 w}{\partial x \partial y}\sin\theta\cos\theta + \frac{\partial^2 w}{\partial y^2}\sin^2\theta\right) + v\nabla^2 w = 0 \tag{18-41}$$

When the plate edge is straight, the two conditions of Eq. 18-40 will be satisfied if the plate is considered continuous with a fictitious symmetrical plate subjected to antisymmetrical loading so that the deflections of the middle surface are equal and opposite at symmetrical points.

If the simply supported edge is parallel to the x axis, as in Fig. 18-11b, the conditions of Eq. 18-40 give $w_D = w_B = w_E = 0$ and $w_C = -w_A$. These equations are substituted in the finite-difference Eq. 18-36 applied at node A in a manner similar to that used for the built-in edge.

(c) *FREE EDGE.* The bending moment at any point on a free edge is zero, that is $M_n = 0$. This means that Eq. 18-41 must be satisfied. Further, at a free edge there should be no transverse forces. Now, the shear V_n and the twisting moment M_{nt} (Fig. 18-10) can be reduced to transverse forces only[5] and, since these vanish at a free edge, we can write

$$V_n - \frac{\partial M_{nt}}{\partial s} = 0 \tag{18-42}$$

Using Eqs. 18-32, 18-34, 18-37, and 18-38, this condition can be represented in terms of the deflection in the form

$$(1 - v)\frac{\partial}{\partial s}\left[\left(\frac{\partial^2 w}{\partial x^2} - \frac{\partial^2 w}{\partial y^2}\right)\sin\theta\cos\theta - \frac{\partial^2 w}{\partial x \partial y}(\cos^2\theta - \sin^2\theta)\right]$$

$$- \left(\frac{\partial^3 w}{\partial x^3} + \frac{\partial^3 w}{\partial x \partial y^2}\right)\cos\theta - \left(\frac{\partial^3 w}{\partial y^3} + \frac{\partial^3 w}{\partial y \partial x^2}\right)\sin\theta = 0 \tag{18-43}$$

[5]See S. Timoshenko and S. Wojnowsky-Krieger, *Theory of Plates and Shells*, 2d ed., McGraw-Hill, New York, 1959.

When the free edge is parallel to the x axis ($\theta = 90°$ and $\partial s = -\partial x$), the conditions at the free edge become

$$\frac{\partial^2 w}{\partial y^2} + v\frac{\partial^2 w}{\partial x^2} = 0 \qquad (18\text{-}44)$$

and

$$\frac{\partial^3 w}{\partial y^3} + (2 - v)\frac{\partial^3 w}{\partial y\partial x^2} = 0 \qquad (18\text{-}45)$$

When the finite-difference Eq. 18-36 is applied at node A on the first mesh line inside the boundary (Fig. 18-11c), the deflection at a fictitious node C can be eliminated using the following finite-difference form of Eq. 18-44 applied at node B:

$$\frac{1}{\lambda_y^2}(w_C - 2w_B + w_A) + \frac{v}{\lambda_x^2}(w_E - 2w_B + w_D) \cong 0 \qquad (18\text{-}46)$$

Application of Eq. 18-36 at node A on a free edge (Fig. 18-11d) includes the deflection at the fictitious nodes H, D, K, and E. The deflection w_E can be expressed in terms of the deflection at the other nodes, using the following finite-difference form of Eq. 18-45 applied at A:

$$\frac{1}{2\lambda_y^3}(w_E - 2w_D + 2w_C - w_B) + \frac{(2 - v)}{2\lambda_y}\left(\frac{w_K - 2w_D + w_H}{\lambda_x^2}\right.$$
$$\left. - \frac{w_I - 2w_C + w_F}{\lambda_x^2}\right) \cong 0 \qquad (18\text{-}47)$$

Three equations similar to Eq. 18-46 can be applied at nodes G, A, and J (Fig. 18-11d) and used to eliminate the fictitious deflections w_H, w_D, and w_K.

(d) *CORNER.* At a corner where two free edges parallel to the x and y axes meet (Fig. 18-11e), the boundary conditions of Eqs. 18-44 and 18-45 apply (as to all the nodes on the edge parallel to the x axis). Two similar equations apply to the nodes on the edge parallel to the y axis. In addition, at a corner,[6] the twisting moment $M_{xy} = 0$; this can be written in finite-difference form (see Fig. 18-9) as

$$\frac{N(1 - v)}{4\lambda_x\lambda_y}(w_B - w_C + w_D - w_E) \cong 0 \qquad (18\text{-}48)$$

Application of Eq. 18-36 at a corner node A (Fig. 18-11e) includes deflections at the fictitious nodes B, D, G, F, H, I, and E. The deflection at E can be eliminated by the use of Eq. 18-48, and the other fictitious deflections

---

[6]For a discussion of this condition, see reference in footnote 5 of this chapter.

are eliminated using the remainder of the boundary conditions in the same manner as for the free edge.

18-9 ANALYSIS OF PLATES IN BENDING

The procedure outlined in the preceding sections leads to linear equations relating the deflections at nodes within the boundary to the external applied load. These equations will have one of the forms given in Fig. 18-12 when

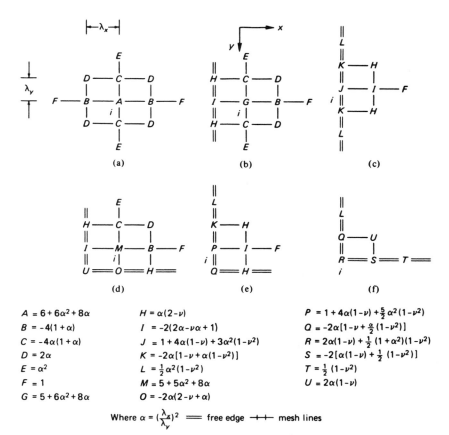

$A = 6 + 6\alpha^2 + 8\alpha$

$B = -4(1 + \alpha)$

$C = -4\alpha(1 + \alpha)$

$D = 2\alpha$

$E = \alpha^2$

$F = 1$

$G = 5 + 6\alpha^2 + 8\alpha$

$H = \alpha(2 - \nu)$

$I = -2(2\alpha - \nu\alpha + 1)$

$J = 1 + 4\alpha(1 - \nu) + 3\alpha^2(1 - \nu^2)$

$K = -2\alpha[1 - \nu + \alpha(1 - \nu^2)]$

$L = \frac{1}{2}\alpha^2(1 - \nu^2)$

$M = 5 + 5\alpha^2 + 8\alpha$

$O = -2\alpha(2 - \nu + \alpha)$

$P = 1 + 4\alpha(1 - \nu) + \frac{5}{2}\alpha^2(1 - \nu^2)$

$Q = -2\alpha[1 - \nu + \frac{\alpha}{2}(1 - \nu^2)]$

$R = 2\alpha(1 - \nu) + \frac{1}{2}(1 + \alpha^2)(1 - \nu^2)$

$S = -2[\alpha(1 - \nu) + \frac{1}{2}(1 - \nu^2)]$

$T = \frac{1}{2}(1 - \nu^2)$

$U = 2\alpha(1 - \nu)$

Where $\alpha = (\frac{\lambda_x}{\lambda_y})^2$ ═══ free edge ┼┼ mesh lines

Fig. 18-12. Finite-difference coefficients in Eq. 18-49 applied at node i on or near the edge of a plate in bending (from reference in footnote 7 of this chapter).

applied at a node i, which may be a general interior node or a node on or near the boundary. Any one of these forms can be expressed as

$$(N\lambda_y/N\lambda_x^3)[\text{coefficients}]\{w\} \cong Q_i \qquad (18\text{-}49)$$

where the coefficients are as given in Fig. 18-12, and Q_i is an equivalent concentrated load at i. When the load is uniform and of intensity q, and i is an interior node, $Q_i = q\lambda_x\lambda_y$. The values of Q when i is on an edge or at a corner are respectively, $q\lambda_x\lambda_y/2$ and $q\lambda_x\lambda_y/4$.

In the case of nonuniform loading, the equivalent concentrated load can be calculated from the equations given in Sec. 9-51. For example, if we use the expression for parabolic variation (see Fig. 9-11) to find the equivalent concentrated load Q_0 at node 0 of the mesh in Fig. 18-1, we write

$$\left.\begin{aligned} p_3 &= \frac{\lambda_x}{12}(q_9 + 10q_3 + q_5) \\ p_0 &= \frac{\lambda_x}{12}(q_1 + 10q_0 + q_2) \\ p_4 &= \frac{\lambda_x}{12}(q_{10} + 10q_4 + q_8) \end{aligned}\right\} \qquad (18\text{-}50)$$

where the q terms are the load intensities at the nodes (force/area) and the p values are line load intensities (force/length) on a line parallel to the y axis through 0. Applying the same expression for the load p, we obtain

$$Q_0 = \frac{\lambda_y}{12}(p_3 + 10p_0 + p_4) \qquad (18\text{-}51)$$

Combining Eqs. 18-50 and 18-51, we find

$$Q_0 = \frac{\lambda_x\lambda_y}{144}\left[100q_0 + 10(q_1 + q_2 + q_3 + q_4) + (q_9 + q_5 + q_{10} + q_8)\right] \qquad (18\text{-}52)$$

For other forms of load variation, or for edge or corner nodes, the other equations for the line loads given in Fig. 9-11 can be used to derive corresponding expressions for the equivalent concentrated load.

Applying the finite-difference equations at all mesh points where the deflection is not known leads to a system of simultaneous equations which may be written in matrix form

$$[K]\{w\} \cong \{Q\} \qquad (18\text{-}53)$$

where $[K]$ is a square matrix formed by the finite-difference coefficients. Examination of the coefficients in Fig. 18-12 shows that the matrix $[K]$ is symmetrical. For example, the finite-difference equation applied at a corner node (Fig. 18-12f) has a coefficient $U = 2\alpha(1 - \nu)$ for the deflection at a diagonally opposite node, and when the equation is applied at this latter node, (see Fig. 18-12d) the coefficient at the corner is also equal to U. We may consider the matrix $[K]$ as an equivalent stiffness matrix, and we shall so use it in the following chapter.

The procedure of the analysis of plates in bending by finite differences can be summarized as follows. A system of linear simultaneous equations is written in the form of Eq. 18-53, in which $\{Q\}$ are the known forces and $\{w\}$ the unknown deflections. The elements of the matrix $[K]$ are determined using Fig. 18-12, which covers the majority of practical cases. The solution of Eq. 18-53 gives the deflections, and these are then substituted in the finite-difference equations in Fig. 18-9 to determine the stress resultants.

The finite-difference equations in Fig. 18-12 can be used for a plate with a simply supported edge by substituting zero for the deflection at nodes on this edge. A figure similar to Fig. 18-12 can be prepared for the finite-difference equations applied at nodes on the first mesh line inside the boundary parallel to a built-in edge. This figure is not given as it is easy to derive from Fig. 18-12a if we observe that the deflections at fictitious nodes outside the boundary are a mirror image of the deflections at corresponding nodes within the boundary. When the free edge is inclined to the x and y directions, the finite-difference equations can also be derived by a similar procedure.[7]

When a plate has a curved boundary, the nodes of a rectangular mesh may not fall on the plate edge. In this case, the finite-difference expressions have to be used for unevenly spaced nodes.[8]

In the analysis of a parallelogram-shaped plate, it may be more suitable to use skew coordinate axes and a skew mesh in the finite-difference equations.[9] In some other cases, a mesh formed by arcs of circles and radial lines may be used. The coefficients of the finite-difference equations for these types of meshes are not given here, but the basic ideas in the analysis are the same in all cases.

18-91   Stiffened Plates

If a plate is stiffened by beams located on mesh lines and assumed to be connected to the plate in the idealized fashion indicated in Figs. 18-13a or b, the effect of the stiffening can be easily included by substituting $[K] = [K]_p + [K]_b$ in Eq. 18-53, where $[K]_p$ is the plate equivalent stiffness matrix for the plate alone, obtained as described before from the finite-difference coefficients, and $[K]_b$ is the equivalent stiffness matrix for the beams, defined by

$$[K]_b \{w\} = \{Q\}_b \qquad (18\text{-}54)$$

---

[7]See A. Ghali and K.-J. Bathe, "Analysis of Plates in Bending Using Large Finite Elements," International Association for Bridges and Structural Engineering, 30/II, Zurich, 1970.

[8]For further discussion see S. H. Crandall, *Engineering Analysis*, McGraw-Hill, New York, 1956.

[9]V. P. Jensen, *Analysis of Skew Slabs*, University of Illinois Engineering Experiment Station, Bulletin No. 332.

Here $\{w\}$ are the nodal deflections of the composite system beam-plate, $\{Q\}_b$ is a vector of the part of $\{Q\}$ carried by the beams alone, and $[K]_b$ is

Beam   Plate

(a)   (b)

Fig. 18-13. Idealized plate-beam connections.

an equivalent stiffness matrix for the beams. The elements of $[K]_b$ can be derived from the finite-difference coefficients given in Figs. 17-4 and 17-5.

Example 18-2   Find the deflection and moment M_y at node 2 and the moments M_x and M_y at node 6 in the plate shown in Fig. 18-14, subjected to

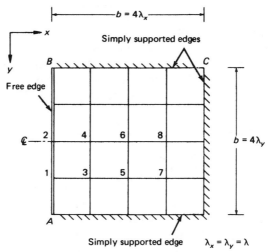

Fig. 18-14. Square plate subjected to a uniform transverse load analyzed by finite difference in Example 18-2.

a uniform load q per unit area. The plate is of constant thickness h, and has three edges simply supported and one edge free. Assume Poisson's ratio $v = 0.3$.

Applying Eq. 18-49 at each of the 8 nodes (see Fig. 18-14) and using the finite-difference patterns of coefficients in Fig. 18-12, the following simultaneous equations can be written:

$$\frac{N}{\lambda^2}\begin{bmatrix} J & K & I & H & F & & & \\ 2K & J & 2H & I & & F & & \\ I & H & G & C & B & D & F & \\ 2H & I & 2C & G & 2D & B & & F \\ F & & B & D & A & C & B & D \\ & F & 2D & B & 2C & A & 2D & B \\ & & F & & B & D & G & C \\ & & & F & 2D & B & 2C & G \end{bmatrix}\begin{Bmatrix} w_1 \\ w_2 \\ w_3 \\ w_4 \\ w_5 \\ w_6 \\ w_7 \\ w_8 \end{Bmatrix} \cong q\lambda^2 \begin{Bmatrix} 0.5 \\ 0.5 \\ 1.0 \\ 1.0 \\ 1.0 \\ 1.0 \\ 1.0 \\ 1.0 \end{Bmatrix}$$

where the coefficients A, B, ... are defined in Fig. 18-12. The square matrix on the left-hand side can be made symmetrical by multiplying by 2 each of the first, third, fifth and seventh equations. The figure 2 represents the number of nodes of the mesh which have the same deflection as each of w_1, w_3, w_5, and w_7. The property of symmetry is useful to detect any mistakes which may have been made in forming the equations.

Substituting the values of the coefficients A, B, ... and solving, we obtain

$$\{w\} \cong \frac{q\lambda^4}{N} \{2.3984, 3.3516, 1.9361, 2.7033, 1.4899, 2.0749, 0.8641, 1.1979\}$$

The required deflection at node 2 is $w_2 \cong 0.1429\, qb^4/Eh^3$. The moment M_y at node 2 on the free edge is (from Eq. 18-32)

$$(M_y)_2 = -N\left(\frac{\partial^2 w}{\partial y^2} + v\frac{\partial^2 w}{\partial x^2}\right)_2$$

and the moment

$$(M_x)_2 = -N\left(\frac{\partial^2 w}{\partial x^2} + v\frac{\partial^2 w}{\partial y^2}\right)_2 = 0$$

Hence

$$(M_y)_2 = -N(1 - v^2)\frac{\partial^2 w}{\partial y^2}$$

Using finite-difference approximation for the derivatives, we obtain

$$(M_y)_2 \cong -N\frac{(1 - v^2)}{\lambda_y^2}(2w_1 - 2w_2) = 0.1084qb^2$$

Substituting for w in the M_x and M_y equations in Fig. 18-9, we obtain

$$(M_x)_6 \cong 0.0374qb^2 \qquad \text{and} \qquad (M_y)_6 \cong 0.0777qb^2$$

Exact answers[10] to this problem are: $w_2 = 0.1404qb^4/Eh^3$, $(M_y)_2 = 0.112qb^2$, $(M_x)_6 = 0.039qb^2$, and $(M_y)_6 = 0.080qb^2$.

[10]Taken from reference in footnote 5 of this chapter.

18-10   BUCKLING OF THIN PLATES

We shall now use the method of finite differences to determine the critical values of a system of in-plane forces (with a given ratio between them) at which the flat form of equilibrium becomes unstable and the plate begins to buckle. This behavior is, of course, similar to buckling of struts.

We shall limit the discussion to the case of isotropic plates with no body forces and no transverse loading. We assume that the plate buckles *slightly* under the action of the in-plane forces and then determine the values that these forces must have in order to keep the plate in this slightly deflected shape.

Under these circumstances, a small block A (see Figs. 18-3 or 18-8b), with sides $dx = dy = 1$ and h, will be subjected to stress resultants indicated in Fig. 18-8b together with the forces $P_x = \sigma_x h$, $P_y = \sigma_y h$, and $P_{xy} = \tau_{xy} h$ in the directions of the stresses indicated in Fig. 18-3. Because of the deflected shape of the element, the forces P_x, P_y, and P_{xy} have a resultant in the z direction equal to

$$p = P_x \frac{\partial^2 w}{\partial x^2} + P_y \frac{\partial^2 w}{\partial y^2} + 2P_{xy} \frac{\partial^2 w}{\partial x \partial y} \tag{18-55}$$

Since there is no transverse loading ($q = 0$), the right-hand side of the equilibrium Eq. 18-28 becomes $-(q + p) = -p$, and consequently the governing differential Eq. 18-35 (which is derived from Eq. 18-28) becomes

$$\frac{\partial^4 w}{\partial x^4} + 2 \frac{\partial^4 w}{\partial x^2 \partial y^2} + \frac{\partial^4 w}{\partial y^4} = \frac{1}{N} \left(P_x \frac{\partial^2 w}{\partial x^2} + P_y \frac{\partial^2 w}{\partial y^2} + 2P_{xy} \frac{\partial^2 w}{\partial x \partial y} \right) \tag{18-56}$$

To solve for the buckling load using finite differences, we first determine the quantities $\{P\} = \{P_x, P_y, P_{xy}\}$ at nodes of a chosen mesh under assumed values of the external applied forces; this is achieved by solving Eq. 18-17, using the procedure discussed in Sec. 18-5. Because the external applied forces occur in a given ratio, we can assume that, at the critical loading, the P-forces have the values $\gamma\{P\}$ at the nodes and the problem then reduces to the determination of the value of the constant γ which corresponds to the critical load.

When Eq. 18-56 is multiplied by $N\lambda_x \lambda_y$ and then applied in finite-difference form at the nodes (satisfying the boundary conditions of the plate in bending, as discussed in Secs. 18-8 and 18-9), a system of simultaneous equations can be written in matrix form

$$[K]\{w\} \cong \gamma [A]\{w\} \tag{18-57}$$

where $[K]$ is a square symmetrical matrix considered previously in connection with Eq. 18-53. The elements of $[A]$ are obtained from the finite-difference coefficients of w in the derivatives

$$\frac{\partial^2 w}{\partial x^2} \qquad \frac{\partial^2 w}{\partial y^2} \qquad \text{and} \qquad \frac{2\partial^2 w}{\partial x \partial y}$$

multiplied by the known P-values.

Multiplying both sides of Eq. 18-57 by $[K]^{-1}$ (which is a flexibility matrix for the plate corresponding to transverse coordinates at the nodes), we obtain

$$[G]\{w\} \cong \beta\{w\} \tag{18-58}$$

where

$$\beta = \frac{1}{\gamma} \tag{18-59}$$

and

$$[G] = [K^{-1}][A] \tag{18-60}$$

The solution of Eq. 18-58 for the largest eigenvalue β gives the smallest value of γ, which corresponds to the critical buckling load.

We can see that the above procedure is similar to that used to determine the buckling load in columns (Sec. 17-10).

Example 18-3 The plate in Fig. 18-15 is simply supported along three edges and is free at one edge, and is of constant thickness h. Find the value σ of the uniform stress on edges CD and EF which causes buckling of the plate. Assume $v = 0.25$.

We assume that the smallest critical value of σ is $\sigma_{cr} = \gamma$ which causes the plate to buckle in a symmetrical form about the line AB. This assumption is made in order to simplify the example, but in a general case we should verify that the symmetrical buckling mode corresponds to the smallest critical value of σ.

At any node, the stresses are: $\sigma_y = -\gamma$ and $\sigma_x = \tau_{xy} = 0$. Thus, $P_y = -\gamma h$, and $P_x = P_{xy} = 0$. Hence Eq. 18-56 becomes

$$N\left(\frac{\partial^4 w}{\partial x^4} + \frac{2\partial^4 w}{\partial x^2 \partial y^2} + \frac{\partial^4 w}{\partial y^4}\right) = -\gamma h \frac{\partial^2 w}{\partial y^2}$$

Applying this equation in finite-difference form at each of the 10 nodes in one half of the symmetrical mesh shown in Fig. 18-15, and multiplying

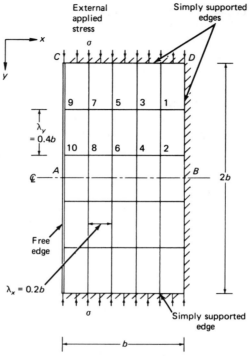

Fig. 18-15. Buckling load for the plate of Example 18-3 calculated by finite differences.

both sides of the first eight equations by $\lambda_x \lambda_y$ and of the last two by $\lambda_x \lambda_y/2$, we obtain (see Eq. 18-57)

$$[K]_{10 \times 10}\{w\}_{10 \times 1} = \gamma[A]_{10 \times 10}\{w\}_{10 \times 1}$$

The equivalent stiffness matrix $[K]$, derived by the use of the pattern of coefficients in Fig. 18-12, is

$$[K] = \frac{N\lambda_y}{\lambda_x^3}
\begin{array}{c}
 \\ 1 \\ 2 \\ 3 \\ 4 \\ 5 \\ 6 \\ 7 \\ 8 \\ 9 \\ 10
\end{array}
\begin{bmatrix}
\overset{1}{A-F-E} & \overset{2}{} & \overset{3}{} & \overset{4}{} & \overset{5}{} & \overset{6}{} & \overset{7}{} & \overset{8}{} & \overset{9}{} & \overset{10}{} \\
C+E & A-F+C & & & & & & & & \\
B & D & A-E & & & \text{symmetrical} & & & & \\
D & B+D & C+E & A+C & & & & & & \\
F & & B & D & A-E & & & & & \\
& F & D & B+D & C+E & A+C & & & & \\
& & F & & B & D & G-E & & & \\
\text{Elements not} & & & F & D & B+D & C+E & G+C & & \\
\text{shown are zero} & & & & F & & B & D & J-L & \\
& & & & & F & & H & I+H & K+L & J+K
\end{bmatrix}$$

where $\lambda_y = 0.4b$ and $\lambda_x = 0.2b$.

The matrix $[A]$ can be easily formed using the finite differences pattern of coefficients given in Fig. 17-2. Thus

$$[A] = -\frac{h\lambda_x}{\lambda_y}\begin{bmatrix} -2 & 1 & & & & & & \\ 1 & -1 & & & & & & \\ & & -2 & 1 & & & & \\ & & 1 & -1 & & & & \\ & & & & -2 & 1 & & \\ & & & & 1 & -1 & & \\ & & & & & & -2 & 1 \\ & & & & & & 1 & -1 \\ & \text{Elements not} & & & & & -1 & \tfrac{1}{2} \\ & \text{shown are zero} & & & & & \tfrac{1}{2} & -\tfrac{1}{2} \end{bmatrix}$$

Performing the matrix operations in Eq. 18-58 and solving for the largest eigenvalue, we obtain

$$\beta = 14.9047 \frac{h\lambda_x^4}{N\lambda_y^2}$$

whence the critical buckling stress (see Eq. 18-59) is

$$\sigma_{\mathrm{cr}} = \gamma = \frac{1}{14.9047}\frac{N\lambda_y^2}{h\lambda_x^4}$$

and, substituting $N = Eh^3/12(1 - v^2)$, we obtain

$$\sigma_{\mathrm{cr}} = 0.596E(h/b)^2$$

Timoshenko and Gere[11] give a value of $\sigma = 0.613E(h/b)^2$, obtained by an energy approach using a trigonometric series to represent the deflected surface of the buckled plate.

18-11   STIFFNESS MATRIX EQUIVALENT

In Chapter 17 and this chapter we have shown that the use of finite differences for solving differential equations relating the transverse load to the deflection in beams and slabs results in a system of simultaneous linear equations of the form

$$[K]\{D\} \cong \{F\} \tag{18-61}$$

where $[K]$ is a matrix formed by the finite-difference coefficients of the transverse deflections at the nodes $\{D\}$, and $\{F\}$ are nodal transverse loads (compare with Eqs. 17-31 and 18-53).

Comparing Eq. 18-61 with the fundamental equation

$$[S]\{D\} = \{F\} \tag{18-62}$$

[11] See S. P. Timoshenko and J. M. Gere, *Theory of Elastic Stability*, 2nd ed., McGraw-Hill, New York, 1961, p. 362.

we see that replacing the stiffness matrix $[S]$ by $[K]$ results in the approximate Eq. 18-61. The matrix $[K]$ can, therefore, be treated as equivalent to the stiffness matrix. The inverse of the equivalent stiffness matrix gives the approximate influence coefficients of transverse deflections. From these, influence coefficients of the stress resultants and reactions can be calculated.

The equivalent stiffness matrix $[K]$ is easily generated by the use of the finite-difference patterns of coefficients given in chapter 17 and this chapter. The matrix $[K]$ which corresponds to coordinates representing transverse deflections can be transformed to stiffness matrices corresponding to other sets of coordinates.

The principles discussed are general and can be applied to a variety of structures, such as continuous beams, frames, grids, slabs, and shells.

18-12 COMPARISON BETWEEN EQUIVALENT STIFFNESS MATRIX AND STIFFNESS MATRIX

Consider a beam on an elastic foundation (Fig. 18-16a) which for simplicity is assumed to have a constant flexural rigidity EI, and a constant foundation modulus k per unit length per unit deflection.

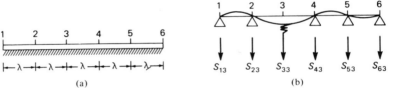

Fig. 18-16. Beam on an elastic foundation considered in Sec. 18-12. (a) Points of division in a beam on an elastic foundation. (b) Forces in the third column of the stiffness matrix $[S]$, Eq. 18-64.

Let the beam be divided into five equal intervals λ. Using the finite-difference pattern of coefficients of Fig. 17-5, we can generate the following equivalent stiffness matrix relating the nodal transverse deflections to nodal transverse forces:

$$[K] = \frac{EI}{\lambda^3} \begin{bmatrix} \left(1 + \dfrac{k\lambda^4}{2EI}\right) & & & & & \text{symmetrical} \\ -2 & \left(5 + \dfrac{k\lambda^4}{EI}\right) & & & & \\ 1 & -4 & \left(6 + \dfrac{k\lambda^4}{EI}\right) & & & \\ & 1 & -4 & \left(6 + \dfrac{k\lambda^4}{EI}\right) & & \\ & \text{Elements not shown} & 1 & -4 & \left(5 + \dfrac{k\lambda^4}{EI}\right) & \\ & \text{are zero} & & 1 & -2 & \left(1 + \dfrac{k^4}{2EI}\right) \end{bmatrix} \quad (18\text{-}63)$$

Replace the elastic foundation by elastic supports (springs) of stiffness equal to $k\lambda$ at nodes 2, 3, 4, and 5, and equal to $k\lambda/2$ at the end nodes 1 and 6. The elements of the stiffness matrix of this beam are the support reactions of the continuous beam in Fig. 18-16b when one support undergoes a unit downward displacement. These can be obtained from Appendix E. Thus,

$$[S] = \frac{EI}{\lambda^3}\begin{bmatrix}
\left(1.6077 + \frac{k\lambda^4}{2EI}\right) & & & & & \\
-3.6459 & \left(9.8756 + \frac{k\lambda^4}{EI}\right) & & \text{symmetrical} & & \\
2.5837 & -9.5024 & \left(14.0096 + \frac{k\lambda^4}{EI}\right) & & & \\
-0.6890 & 4.1340 & -10.5359 & \left(14.0096 + \frac{k\lambda^4}{EI}\right) & & \\
0.1723 & -1.0335 & 4.1340 & -9.5024 & \left(9.8756 + \frac{k\lambda^4}{EI}\right) & \\
-0.0287 & 0.1723 & -0.6890 & 2.5837 & -3.6459 & \left(1.6077 + \frac{k\lambda^4}{2EI}\right)
\end{bmatrix}$$

$$(18\text{-}64)$$

We can now verify that the equivalent stiffness matrix $[K]$ in Eq. 18-63 has all the general properties of stiffness matrices discussed in Sec. 4-6. The sum of the elements of any column (or row) of $[K]$ or $[S]$ is equal to the value $k\lambda$ (or $k\lambda/2$). This value is the additional force required at the displaced support to overcome the stiffness of the spring below this support, so that either matrix satisfies the equilibrium condition that the sum of the reactions produced by the displacement of one support of a continuous beam is zero. If k is zero, that is if the elastic foundation is removed, the beam becomes unstable. This is reflected in both the stiffness matrix and the equivalent stiffness matrix: their determinants vanish and both $[S]$ and $[K]$ become singular.

The two matrices $[K]$ and $[S]$, when inverted, give the flexibility matrices $[\bar{f}]$ and $[f]$, in which any element $\bar{f}_{ij} \cong f_{ij}$, the degree of approximation depending on the chosen node spacing. If, for example, we take $k\lambda^4/(EI) = 0.1$ in Eqs. 18-63 and 18-64 and invert, we obtain the following flexibility matrices, which can be compared

$$[\bar{f}] = [K]^{-1} = \frac{\lambda^3}{EI}\begin{bmatrix}
8.569 & & & & & \\
5.285 & 4.095 & & \text{symmetrical} & & \\
2.572 & 2.642 & 2.583 & & & \\
0.474 & 1.250 & 2.003 & 2.583 & & \\
-1.224 & -0.017 & 1.250 & 2.642 & 4.095 & \\
-2.782 & -1.224 & 0.474 & 2.572 & 5.284 & 8.569
\end{bmatrix}$$

$$(18\text{-}65)$$

and

$$[f] = [S]^{-1} = \frac{\lambda^3}{EI} \begin{bmatrix} 8.489 \\ 5.304 & 4.062 & & & \text{symmetrical} \\ 2.605 & 2.654 & 2.529 \\ 0.483 & 1.265 & 2.001 & 2.529 \\ -1.242 & -0.020 & 1.265 & 2.654 & 4.062 \\ -2.807 & -1.242 & 0.483 & 2.605 & 5.304 & 8.489 \end{bmatrix}$$

(18-66)

The methods of condensation of the stiffness matrix discussed in Sec. 4-5 can also be used with the equivalent stiffness matrix. For example, if in the beam of Fig. 18-16a the displacement at node 6 is prevented by the introduction of a support, the deflection $D_6 = 0$, and the stiffness matrix of the resulting structure is obtained by deletion of the 6th row and column from $[K]$. The last row of $[K]$ can then be used to find the force F_6 at the support introduced:

$$F_6 = [K_{61} K_{62} \ldots K_{65}] \begin{Bmatrix} D_1 \\ D_2 \\ \ldots \\ D_5 \end{Bmatrix}$$

(18-67)

18-13   GENERAL

The finite-difference method is suitable for the analysis of plates subjected to in-plane or transverse loading. In either case, the analysis involves the solution of a system of simultaneous linear equations formed following a pattern of finite-difference coefficients. Hence, we obtain the value of a stress function or of transverse deflections at node points defined by a mesh on the plate. The stresses in the plane-stress problem or the stress resultants in the bending problem are determined using finite-difference approximations of the derivatives of the stress function or deflections.

When the pattern of the finite-difference coefficients is known, the formation of the simultaneous equations is an easy process and the finite-difference solution can be used to advantage. All the steps of the analysis can be easily programmed for the use of computers. However, in some cases, such as with irregular boundaries, it is rather difficult to derive the finite-difference equations which suit the boundary conditions.

The stresses arising from in-plane forces and those due to bending can be treated separately when the transverse deflections are small compared with the dimensions of the plate and the in-plane forces are small. However,

---

[11]See p. 362 of reference in footnote 1 of this chapter.

large in-plane forces can cause buckling: the critical load can be found using the finite-difference method.

The finite-difference method can be used to solve also structural problems not included in this chapter, for example a stretched membrane, or torsion, shell, and vibration problems. In all these cases we find solution of a partial differential equation using finite differences in a manner similar to that used in the plate problems.

PROBLEMS

A finite-difference mesh is suggested in each of the following problems. This mesh was used to find the answers given at the end of this book. Coarse meshes were chosen in order to reduce the amount of computation for the solution. A high accuracy cannot be expected in all cases but this does not reduce the instructive value of the problems. If the calculations are performed by hand, the reader may use the given answers to check the simultaneous equations involved in the solution without actually solving them.

18-1   Find the variation in stress σ_x in section GH of the deep beam in Fig. 18-6c.

18-2   Solve Prob. 18-1 with the downward load P distributed on DC instead of AB.

18-3   Using finite differences with the mesh shown in the figure, find the stress σ_x and σ_y at node 4. What is the shear stress at A, the center of the rectangle 1-2-4-3?

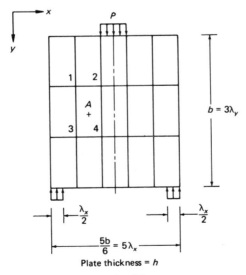

Prob. 18-3

18-4   Using finite differences with the mesh shown in the figure, find the deflection at the nodes and the bending moment M_y at nodes 1 and 2 of the plate shown.

The plate is subjected to a uniform transverse load q, has a constant thickness h, and Poisson's ratio $v = 0$.

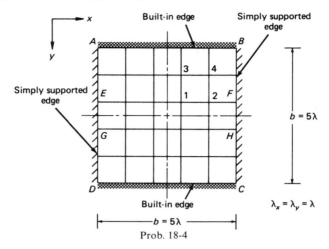

Prob. 18-4

18-5   Solve Prob. 18-4 with the width of the plate in the x direction of $2b$, and use the same mesh with $\lambda_x = 2\lambda_y$.

18-6   Solve Prob. 18-4 with the plate subjected (in lieu of q) to a uniform positive bending moment M on the two edges AD and BC. (*Hint:* Replace the edge moments by two equal and opposite transverse loads on the edge nodes and at nodes on a mesh line at distance λ_x from the edge.)

18-7   Solve Prob. 18-4 with the plate reinforced by beams of flexural rigidity $EI = 2 N\lambda$ along mesh lines EF and GH. The beams are assumed to be attached to the plate in the idealized fashion indicated in Fig. 18-13b and have the ends simply supported. What is the bending moment in beam EF at node 1?

18-8   Using finite differences with the mesh shown in the figure, find the deflections at the nodes and M_y at points A, B, and C of the plate in the figure. The plate has a constant thickness h, Poisson's ratio $v = 0$, and is loaded by one concentrated load P on the free edge DE, as shown.

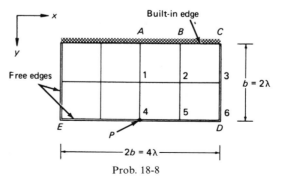

Prob. 18-8

18-9 Solve Prob. 18-8 with the plate reinforced by a beam along DE with free ends and a flexural rigidity $EI = 2N\lambda$. The beam is assumed to be attached to the plate in the idealized fashion indicated in Fig. 18-13a.

18-10 Using finite differences with the mesh shown in the figure, find the deflections at the nodes and the reaction at the support A of the plate in the figure. The plate has a constant thickness h, Poisson's ratio $v = 0$, is built-in along two edges, free at the other two and has a support at A which can provide a transverse reaction only. The load intensity on the plate varies as shown. (*Hint:* Apply the finite-difference equations in Fig. 18-12 at nodes 1, 2, and 3, solve for the deflections, and then use the equation in Fig. 18-12f to calculate the reaction at A.)

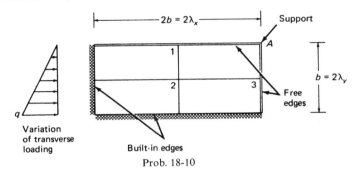

Prob. 18-10

18-11 Using a 4 × 4 mesh, find the buckling stress of a square plate subjected to a uniform compressive stress σ in the x and y directions, that is $\sigma_x = \sigma_y = -\sigma$ at all nodes. x and y being parallel to the plate edge. The plate is simply supported at its 4 edges, has side length b, thickness h, and $v = 0.3$. Assume that the plate buckles in a form symmetrical about central axes parallel to the sides.

CHAPTER 19

Finite-element method

19-1 INTRODUCTION

The finite-element method is widely used in structural analysis. The method is also used in a wide range of physical problems[1] including heat transfer, seepage, flow of fluids, and electrical and magnetic potential. In the finite-element method, a continuum is idealized as an assemblage of finite elements with specified nodes. The infinite number of degrees of freedom of the continuum is replaced by specified unknowns at the nodes.

In essence, the analysis of a structure by the finite-element method is an application of the displacement method. In frames, trusses and grids, the elements are bars connected at the nodes; these elements are considered to be one-dimensional. Two-dimensional or three-dimensional finite elements are used in the analysis of walls, slabs, shells and mass structures. The finite elements can have many shapes with nodes at the corners or on the sides (Fig. 19-1). The unknown displacements are nodal translations or rotations or derivatives of these.

The use of a computer is essential in the finite-element method because of the large number of degrees of freedom commonly involved. Chapters 24 and 25 discuss the computer analysis of structures; the approach there applies to structures which are composed of finite elements of any type. However, the matrices for individual elements, given explicitly in earlier chapters (e.g. Eqs. 4-5 to 4-7) and in Chapters 24 and 25, are for one-dimensional bar elements. The present chapter is mainly concerned with the generation of matrices for finite elements other than straight bars. The element matrices required are: the stiffness matrix, relating nodal forces to nodal displacements; the stress matrix, relating the stress or internal forces at any point within an element to its nodal displacements; and vectors of restraining nodal forces for use when the external forces are applied away from the nodes or when the element is subjected to temperature variation.

For finite elements other than bars, "exact" element matrices cannot be

---

[1]See Zienkiewicz, O. C., *The Finite Element Method in Engineering Science*, McGraw-Hill, London, 1977.

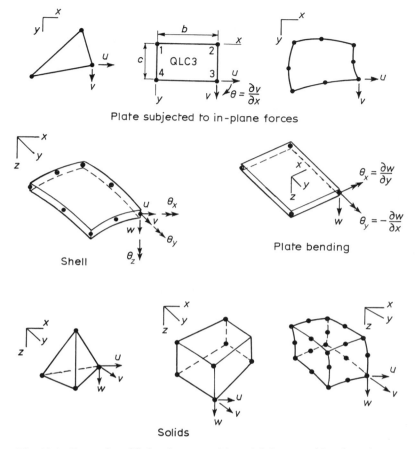

Fig. 19-1. Examples of finite elements with nodal degrees of freedom shown at a typical node only.

generated. The displacements (e.g. u and v) within an element are expressed in terms of the nodal displacements. *Assumed* displacement fields (e.g. a polynomial in x and y) are used. The corresponding strains are determined by differentiation and the stress by using Hooke's law. Use of the principle of virtual work or minimization of the total potential energy (Secs. 5-5, 5-6 and 8-10) with respect to the element nodal displacements gives the desired element matrices.

Use of displacement fields as described above is similar to the Rayleigh-Ritz method presented in Sec. 9-8, where the deflection of a beam is represented by a series of assumed functions with indeterminate parameters. The unknown parameters are derived by the principle of virtual work. They can also be derived by minimization of total potential energy. In the finite-element formulation, using assumed displacement fields, the indeterminate parameters are the nodal displacements. As in the Rayleigh-Ritz method, the solution obtained by the finite-element method is approximate; however, convergence to the exact solution is achieved as the number of unknown parameters is increased. In other words, when a finer finite-element mesh is used, more unknown displacements are involved and a greater accuracy is achieved.

The use of smaller elements for convergence is not required when the true displacement shapes can be derived and used to generate the element matrices. This is the case with prismatic bar elements.

This chapter gives the general procedures to derive matrices for finite elements of any type. This is explained by reference to bar elements and to plate elements subjected to in-plane forces or to bending. Other element types are discussed in Chapter 20.

It is most common to derive the matrices for finite elements using assumed displacement fields, and this method will be mainly used in this chapter. It is also possible to generate the element matrices using assumed fields for displacements and for stresses (or strains); this results in what are known as *hybrid elements*. The unknowns will then be the stress (or strain) parameters, in addition to the nodal displacements. However, all the unknown parameters other than the nodal displacements are eliminated at the element level; the global system of equations for the structure involves nodal displacements only. When the same number of equations is solved, a more accurate analysis is obtained with hybrid elements compared with the analysis using displacement-based finite elements. Thus the use of hybrid elements represents a convenient practical approach. Section 20-13 discusses the generation of stiffness and stress matrices for hybrid elements.

19-2 APPLICATION OF THE FIVE STEPS OF DISPLACEMENT METHOD

In essence, the analysis of a structure by the finite-element method is an application of the five steps of the displacement method summarized in Sec. 3-6. This analysis is explained below by reference to a plate subjected to in-plane forces (Fig. 19-2) which is idealized as an assemblage of rectangular finite elements. Each element has four nodes with two degrees of freedom per node, that is translations u and v in the x and y directions, respectively. The purpose of the analysis is to determine the stress components $\{\sigma\} = \{\sigma_x, \sigma_y, \tau_{xy}\}$ at O, the center of each element. The external loads are nodal forces $\{F_x, F_y\}_i$ at any node i, and body forces with intensities per unit volume of $\{p_x, p_y\}_m$ distributed over any element m. Other loadings can also be considered, such as the effects of temperature variation or of shrinkage (or swelling).

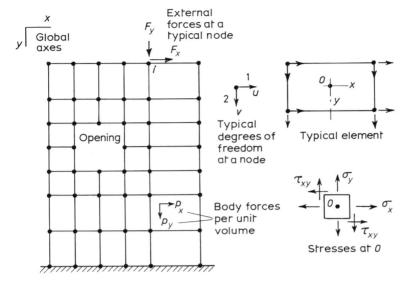

Fig. 19-2. Example of a finite-element model for the analysis of stresses in a wall subjected to in-plane forces.

The five steps in the analysis are as follows:

Step 1 Define the unknown degrees of freedom by two coordinates u and v at each node. The actions to be determined for any element m are $\{A\}_m \equiv \{\sigma\}_m = \{\sigma_x, \sigma_y, \tau_{xy}\}_m$.

Step 2 With the loading applied, determine the restraining forces $\{F\}$ to prevent the displacements at all coordinates. Also, for any element, determine $\{A_r\}_m \equiv \{\sigma_r\}_m$, which represents the values of the actions (the stresses) with the nodal displacements prevented. The stresses $\{\sigma_r\}$ are produced only when effects of temperature are considered; $\{\sigma_r\}$ due to body forces is commonly ignored.

The vector $\{F\}$ is considered equal to the sum of two vectors:

$$\{F\} = \{F_a\} + \{F_b\} \tag{19-1}$$

The vector $\{F_a\}$ is composed of the external nodal forces reversed in sign; $\{F_b\}$ is generated by assemblage of $\{F_b^*\}_m$ for individual elements. The vector $\{F_b^*\}_m$ is composed of forces at the nodes of element m in equilibrium with the external forces on the element body away from the nodes; in the case of temperature variation, $\{F_b^*\}_m$ represents a system of nodal forces in equilibrium producing stresses $\{\sigma_r\}$. (Eq. 4-42 gives the nodal forces due to temperature variation for a member of a plane frame; for other finite elements, see Sec. 19-61.)

Step 3   Generate the structure stiffness matrix $[S]$ by assemblage of the stiffness matrices $[S]_m$ of individual elements. Also, generate $[A_u]_m \equiv [\sigma_u]_m$, which represents the stress components at O in any element due to unit displacement introduced separately at the element nodal coordinates. For the example considered in Fig. 19-2, $[\sigma_u]_m$ will be a 3×8 matrix.

Step 4   Solve the equilibrium equations

$$[S]\{D\} = -\{F\} \tag{19-2}$$

This gives the structure nodal displacements $\{D\}$. In the example considered, the number of elements in $\{D\}$ is twice the number of nodes.

Step 5   Calculate the required stress components for each element:

$$\{A\}_m = \{A_r\}_m + [A_u]_m\{D\}_m \tag{19-3}$$

or

$$\{\sigma\}_m = \{\sigma_r\}_m + \{\sigma_u\}_m\{D\}_m \tag{19-4}$$

The values $\{D\}_m$ are the nodal displacements for the element m; in the example considered (Fig. 19-2), $\{D\}_m$ has eight values (subset of the structure displacement vector $\{D\}$).

Ignoring $\{\sigma_r\}$ caused by body forces (steps 2 and 5) produces an error which diminishes as the size of the finite elements is reduced. However, when the elements are bars (in framed structures), $\{\sigma_r\} \equiv \{A_r\}$ and the other matrices for individual members can be determined exactly; for this reason, $\{\sigma_r\}$ is commonly not ignored and the exact answers can be obtained without the need to reduce the size of the elements for convergence.

Assemblage of the structure load vector and of the stiffness matrix may be done by Eqs. 24-34 and 24-31. The nonzero elements of $[S]$ are generally limited to a band adjacent to the diagonal. This property, combined with the symmetry of $[S]$, is used to conserve computer storage and to reduce the number of computations. These topics and the methods of solution of Eq. 19-2 to satisfy displacement constraints are discussed in Chapters 24 and 25. Examples of displacement constraints are a zero or a prescribed value for the displacement at a support.

19-3   BASIC EQUATIONS OF ELASTICITY

The stresses and strains in an elastic body are related by Hooke's law, which can be written in the generalized form

$$\{\sigma\} = [d]\{\varepsilon\} \tag{19-5}$$

where $\{\sigma\}$ and $\{\varepsilon\}$ are generalized stress and strain vectors, respectively, and $[d]$ is a square symmetrical matrix referred to as the *elasticity matrix*.

The strain components are defined as derivatives of the displacement component by the generalized equation

$$\{\varepsilon\} = [\partial]\{f\} \qquad (19\text{-}6)$$

where $[\partial]$ is a matrix of the differential operator, and $\{f\}$ is a vector of functions describing the displacement field.

The symbols $\{\sigma\}$ and $\{\varepsilon\}$ will be used to represent stress or strain components in one-, two- and three-dimensional bodies. The displacement field $\{f\}$ will have one, two or three components: u, v and w in the direction of orthogonal axes x, y and z. The differential operator matrix $[\partial]$ will represent derivatives with respect to one, two or three of the variables x, y and z.

In a bar subjected to an axial force (Fig. 5-5a), each of $\{\sigma\}$ and $\{\varepsilon\}$ has one component and $[d]$ has one element equal to E, the modulus of elasticity. The strain ε is equal to du/dx, where u is displacement along the beam axis and x is distance measured in the same direction. Thus, we can use Eqs. 19-5 and 19-6 for the uniaxial stress state, with the symbols having the following meanings:

$$\{\sigma\} \equiv \sigma \quad \{\varepsilon\} \equiv \varepsilon \quad [d] \equiv E \quad \{f\} \equiv u \quad [\partial] = d/dx \quad (19\text{-}7)$$

We shall also use the symbol $\{\sigma\}$ to represent a vector of stress resultants. For the bar considered above, we can take $\{\sigma\} \equiv N$, the axial force on the bar cross section, and $[d] \equiv Ea$, where a is the cross-sectional area. Again Eqs. 19-5 and 19-6 apply, with the symbols having the following meanings:

$$\{\sigma\} \equiv N \quad \{\varepsilon\} \equiv \varepsilon \quad [d] \equiv Ea \quad \{f\} \equiv u \quad [\partial] = d/dx \quad (19\text{-}8)$$

The generalized Eqs. 19-5 and 19-6 apply to a bar in bending (Fig. 5-5b), with the symbols having the following meanings:

$$\{\sigma\} \equiv M \quad \{\varepsilon\} \equiv \varepsilon \quad [d] \equiv EI \quad \{f\} \equiv v \quad [\partial] = d^2/dx^2 \quad (19\text{-}9)$$

where M is the bending moment, ε is the curvature, I is the second moment of area about the centroidal axis, and v is the displacement in the y direction.

The product $\{\sigma\}^T\{\varepsilon\}$ integrated over the volume of an element appears in the strain energy Eq. 5-12 and in the virtual work Eq. 5-35, both of which will be frequently used. When $\{\sigma\}$ represents stress resultants over a cross section of a bar the integral over the volume has to be replaced by an integral over the length (see Eq. 5-28). For plates in bending, we shall use $\{\sigma\}$ to represent bending and twisting moments $\{M_x, M_y, M_{xy}\}$ and the integral will be over the area.

In the following subsections we shall apply the generalized Eqs. 19-5 and 19-6 in three stress states.

19-31   Plane Stress and Plane Strain

Consider a plate subjected to in-plane forces (Fig. 18-3). At any point, the stress, strain and displacement components are

$$\{\sigma\} = \{\sigma_x, \sigma_y, \tau_{xy}\} \qquad \{\varepsilon\} = \{\varepsilon_x, \varepsilon_y, \gamma_{xy}\} \qquad \{f\} = \{u, v\} \qquad (19\text{-}10)$$

The strains are defined as derivatives of $\{f\}$ by the generalized Eq. 19-6, with the differential operator matrix

$$[\partial] = \begin{bmatrix} \partial/\partial x & 0 \\ 0 & \partial/\partial y \\ \partial/\partial y & \partial/\partial x \end{bmatrix} \qquad (19\text{-}11)$$

The stress and strain vectors are related by generalized Hooke's law (Eq. 19-5) with the elasticity matrix $[d]$ given by one of the Eqs. 19-12 or 19-13.

When strain in the z direction is free to occur, $\sigma_z = 0$ and we have the state of plane stress. Deep beams and shear walls are examples of structures in a state of plane stress. When strain in the z direction cannot occur, $\varepsilon_z = 0$ and we have the state of plane strain. The state of plane strain occurs in structures which have a constant cross section perpendicular to the z direction and also have the dimension in the z direction much larger than those in the x and y directions. Concrete gravity dams and earth embankments are examples of structures in this category. The analysis of these structures may be performed for a slice of unit thickness in a state of plane strain.

For an isotropic material; the elasticity matrix in a plane-stress state is

$$[d] = \frac{E}{1 - v^2} \begin{bmatrix} 1 & v & 0 \\ v & 1 & 0 \\ 0 & 0 & (1-v)/2 \end{bmatrix} \qquad (19\text{-}12)$$

where E is the modulus of elasticity in tension or in compression and v is Poisson's ratio.

The elasticity matrix for a plane-strain state is

$$[d] = \frac{E(1-v)}{(1+v)(1-2v)} \begin{bmatrix} 1 & v/(1-v) & 0 \\ v/(1-v) & 1 & 0 \\ 0 & 0 & (1-2v)/2(1-v) \end{bmatrix} \qquad (19\text{-}13)$$

Equation 19-12 can be derived from Eq. 5-6 or by inversion of the square matrix in Eq. 18-10. Equation 19-13 can also be derived from Eq. 5-6 by setting $\varepsilon_z = 0$ (in addition to $\tau_{xz} = \tau_{yz} = 0$). The same equations give the normal stress in the z direction in the plane-strain state

$$\sigma_z = v(\sigma_x + \sigma_y) \qquad (19\text{-}14)$$

19-32 Bending of Plates

For a plate in bending (Fig. 18-8), the generalized stress and strain vectors are defined as

$$\{\sigma\} = \{M_x, M_y, M_{xy}\} \qquad (19\text{-}15)$$

$$\{\varepsilon\} = \{-\partial^2 w/\partial x^2, -\partial^2 w/\partial y^2, 2\partial^2 w/(\partial x\partial y)\} \qquad (19\text{-}16)$$

One component of body force and one component of displacement exist:

$$\{p\} \equiv \{q\} \qquad (19\text{-}17)$$

$$\{f\} \equiv \{w\} \qquad (19\text{-}18)$$

where q is force in the z-direction per unit area and w is deflection in the same direction. The generalized Eqs. 19-5 and 19-6 apply to a plate in bending, with

$$[\partial] = \left\{ \begin{array}{c} -\partial^2/\partial x^2 \\ -\partial^2/\partial y^2 \\ 2\partial^2/(\partial x\partial y) \end{array} \right\} \qquad (19\text{-}19)$$

For an orthotropic plate in bending, the elasticity matrix is

$$[d] = \frac{h^3}{12} \begin{bmatrix} E_x/(1-v_x v_y) & v_x E_y/(1-v_x v_y) & 0 \\ v_x E_y/(1-v_x v_y) & E_y/(1-v_x v_y) & 0 \\ 0 & 0 & G \end{bmatrix} \qquad (19\text{-}20)$$

where E_x, E_y are moduli of elasticity in tension or in compression in the x and y directions; v_x and v_y are Poisson's ratios; G is shear modulus of elasticity; and h is the plate thickness. When the plate is isotropic we set $E = E_x = E_y$ and $v = v_x = v_y$ in Eq. 19-20:

$$[d] = \frac{Eh^3}{12(1-v^2)} \begin{bmatrix} 1 & v & 0 \\ v & 1 & 0 \\ 0 & 0 & (1-v)/2 \end{bmatrix} \qquad (19\text{-}21)$$

It can be noted that, for a plate in bending, the generalized Eq. 19-5 is simply a condensed form of Eq. 18-32.

19-33 Three-Dimensional Solid

For a three-dimensional body, the generalized Eqs. 19-5 and 19-6 apply again, with the vectors $\{p\}$ and $\{f\}$ having the following meaning:

$$\{p\} = \{p_x, p_y, p_z\} \qquad (19\text{-}22)$$

$$\{f\} = \{u, v, w\} \qquad (19\text{-}23)$$

where $\{p_x, p_y, p_z\}$ are body forces per unit volume, and $\{u, v, w\}$ are translations in the x, y and z directions (Fig. 5-4). The stress and strain vectors $\{\sigma\}$ and $\{\varepsilon\}$ are defined by Eqs. 5-7 and 5-8, and the $[d]$ matrix is given by Eq. 5-11. The differential operator is

$$[\partial]^T = \begin{bmatrix} \partial/\partial x & 0 & 0 & \partial/\partial y & 0 & \partial/\partial z \\ 0 & \partial/\partial y & 0 & \partial/\partial x & \partial/\partial z & 0 \\ 0 & 0 & \partial/\partial z & 0 & \partial/\partial y & \partial/\partial x \end{bmatrix} \qquad (19\text{-}24)$$

19-4 DISPLACEMENT INTERPOLATION

In the derivation of element matrices, interpolation functions are required to define the deformed shape of the element. For a finite element of any type, the displacements at any point within the element can be related to the nodal displacement by the equation

$$\{f\} = [L]\{D^*\} \tag{19-25}$$

where $\{f\}$ is a vector of displacement components at any point; $\{D^*\}$ is a vector of nodal displacements; and $[L]$ is a matrix of functions of coordinates defining the position of the point considered within the element (e.g. x and y or ξ and η in two-dimensional finite elements).

The interpolation functions $[L]$ are also called shape functions; they describe the deformed shape of the element due to unit displacements introduced separately at each coordinate. Any interpolation function L_i represents the deformed shape when $D_i^* = 1$ while the other nodal displacements are zero.

The accuracy of a finite element depends upon the choice of the shape functions. These should satisfy conditions which will ensure convergence to correct answers when a finer finite-element mesh is used. The derivation of the shape functions and the conditions which should be satisfied are discussed in Secs. 19-8 and 19-9. Examples of shape functions with various types of elements are given below.

19-41 Straight Bar Element

For the axial deformation of a bar (Fig. 19-3), $\{f\} \equiv \{u\}$ is the translation in the x direction at any section, and $\{D^*\} = \{u_1, u_2\}$ is the translation at the two ends. The matrix $[L]$ may be composed of two linear interpolation functions:

$$[L] = [1 - \xi, \ \xi] \tag{19-26}$$

where $\xi = x/l$ and l is the bar length.

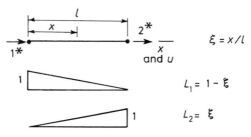

Fig. 19-3. Linear interpolation functions. Shape functions for a bar subjected to an axial force.

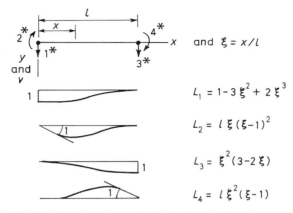

Fig. 19-4. Shape functions for deflection of a bar in bending.

The shape functions which can be used for a bar in bending are given in Fig. 19-4. For this element, $\{f\} \equiv \{v\}$ is the transverse deflection in the y direction at any section; the nodal displacements are defined as

$$\{D^*\} = \left\{ v_{x=0}, \left(\frac{dv}{dx}\right)_{x=0}, v_{x=l}, \left(\frac{dv}{dx}\right)_{x=l} \right\} \tag{19-27}$$

The shape functions $[L]$ can be the four cubic polynomials

$$[L] = [1 - 3\xi^2 + 2\xi^3 \,|\, l\xi(\xi - 1)^2 \,|\, \xi^2(3 - 2\xi) \,|\, l\xi^2(\xi - 1)] \tag{19-28}$$

Each shape function L_i in Eqs. 19-26 and 19-28 satisfies the requirement that $D_i^* = 1$ with other displacements zero. This requirement is sufficient to derive the functions (see Sec. 19-8).

The shape functions in Eqs. 19-26 and 19-28 correspond to the true deformed shapes of a prismatic bar (with shear deformation ignored). The same shape functions may be used to derive matrices for nonprismatic bars (see Examples 19-4 and 19-5).

We can recognize that the shape functions in Figs. 19-3 and 19-4 are the same as the influence lines for the nodal forces, reversed in sign. (Compare, for example, L_2 and L_4 in Fig. 19-4 with the influence lines of the member end-moments in Fig. 13-7d.)

The shape functions L_1 in Fig. 19-4 can be considered to be equal to the sum of the straight lines $1 - \xi$ and $(L_2 + L_4)/l$; the latter term represents the deflected shape corresponding to clockwise rotations, each equal to $1/l$ at the two ends. By similar reasoning, we can verify that $L_3 = \xi - (L_2 + L_4)/l$.

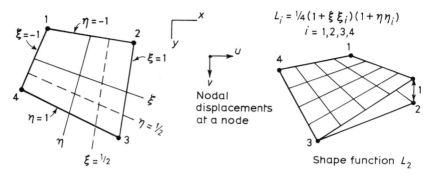

Fig. 19-5. Plane-stress or plane-strain quadrilateral element. Natural coordinates ξ and η define the location of any point. Pictorial view of a shape function (hyperbolic paraboloid).

19-42　Quadrilateral Element Subjected to In-Plane Forces

Figure 19-5 shows a quadrilateral element with corner nodes and two degrees of freedom per node. The element may be used in plane-stress and plane-strain analyses. In this case, the symbols in the generalized Eq. 19-25 have the following meaning:

$$[f] = \{u, v\} \tag{19-29}$$

where u and v are translations in the x and y directions, and

$$\{D^*\} = \{u_1, v_1, u_2, v_2, u_3, v_3, u_4, v_4\} \tag{19-30}$$

where u_i and v_i are translations at node i in the x and y directions. The $[L]$ matrix for the element in Fig. 19-5 is

$$[L] = \begin{bmatrix} L_1 & 0 & L_2 & 0 & L_3 & 0 & L_4 & 0 \\ 0 & L_1 & 0 & L_2 & 0 & L_3 & 0 & L_4 \end{bmatrix} \tag{19-31}$$

The function L_i is a sum of bilinear functions in the natural coordinates ξ and η, defined in Fig. 19-5. The value of L_i is unity at node i and zero at the other three nodes; any of the four shape functions may be expressed as

$$L_i = \frac{1}{4}(1 + \xi\xi_i)(1 + \eta\eta_i) \quad \text{with } i = 1, 2, 3, 4 \tag{19-32}$$

If the value of L_i is plotted perpendicular to the surface of the element, a hypersurface is obtained. Along the lines $\xi = \text{constant}$ or $\eta = \text{constant}$, the surface follows straight lines. The shape function L_2 is plotted in pictorial view in Fig. 19-5; at node 2, $\xi_i = 1$ and $\eta_i = -1$, and the function L_2 defining the hypersurface is obtained by substitution of the two values in Eq. 19-32.

Equation 19-32 represents one of a family of functions used for interpolation in the *isoparametric elements*, discussed in Sec. 20-2. We should note that the sum of the values of the four L_i functions at any point is unity.

19-43   Rectangular Plate-Bending Element

The rectangular element in Fig. 19-6, used in the analysis of plates in bending, has twelve degrees of freedom (three at each corner), defined as

$$\{D^*\} = \{(w, \theta_x, \theta_y)_1 \,\vdots\, (w, \theta_x, \theta_y)_2 \,\vdots\, (w, \theta_x, \theta_y)_3 \,\vdots\, (w, \theta_x, \theta_y)_4\} \qquad (19\text{-}33)$$

Here, the displacement $f \equiv w$ is the deflection at any point. The rotations θ are treated as derivatives of w:

$$\theta_x = \partial w/\partial y \qquad \theta_y = -\partial w/\partial x \qquad (19\text{-}34)$$

The shape functions for the rectangular bending element (Fig. 19-6) are:

$$[L] = [(1 - \xi)(1 - \eta) - \frac{1}{b}(L_3 + L_6) + \frac{1}{c}(L_2 + L_{11}) \qquad 1$$

$$c\eta(\eta - 1)^2(1 - \xi) \qquad 2$$

$$-b\xi(\xi - 1)^2(1 - \eta) \qquad 3$$

$$\xi(1 - \eta) + \frac{1}{b}(L_3 + L_6) + \frac{1}{c}(L_5 + L_8) \qquad 4$$

$$c\eta(\eta - 1)^2\xi \qquad 5$$

$$-b\xi^2(\xi - 1)(1 - \eta) \qquad 6 \qquad\qquad (19\text{-}35)$$

$$\xi\eta + \frac{1}{b}(L_9 + L_{12}) - \frac{1}{c}(L_5 + L_8) \qquad 7$$

$$c\eta^2(\eta - 1)\xi \qquad 8$$

$$-b\xi^2(\xi - 1)\eta \qquad 9$$

$$(1 - \xi)\eta - \frac{1}{b}(L_9 + L_{12}) - \frac{1}{c}(L_2 + L_{11}) \qquad 10$$

$$c\eta^2(\eta - 1)(1 - \xi) \qquad 11$$

$$-b\xi(\xi - 1)^2\eta] \qquad 12$$

where b and c are lengths of element sides; $\xi = x/b$ and $\eta = y/c$. The functions $L_2, L_3, L_5, L_6, L_8, L_9, L_{11}, L_{12}$ are shape functions given explicitly in Eq. 19-35 on lines 2, 3, 5, 6, 8, 9, 11 and 12, respectively; they correspond to unit rotations. Pictorial views of three deflected shapes, L_4, L_5 and L_6, corresponding to unit displacements at node 2, are included in Fig. 19-6. It can be seen that L_6 is zero along three edges, while along the fourth edge (1–2) the function is the same as

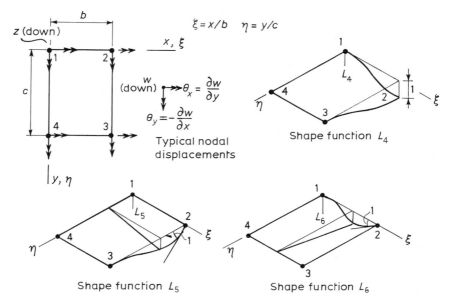

$\xi = x/b \quad \eta = y/c$

$\theta_x = \dfrac{\partial w}{\partial y}$

$\theta_y = -\dfrac{\partial w}{\partial x}$

Typical nodal displacements

Shape function L_4

Shape function L_5

Shape function L_6

Fig. 19-6. Plate-bending element.

the shape function for a beam (compare with L_4 in Fig. 19-4, reversed in sign). Along any line $\xi = $ constant, L_6 varies linearly, as shown in Fig. 19-6. Similarly, L_5 has the same shape as a deflected bar along the edge 3–2 and varies linearly along any line $\eta = $ constant.

The shape functions L_1, L_4, L_7, L_{10}, corresponding to $w = 1$ at the corners, are expressed as the sum of bilinear shape functions (similar to the one shown in Fig. 19-5) and the shape functions corresponding to rotations at the nodes equal to $\pm (1/b)$ or $\pm (1/c)$. Each of the four functions has values along two edges which are the same as the shape functions of a bar (L_1 or L_3 in Fig. 19-4). At any point, the sum $L_1 + L_4 + L_7 + L_{10}$ is equal to unity.

We should note that the deflected surface defined by any of the twelve shape functions does not generally have a zero slope normal to the element edges. This can produce incompatibility of slopes in adjacent elements; the effects of these incompatibilities will be discussed in Sec. 19-9.

19-5 STIFFNESS AND STRESS MATRICES FOR DISPLACEMENT-BASED ELEMENTS

Equation 19-25 defines the element displacement field in terms of the nodal displacement. By appropriate differentiation, we can derive the strain

(Eq. 19-6):

$$\{f\} = [L]\{D^*\} \tag{19-36}$$

$$\{\varepsilon\} = [\partial][L]\{D^*\} \tag{19-37}$$

Thus the strain at any point in a displacement-based element is

$$\{\varepsilon\} = [B]\{D^*\} \tag{19-38}$$

where

$$[B] = [\partial][L] \tag{19-39}$$

The matrix $[B]$ may be referred to as the nodal displacement-strain transformation matrix. Any column j of $[B]$ represents the strain components $\{\varepsilon_{uj}\}$ due to $D_j^* = 1$. Use of Hooke's law (Eq. 19-5) gives the stress at any point in a displacement-based element:

$$\{\sigma\} = [d][B]\{D^*\} \tag{19-40}$$

or

$$\{\sigma\} = [\sigma_u]\{D^*\} \tag{19-41}$$

where $[\sigma_u]$ is the *stress matrix* for the element:

$$[\sigma_u] = [d][B] \tag{19-42}$$

The elements of any column j of $[\sigma_u]$ are the stress components at any point due to $D_j^* = 1$.

An element S_{ij}^* of the stiffness matrix is the force at coordinate i corresponding to unit displacement at j. S_{ij}^* can be determined by the unit-displacement theorem (Eq. 5-39):

$$S_{ij}^* = \int_v \{\sigma_{uj}\}^T \{\varepsilon_{ui}\} \, dv \tag{19-43}$$

where $\{\sigma_{uj}\}$ represents the "actual" stresses at any point due to unit displacement at j; $\{\varepsilon_{ui}\}$ represents the strains at the same point corresponding to unit virtual displacement at i; and dv is an elemental volume.[2] The integral over the volume is replaced by an integral over the length in the case of a bar and over the area in the case of a plate. For this purpose, the symbols $\{\sigma\}$ and $\{\varepsilon\}$ in Eq. 19-43 represent generalized stress and strain, respectively. For example, in a bar, $\{\sigma\}$ represents internal forces at a section; in a plate in bending, $\{\varepsilon\}$ represents curvatures (Eq. 19-16).

Using the shape functions $[L]$ to determine the actual stresses and the virtual strains via Eqs. 19-39 and 19-42, substitution in Eq. 19-43 gives any

---

[2]The symbol dv representing elemental volume should not be confused with v, which represents a translational displacement.

element of the stiffness matrix:

$$S_{ij}^* = \int_v \{B\}_j^T [d] \{B\}_i \, dv \qquad (19\text{-}44)$$

where $\{B\}_i$ and $\{B\}_j$ are the ith and jth columns of $[B]$. The stiffness matrix of a finite element is given by:

$$[S^*] = \int_v [B]^T [d][B] \, dv \qquad (19\text{-}45)$$

In Eqs. 19-43 and 19-44 we are accepting an assumed displacement field, namely L_j, as actual. However, in general the assumed shape is different from the actual. What we are doing then is tantamount to imposing the assumed configuration by the application of small distributed forces on the element body in addition to the nodal forces. The distributed forces have the effect of changing the actual configuration to the assumed shape; these forces, not accounted for, cause the stiffness calculated by Eq. 19-45 to be an overestimate. In other words, a finite-element analysis in which the element stiffness matrices are derived by the above procedure is expected to give smaller displacements than the actual ones.

19-6   ELEMENT LOAD VECTORS

The vector of restraining forces, to be used in the equilibrium Eq. 19-2, includes a component $\{F_b\}$ representing the forces in equilibrium with the external loads applied on the body of the elements away from the nodes (Eq. 19-1). Considering a single element, the equilibrant at node j to the body forces can be determined by

$$F_{bj} = - \int_v \{L_j\}^T \{p\} \, dv \qquad (19\text{-}46)$$

where $\{p\}$ represents the magnitudes per unit volume of forces applied in the same directions as the displacements $\{f\}$.

Equation 19-46 can be explained by the principle of virtual work (Eq. 5-35) and also, succinctly, by Betti's theorem (Sec. 8-2). Here, the body forces and the nodal equilibrants form one system; the element subjected only to those nodal forces which produce the displacement configuration L_j represents the second system. According to Betti's theorem, the work of the forces of the first system during displacements by the second system is equal to the work of the second system during displacements by the first system. Now, the second quantity is zero because the second system has forces at the nodes only, and the nodal displacements in the first system are all zero.

By the use of Eq. 19-46, we are treating the shape function L_j as the influence line (or influence surface) of the nodal force at j, reversed in sign (see Sec. 13-3). We should remember that an approximation is involved in Eq. 19-46 by the acceptance of an assumed deflected shape as the actual displacement field.

The vector of nodal forces in equilibrium with the forces applied on the element away from the nodes is

$$\{F_b^*\} = -\int_v [L]^T\{p\}\,dv \qquad (19\text{-}47)$$

This vector is referred to as the element *consistent load vector* because the same shape functions $[L]$ are used to generate $[S^*]$ and $\{F_b^*\}$. The superscript $*$ is used here to refer to local coordinates of an individual element.

When the external forces are applied to the surface of the element, the integral in Eq. 19-47 should be taken over the area of the element. When concentrated forces act, the integral is replaced by a summation of the forces multiplied by the values of $[L]^T$ at the load positions.

19-61   Analysis of Effects of Temperature Variation

When an element is subjected to temperature variation (or to shrinkage), with the displacements restrained, the stresses at any point are given by (Eq. 19-9):

$$\{\sigma_r\} = -[d]\{\varepsilon_0\} \qquad (19\text{-}48)$$

where $\{\varepsilon_0\}$ represents the strains which would exist if the change in volume were free to occur. In a two-dimensional plane-stress or plane-strain state, a rise of temperature of T degrees produces the free strain

$$\{\varepsilon_0\} = \alpha T \begin{Bmatrix} 1 \\ 1 \\ 0 \end{Bmatrix} \qquad (19\text{-}49)$$

where α is the coefficient of thermal expansion.

For an element subjected to volume change, the consistent vector of restraining forces is given by

$$\{F_b^*\} = -\int_v [B]^T[d]\{\varepsilon_0\}\,dv \qquad (19\text{-}50)$$

Again, the unit-displacement theory may be used to derive Eq. 19-50. With the actual stress being $\{\sigma_r\} = -[d]\{\varepsilon_0\}$ and the virtual strain being $\{\varepsilon_{uj}\} = \{B\}_j$, Eq. 5-39 gives the jth element of the consistent load vector.

In most cases, the integrals involved in generating the stiffness matrix and the load vectors for individual elements are evaluated numerically using Gaussian quadrature (see Sec. 20-9).

19-7 DERIVATION OF ELEMENT MATRICES BY MINIMIZATION OF TOTAL POTENTIAL ENERGY

The element stiffness matrix $[S^*]$ and the consistent load vector $\{F_b^*\}$ (Eqs. 19-45 and 19-47) can be derived by the principle of total potential energy (Sec. 8-10). Consider a finite element subjected to body forces $\{p\}$ and nodal forces $\{Q^*\}$. The total potential energy is defined as the sum of potential energy and strain energy (Eq. 8-51):

$$\Phi = -\{D^*\}^T[Q^*] - \int_v \{f\}^T\{p\}\,dv + \frac{1}{2}\int_v \{\sigma\}^T\{\varepsilon\}\,dv \qquad (19\text{-}51)$$

where $\{f\}$ are displacement components at any point; $\{p\}$ are body forces per unit volume applied in the same directions as $\{f\}$; $\{\sigma\}$ are stresses; $\{\varepsilon\}$ are strains; and $\{D^*\}$ are nodal displacements. Substitution of Eqs. 19-36, 19-38 and 19-40 in the above equation gives

$$\Phi = -\{D^*\}^T\{Q^*\} - \int_v \{D^*\}^T[L]^T\{p\}\,dv$$

$$+\frac{1}{2}\int_v \{D^*\}^T[B]^T[d][B]\{D^*\}\,dv \qquad (19\text{-}52)$$

The principle of minimum total potential energy can be expressed as (Eq. 8-54)

$$\frac{\partial\Phi}{\partial D_i^*} = 0 \qquad (19\text{-}53)$$

where the subscript i refers to any of the nodal displacements. Partial differentiation[3] with respect to each nodal displacement gives

$$\frac{\partial\Phi}{\partial\{D^*\}} = -\{Q^*\} - \int_v [L]^T\{p\}\,dv + \int_v [B]^T[d][B]\{D^*\}\,dv = 0 \quad (19\text{-}54)$$

Equation 19-54 can be rewritten in the form

$$[S^*]\{D^*\} = -\{F^*\} \qquad (19\text{-}55)$$

where $\{F^*\}$ are nodal forces which would prevent the nodal displacements. The restraining forces are the sum of the nodal forces $\{Q^*\}$ in a reversed direction and of the nodal equilibrants of the body forces. Thus

$$\{F^*\} = \{F_a^*\} + \{F_b^*\} \qquad (19\text{-}56)$$

where $\{F_a^*\} = -\{Q^*\}$, and $\{F_b^*\}$ is the element consistent load vector. The

---

[3]It can be shown that if a scalar quantity y is expressed as a sum of products of matrices $y = (1/2)\{x\}^T[a]\{x\} + \{x\}^T\{b\}$, where $[a]$ and $\{b\}$ are constants and $[a]$ is symmetrical, differentiation of y with respect to x_i for $i = 1, 2, \ldots$ gives $\partial y/\partial x = \{\partial y/\partial x_1, \partial y/\partial x_2, \ldots\} = [a]\{x\} + \{b\}$

matrix $[S^*]$ is the element stiffness matrix. Combining Eqs. 19-55 and 19-56 and comparing with Eq. 19-54, we obtain, by analogy, Eqs. 19-47 and 19-45.

19-8   DERIVATION OF SHAPE FUNCTIONS

The displacement field $\{f\}$ may be expressed as polynomials of the coordinates x and y (or ξ and η) defining the position of any point. For example, the deflection w in a plate-bending element or the translations u and v in a plane-stress or a plane-strain element may be expressed as

$$f(x, y) = [1, x, y, x^2, \ldots]\{A\} = [P]\{A\} \qquad (19\text{-}57)$$

where $\{A\}$ is a vector of constants, yet to be determined; and $[P]$ is a matrix of polynomial terms, the number of which equals the number of nodal degrees of freedom. Pascal's triangle (Fig. 19-7) can be used to select the polynomial terms to be included in $[P]$. In general, the lower-degree terms are used. Examples of the polynomial terms used in several elements are given later in this section.

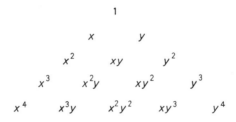

Fig. 19-7. Pascal's triangle.

The nodal displacements $\{D^*\}$ can be related to the constants $\{A\}$ by substituting x and y (or ξ and η) values at the nodes in Eq. 19-57 (or its derivatives). This gives

$$\{D^*\} = [C]\{A\} \qquad (19\text{-}58)$$

The elements of $[C]$ are known values depending upon (x_i, y_i), with $i = 1, 2, \ldots$ referring to the node numbers. The undetermined constants $\{A\}$ can now be expressed in terms of $\{D^*\}$ by inversion:

$$\{A\} = [C]^{-1}\{D^*\} \qquad (19\text{-}59)$$

Substituting Eq. 19-59 into Eq. 19-57, and by analogy of the resulting equation with Eq. 19-36, we obtain the shape functions

$$[L] = [P][C]^{-1} \qquad (19\text{-}60)$$

As an example of polynomial selection, let us consider the bar element shown in Fig. 19-3, which is subjected to an axial force. With two degrees of freedom, $[P]$ has only two terms:

$$[P] = [1, \quad \xi] \tag{19-61}$$

For a bar in bending (Fig. 19-4),

$$[P] = [1, \quad \xi, \quad \xi^2, \quad \xi^3] \tag{19-62}$$

For a plate element subjected to in-plane forces (Fig. 19-5), each of the displacements u and v is associated with four nodal displacements. A polynomial with four terms is used for each of u and v:

$$[P] = [1 \quad \xi \quad \eta \quad \xi\eta] \tag{19-63}$$

For the plate-bending element in Fig. 19-6, with twelve degrees of freedom, we express the deflection as $w = [P]\{A\}$, with

$$[P] = [1 \quad \xi \quad \eta \quad \xi^2 \quad \xi\eta \quad \eta^2 \quad \xi^3 \quad \xi^2\eta \quad \xi\eta^2 \quad \eta^3 \quad \xi^3\eta \quad \xi\eta^3] \tag{19-64}$$

For elements with two or more variables, such as x and y (or x, y and z), the terms included in $[P]$ should be invariant if the reference axes x and y (or x, y and z) are interchanged. Thus in Eq. 19-63 we should not replace the terms $\xi\eta$ by ξ^2 or by η^2. Similarly, in Eq. 19-64 we should not replace $\xi^3\eta$ or $\xi\eta^3$ by $\xi^2\eta^2$. In other words, $[P]$ includes symmetrical terms from Pascal's triangle (Fig. 19-7). When this requirement is satisfied, the element does not have a "preferred" direction. In consequence, the use of such an element in the analysis of the structure shown in Fig. 19-2 will give the same answers regardless of whether the global axes x and y are as shown or are rotated through 90° so that x becomes vertical.

The invariance requirement is relaxed for the rectangular plate element subjected to in-plane forces included in Fig. 19-1. The element has four nodes with three nodal displacements per node: $u, v, \partial v/\partial x$. The inclusion of $\partial v/\partial x$ (but not of other derivatives of u and v) makes the element behave differently in the x and y directions. The polynomials used for u and for v are different:

$$u(x, y) = [1 \ x \ y \ xy]\{A\}_u \tag{19-65}$$

$$v(x, y) = [1 \ x \ y \ x^2 \ xy \ x^3 \ x^2y \ x^3y]\{A\}_v \tag{19-66}$$

The variation of v in the x direction is cubic, while a linear variation is used for v in the y direction and for u in both the x and y directions. The shape functions for the element are given in Eq. 19-67 (see Example 19-3).

This element[4] gives excellent accuracy when used for structures which

---

[4]The rectangular element considered here is a special case of the quadrilateral element referred to as QLC3. See Sisodiya, R., Cheung, Y. K. and Ghali, A. "New Finite Element with Application to Box Girder Bridges," *Proceedings, The Institution of Civil Engineers*, London, Supplement 1972, Paper 7495, pp. 207–225. The combination of QLC3 with the rectangular bending element shown in Fig. 19-6 gives a good shell element with three translations and three rotations per node for the analysis of spatial structures (Sec. 20-101).

have beam-like behavior, e.g. folded plates and box girders. For this use, the element local x axis must be in the direction of the "beam". For comparison of accuracy of results of this element with other elements, see Prob. 20-18.

Example 19-1   Derive the shape functions for the bar element in Fig. 19-4.

The deflection v is expressed as a cubic polynomial of ξ, where $\xi = x/l$ (Eqs. 19-57 and 19-62):

$$v = [1 \; \xi \; \xi^2 \; \xi^3]\{A\} = [P]\{A\}$$

The nodal displacements defined by Eq. 19-27 are substituted in the above equations to give

$$\{D^*\} = \begin{bmatrix} 1 & 0 & 0 & 0 \\ 0 & 1/l & 0 & 0 \\ 1 & 1 & 1 & 1 \\ 0 & 1/l & 2/l & 3/l \end{bmatrix}\{A\}$$

Inversion of the square matrix in the above equation gives

$$[C]^{-1} = \begin{bmatrix} 1 & 0 & 0 & 0 \\ 0 & l & 0 & 0 \\ -3 & -2l & 3 & -l \\ 2 & l & -2 & l \end{bmatrix}$$

The product $[P][C]^{-1}$ gives the shape functions $[L]$ (Eq. 19-28).

Example 19-2   Derive the shape functions for the plane-stress or plane-strain quadrilateral element shown in Fig. 19-5.

We have u or $v = [P]\{A\}$, with $[P]$ given in Eq. 19-63. Substituting for ξ and η by their values at the four corners, we write for u:

$$\begin{Bmatrix} u_1 \\ u_2 \\ u_3 \\ u_4 \end{Bmatrix} = \begin{bmatrix} 1 & -1 & -1 & 1 \\ 1 & 1 & -1 & -1 \\ 1 & 1 & 1 & 1 \\ 1 & -1 & 1 & -1 \end{bmatrix}\{A\}$$

Inversion of $[C]$, which is the square matrix in this equation, gives

$$[C]^{-1} = (1/4)\begin{bmatrix} 1 & 1 & 1 & 1 \\ -1 & 1 & 1 & -1 \\ -1 & -1 & 1 & 1 \\ 1 & -1 & 1 & -1 \end{bmatrix}$$

Substitution in Eq. 19-60 gives

$$[L_1, L_2, L_3, L_4] = \tfrac{1}{4}[1 - \xi - \eta + \xi\eta, \quad 1 + \xi - \eta - \xi\eta, \quad 1 + \xi + \eta + \xi\eta,$$
$$1 - \xi + \eta - \xi\eta]$$

which is the same as Eq. 19-32. The same shape functions apply to v.

Example 19-3   Using the polynomials in Eqs. 19-66, derive the shape functions for the displacement v in a rectangular plate element QLC3 subjected to in-plane forces (Fig. 19-1).

The nodal displacements are defined as:

$$\{D^*\} = \left\{ \left(u, v, \frac{\partial v}{\partial x} \right)_1, \left(u, v, \frac{\partial v}{\partial x} \right)_2, \left(u, v, \frac{\partial v}{\partial x} \right)_3, \left(u, v, \frac{\partial v}{\partial x} \right)_4 \right\}$$

Using the symbols $\xi = x/b$ and $\eta = y/c$, Eq. 19-66 can be written as $v = [P]\{\overline{A}\}$, with

$$[P] = [1 \ \xi \ \eta \ \xi^2 \ \xi\eta \ \xi^3 \ \xi^2\eta \ \xi^3\eta]$$

The nodal displacements associated with v are:

$$\{D_v^*\} = \{D_2^*, D_3^*, D_5^*, D_6^*, D_8^*, D_9^*, D_{11}^*, D_{12}^*\}$$

Substituting for ξ and η by their values at the nodes, we can write $\{D_v^*\} = [C]\{\overline{A}\}$, with

$$[C] = \begin{array}{c} 2 \\ 3 \\ 5 \\ 6 \\ 8 \\ 9 \\ 11 \\ 12 \end{array} \left[\begin{array}{cccccccc} 1 & 0 & 0 & 0 & 0 & 0 & 0 & 0 \\ 0 & 1/b & 0 & 0 & 0 & 0 & 0 & 0 \\ 1 & 1 & 0 & 1 & 0 & 1 & 0 & 0 \\ 0 & 1/b & 0 & 2/b & 0 & 3/b & 0 & 0 \\ 1 & 1 & 1 & 1 & 1 & 1 & 1 & 1 \\ 0 & 1/b & 0 & 2/b & 1/b & 3/b & 2/b & 3/b \\ 1 & 0 & 1 & 0 & 0 & 0 & 0 & 0 \\ 0 & 1/b & 0 & 0 & 1/b & 0 & 0 & 0 \end{array} \right]$$

Inversion of $[C]$ and substitution in Eq. 19-60 gives the eight shape functions associated with v: $L_2, L_3, L_5, L_6, L_8, L_9, L_{11}, L_{12}$. For reference, we give the complete shape functions for the element as follows:

$$\begin{array}{lll} L_1 = (1 - \xi)(1 - \eta) & L_2 = (1 - 3\xi^2 + 2\xi^3)(1 - \eta) & L_3 = b\xi(\xi - 1)^2(1 - \eta) \\ L_4 = \xi(1 - \eta) & L_5 = \xi^2(3 - 2\xi)(1 - \eta) & L_6 = b\xi^2(\xi - 1)(1 - \eta) \\ L_7 = \xi\eta & L_8 = \xi^2(3 - 2\xi)\eta & L_9 = b\xi^2(\xi - 1)\eta \\ L_{10} = (1 - \xi)\eta & L_{11} = (1 - 3\xi^2 + 2\xi^3)\eta & L_{12} = b\xi(\xi - 1)^2\eta \end{array}$$

$$(19\text{-}67)$$

These functions can be used to express u and v by the equation $\{u, v\} = [L]\{D^*\}$ (see Eq. 19-25), with

$$[L] = \begin{bmatrix} L_1 & 0 & 0 & L_4 & 0 & 0 & L_7 & 0 & 0 & L_{10} & 0 & 0 \\ 0 & L_2 & L_3 & 0 & L_5 & L_6 & 0 & L_8 & L_9 & 0 & L_{11} & L_{12} \end{bmatrix} \quad (19\text{-}68)$$

19-9 CONVERGENCE CONDITIONS

In principle, it is possible to use any continuous shape function for the displacement field of a finite element; however, polynomials are commonly used. For monotonous convergence to the correct answer as smaller elements are used, the shape functions should satisfy the following three requirements:

1 The displacements of adjacent elements along a common boundary must be identical.

It can be seen that this condition is satisfied in the plane-stress or plane-strain element in Fig. 19-5 and in the plate-bending element in Fig. 19-6. Along any side 1–2, the translations u and v in Fig. 19-5 or the deflection w in Fig. 19-6 are functions of the nodal displacements at 1 and 2 only. It thus follows that two adjacent elements sharing nodes 1 and 2 will have the same nodal displacements at the two nodes and the same u and v or w along the line 1–2.

In some cases, the first partial derivatives of the element should also be compatible. This condition needs to be satisfied in plate-bending elements but not in plane-stress or plane-strain elements.

Two adjacent elements of the type shown in Fig. 19-6 have the same deflection and hence the same slope along the common edge. However, normal to the common edge, the tangents of the deflected surfaces of the two elements have the same slope only at the nodes. Away from the nodes, the tangents normal to a common edge can have slopes differing by an angle α. The quantity $(1/2)\int M_n \alpha \, ds$ represents, for the assembled structure, strain energy not accounted for in the process of minimizing the total potential energy (see Eq. 19-52); here M_n is the resultant of stresses normal to the common edge and ds is an elemental length of the edge.

The rule to ensure convergence is that the compatibility be satisfied for $[L]$ and its derivatives of order one less than the derivatives included in $[B]$. For a plate-bending element, $[B]$ includes second derivatives (Eqs. 19-39 and 19-19): $-\partial^2 w/\partial x^2, -\partial^2 w/\partial y^2, 2\partial^2 w/(\partial x \partial y)$. Thus compatibility is required for $\partial w/\partial x$ and $\partial w/\partial y$ in addition to w.

Several elements, said to be incompatible or nonconforming, such as the element in Fig. 19-6, do not satisfy the requirement of compatibility of displacement derivatives, and yet some of these elements give excellent results. An explanation of this behavior is that the excess stiffness, which is a characteristic of displacement-based finite elements, is compensated by the increase in flexibility resulting from a lack of compatibility of slopes.

Nonconforming elements converge towards the correct answer when the incompatibilities disappear as the mesh becomes finer and the strains within the element tend to constants.

2 When the nodal displacements $\{D^*\}$ correspond to rigid-body motion, the strains $[B]\{D^*\}$ must be equal to zero.

This condition is easily satisfied when the polynomial matrix $[P]$ in Eq. 19-57 includes the lower-order terms of Pascal's triangle. For example, for the plate-bending element (Fig. 19-6), inclusion of the terms 1, x, y would allow w to be represented by the equation of an inclined plane.

We can verify that the shape functions in Eq. 19-35 for the element shown in Fig. 19-6 allow translation as a rigid body by setting $w = 1$ at the four corners and $\theta_x = \theta_y = 0$ at all nodes; Eq. 19-25 will then give $w = 1$, representing unit downward translation. (This is because $L_1 + L_4 + L_7 + L_{10} = 1$, as noted earlier.) To check that the shape functions as allow rigid-body rotation, let $w = 1$ at nodes 1 and 2 and $\theta_x = -1/c$ at the four nodes of the element, while all other nodal displacements are zero. It can be seen that substitution of these nodal displacements and of Eq. 19-35 in Eq. 19-25 gives $w = 1 - \eta$, which is the equation of the plane obtained by rotation of the element through an angle $\theta_x = -1/c$ about the edge 4–3. In a similar way, we can verify that the shape functions allow a rigid-body rotation $\theta_y = $ constant.

3 The shape functions must allow the element to be in a state of constant strain.

This is required because, as the elements become smaller, the strains within individual elements tend to constants. Thus a smooth curve (or surface) representing the strain variation can be approximated by step variation.

This requirement will be satisfied when the polynomial $[P]$ in Eq. 19-57 includes the lower terms which contribute to the strain. For example, for the rectangular bending element shown in Fig. 19-6, the strains are $\{\varepsilon\} = \{-\partial^2 w/\partial x^2, -\partial^2 w/\partial y^2, 2\partial^2 w/(\partial x \partial y)\}$; the terms x^2, xy and y^2 of Pascal's triangle must be included in $[P]$.

We can verify that the shape functions in Eq. 19-35 allow a constant curvature, i.e. $-\partial^2 w/\partial x^2 = $ constant, by setting $\theta_y = 1$ at nodes 1 and 4 and $\theta_y = -1$ at nodes 2 and 3 while the remaining nodal displacements are equal to zero. Substitution in Eq. 19-36 gives $w = b\xi(1 - \xi)$, which represents the surface of a cylinder with a constant curvature of $2/b$.

For the same element, $2\partial^2 w/(\partial x \partial y)$ will be constant $(= -4/bc)$ when the element is twisted so that $w = 1$ at nodes 2 and 4, with the edges remaining straight. Thus the nodal displacements will be: $w = 0$ at 1 and 3; $w = 1$ at 2 and 4; $\theta_x = 1/c$ at 1 and 4; $\theta_x = -1/c$ at 2 and 3; $\theta_y = -1/b$ at 1 and 2; and $\theta_y = 1/b$ at 3 and 4.

Requirement 2 can be considered to be a special case of requirement 3 when the constant strain is zero. The "patch test" (Sec. 19-10) is a numerical method for nonconforming elements to verify that an assemblage of elements can assume a constant-strain state.

19-10   THE PATCH TEST FOR CONVERGENCE

It is advisable to perform a patch test before a new computer program or a new element is adopted.[5] The test is conducted on a patch of several elements (Fig. 19-8). Prescribed displacements are introduced at boundary nodes which correspond to constant-strain conditions. If the strains determined by the analysis are constant, the patch test is passed. If the test is not passed, convergence with an arbitrary fine mesh is not ensured. Passing the patch test for conforming or nonconforming elements means that convergence requirements 2 and 3 in the preceding section are satisfied.

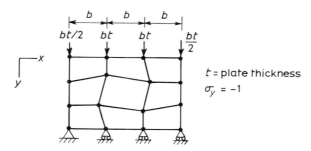

Fig. 19-8. Model for the patch test for the plane-stress or plane-strain element of Fig. 19-5. Consistent nodal forces correspond to a uniform stress $\sigma_y = -1$ on the top edge of the patch.

In the patch of plane-stress or plane-strain elements shown in Fig. 19-8, a constant-strain state should exist for all elements of the patch when the displacements at the boundary nodes are prescribed by

$$u = a_1 x + a_2 y + a_3 \qquad v = a_4 x + a_5 y + a_6$$

where a_1 to a_6 are arbitrary constants. The inner nodes should be left free. The expected strains are (Eq. 19-6): $\{\varepsilon_x, \varepsilon_y, \gamma_{xy}\} = \{a_1, a_5, (a_2 + a_4)\}$. When the patch test is passed, the computed strains agree with the exact values to the limit of computer accuracy.

Alternatively, appropriate supports and consistent nodal forces (calculated by Eq. 19-47) may be introduced to represent the state of constant stress. An example of this is shown in Fig. 19-8, in which is represented a constant stress $\sigma_y = -1$. Other constant-stress (or constant-strain) states and more than one geometry should be tested to guarantee convergence.

---

[5]The importance of the patch test, what it achieves and the philosophy behind it are the subject of a chapter in Irons, B. M. and Ahmad, S., *Techniques of Finite Elements*, Ellis Horwood, Chichester, England, and Halsted Press (Wiley), New York, 1980, (see pp. 149–162).

Example 19-4   Generate the stiffness matrix and the vector of restraining forces due to a uniform rise of T degrees for the bar shown in Fig. 19.3. Assume that the cross section varies as $a = a_0(2 - \xi)$, where a_0 is constant. The coefficient of thermal expansion is α and the modulus of elasticity is E.

Using the shape functions of Eq. 19-26, the $[B]$ matrix is (Eq. 19-39)

$$[B] = \frac{d}{dx} [1 - \xi \quad \xi] = \frac{d}{ld\xi} [1 - \xi \quad \xi] = \frac{1}{l} [-1 \quad 1]$$

The elasticity matrix in this case has one element $[d] \equiv [E]$. Substitution in Eq. 19-45 gives the stiffness matrix:

$$[S^*] = l \int_0^1 \frac{1}{l} \begin{Bmatrix} -1 \\ 1 \end{Bmatrix} [E] \frac{1}{l} [-1, 1] a_0(2 - \xi) \, d\xi$$

or

$$[S^*] = \frac{Ea_0}{l} \begin{bmatrix} 1 & -1 \\ -1 & 1 \end{bmatrix} \left[2\xi - \frac{\xi^2}{2} \right]_0^1 = \frac{1.5Ea_0}{l} \begin{bmatrix} 1 & -1 \\ -1 & 1 \end{bmatrix}$$

If the effect of the temperature change is not restrained, the strain is $\varepsilon_0 = \alpha T$. The nodal forces to restrain nodal displacements (Eq. 19-50) are then

$$\{F_b\}_m = -l \int_0^1 \frac{1}{l} \begin{Bmatrix} -1 \\ 1 \end{Bmatrix} E\alpha T a_0(2 - \xi) \, d\xi = 1.5E\alpha T a_0 \begin{Bmatrix} 1 \\ -1 \end{Bmatrix}$$

The same results would be obtained if the member were treated as a prismatic bar with a constant cross-sectional area equal to the average of the values at the two ends. The exact answer for the stiffness matrix is the same as above with the constant 1.5 replaced by $(1/\ln 2) = 1.443$ (obtained by considering the true deformed shape). As expected, the use of assumed shape functions resulted in an overestimate of stiffness.

Example 19-5   Determine element S^*_{12} of the stiffness matrix for the bar shown in Fig. 19-4, assuming the second moment of the cross section to vary as $I = I_0(1 + \xi)$, where I_0 is constant. Consider bending deformations only; $E =$ constant. Also, generate the vector of nodal equilibrants of a uniform load q per unit length covering the entire length.

For a beam in bending, $[d] \equiv [EI]$, $\{\sigma\} \equiv \{M\}$ and $\{\varepsilon\} \equiv -d^2v/dx^2$. Using the shape functions in Fig. 19-4, the $[B]$ matrix is (Eq. 19-39)

$$[B] = -\frac{d^2}{dx^2} [1 - 3\xi^2 + 2\xi^3 \quad l\xi(\xi - 1)^2 \quad \xi^2(3 - 2\xi) \quad l\xi^2(\xi - 1)]$$

$$[B] = -\frac{1}{l^2} [-6 + 12\xi \quad l(6\xi - 4) \quad 6 - 12\xi \quad l(6\xi - 2)]$$

The required element of the stiffness matrix (Eq. 19-44) is

$$S^*_{12} = \int_0^1 \left(\frac{6 - 12\xi}{l^2} \right) EI_0(1 + \xi) \left(\frac{-6\xi + 4}{l} \right) ld\xi = 8\frac{EI_0}{l^2}$$

The exact answer can be calculated by Eq. 10-20, giving $S_{12}^* = 7.72\,EI_0/l^2$. As expected, the stiffness is overestimated by the use of the assumed shape function L_2 instead of the true deflected shape due to $D_2^* = 1$.

The entire stiffness $[S^*]$ derived by Eq. 19-45 may be compared with the exact stiffness matrix (by Eq. 10-20) in the answers to Prob. 19-1.

The nodal forces in equilibrium with the uniform load q, with nodal displacements prevented, are (Eq. 19-47)

$$
\{F_b\}_m = -\int_0^1
\begin{Bmatrix}
1 - 3\xi^2 + 2\xi^3 \\
l\xi(\xi - 1)^2 \\
\xi^2(3 - 2\xi) \\
l\xi^2(\xi - 1)
\end{Bmatrix}
ql\,d\xi =
\begin{Bmatrix}
-ql/2 \\
-ql^2/12 \\
-ql/2 \\
ql^2/12
\end{Bmatrix}
$$

These are the same forces as for a prismatic beam; again, an approximation is involved in accepting the deflected shapes of a prismatic bar for a bar with a variable I.

Example 19-6   Determine element S_{22}^* of the stiffness matrix for a rectangular plane-stress element of constant thickness h (Fig. 19-1). The shape functions for this element are derived in Example 19-3. Determine also the stresses at any point due to $D_2^* = 1$.

Due to $D_2^* = 1$, the displacements at any point are given by (Eqs. 19-67 and 19-68)

$$
\begin{Bmatrix} u \\ v \end{Bmatrix} = \begin{Bmatrix} 0 \\ L_2 \end{Bmatrix} = \begin{Bmatrix} 0 \\ (1 - 3\xi^2 + 2\xi^3)(1 - \eta) \end{Bmatrix}
$$

The strain at any point (Eqs. 19-39 and 19-11) is

$$
\{B\}_2 = \begin{Bmatrix} 0 \\ (1/c)(-1 + 3\xi^2 - 2\xi^3) \\ (1/b)(-6\xi + 6\xi^2)(1 - \eta) \end{Bmatrix}
$$

The required element of the stiffness matrix is given by Eq. 19-44 as

$$
S_{22}^* = bch \int_0^1 \int_0^1 \{B\}_2^T [d] \{B\}_2 \, d\xi \, d\eta
$$

Substituting for $[d]$ from Eq. 19-12 and performing the integral gives

$$
S_{22}^* = \frac{Eh}{1 - v^2} \left[\frac{13}{35}\left(\frac{b}{c}\right) + \frac{1}{5}\left(\frac{c}{b}\right)(1 - v) \right]
$$

The stress at any point due to $D_2^* = 1$ is (Eq. 19-42)

$$
\{\sigma_u\}_2 = [d]\{B\}_2
$$

$$
= \frac{E}{1 - v^2} \left\{ \frac{v}{c}(-1 + 3\xi^2 - 2\xi^3), \frac{1}{c}(-1 + 3\xi^2 - 2\xi^3), \frac{3(1 - v)}{b}(1 - \eta)(-\xi + \xi^2) \right\}
$$

19-11 CONSTANT-STRAIN TRIANGLE

The triangular element shown in Fig. 19-9 may be used in a plane-stress or plane-strain analysis. The element has three nodes at the corners, with two nodal displacements u and v. The element is called a constant-strain triangle because the strain, and hence the stress, within the element is constant.

Each of u and v is associated with three of the six nodal displacements. The same polynomial may therefore be used for the two variables:

$$[P] = [1, x, y] \tag{19-69}$$

We shall now derive the shape functions associated with u, which are the same as the shape functions associated with v. At the three nodes we have

$$\begin{Bmatrix} u_i \\ u_j \\ u_k \end{Bmatrix} = \begin{Bmatrix} 1 & x_i & y_i \\ 1 & x_j & y_j \\ 1 & x_k & y_k \end{Bmatrix} \{A\} \tag{19-70}$$

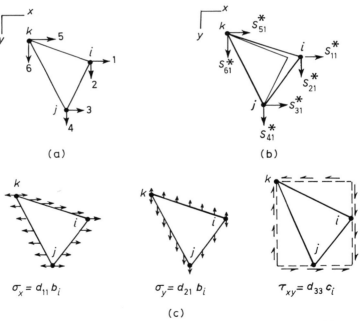

$$\sigma_x = d_{11} b_i \qquad \sigma_y = d_{21} b_i \qquad \tau_{xy} = d_{33} c_i$$

(c)

Fig. 19-9. Triangular plane-stress or plane-strain element. (a) Nodal coordinates. (b) and (c) Stresses on element edges lumped in equivalent forces at the nodes.

or

$$\{u\} = [C]\{A\} \tag{19-71}$$

Inversion of $[C]$, using Cramer's rule (Sec. A-9) or other methods, gives:

$$[C]^{-1} = \frac{1}{2\Delta} \begin{bmatrix} a_i & a_j & a_k \\ b_i & b_j & b_k \\ c_i & c_j & c_k \end{bmatrix} \tag{19-72}$$

where 2Δ is the determinant of $[C] = 2 \times$ area of triangle. Here,

$$a_i = x_j y_k - y_j x_k \qquad b_i = y_j - y_k \qquad c_i = x_k - x_j \tag{19-73}$$

By cyclic permutation of the subscripts i, j and k, similar equations can be written for $\{a_j, b_j, c_j\}$ and for $\{a_k, b_k, c_k\}$.

Substitution in Eq. 19-60 gives the shape functions:

$$[L] = \frac{1}{2\Delta}[(a_i + b_i x + c_i y) \quad (a_j + b_j x + c_j y) \quad (a_k + b_k x + c_k y)] \tag{19-74}$$

The displacements $\{u, v\}$ at any point may be expressed in terms of the nodal displacements (Eq. 19-25):

$$\begin{Bmatrix} u \\ v \end{Bmatrix} = \begin{bmatrix} L_1 & 0 & L_2 & 0 & L_3 & 0 \\ 0 & L_1 & 0 & L_2 & 0 & L_3 \end{bmatrix} \{D^*\} \tag{19-75}$$

where

$$\{D^*\} = \{u_i, v_i, u_j, v_j, u_k, v_k\}$$

The matrix $[B]$, relating strains $\{\varepsilon\}$ to $\{D^*\}$, is (Eqs. 19-39 and 19-11)

$$[B] = \frac{1}{2\Delta} \begin{bmatrix} b_i & 0 & b_j & 0 & b_k & 0 \\ 0 & c_i & 0 & c_j & 0 & c_k \\ c_i & b_i & c_j & b_j & c_k & b_k \end{bmatrix} \tag{19-76}$$

The product $[d][B]$ gives the stresses due to unit nodal displacements (Eq. 19-42):

$$[\sigma_u] = \frac{1}{2\Delta} \begin{bmatrix} d_{11}b_i & d_{21}c_i & d_{11}b_j & d_{21}c_j & d_{11}b_k & d_{21}c_k \\ d_{21}b_i & d_{22}c_i & d_{21}b_j & d_{22}c_j & d_{21}b_k & d_{22}c_k \\ d_{33}c_i & d_{33}b_i & d_{33}c_j & d_{33}b_j & d_{33}c_k & d_{33}b_k \end{bmatrix} \tag{19-77}$$

where d_{ij} are elements of the elasticity matrix $[d]$, given by Eqs. 19-12 and 19-13 for the states of plane stress and plane strain, respectively. It can be seen that all the elements of $[B]$ and $[\sigma_u]$ are constants, indicating constant strain and stress. If the thickness h is constant, the stiffness matrix of the element is (Eq. 19-45)

$$[S^*] = h\Delta[B]^T[d][B] = h\Delta[B]^T[\sigma_u] \tag{19-78}$$

That is,

$$[S^*] = \frac{h}{4\Delta}
\begin{bmatrix}
d_{11}b_i^2 + d_{33}c_i^2 & & & \text{symmetrical} \\
d_{21}b_ic_i + d_{33}b_ic_i & d_{22}c_i^2 + d_{33}b_i^2 \\
d_{11}b_ib_j + d_{33}c_ic_j & d_{21}c_ib_j + d_{33}c_jb_i & d_{11}b_j^2 + d_{33}c_j^2 \\
d_{21}b_ic_j + d_{33}b_jc_i & d_{22}c_ic_j + d_{33}b_ib_j & d_{21}b_jc_j + d_{33}c_jb_j \\
d_{11}b_ib_k + d_{33}c_ic_k & d_{21}c_ib_kd_{33}b_ic_k & d_{11}b_jb_k + d_{33}c_jc_k \\
d_{21}b_ic_k + d_{33}b_kc_i & d_{22}c_ic_k + d_{33}b_ib_k & d_{21}b_jc_k + d_{33}c_jb_k
\end{bmatrix}$$

$$\left.
\begin{matrix}
d_{22}c_j^2 + d_{33}b_j^2 \\
d_{21}c_jb_k + d_{33}b_jc_k & d_{11}b_k^2 + d_{33}c_k^2 \\
d_{22}c_jc_k + d_{33}b_jb_k & d_{21}b_kc_k + d_{33}c_kb_k & d_{22}c_k^2 + d_{33}b_k^2
\end{matrix}
\right] \quad (19\text{-}79)$$

If the element is subjected to uniform body forces with intensities per unit volume of $\{p\} = \{q_x, q_y\}$, the equilibrants at the nodes, when the nodal displacements are prevented, are (Eq. 10-47)

$$\{F_b^*\} = -h \iint
\begin{bmatrix}
L_1 & 0 & L_2 & 0 & L_3 & 0 \\
0 & L_1 & 0 & L_2 & 0 & L_3
\end{bmatrix}^T
\begin{Bmatrix} q_x \\ q_y \end{Bmatrix} dx\,dy \quad (19\text{-}80)$$

Evaluation of the integrals is simplified by noting that $\iint dx\,dy = \Delta$; $\iint x\,dx\,dy$ is the first moment of area about the y axis, so that $\iint y\,dx\,dy = (\Delta/3)(y_i + y_j + y_k)$. The consistent vector of forces in equilibrium with the body forces thus becomes

$$\{F_b^*\} = -\frac{\Delta h}{3}\{q_x, q_y, q_x, q_y, q_x, q_y\} \quad (19\text{-}81)$$

This means that one-third of the load on the triangle is assigned to each node. The same distribution could have been suggested intuitively. (However, it is not always possible to determine $\{F_b^*\}$ intuitively; see Fig. 20-7.)

The consistent vector of restraining forces when the element is subjected to a rise in temperature of T degrees is obtained by substitution of Eqs. 19-49 and 19-76 into Eq. 19-50:

$$\{F_b\}_m = -\frac{\alpha T h}{2}(d_{11} + d_{21})\{b_i, c_i, b_j, c_j, b_k, c_k\} \quad (19\text{-}82)$$

Here, we have assumed that the material is isotropic ($d_{22} = d_{11}$).

19-12 INTERPRETATION OF NODAL FORCES

In the derivation of the stiffness matrix of individual elements, the forces distributed along the edges of the elements are replaced by equivalent forces lumped at the nodes. The equivalent forces are determined by the use of the principle of virtual work.

This approach can be seen in the example of the constant-strain triangular element shown in Fig. 19-9. For a unit nodal displacement, say $D_1^* = 1$, while other nodal displacements are zero, the stresses are (first column of $[\sigma_u]$, Eq. 19-77)

$$\{\sigma\} = \frac{1}{2\Delta}\{d_{11}b_i, d_{21}b_i, d_{33}c_i\} \tag{19-83}$$

where d_{ij} are elements of the elasticity matrix $[d]$ given in Eqs. 19-12 and 19-13 for plane-stress or plane-strain states, respectively; $b_i = y_j - y_k$; $c_i = x_k - x_j$; and Δ = area of the triangle.

The stresses are represented by uniform forces on the edges of the element in Fig. 19-9c.

The distributed forces in Fig. 19-9c are lumped at the nodes in Fig. 19-9b. The force in the x or y direction at a node is equal to the sum of one-half of the distributed load on each of the two sides connected to the node. For example, the horizontal force at node k is one-half of the load on edges ki and kj; thus

$$S_{51}^* = h\left\{\frac{\sigma_x}{2}[(y_i - y_k) - (y_j - y_k)] + \frac{\tau_{xy}}{2}[(x_j - x_k) - (x_i - x_k)]\right\} \tag{19-84}$$

We can now verify that this is the same as element S_{51}^* of the stiffness matrix in Eq. 19-79. Any other element of the stiffness matrix can be verified in a similar way.

Equation 19-84 means that the virtual work of force S_{51}^* during nodal displacement $D_1^* = 1$ is the same as the work done by the forces distributed over edges ki and kj during their corresponding virtual displacements (which, in this example, vary linearly).

The consistent load vector also represents the distributed forces lumped at the nodes. If the element shown in Fig. 19-9b is subjected to a rise in temperature of T degrees and the element expansion is restrained, the stresses in an isotropic material will be (Eq. 19-48 and 19-49)

$$\{\sigma_r\} = -[d]\{\varepsilon_0\} = -\alpha T(d_{11} + d_{21})\begin{Bmatrix} 1 \\ 1 \\ 0 \end{Bmatrix} \tag{19-85}$$

where α is the coefficient of thermal expansion.

Under the conditions of restraint, uniform forces act on the edges so as to produce $\sigma_x = \sigma_y = -\alpha T(d_{11} + d_{21})$. Lumping one-half of the distributed load on any edge at the two end nodes gives the nodal forces of Eq. 19-82.

19-13   GENERAL

The displacement method of analysis is applicable to structures composed of finite elements which may be one-, two- or three-dimensional. The analysis requires the generation of stiffness and stress matrices and of load vectors for individual elements. Exact element matrices can be generated only for bars. Procedures to generate approximate matrices for elements of any type have been presented in this chapter. Convergence to the exact solution can be ensured as a finer finite-element mesh is used. The general procedures have been applied to bar elements and to plate elements subjected to in-plane forces and to bending. Other finite elements are discussed in Chapter 20.

Assemblage of stiffness matrices and load vectors of individual elements, generation of structure equilibrium equations and their solutions, and the use of computers are discussed in Chapters 24 and 25. This is done mainly for structures composed of bar elements (framed structures). However, the same techniques apply when any type of finite-element is used.

PROBLEMS

19-1   Determine any element S_{ij}^* for the bar of Example 19-5, using the shape functions in Fig. 19-4. Compare the answer with an exact value as given by Eq. 10-20.

19-2   Consider the member of Prob. 19-1 as a cantilever fixed at the end $x = l$ and subjected to a transverse concentrated load P at the free end. Find the deflection at the tip of the cantilever, using the two matrices given in the answers to Prob. 19-1.

19-3   For the prismatic bar shown, generate the stiffness matrix corresponding to the three coordinates indicated. Use the following shape functions: $L_1 = -(1/2)\xi(1 - \xi)$; $L_2 = (1/2)\xi(1 + \xi)$; $L_3 = 1 - \xi^2$ (Lagrange polynomials, Fig. 20-3). Condense the stiffness matrix by elimination of node 3.

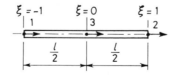

Prob. 19-3

19-4   Assume that the displacement at coordinate 1 of the bar element of Prob. 19-3 is prevented. A spring support is provided at coordinate 2 so that $F_2 = -Ku_2$, where F_2 and u_2 are, respectively, the force and the displacement at coordinate 2, and K is the spring constant equal to $Ea/(3l)$. What will be the displacements at coordinates 2 and 3 and the stress in the bar when it is subjected to a rise in temperature of $T = (1/2)T_0(1 + \xi)$, where T_0 is a constant? Use the shape functions given in Prob. 19-3. Do you expect exact answers?

19-5   Determine $S_{11}^*, S_{21}^*, S_{31}^*, S_{41}^*$ and S_{51}^* of the stiffness matrix of the plane-stress finite element in Fig. 19-5. Assume that the element is rectangular with sides b and c parallel to the x and y axes, respectively, and is made of an isotropic material with $v = 0.2$. Element thickness is constant and equal to h. The nodal displacement vector and shape functions are defined by Eqs. 19-30 to 19-32.

19-6   Determine the consistent vector of restraining forces $\{F_b^*\}$ for a rectangular plane-stress element (Example 19-2) subjected to a uniform temperature rise of T degrees. Assume an isotropic material with $v = 0.2$. Only the first two elements of the vector need to be determined by Eq. 19-50; the remaining elements can be generated by consideration of equilibrium and symmetry.

19-7   Considering symmetry and equilibrium, use the results of Prob. 19-5 to generate $[S^*]$ for the element when $b = c$. For a comparison of the accuracy of this element with other elements, see Prob. 20-18.

19-8   Use the results of Prob. 19-5 to calculate the deflection at the middle of a beam idealized by two elements as shown. Take the beam width as h and Poisson's ratio as 0.2. Compare the result with that obtained by beam theory, considering bending and shear deformations. Note that the deflection calculated by the finite-element method is smaller than the more accurate value obtained by beam theory and that the percentage error increases with an increase in the ratio b/c.

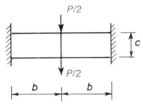

Prob. 19-8

19-9   Use the displacement determined in Prob. 19-8 to calculate the stresses in the top fiber at the fixed end.

19-10   Determine the consistent vector of restraining forces $\{F_b^*\}$ for the rectangular plane-stress element considered in Example 19-3 when it is subjected to a uniform temperature rise of T degrees. Assume an isotropic material. Only the first three elements of the vector need to be determined by Eq. 19-50; the remaining elements can be obtained by considering equilibrium and symmetry.

19-11   Determine S_{11}^* for the rectangular plate-bending element shown in Fig. 19-6. The nodal displacement vector and the shape functions are defined in Eqs. 19-33 to 19-35. Use the result to calculate the central deflection of a rectangular plate $2b \times 2c$ with built-in edges and subjected to a concentrated load P at midpoint. Consider an isotropic material with $v = 0.3$. Idealize the plate by four elements and perform the analysis for one element only, taking advantage of symmetry. Use Fig. 19-6 to represent the element analyzed with node 1 at the center of the plate. Determine also M_x at node 2.

The answers to this problem are given in terms of the ratio b/c. For comparison, we give here the exact[6] answers for a square plate ($b/c = 1$): deflection $= 0.244Pb^2/Eh^3$; M_x at node 2 $= -0.126P$. The exact deflection is smaller than the finite-element solution; why?

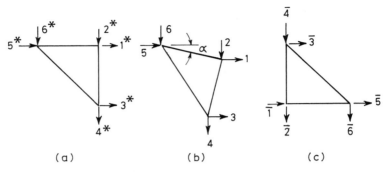

Prob. 19-12

19-12   The figure shows identical plane-stress elements with three coordinate systems. For the system in (a), calculate S^*_{11}, S^*_{21}, S^*_{31}, S^*_{33}, S^*_{34}, and S^*_{44} and, by considering equilibrium and symmetry, generate the remaining elements of $[S^*]$. What are the transformation matrix $[T]$ and the equation to be used to transform $[S^*]$ into $[S]$ for the coordinate system in (b)? What is the value of the angle α to be substituted in the equations to give $[\bar{S}]$ corresponding to the coordinates in (c)? Consider an isotropic material with Poisson's ratio $v = 0$. Element thickness is h; element sides are b, b, $b\sqrt{2}$.

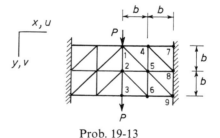

Prob. 19-13

19-13   Using the answers of Prob. 19-12 and taking advantage of symmetry, determine, for the beam shown, the nodal displacements and the stress in element 2–6–3. The beam is of isotropic material; $v = 0$; beam width is h. The structure idealization is composed of isosceles right-angle triangles of the same size. The equilibrium equations can be checked by substitution of $\{D\}$ given in the answers.

19-14   Generate columns 1, 2 and 3 of the $[B]$ matrix for element QLC3 (Fig. 19-1).

[6]From the reference in Chapter 18, footnote 2.

19-15. Calculate any stiffness coefficient S_{ij}^* for a rectangular plane-stress element QLC3 (Fig. 19-1 and Example 19-3). Consider an isotropic square element of side b, constant thickness h and Poisson's ratio $v = 0.2$. The stiffness matrix is given in the answers and is used in Prob. 20-18 to compare the accuracy of results when using this approach and two other types of element (the element in Fig. 19-5 and the hybrid element of Example 20-6).

CHAPTER 20

Further development of finite-element method

20-1 INTRODUCTION

This chapter is a continuation of Chapter 19, introducing various types of finite elements which are frequently used in practice.

The isoparametric formulation[1] is widely used because of its effectiveness and because it allows the elements to have curved shapes. In this chapter it is used for one-, two- and three-dimensional elements. For this purpose, functions in natural coordinates ξ, (ξ, η) or (ξ, η, ζ) are used to describe the geometry of the element, and the same functions are used as shape functions describing the variation of the displacement components over the elements.

The integration involved in the generation of the finite-element matrices is, in the majority of cases, evaluated numerically. The Gauss quadrature method, which is frequently used, is discussed in Sec. 20-9.

Sections 20-10 and 20-11 present elements suitable for the analysis of shell structures and solids of revolution.

Instead of the division of a structure into elements by a mesh in two or three directions, it is possible to perform the analysis on an assemblage of finite strips or finite prisms. For this purpose the structure is divided in one dimension or in two dimensions only, instead of two or three dimensions, respectively. The division results in finite strips or finite prisms running the full length of the structure. The analysis using this structural idealization is discussed in Sec. 20-12.

Section 20-13 is an introduction to the hybrid elements which were briefly discussed in Sec. 19-1.

20-2 ISOPARAMETRIC ELEMENTS

Isoparametric elements can be in the form of a curved bar, a triangular or quadrilateral plate with curved edges, or a three-dimensional brick with

---

[1]The first to introduce the technique (in 1966) was B. M. Irons, who made important contributions to the method of finite elements. He is a coauthor of three books included in the General References to this book.

curved edges. The quadrilateral plane-stress or plane-strain element in Fig. 19-5 is an example of an isoparametric element. The nodal displacements for the element are

$$\{D^*\} = \{u_1, v_1, u_2, v_2, u_3, v_3, u_4, v_4\} \tag{20-1}$$

The displacements u and v at any point are determined from the nodal displacements, using shape functions:

$$u = \sum L_i u_i \qquad v = \sum L_i v_i \tag{20-2}$$

where L_i with $i = 1, 2, 3, 4$ represents shape functions of the natural coordinates ξ and η. The shape functions for this element are given by Eq. 19-32, which is repeated here:

$$L_i = \tfrac{1}{4}(1 + \xi\xi_i)(1 + \eta_i) \tag{20-3}$$

The natural coordinates varying between -1 and 1 (Fig. 19-5) define the relative position of any point with respect to the corner nodes. The same shape functions can be used to determine the (x, y) coordinates of any point in terms of ξ and η:

$$x = \sum L_i x_i \qquad y = \sum L_i y_i \tag{20-4}$$

where (x_i, y_i) with $i = 1, 2, 3, 4$ represent cartesian coordinates at the nodes. The element is called isoparametric because the same interpolation functions are used to express the location of a point and the displacement components in terms of ξ and η.

The strain components at any point involve derivatives $\partial/\partial x$ and $\partial/\partial y$ with respect to the cartesian coordinates. Derivatives of any variable g with respect to the natural coordinates are obtained by the chain rule:

$$\frac{\partial g}{\partial \xi} = \frac{\partial g}{\partial x}\frac{\partial x}{\partial \xi} + \frac{\partial g}{\partial y}\frac{\partial y}{\partial \xi} \qquad \frac{\partial g}{\partial \eta} = \frac{\partial g}{\partial x}\frac{\partial x}{\partial \eta} + \frac{\partial g}{\partial y}\frac{\partial y}{\partial \eta} \tag{20-5}$$

We can rewrite Eq. 20-5 in matrix form:

$$\begin{Bmatrix} \partial g/\partial \xi \\ \partial g/\partial \eta \end{Bmatrix} = [J] \begin{Bmatrix} \partial g/\partial x \\ \partial g/\partial y \end{Bmatrix} \qquad \begin{Bmatrix} \partial g/\partial x \\ \partial g/\partial y \end{Bmatrix} = [J]^{-1} \begin{Bmatrix} \partial g/\partial \xi \\ \partial g/\partial \eta \end{Bmatrix} \tag{20-6}$$

Here, $[J]$ is the *Jacobian matrix*; it serves to transform the derivatives of any variable with respect to ξ and η into derivatives with respect to x and y, and vice versa. The elements of $[J]$ are given by

$$[J] = \begin{bmatrix} \partial x/\partial \xi & \partial y/\partial \xi \\ \partial x/\partial \eta & \partial y/\partial \eta \end{bmatrix} \tag{20-7}$$

When Eq. 20-4 applies, the Jacobian can be expressed as

$$[J] = \sum \begin{bmatrix} x_i(\partial L_i/\partial \xi) & y_i(\partial L_i/\partial \xi) \\ x_i(\partial L_i/\partial \eta) & y_i(\partial L_i/\partial \eta) \end{bmatrix} \tag{20-8}$$

The element geometry is defined by the cartesian coordinates (x, y) at the four nodes. For a given quadrilateral the Jacobian varies, with ξ and η defining the position of any point.

The strains at any point are given by $\{\varepsilon\} = [\partial]\{u, v\}$; the derivative operator $[\partial]$ is defined by Eq. 19-11. Substitution of Eq. 20-2 gives

$$\{\varepsilon\} = \sum_{i=1}^{n} \left\{ [B]_i \begin{Bmatrix} u_i \\ v_i \end{Bmatrix} \right\} \tag{20-9}$$

where n is the number of nodes; for the quadrilateral element in Fig. 19-5, $n = 4$. Here,

$$[B]_i = \begin{bmatrix} \dfrac{\partial L_i}{\partial x} & 0 \\ 0 & \dfrac{\partial L_i}{\partial y} \\ \dfrac{\partial L_i}{\partial y} & \dfrac{\partial L_i}{\partial x} \end{bmatrix} \qquad [B] = [[B]_1 [B]_2 \cdots [B]_n] \tag{20-10}$$

When $[J]$ and its inverse are determined at a particular point, numerical values of the derivatives $(\partial L_i/\partial x)$ and $(\partial L_i/\partial y)$ required for generating $[B]_i$ are calculated from $\partial L_i/\partial \xi$ and $\partial L_i/\partial \eta$ via Eq. 20-6. The integrals involved in the generation of the element stiffness matrix and of the consistent load vector are obtained numerically.

The element stress matrix relating the stresses at any point to nodal displacements (Eq. 19-42) is

$$[\sigma_u] = [d][B] \tag{20-11}$$

The element stiffness matrix is given by (Eq. 19-45)

$$[S^*] = h \int_{-1}^{1} \int_{-1}^{1} [B]^T [d][B]|J| \, d\xi \, d\eta \tag{20-12}$$

where h is the element thickness, which is assumed constant, and

$$|J| \, d\xi \, d\eta = da \tag{20-13}$$

Here, da is the elemental area (parallelogram) shown in Fig. 20-1. The determinant of the Jacobian serves as a scaling factor (length^2) to transform the dimensionless product $d\xi \, d\eta$ into an elemental area. The validity of Eq. 20-13 can be verified by considering the following areas:

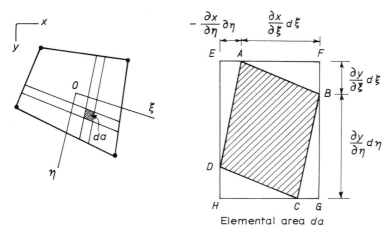

Fig. 20-1. Verification of Eq. 20-12.

parallelogram $ABCD$ = rectangle $EFGH$ − 2(triangle EAD + triangle AFB)

If the element is subjected to a rise in temperature of T degrees and the expansion is restrained, the stress $\{\sigma_r\}$ will be the same as given by Eq. 19-48, and the consistent vector of restraining forces will be given by (Eq. 19-50)

$$\{F_b^*\} = -h \int_{-1}^{1} \int_{-1}^{1} [B]^T [d] \{\varepsilon_0\} |J| \, d\xi \, d\eta \qquad (20\text{-}14)$$

where $\{\varepsilon_0\}$ represents the change in strains if the expansion is free to occur (Eq. 19-49).

For a quadrilateral element, the determinant $|J|$ at the center of the element is equal to one-quarter of its area. For a rectangle, $[J]$ is constant. In general, the determinant $|J|$ is positive, the matrix $[J]$ is nonsingular, and the matrix inversion in Eq. 20-6 is possible. If two corners of the quadrilateral coincide, the element becomes triangular. At the point where they coincide, $[J]$ is singular, indicating that the strains cannot be determined at this point. When the internal angle at each of the four corners of a quadrilateral is smaller than 180°, $[J]$ is nonsingular at all points and its determinant is positive.

Equations 20-2 and Eqs. 20-4 to 20-14 are general for any isoparametric plane-stress or plane-strain element, with two degrees of freedom per node. The shape functions $[L]$ differ depending on the number of nodes and the arrangement of the nodes in each element. Methods of derivation of shape functions for various isoparametric elements are given in Sects. 20-4 and 20-5.

Example 20-1 Determine the Jacobian matrix at the center of the quadrilateral element shown in Fig. 19-5, using the following (x, y) coordinates: node

1 (0,0); node 2 (7, 1); node 3 (6, 9); node 4 ($-$ 2, 5). Use the determinant of the Jacobian to calculate the area of the quadrilateral.

The derivatives of the shape function in Eq. 20-3 at $\xi = \eta = 0$ are

$$\frac{\partial L_i}{\partial \xi} = \frac{\xi_i}{4} \qquad \frac{\partial L_i}{\partial \eta} = \frac{\eta_i}{4}$$

The Jacobian is (Eq. 20-8)

$$[J] = \begin{bmatrix} 0\left(\dfrac{-1}{4}\right) & 0\left(\dfrac{-1}{4}\right) \\ 0\left(\dfrac{-1}{4}\right) & 0\left(\dfrac{-1}{4}\right) \end{bmatrix} + \begin{bmatrix} 7\left(\dfrac{1}{4}\right) & 1\left(\dfrac{1}{4}\right) \\ 7\left(\dfrac{-1}{4}\right) & 1\left(\dfrac{-1}{4}\right) \end{bmatrix}$$

$$+ \begin{bmatrix} 6\left(\dfrac{1}{4}\right) & 9\left(\dfrac{1}{4}\right) \\ 6\left(\dfrac{1}{4}\right) & 9\left(\dfrac{1}{4}\right) \end{bmatrix} + \begin{bmatrix} -2\left(\dfrac{-1}{4}\right) & 5\left(\dfrac{-1}{4}\right) \\ -2\left(\dfrac{1}{4}\right) & 5\left(\dfrac{1}{4}\right) \end{bmatrix}$$

$$[J] = \begin{bmatrix} \dfrac{15}{4} & \dfrac{5}{4} \\ -\dfrac{3}{4} & \dfrac{13}{4} \end{bmatrix}$$

Area $= 4|J| = 4(13.125) = 52.5$

20-3   CONVERGENCE OF ISOPARAMETRIC ELEMENTS

One of the requirements for the shape functions for a finite element is that they allow a constant-strain state (see Sec. 19-9). To verify that the shape functions of isoparametric elements satisfy this requirement, let the element acquire nodal displacements given by

$$u_i = a_1 x_i + a_2 y_i + a_3 \qquad v_i = a_4 x_i + a_5 y_i + a_6 \qquad (20\text{-}15)$$

where a_1 to a_6 are arbitrary constants, and i is the node number. The displacement at any point (ξ, η) is given by substitution of Eq. 20-15 in Eq. 20-2:

$$u = a_1 \sum L_i x_i + a_2 \sum L_i y_i + a_3 \sum L_i$$
$$v = a_4 \sum L_i x_i + a_5 \sum L_i y_i + a_6 \sum L_i \qquad (20\text{-}16)$$

We note that $\sum L_i x_i = x$ and $\sum L_i y_i = y$ (Eq. 20-4). If $\sum L_i = 1$, Eq. 20-16 can be rewritten as

$$u = a_1 x + a_2 y + a_3 \qquad v = a_4 x + a_5 y + a_6 \qquad (20\text{-}17)$$

Equation 20-17 indicates that the strains are constant when the sum $\sum L_i = 1$. This condition is satisfied for the shape functions in Eq. 20-3. It is also satisfied for the shape functions of other isoparametric elements discussed below.

Another requirement for convergence is that the shape functions allow rigid-body motion with zero strains. This represents a special case of the state of constant strain. This then verifies the fact that the shape functions allow the element to move as a rigid body.

20-4   LAGRANGE INTERPOLATION

Consider a function $g(\xi)$ for which the n values $g_1, g_2, \ldots, g_n$ are known at $\xi_1, \xi_2, \ldots, \xi_n$ (Fig. 20-2). The Lagrange equation gives a polynomial $g(\xi)$ which passes through the n points. The equation takes the form of a summation of n polynomials:

$$g(\xi) = \sum_{i=1}^{n} g_i L_i \tag{20-18}$$

where L_i is a polynomial in ξ of degree $n - 1$. Equation 20-18 can be used to interpolate between $g_1, g_2, \ldots, g_n$ to give the g value at any intermediate ξ.

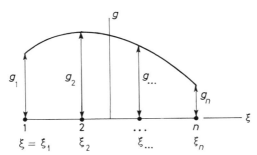

Fig. 20-2.   Function $g(\xi)$ with n known values $g_1, g_2, \ldots, g_n$ at $\xi = \xi_1, \xi_2, \ldots, \xi_n$.

Figure 20-3 shows *Lagrange interpolation functions* for $n = 2$, 3 and 4. It can be seen that any function L_i has a unit value at ξ_i and a zero value at ξ_j, where $j \neq i$.

Lagrange interpolation functions are expressed as a product of $n - 1$ terms:

$$L_i = \frac{\xi - \xi_1}{\xi_i - \xi_1} \frac{\xi - \xi_2}{\xi_i - \xi_2} \cdots \frac{\xi - \xi_{i-1}}{\xi_i - \xi_{i-1}} \frac{\xi - \xi_{i+1}}{\xi_i - \xi_{i+1}} \cdots \frac{\xi - \xi_n}{\xi_i - \xi_n} \tag{20-19}$$

which gives polynomials of order $n - 1$.

Equation 20-19 can be used to derive any of the functions L_i in Fig. 20-3, which are linear, quadratic and cubic corresponding to $n = 2$, 3 and 4,

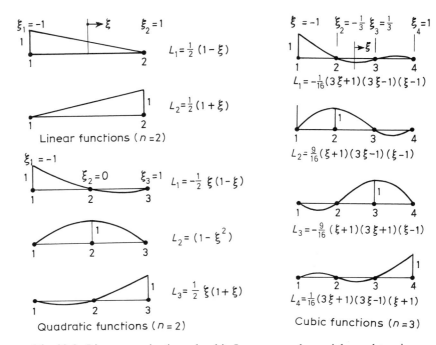

Fig. 20-3. Linear, quadratic and cubic Lagrange polynomials used to give the value of a function at any point in terms of n known values, with $n = 2, 3$ and 4, respectively (Eq. 20-18).

respectively. It can be verified that, for any n, the sum $(L_1 + L_2 + \cdots + L_n)$ is equal to 1 for any value of ξ.

20-5   SHAPE FUNCTIONS FOR TWO- AND THREE-DIMENSIONAL ISOPARAMETRIC ELEMENTS

Figure 20-4 shows a quadrilateral element with corner nodes and also intermediate nodes on the edges. This element can be used for plane-stress or plane-strain analysis; the degrees of freedom at each node are u and v, representing translations in x and y directions. The nodes are equally spaced on each edge. The number of intermediate nodes on any edge can be $0, 1, 2, \ldots$, but the most commonly used elements have one or two intermediate nodes on each edge.

The shape functions are represented by Lagrange polynomials at one edge (or at two adjacent edges) with linear interpolation between each pair of opposite edges. As an example, the shape functions L_5 and L_6 for the element in Fig. 20-4 are

$$L_5 = \tfrac{1}{2}(1 - \xi^2)(1 - \eta) \tag{20-20}$$

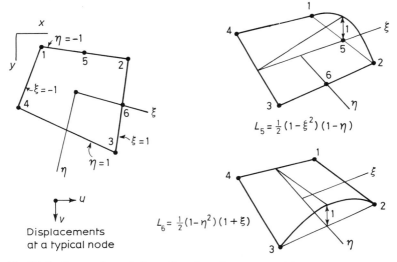

Fig. 20-4. Generation of shape functions by superposition of shapes. The function L_2 is the shape function shown in Fig. 19-5 minus one-half of $(L_5 + L_6)$ of this figure.

$$L_6 = \tfrac{1}{2}(1 - \eta^2)(1 + \xi) \tag{20-21}$$

The shape functions L are used for interpolation of u or v and also for the coordinates x and y between nodal values of the same parameters. Figure 20-4 includes pictorial views of L_5 and L_6 plotted perpendicular to the surface of the element.

For corner node 2, the shape function is

$$L_2 = \tfrac{1}{4}(1 + \xi)(1 - \eta) - \frac{L_5}{2} - \frac{L_6}{2} \tag{20-22}$$

The first term in this equation is the shape function for the same corner node in an element without mid-side nodes (Fig. 19-5). The terms $L_5/2$ and $L_6/2$ are subtracted to make L_2 equal to zero at the two added mid-side nodes 5 and 6.

The use of Lagrange polynomials at the edges and linear interpolation between opposite sides, in the manner described above, can be used to write directly (by intuition) the shape functions for any corner or intermediate node on the element edges. Figure 20-5 gives the shape functions for an eight-node element; such an element is widely used in plane-stress or plane-strain analysis because it gives accurate results. In its general form, the edges are curved (second-degree parabola) and the coordinates ξ and η are curvilinear. The shape functions given in this figure can be used with Eq. 20-2 and Eqs. 20-4 to

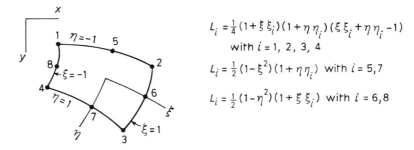

$$L_i = \tfrac{1}{4}(1 + \xi \xi_i)(1 + \eta \eta_i)(\xi \xi_i + \eta \eta_i - 1)$$
with $i = 1, 2, 3, 4$

$$L_i = \tfrac{1}{2}(1 - \xi^2)(1 + \eta \eta_i) \text{ with } i = 5,7$$

$$L_i = \tfrac{1}{2}(1 - \eta^2)(1 + \xi \xi_i) \text{ with } i = 6,8$$

Fig. 20-5. Shape functions for a two-dimensional element with corner and mid-side nodes.

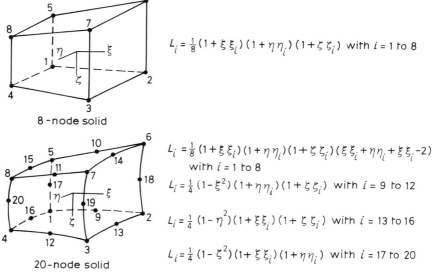

$$L_i = \tfrac{1}{8}(1 + \xi \xi_i)(1 + \eta \eta_i)(1 + \zeta \zeta_i) \text{ with } i = 1 \text{ to } 8$$

8-node solid

$$L_i = \tfrac{1}{8}(1 + \xi \xi_i)(1 + \eta \eta_i)(1 + \zeta \zeta_i)(\xi \xi_i + \eta \eta_i + \xi \xi_i - 2)$$
with $i = 1$ to 8

$$L_i = \tfrac{1}{4}(1 - \xi^2)(1 + \eta \eta_i)(1 + \zeta \zeta_i) \text{ with } i = 9 \text{ to } 12$$

$$L_i = \tfrac{1}{4}(1 - \eta^2)(1 + \xi \xi_i)(1 + \zeta \zeta_i) \text{ with } i = 13 \text{ to } 16$$

$$L_i = \tfrac{1}{4}(1 - \zeta^2)(1 + \xi \xi_i)(1 + \eta \eta_i) \text{ with } i = 17 \text{ to } 20$$

20-node solid

Fig. 20-6. Shape functions for isoparametric solid elements with eight and twenty nodes.

20-14 to generate a stiffness matrix and a load vector without any complication caused by the curvilinear coordinates.

The above approach can be extended directly to three-dimensional solid elements with three degrees of freedom per node, u, v and w, in the global x, y and z directions. The shape functions are given in Fig. 20-6 for an element with corner nodes and for an element with corner as well as mid-edge nodes. With the addition of mid-edge nodes, the accuracy is increased and the element can have curved edges.

For the three-dimensional isoparametric elements shown in Fig. 20-6, the shape functions $L(\xi, \eta, \zeta)$ are used for interpolation of displacements u, v, w and coordinates x, y, z as follows:

$$u = \sum L_i u_i \qquad v = \sum L_i v_i \qquad w = \sum L_i w_i \qquad (20\text{-}23)$$

$$x = \sum L_i x_i \qquad y = \sum L_i y_i \qquad z_i = \sum L_i z_i \qquad (20\text{-}24)$$

with $i = 1$ to 8 or 1 to 20.

It is of interest to note that, by use of Lagrange polynomials, the sum of the interpolation functions at any point equals unity. This can be verified for the two- and three-dimensional elements discussed above.

Example 20-2   Construct shape functions for the isoparametric element shown in Fig. 20-5 with four corner nodes and only one mid-edge node 5 (nodes 6, 7 and 8 are deleted).

The shape functions can be generated by intuition as explained in Sec. 20-5. The shape function L_5 is

$$L_5 = \tfrac{1}{2}(1 - \xi^2)(1 - \eta)$$

For L_1 and L_2, combine the shape functions in Fig. 19-5 with L_5:

$$L_1 = \tfrac{1}{4}(1 - \xi)(1 - \eta) - \frac{L_5}{2} \qquad L_2 = \tfrac{1}{4}(1 + \xi)(1 - \eta) - \frac{L_5}{2}$$

The shape functions L_3 and L_4 are the same as in Fig. 19-5:

$$L_3 = \tfrac{1}{4}(1 + \xi)(1 + \eta) \qquad L_4 = \tfrac{1}{4}(1 - \xi)(1 + \eta)$$

20-6   CONSISTENT LOAD VECTORS FOR RECTANGULAR PLANE ELEMENT

Equations 19-47 and 19-50 can be used to determine the consistent load vectors due to body forces or due to temperature variation for the plane-stress or plane-strain element shown in Fig. 20-5. Consider a special case when the element is a rectangle subjected to a uniform load of q_x per unit volume (Fig. 20-7). The consistent vector of restraining forces is (Eq. 19-47)

$$\{F_b^*\} = -\frac{hbc}{4} \int_{-1}^{1} \int_{-1}^{1} \begin{bmatrix} L_1 & 0 & L_2 & 0 & \cdots & L_8 & 0 \\ 0 & L_1 & 0 & L_2 & \cdots & 0 & L_8 \end{bmatrix}^T \begin{Bmatrix} q_x \\ 0 \end{Bmatrix} d\xi \, d\eta$$
$$(20\text{-}25)$$

where b and c are the element sides and h is its thickness; L_1 to L_8 are given in Fig. 20-6. The integrals in Eq. 20-25 give the consistent nodal forces shown in Fig. 20-7.

Note the unexpected result: the forces at corner nodes are in the opposite direction to the forces at mid-side nodes. But, as expected, the sum of the forces

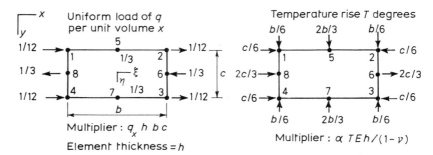

Fig. 20-7. Rectangular plane-stress element. Consistent nodal forces due to a uniform load q_x and due to a rise in temperature of T degrees, with the nodal displacements prevented.

at the eight nodes is equal to $q_x hbc$. Apportionment of the load on the nodal forces intuitively may be sufficient when the finite-element mesh is fine, but greater accuracy is obtained using the consistent load vector, particularly when the mesh is coarse.

Let the element in Fig. 20-7 be subjected to a temperature rise of T degrees; if the expansion is restrained, the stresses are

$$\{\sigma_r\} = -[E\alpha T/(1-v)]\{1,1,0\} \tag{20-26}$$

where E, v and α are modulus of elasticity, Poisson's ratio and coefficient of thermal expansion of the material (assumed isotropic). The corresponding consistent nodal forces can be determined by substitution of Eqs. 20-10, 20-26 and 19-48 into 19-50, using the shape functions in Fig. 20-5. The resulting nodal forces are included in Fig. 20-7.

The same results can be obtained by integration, over the length of the edges, of the product of σ_{rx} or σ_{ry} and the shape functions, with the result multiplied by the thickness h. For example, the force at node 8 is

$$F_8 = \sigma_{rx} h \int_{-c/2}^{c/2} (L_8)_{\xi=-1}\, dy = \sigma_{rx}\frac{hc}{2}\int_{-1}^{1}(1-\eta^2)\,d\eta = \tfrac{2}{3}\sigma_{rx}hc$$

$$\tag{20-27}$$

Here again, the consistent load vector cannot be generated by intuition.

20-7 TRIANGULAR PLANE-STRESS AND PLANE-STRAIN ELEMENTS

Consider the triangular element 1–2–3 shown in Fig. 20-8a, which has an area Δ. Joining any point (x, y) inside the triangle to its three corners divides Δ into three areas, A_1, A_2 and A_3. The position of the point can be

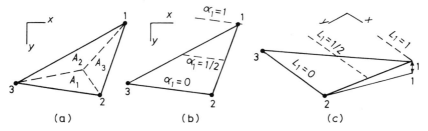

Fig. 20-8. Constant-strain triangle. (a) Area coordinates $\{\alpha_1, \alpha_2, \alpha_3\} = (1/\Delta)\{A_1, A_2, A_3\}$. (b) Lines of equal α_1. (c) Pictorial view of shape function $L_1 = \alpha_1$.

defined by the dimensionless parameters α_1, α_2 and α_3, where

$$\alpha_1 = A_1/\Delta \qquad \alpha_2 = A_2/\Delta \qquad \alpha_3 = A_3/\Delta \qquad (20\text{-}28)$$

and

$$\alpha_1 + \alpha_2 + \alpha_3 = 1 \qquad (20\text{-}29)$$

The parameters α_1, α_2 and α_3 are called *area coordinates* or *areal coordinates*. Any two of them are sufficient to define the position of a point within the element. Lines parallel to the sides of the triangle are lines of equal α_i (Fig. 20-8b).

Figure 20-8c is a pictorial view of the shape function L_1 for the constant-strain triangle discussed in Sec. 19-11. The element has three corner nodes with degrees of freedom u and v per node, representing translations in the x and y directions. The ordinates L_1, which are plotted perpendicular to the surface of the triangle, represent the variation in u or v when the nodal displacement u_1 or v_1 equals unity. If we plot lines of equal L_1, they will be identical to the lines of equal α_1, indicating that α_1 is equal to L_1. Thus the shape functions α_1, α_2 and α_3 can serve as shape functions L_1, L_2 and L_3 for the constant-strain triangle.

Therefore we can express the displacement at any point by

$$u = \sum L_i u_i \qquad v = \sum L_i v_i \qquad (20\text{-}30)$$

It can be shown (by linear interpolation) that

$$x = \sum L_i x_i \qquad y = \sum L_i y_i \qquad (20\text{-}31)$$

Equations 20-30 and 20-31 are the same as Eqs. 20-2 and 20-4, indicating that the constant-strain triangle is an isoparametric element in which the area coordinates serve as shape functions:

$$L_1 = \alpha_1 \qquad L_2 = \alpha_2 \qquad L_3 = \alpha_3 \qquad (20\text{-}32)$$

For a triangle with *straight* edges (Fig. 20-8a), the area coordinates can be expressed in terms of x and y by combining Eqs. 20-29, 20-31 and 20-32, giving

$$\{\alpha_1, \alpha_2, \alpha_3\} = \frac{1}{2\Delta}\{a_1 + b_1 x + c_1 y, \quad a_2 + b_2 x + c_2 y, \quad a_3 + b_3 x + c_3 y\}$$

(20-33)

where

$$a_1 = x_2 y_3 - y_2 x_3 \qquad b_1 = y_2 - y_3 \qquad c_1 = x_3 - x_2 \qquad (20\text{-}34)$$

By cyclic permutation of the subscripts 1, 2 and 3, similar equations can be written for a_2, b_2, c_2 and a_3, b_3, c_3.

20-71 Linear-Strain Triangle

An isoparametric triangular element with corner and mid-side nodes is shown in Fig. 20-9a. This element is widely used in practice for plane-stress or plane-strain analysis because it gives accurate results. The nodal displacements are u_i and v_i with $i = 1, 2, \dots, 6$. Equations 20-30 and 20-31 apply to this element, with the shape functions L_1 to L_6 derived by superposition for various shapes, in a way similar to that used for the quadrilateral element in Fig. 20-4.

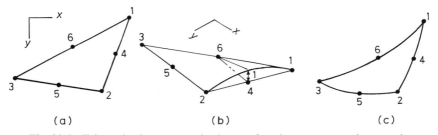

(a) (b) (c)

Fig. 20-9. Triangular isoparametric element for plane-stress or plane-strain analysis. Two degrees of freedom per node: u and v in x and y directions. (a) Node numbering. (b) Pictorial view of shape function L_4. (c) General quadratic triangle.

The shape functions for the mid-side nodes in Fig. 20-9a are

$$L_4 = 4\alpha_1 \alpha_2 \qquad L_5 = 4\alpha_2 \alpha_3 \qquad L_6 = 4\alpha_3 \alpha_1 \qquad (20\text{-}35)$$

A pictorial view of L_4, plotted perpendicular to the plane of the element, is shown in Fig. 20-9b. The function is quadratic along edge 1–2, but linear along lines α_1 or $\alpha_2 = $ constant.

The shape functions for corner nodes are obtained by an appropriate combination of L_4, L_5 and L_6 with the shape functions for the three-node element. For example, combining the shape function of Fig. 20-8 with $- L_4/2$ and $- L_6/2$ gives shape function L_1 for the six-node element. The function obtained in this way has a unit value at 1 and a zero value at all other nodes, as it should. The shape functions for the three corner nodes for the element in Fig. 20-9a are

$$L_1 = \alpha_1(2\alpha_1 - 1) \qquad L_2 = \alpha_2(2\alpha_2 - 1) \qquad L_3 = \alpha_3(2\alpha_3 - 1) \quad (20\text{-}36)$$

It can be seen from Eqs. 20-35 and 20-36 that the shape functions are quadratic, but the first derivatives, which give the strains, are linear. The six-node triangular element is sometimes referred to as a *linear-strain* triangle.

The six-node element can have curved sides. The x, y coordinates of any point are given by Eq. 20-31, using the shape functions in Eqs. 20-35 and 20-36. The lines of equal α will not be straight in this case, and Eq. 20-33 does not apply.

The stiffness matrix and the consistent load vectors for the six-node isoparametric triangular element can be generated using the equations given in Sec. 20-2 for a quadrilateral element, replacing ξ and η by α_1 and α_2, respectively. The third parameter α_3 is eliminated from the shape functions by substituting $\alpha_3 = 1 - \alpha_1 - \alpha_2$.

The integrals involved in the derivation of the matrices of triangular elements can be put in the form

$$\int_a f(\alpha_1, \alpha_2)\, da = \int_0^1 \int_0^{1-\alpha_1} f(\alpha_1, \alpha_2)|J|\, d\alpha_2\, d\alpha_1 \qquad (20\text{-}37)$$

where $f(\alpha_1, \alpha_2)$ is a function of α_1 and α_2, and $|J|$ is the determinant of the Jacobian. The inner integral on the right-hand side represents the value of the integral over a line of constant α_1; over such a line, the parameter α_2 varies between 0 and $1 - \alpha_1$.

The Jacobian can be determined by Eq. 20-8, replacing ξ and η as mentioned above. We can verify that, when the sides of the triangle are straight, $[J]$ is constant and its determinant is equal to twice the area.

The integral on the right-hand side of Eq. 20-37 is commonly evaluated numerically, using Eq. 20-48. However, in the case of a triangle with *straight* edges, the determinant $|J|$ can be taken out of the integral and the following closed-form equation can be used to evaluate the functions in the form:

$$\int_a \alpha_1^{n_1}\alpha_2^{n_2}\alpha_3^{n_3}\, da = 2\Delta\, \frac{n_1! n_2! n_3!}{(n_1 + n_2 + n_3 + 2)!} \qquad (20\text{-}38)$$

where n_1, n_2 and n_3 are any integers; the symbol ! indicates a factorial.

20-8 TRIANGULAR PLATE-BENDING ELEMENTS

A desirable triangular element which can be used for the analysis of plates in bending and for shell structures (when combined with plane-stress elements) should have three degrees of freedom per node, namely w, θ_x and θ_y, representing deflection in a global z direction and two rotations θ_x and θ_y

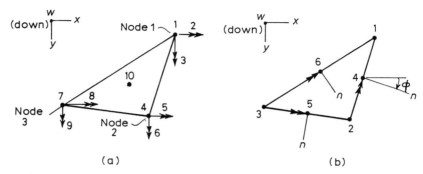

Fig. 20-10. Triangular bending elements. (a) Number of degrees of freedom 10: w, θ_x, θ_y at each corner and w at centroid. (b) Number of degrees of freedom 6: w at corners and $\partial w/\partial n$ at mid-side nodes.

(Fig. 20-10a). With nine degrees of freedom, the deflection w should be expressed by a polynomial with nine terms. Pascal's triangle (Fig. 19-7) indicates that a complete quadratic polynomial has six terms and a cubic polynomial has ten. Leaving out one of the terms x^2y or xy^2 gives an element for which the outcome is dependent upon the choice of x and y axes.

It is possible to express w using area coordinates and a polynomial of nine terms (which are not a simple selection). This gives a nonconforming element[2] which is widely used in practice because its results are accurate (although it does not pass the patch test, Sec. 19-10).

The two triangular elements shown in Fig. 20-10b and 20-10a have six and ten degrees of freedom, respectively. The deflection w for the two elements can be represented by complete quadratic and cubic polynomials, respectively, containing the top six or ten terms of Pascal's triangle (Fig. 19-7).

The six degrees of freedom for the element in Fig. 20-10b are

$$\{D^*\} = \left\{ w_1, w_2, w_3, \left(\frac{\partial w}{\partial n}\right)_4, \left(\frac{\partial w}{\partial n}\right)_5, \left(\frac{\partial w}{\partial n}\right)_6 \right\} \qquad (20\text{-}39)$$

where $\partial w/\partial n$ is the slope of a tangent normal to the sides half-way between the nodes. The positive direction of the normal vector is pointing either inwards or outwards, so that the angle ϕ, measured in the clockwise direction from the global x axis to the normal, is smaller than π. Thus, for the triangle in Fig. 20-10b, the normal at node 6 points inwards. However, for an adjacent triangle (not shown) sharing side 1–3, the normal points outwards; the two triangles

[2]See Bazeley, G. P., Cheung, Y. K., Irons, B. M. and Zienkiewicz, O. C., "Triangular Elements in Plate Bending: Conforming and Non-Conforming Solutions," Proceedings of the Conference on Matrix Methods in Structural Mechanics, Air Force Institute of Technology, Wright Patterson Base, Ohio, 1965.

share the same degree of freedom $\partial w/\partial n$, which is indicated in a positive direction (Fig. 20-10b). The derivative $\partial w/\partial n$ can be expressed as

$$\frac{\partial w}{\partial n} = \frac{\partial w}{\partial x}\cos\phi + \frac{\partial w}{\partial y}\sin\phi \tag{20-40}$$

The deflection is expressed by the upper six terms of Pascal's triangle (Fig. 19-7):

$$w = [1 \quad x \quad y \quad x^2 \quad xy \quad y^2]\{A\} = [P]\{A\} \tag{20-41}$$

Substituting for x and y by their values at the nodes, the nodal displacements can be expressed by $[D^*] = [C]\{A\}$, with

$$[C] = \begin{bmatrix} 1 & x_1 & y_1 & x_1^2 & x_1y_1 & y_1^2 \\ 1 & x_2 & y_2 & x_2^2 & x_2y_2 & y_2^2 \\ 1 & x_3 & y_3 & x_3^2 & x_3y_3 & y_3^2 \\ 0 & c_4 & s_4 & 2x_4c_4 & y_4c_4 + x_4s_4 & 2y_4s_4 \\ 0 & c_5 & s_5 & 2x_5c_5 & y_5c_5 + x_5s_5 & 2y_5s_5 \\ 0 & c_6 & s_6 & 2x_6c_6 & y_6c_6 + x_6s_6 & 2y_6s_6 \end{bmatrix} \tag{20-42}$$

where $s_i = \sin\phi_i$, $c_i = \cos\phi_i$.

The shape functions can be derived by Eq. 19-60, the $[B]$ matrix by Eqs. 19-39 and 19-19, and the stiffness matrix and consistent load vectors by Eqs. 19-45, 19-47 and 19-50. The strains for this element are constant and the element matrices are relatively simple to derive (see Prob. 20-8).

Despite its simplicity, the element with six degrees of freedom (Fig. 20-10b) gives fairly accurate results. It can be shown that the element is nonconforming because the inter-element deflections and their derivatives are not compatible. However, the element passes the patch test and the results converge quickly to the exact solution as the finite-element mesh is refined.[3]

The element in Fig. 20-10a has ten degrees of freedom $\{D^*\}$, as shown. The tenth displacement is a downward deflection at the centroid. The deflection can be expressed as $w = [L]\{D^*\}$, where $[L]$ is composed of ten shape functions given in terms of area coordinates α_1, α_2 and α_3 (Eq. 20-28) as follows:

$$\begin{aligned} L_1 &= \alpha_1^2(\alpha_1 + 3\alpha_2 + 3\alpha_3) - 7\alpha_1\alpha_2\alpha_3 \\ L_2 &= \alpha_1^2(b_2\alpha_3 - b_3\alpha_2) + (b_3 - b_2)\alpha_1\alpha_2\alpha_3 \\ L_3 &= \alpha_1^2(c_2\alpha_3 - c_3\alpha_2) + (c_3 - c_2)\alpha_1\alpha_2\alpha_3 \\ L_{10} &= 27\alpha_1\alpha_2\alpha_3 \end{aligned} \tag{20-43}$$

---

[3]Numerical comparison of results obtained by various types of triangular bending elements and convergence as the mesh is refined can be seen in Gallagher, R. H., *Finite Element Analysis Fundamentals*, Prentice-Hall, Englewood Cliffs, NJ, 1975, p. 350. The same book includes other triangular elements and a list of references on this topic.

where b_i and c_i are defined by Eq. 20-34. The equation for L_1 can be used for L_4 and L_7 by cyclic permutation of the subscripts 1, 2 and 3. Similarly, the equation for L_2 can be used for L_5 and L_8 and the equation for L_3 can be used for L_6 and L_9. We can verify that the ten shape functions and their derivatives take unit or zero values at the nodes, as they should. This involves the derivatives $\partial \alpha_i/\partial x$ and $\partial \alpha_i/\partial y$ which are, respectively, equal to $b_i/2\Delta$ and $c_i/2\Delta$, where Δ is the area of the triangle (see Eq. 20-33).

The inter-element displacement compatibility for $\partial w/\partial n$ is violated with the shape functions in Eq. 20-43, but the values of w of two adjacent elements are identical along a common side. This element does not give good accuracy.

However, the accuracy of the element shown in Fig. 20-10a can be improved by imposing constraints in the solution of the equilibrium equations of the assembled structure (Sec. 25-4). The constraints specify continuity of normal slopes $\partial w/\partial n$ at mid-points of element sides. With such treatment, the accuracy of this element will be better than that of the element shown in Fig. 20-10b.

20-9   NUMERICAL INTEGRATION

The Gauss method of integration is used extensively in generating the matrices of finite elements. The method is given below without derivation[4] for one-, two- and three-dimensional integration.

A definite integral of a one-dimensional function $g(\xi)$ (Fig. 20-11) can be

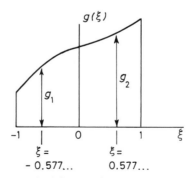

Fig. 20-11. Gauss numerical integration of function $g(\xi)$ by Eq. 20-44. Example for a number of sampling points $n = 2$.

---

[4]Derivation can be found in Irons, B. M. and Shrive, N. G., *Numerical Methods in Engineering and Applied Science*, Ellis Horwood, Chichester, England and Halsted Press (Wiley), New York, 1987.

calculated by

$$\int_{-1}^{1} g(\xi)d\xi \simeq \sum_{i=1}^{n} W_i g_i \qquad (20\text{-}44)$$

where g_i represents the values of the function at n sampling points $\xi_1, \xi_2, \ldots, \xi_n$, and W_i represents the weight factors. The sampling points are sometimes referred to as *Gauss points*. The values ξ_i and W_i are given to 15 decimal places in Table 20-1 for n between 1 and 6.[5] The values of ξ_i and W_i used in computer programs should include as many digits as possible.

Table 20-1. Location of Sampling Points and Weight Factors for Gauss Numerical Integration by Eqs. 20-44 to 20-46

n	ξ_i, η_i or ζ_i	W_i
1	0.000 000 000 000 000	2.000 000 000 000 000
2	± 0.577 350 269 189 626	1.000 000 000 000 000
3	0.000 000 000 000 000	0.888 888 888 888 888
	± 0.774 596 669 241 483	0.555 555 555 555 556
4	± 0.339 981 043 584 856	0.652 145 154 862 546
	± 0.861 136 311 594 053	0.347 858 485 137 454
5	0.000 000 000 000 000	0.568 888 888 888 889
	± 0.538 469 310 105 683	0.478 628 670 499 366
	± 0.906 179 845 938 664	0.236 926 885 056 189
6	± 0.238 619 186 083 197	0.467 913 934 572 691
	± 0.661 209 386 466 265	0.360 761 573 048 139
	± 0.932 469 514 203 152	0.171 324 492 379 170

The location of the sampling points, specified by the values of ξ_i in the table, is chosen to achieve the maximum accuracy for any given n. The sampling points are located symmetrically with respect to the origin.

If the function is a polynomial, $g = a_0 + a_1\xi + \cdots + a_n\xi^m$, Eq. 20-44 is exact when the number of sampling points is $n \geqslant (m+1)/2$. In other words, n Gauss points are sufficient to integrate exactly a polynomial of order $2n - 1$. For example, we can verify that using $n = 2$ (Fig. 20-11) gives exact integrals for each of the functions $g = a_0 + a_1\xi$, $g = a_0 + a_1\xi + a_2\xi^2$ and $g = a_0 + a_1\xi + a_2\xi^2 + a_3\xi^3$. The two sampling points are at $\xi = \pm 0.577\ldots$ and the weight factors are equal to unity.

We should note that the sum of the weight factors in Eq. 20-44 is 2.0.

A definite integral for a two-dimensional function $g(\xi, \eta)$ can be calculated by

$$\int_{-1}^{1} \int_{-1}^{1} g(\xi, \eta)\,d\xi\,d\eta \simeq \sum_{j}^{n_\eta} \sum_{i}^{n_\xi} W_j W_i\, g(\xi_i, \eta_j) \qquad (20\text{-}45)$$

[5]Values of ξ_i and W_i for n between 1 and 10 can be found on p. 79 of the reference in footnote 4 of this chapter. See also Kopal, Z., *Numerical Analysis*, 2nd ed., Chapman and Hall, 1961.

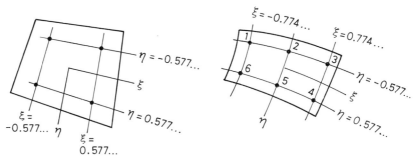

Fig. 20-12. Examples of sampling points for two-dimensional Gauss numerical integration.

where n_ξ and n_η are the number of sampling points on ξ and η axes, respectively. Figure 20-12 shows two examples of sampling points. In most cases, n_ξ and n_η are the same (equal to n); then $n \times n$ Gauss points integrate exactly a polynomial $g(\xi, \eta)$ of order $2n - 1$.

Equation 20-45 can be derived from Eq. 20-44 by integrating first with respect to ξ and then with respect to η. For 3×2 Gauss points ($n_\xi = 3, n_\eta = 2$), Eq. 20-45 can be rewritten (Fig. 20-12) as

$$\int_{-1}^{1} \int_{-1}^{1} g(\xi, \eta)\, d\xi\, d\eta$$

$$\simeq (1.0)(0.555\cdots)g_1 + (1.0)(0.888\cdots)g_2 + (1.0)(0.555\cdots)g_3$$

$$+ (1.0)(0.555\cdots)g_4 + (1.0)(0.888\cdots)g_5 + (1.0)(0.555\cdots)g_6 \quad (20\text{-}46)$$

We can note that the sum of the multipliers of g_1 to g_6 is 4.0.

For a function in three dimensions, $g(\xi, \eta, \zeta)$, the integral is

$$\int_{-1}^{1} \int_{-1}^{1} \int_{-1}^{1} g(\xi, \eta, \zeta)\, d\xi\, d\eta\, d\zeta = \sum_k^{n_\zeta} \sum_j^{n_\eta} \sum_i^{n_\xi} W_k W_j W_i g(\xi_i, \eta_j, \zeta_k) \quad (20\text{-}47)$$

Equations 20-44, 20-45 and 20-47 can be used for one-, two- and three-dimensional elements with curvilinear axes.

When the element is triangular and the area coordinates α_1, α_2 and α_3 are used (Fig. 20-8a), any variable over the area of the element can be expressed as a function $f(\alpha_1, \alpha_2)$, in which α_3 is eliminated by substituting $\alpha_3 = 1 - \alpha_1 - \alpha_2$. The integral $\int\int f\, da$ over the area of the triangle is expressed by Eq. 20-37, and can be evaluated numerically using

$$\int_0^1 \int_0^{1-\alpha_1} g(\alpha_1, \alpha_2)\, d\alpha_2\, d\alpha_1 \simeq \sum_{i=1}^n W_i g_i \quad (20\text{-}48)$$

In this equation, the symbol $g(\alpha_1, \alpha_2)$ stands for the product $f(\alpha_1, \alpha_2)|J|$ in

	Point	α_1 or α_2	Point	Weight factor
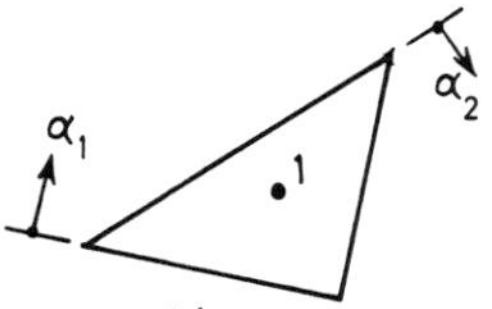 **Linear**	1	0.333 333 333 3	1	W_1 = 0.500 000 000 0
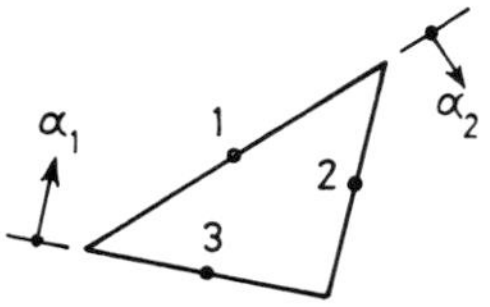 **Quadratic**	3 1,2	0.000 000 000 0 0.500 000 000 0	1 2,3	$W_{1,2,3}$ = 0.166 666 666 7
Cubic	4,6,7 1 2,3 5	0.000 000 000 0 0.333 333 333 3 0.500 000 000 0 1.000 000 000 0	2,5,7 1 3,4 6	W_1 = 0.225 000 000 0 $W_{2,3,4}$ = 0.066 666 666 7 $W_{5,6,7}$ = 0.025 000 000 0
Quintic	4 6,7 1 2,3 5	0.059 715 871 8 0.101 286 507 3 0.333 333 333 3 0.470 142 064 1 0.797 426 985 4	2 5,7 1 3,4 6	W_1 = 0.112 500 000 0 $W_{2,3,4}$ = 0.066 197 076 4 $W_{5,6,7}$ = 0.062 969 590 3

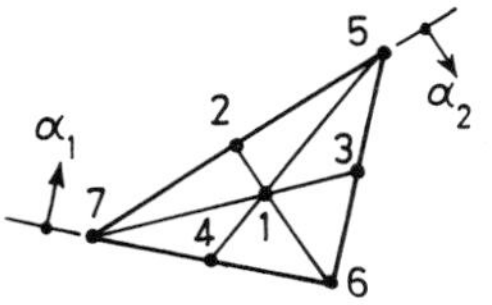

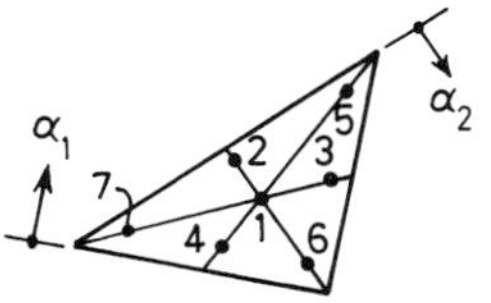

Fig. 20-13. Numerical integration of function $g(\alpha_1, \alpha_2, \alpha_3)$ over the area of a triangle. Values of α_1, α_2 and α_3 defining sampling points and weight factors W_i for use in Eq. 20-48 ($\alpha_3 = 1 - \alpha_1 - \alpha_2$).

Eq. 20-37, where $|J|$ is the determinant of the Jacobian.

The location of the sampling points and the corresponding weight factors for use in Eq. 20-48 are given in Fig. 20-13*. These values give the exact integrals when g is linear, quadratic, cubic and quintic, as indicated.

We should note that the sum of the weight factors for all the sampling points of any triangle is equal to one-half. For a triangle with straight edges,

*See Cowper, G. R. (1973) Gaussion quadrature formulas for triangles. *Int. J. Num. Math. Eng.*, **7**, 405–8 and Zienkiewicz, O. C., *The Finite Element Method in Engineering Science*, McGraw-Hill, London, 1977.

$|J| = 2\Delta$, where Δ is the area. If we set $f(\alpha_1, \alpha_2) = 1$, the function $g(\alpha_1, \alpha_2) = 2\Delta$, and the value of the integral given by Eq. 20-48 is equal to $(2\Delta)\sum W_i = \Delta$, as it should.

The number of sampling points in the numerical integration to generate the stiffness matrices of finite elements influences the accuracy and the convergence. The appropriate choice of sampling points is discussed in more detail in references devoted specifically to the finite-element method.[6]

A small number of points is generally used in order to reduce computation. Furthermore, the use of a small number of points tends to reduce the stiffness, thus compensating for the excess in stiffness associated with displacement-based finite elements (see Sec. 19-5). However, the number of sampling points cannot be reduced without limit. Some elements have a *spurious mechanism* when a specified pattern of sampling points is used in the integration for the derivation of the stiffness matrix. This mechanism occurs when the element can deform in such a way that the strains at the sampling points are zero.

Two-by-two sampling points are frequently used in quadrilateral plane linear elements with corner nodes only or in quadratic elements with corner and mid-side nodes. Three or four sampling points in each direction are used when the elements are elongated or in cubic[7] elements.

Example 20-3 Determine coefficient S^*_{11} of the stiffness matrix of the quadrilateral element of Example 20-1 and Fig. 19-5. The element is of isotropic material; Poisson's ratio $v = 0.2$. Also find the consistent restraining force at coordinate 1 when the element is subjected to a constant rise of temperature of T degrees. Use only one Gauss integration point.

The displacement field corresponding to $D^*_1 = 1$ is $\{u, v\} = \{L_1, 0\}$, where

$$L_1 = \tfrac{1}{4}(1 - \xi)(1 - \eta)$$

Derivatives of L_1 with respect to x and y at the center ($\xi = \eta = 0$) are (using Eq. 20-6, with $[J]$ from Example 20-1)

$$\begin{Bmatrix} \partial L_1/\partial x \\ \partial L_1/\partial y \end{Bmatrix} = \begin{bmatrix} 15/4 & 5/4 \\ -3/4 & 13/4 \end{bmatrix}^{-1} \begin{Bmatrix} -(1-\eta)/4 \\ -(1-\xi)/4 \end{Bmatrix} = \frac{1}{210} \begin{Bmatrix} -8 \\ -18 \end{Bmatrix}$$

When $D^* = 1$ the strains at the same point are (Eq. 20-10)

$$\{B\}_1 = \tfrac{1}{210}\{-8, 0, -18\}$$

The value of the product in the stiffness matrix of Eq. 20-12 at the center of the

---

[6]See, for example, the reference in footnote 5 of Chapter 19. Other selected books are included in the General References to this book.

[7]For example, element QLC3; see footnote 4 of Chapter 19, and Example 19-3.

element (Eq. 19-12) is

$$g_1 = h\{B\}_1^T[d]\{B\}_1|J|$$

$$= \frac{1}{(210)^2}[-8, \ 0, \ -18]\frac{Eh}{1-(0.2)^2}\begin{bmatrix}1 & 0.2 & 0 \\ 0.2 & 1 & 0 \\ 0 & 0 & (1-0.2)/2\end{bmatrix}\begin{Bmatrix}-8 \\ 0 \\ -18\end{Bmatrix}|J|$$

$$= 4.57 \times 10^{-3}\, Eh|J|$$

The determinant of the Jacobian is 13.125 (see Example 20-1). Substitution in Eq. 20-45 with $W_1 = 2$ (from Table 20-1) gives

$$S_{11}^* = 2 \times 2(4.57 \times 10^{-3})Eh(13.125) = 0.240Eh$$

The free strains due to the temperature rise are $\alpha T\{1, 1, 0\}$. Substitution in Eq. 20-14 gives

$$F_{b1}^* = 2 \times 2(13.125)(-h)\left(\frac{1}{210}\right)[-8, 0, -18]$$

$$\times \frac{E}{1-(0.2)^2}\begin{bmatrix}1 & 0.2 & 0 \\ 0.2 & 1 & 0 \\ 0 & 0 & (1-0.2)/2\end{bmatrix}\begin{Bmatrix}1 \\ 1 \\ 0\end{Bmatrix}\alpha T$$

$$= 2.5E\alpha Th$$

Example 20-4 Determine the Jacobian at $(\alpha_1, \alpha_2) = (1/3, 1/3)$ in a triangular isoparametric element with corner and mid-side nodes, where α_1 and α_2 are area coordinates. The (x, y) coordinates of the nodes are: $(x_1, y_1) = (0, 0)$; $(x_2, y_2) = (1, 1)$; $(x_3, y_3) = (0, 1)$; $(x_4, y_4) = ((\sqrt{2})/2, 1, -(\sqrt{2})/2)$; $(x_5, y_5) = (0.5, 1)$; $(x_6, y_6) = (0, 0.5)$. These correspond to a quarter of a circle with its center at node 3 and a radius of unity. By numerical integration of Eq. 20-48, determine the area of the element, using only one sampling point.

The shape functions L_1 to L_6 are given by Eqs. 20-35 and 20-36. We eliminate α_3 by substituting $\alpha_3 = 1 - \alpha_1 - \alpha_2$, and then determine the derivatives $\partial L_i/\partial \alpha_1$ and $\partial L_i/\partial \alpha_2$ at $(\alpha_1, \alpha_2) = (1/3, 1/3)$. The Jacobian at this point is (from Eq. 20-8, replacing ξ and η by α_1 and α_2)

$$[J] = \begin{bmatrix}0\left(\dfrac{1}{3}\right) & 0\left(\dfrac{1}{3}\right) \\ 0(0) & 0(0)\end{bmatrix} + \begin{bmatrix}1(0) & 1(0) \\ 1\left(\dfrac{1}{3}\right) & 1\left(\dfrac{1}{3}\right)\end{bmatrix} + \begin{bmatrix}0\left(-\dfrac{1}{3}\right) & 1\left(-\dfrac{1}{3}\right) \\ 0\left(-\dfrac{1}{3}\right) & 1\left(-\dfrac{1}{3}\right)\end{bmatrix}$$

$$+ \begin{bmatrix}\dfrac{\sqrt{2}}{2}\left(\dfrac{4}{3}\right) & \left(1-\dfrac{\sqrt{2}}{2}\right)\left(\dfrac{4}{3}\right) \\ \dfrac{\sqrt{2}}{2}\left(\dfrac{4}{3}\right) & \left(1-\dfrac{\sqrt{2}}{2}\right)\left(\dfrac{4}{3}\right)\end{bmatrix} + \begin{bmatrix}0.5\left(-\dfrac{4}{3}\right) & 1\left(-\dfrac{4}{3}\right) \\ 0.5(0) & 1(0)\end{bmatrix} +$$

$$+ \begin{bmatrix} 0(0) & 0.5(0) \\ 0\left(-\dfrac{4}{3}\right) & 0.5\left(-\dfrac{4}{3}\right) \end{bmatrix}$$

$$= \begin{bmatrix} 0.2761 & -1.2761 \\ 1.2761 & -0.2761 \end{bmatrix}$$

The area of the element is (from Eq. 20-48 with W_1 from Fig. 20-13)

$$\text{area} = \int_0^1 \int_0^{1-\alpha_2} |J| \, d\alpha_2 \, d\alpha_1 = 0.5|J| = 0.7761$$

20-10 SHELLS AS ASSEMBLAGE OF FLAT ELEMENTS

Finite elements in the form of flat quadrilateral or triangular plates can be used to idealize a shell (Fig. 20-14). In general, the elements will be subjected to in-plane forces and to bending. The element matrices derived separately in earlier sections for elements in a state of plane stress and for bending elements can be combined for a shell element as discussed below.

Fig. 20-14. Cylindrical shells idealized as an assemblage of flat elements.

Idealization of a shell, using curved elements, may be necessary if large elements are employed, particularly in double-curved shells. However, in practice many shells, particularly those of cylindrical shape, have been analyzed successfully using triangular, quadrilateral or rectangular elements.

Flat shell elements can be easily combined with beam elements to idealize edge beams or ribs, which are common in practice.

20-101 Rectangular Shell Element

Figure 20-15 represents a rectangular shell element with six nodal displacements at each corner: three translations $\{u, v, w\}$ and three rotations $\{\theta_x, \theta_y, \theta_z\}$. The derivation of the matrices for a plane-stress rectangular element with nodal displacements u, v and θ_z at each corner was discussed in Sec. 19-8 (see Example 19-3). A rectangular bending element with nodal displacements w, θ_x and θ_y at each corner was discussed in Sec. 19-43.

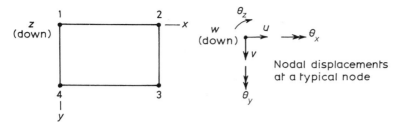

Fig. 20-15. Rectangular flat shell element.

The stiffness matrix for the shell element in Fig. 20-15 can be written in the form

$$[S^*] = \begin{bmatrix} [S_m^*] & [0] \\ [0] & [S_b^*] \end{bmatrix}$$ (20-49)

where $[S_m^*]$ is a 12×12 membrane stiffness matrix relating forces to the displacements u, v and θ_z, and $[S_b^*]$ is a 12×12 bending stiffness matrix relating forces to the displacements w, θ_x and θ_y. The off-diagonal submatrices in Eq. 20-49 are null because the first set of displacements produces no forces in the directions of the second set, and vice versa. The two sets of displacements are said to be *uncoupled*.

For convenience in coding, the degrees of freedom at each node are commonly arranged in the order $\{u, v, w, \theta_x, \theta_y, \theta_z\}$. The columns and rows in the 24×24 stiffness matrix in Eq. 20-49 are rearranged so that the nodal coordinates are numbered in the following sequence: six coordinates at node 1, followed by six coordinates at node 2, and so on.

20-102 Fictitious Stiffness Coefficients

The majority of plane-stress elements have two degrees of freedom per node, u and v. Combining such an element with a bending element (with nodal displacements w, θ_x and θ_y) so as to form a shell element will leave the θ_z coordinate unused. The five degrees of freedom, u, v, w, θ_x and θ_y, are not sufficient to analyze a spatial structure. This can be seen by considering elements in two intersecting planes; no matter how the global axes are chosen, the structure will, in general, have three rotation components in the three global directions at the nodes on the line of intersection.

We may think of generating a stiffness matrix for the shell element with six degrees of freedom per node, with zero columns and rows corresponding to the θ_z coordinates. However, when elements situated in one plane meet at a node, a zero will occur at the diagonal of the stiffness matrix of the assembled structure; this will result in an error message with common computer equation solvers (Secs. 24-11 and A-9) when attempting to divide a number by the stiffness coefficient on the diagonal. To avoid this difficulty Zienkiewicz[8] assigns fictitious stiffness coefficients, instead of zeros, in the columns and rows corresponding to θ_z. For a triangular element with three nodes, the fictitious coefficients are

$$[S]_{fict} = \beta E \begin{bmatrix} 1 & -0.5 & -0.5 \\ -0.5 & 1 & -0.5 \\ -0.5 & -0.5 & 1 \end{bmatrix} \text{(volume)} \qquad (20\text{-}50)$$

where E is the modulus of elasticity, the multiplier at the end is the volume of the element, and β is an arbitrarily chosen coefficient. Zienkiewicz suggested a value $\beta = 0.03$ or less. We can note that the sum of the fictitious coefficients in any column is zero, which is necessary for equilibrium.

20-11   SOLIDS OF REVOLUTION

Figure 20-16 represents a solid of revolution which is subjected to axisymmetrical loads (not shown). For the analysis of stresses in this body, it is divided into finite elements in the form of rings, the cross sections of which are triangles or quadrilaterals. The element has nodal circular lines, rather than point nodes. The displacements at the nodes are u and w in the radial direction r and the z direction, respectively.

The arrangement of the nodes and the shape functions used for plane-stress and plane-strain elements (Secs. 19-42, 19-11, 20-2, 20-5 and 20-7) can be used for axisymmetric solid elements. The parameters x, y, u and v used for plane elements have to be replaced by r, z, u and w, respectively.

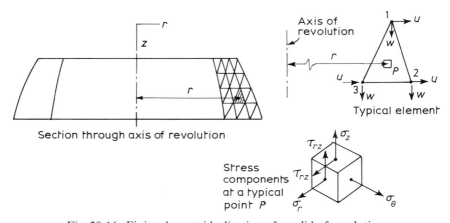

Fig. 20-16. Finite-element idealization of a solid of revolution.

[8]See reference mentioned in footnote 1 of Chapter 19.

One additional strain component needs to be considered in solids of revolution, namely the hoop strain ε_θ, which represents elongation in the tangential direction, normal to the plane of the element. Thus the strain vector has four components:

$$\{\varepsilon\} = \begin{Bmatrix} \varepsilon_r \\ \varepsilon_w \\ \varepsilon_\theta \\ \gamma_{rz} \end{Bmatrix} = \begin{bmatrix} \dfrac{\partial}{\partial r} & 0 \\ 0 & \dfrac{\partial}{\partial w} \\ \dfrac{1}{r} & 0 \\ \dfrac{\partial}{\partial w} & \dfrac{\partial}{\partial r} \end{bmatrix} \begin{Bmatrix} u \\ w \end{Bmatrix} \tag{20-51}$$

The stress vector has also one additional component, the hoop stress σ_θ (Fig. 20-16). The stress–strain relation for solids of revolution of isotropic material is

$$\{\sigma\} = [d]\{\varepsilon\} \tag{20-52}$$

where

$$\{\sigma\} = \{\sigma_r, \sigma_w, \sigma_\theta, \gamma_{rz}\} \tag{20-53}$$

and

$$[d] = \frac{E}{(1+v)(1-2v)} \begin{bmatrix} 1-v & v & v & 0 \\ v & 1-v & v & 0 \\ v & v & 1-v & 0 \\ 0 & 0 & 0 & \dfrac{1-2v}{2} \end{bmatrix} \tag{20-54}$$

The element stiffness matrix for an axisymmetric solid element is given by (see Eq. 19-45)

$$[S^*] = 2\pi \iint [B]^T[d][B] r\, dr\, dz \tag{20-55}$$

When the natural coordinates ξ and η are used, the elemental area $dr\, dz$ is replaced by $|J|\, d\xi\, d\eta$, where $|J|$ is the determinant of the Jacobian (Eq. 20-7 or 20-8). The limits for each of the two integrals become -1 to 1.

Any element S_{ij}^* of the stiffness matrix represents $2\pi r_i$ multiplied by the intensity of a uniform load distributed on a nodal line at node i when the displacement at j is 1. Thus S_{ij}^* has the units of force per length.

If the element is subjected to a rise in temperature of T degrees and the expansion is restrained, the stress at any point is

$$\{\sigma_0\} = -E\alpha T[d]\{1, 1, 1, 0\} \tag{20-56}$$

and the consistent vector of restraining forces is

$$\{F_b\} = 2\pi \iint [B]^T\{\sigma_0\} r\, dr\, dz \tag{20-57}$$

20-12   FINITE STRIP AND FINITE PRISM

Figure 20-17 shows a top view of a simply supported bridge with three types of cross section. Analysis of stress and strain can be performed by treating the structure as an assemblage of finite strips or finite prisms as shown.[9] Similarly to the elements used in the analysis of solids of revolution (Sec. 20-11), the finite strips and finite prisms have nodal lines rather than node points.

Over a nodal line, any displacement component is expressed as a sum of a series $\sum_{k=1}^{n} a_k Y_k$, where the values a are unknown amplitudes. The terms Y are functions of y which satisfy *a priori* the displacement conditions at the extremities of the strip.

An example of Y functions suitable for the determination of the deflection of a simply supported strip is

$$Y_k = \sin\frac{\mu_k y}{l} \quad \text{with } \mu_k = k\pi \qquad (20\text{-}58)$$

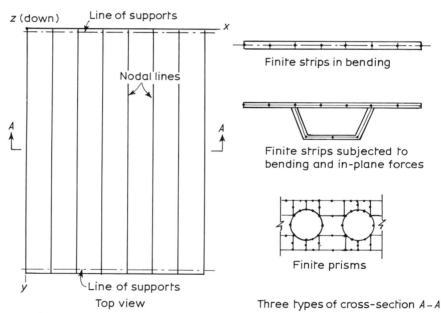

Fig. 20-17. Example of the use of finite strips and finite prisms for the analysis of a simply supported bridge with three types of cross section: slab, box and voided slab.

[9]The theory and applications of finite strips and finite prisms are discussed in more detail in Cheung, Y. K., *Finite Strip Method in Structural Analysis*, Pergamon Press, Oxford, New York, Toronto, 1976.

Over the width of the strip, or over the cross-sectional area of the prism, the displacement components are assumed to vary as polynomials of x or of x and z. The same polynomials as used for one-dimensional or two-dimensional elements are generally used for the strip and the prism, respectively.

In general, the displacement components in a finite strip or a finite prism are expressed as

$$\{f\} = \sum_{k=1}^{n} [L]_k \{D^*\}_k \qquad (20\text{-}59)$$

where $\{D^*\}_k$ is a vector of nodal parameters (displacement amplitudes, i.e. the a values), and $[L]_k$ is a matrix of shape functions pertaining to the kth term of the series. The nodal displacement parameters for a strip or a prism are

$$\{D^*\} = \{\{D^*\}_1, \{D^*\}_2, \ldots, \{D^*\}_n\} \qquad (20\text{-}60)$$

For the finite strip in Fig. 20-18a, usable for the analysis of plates in bending,

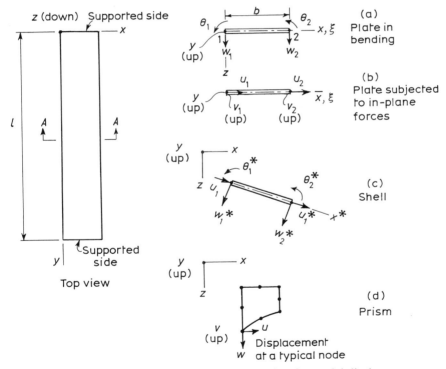

Fig. 20-18. Top view and cross section $A - A$ showing nodal displacement parameters for (a) finite strip in bending (b) finite strip subjected to in-plane forces (c) finite strip subjected to bending combined with in-plane forces (a shell) (d) finite prism.

the displacement field $\{f\} \equiv \{w\}$ represents deflection, and the nodal parameters are defined as

$$\{D^*\}_k = \{w_1, \theta_1, w_2, \theta_2\}_k \tag{20-61}$$

where the subscripts 1 and 2 refer to nodal lines along the sides of the strip; $\theta = \theta_y = -\partial w/\partial x$. The strip has $4n$ degrees of freedom.

The shape functions for a finite strip in bending are (Fig. 20-18a)

$$[L]_k = [1 - 3\xi^2 + 2\xi^3, -b\xi(\xi - 1)^2, \xi^2(3 - 2\xi), -b\xi^2(\xi - 1)]Y_k \tag{20-62}$$

We recognize that the functions in the 1×4 matrix are the same as the shape functions for a bar in bending (Fig. 19-4); here, b is the width of the strip and $\xi = x/b$.

A finite strip suitable for the analysis of plates subjected to in-plane forces is shown in Fig. 20-18b. The displacement components at any point are $\{f\} \equiv \{u, v\}$ and the nodal displacement parameters are

$$\{D^*\}_k = \{u_1, v_1, u_2, v_2\} \tag{20-63}$$

The same Y functions which are used for plates in bending are suitable to describe the variation of u in the y direction. However, for the variation of v it is suitable to use the derivative $Y' = dY/dy$. This choice can represent the behavior of a beam in bending with the same cross section as the strip and subjected to forces in the x direction.

The u displacement represents the deflection in the direction of the applied load. If the beam is simply supported, the sinusoidal Y function (Eq. 20-58) can be used to represent the variation of u in the y direction. However, the displacement in the v direction (caused by the shortening or elongation of the fibers due to bending) is equal to Y' multiplied by the distance from the neutral axis of the beam.

In general, to describe the variation of the displacement components in the y direction, we use the Y_k functions for u and w, and $Y'_k l/\mu_k$ for v. Thus, when Y is a sine series (Eq. 20-58), the variation of v is described by a cosine series.

The shape functions for a finite strip subjected to in-plane forces are (Fig. 20-18b)

$$[L]_k = \begin{bmatrix} (1 - \xi)Y_k & 0 & \xi & 0 \\ 0 & (1 - \xi)\dfrac{l}{\mu_k}Y'_k & 0 & \xi\dfrac{l}{\mu_k}Y'_k \end{bmatrix} \tag{20-64}$$

For the variation of each of u and w over the width of the strip, we use linear functions (Fig. 20-3).

When a strip is subjected to bending combined with in-plane forces, the nodal displacement parameters will be those defined by Eqs. 20-61 and 20-63 combined. We should note that the two sets of displacements are uncoupled.

Thus the stiffness matrix of a finite strip, suitable for the analysis of shells, folded plates or box girders, is obtained by combining the stiffness matrices of a bending strip and a plane-stress strip as shown in Eq. 20-49.

The finite strip for shell analysis is shown in Fig. 20-18c in a general inclined position. The nodal displacement parameters are indicated in local directions. The stiffness matrix has to be derived with respect to these local directions and then transformed to correspond to the global directions before the assemblage of strip matrices can be performed (see Secs. 24-8 and 24-9).

The nodal displacement parameters for the finite prism in Fig. 20-18d are

$$\{D^*\}_k = \{u_1, v_1, w_1, u_2, v_2, w_2, \ldots, u_8, v_8, w_8\}_k \qquad (20\text{-}65)$$

The shape functions for the finite prism can be expressed by the shape functions for the plane-stress element shown in Fig. 20-5 multiplied by Y_k or by $(l/\mu_k)Y'_k$.

20-121   Stiffness Matrix

The strains at any point in a finite strip or a finite prism are given by Eq. 19-6:

$$\{\varepsilon\} = [\partial]\{f\} \qquad (20\text{-}66)$$

The differential operator $[\partial]$ is given by Eqs. 19-19, 19-11 and 19-24 for the elements in Figs. 20-18a,b and d, respectively.

Substitution of Eq. 20-59 into Eq. 20-66 gives

$$\{\varepsilon\} = [B]\{D^*\} = \sum_{k=1}^{n} [B]_k \{D^*\}_k \qquad (20\text{-}67)$$

The $[B]$ matrix is thus partitioned into n submatrices, given by

$$[B]_k = [\partial][L]_k \qquad (20\text{-}68)$$

The stiffness matrix of a finite strip or a finite prism can be generated by Eq. 19-45, which is repeated here:

$$[S^*] = \int_v [B]^T[d][B] \, dv \qquad (20\text{-}69)$$

The number of rows or columns in $[S^*]$ is $n_1 n_2 n$, where n_1 is the number of nodal lines in the finite strip or prism, n_2 is the number of degrees of freedom per node, and n is the number of terms in the Y series.

The matrix $[S^*]$ can be partitioned into $n \times n$ submatrices, of which a typical submatrix is

$$[S^*]_{kr} = \int_v [B]_k^T[d][B]_r \, dv \qquad (20\text{-}70)$$

Generation of the stiffness submatrix $[S^*]_{kr}$ by Eq. 20-70 involves integration over the length of nodal lines of the products $Y_k Y_r$, $Y_k'' Y_r''$, $Y_k' Y_r'$ and $Y_k Y_r''$. For $k \neq r$, the integrals for the first two products are zero when Y is given by any of Eqs. 20-58, 20-79, 20-80 or 20-81. However, the integrals of the remaining two products are also zero only when Y is a sine series (Eq. 20-58). Because of these properties, with $k \neq r$, Eq. 20-70 gives submatrices $[S^*]_{kr}$ equal to zero only when Y is a sine series. In this case, the stiffness matrix of the finite strip or finite prism takes the form

$$[S^*] = \begin{bmatrix} [S^*]_{11} & & \text{submatrices not} \\ & [S^*]_{22} & \text{shown are null} \\ & & \cdots \\ & & [S^*]_{nn} \end{bmatrix} \qquad (20\text{-}71)$$

The stiffness matrices of individual strips or prisms have to be assembled to obtain the stiffness matrix of the structure (see Sec. 24-9).

The stiffness matrix of the assembled structure will also have the form of Eq. 20-71 when Y is a sine series. Thus the equilibrium equations $[S]\{D\} = -\{F\}$ will uncouple, which means that, instead of a large set of simultaneous equations having to be solved, n subsets (of much smaller band width) are solved separately. Each subset is of the form $[S]_k\{D\}_k = -\{F\}_k$ and its solution gives the nodal displacement parameters for the kth term of the Y series. This results in a substantial reduction of computing.

The variation of a displacement component over a nodal line is given by

$$D_i = \sum_{k=1}^{n} D_{ik} Y_k \qquad (20\text{-}72)$$

where $D_{i1}, D_{i2}, \ldots, D_{in}$ are the nodal parameters for D_i obtained by the solution of the equilibrium equations.

20-122 Consistent Load Vector

When a strip or a prism is subjected to body forces $\{p\}$ per unit volume, the consistent vector of restraining forces is given by Eq. 19-47, which is repeated here:

$$\{F_b^*\} = -\int_v [L]^T \{p\} \, dv \qquad (20\text{-}73)$$

The vector $\{F_b^*\}$ can be partitioned into n subvectors, given by

$$\{F_b^*\}_k = -\int_v [L]_k^T \{p\} \, dv \qquad (20\text{-}74)$$

Similarly, the consistent vector of forces to restrain a temperature expansion is given by Eq. 19-50. The vector can be partitioned into subvectors, given by

$$\{F_b^*\}_k = - \int_v [B]_k^T [d] \{\varepsilon_0\} \, dv \qquad (20\text{-}75)$$

where $\{\varepsilon_0\}$ represents the strains if the expansion were free to occur.

Element matrices for a finite strip for the analysis of plates in bending (Fig. 20-18a) are given explicitly in Sec. 20-124.

We shall now consider the case when an external load of intensity q per unit length is applied on a nodal line in the direction of one of the nodal coordinates. The displacement due to this loading can be restrained by a consistent force equal to the sum of n terms of the form

$$F_{ak}^* = - \int_0^l q Y_k \, dy \qquad (20\text{-}76)$$

When a concentrated force P is applied on the nodal line, the integral in Eq. 20-74 is replaced by the product $P \, Y_k(y_p)$, where $Y_k(y_p)$ is the value of Y_k at y_p defining the location of P:

$$F_{ak}^* = - P \, Y_k(y_p) \qquad (20\text{-}77)$$

The vector of restraining forces to prevent the nodal displacements is (Eq. 24-32)

$$\{F\} = \{F_a\} + \{F_b\} \qquad (20\text{-}78)$$

where $\{F_a\}$ accounts for external forces applied on the nodal lines, calculated by Eqs. 20-76 and 20-77 and transformed to global directions. The vector $\{F_b\}$ accounts for body forces. Consistent force vectors for individual strips or prisms are calculated by Eqs. 20-73 to 20-75, transformed to global directions and assembled into one vector by Eq. 24-34.

20-123 Displacement Variation over a Nodal Line

The variation of all displacement components in the y direction is described by the Y functions which satisfy the boundary conditions at $y = 0$ and $y = l$. Functions of the shape of free vibration modes of prismatic beams are suitable.[10]

For simply supported finite strips or prisms, Y and Y'' are zero at $y = 0$ and $y = l$; these conditions are satisfied when Y is a sine series (Eq. 20-58). When one end is simply supported and the other end is built in, Y and Y'' are

[10]Derivations of the Y functions used here are given in Vlasov, V. Z., *General Theory of Shells and its Applications in Engineering*; Translated from Russian by NASA; distributed by Office of Technical Services, Department of Commerce, Washington, DC, 1949, pp. 699–707. This reference gives Y functions suitable for other boundary conditions not included above.

zero at $y = 0$ while Y and Y' are zero at $y = l$. These conditions are satisfied by

$$Y_k = \sin \frac{\mu_k y}{l} - \alpha_k \sinh \frac{\mu_k y}{l} \tag{20-79}$$

where

$$\mu_k = 3.927, 7.068, \ldots, \frac{4k + 1}{4}\pi \qquad \alpha_k = \frac{\sin \mu_k}{\sinh \mu_k}$$

When both ends of a finite strip or prism are built in, Y and Y' are zero there:

$$Y_k = \sin \frac{\mu_k y}{l} - \sinh \frac{\mu_k y}{l} + \alpha_k \left(\cosh \frac{\mu_k y}{l} - \cos \frac{\mu_k y}{l} \right) \tag{20-80}$$

where

$$\mu_k = 4.730, 7.853, \ldots, \frac{2k + 1}{2}\pi \qquad \alpha_k = \frac{\sinh \mu_k - \sin \mu_k}{\cosh \mu_k - \cos \mu_k}$$

When the end $y = 0$ is built in and the other end is free, Y and Y' are zero at $y = 0$ while Y'' and Y''' are zero at $y = l$. These conditions are satisfied by

$$Y_k = \sin \frac{\mu_k y}{l} + \sinh \frac{\mu_k y}{l} + \alpha_k \left(\cosh \frac{\mu_k y}{l} - \cos \frac{\mu_k y}{l} \right) \tag{20-81}$$

where

$$\mu_k = 1.875, 4.694, \ldots, \frac{2k - 1}{2}\pi \qquad \alpha_k = \frac{\sinh \mu_k + \sin \mu_k}{\cosh \mu_k + \cos \mu_k}$$

20-124   Plate-bending Finite Strip

The strains at any point in the finite strip shown in Fig. 20-18a are given by Eq. 20-67, in which the matrix $[B]_k$ is (Eqs. 20-68, 19-19 and 20-62)

$$[B]_k = \begin{bmatrix} \frac{1}{b^2}(6 - 12\xi)Y_k & \frac{1}{b}(-4 + 6\xi)Y_k & \frac{1}{b^2}(12\xi - 6)Y_k & \frac{1}{b}(-2 + 6\xi)Y_k \\ -(1 - 3\xi^2 + 2\xi^3)Y_k'' & b\xi(\xi - 1)^2 Y_k'' & -\xi^2(3 - 2\xi)Y_k'' & 6\xi^2(\xi - 1)Y_k'' \\ \frac{2}{b}(-6\xi + 6\xi^2)Y_k' & 2(-3\xi^2 + 4\xi - 1)Y_k' & \frac{1}{b}(6\xi - 6\xi^2)Y_k' & 2(-3\xi^2 + 2\xi)Y_k' \end{bmatrix} \tag{20-82}$$

For a simply supported strip, $Y_k = \sin(k\pi y/l)$, and the stiffness matrix submatrices $[S^*]_{kr}$ are nonzero only when $k = r$.

Equation 20-70 can be rewritten in the form

$$[S^*]_{kk} = b \int_0^l \int_0^1 [B]_k^T [d][B]_k \, d\xi \, dy \tag{20-83}$$

The elasticity matrix $[d]$ for a plate in bending is given by Eq. 19-20 or Eq. 19-21. For a strip of constant thickness, the integrals involved in Eq. 20-83 can be expressed explicitly, noting that

$$\int_0^l Y_k^2 \, dy = \frac{l}{2} \qquad \int_0^l Y_k Y_k'' \, dy = -\frac{k^2\pi^2}{2l}$$

$$\int_0^l (Y_k'')^2 \, dy = \frac{k^4\pi^4}{2l^3} \qquad \int_0^l (Y_k')^2 \, dy = \frac{k^2\pi^2}{2l} \qquad (20\text{-}84)$$

Thus the submatrix $[S^*]_{kk}$ of a simply supported finite strip is (Fig. 20-18a)

$$= b \begin{bmatrix}
\dfrac{6l}{b^4}d_{11} + \dfrac{13k^4\pi^4}{70l^3}d_{22} & & & & \text{symmetrical} \\[2mm]
+\dfrac{6k^2\pi^2}{5b^2l}d_{21} + \dfrac{12k^2\pi^2}{5b^2l}d_{33} & & & & \\[4mm]
\dfrac{3l}{b^3}d_{11} - \dfrac{11bk^4\pi^4}{420l^3}d_{22} & \dfrac{2l}{b^2}d_{11} + \dfrac{b^2k^4\pi^4}{210l^3}d_{22} & & \\[2mm]
-\dfrac{3k^2\pi^2}{5bl}d_{21} - \dfrac{k^2\pi^2}{5bl}d_{33} & +\dfrac{2k^2\pi^2}{15l}d_{21} & & \\[2mm]
& +\dfrac{4k^2\pi^2}{15l}d_{33} & & \\[4mm]
-\dfrac{6l}{b^4}d_{11} + \dfrac{9k^4\pi^4}{140l^3}d_{22} & \dfrac{3l}{b^3}d_{11} - \dfrac{13bk^4\pi^4}{840l^3}d_{22} & \dfrac{6l}{b^4}d_{11} + \dfrac{13k^4\pi^4}{70l^3}d_{22} & \\[2mm]
-\dfrac{6k^2\pi^2}{5b^2l}d_{21} - \dfrac{12k^2\pi^2}{5b^2l}d_{33} & +\dfrac{k^2\pi^2}{10bl}d_{21} + \dfrac{k^2\pi^2}{5bl}d_{33} & +\dfrac{6k^2\pi^2}{5b^2l}d_{21} + \dfrac{12k^2\pi^2}{5b^2l}d_{33} & \\[4mm]
-\dfrac{3l}{b^3}d_{11} + \dfrac{13bk^4\pi^4}{840l^3}d_{22} & \dfrac{l}{b^2}d_{11} - \dfrac{b^2k^4\pi^4}{280l^3}d_{22} & \dfrac{3l}{b^3}d_{11} + \dfrac{11bk^4\pi^4}{420l^3}d_{22} & \dfrac{2l}{b^2}d_{11} \\[2mm]
& & & +\dfrac{b^2k^4\pi^4}{210l^3}d_{22} \\[2mm]
-\dfrac{k^2\pi^2}{10bl}d_{21} - \dfrac{k^2\pi^2}{5bl}d_{33} & -\dfrac{k^2\pi^2}{30l}d_{21} - \dfrac{k^2\pi^2}{15l}d_{33} & +\dfrac{3k^2\pi^2}{5bl}d_{21} + \dfrac{k^2\pi^2}{5bl}d_{33} & +\dfrac{2k^2\pi^2}{15l}d_{21} \\[2mm]
& & & +\dfrac{4k^2\pi^2}{15l}d_{33}
\end{bmatrix}$$

$$(20\text{-}85)$$

where d_{ij} represents elements of the elasticity matrix (Eq. 19-20 or Eq. 19-21).
When a strip is subjected to a load in the z direction, distributed over a

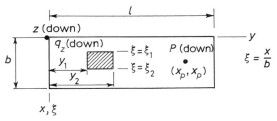

Fig. 20-19. Finite strip loads considered in Eqs. 20-86 and 20-87.

rectangle (Fig. 20-19), the consistent subvector of restraining forces is (see Eq. 20-74)

$$\{F_b^*\}_k = -b \int_{y_1}^{y_2} \int_{\xi_1}^{\xi_2} q_z [L]_k^T d\xi \, dy \qquad (20\text{-}86)$$

where q_z is the load intensity.

When a concentrated load P is applied at (x_p, y_p) (Fig. 20-19), Eq. 20-86 becomes

$$\{F_b^*\}_k = -P[L(\xi_p, y_p)]_k^T \qquad \text{with } \xi_P = x_P/b \qquad (20\text{-}87)$$

When the load q_z is constant over the whole area of the strip, Eq. 20-86 becomes

$$\{F_b^*\}_k = -q_z \left\{ \frac{b}{2}, -\frac{b^2}{12}, \frac{b}{2}, \frac{b^2}{12} \right\} \int_0^l Y_k \, dy \qquad (20\text{-}88)$$

When a sine series is used for Y_k (Eq. 20-58), the integral in Eq. 20-88 can be replaced by $2l/k\pi$ or by zero when k is odd or even, respectively.

When a plate in bending is stiffened by a beam running along a nodal line, compatibility of displacement will be ensured by assuming that the deflection w and the angle of twist θ vary over the beam length according to the same Y function as used for the finite strips. To account for the contribution of the beam to the stiffness of the system, we can derive flexural and torsional stiffnesses for the beam (see answers to Prob. 20-15) and add them to the diagonal coefficients, corresponding to w and θ, in the stiffness matrix of the finite strips assemblage.

Example 20-5   For a horizontal rectangular plate with sides of length l and c, determine the deflection at the center due to (a) a downward line load of p per unit length along the axis of symmetry parallel to the dimension l, and (b) a uniform downward load of q per unit area. The plate is simply supported at its four sides, has a constant thickness h, and is of isotropic material with a modulus of elasticity E and Poisson's ratio $v = 0.3$. The aspect ratio is $l/c = 2$.

Divide the slab into two strips parallel to the l dimension. Because of symmetry, only one strip needs to be considered; this strip can be represented by Fig. 20-18a, with nodal lines 1 and 2 running along the supported edge and the slab center line, respectively. The width of the strip is $b = c/2 = l/4$. The displacement parameters w_1 and θ_2 are zero. This leaves only two unknowns:

$$\{D\} = \{\theta_1, w_2\}$$

In case (a), only one-half of the line load intensity $(p/2)$ is applied to the half-slab considered.

For an isotropic slab with thickness h, the elements of the elasticity matrix are (Eq. 19-21)

$$d_{11} = d_{22} = \frac{Eh^3}{12(1 - v^2)} = 91.58 \times 10^{-3} Eh^3$$

$$d_{21} = vd_{11} = 27.47 \times 10^{-3} Eh^3$$

$$d_{33} = \frac{Eh^3}{24(1 + v)} = 32.05 \times 10^{-3} Eh^3$$

The submatrix $[S]_{kk}$ for the half-slab is given by Eq. 20-85 after deletion of the first and the fourth rows and columns:

$$[S]_{kk}$$
$$= 10^{-3} Eh^3 \begin{bmatrix} 732.6 + 0.6637k^4 + 30.13k^2 & \text{symmetrical} \\ \dfrac{1}{c}(2198 - 4.314k^4 + 45.19k^2) & \dfrac{1}{c^2}(8791 + 103.5k^4 + 1085k^2) \end{bmatrix}$$

The consistent subvectors of restraining forces for the two loading cases are null for values of k which are even numbers. For $k = 1, 3, \ldots$, the subvectors are (Eqs. 20-76 and 20-88)

$$[F]_k = - \begin{bmatrix} 0 & -\dfrac{b^2}{12}\left(\dfrac{2l}{k\pi}\right) \\ \dfrac{p}{2}\dfrac{2l}{k\pi} & q\dfrac{b}{2}\left(\dfrac{2l}{k\pi}\right) \end{bmatrix} = \begin{bmatrix} 0 & q\dfrac{c^3}{12k\pi} \\ -\dfrac{2pc}{k\pi} & -\dfrac{qc^2}{k\pi} \end{bmatrix}$$

The equilibrium equations are

$$[S]_k\{D\}_k = -\{F\}_k$$

Substitution and solution with $k = 1$ and $k = 3$ gives the nodal displacement parameters in the two cases of loading:

$$[D]_1 = \frac{c^2}{Eh^3}\begin{bmatrix} -0.5470p & -0.3751qc \\ 0.1865pc & 0.1161qc^2 \end{bmatrix}$$

$$[D]_3 = \frac{c^2}{Eh^3}\begin{bmatrix} -0.0205p & -0.0205qc \\ 0.0096pc & 0.0057qc^2 \end{bmatrix}$$

It can be seen that the contribution of the third term ($k = 3$) is much smaller than the contribution of the first term. Using $k = 1$ and $k = 3$, the deflection at the center of the slab ($\xi = 1$; $y = l/2$) is as follows (Eq. 20-59):

Case (a):

$$w_{center} = \frac{pc^3}{Eh^3} \left(0.1865 \sin \frac{\pi}{2} + 0.0096 \sin \frac{3\pi}{2} \right) = 0.1769 \frac{pc^3}{Eh^3}$$

Case (b):

$$w_{center} = \frac{qc^4}{Eh^3} \left(0.1161 \sin \frac{\pi}{2} + 0.0057 \sin \frac{3\pi}{2} \right) = 0.1104 \frac{qc^4}{Eh^3}$$

The "exact" answers[11] to the same problem are $0.1779pc^3/Eh^3$ and $0.1106qc^4/Eh^3$, respectively.

While high accuracy in the calculation of deflection is achieved with only two strips and two nonzero terms, for the same accuracy in the calculation of moments a larger number of strips and of terms may be necessary, particularly with concentrated loads.

Finally, the displacement parameters can be used to calculate the strains (curvatures) and the stresses (moments) at any point by Eqs. 20-82, 19-15 and 19-40.

20-13 HYBRID FINITE ELEMENTS

In displacement-based finite elements, the displacement components are assumed to vary as (Eq. 19-25)

$$\{f\} = [L]\{D^*\} \tag{20-89}$$

where $\{D^*\}$ are nodal displacements and $[L]$ are displacement shape functions. The strains and the stress components are determined from the displacements by Eqs. 19-37 to 19-40.

Two techniques of formulating hybrid finite elements will be discussed: *hybrid stress* and *hybrid strain* formulations.[12] In the first of these, the displacements are assumed to vary according to Eq. 20-89 and the stresses according to an assumed polynomial

$$\{\sigma\} = [P]\{\beta\} \tag{20-90}$$

In the second technique, in addition to the displacement field (Eq. 20-89), a strain field is assumed:

$$\{\varepsilon\} = [P]\{\alpha\} \tag{20-91}$$

---

[11]See Timoshenko S. and Wojnowsky-Krieger S., *Theory of Plates and Shells* (2nd edn), McGraw-Hill, New York, 1959.

[12]The material in this section is based on Ghali, A. and Chieslar, J., "Hybrid Finite Elements," *Journal of Structural Engineering*, 112 (11) (November 1986), American Society of Civil Engineers, pp. 2478–2493.

In the last two equations, $[P]$ represents assumed polynomials; $\{\beta\}$ and $\{\alpha\}$ are unknown multipliers referred to as the stress parameter and the strain parameter, respectively.

In the hybrid stress or hybrid strain formulation, a finite element has two sets of unknown parameters, $\{D^*\}$ with $\{\beta\}$ or $\{D^*\}$ with $\{\alpha\}$. However, the parameters $\{\beta\}$ or $\{\alpha\}$ are eliminated, by condensation, before assemblage of the element matrices in order to obtain the stiffness matrix of the structure. The equilibrium equations $[S]\{D\} = -\{F\}$ are generated and solved by the general techniques discussed in Secs. 24-9 to 24-11.

In the formulation of displacement-based elements and hybrid elements, the forces applied on the body of the element are represented by a vector of consistent restraining nodal forces (Eq. 19-47)

$$\{F_b^*\} = -\int_v [L]^T\{p\}\, dv \tag{20-92}$$

where $\{p\}$ are body forces per unit volume.

20-131 Stress and Strain Fields

We give here, as an example, the polynomials $[P]$ selected successfully[13] to describe the stress field by Eq. 20-90 in a plane-stress or plane-strain hybrid stress element, with corner nodes and two degrees of freedom per node (Fig. 20-20):

$$[P] = \begin{bmatrix} 1 & y & 0 & 0 & 0 \\ 0 & 0 & 1 & x & 0 \\ 0 & 0 & 0 & 0 & 1 \end{bmatrix} \tag{20-93}$$

The stress vector in this example is $\{\sigma\} = \{\sigma_x, \sigma_y, \tau_{xy}\}$.

For a general quadrilateral, x and y can be replaced by the natural coordinates ξ and η, and the use of the Jacobian matrix will be necessary in the

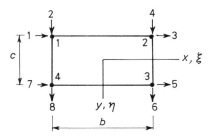

Fig. 20-20. Hybrid finite element for plane-stress and plane-strain analysis.

[13]Pian, T. H. H. and Sumihara, K., "Rational Approach for Assumed Stress Finite Elements," *International Journal for Numerical Methods in Engineering*, 20 (1984), pp. 1685–1695.

derivation of element matrices (Sec. 20-2). The same matrix $[P]$ can be used for a hybrid strain element.

The number of stress or strain parameters, $\{\beta\}$ or $\{\alpha\}$, must be at least equal to the difference between the number of degrees of freedom of the element and the minimum number of degrees of freedom at which displacements have to be restrained in order to prevent rigid-body motion. The element shown in Fig. 20-20 has eight degrees of freedom, and rigid-body motion will not occur if the displacements are prevented at three coordinates (e.g. two in the x direction and one in the y direction). Thus, for this element, the minimum number of parameters in $\{\beta\}$ or $\{\alpha\}$ is five.

20-132 Hybrid Stress Formulation

For this formulation we shall use the principle of virtual work (Eq. 5-35), which is rewritten here as

$$\{F^*\}^T\{D^*\} = \int \{\sigma\}^T\{\varepsilon\}\, dv \qquad (20\text{-}94)$$

We recall that two interpretations are possible for the symbols in this equation. In one, $\{F^*\}$ and $\{\sigma\}$ represent actual nodal forces and corresponding stresses, while $\{D^*\}$ and $\{\varepsilon\}$ represent virtual nodal displacements and the corresponding strains. In the second interpretation, $\{D^*\}$ and $\{\varepsilon\}$ are real while $\{F^*\}$ and $\{\sigma\}$ are virtual.

The two interpretations will be used in the formulation of hybrid elements. Equations 20-90 and 20-91 will be considered to represent real stresses and strains, but the virtual strains are chosen as the derivatives of the displacement shape functions (Eq. 19-39)

$$[B] = [\partial][L] = [\{B\}_1 \{B\}_2 \cdots \{B\}_n] \qquad (20\text{-}95)$$

where $[\partial]$ is a matrix of differential operators (see, for example, the 3×2 matrix in Eq. 19-11), and n is the number of nodal displacements. Each column of $[B]$ may be used to represent a virtual strain field.

We apply the principle of virtual work (Eq. 20-94) to a finite element subjected to nodal forces $\{F^*\}$ producing stresses $\{\sigma\} = [P]\{\beta\}$. The virtual displacement field is assumed to correspond to $D_1^* = 1$, while the other nodal displacements are zero. The corresponding virtual strains are

$$\{\varepsilon\}_1 = \{B\}_1 = [\partial]\{L\}_1 \qquad (20\text{-}96)$$

For this application, Eq. 20-94 gives

$$\{F^*\}^T \left\{ \begin{matrix} 1 \\ 0 \\ \ldots \end{matrix} \right\} = \int_v \{\sigma\}^T\{\varepsilon\}_1\, dv \qquad (20\text{-}97)$$

Substitution of Eq. 20-90 produces

$$\{F^*\}^T \left\{ \begin{matrix} 1 \\ 0 \\ \cdots \end{matrix} \right\} = \int_v ([P]\{\beta\})^T \{B\}_1 \, dv \tag{20-98}$$

By repeated application of Eq. 20-94, using pairs of virtual vectors, $\{0, 1, 0, \ldots\}$ and $\{B\}_2$, and so on, and combining the resulting equation, we obtain

$$\{F^*\}^T [I] = \{\beta\}^T \int_v [P]^T [B] \, dv \tag{20-99}$$

We can rewrite Eq. 20-99 as

$$\{F^*\} = [G]^T \{\beta\} \tag{20-100}$$

where

$$[G] = \int_v [P]^T [B] \, dv \tag{20-101}$$

The elements in any row i of $[G]$ are forces at the nodes when the stresses are represented by the ith column of $[P]$. The elements in any row of $[G]$ form a set of self-equilibrating forces (see Example 20-6).

The procedure to generate $[G]$, used above, is tantamount to a repeated application of the unit-displacement theorem (Eq. 5-39) in order to determine nodal forces corresponding to given stress fields $\{\sigma\} = [P]\{\beta\}$.

The strains in the element considered above can be determined from the stresses by

$$\{\varepsilon\} = [e]\{\sigma\} = [e][P]\{\beta\} \tag{102}$$

where

$$[e] = [d]^{-1}$$

Here, $[d]$ is the material elasticity matrix (Eqs. 5-11, 19-12, 19-13, 19-20 and 19-21); $[e]$ is given in Eq. 5-9 for an isotropic three-dimensional solid.

Let us now apply again the principle of virtual work (Eq. 20-94) but this time with $\{D^*\}$ and $\{[e][P]\{\beta\}\}$ representing real displacements and strains, and $[G]^T$ and $[P]$ representing virtual forces and stresses:

$$[G]\{D^*\} = \int_v [P]^T [e][P]\{\beta\} \, dv \tag{20-103}$$

or

$$[G]\{D^*\} = [H]\{\beta\} \tag{20-104}$$

where

$$[H] = \int [P]^T [e][P] \, dv \tag{20-105}$$

The matrix $[H]$ is a symmetrical matrix which we refer to as the *quasi-*

flexibility matrix. The explanation of this term is as follows. For a regular flexibility matrix, we should integrate the product of stress and strain produced by unit forces at the nodes. For the quasi-flexibility matrix, we also integrate the product of stress and strain, but these correspond to nodal forces $[G]$. We can obtain the stiffness matrix of the hybrid stress element by inverting $[H]$ and pre- and post-multiplying by $[G]^T$ and $[G]$, respectively:

$$[S^*]_{\text{hybrid stress}} = [G]^T[H]^{-1}[G] \tag{20-106}$$

This equation can be derived from Eq. 20-100 by elimination of $\{\beta\}$ using Eq. 20-104:

$$[G]^T[H]^{-1}[G]\{D^*\} = \{F^*\} \tag{20-107}$$

or

$$[S^*]\{D^*\} = \{F^*\} \tag{20-108}$$

A comparison of the last two equations gives Eq. 20-106.

When the nodal displacements have been determined (by the solution of the equilibrium equations for the structure) the stress parameters for individual hybrid stress elements can be determined by Eq. 20-104:

$$\{\beta\} = [H]^{-1}[G]\{D^*\} \tag{20-109}$$

The stresses due to unit nodal displacements in the hybrid stress element are:

$$[\sigma_u]_{\text{hybrid stress}} = [P][H]^{-1}[G] \tag{20-110}$$

Here, $[\sigma_u]$ is referred to as the *stress matrix*. The product $[P][H]^{-1}[G]$ is the equivalent of the product $[d][B]$ in displacement-based finite elements (Eqs. 19-41 and 19-42).

Example 20-6   Generate the matrices $[G]$ and $[H]$ for a hybrid stress rectangular element for plane-stress analysis (Fig. 20-20). Use the polynomial matrix of Eq. 20-93 and the shape functions given by Eqs. 19-31 and 19-32. The element has a constant thickness h, modulus of elasticity E, and Poisson's ratio v.

$$[B] = \begin{bmatrix} \partial/\partial x & 0 \\ 0 & \partial/\partial y \\ \partial/\partial y & \partial/\partial x \end{bmatrix} [L]$$

$$= \frac{1}{4} \begin{bmatrix} -\dfrac{2}{b}(1-\eta) & 0 & \dfrac{2}{b}(1-\eta) & 0 & \dfrac{2}{b}(1+\eta) & 0 & -\dfrac{2}{b}(1+\eta) & 0 \\[2mm] 0 & -\dfrac{2}{c}(1-\xi) & 0 & -\dfrac{2}{c}(1+\xi) & 0 & \dfrac{2}{c}(1+\xi) & 0 & \dfrac{2}{c}(1-\xi) \\[2mm] -\dfrac{2}{c}(1-\xi) & -\dfrac{2}{b}(1-\eta) & -\dfrac{2}{c}(1+\xi) & \dfrac{2}{b}(1-\eta) & \dfrac{2}{c}(1+\xi) & \dfrac{2}{b}(1+\eta) & \dfrac{2}{c}(1-\xi) & -\dfrac{2}{b}(1+\eta) \end{bmatrix}$$

Substitution in Eq. 20-101 gives

$$[G] = \int_v [P]^T[B]\,dv = h\frac{bc}{4}\int_{-1}^{1} [P]^T[B]\,d\xi\,d\eta$$

$$= hbc \begin{bmatrix}
-1/(2b) & 0 & 1/(2b) & 0 & 1/(2b) & 0 & -1/(2b) & 0 \\
c/(12b) & 0 & -c/(12b) & 0 & c/(12b) & 0 & -c/(12b) & 0 \\
0 & -1/(2c) & 0 & -1/(2c) & 0 & 1/(2c) & 0 & 1/(2c) \\
0 & b/(12c) & 0 & -b/(12c) & 0 & b/(12c) & 0 & -b/(12c) \\
-1/(2c) & -1/(2b) & -1/(2c) & 1/(2b) & 1/(2c) & 1/(2b) & 1/(2c) & -1/(2b)
\end{bmatrix}$$

The inverse of the elasticity matrix for the plane-stress state is (Eq. 19-12)

$$[e] = [d]^{-1} = \frac{1}{E}\begin{bmatrix} 1 & -v & 0 \\ -v & 0 & 0 \\ 0 & 0 & 2(1+v) \end{bmatrix}$$

Substitution of this matrix and of Eq. 20-93 in Eq. 20-105 gives

$$[H] = \frac{hbc}{E}\begin{bmatrix}
1 & & & & \\
0 & c^2/12 & & \text{symmetrical} & \\
-v & 0 & 1 & & \\
0 & 0 & 0 & b^2/12 & \\
0 & 0 & 0 & 0 & 2(1+v)
\end{bmatrix}$$

The accuracy of the results obtained by this element is compared with two displacement-based elements in Prob. 20-18.

20-133   Hybrid Strain Formulation

We shall now consider an element in which the strain field is assumed to be $\{\varepsilon\}$ $= [P]\{\alpha\}$ (Eq. 20-91) while the corresponding stress field is determined from

$$\{\sigma\} = [d]\{\varepsilon\} = [d][P]\{\alpha\} \tag{20-111}$$

Application of the principle of virtual work (Eq. 20-94) in two different ways, as was done in Sec. 20-132, gives

$$\{F^*\} = [\bar{G}]\{\alpha\} \tag{20.112}$$

and

$$[\bar{G}]\{D^*\} = [\bar{H}]\{\alpha\} \tag{20-113}$$

where

$$[\bar{G}] = \int_v [P]^T[d][B]\,dv \tag{20.114}$$

and

$$[\bar{H}] = \int_v [P]^T[d][P]\,dv \tag{20.115}$$

The elements in any row i of $[\bar{G}]$ are the nodal forces which produce the strain vector represented by the ith column of $[P]$. The symmetrical matrix $[\bar{H}]$ is referred to as the *quasi-stiffness* matrix.

The stiffness matrix of a hybrid strain element is

$$[S^*]_{\text{hybrid strain}} = [\bar{G}]^T [\bar{H}]^{-1} [\bar{G}] \qquad (20.116)$$

The strain parameters are given by (Eq. 20.113)

$$\{\alpha\} = [\bar{H}]^{-1} [\bar{G}] \{D^*\} \qquad (20.117)$$

The stress matrix for a hybrid strain element is

$$[\sigma_u]_{\text{hybrid strain}} = [d][P][\bar{H}]^{-1}[\bar{G}] \qquad (20.118)$$

The product of the four matrices in this equation can be used to replace the product $[d]$ $[B]$ in displacement-based finite elements.

The assumption that the strain field is continuous over the element (rather than the stresses being continuous) is more representative of actual conditions in some applications. These include composite materials and materials with a nonlinear stress–strain relationship.

20.14 GENERAL

In this chapter and in Chapter 19 we have discussed the main concepts and techniques widely used in the analysis of structures by the finite-element method. This was possible because the finite-element method is an application of the displacement method, which is extensively discussed in earlier chapters and in Chapters 24 and 25.

Selected books on the finite-element method are included in the General References to this book. Some of these books contain extensive lists of references on the subject.

PROBLEMS

20-1   Generate $[S]$ for the isoparametric bar element of Prob. 19-3, assuming that the cross-sectional area equals $2a_0$, a_0 and $1.25a_0$ at nodes 1, 2 and 3, respectively. For interpolation between these values, use the displacement shape functions. Evaluate the integrals numerically with two sampling points. Would you expect a change in the answer if three or four sampling points were used?

20-2   The figure shows two load distributions on an element boundary. Determine the consistent vectors of restraining forces, assuming that the variation of displacement is (a) linear between each two adjacent nodes (b) quadratic polynomial. Use Lagrange interpolation (Fig. 20-3). The answers to (a) are included in Fig. 9-11. Compare the answers to (b) when $q_1 = q_2 = q_3 = q$ with the consistent nodal forces due to a rise in temperature as given in Fig. 20-7.

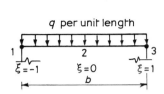 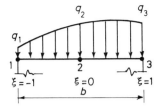

Prob. 20-2

20-3   Use the shape functions in Fig. 20-5 to verify the consistent vectors of restraining nodal forces given in Fig. 20-7.

20-4   Using Lagrange polynomials (Fig. 20-3), construct shape functions L_1, L_2, L_{10} and L_{11} for the element shown. Using these shape functions and considering symmetry and equilibrium, determine the consistent vector of nodal forces in the y direction when the element is subjected to a uniform load of q_y per unit volume.

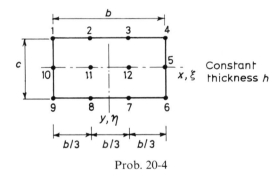

Prob. 20-4

20-5   Determine the strains at nodes 1, 2 and 3 in the linear-strain triangle (Fig. 20-9) due to $D_1^* \equiv u_1 = 1$ and to $D_2^* \equiv v_1 = 1$.

20-6   Show that the shape functions of Eqs. 20-35 and 20-36 give linear strains in the element shown in Fig. 20-9. The answers to Prob. 20-5 give the strains at the three corners of the triangle. By linear interpolation, using area coordinates, determine the strains at any point in the element. These give the first two columns of the $[B]$ matrix.

20-7   Use the answers to Prob. 20-6 to determine S_{11}^* and S_{12}^* of the linear-strain triangle (Fig. 20-9).

20-8   Derive $[B]$ and $[S^*]$ for the triangular bending element with equal sides l shown. Assume a constant thickness h and an isotropic material with $v = 0.3$.

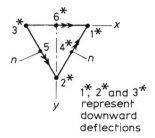

1*, 2* and 3* represent downward deflections

Prob. 20-8

20-9   The hexagonal horizontal plate shown is simply supported at the corners and carries a downward concentrated load P at the center. Calculate the deflection at this point and the moments at A by idealizing the plate as an assemblage of six triangular elements identical to the element of Prob. 20-8. Check the value of M_x by taking the moment of the forces situated to the left or to the right of the y axis.

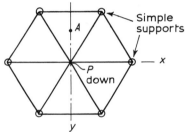

Prob. 20-9

20-10   If the plate in Prob. 20-9 is built-in along the edges, find the deflection at the center and the moments M_x, M_y and M_{xy} at A caused by a change in temperature of $T/2$ and $-T/2$ degrees at the bottom and top surfaces, respectively, with a linear variation over the thickness. Take the coefficient of thermal expansion as α.

20-11   Using one strip and one term of the series, calculate the deflection at A and also M_x at B in the plate shown. The plate is subjected to a uniform load of q per unit area. Assume a constant thickness h and Poisson's ratio $v = 0.2$.

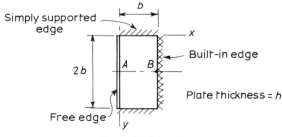

Prob. 20-11

20-12   Solve Prob. 20-11 for a uniform line load of p per unit length applied along the plate center line parallel to the x axis.

20-13   Find the deflection at the center of a square, simply supported plate due to a uniform load of q per unit area. Assume a constant thickness h, a length of side l, and an isotropic material with $v = 0.3$. Divide the slab into two finite strips and use one term of the series.

20-14   Find the variation in the deflection of the plate in Example 20-5 along the center line parallel to the l dimension, due to a hydrostatic pressure $q_z = q_0 y/l$. Divide the plate into two strips of length l and width $c/2$, and use three terms of the series.

20-15   The figure shows a top view of a rectangular plate simply supported at two edges, with each of the other two edges stiffened by a beam. The centroid of the beam cross section is in the plane of middle surface of the plate; the cross-sectional area properties of the beam are $I = 3lh^3/100$ and $J = 4lh^3/100$. Poisson's ratio $v = 0$. The plate is subjected to a concentrated load P at the center.
   (a) Derive the flexural and torsional stiffnesses for the beam which can be combined with the finite strip stiffness. Use a sine series (Eq. 20-58) to describe the variation in the deflection of the beam, w, and in the angle of twist, θ.
   (b) Find the deflection at the center of the slab and the bending moment in the edge beams at mid-span. Divide the plate into two strips and use one term of the series.

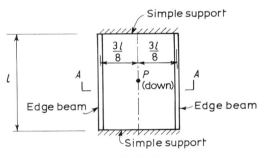

Prob. 20-15

20-16   Derive $S_{11}^*, S_{21}^*, S_{31}^*, S_{41}^*$ and S_{51}^* for the hybrid plane-stress element in Fig. 20-20. Assume a constant thickness h and an isotropic material with $v = 0.2$.

20-17   Use the results of Prob. 20-16 and consider symmetry and equilibrium to generate $[S^*]$ for the hybrid rectangular element when $b = c$.

20-18   Use the stiffness matrices generated in Probs. 19-7, 19-15 and 20-17 to calculate the deflection at A, the stress σ_x at B, and τ_{xy} at C for one or more of the cantilevers shown. The answers to this problem are given in the figure, which indicates the degree of accuracy when we use two displacement-based elements and a hybrid element. The stresses are determined at two points on a vertical section at the middle of an element rather than at the corners because, at the chosen locations, the stresses are more accurate.

Width of cantilevers = h ; $\nu = 0.2$

Element type	Deflection at A	Normal stress σ_x at B	Shear stress τ_{xy} at C
1	4.46	2.14	1.00
2	5.01	3.00	1.17
3	5.40	3.00	1.00
Exact	6.88	3.00	1.50
1	25.4	6.43	1.00
2	33.3	9.00	1.15
3	34.8	9.00	1.00
Exact	37.8	9.00	1.50
1	79.2	10.7	1.00
2	107.7	15.0	1.15
3	112.2	15.0	1.00
Exact	116.6	15.0	1.50
1	182.4	15.0	1.00
2	251.2	21.0	1.15
3	261.6	21.0	1.00
Exact	267.5	21.0	1.50
Multipliers	P/Eh	P/hb	

Element types:

1 Displacement based, 8 degrees of freedom (Example 19-2)

2 Displacement based, 12 degrees of freedom (element QLC3, Example 19-3)

3 Hybrid stress, 8 degrees of freedom (Example 20-6)

Prob. 20-18

CHAPTER 21

Plastic analysis of continuous beams and frames

21-1 INTRODUCTION

An elastic analysis of a structure is important to study its performance, especially with regard to serviceability, under the loading for which the structure is designed. However, if the load is increased until yielding occurs at some locations, the structure undergoes elastic-plastic deformations, and on further increase a fully plastic condition is reached, at which a sufficient number of *plastic hinges* are formed to transform the structure into a mechanism. This mechanism would collapse under any additional loading. A study of the mechanism of failure and the knowledge of the magnitude of the collapse load are necessary to determine the load factor in analysis. Alternatively, if the load factor is specified, the structure can be designed so that its collapse load is equal to, or higher than, the product of the load factor and the service loading.

Design of structures based on the plastic approach (referred to as *limit design*) is increasingly used and accepted by various codes of practice, particularly for steel construction. The material is assumed to deform in the idealized manner shown in Fig. 21-1. The strain and stress are proportional to one another up to the yield stress, at which the strain increases indefinitely without any further increase in stress. This type of stress-strain relation is

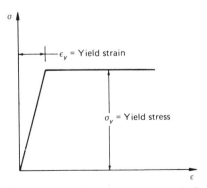

Fig. 21-1. Idealized stress-strain relation.

not greatly removed from that existing in mild steel. However, there exists a reserve of strength due to strain hardening but this will not be allowed for in the analysis in this chapter.

We shall now consider the principles of plastic analysis of plane frames in which buckling instability is prevented and fatigue or brittle failure is not considered possible. In most cases, the calculation of the collapse load involves trial and error, which may become tedious in large structures. A general procedure of limit analysis by the displacement method suitable for use with computers will be discussed in Chapter 25.

21-2   ULTIMATE MOMENT

Consider a beam whose cross section has an axis of symmetry as shown in Fig. 21-2a. Let the beam be subjected to bending in the plane of symmetry. If the bending moment is small, the stress and the strain vary linearly across the section as shown in Fig. 21-2b. When the moment is increased, yield stress is attained in the top fiber (Fig. 21-2c), and with a further increase the yield stress is reached in the bottom fiber as well, as shown in Fig. 21-2d. If the bending moment continues to increase, yield will spread from the outer fibers inward until the two zones of yield meet (Fig. 21-2e); the cross section in this state is said to be *fully plastic*.

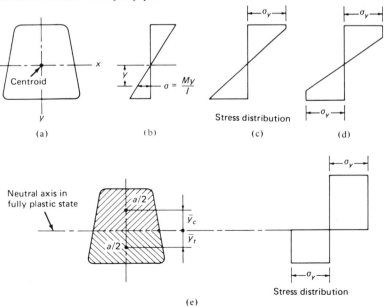

Fig. 21.2. Stress distribution in a symmetrical cross section subjected to a bending moment of increasing magnitude. (a) Beam cross section. (b) Elastic. (c) Plastic at top fiber. (d) Plastic at top and bottom fibers.

The value of the ultimate moment in the fully plastic condition can be calculated in terms of the yield stress σ_y. Since the axial force is zero in the case considered, the neutral axis in the fully plastic condition divides the section into two equal areas, and the resultant tension and compression are each equal to $(a\,\sigma_y/2)$, forming a couple equal to the ultimate moment

$$M_p = \frac{1}{2} a\,\sigma_y(\bar{y}_c + \bar{y}_t) \qquad (21\text{-}1)$$

where $\bar{y}_c$ and $\bar{y}_t$ are respectively the distance of the centroid of the compression and tension area from the neutral axis in the fully plastic condition.

The maximum moment which a section can carry without exceeding the yield stress is $M_y = \sigma_y Z$, where Z is the section modulus. The ratio $\alpha = M_p/M_y$ depends on the shape of the cross section and is referred to as the *shape factor*; it is always greater than unity. For a rectangular section of breadth b and depth d, $Z = bd^2/6$, $M_p = \sigma_y bd^2/4$; hence $\alpha = 1.5$. For a solid circular cross section $\alpha = 1.7$, while for I-beams and channels α varies within the small range of 1.15 to 1.17.

21-3 PLASTIC BEHAVIOR OF A SIMPLE BEAM

To consider displacements let us assume an idealized relation between the bending moment and curvature at a section, as shown in Fig. 21-3.

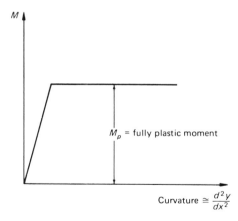

Fig. 21-3. Idealized moment-curvature relation.

If a load P at the mid-span of a simple beam (Fig. 21-4a) is increased until the bending at the mid-span cross section reaches the fully plastic moment M_p, a plastic hinge is formed at this section and collapse will occur under any further load increase. According to the assumed idealized relation, the curvature, and hence the rotation, at the plastic hinge increase at a constant load, so does deflection. The collapse load P_c can be easily calculated from statics

$$P_c = 4M_p/l \qquad (21\text{-}2)$$

The bending moment at sections other than mid-span is less than M_p, and by virtue of the assumed idealized relations the beam remains elastic away from this section. The deflected configuration of the beam in the elastic and plastic stages are shown in Fig. 21-4b. The increase in deflection during collapse is caused by the rotation at the central hinge without a concurrent change in curvature of the two halves of the beam. Figure 21-4c represents the change in deflection during collapse; this is a straight line for each half of the beam. The same figure shows also the collapse mechanism of the beam.

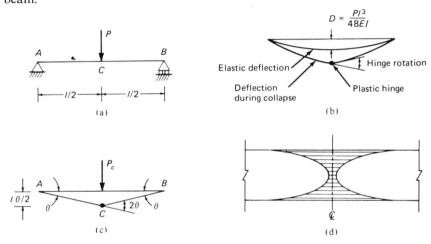

Fig. 21-4. Plastic behavior of a simple beam. (a) Beam. (b) Deflection lines. (c) Change in deflection during collapse. (d) Elevation of beam showing yielding near the mid-span section.

The collapse load of the beam (and this applies also to statically indeterminate structures) can be calculated by equating the external and internal work during a virtual movement of the collapse mechanism. Let each half of the beam in Fig. 21-4c acquire a virtual rotation θ, so that the corresponding rotation at the hinge is 2θ, and the downward displacement of the load P_c is $l\theta/2$. Equating the work done by P_c to the work of the moment M_p at the plastic hinge, we obtain

$$P_c \frac{l\theta}{2} = M_p \, 2\theta \qquad (21\text{-}3)$$

which gives the same result as Eq. 21-2.

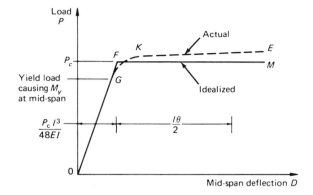

Fig. 21-5. Load–deflection relation for the beam in Fig. 21-4.

The idealized relation between load and central deflection for this beam is represented by the line OFM in Fig. 21-5. When the collapse load corresponding to point F in Fig. 21-5 is reached, the elastic deflection at mid-span is

$$D_F = P_c \frac{l^3}{48EI} \tag{21-4}$$

However, the actual load-deflection relation follows the dotted curve GKE. When the yield moment $M_y(= \sigma_y Z)$ is reached at the mid-span section, the upper or the lower fiber, or both, yield and the elastic behavior comes to an end. If the load is increased further, the yield spreads inward at this section and also laterally to other nearby sections. Figure 21-4d illustrates this spread of yield. After M_y has been reached, the deflection increases at a greater rate per unit increase of load until M_p is reached, as indicated by the curve GK in Fig. 21-5. In practice, rolled steel sections continue to show a small rise in the load-deflection curve during collapse (line KE); this is due to strain hardening which is generally not considered in ordinary plastic analysis.

21-4   ULTIMATE STRENGTH OF FIXED-ENDED AND CONTINUOUS BEAMS

Consider a prismatic fixed-ended beam subjected to a uniform load of intensity q (Fig. 21-6a). The resulting bending moments are $M_A = M_C = -ql^2/12$ and $M_B = ql^2/24$. When the load intensity is increased to q_1 such that the moments at the supports reach the fully plastic moment $M_p = q_1 l^2/12$, hinges are formed at A and C. If q is further increased, the moment at the supports will remain constant at M_p; free rotation will take place there so that the deflection due to the load in excess of q_1 will be the

same as in a simply supported beam. The collapse will occur at a load
intensity q_c which produces the moment at mid-span of magnitude M_p, so
that a third hinge is formed at B. The bending-moment diagrams due to
load intensity q for the cases $q = q_1$, $q_1 < q < q_c$, and $q = q_c$ are shown in
Fig. 21-6b and the collapse mechanism in Fig. 21-6c. The collapse load q_c
is calculated by the virtual-work equation

$$M_p(\theta + 2\theta + \theta) = 2\left(\frac{q_c l}{2}\right)\frac{\theta l}{4} \tag{21-5}$$

where 0, 2θ, and 0 are the virtual rotations at the plastic hinges A, B, and C

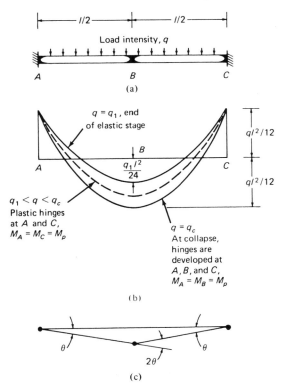

Fig. 21-6. Collapse of a beam with fixed ends under a uniformly distributed
load. (a) Beam. (b) Bending-moment diagrams for three load intensities.
(c) Collapse mechanism.

respectively, and $\theta l/4$ is the corresponding downward displacement of the
resultant load on one half of the beam. Equation 21-5 gives the intensity of
the collapse load:

$$q_c = \frac{16M_p}{l^2} \tag{21-6}$$

If the beam is of solid rectangular section, $M_p = 1.5M_y$ and the maximum load intensity computed by elastic theory with the maximum fiber stress σ_y is $q_E = 12M_y/l^2$. Thus the ratio $q_c/q_E = 2$, which clearly indicates that the design of the beam considered by elastic theory is conservative.

In plastic design of continuous beams, we draw the bending-moment diagram for each span as a simple beam loaded with the design load multiplied by the load factor. Arbitrary values may be chosen for the bending moment at the supports and a closing line such as $AB'C'D$ in Fig. 21-7d is drawn. The value of the bending moment at any section will then be the ordinate between the closing line and the simple-beam moment diagram. The beam will have the required ultimate capacity if the sections are then selected so that the plastic moment of resistance is everywhere equal to or in excess of

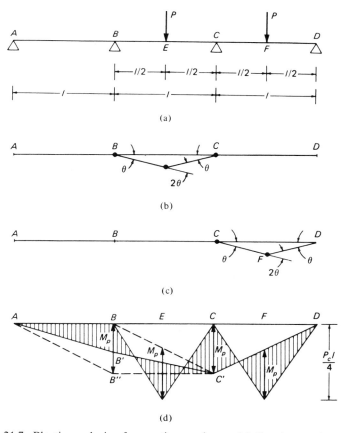

(a)

(b)

(c)

(d)

Fig. 21.7. Plastic analysis of a continuous beam. (a) Continuous beam of constant section and of plastic moment of resistance M_p. (b) Collapse mechanism 1. (c) Collapse mechanism 2. (d) Bending-moment diagram.

the bending moment. However, the most economical design is generally attained when a regular section is used of size such that a collapse mechanism develops.

If the beam sections are given or assumed, the values of the collapse load corresponding to all possible mechanisms are determined; the actual collapse load is the smallest one of these. Consider, for example, the continuous beam in Fig. 21-7a which has constant section with a plastic moment of resistance M_p. We want to find the value of the two equal loads P which causes collapse; denote this value by P_c. Failure can occur only by one of the two mechanisms shown in Figs. 21-7b and c. By a virtual-work equation for each of the two mechanisms, we obtain

$$P_{c1}\left(\frac{l\theta}{2}\right) = M_p(\theta + 2\theta + \theta) \qquad \text{(see Fig. 21-7b)}$$

whence

$$P_{c1} = 8M_p/l \tag{21-7}$$

and

$$P_{c2}\left(\frac{l\theta}{2}\right) = M_p(\theta + 2\theta) \qquad \text{(see Fig. 21-7c)}$$

whence

$$P_{c2} = 6M_p/l \tag{21-8}$$

The smaller of these two values is the true collapse load, $P_c = 6M_p/l$. The corresponding bending-moment diagram is shown in Fig. 21-7d, in which the values of the bending-moment at C and F are equal to M_p. When the collapse occurs, the part of the beam between A and C is still in the elastic stage, and the value of the bending moment at B can be calculated by analyzing a continuous beam ABC, hinged at A and C, and subjected to a clockwise couple of magnitude M_p at C and a vertical load $P_c = 6M_p/l$ at E. However, this calculation has no practical value in limit design. It is clear that the closing line $AB'C'$ of the bending-moment diagram can take any position between the limiting lines $AB''C'$ and ABC' with the bending moment at no section exceeding M_p.

21-5 RECTANGULAR PORTAL FRAME

Let us determine the collapse load for the frame shown in Fig. 21-8a, assuming the plastic moment of resistance $2M_p$ for the beam BC and M_p for the columns. There are only three possible collapse mechanisms, which are shown in Figs. 21-8b, c, and d.

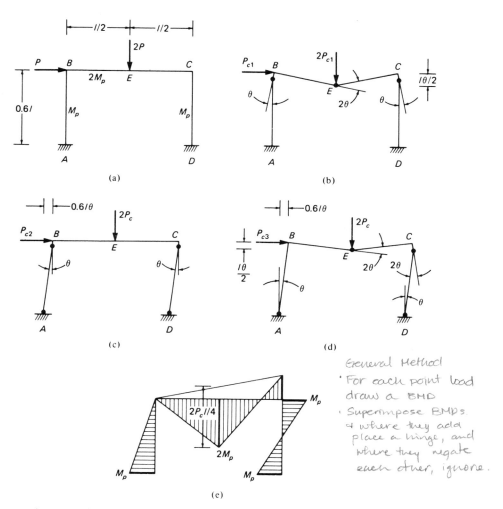

Fig. 21-8. Plastic analysis of a rectangular portal frame. (a) Frame loading and properties. (b) Mechanism corresponding to collapse load P_{c1}. (c) Mechanism corresponding to collapse load P_{c2}. (d) Mechanism corresponding to collapse load P_{c3}. (e) Bending-moment diagram at collapse.

A virtual-work equation for each of these mechanisms gives

$$M_p(\theta + \theta) + 2M_p(2\theta) = 2P_{c1}\left(\frac{l\theta}{2}\right) \qquad \text{(see Fig. 21-8b)}$$

or

$$P_{c1} = \frac{6M_p}{l}, \tag{21-9}$$

$$M_p(\theta + \theta + \theta + \theta) = P_{c2}(0.6\,l\theta) \qquad \text{(see Fig. 21-8c)}$$

or
$$P_{c2} = 6.67 \frac{M_p}{l} \tag{21-10}$$

and

$$M_p(\theta + \theta + 2\theta) + 2M_p(2\theta) = P_{c3}(0.6\ l\theta) + 2P_{c3}\left(\frac{l\theta}{2}\right) \qquad \text{(see Fig. 21-8d)}$$

or

$$P_{c3} = 5\frac{M_p}{l} \tag{21-11}$$

The collapse load is the smallest of P_{c1}, P_{c2}, and P_{c3}; thus, $P_c = 5M_p/l$, and failure of the frame will occur with the mechanism of Fig. 21-8d. The corresponding bending-moment diagram in Fig. 21-8e has an ordinate M_p at the plastic hinges A, C, and D and $2M_p$ at E, with the plastic moment of resistance exceeded nowhere. A check on the calculations can be made by verifying that the bending-moment diagram in Fig. 21-8e satisfies statical equilibrium.

21-51   Location of Plastic Hinges under Distributed Loads

Consider the frame analyzed in the previous section but with a vertical load $4P$ distributed over the beam BC, as shown in Fig. 21-9a; the horizontal load P is unchanged. In this case, the position of the maximum positive bending moment in BC is not known, so that the location of the plastic hinge has to be determined.

Let us apply the virtual-work equation to the mechanism in Fig. 21-8d, loaded as in Fig. 21-9a, with the hinge in the beam assumed at mid-span. Each half of the beam is subjected to a vertical load whose resultant $2P_{c3}$ moves through a vertical distance $l\theta/4$. The internal virtual work and the external virtual work of the horizontal force are the same as before (Eq. 21-11); thus

$$M_p(\theta + \theta + 2\theta) + 2M_p(2\theta) = P_{c3}(0.6\ l\theta) + 2 \times 2P_{c3}\left(\frac{l\theta}{4}\right) \tag{21-12}$$

This equation gives the same value of the collapse load as Eq. 21-11.

The bending-moment diagram for the mechanism in Fig. 21-8d with the loads in Fig. 21-9a is shown in Fig. 21-9b, from which it can be seen that the fully plastic moment $2M_p$ is slightly exceeded in the left-hand half of the beam.

This maximum moment occurs at a distance $x = 0.45l$ from B, and its value is $2.025M_p$. It follows that the assumed collapse mechanism is not correct, the reason for this being that the plastic hinge in the beam should not be located at mid-span. The calculated value of $P_c = 5M_p/l$ is an upper bound on the value of the collapse load. If the load P_c is given, the structure

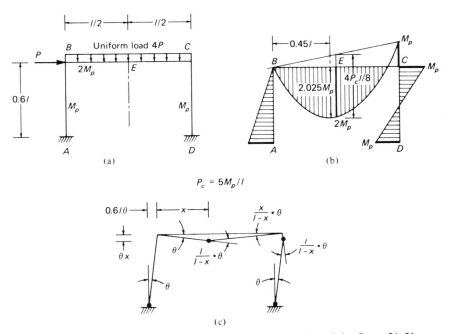

Fig. 21-9. Frame subjected to distributed load analyzed in Sect. 21-51. (a) Frame loading and properties. (b) Bending-moment diagram due to loading in part (a) on the mechanism in Fig. 21-8d. (c) Mechanism corresponding to collapse load P_c in Eq. 21-14.

must be designed for a plastic moment M_p slightly greater than $P_c l/5$, and this value is to be considered a lower bound on the required plastic moment. It is clear that the structure will be safe if designed for

$$M_p = \frac{2.025}{2}\left(\frac{P_c l}{5}\right) = 1.0125\,\frac{P_c l}{5}$$

Therefore, the required value of M_p is

$$\frac{P_c l}{5} < M_p < 1.0125\,\frac{P_c l}{5} \tag{21-13}$$

These limits define the value of M_p to within 1.25 percent and are accurate enough for practical purposes. However, the precise value of M_p can be calculated considering the mechanism in Fig. 21-9c, with a plastic hinge F at a distance x from B. By a virtual-work equation M_p can be derived in terms of x. The value of x is then chosen so that M_p is a maximum, that is $dM_p/dx = 0$.

The rotation at the hinges and the translation of the loads are shown in Fig. 21-9c, and the virtual-work equation is

$$M_p\left(\theta + \theta + \frac{l}{l-x}\theta\right) + 2M_p\left(\frac{l}{l-x}\theta\right) = P_c(0.6l\theta) + \frac{4P_c}{l}x\left(\frac{x\theta}{2}\right) +$$
$$+ \frac{4P_c}{l}(l-x)\left(\frac{x\theta}{2}\right)$$

or

$$M_p = P_c\frac{(0.6l^2 + 1.4lx - 2x^2)}{(5l - 2x)} \tag{21-14}$$

Putting $dM_p/dx = 0$, we obtain

$$4x^2 - 20lx + 8.2l^2 = 0 \tag{21-15}$$

whence $x = 0.4505l$. Substituting in Eq. 21-14, we obtain the maximum value of M_p

$$M_p = 1.0061\frac{P_cl}{5} \tag{21-16}$$

The values of x and M_p differ slightly from the approximate value $x = 0.45l$ and the conservative value of $M_p = 1.0125\,(P_cl/5)$ obtained from the bending-moment diagram in Fig. 21-9b.

21-6 COMBINATION OF ELEMENTARY MECHANISMS

We can now consider the plastic analysis of plane frames in general. A sufficient number of plastic hinges are introduced at assumed locations to form a mechanism and the corresponding collapse load is calculated by virtual work. The value determined in this way is an upper bound on the correct load capacity, that is the load indicated is greater than the correct load. In other words, if the plastic moment M_p required for any specified collapse is calculated according to any mechanism, then the resulting value of M_p is a lower bound on the necessary plastic resisting moment. In practical calculation, the correct mechanism may be reached by considering elementary mechanisms and combining them to obtain the lowest that is the correct load capacity.

Consider the frame of Fig. 21-10a with prismatic members; the relative values of the fully plastic moment are indicated. Three elementary mechanisms are shown in Figs. 21-10b, c, and d and the work equation for each is given alongside.

Mechanism 2 gives the lowest P_c and we combine it with the other mechanisms to reach a smaller value of P_c. Two combinations are shown in Figs. 21-10e and f. Mechanism 4 is a combination of mechanisms 2 and 3

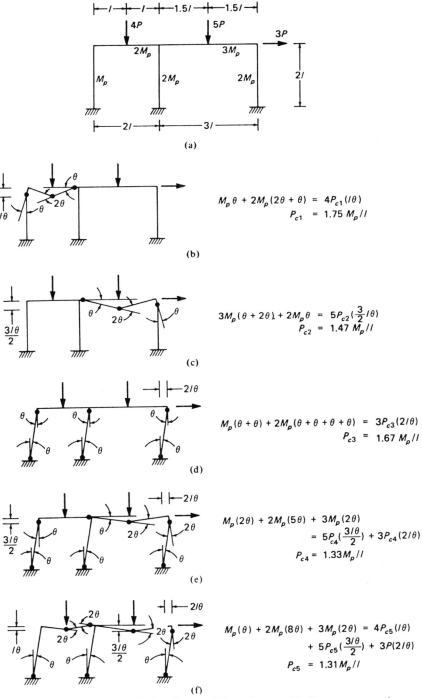

Fig. 21-10. Plastic analysis of a multibay frame. (a) Frame properties and loading. (b) Mechanism 1. (c) Mechanism 2. (d) Mechanism 3. (e) Mechanism 4. (f) Mechanism 5.

with a modification of the location of the hinge above the central column. The resulting collapse load is $P_{c4} = 1.33 M_p/l$. Mechanism 5 is a combination of mechanisms 1, 2, and 3, again with the same modification, giving a value $P_{c5} = 1.31 M_p/l$, which is lower than all the previous values.

To find whether or not P_{c5} is the lower bound on P_c the corresponding bending-moment diagram is drawn. If the moment at no section is greater than its plastic moment, P_{c5} is the lower bound and the solution is correct. The bending-moment diagram of Fig. 21-11 shows that mechanism 5 is the correct mechanism. In the construction of this diagram, the known values of moments at the plastic hinges were first plotted, and the other ordinates were then derived by simple statics.

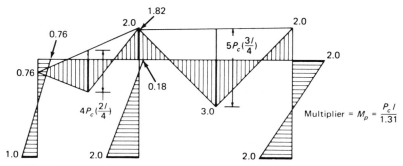

Fig. 21-11. Bending-moment diagram for mechanism 5 in Fig. 21-10f.

It is possible that on plotting the bending-moment diagram we find that the moment at some section exceeds the plastic moment of resistance. If the excess is small, say 5 percent, then it is safe to consider that the frame can carry the loads previously calculated, each reduced by 1/1.05. Some publications[1] suggest that if for a specified collapse load calculated for a wrong mechanism the plastic moment of resistance is nowhere exceeded by more than 30 percent, a fairly close estimate of the required bending strength can be obtained by increasing the computed value of M_p by one-half of the largest excess moment.

21-7   FRAMES WITH INCLINED MEMBERS

In computing the angle between the ends of members at a plastic hinge in a nonrectangular frame, it may be convenient to locate the instantaneous center about which a part of the frame rotates during collapse.

Consider the gable frame in Fig. 21-12a, for which a trial mechanism 1 is given in Fig. 21-12b. Joining the plastic hinges at A and C and extending the

[1]See *Plastic Design of Steel*, The American Institute of Steel Construction, New York, 1959, p. 7.

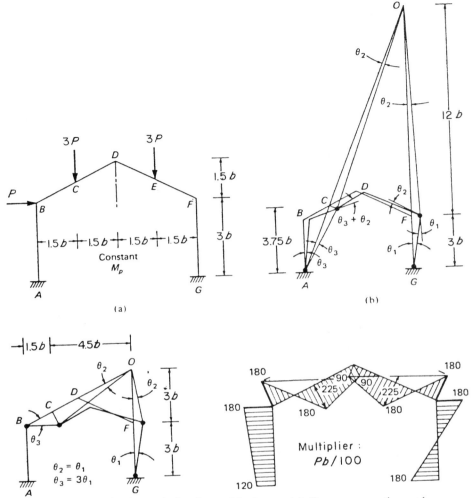

Fig. 21-12. Plastic analysis of a gable frame. (a) Frame properties and loading. (b) Mechanism 1. (c) Mechanism 2. (d) Bending-moment diagram for Mechanism 2.

line AC to meet GF produced, locates O, the center of rotation of the part CDF. Parts ABC and GF rotate about A and G, respectively. During collapse points C and F move along the normals to the lines AO and GO respectively, that is along the common tangent to circles having their centers located at the centers of rotation A, O, and G. Assuming the angle of rotation of GF to be θ_1, the angles of rotation of the frame parts about their instantaneous centers can be determined easily by geometry: $\theta_2 = \theta_1/4$ and $\theta_3 = 3\theta_2 = 3\theta_1/4$. The relative rotations of frame parts at the plastic hinges are indicated in Fig. 21-12b, and the corresponding internal virtual work W is the sum of the work done at the four hinges A, C, F, and G; thus

$$W = M_p[\theta_3 + (\theta_3 + \theta_2) + (\theta_2 + \theta_1) + \theta_1]$$
$$= 4M_p\theta_1$$

The translation of the external loads corresponding to the above rotations is calculated from the geometry of the mechanism, and the external virtual work is obtained as follows.

Load point	Displacement	Load	External work
B	$0.75\theta_1 \times 3b = 2.25\theta_1 b$	P	$2.25Pb\theta_1$
C	$0.75\theta_1 \times 1.5b = 1.125\theta_1 b$	$3P$	$3.375Pb\theta_1$
E	$0.25\theta_1 \times 1.5b = 0.375\theta_1 b$	$3P$	$1.125Pb\theta_1$

$$\text{Total} = 6.75Pb\theta_1$$

Equating the external and internal work, we obtain

$$M_p = 1.6875Pb$$

If the bending-moment corresponding to this mechanism is sketched, it will be clear that the value $M_p = 1.6875Pb$ is exceeded at several points, which means that mechanism 1 is not the correct collapse mechanism. A second mechanism is shown in Fig. 21-12c, for which the instantaneous centers are B, O, and G for frame parts BC, CF, and FG, respectively. An arbitrary small virtual rotation θ_1 is assumed for FG, and the corresponding rotations for the other parts are determined in terms of θ_1 by considering the geometry of the mechanism. The work equation for this mechanism gives $M_p = 1.8Pb$. The corresponding bending-moment diagram is shown in Fig. 21-12d: the value of M_p is not exceeded, indicating that collapse takes place according to mechanism 2.

21-8 EFFECT OF AXIAL FORCES ON PLASTIC MOMENT CAPACITY

In previous sections we assumed that the fully plastic state at a hinge is induced solely by the bending-moment M_p. However, in the presence of a high axial compressive or tensile force, a plastic hinge can be formed at a moment M_{pc} lower than the value M_p. We shall now derive an expression for M_{pc} for a rectangular cross section subjected to a tensile or compressive axial force; any buckling effect will be ignored.

Figures 21-13a to e show the changes in stress distribution in a rectangular cross section $b \times d$ subjected to an axial compressive force P together with a bending moment M in the vertical plane, as the magnitude of M and P is increased at a constant value of M/P until the fully plastic stage is reached. The values of the axial force and the moment at this stage are

$$P = 2\sigma_y b y_0 \qquad (21\text{-}17)$$

and

$$M_{pc} = \frac{\sigma_y b}{4} (d^2 - 4y_0^2) \qquad (21\text{-}18)$$

where σ_y is the yield stress and y_0 is the distance from the centroid of the section to the neutral axis.

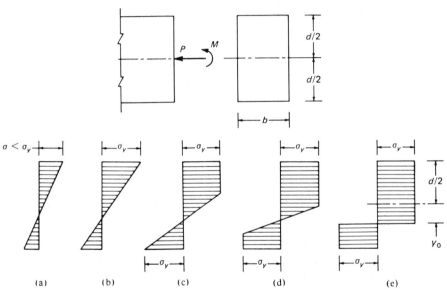

Fig. 21-13. Stress distribution in a rectangular section subjected to an increasing bending moment M and axial force P at a constant value of M/P.

In the absence of the axial force, $y_0 = 0$, and the fully plastic moment is then

$$M_p = \sigma_y \frac{bd^2}{4} \qquad (21\text{-}19)$$

On the other hand, if the axial force alone causes a fully plastic state, its magnitude is

$$P_y = \sigma_y bd \qquad (21\text{-}20)$$

From Eqs. 21-17–21-20 we can obtain the interaction equation

$$\frac{M_{pc}}{M_p} = 1 - \left(\frac{P}{P_y}\right)^2 \tag{21-21}$$

The interaction curve of M_{pc}/M_p plotted against P/P_y in Fig. 21-14 can be used to determine the strength of a section under combined loading of an axial force and a bending moment from the strengths in two simple types of loading: axial force only and bending moment only.

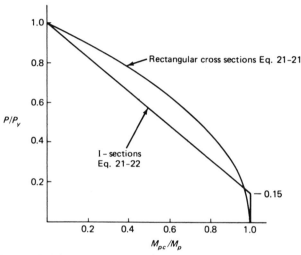

Fig. 21-14. Axial-force/bending-moment interaction curves for rectangular cross sections and for wide-flange I-sections.

The shape of the interaction curve depends on the geometry of the cross section. For all wide flange I-sections used in steel construction, the interaction curves fall within a narrow band and may be approximated by two straight lines[2] shown in Fig. 21-14. From this figure it is apparent that when $P/P_y \leq 0.15$, the effect of the axial load is neglected and when $P/P_0 \geq 0.15$, we use the equation

$$\frac{M_{pc}}{M_p} = 1.18\left(1 - \frac{P}{P_y}\right) \tag{21-22}$$

21-9 EFFECT OF SHEAR ON PLASTIC MOMENT CAPACITY

The presence of shear at a section at which the bending moment is high may limit the ultimate load capacity of a structure to a value below that

[2]Joint Committee of the Welding Research Council and the American Society of Civil Engineers, Commentary on Plastic Design in Steel, *ASCE Manual of Engineering Practice*, No. 41, New York, 1961.

which produces the fully plastic moment M_p at the location of the plastic hinges. Although in the majority of practical cases the influence of shear is small, there are circumstances when shear may produce a plastic hinge at a bending moment M_{ps} appreciably smaller than M_p.

The main assumptions made in deriving an interaction equation for M_{ps}/M_p for rectangular and wide-flange I-sections are:

1. The outer portions of the beam (the flanges and a part of the web) are in a yield state and resist bending only, while the middle portion is in an elastic state and resists shear.

2. The shearing stress in the (rectangular) center portion is calculated by ordinary elastic equations.

The resistance of the section is assumed to be exhausted when the shearing stress in the center portion reaches yield under pure shear, which, according to von Mises-Hencky yield criterion, equals $\sigma_y/\sqrt{3}$. With these assumptions, the following equation can be derived[3]

$$\frac{M_{ps}}{M_p} = \frac{8bc^2}{9\alpha Z}\left(\sqrt{1 + \frac{9}{4}\frac{\alpha Z}{bc^2}} - 1\right) \tag{21-23}$$

where b is the width of the rectangular section or of the web of an I-section, Z is the section modulus, α is as defined at the end of Sec. 21-2, and c is the shear span ($=$ bending moment/shearing force).

21-10 GENERAL

This chapter is no more than an introduction to the structural analysis required in the plastic design of steel structures. For safe and efficient use of plastic design of continuous framed structures, it is important to understand the restrictions and limitations of this design method, such as the effect of repeated loading, instability, and also to be able to estimate deflections at working and ultimate loads.[4]

PROBLEMS

21-1   Find the required plastic moment of resistance of the cross section(s) for the beam in the figure, which is to be designed to carry the given loads with a load factor of 1.7. Assume that the beam has: (a) a constant cross section, (b) two different cross sections one from A to C, and the other from C to D.

---

[3] See pp. 35–40 of the reference in footnote 2 of this chapter for proof and limitations of this equation, and also for the combined effect of shear and axial force.

[4] See, for instance, B. G. Neal, *The Plastic Methods of Structural Analysis*, 2d ed., Chapman and Hall, London, 1963. A list of other references can be found at the end of the manual referred to in footnote 2 of this chapter.

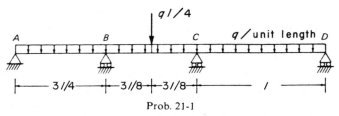

Prob. 21-1

21-2   Determine the fully plastic moment for the frames shown, with the collapse
loads indicated. Ignore the effects of shear and axial forces.

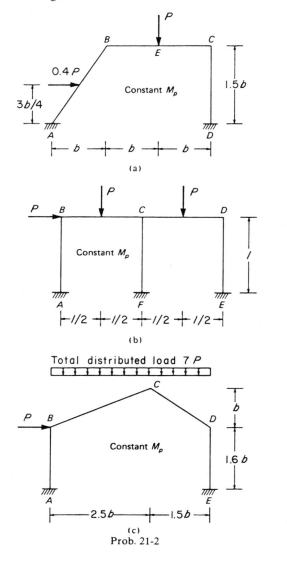

Prob. 21-2

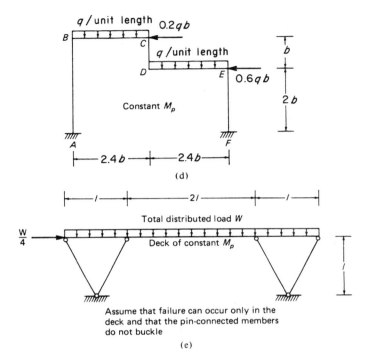

Prob. 21-2 (*continued*)

21-3 What is the value of M_p for the frame in Prob. 21-2b, if the axial force effect (excluding buckling) is taken into account. Assume that the frame has a **constant rectangular section** $b \times d$, with $d = l/15$.

21-4 What is the value of M_p for the beam in Prob. 21-1, if the shear effect is taken into account. Assume that the beam has a constant rectangular section $b \times d$, with $d = l/30$.

CHAPTER 22

Yield-line and strip methods
for slabs

22-1 INTRODUCTION[1]

This chapter deals in a general way with plates but many of the applications are virtually limited to reinforced concrete slabs. For this reason, the term slab will be generally used.

An elastic analysis of a reinforced concrete slab gives no indication of its ultimate load-carrying capacity and further analyses have to be made for this condition. An exact solution for the ultimate flexural strength of a slab can be found only rarely, but it is possible to determine upper and lower bounds to the true collapse load.

The yield-line method of analysis gives an upper bound to the ultimate load capacity of a reinforced concrete slab by a study of assumed mechanisms of collapse. This method, developed by Johansen,[2] is a powerful tool for estimating the required bending resistance and hence the necessary reinforcement, especially for slabs of nonregular geometry or loading. Two approaches are possible in yield-line theory. The first one is an energy method in which the external work done by the loads during a small virtual movement of the collapse mechanism is equated to the internal work. The alternative approach is by the study of the equilibrium of the various parts of the slab into which the slab is divided by the yield lines. We may note that it is the equilibrium of slab parts that is considered and not the equilibrium of forces at all points of the yield line.

In contrast to the above, lower-bound solutions to the collapse load are obtained by satisfying equilibrium at *all* points in the slab, and necessitate the determination of a complete bending-moment field in equilibrium with the applied loading. We shall restrict our discussion of these lower-bound solutions to the special case where twisting moments are absent — the so-called strip method. This strip method is more of a direct design procedure than the yield-line method as the designer chooses the layout of reinforcement as the calculation progresses.

Ultimate load designs according to the yield-line or strip methods do not guarantee safety against cracking or excessive deformations. Therefore, an

---

[1] Sections 22-1, 22-5, 22-6, 22-8, 22-9, and 22-10 were written in collaboration with Dr. J. Harrop, University of Leeds.
[2] K. W. Johansen, *Yield-Line Theory*, Cement and Concrete Association, London 1962, pp. 181.

understanding of elastic behavior is necessary for the effective distribution of reinforcement when an ultimate load design is made.

22-2 FUNDAMENTALS OF YIELD-LINE THEORY

The slab is assumed to collapse at a certain ultimate load through a system of yield lines or fracture lines, called the pattern of fracture. The working load is obtained by dividing this ultimate load by the required load factor. For design, the working load is multiplied by the load factor, and the required ultimate moment of resistance is determined.

The basic fundamentals and main assumptions of the yield-line theory are as follows:

1. At fracture, the bending moment per unit length along all the fracture lines is constant and equal to the yield value corresponding to the steel reinforcement. The fracture is assumed to occur due to the yield of the steel.

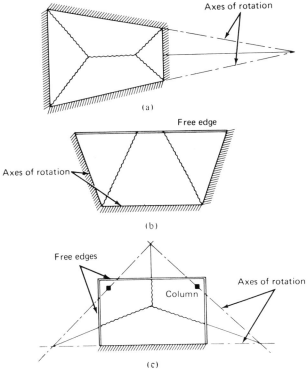

Fig. 22-1. Typical fracture patterns in slabs. (a) Slab simply supported on four sides. (b) Slab simply supported on three sides and free along the fourth side. (c) Slab simply supported on two columns and on one side.

2. The slab parts rotate about axes along the supported edges. In a slab supported directly on columns, the axes of rotation pass through the columns. Figure 22-1 shows some typical fracture patterns.

3. At fracture, elastic deformations are small compared with the plastic deformations and are therefore ignored. From this assumption and the previous one it follows that fractured slab parts are plane and therefore they intersect in straight lines. In other words, the yield lines are straight.

4. The lines of fracture on the sides of two adjacent slab parts pass through the point of intersection of their axes of rotation.

Figure 22-2 shows the fracture pattern of a uniformly loaded slab simply supported on three sides. It is readily seen that the pattern satisfies the requirements given above. Each of the three slab parts rotates about its axis of rotation by an angle θ_i which is related to the rotation of the other parts. Let points E and F have a virtual downward displacement $w = 1$. The rotations of the slab parts are then

$$\theta_1 = \frac{1}{c_1} \qquad \theta_2 = \frac{1}{y} \qquad \text{and} \qquad \theta_3 = \frac{1}{c_2}$$

22-21 Convention of Representation

The different conditions of supports will be indicated thus

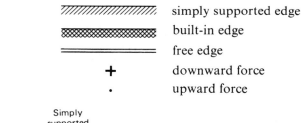

simply supported edge
built-in edge
free edge

+   downward force
•   upward force

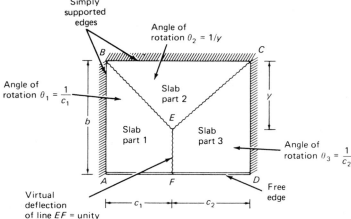

Fig. 22-2. Angular rotation of slab parts.

A positive ultimate moment m per unit length causes yield of the bottom reinforcement. A yield line formed by a positive moment is referred to as a positive yield line. Negative ultimate moment m' per unit length causes yield of the top reinforcement along a negative yield line.[3]

22-22   Ultimate Moment of a Slab Equally Reinforced in Two Perpendicular Directions

Consider a slab reinforced in two perpendicular directions, x and y, with different reinforcement corresponding to ultimate positive moments m_1 and m_2 (Fig. 22-3a). For equilibrium of the element shown in Fig. 22-3b,

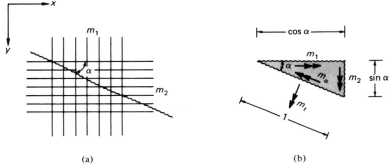

(a)   (b)

Fig. 22-3. Moments at a fracture line inclined to the direction of reinforcement.

the bending (m) and twisting (m_t) moments at a fracture line making an angle α with the x axis are

$$m_\alpha = m_1 \cos^2\alpha + m_2 \sin^2\alpha$$
$$(m_t)_\alpha = (m_1 - m_2) \sin\alpha \cos\alpha \qquad (22\text{-}1)$$

The moments are represented in Fig. 22-3b (as usual) by double-headed arrows in the direction of the progress of a right-hand screw rotating in the same direction as the moment. In an isotropic slab — that is, one equally reinforced in two perpendicular directions — $m_1 = m_2 = m$, and the moment on any inclined fracture line is

$$m_\alpha = m(\cos^2\alpha + \sin^2\alpha) = m$$
and
$$m_t = 0 \qquad (22\text{-}2)$$

[3]There are several methods of calculation of the value of the ultimate moment in terms of the depth, area of reinforcement and the strengths of concrete and steel. Refer, for example, to the American Concrete Institute Standard 318–89 Building Code Requirements for Reinforced Concrete.

The values of m_x and m_t can also be determined by Mohr's circle as shown in Fig. 22-4.

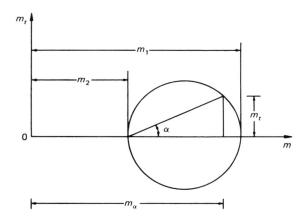

Fig. 22-4. Moments on an inclined fracture line by Mohr's circle.

22-3   ENERGY METHOD

In this method the pattern of fracture is assumed and the slab is allowed to deflect in the fractured state as a mechanism. Each slab part will rotate a small virtual angle θ about its axis of rotation. The relation between the rotation of the slab parts is defined by the choice of the fracture pattern. The internal energy dissipated on the yield lines during the virtual rotation is equated to the external virtual work done in deflecting the slab. From this equation, the value of the ultimate moment is obtained. We should note, when calculating the internal energy, that only the ultimate moments in the yield lines do work during rotation.

The virtual-work equation (similar to the equation used for plastic analysis of frames, Chapter 21) gives either the correct ultimate moment or a value smaller than the correct value. In other words, if the virtual-work equation is used to find the ultimate load for a slab with an assumed bending resistance then the value obtained will be an upper bound on the carrying capacity of the slab. This means that the solution obtained is either correct or unsafe. In practical calculations, one or two fracture patterns are assumed, and the value obtained is usually within 10 percent of the correct value. It seems to be a reasonable design procedure to increase the moment obtained by the work equation by a small percentage, depending on the number of trials and on the uncertainty of the chosen fracture pattern. The theoretical exact pattern is that for which the ultimate moment is a maximum. This can be reached, if we define the fracture pattern by certain parameters

$x_1, x_2, ...$; the work equation will then give the value of m as a function of these parameters, i.e., $m = f(x_1, x_2, ...)$. The value of the parameters corresponding to the maximum moment is determined by partial differentiation: $(\partial f/\partial x_1) = 0$, $(\partial f/\partial x_2) = 0$, etc. This process can become laborious except for simple slabs in which the designer can define a reasonable pattern and proceed as suggested above.

The internal work done during a virtual rotation θ of a slab part is equal to the scalar product of a vector $\vec{M} = \overrightarrow{ml}$ and a vector $\vec{\theta}$ along the axis of rotation (Fig. 22-5). The internal work for this slab part is then $\vec{M}.\vec{\theta} = ml$ $(\cos \alpha)\theta$, where α is the angle between the two vectors. This means that, for any part of the slab, the internal work is equal to the rotation of that part multiplied by the projection of the ultimate moment upon the axis of rotation. It is sometimes convenient to consider the components of the moments in two perpendicular directions x and y in the plane of the slab. The total internal virtual work for all the slab parts is then

$$U = \sum \vec{M}.\vec{\theta} = \Sigma M_x.\theta_x + \Sigma M_y.\theta_y \qquad (22\text{-}3)$$

where θ_x and θ_y are the x and y components of the rotation vector, and M_x and M_y are the x and y components of the vector $\overline{ml}$.

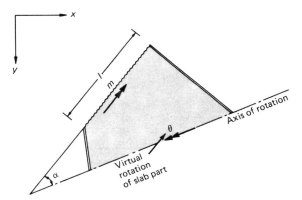

Fig. 22-5. Data for calculation of internal virtual work.

If x and y are the coordinates of any point on the slab and w is the vertical displacement corresponding to the virtual rotation of the slab parts, then the total external virtual work is

$$W = \int\int qw\,dx\,dy \qquad (22\text{-}4)$$

where q is the load intensity. The virtual-work equation is

$$W = U \qquad (22\text{-}5)$$

whence

$$\sum (M_x.\theta_x + M_y.\theta_y) = \int \int qw\, dx\, dy \qquad (22\text{-}6)$$

Example 22-1 Determine the ultimate moment of a square isotropic slab simply supported on three sides and subjected to a uniform load q per unit area.

Because of symmetry the fracture pattern is fully determined by one parameter x, as shown in Fig. 22-6. Let the junction of the three fracture lines have a virtual displacement $w = 1$. The rotations of the slab parts are

$$\theta_1 = \frac{1}{x} \qquad \theta_2 = \theta_3 = \frac{2}{l}$$

The internal virtual work is

$$U = ml\,\frac{1}{x} + 2ml\,\frac{2}{l}$$

where the first term applies to part 1 and the second to parts 2 and 3.

The work done by the distributed load is equal to the resultant on each part multiplied by the vertical displacement of its point of application. Thus

$$W = q\left\{ l\frac{x}{2}\frac{1}{3} + 2\left[(l - x)\,\frac{l}{2}\frac{1}{2} + \frac{xl}{4}\frac{1}{3} \right] \right\}$$

Here again, the first term is for part 1 and the second term parts 2 and 3.

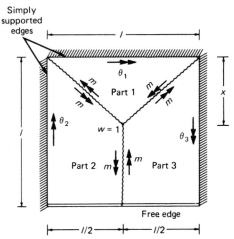

Fig. 22-6. Slab considered in Example 22-1.

Equating the internal and external work,

$$m\left(\frac{l}{x} + 4\right) = ql^2\left(\frac{1}{2} - \frac{x}{6l}\right)$$

$$m = ql^2 \left[\frac{3 - \dfrac{x}{l}}{6\left(\dfrac{l}{x} + 4\right)}\right]$$

The maximum value of m is obtained when $dm/dx = 0$, which gives $x/l = 0.65$. The corresponding ultimate moment is

$$m = \frac{ql^2}{14.1}$$

22-4 ORTHOTROPIC SLABS

The analysis of certain types of orthotropic slabs was simplified by Johansen to that of an isotropic affine slab for which the length of the sides and the loading are altered in certain ratios depending on the ratio of the ultimate resistance of the orthotropic slab in the two perpendicular directions.

Consider a part of a slab $ABCDEF$ shown in Fig. 22-7, limited by positive and negative yield lines and a free edge, assumed to rotate through a virtual angle θ about an axis of rotation $R-R$. Assume that the bottom and top reinforcement are placed in the x and y directions. Let the reinforcement in the y direction[4] provide ultimate moments of m and m', and let the corresponding values in the x direction be ϕm and $\phi m'$; this means that the ratio of the top to the bottom reinforcement is the same in both directions. The vectors $\vec{c}$ and $\vec{b}$ represent the resultants of the positive and negative moments, respectively.

The internal virtual work for this slab part is

$$U = (mc_x + m'b_x)\theta_x + \phi(mc_y + m'b_y)\theta_y \qquad (22\text{-}7)$$

where c_x, c_y and b_x, b_y are the projections in the x and y directions of the lengths c and b, and θ_x and θ_y are the x and y components of the rotation vector $\vec{\theta}$.

Assuming that the virtual deflection at a point n, distance r from the axis $R-R$, is unity, the rotation θ and its components can be written as

$$\theta = \frac{1}{r} \qquad \theta_x = \theta \cos \alpha = \frac{1}{r_y} \qquad \theta_y = \theta \sin \alpha = \frac{1}{r_x} \qquad (22\text{-}8)$$

[4] The ultimate moments in the x and y directions are indicated by vectors in the right-hand top corner of Fig. 22-7.

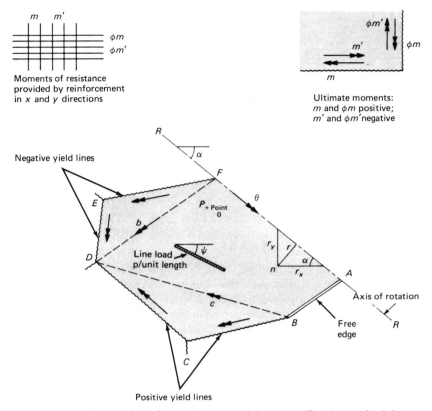

Fig. 22.7. Conversion of an orthotropic slab to an affine isotropic slab.

α being the angle between the x axis and the axis of rotation. Substituting Eq. 22-8 into Eq. 22-7, we obtain

$$U = \theta[(mc_x + m'b_x)\cos\alpha + \phi(mc_y + m'b_y)\sin\alpha] \qquad (22\text{-}9)$$

Let us assume now that the loading on the slab part consists of a uniformly distributed load q per unit area, a line load p per unit length on a length l in a direction making an angle ψ with the x axis and a concentrated load P at point O. The external virtual work of the loads on this slab part is

$$W = \int\int qw\,dx\,dy + \int pw_l\,dl + Pw_P \qquad (22\text{-}10)$$

where w is the deflection at any point (x, y), w_l deflection of any point below the line load, and w_P the deflection below the concentrated load. The integration is carried out for the whole area and for the loaded length. The virtual-work equation is

$$\sum \left[(mc_x + m'b_x)\frac{1}{r_y} + \phi(mc_y + m'b_y)\frac{1}{r_x} \right]$$

$$= \sum \left(\int\int qw\,dx\,dy + \int pw_l\,dl + Pw_P \right) \qquad (22\text{-}11)$$

where the summation is for all slab parts.

Consider an affine slab equally reinforced in the x and y directions so that the ultimate positive and negative moments are m and m' respectively. Suppose that this affine slab has all its dimensions in the x direction equal to those of the actual slab multiplied by a factor λ. The pattern of fracture remains similar and the corresponding points can still have the same vertical displacements. The internal virtual work for the part of the affine slab is

$$U' = (m\lambda c_x + m'\lambda b_x)\frac{1}{r_y} + (mc_y + m'b_y)\frac{1}{\lambda r_x} \qquad (22\text{-}12)$$

Let the loading on the affine slab be a distributed load of q' per unit area, a line load p' per unit length, and a concentrated load P'. The external work for this part of the affine slab is

$$W' = \int\int q'w\lambda\,dx\cdot dy + \int p'w_l\sqrt{(dy)^2 + \lambda^2(dx)^2} + P'w_P \qquad (22\text{-}13)$$

Dividing both the internal and external work by λ will not change the work equation, which then becomes

$$\sum(mc_x + m'b_x)\frac{1}{r_y} + \frac{1}{\lambda^2}\sum(mc_y + m'b_y)\frac{1}{r_x}$$

$$= \sum\left(\int\int q'w\,dx\,dy + \int p'w_l\sqrt{\frac{(dy)^2}{\lambda^2} + (dx)^2} + \frac{P'}{\lambda}w_P \right) \qquad (22\text{-}14)$$

All terms of the virtual-work Eqs. 22-11 and 22-14 are identical provided that

$$\phi = \frac{1}{\lambda^2} \quad \text{or} \quad \lambda = \sqrt{\frac{1}{\phi}} \qquad (22\text{-}15)$$

$$q' = q \qquad (22\text{-}16)$$

$$p'\sqrt{\frac{(dy)^2}{\lambda^2} + (dx)^2} = pdl$$

or

$$p' = \frac{p}{\sqrt{\phi \sin^2 \psi + \cos^2 \psi}} \qquad (22\text{-}17)$$

and

$$P' = P\sqrt{\frac{1}{\phi}} \qquad (22\text{-}18)$$

It follows that an orthotropic slab with positive and negative ultimate moments m and m' in the x direction and ϕm and $\phi m'$ in the y direction, can be analyzed as an isotropic slab with moments m and m' but the linear dimensions in the x direction multiplied by $\sqrt{1/\phi}$. The intensity of a uniformly distributed load remains the same. A linear load has to be multiplied by $1/\sqrt{\phi \sin^2 \psi + \cos^2 \psi}$, ψ being the angle between the load line and the x axis. A concentrated load has to be multiplied by $\sqrt{1/\phi}$.

Example 22-2   A rectangular orthotropically reinforced slab is shown in Fig. 22-8a. Find the dimensions of an isotropic affine slab.

We have $\phi = \frac{1}{2}$, so that side AB is to be changed to $l\sqrt{1/\phi} = 1.414l$ for an isotropic affine slab with ultimate moments of $2m$ and $2m'$. The same orthotropic slab can also be analyzed as an isotropic slab for which side AB remains unchanged and side AD is changed to $(3l/4)/\sqrt{2} = 0.530l$, the ultimate moments being m and m' (Fig. 22-8c).

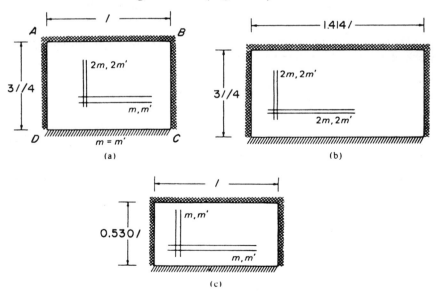

Fig. 22.8. Orthotropic slab and affine slabs considered in Example 22-2. (a) Orthotropic slab. (b)Affine slab. (c) Alternative affine slab.

22-5 EQUILIBRIUM OF SLAB PARTS

The energy method of the last two sections gives, it will be remembered an upper-bound value to the collapse load and can always be used for any assumed mechanism of collapse. Where the mechanism is complex and its layout is defined by several initially unknown dimensions the algebraic manipulation necessary to obtain a solution can be long and tedious. There can, however, be a saving of work in many cases by considering the equilibrium of the slab parts.

In this approach, we abandon the virtual-work equations of the energy method and consider instead the equilibrium of each slab part when acted upon by the external applied load, and by the forces acting at a fracture line. In general, these are bending moment, shearing force acting perpendicular to the slab plane, and twisting moment.

To establish the equilibrium conditions it is not necessary to know the precise distribution of the shear and of the twisting moment: they can be replaced by two forces perpendicular to the plane of the slab, one at each end of the fracture line. These two forces are referred to as nodal forces and are denoted by V; they are considered positive when acting upwards.

22-51 Nodal Forces

The formulas given here follow the original theory of Johansen. There are some restrictions on their use, although in the majority of cases they give a satisfactory solution. These restrictions have been studied by Wood and Jones.[5]

In Fig. 22-9a the shears and twists on fracture lines (1), (2), and (3) are represented by the equivalent nodal forces V_1 and V_1', etc., on ends of lines (1), (2), and (3). The nodal forces are equal and opposite on the two sides of each fracture line. It is clear that a summation of all the nodal forces at any junction of the fracture lines is zero.

Consider an elemental triangle ABC of area ΔA limited by positive fracture lines (2) and (3) (Fig. 22-9b) and any adjacent line at a small angle $d\alpha$ to fracture line (2). This adjacent line is assumed to have the same bending moment m_2 as line (2) to a first order approximation. The resultant moment on the triangle ΔA is $(m_3 - m_2) ds$ directed from C to A. For equilibrium of the triangle ABC, the moments about BC vanish:

$$V_{\Delta A} ds \sin \gamma + (m_3 - m_2) ds \cos (180° - \gamma) - dP \frac{ds \sin \gamma}{3} = 0$$

where $V_{\Delta A}$ is the nodal force at A replacing the twisting moment and shearing force on the fracture line (2) and the length ds of the fracture line (3), and dP

[5]"Recent Developments in Yield Line Theory," *Magazine of Concrete Research*, May 1965. See also L. L. Jones and R. H. Wood, "Yield Line Analysis of Slabs," Thames and Hudson, Chatto and Windus, London, 1967, p. 405.

is the external load on the triangle ABC, assumed to be uniformly distributed. As the triangle ΔA tends to zero $\gamma \to \beta$ and $dP \to 0$, so that

$$V_{\Delta A} = (m_3 - m_2) \cot \beta \qquad (22\text{-}19)$$

The general form of this equation is

$$V_{\Delta A} = (m_{ds} - m_{\text{long side}}) \cot \beta \ (\text{acting upward}) \qquad (22\text{-}20)$$

where the subscripts of m indicate the bending moment per unit length on the sides ds and the long side of the infinitesimal triangle.

Equation 22-20 can be used to find the value of the nodal force V between any two lines at the junction of fracture lines. The nodal force between lines (1) and (3) in Fig. 22-9c is equal and opposite to the sum of the two

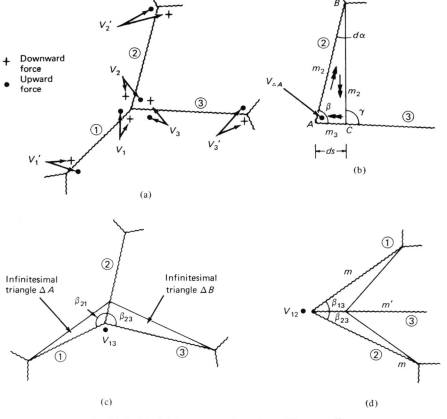

Fig. 22-9. Nodal forces at a junction of fracture lines.

nodal forces $V_{\Delta A}$ and $V_{\Delta B}$ of the infinitesimal triangles ΔA and ΔB — that is,

$$V_{13} = -V_{\Delta A} - V_{\Delta B}$$

or

$$V_{13} = -(m_2 - m_1)\cot\beta_{21} - (m_2 - m_3)\cot\beta_{23} \qquad (22\text{-}21)$$

It follows from the above that, when the reinforcement is equal in two orthogonal directions, the moments are $m_1 = m_2 = m_3 = m$, and all the nodal forces are zero at the junction of fracture lines of the same sign.

At a junction where two positive fracture lines (1) and (2) meet one negative fracture line (3) in an isotropic slab (Fig. 22-9d) the nodal forces are:

$$V_{12} = -(-m' - m)\cot\beta_{13} - (-m' - m)\cot\beta_{23}$$

or

$$V_{12} = (m' + m)(\cot\beta_{13} + \cot\beta_{23}) \qquad (22\text{-}22)$$

where m' is the absolute value of the bending moment on the negative fracture line. Similar equations can be written for the forces between other lines. When a fracture line meets a free or simply supported edge (Fig. 22-10a), we have $V_{12} = m\cot\beta$. When the fracture line is positive the nodal force V is downward in the acute angle.

If the edge rotation is restrained or otherwise subjected to a negative moment m' (Fig. 22-10b), the nodal force is $V_{12} = (m + m')\cot\beta$, acting downward in the acute angle.

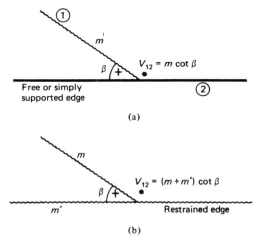

Fig. 22-10. Nodal force at a free or simply supported edge and at a restrained edge.

22-6 EQUILIBRIUM METHOD

As mentioned earlier, the slab parts are in equilibrium under the effect of the external loading, the moments on the yield lines, the nodal forces and the support reactions. For each slab part, three equations of equilibrium can be written, viz., two moment equations about two axes in the plane of the slab and an equation for the forces perpendicular to the plane of the slab which add up to zero.

The fracture pattern for a slab is completely defined if the axes of rotation are known, together with the ratios of the rotations θ_1, θ_2, ..., θ_n of slab parts when the mechanism acquires a small virtual deflection. For n parts, we require $(n - 1)$ ratios. The fracture pattern in Fig. 22-11a is determined by drawing contour lines of the deflected mechanism. The contour line of deflection w is composed of n straight segments parallel to the axes of rotation

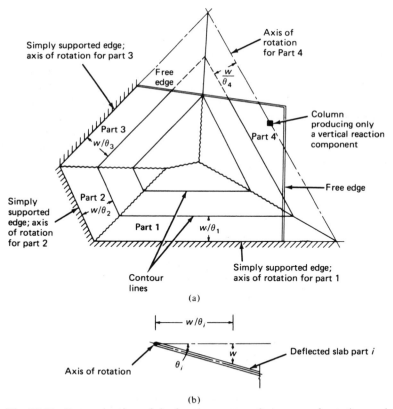

Fig. 22.11. Determination of the fraction pattern from axes of rotation and the ratios between the virtual rotations of slab parts. (a) Fracture pattern. (b) Cross section through the ith slab part in a direction perpendicular to its axis of rotation.

and distant from them $w/\theta_1, w/\theta_2, \ldots, w/\theta_n$ (see Fig. 22-11b). The intersections of the segments define points on the fracture lines.

For a part supported on one side, the position and magnitude of the reaction are unknown, thus representing two unknowns. For a part supported on a column, the axis of rotation passes through the column, but its direction is unknown and so is of course the magnitude of the reaction; hence, again there are two unknowns for the part of the slab. For a nonsupported part, the direction and position of the axis of rotation are unknown, so that once again we have two unknowns.

For n parts of a slab the unknowns are: the value of the ultimate moment m, $(n-1)$ relations between the rotations of the parts, and two unknowns for each part. Hence the total number of unknowns is $3n$ — the same as the number of equations of equilibrium (three for each part).

The formulation of the equilibrium equations becomes complicated except in simple cases such as the slabs considered below.

(a) Isotropic square slab, simply supported on four edges and carrying distributed load q per unit area. The fracture pattern is shown in Fig. 22-12.

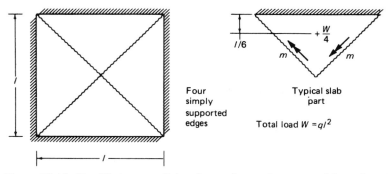

Figure 22-12. Equilibrium condition for an isotropic square slab under a uniformly distributed load of total magnitude W.

Taking the moment about a supported edge for one of the parts, we obtain

$$lm = \frac{ql^2 l}{4 \times 6}$$

whence

$$m = \frac{ql^2}{24} \qquad (22\text{-}23)$$

or

$$m = \frac{W}{24} \qquad (22\text{-}23a)$$

where W is the total load.

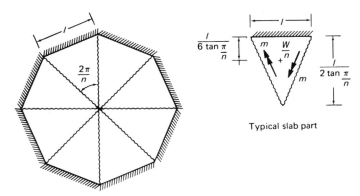

Typical slab part

Fig. 22-13. Equilibrium condition for an isotropic polygonal slab under uniformly distributed load of total magnitude W.

(b) Isotropic polygonal slab simply supported on n equal sides, carrying a uniform load whose total magnitude is W. The fracture pattern is shown in Fig. 22-13. Taking moments about the edge for any slab part, it can be shown that

$$m = \frac{W}{6n \tan (\pi/n)} \qquad (22\text{-}24)$$

The values of the ultimate moment for simply supported slabs in the form of different regular polygons are given in Fig. 22-14. They are calculated by Eq. 22-24 with n tending to infinity for the circular slab.

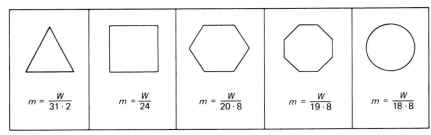

Fig. 22-14. Ultimate moment for simply supported polygonal slabs under a uniform load of total magnitude W.

(c) Isotropic square slab simply supported at corners, carrying a uniform load whose total magnitude is W. Taking moment for one slab part (Fig. 22-15) about the axis of rotation passing through the corner support and inclined at $45°$ to the edges, we find

$$m = \frac{W}{8}$$

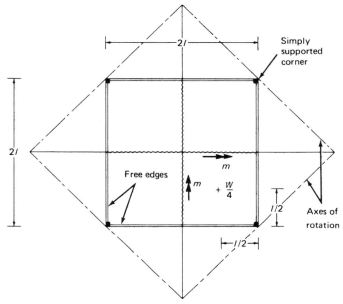

Fig. 22-15. Equilibrium condition for an isotropic square slab supported at corners and carrying a uniformly distributed load of total magnitude W.

22-7 NONREGULAR SLABS

In the case of slabs of nonregular shape or loading, the calculation of the ultimate moment is as follows.

(a) A fracture pattern is assumed, and the corresponding ultimate moment is computed by considering the equilibrium of each slab part. A moment equation about the axis of rotation of each part will give different values of the ultimate moment. For the correct yield pattern, all yield moment values must be equal.

(b) In general, the ultimate moments computed for the first assumed fracture pattern are not equal and the values will indicate how the pattern should be corrected. The procedure is then repeated with a "more correct" pattern until the exact pattern is obtained. This then is an iteration method.

(c) If for an assumed pattern the values of the ultimate moment obtained by equilibrium considerations do not differ much from one another, the application of the work equation for this fracture mechanism will give a value of the ultimate moment very close to the correct answer.

Example 22-3   Find the ultimate moment for the isotropic slab shown in Fig. 22-16a. The ultimate positive and negative moments are equal.

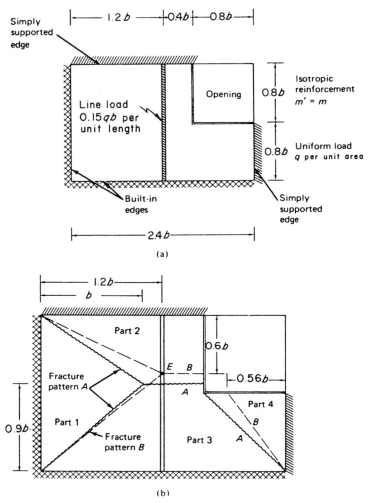

Fig. 22-16. Slab analyzed in Example 22-3. (a) Slab dimensions and loading.
(b) Fracture patterns.

For the first trial, we assume the fracture pattern A of Fig. 22-16b. The only nodal forces that need to be considered are those at the intersection of the inclined fracture line from the slab corner with the free edge. A moment equation about the axis of rotation of each slab part gives

Part 1:

$$2m(1.6b) = \frac{q(1.6b)b^2}{6} \qquad \therefore \ m = 0.0833qb^2$$

Part 2:

$$m(1.6b) = q\left[\frac{b(0.7b)^2}{6} + \frac{0.6b(0.7b)^2}{2}\right] + 0.15qb\frac{(0.7b)^2}{2} \quad \therefore \ m = 0.1659qb^2$$

Part 3:

$$2m(2.4b) = q\left[\frac{b(0.9b)^2}{6} + \frac{0.6b(0.9b)^2}{2} + \frac{0.8b(0.8b)^2}{6}\right]$$

$$+ 0.15qb\frac{(0.9b)^2}{2} - m\frac{0.8b}{0.8b}(0.8b) \quad \therefore \ m = 0.0936qb^2$$

Part 4:

$$m(0.8b) = q\frac{(0.8b)(0.8b)^2}{6} + m\frac{0.8b}{0.8b}(0.8b) \quad \therefore \ m = \infty$$

It is clear that the chosen pattern is not the correct one. However, the moment values for the various slab parts indicate the way in which we should move the yield lines to achieve a better result. Specifically, we can see that parts 1 and 3 should be increased in size and parts 2 and 4 decreased. The amended pattern B is therefore chosen for the second trial. The moment equation gives

Part 1:

$$2m(1.6b) = q\frac{(1.6b)(1.2b)^2}{6} \quad \therefore \ m = 0.1200qb^2$$

Part 2:

$$m(1.6b) = q\left[\frac{1.2b(0.6b)^2}{6} + \frac{0.4b(0.6b)^2}{2}\right] + 0.15qb\frac{(0.6b)^2}{2} \quad \therefore \ m = 0.1069qb^2$$

Part 3:

$$m(2.4b + 2.16b) = q\left[\frac{1.2bb^2}{6} + \frac{0.4bb^2}{2} + \frac{0.24b(0.8b)^2}{2} + \frac{0.56b(0.8b)^2}{6}\right]$$

$$+ 0.15qb\frac{b^2}{2} - m\frac{0.56b}{0.8b}(0.8b) \quad \therefore \ m = 0.1194qb^2$$

Part 4:

$$m(0.8b) = q\frac{(0.8b)(0.56b)^2}{6} + m\frac{0.56b}{0.8b}(0.56b) \quad \therefore \ m = 0.1025qb^2$$

Assuming that the mechanism corresponding to pattern B acquires a unit virtual deflection at point E, the corresponding rotation of slab parts will be $1/(1.2b)$, $1/(0.6b)$, $1/b$, and $1/(0.7b)$ for slab parts 1, 2, 3, and 4 respectively. Equating the internal and external virtual work gives

$$2m(1.6b)\frac{1}{1.2b} + m(1.6b)\frac{1}{0.6b} + (2.4b + 2.16b)m\frac{1}{b}$$

$$+ m(0.8b)\frac{1}{0.7b} = q\left[1.2b(1.6b)\frac{1}{3} + 0.4b(1.6b)\frac{1}{2}\right.$$

$$\left. + 0.24b(0.8b)\left(\frac{0.8b}{b}\right)\frac{1}{2} + 0.56b(0.8b)\left(\frac{0.8b}{b}\right)\frac{1}{3}\right] + 0.15qb(1.6b)\frac{1}{2}$$

whence $m = 0.1156qb^2$.

22-8 STRIP METHOD

We remember that the yield-line method of analysis for slabs always gives an upper bound to the true collapse load, though for certain simple cases the exact collapse load can be intuitively achieved by guessing the correct mechanism of failure. It is clear that for design purposes we might justifiably consider a lower bound to the true collapse load to be preferable.

Any lower-bound solution for a slab with given loading must have a moment field which satisfies the governing equilibrium equation at all points, and must not violate the particular yield criterion anywhere. The equilibrium equation in rectangular coordinates is (see Eq. 18-28)

$$\frac{\partial^2 M_x}{\partial x^2} + \frac{\partial^2 M_y}{\partial y^2} - 2\frac{\partial^2 M_{xy}}{\partial x \partial y} = -q \tag{22-25}$$

This equation must hold for both lower-bound and exact solutions regardless of the material properties of the slab, and it is evident that there is an infinite number of moment fields which satisfy Eq. 22-25. If we derive a complete moment field in equilibrium with the desired ultimate load and then provide reinforcement such that the ultimate moments of resistance at all points exceed or are equal to the equilibrium moments then a lower-bound solution will be achieved. A yield-line analysis of the completed design will of course give an upper-bound to the collapse load.

Because of the infinite number of possible equilibrium moment fields for a given slab and loading and the difficulties of proportioning reinforcement in the cases where twisting moments are present, Hillerborg[6] suggested

[6]A. Hillerborg, "A Plastic Theory for the Design of Reinforced Concrete Slabs," *Proc. 6th Congr. Int. Ass. Bridg. Struct. Eng.*, Stockholm, 1960. A. Hillerborg, "Jamviktsteori för Armerade Betongplattor," *Betong* 41(4) (1956), pp. 171–182.

that the twisting moments M_{xy} be made zero at the outset of the analysis. In this case, the equilibrium Eq. 22-25 reduces to

$$\frac{\partial^2 M_x}{\partial x^2} + \frac{\partial^2 M_y}{\partial y^2} = -q \tag{22-26}$$

and the load q is carried by strips running in the x and y directions. This strip method has been critically examined by Wood,[7] who found that if the reinforcement in the slab is curtailed so that precise correspondence between the ultimate moment field and the equilibrium moment field is obtained, then the strip method gives an exact correspondence between the design load and the collapse load.

To determine how the load is shared between the strips in the x and y directions, we partition Eq. 22-26 as follows:

$$\left. \begin{array}{l} \dfrac{\partial^2 M_x}{\partial x^2} = -\alpha q \\[2ex] \dfrac{\partial^2 M_y}{\partial y^2} = -(1 - \alpha)q \end{array} \right\} \tag{22-27}$$

where the parameter α governs the way in which load is dispersed at all points in the slab. This load dispersion parameter may have any value between 0 and 1, and can vary from point to point in the slab. In practice, for simplicity of design, α will normally have a constant value in specified zones within the slab. Because we are now dealing with beam equations of the type of Eq. 22-27, the moment profiles are easily obtained.

Consider the design of a simply supported slab shown in Fig. 22-17a, which carries a uniformly distributed load q per unit area. We can assume the load dispersion discontinuity lines as shown in the figure: they are consistent with the major part of the slab spanning one-way in the short-span direction. The slab strip 1-1 carries then the full load q over its full length and the maximum moment will be $ql^2/8$. Slab strips 2-2 and 3-3 are loaded over the end regions only as shown in Fig. 22-17b. In these cases the maximum moment is $qc^2/2$, the moment being constant in the unloaded region.

Since the full bending-moment pattern for the slab is known, the reinforcement can be provided at all points to ensure that the ultimate moment of resistance is greater than the calculated value. For efficiency in design, variable reinforcement in the end-loaded strips (Fig. 22-17b) will be necessary. This is difficult to handle in practical cases and it is preferable to use reinforcement in distinct uniform bands.

[7] R. H. Wood and G. S. T. Armer, "The Theory of the Strip Method for Design of Slabs," *Proc. I.C.E.* 41 (1968), pp. 285–311.

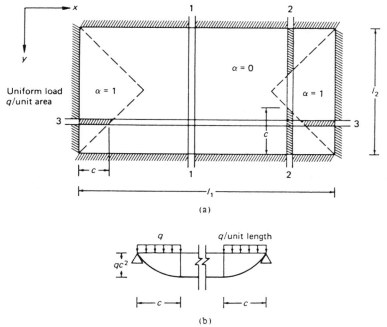

Fig. 22-17. Loading in the strip method. (a) Assumed α values. (b) Loading and bending moment diagrams for strips of unit width, 2–2 and 3–3.

22-9   USE OF BANDED REINFORCEMENT

Wherever the load dispersion discontinuity lines are not parallel to the support lines of the particular strip, the maximum moment varies from strip to strip. We can, however, ensure that the moment profiles for strips within a given bandwidth are the same by making the load-dispersion discontinuity lines parallel to the support lines of the band.

Consider again the simply supported slab in Fig. 22-17a. We know that the elastic curvature of the slab parallel to and near the edges is small and only small amounts of reinforcement need to be placed there. We choose then the load-dispersion zones of Fig. 22-18a. The size of the various zones is quite arbitrary, as are the load-dispersion proportions in the corner regions.

The maximum moments in the four different bands are then

$$
\begin{array}{lll}
\text{Strip 1–1} & M_{\max} = ql^2/8 \\
\text{Strip 2–2} & M_{\max} = qc^2/4 \\
\text{Strip 3–3} & M_{\max} = qb^2/2 \\
\text{Strip 4–4} & M_{\max} = qb^2/4
\end{array}
$$

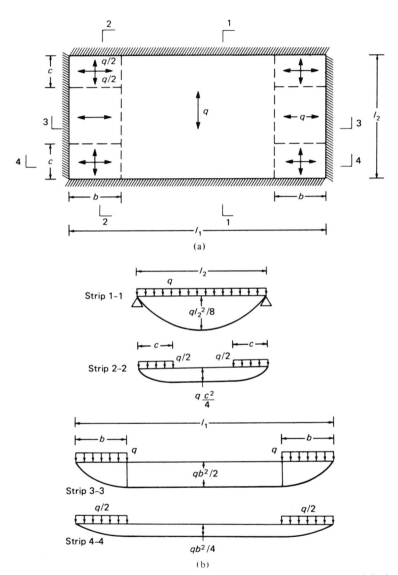

Fig. 22.18. Load-dispersion zones and moment profiles for the slab in Fig. 22-17a. (a) Load-dispersion zones. (b) Loading and bending moment diagrams for strips of unit width.

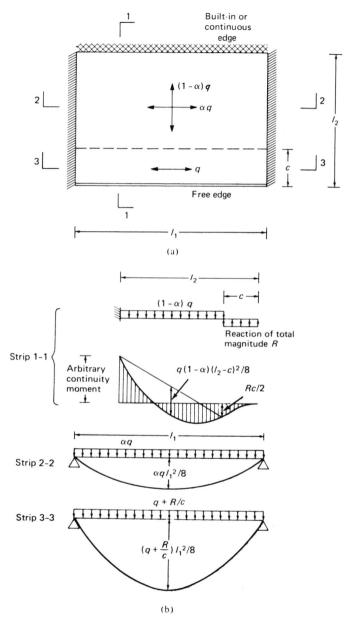

Fig. 22-19. Strip method for a slab with a free edge. (a) Slab details and load dispersion. (b) Loading and bending-moment diagrams for strips of unit width.

and reinforcement is provided accordingly. The reinforcement can be curtailed, if desired for economy reasons, in accordance with the moment profiles of Fig. 22-18b.

One advantage of the strip method is that the reactions to strips are known and hence the loading on any supporting beam is immediately established. Continuity at the supports poses no difficulty; all that is necessary is to provide an arbitrary value of moment at these points, with consequent reduction in the interior slab moments. In the case of slabs having a free edge, all load dispersion adjacent to the free edge must take place parallel to the edge. This will necessitate in general a strong band of reinforcement parallel to the edge. Strips normal to the edge are supported on this strong band and on the remote support. Figure 22-19 shows an example of a slab with one free edge, the opposite edge continuous, and the remaining edges simply supported, and gives the moment profiles in the various bands of the slab. The uniformly distributed reaction, R, to strip 1–1 can be calculated once the continuity moment value has been fixed. The loading on strip 3–3 includes the reaction R in addition to the uniformly distributed external loading q.

Because of the large measure of freedom in using the strip method, the designer will find an appreciation of the elastic moment fields particularly useful. Provided the equilibrium moment field derived is not too far removed from that expected in elastic design, the working load behavior in terms of cracking and deflections will generally be satisfactory.

22-10 GENERAL

The methods considered in this chapter are of two basic types: the energy and equilibrium methods of yield-line theory are, strictly speaking, methods of analysis, while the strip method is a direct design procedure. For simple cases of slab geometry and loading, the yield-line method can be safely used as a design method since the fracture pattern to give an upper bound close to the correct collapse load can be readily obtained. For complex cases of geometry and loading, care must be exercised in the choice of fracture patterns, particularly where concentrated loads occur, since local modes of collapse, called fan modes, can occur. A fracture pattern involving fan modes will generally give lower collapse load than that obtained from the corresponding straight line pattern.

Although the collapse loads given by the yield-line method are theoretically upper-bound values, in practice the actual collapse load of a reinforced concrete slab may be above the calculated value because of the presence of various secondary effects.

It has been assumed in the analysis that Eq. 22-1 can be used for calculating the moment of resistance in a yield line at any angle to a set of orthogonal reinforcement. This method tends to underestimate the true moment of resistance which can be increased by up to 14 percent for the case of a yield line at 45° to an isotropic orthogonal set of reinforcement. The reason for this is that the analysis takes no account of kinking of reinforcement, which places the reinforcement almost at right angles to the fracture line and therefore increases the moment of resistance.

Where the boundaries of any slab are restrained from horizontal movement (as might occur in the interior panels of a continuous slab) the formation of the collapse mechanism develops high compressive forces in the plane of the slab with a consequent increase in carrying capacity. At very large deflections of the slab, it is possible finally to develop tensile membrane action where cracks go right through the slab so that the load is supported on the net of reinforcement.

These secondary effects will also occur in the actual behavior of slabs designed by the strip method. The strip method will always give a lower-bound or correct value for the collapse load depending on the degree of curtailment of reinforcement that the designer imposes. An advantage of the strip method is that the reactions on the supports to the slab are precisely determined.

Whenever variable reinforcement is required in a slab as a move toward efficiency, then the strip method is preferable, since economy can be readily achieved by suitable choice of the load-dispersion parameter. The use of the yield-line method with variable reinforcement involves the consideration of a large number of differing yield-line patterns.

Neither the yield-line method nor the strip method of ultimate load design guarantees satisfactory deflection and cracking behavior at working loads, though tests carried out on slabs designed by both methods generally exhibit satisfactory behavior. Distribution of reinforcement within the slab which is not too far removed from that expected with an elastic distribution of bending moments will generally ensure satisfactory working-load behavior. In the context of elastic analyses we should also remember that the bending-moment distribution can change markedly for changes in the relative stiffnesses of the slab and its supporting beams and that many of the design tables based on elastic analyses are consistent only with nondeflecting supporting beams.

PROBLEMS

The representation of the different edge conditions of the slabs in the following problems are made according to the convention indicated in Sec. 22-21.

22-1   Using the yield-line theory, find the ultimate moment for the isotropic slabs shown under the action of uniformly distributed load. Show in a pictorial view the forces on the fractured slab parts in Prob. 22-1a. The columns in Prob. 22-1d and e are assumed to produce only a normal concentrated reaction component.

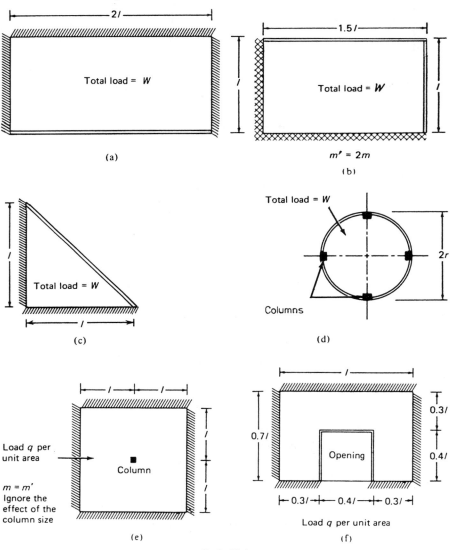

Prob. 22-1

22-2 Using the yield-line theory, find the ultimate moment for the slabs shown assuming the given ratio of reinforcement in each case. Assume that the column in Prob. 22-b produce only a normal concentrated reaction component.

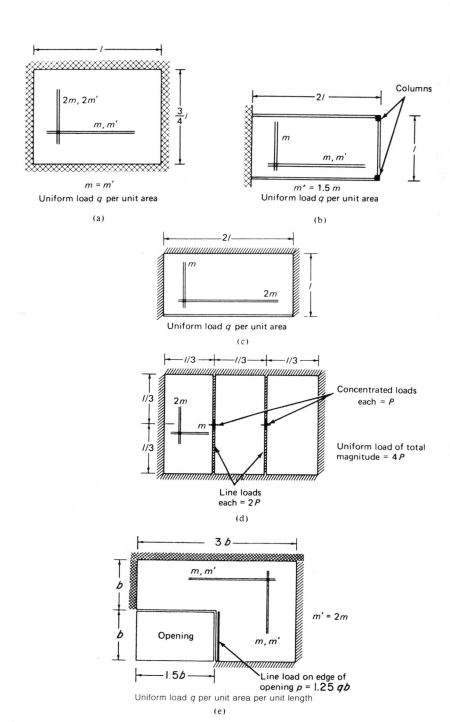

$2m, 2m'$

m, m'

$\frac{3}{4}l$

l

$m = m'$

Uniform load q per unit area

(a)

Columns

$2l$

m

m, m'

l

$m' = 1.5 m$

Uniform load q per unit area

(b)

$2l$

m

$2m$

l

Uniform load q per unit area

(c)

$l/3$ $l/3$ $l/3$

$l/3$

$2m$

m

$l/3$

Concentrated loads
each $= P$

Uniform load of total
magnitude $= 4P$

Line loads
each $= 2P$

(d)

$3b$

b

m, m'

b

Opening

m, m'

$m' = 2m$

$1.5b$

Line load on edge of
opening $p = 1.25\, qb$

Uniform load q per unit area per unit length

(e)

Prob. 22-2

CHAPTER 23

Structural dynamics

23-1 INTRODUCTION

The previous chapters dealt with structures subjected to static forces producing displacements which do not vary with time. We shall now consider dynamic problems, in which the forces are time-dependent and cause vibration of the structure: hence, it is necessary to take into account the forces produced by the inertia of the accelerating masses. For this purpose, we use Newton's second law of motion which states that the product of the mass and its acceleration is equal to force. In practice, dynamic loading is produced by seismic forces, nonsteady wind, blast, reciprocating machinery, or impact of moving loads.

An elastic structure disturbed from its equilibrium condition by the application and removal of forces will oscillate about its position of static equilibrium. Thus the displacement at any point on the structure will vary periodically between specific limits in either direction. The distance of either of these limits from the position of equilibrium is called the amplitude of the vibration. In the absence of external forces, the motion is called *free vibration*, and may continue with the same amplitude for an indefinitely long time. In practice, there are always forces tending to oppose the motion, such as friction, air resistance, or imperfect elasticity, which cause the amplitude to diminish gradually until the motion ceases. This type of motion is called *damped free vibration*.

During a period over which external forces are applied we have *forced motion*. This may be *damped* or *undamped* forced motion, depending on the presence or absence of resisting forces.

23-2 COORDINATES AND LUMPED MASSES

In a vibrating structure, the product of the mass of all the members and of their acceleration represents distributed forces. This causes some difficulties. For instance, for the beam in Fig. 23-1a, with a distributed mass, an infinite number of coordinates are required to define the displacement configuration. If we imagine that the beam mass is lumped into two bodies, with the force-displacement properties of the beam unchanged (Fig. 23-1b), and assume that the external forces causing the motion are applied at these two masses, the deflected shape of the beam at any time can be completely defined by 12 coordinates, 6 at each mass (Fig. 23-1c). The idealized model of Fig. 23-1b which has 12 degrees of freedom may therefore conveniently be considered in dynamic problems in place of the actual structure. A more accurate representation can be obtained by using a larger number of masses at closer intervals but this requires a greater number of coordinates.

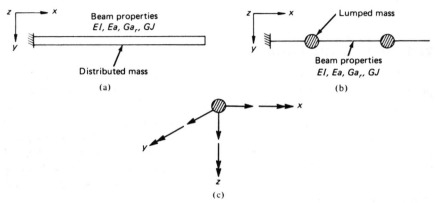

Fig. 23-1. Idealization of a distributed mass by a lumped-mass system. (a) A vibrating beam has infinite degrees of freedom. (b) Idealization of the beam in part (a) into a lumped-mass system. (c) Typical degrees of freedom at a lumped mass.

To demonstrate the use of lumped mass idealization for structures, consider the free vibration of a multistorey frame in its own plane (Fig. 23-2a). For the idealized model, we ignore the axial deformations of the members, add the mass of the columns to the mass of the floors, and assume that the total mass is mounted on a simple beam (Fig. 23-2b), so that the mass can sidesway only (i.e., it is not subject to rotation).

Let the horizontal displacements at floor levels at any time be D_1, D_2, ..., D_n, and let the masses of the floors be m_1, m_2, ..., m_n. Writing Newton's second law for the ith mass, (see Fig. 23-2d), we have

$$Q_i = m_i \ddot{D}_i \tag{23-1}$$

and for n masses we can write in matrix form

$$\{Q\} = [m]\{\ddot{D}\} \tag{23-2}$$

where $[m]$ is a diagonal matrix

$$[m] = \begin{bmatrix} m_1 & & & \\ & m_2 & & \\ & & \cdots & \\ & & & m_n \end{bmatrix} \tag{23-3}$$

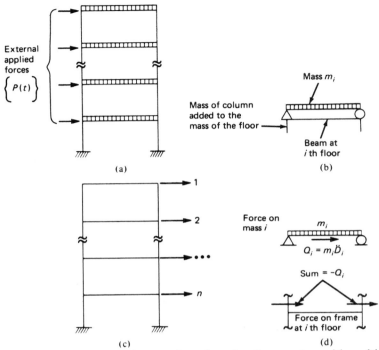

External
applied
forces
$\{P(t)\}$

Mass of column
added to the
mass of the floor

Mass m_i

Beam at
i th floor

(a)

(b)

1

2

•••

n

Force on
mass i

m_i

$Q_i = m_i \ddot{D}_i$

Sum = $-Q_i$

Force on frame
at i th floor

(c)

(d)

Fig. 32-2. Derivation of the equation of motion for a system with multi-degrees of freedom. (a) Frame. (b) Idealization of the mass of a typical floor. (c) Coordinate system representing positive directions of forces $\{F\}$, $\{Q\}$, and $\{P\}$ and of displacements $\{D\}$. (d) Application of Newton's second law to the mass at the ith floor.

The acceleration $\ddot{D}_i$ of the ith mass is equal to the second derivative of the displacement D_i with respect to time t, $(d^2 D_i/dt^2)$. The positive direction of forces and displacements are arbitrarily chosen in Fig. 23-2c. The force Q_i acts in the positive direction on the mass i with an equal and opposite force acting on the frame (see Fig. 23-2d). If, in addition, the frame is subjected at the n coordinates to forces $\{P\}$ (generally varying with time), the net forces on the frame at any time will be

$$\{F\} = \{P\} - \{Q\} \tag{23-4}$$

The forces $\{F\}$ and the displacements $\{D\}$ are related by

$$[S]\{D\} = \{F\} \tag{23-5}$$

where $[S]$ is the stiffness matrix of the frame[1] corresponding to the coordinates in Fig. 23-2c.

[1]This matrix may be derived by considering first a stiffness matrix of the order $3n \times 3n$ corresponding to three coordinates per floor (representing two rotations at beam ends and one sidesway). This matrix is then condensed by the method of Sec. 4-5. (See also Chap. 16).

Combining Eqs. 23-2, 23-4, and 23-5, we obtain the following *equation of motion* of an *n*-degree-of-freedom system subjected to dynamic forces (undamped forced motion).

$$[m]\{\ddot{D}\} + [S]\{D\} = \{P\}$$ (23-6)

When the system is vibrating freely, $\{P\} = \{0\}$ and the equation of motion becomes (undamped free vibration)

$$[m]\{\ddot{D}\} + [S]\{D\} = \{0\}$$ (23-7)

23-3 CONSISTENT MASS MATRIX

From the preceding section, we see that when a structure is vibrating, the product of the mass and acceleration with a negative sign is the inertia force in the direction of the assumed degrees of freedom (coordinates). The inertia force $-m\ddot{D}$ may represent a load or a couple, and thus the lumped masses refer to translational or rotational inertia of the elements of a structure. As we have seen, the mass of the various structural elements is arbitrarily lumped at nodes, resulting in a diagonal mass matrix $[m]$ in the equation of motion, and hence in some simplification in the analysis. A more accurate representation of the distributed mass can be achieved by the use of the consistent mass matrix,[2] which will be derived here in a general form.

If at time *t* the displacements at the assumed degrees of freedom are $\{D\}$, the displacement *f* at any other point (x, y, z) may be described by a set of independent functions $L_i(x, y, z)$ such that

$$f = [L_1 \quad L_2 \quad \dots \quad L_n]\{D\}$$ (23-8)

The choice of the displacement functions was discussed in Chapter 19 (see Eq. 19-25).

The element m_{ij} of the consistent mass matrix is defined as the inertia force acting (in the negative direction) at coordinate *i* associated with a unit acceleration of coordinate *j*. This quantity is referred to as the *mass coupling* between coordinates *i* and *j*. The element m_{ij} is given by

$$m_{ij} = \int_v \gamma L_i L_j \, dv$$ (23-9)

where $\gamma = \gamma(x, y, z)$ is the mass per unit volume and *dv* is elemental volume.

Equation 23-9 can be proved as follows. If the structure is restrained so that the motion is prevented at all the coordinates except *j*, which is given

---

[2]This was first recognized by J. S. Archer in his paper Consistent Mass Matrix for Distributed Mass Systems, *Proc. ASCE.*, Vol. 89, ST4, pp. 161-178. O. C. Zienkiewicz and Y. K. Cheung extended the use of consistent mass matrices to represent distributed mass of plates in bending, and R. W. Clough used them for plates subjected to in-plane forces. See O. C. Zienkiewicz and Y. K. Cheung, *The Finite Element Method in Structural and Continuum Mechanics*, McGraw-Hill, New York, 1967.

a unit acceleration $\ddot{D}_j = 1$, the mass element $\gamma\,dv$ will have an inertia force $-\gamma\,dvL_j$. The restraining forces (reactions) at any coordinate i are readily obtained by the Müller-Breslau principle (see Sec. 13-3): $-L_i$ represents the influence line (or surface) of the reaction at i. Thus, due to the force $-\gamma\,dvL_j$, the reaction at i is $(-L_i)(-\gamma\,dvL_j)$. Integrating over the volume of the structure, we obtain the total reaction at i due to the distributed inertia forces:

$$R_{ij} = \int_v \gamma L_i L_j\,dv \qquad (23\text{-}10)$$

The distribution inertia forces can be replaced by forces at the coordinates equal and opposite to the reaction. Therefore it follows that $R_{ij} = m_{ij}$, and hence substitution of m_{ij} for R_{ij} in Eq. 23-10 gives Eq. 23-9.

It is clear from Eq. 23-9 that $m_{ij} = m_{ji}$, which means that the consistent mass matrix is symmetrical.

As an example of the use of Eq. 23-9, we can derive the consistent mass matrix for a prismatic beam corresponding to the four coordinates in Fig. 4-4c. For convenience, the beam is shown again in Fig. 23-3a. The mass per unit volume γ is assumed constant.

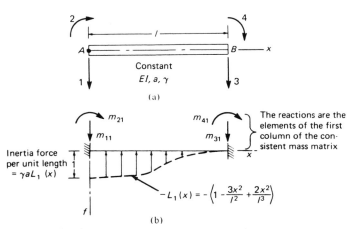

(a)

(b)

Fig. 23-3. Derivation of the consistent mass matrix for a prismatic beam. (a) Prismatic beam of total mass $\gamma a l$. (b) Distributed inertia forces associated with acceleration $\ddot{D}_1 = 1$.

The transverse deflection of the beam (neglecting shear deformation) is described by Eq. 23-8 with the shape functions given in Eq. 19-28, which can be written as

$$[L] = \left[\left(1 - \frac{3x^2}{l^2} + \frac{2x^3}{l^3}\right)\left(x - \frac{2x^2}{l} + \frac{x^3}{l^2}\right)\left(\frac{3x^2}{l^2} - \frac{2x^3}{l^3}\right)\left(-\frac{x^2}{l} + \frac{x^3}{l^2}\right)\right]$$

$$(23\text{-}11)$$

Figure 23-3b shows the distributed inertia forces associated with $\ddot{D}_1 = 1$ with the motion prevented at the other coordinates. Equation 23-9 for the present example can be written in the form

$$m_{ij} = \gamma a \int_0^l L_i L_j \, dx \tag{23-12}$$

where a is the beam cross-sectional area.

Substituting for L_r from Eq. 23-11 into Eq. 23-12, the following consistent mass matrix for the beam in Fig. 23-3a is generated:

$$[m] = \frac{\gamma a l}{420} \begin{bmatrix} 156 & & \text{symmetrical} & \\ 22l & 4l^2 & & \\ 54 & 13l & 156 & \\ -13l & -3l^2 & -22l & 4l^2 \end{bmatrix} \tag{23-13}$$

23-4 UNDAMPED FREE VIBRATION OF A SYSTEM WITH ONE DEGREE OF FREEDOM

For a single-degree-of-freedom system vibrating freely, the equation of motion (Eq. 23-7) reduces to

$$m\ddot{D} + SD = 0 \tag{23-14}$$

Examples of such a system are the frame in Fig. 23-2 with the number of floors reduced to one and the cantilever in Fig. 23-1 with one lumped mass. The quantity S in this case is the static force necessary to produce a unit displacement. Equation 23-14 can be written as

$$\ddot{D} + \omega^2 D = 0 \tag{23-15}$$

with

$$\omega = \sqrt{\frac{S}{m}} \tag{23-16}$$

The solution of Eq. 23-15 is

$$D = C_1 \sin \omega t + C_2 \cos \omega t \tag{23-17}$$

This equation is harmonic — that is, D has the same value at time t as time $t + T$ such that

$$T = \frac{2\pi}{\omega} \tag{23-18}$$

The time interval T is the *natural period of vibration*, and the reciprocal of T is the *natural frequency*. The quantity ω is called *natural angular frequency*. The velocity of the vibrating mass $\dot{D} = dD/dt$ is

$$\dot{D} = C_1 \omega \cos \omega t - C_2 \omega \sin \omega t \qquad (23\text{-}19)$$

The integration constants C_1 and C_2 in Eqs. 23-17 and 23-19 can be determined from the initial conditions of motion. If at time $t = 0$, the mass has a displacement D_0 from the equilibrium position and an initial velocity $\dot{D}_0$, then by substitution in Eqs. 23-17 and 23-19, we obtain $C_2 = D_0$ and $C_1 = \dot{D}_0/\omega$, so that Eq. 23-17 becomes

$$D = \frac{\dot{D}_0}{\omega} \sin \omega t + D_0 \cos \omega t \qquad (23\text{-}20)$$

The amplitude of vibration is the maximum value of displacement, which from Eq. 23-20 is

$$D_{max} = \bar{a} = \sqrt{D_0^2 + \left(\frac{\dot{D}_0}{\omega}\right)^2} \qquad (23\text{-}21)$$

Equation 23-20, can be written in the form

$$D = \bar{a} \sin(\omega t + \alpha) \qquad (23\text{-}22)$$

where

$$\alpha = \tan^{-1}(D_0\omega/\dot{D}_0) \qquad (23\text{-}23)$$

A graph of D against t is shown in Fig. 23-4, from which it can be seen

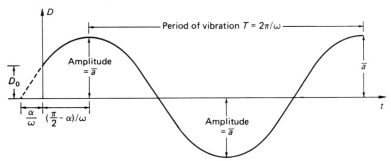

Fig. 23-4. Time-displacement relationships for a single-degree-of-freedom system.

that the vibrating mass reaches its first extreme position at time

$$t = \left(\frac{\pi}{2} - \alpha\right)\bigg/\omega$$

Example 23-1 A weightless cantilever of length l and constant flexural rigidity EI carries a weight W at its end. Neglecting the moment of inertia $\bar{I}$ of the mass about its center,

    (a) Find the natural angular frequency and the natural period of vibration of the system.

(b) If the motion is initiated by displacing the mass in a direction perpendicular to the cantilever by a distance of $Wl^3/3EI$ and then leaving the system to vibrate freely, what is the maximum displacement? What is the displacement at any time t?

The static force required to produce a unit transverse deflection at the end of the cantilever is

$$S = 3EI/l^3$$

Equations 23-16 and 23-18 give

$$\omega = \sqrt{\frac{3EIg}{Wl^3}} \quad \text{and} \quad T = 2\pi\sqrt{\frac{Wl^3}{3EIg}}$$

where g is the acceleration due to gravity.

The initial conditions are

$$D_0 = \frac{Wl^3}{3EI} \quad \text{and} \quad \dot{D}_0 = 0$$

Substitution in Eqs. 23-20 and 23-21 gives

$$D_{max} = D_0 = \frac{Wl^3}{3EI}$$

$$D = \frac{Wl^3}{3EI}\cos \omega t$$

23-5   RESPONSE OF A SINGLE-DEGREE-OF-FREEDOM UNDAMPED SYSTEM

For a single-degree-of-freedom system, the equation of undamped motion (see Eq. 23-6) is

$$m\ddot{D} + SD = P \tag{23-24}$$

23-51   Special Case: Harmonic Force

We can now distinguish the special case when the externally applied force P varies harmonically:

$$P = P_0 \sin \Omega t \tag{23-25}$$

where P_0 is the maximum value of the force and Ω is the angular frequency of the disturbing force, called the *impressed frequency*. This type of loading is produced in practice by centrifugal forces caused by some imbalance in rotating machines. With this loading, the equation of motion becomes

$$\ddot{D} + \omega^2 D = \frac{P_0}{m}\sin \Omega t \tag{23-26}$$

where, as before, $\omega^2 = S/m$, and ω is the natural angular frequency of the system.

The solution of this differential equation is

$$D = C_1 \cos \omega t + C_2 \sin \omega t + \frac{P_0}{S}\left(\frac{1}{1 - \Omega^2/\omega^2}\right) \sin \Omega t \qquad (23\text{-}27)$$

The first two terms represent free vibrations (see Eq. 23-17) and the last term represents the forced vibrations. In practical applications of Eq. 23-27, the forced vibration term becomes large compared to the free vibration terms, which may therefore be neglected. The motion defined by the third term in Eq. 23-27 is a *steady-state* forced vibration. Considering only the forced vibration, Eq. 23-27 becomes

$$D = \frac{P_0}{S}\left(\frac{1}{1 - \Omega^2/\omega^2}\right) \sin \Omega t \qquad (23\text{-}28)$$

The quantity $(P_0/S) \sin \Omega t$ is the displacement which would occur if the external force were applied statically and the term $1/(1 - \Omega^2/\omega^2)$ accounts for the dynamic effect of this force. The absolute value of this term is called the *magnification factor*. It is nearly equal to unity when the impressed frequency Ω is small compared with the natural frequency ω, and reaches infinity when $\Omega = \omega$. This is the *resonance* condition in which the amplitude of vibration increases indefinitely. However, in practice, damping forces always exist and thus limit the amplitude to finite values but these may be large enough to cause collapse. At higher frequencies, the magnification factor decreases and approaches zero when $\Omega \gg \omega$, resulting in vibrations of very small amplitude.

23-52 General Case: Any Disturbing Force

The dynamic loading resulting from blast, gusts of wind, or seismic forces is generally not harmonic. In such case the equation of motion has to be solved numerically, and a solution in exact form can be obtained only if some idealized loading is used to represent the true loading.

Consider an undamped system with a single degree of freedom subjected to a disturbing force $P(t)$ which is a function of time (Fig. 23-5). The disturbing force is assumed to begin to act at time $t = 0$, when the mass is in equilibrium position, so that $D_0 = 0$, and $\dot{D}_0 = 0$. We want to find the displacement D_τ at a later instant τ.

The increase in velocity of the mass in any time interval dt is

$$d\dot{D} = \frac{P}{m} dt \qquad (23\text{-}29)$$

This equation states that the change in momentum is equal to the impulse. If we consider the increment $d\dot{D}$ as an initial velocity acquired by the mass at time t, the corresponding displacement at time τ can be expressed by Eq. 23-20

$$dD_\tau = \frac{d\dot{D}}{\omega} \sin(\omega\tau - \omega t) \tag{23-30}$$

Substituting Eq. 23-29 into Eq. 23-30 and integrating over the interval $t = 0$ to $t = \tau$, we obtain

$$D_\tau = \frac{1}{\omega m} \int_0^\tau P \sin(\omega\tau - \omega t)\, dt \tag{23-31}$$

This is Duhamel's superposition integral (or convolution integral); it can be easily evaluated numerically, particularly when a computer is used.

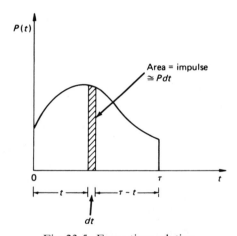

Fig. 23-5. Force-time relation.

To find the displacement during a loading period, we divide the period t into a number of intervals Δt and proceed from one interval to the other. Using the trigonometric identity

$$\sin(\omega\tau - \omega t) = \sin \omega\tau \cos \omega t - \cos \omega\tau \sin \omega t$$

Eq. 23-31 can be written in the form:

$$D_\tau = \frac{1}{\omega m} [\psi_1(\tau) \sin \omega\tau - \psi_2(\tau) \cos \omega\tau] \tag{23-31a}$$

where

$$\psi_1(\tau) = \int_0^\tau P \cos \omega t\, dt \qquad \psi_2(\tau) = \int_0^\tau P \sin \omega t\, dt$$

The value of any of these two integrals for time τ is calculated from its value for time $\tau - \Delta t$ as follows:

$$\left.\begin{aligned}\psi_1(\tau) &= \psi_1(\tau - \Delta t) + \Delta t \, P_{\tau - \Delta t} \cos \omega(\tau - \Delta \tau)\\ \psi_2(\tau) &= \psi_2(\tau - \Delta t) + \Delta t \, P_{\tau - \Delta t} \sin \omega(\tau - \Delta \tau)\end{aligned}\right\} \quad (23\text{-}32)$$

The displacement D_τ is then obtained by substitution in Eq. 23-31a.

If at $t = 0$ the mass has an initial displacement D_0 and velocity $\dot{D}_0$, the displacement at time τ is (see Eq. 23-20)

$$D_\tau^* = \frac{\dot{D}_0}{\omega} \sin \omega\tau + D_0 \cos \omega\tau + \frac{1}{\omega m} \int_0^\tau P \sin(\omega\tau - \omega t) \, dt \quad (23\text{-}33)$$

For the special case when P varies harmonically according to Eq. 23-25, it can be shown that Eq. 23-31 gives the same result as Eq. 23-28.

The solution of the equation of motion can also be performed numerically. Step-by-step methods[3] of integration can be used to give the displacement, the velocity, and the acceleration at time intervals.

23-6 VISCOUSLY DAMPED VIBRATION OF A SINGLE-DEGREE-OF-FREEDOM SYSTEM

Damping forces should not be neglected in cases such as the response of structures to earthquakes. In analysis, it may be convenient to replace the actual damping forces by *viscous damping* forces proportional to the velocity. Thus a viscous damping force is

$$F = c\dot{D} \quad (23\text{-}34)$$

where c is the damping coefficient. With this damping force included, the equation of motion (23-24) of a single-degree-of-freedom system becomes

$$m\ddot{D} + c\dot{D} + SD = P \quad (23\text{-}35)$$

or

$$\ddot{D} + 2\beta\dot{D} + \omega^2 D = P/m \quad (23\text{-}36)$$

where

$$\beta = \frac{c}{2m} \quad (23\text{-}37)$$

and, as before, $\omega = \sqrt{S/m}$ is the natural angular frequency (see Eq. 23-16).

---

[3]See M. G. Salvadori and M. L. Baron, *Numerical Methods in Engineering*, 2d ed., Prentice-Hall, Englewood Cliffs, N.J., 1961, Chapter III, "The Numerical Integration of Initial Value Problems," pp. 113–147.

23-61   Free, Viscously Damped Vibration

For a free, viscously damped vibration, Eq. 23-36 becomes

$$\ddot{D} + 2\beta\dot{D} + \omega^2 D = 0 \qquad (23\text{-}38)$$

Substituting $D = e^{\lambda t}$, we obtain the auxiliary equation

$$\lambda^2 + 2\beta\lambda + \omega^2 = 0 \qquad (23\text{-}39)$$

which has the roots

$$\lambda_1 = -\beta + \sqrt{\beta^2 - \omega^2} \quad \text{and} \quad \lambda_2 = -\beta - \sqrt{\beta^2 - \omega^2} \qquad (23\text{-}40)$$

The general solution of Eq. 23-38 takes the form

$$D = C_1 e^{\lambda_1 t} + C_2 e^{\lambda_2 t} \qquad (23\text{-}41)$$

When $\beta^2 = \omega^2$, we have the case of *critical damping*

$$c_{\text{cr}} = 2m\omega \qquad (23\text{-}42)$$

and the auxiliary equation has repeated roots $\lambda_1 = \lambda_2 = -\beta = -\omega$. The solution of Eq. 23-38 is then

$$D = e^{-\omega t}(C_1 + C_2 t) \qquad (23\text{-}43)$$

If the mass is displaced a distance $(D)_{t=0} = D_0$ and released ($\dot{D}_{t=0} = 0$), the integration constants are $C_1 = D_0$ and $C_2 = \omega D_0$. Eq. 23-43 then becomes

$$D = e^{-\omega t}D_0(1 + \omega t) \qquad (23\text{-}44)$$

This equation indicates no vibratory motion: the mass will return slowly to its static position. The critical value of c is thus that value which will just cause the mass to return to its original position without oscillation.

In most practical cases, $\beta < \omega$, or $c < 2m\omega$, and the damping coefficient is expressed as the ratio ζ:

$$\zeta = \frac{c}{c_{\text{cr}}} = \frac{c}{2m\omega} = \frac{\beta}{\omega} \qquad (23\text{-}45)$$

We limit our discussion to the case when $\beta < \omega$ and introduce the symbol

$$\omega_d^2 = \omega^2 - \beta^2 \qquad (23\text{-}46)$$

ω_d represents the *damped natural circular frequency*. Equation 23-40 can then be written as

$$\lambda_1 = -\beta + i\omega_d \quad \text{and} \quad \lambda_2 = -\beta - i\omega_d \qquad (23\text{-}47)$$

where $i = \sqrt{-1}$.

The general solution of Eq. 23-38 then becomes

$$D = e^{-\beta t}(\bar{C}_1 \sin \omega_d t + \bar{C}_2 \cos \omega_d t) \tag{23-48}$$

In terms of the initial displacement D_0 and the initial velocity $\dot{D}_0$ Eq. 23-48 can be written

$$D = e^{-\beta t}\left[\left(\frac{\dot{D}_0 + \beta D_0}{\omega_d}\right)\sin \omega_d t + D_0 \cos \omega_d t\right] \tag{23-49}$$

The graph of Fig. 23-6 represents the displacement-time relation when $\dot{D}_0 = 0$.

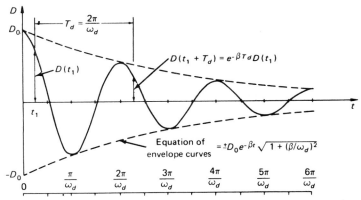

Fig. 23-6. Representation of damped free vibration Eq. 23-49 with $\dot{D}_0 = 0$.

The *natural period of damped vibration* is

$$T_d = \frac{2\pi}{\omega_d} \tag{23-50}$$

The ratio of the displacement at time t to that at time $t + T_d$ is constant (with a value $e^{\beta T_d}$) and the natural logarithm of this ratio is called the *logarithmic decrement*:

$$\delta = \beta T_d \tag{23-51}$$

The logarithmic decrement δ can be determined experimentally.

From Eqs. 23-45, 23-46, 23-50, and 23-51, it can be shown that δ and ζ are related by

$$\delta = \frac{2\pi\zeta}{\sqrt{1 - \zeta^2}} \cong 2\pi\zeta \tag{23-52}$$

In buildings, ζ varies between 0.03 and 0.15, which means that δ varies between 0.19 and 0.95.

23-62 Harmonically Forced, Viscously Damped Vibration

When the disturbing force varies according to Eq. 23-25, the equation of motion (23-36) becomes

$$\ddot{D} + 2\beta\dot{D} + \omega^2 D = \frac{P_0}{m}\sin\Omega t \tag{23-53}$$

where P_0, Ω, and β are as defined in Secs. 23-5 and 23-6. The general solution of Eq. 23-53 has two parts: the first represents the free damped vibration expressed by Eq. 23-48; the second part (the particular solution) represents the forced vibration, and is given by

$$D = \frac{P_0}{S} \frac{\left(1 - \dfrac{\Omega^2}{\omega^2}\right)\sin\Omega t - \dfrac{2\beta\Omega}{\omega^2}\cos\Omega t}{\left[\left(1 - \dfrac{\Omega^2}{\omega^2}\right)^2 + \dfrac{4\beta^2\Omega^2}{\omega^4}\right]} \tag{23-54}$$

If some damping is present, the free-vibration part (Eq. 23-48) is damped out after a few cycles and the total solution may be approximated by the particular solution of Eq. 23-54, which represents a steady-state forced vibration.

For the case of no damping, $\beta = 0$, and Eq. 23-54 becomes identical with Eq. 23-28.

Putting $\bar{A}_1 = 1 - \Omega^2/\omega^2$ and $\bar{A}_2 = -2\beta\Omega/\omega^2$, Eq. 23-54 can be rewritten in the form

$$D = \frac{P_0}{S} \frac{\bar{a}\sin(\Omega t + \bar{\alpha})}{\left[\left(1 - \dfrac{\Omega^2}{\omega^2}\right)^2 + \dfrac{4\beta^2\Omega^2}{\omega^4}\right]} \tag{23-55}$$

where $\bar{a} = \sqrt{\bar{A}_1^2 + \bar{A}_2^2}$ and $\bar{\alpha} = \tan^{-1}(\bar{A}_2/\bar{A}_1)$. Substituting for $\bar{a}$, Eq. 23-55 becomes

$$D = \chi\left[\frac{P_0}{S}\sin(\Omega t + \bar{\alpha})\right] \tag{23-56}$$

The quantity P_0/S is the maximum displacement which would occur if the force $P_0\sin\Omega t$ were applied statically, and χ is the magnification factor which accounts for the dynamic effect. The value of χ is

$$\chi = \left[\left(1 - \frac{\Omega^2}{\omega^2}\right)^2 + \frac{4\beta^2\Omega^2}{\omega^4}\right]^{-\frac{1}{2}} \tag{23-57}$$

and is plotted in Fig. 23-7 for various values of the ratio between the actual and critical damping coefficients, $\zeta = \beta/\omega$. In the resonant condition

$\Omega = \omega$, and the magnification factor is

$$(\chi)_{\Omega = \omega} = \frac{\omega}{2\beta} \tag{23-58}$$

The dotted curve in Fig. 23-7 is for the theoretical case of undamped

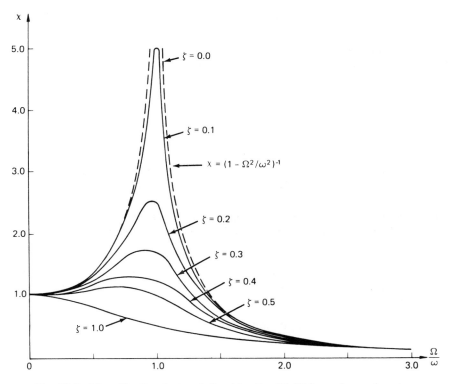

Fig. 23-7. Magnification factor defined by Eq. 23-57 for a damped system subjected to harmonic disturbing forces.

motion (Eq. 23-28). From the other curves it can be seen that even with only small damping, finite amplitudes occur at resonance.

23-63 Response to any Disturbing Force

We can use here the procedure of Sec. 23-52 to find a general solution of Eq. 23-36 when the disturbing force P varies in a random fashion with respect to time. The displacement at time τ caused by an element of impulse applied at an earlier time t (see Fig. 23-5) can be obtained from Eq. 23-49 if we consider the increment $d\dot{D}$ (Eq. 23-29) as the initial velocity D_0 with

$D_0 = 0$. Thus

$$dD_\tau = \frac{1}{\omega_d m} P e^{-\beta(\tau - t)} \sin(\omega_d \tau - \omega_d t)\, dt \qquad (23\text{-}59)$$

and, integrating, we obtain the displacement

$$D_\tau = \frac{1}{\omega_d m} \int_0^\tau P e^{-\beta(\tau - t)} \sin(\omega_d \tau - \omega_d t)\, dt \qquad (23\text{-}60)$$

If at time $t = 0$ there is an initial displacement D_0 and a velocity $\dot{D}_0$, the displacement at time τ will be (see Eq. 23-49)

$$D_\tau = e^{-\beta\tau}\left[\left(\frac{\dot{D}_0 + \beta D_0}{\omega_d}\right)\sin\omega_d\tau + D_0\cos\omega_d\tau\right]$$

$$+ \frac{1}{\omega_d m}\int_0^\tau P e^{-\beta(\tau - t)}\sin(\omega_d\tau - \omega_d t)\, dt \qquad (23\text{-}61)$$

Except in special cases where P is expressed by an idealized function of t, the integrals in Eqs. 23-60 and 23-61 are evaluated numerically.

23-7 UNDAMPED FREE VIBRATION OF A MULTIDEGREE-OF-FREEDOM SYSTEM

In Sec. 23-2 we derived the equation of motion for the free vibration of a system with n degrees of freedom (Eq. 23-7). This is a system of second-order differential equations for which the solution is a set of harmonic motions expressed in the form

$$D_i = \bar{a}_i \sin(\omega t + \alpha) \qquad i = 1, 2, \ldots, n \qquad (23\text{-}62)$$

or

$$\{D\} = \sin(\omega t + \alpha)\{\bar{a}\} \qquad (23\text{-}63)$$

The elements of $\{\bar{a}\}$ are the amplitudes at the n coordinates.

Differentiation of Eq. 23-63 with respect to time yields

$$\{\ddot{D}\} = -\omega^2\{D\} \qquad (23\text{-}64)$$

Substituting Eq. 23-64 into Eq. 23-7, we obtain

$$[S]\{D\} = \omega^2[m]\{D\} \qquad (23\text{-}65)$$

and premultiplication by $[m]^{-1}$ gives

$$[B]\{D\} = \omega^2\{D\} \qquad (23\text{-}66)$$

where

$$[B] = [m]^{-1}[S] \qquad (23\text{-}67)$$

Equation 23-66 is in the form of an eigenvalue problem (see Sec. A-10 of Appendix A) and its solution gives n characteristic values of the quantity ω^2. The smallest characteristic value corresponds to the lowest natural angular frequency, which is referred to as the *first mode* frequency.

For each characteristic value, a mode configuration defined by a *modal vector* $\{D\}$ can be found. Only the ratios between the displacements but not their values can be obtained. A value is selected (often unity) for one of the displacements, say D_i, and we then solve for the remaining D values. From Eq. 23-63, it is apparent that the modal vector $\{D\}$ also defines the ratios between the amplitudes $\{\bar{a}\}$ at various coordinates. These ratios define the natural *mode characteristic shapes*.

If we premultiply Eq. 23-65 by the flexibility matrix $[f]$, where $[f] = [S]^{-1}$, we obtain

$$[G]\{D\} = \frac{1}{\omega^2}\{D\} \tag{23-68}$$

where

$$[G] = [f][m] \tag{23-69}$$

Example 23-2 Find the angular frequencies and the normal mode characteristic shapes for the cantilever in Fig. 23-8a vibrating freely in the plane of the figure. Neglect the axial deformations and consider only the lumped masses.

The system has three degrees of freedom indicated by the three coordinates in Fig. 23-8b. The corresponding stiffness matrix is (using Appendix E):

$$[S] = \frac{EI}{l^3} \begin{bmatrix} 1.6154 & & \text{symmetrical} \\ -3.6923 & 10.1538 & \\ 2.7692 & -10.6154 & 18.4615 \end{bmatrix} \tag{23-70}$$

The mass matrix is

$$[m] = \frac{W}{g} \begin{bmatrix} 4 & 0 & 0 \\ 0 & 1 & 0 \\ 0 & 0 & 1 \end{bmatrix} \tag{23-71}$$

where g is the acceleration due to gravity.

Substituting in Eq. 23-67, we obtain

$$[B] = \frac{gEI}{Wl^3} \begin{bmatrix} 0.4039 & -0.9231 & 0.6923 \\ -3.6923 & 10.1538 & -10.6154 \\ 2.7692 & -10.6154 & 18.4615 \end{bmatrix} \tag{23-72}$$

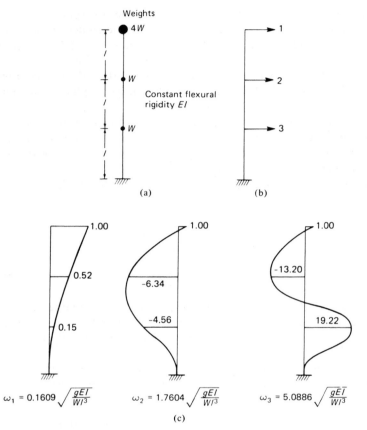

Fig. 23-8. Free vibration of the beam of Example 23-2. (a) System properties.
(b) Coordinate system. (c) Normal mode characteristic shapes.

Substituting in Eq. 23-66 and solving for the eigenvalues ω^2, (or solving the eigenvalue problem in the form of Eq. 23-65), we obtain

$$\omega_1^2 = 0.02588 \frac{gEI}{Wl^3} \quad \omega_2^2 = 3.09908 \frac{gEI}{Wl^3} \quad \text{and} \quad \omega_3^2 = 25.89415 \frac{gEI}{Wl^3}$$

(23-73)

or

$$\omega_1 = 0.1609 \sqrt{\frac{gEI}{Wl^3}} \quad \omega_2 = 1.7604 \sqrt{\frac{gEI}{Wl^3}} \quad \text{and} \quad \omega_3 = 5.0886 \sqrt{\frac{gEI}{Wl^3}}$$

(23-74)

If we choose $D_1 = 1$, the modal vectors corresponding to the above angular frequencies are:

$$\{D^{(1)}\} = \begin{Bmatrix} 1.0000 \\ 0.5224 \\ 0.1506 \end{Bmatrix} \qquad \{D^{(2)}\} = \begin{Bmatrix} 1.0000 \\ -6.3414 \\ -4.5622 \end{Bmatrix} \qquad \{D^{(3)}\} = \begin{Bmatrix} 1.0000 \\ -13.1981 \\ 19.2222 \end{Bmatrix}$$

$$(23\text{-}75)$$

The three modes are shown in Fig. 23-8c.

23-8 ORTHOGONALITY OF THE NATURAL MODES

In the previous section we have seen that the solution of the equation of motion for free undamped vibration of a system with n degrees of freedom gives n values for the natural frequency and a corresponding modal vector (eigenvector) for each. We shall now show that there exist orthogonality relationships between any two modal vectors (with different frequencies).

Consider two eigenvectors $\{D^{(r)}\}$ and $\{D^{(s)}\}$ for which the corresponding angular frequencies are ω_r and ω_s. Each vector and its eigenvalue satisfy the equation of motion 23-65, thus we have

$$[S]\{D^{(r)}\} = \omega_r^2[m]\{D^{(r)}\} \tag{23-76}$$

and

$$[S]\{D^{(s)}\} = \omega_s^2[m]\{D^{(s)}\} \tag{23-77}$$

Premultiplying Eq. 23-76 by $\{D^{(s)}\}^T$, we obtain

$$\{D^{(s)}\}^T[S]\{D^{(r)}\} = \omega_r^2\{D^{(s)}\}^T[m]\{D^{(r)}\} \tag{23-78}$$

Transposing the product of the matrices on both sides of this equation and noting that $[S]$ and $[m]$ are symmetrical, we obtain

$$\{D^{(r)}\}^T[S]\{D^{(s)}\} = \omega_r^2\{D^{(r)}\}^T[m]\{D^{(s)}\} \tag{23-79}$$

Premultiplication of Eq. 23-77 by $\{D^{(r)}\}^T$ gives

$$\{D^{(r)}\}^T[S]\{D^{(s)}\} = \omega_s^2\{D^{(r)}\}^T[m]\{D^{(s)}\} \tag{23-80}$$

The left-hand side of this equation is the same as in Eq. 23-79. Therefore, subtracting Eq. 23-80 from Eq. 23-79, we obtain

$$(\omega_r^2 - \omega_s^2)\{D^{(r)}\}^T[m]\{D^{(s)}\} = 0 \tag{23-81}$$

When the frequencies are different, i.e. $\omega_r \neq \omega_s$, the product of the matrices in Eq. 23-81 vanishes, i.e.,

$$\{D^{(r)}\}^T[m]\{D^{(s)}\} = 0 \qquad r \neq s \tag{23-82}$$

and so does the right-hand side of Eq. 23-79 (or Eq. 23-80). Thus

$$\{D^{(r)}\}^T[S]\{D^{(s)}\} = 0 \qquad r \neq s \qquad (23\text{-}83)$$

Equations 23-82 and 23-83 are the orthogonal relationships between the natural modes, which means that the natural modes are orthogonal with respect to the matrices $[m]$ or $[S]$.

The orthogonality Eq. 23-82 can be illustrated for the structure of Example 23-2 as follows. Substituting $r = 1$ and $s = 2$ in Eq. 23-82 and using the results of the example, we write

$$\{D^{(1)}\}^T[m]\{D^{(2)}\} = [1.00 \quad 0.52 \quad 0.15]\frac{W}{g}\begin{bmatrix} 4 & 0 & 0 \\ 0 & 1 & 0 \\ 0 & 0 & 1 \end{bmatrix}\begin{Bmatrix} 1.000 \\ -6.34 \\ -4.56 \end{Bmatrix}$$

$$= \frac{W}{g}[1.00 \times 4 \times 1.000 + 0.52 \times 1 \times (-6.34)$$

$$+ 0.15 \times 1 \times (-4.56)] = 0$$

23-9 NORMAL COORDINATES

In Sec. 23-2 we derived the equation of motion of a forced undamped system in coordinates $\{D\}$ (Eq. 23-6). The matrix $[S]$ in this equation is generally not a diagonal matrix, and Eq. 23-6 represents a system of simultaneous equations. The equations become uncoupled if both $[m]$ and $[S]$ are diagonal matrices. Transformation of the n coordinates $\{D\}$ in Eq. 23-6 into another system with the same number of coordinates $\{\eta\}$, such that

$$\{D\} = [\phi]\{\eta\} \qquad (23\text{-}84)$$

where $[\phi]_{n \times n}$ is the transformation matrix, gives an equation in which both $[m]$ and $[S]$ are transformed into $[\mathcal{M}]$ and $[K]$ which are generally not diagonal matrices. We shall now derive a transformation matrix $[\phi]$ which results in diagonal mass and stiffness matrices, and hence in uncoupled equations of motion. The corresponding coordinates $\{\eta\}$ are called *normal coordinates*.

Substituting Eq. 23-84 into Eq. 23-6 and premultiplying by $[\phi]^T$, we obtain

$$[\phi]^T[m][\phi]\{\ddot{\eta}\} + [\phi]^T[S][\phi]\{\eta\} = [\phi]^T\{P\} \qquad (23\text{-}85)$$

or

$$[\mathcal{M}]\{\ddot{\eta}\} + [K]\{\eta\} = \{\mathcal{L}\} \qquad (23\text{-}86)$$

where

$$[\mathcal{M}] = [\phi]^T [m] [\phi] \qquad (23\text{-}87)$$

$$[K] = [\phi]^T [S] [\phi] \qquad (23\text{-}88)$$

and

$$\{\mathcal{L}\} = [\phi]^T \{P\} \qquad (23\text{-}89)$$

Performing the matrix multiplications in Eqs. 23-87 and 23-88 will show that any element of the product matrix is given by

$$\mathcal{M}_{rs} = \{\phi^{(r)}\}^T [m] \{\phi^{(s)}\} \qquad (23\text{-}90)$$

or by

$$K_{rs} = \{\phi^{(r)}\}^T [S] \{\phi^{(s)}\} \qquad (23\text{-}91)$$

where $\{\phi^{(j)}\}$ is the jth column of $[\phi]$.

Comparison of Eqs. 23-90 and 23-91 with the orthogonality relations (Eqs. 23-82 and 23-83) shows that for $r \neq s$, $\mathcal{M}_{rs}$ and K_{rs} vanish if $\{\phi^{(j)}\} = \{D^{(j)}\}$. In other words, $[\mathcal{M}]$ and $[K]$ become diagonal matrices if the transformation matrix $[\phi]$ is formed by the eigenvectors $\{D^{(j)}\}$ as follows:

$$[\phi] = [\{D^{(1)}\} \mid \{D^{(2)}\} \mid \ldots \mid \{D^{(n)}\}] \qquad (23\text{-}92)$$

Putting $r = s$ in Eqs. 23-90 or 23-91 gives diagonal elements of $[\mathcal{M}]$ or $[K]$:

$$\mathcal{M}_{rr} = \mathcal{M}_r = \{\phi^{(r)}\}^T [m] \{\phi^{(r)}\} \qquad (23\text{-}93)$$

and

$$K_{rr} = K_r = \{\phi^{(r)}\}^T [S] \{\phi^{(r)}\} \qquad (23\text{-}94)$$

Using Eq. 23-76, we can write

$$[S] \{\phi^{(r)}\} = \omega_r^2 [m] \{\phi^{(r)}\} \qquad (23\text{-}95)$$

Premultiplying this equation by $\{\phi^{(r)}\}^T$ and combining the result with Eqs. 23-93 and 23-94 gives

$$K_r = \omega_r^2 \mathcal{M}_r \qquad (23\text{-}96)$$

For forced undamped vibration, the equation of motion in normal coordinates (Eq. 23-86) now becomes

$$[\mathcal{M}] \{\ddot{\eta}\} + [\Omega] [\mathcal{M}] \{\eta\} = \{\mathcal{L}\} \qquad (23\text{-}97)$$

where

$$[\Omega] = \begin{bmatrix} \omega_1^2 & & & \\ & \omega_2^2 & & \\ & & \cdots & \\ \text{Elements not} & & \omega_n^2 \\ \text{shown are zero} & & \end{bmatrix} \qquad (23\text{-}98)$$

Equation 23-97 is a set of n uncoupled differential equations of which the rth equation is

$$\ddot{\eta}_r + \omega_r^2 \eta_r = \frac{\mathscr{L}_r}{\mathscr{M}_r} \qquad (23\text{-}99)$$

where

$$\mathscr{L}_r = \{\phi^{(r)}\}^T \{P\} \qquad (23\text{-}100)$$

The uncoupled form of the equation of motion is useful when considering response to time-dependent forces. It makes it possible to determine the response in each normal mode separately as an independent system with one degree of freedom. The displacements $\{\eta\}$ are then transformed to the displacement $\{D\}$ by Eq. 23-84. This equation, in fact, superimposes the modes to obtain the total displacements.

Example 23-3   Find the response of the system shown in Fig. 23-8a to a set of harmonic forces at the three coordinates in Fig. 23-8b: $P_1 = 2P_0 \sin \Omega t$, $P_2 = P_0 \sin \Omega t$, and $P_3 = P_0 \sin \Omega t$.

Substituting Eq. 23-75 in Eq. 23-92, we obtain

$$[\phi] = \begin{bmatrix} 1.0000 & 1.0000 & 1.0000 \\ 0.5224 & -6.3414 & -13.1981 \\ 0.1506 & -4.5622 & 19.2222 \end{bmatrix} \qquad (23\text{-}101)$$

Substitution of Eqs. 23-71 and 23-101 into Eq. 23-93 gives

$$\mathscr{M}_1 = 4.296 \frac{W}{g} \qquad \mathscr{M}_2 = 65.027 \frac{W}{g} \qquad \text{and} \qquad \mathscr{M}_3 = 547.680 \frac{W}{g} \quad (23\text{-}102)$$

The external applied load vector is

$$\{P\} = P_0 \begin{Bmatrix} 2 \\ 1 \\ 1 \end{Bmatrix} \sin \Omega t \qquad (23\text{-}103)$$

and using Eq. 23-100 we obtain

$$\left.\begin{array}{l} \mathcal{L}_1 = 2.673\, P_0 \sin \Omega t \\ \mathcal{L}_2 = -8.904\, P_0 \sin \Omega t \\ \mathcal{L}_3 = 8.024\, P_0 \sin \Omega t \end{array}\right\} \qquad (23\text{-}104)$$

Substituting Eqs. 23-73, 23-102, and 23-104 into Eq. 23-99, we obtain the uncoupled equations of motion:

$$\left.\begin{array}{l} \ddot{\eta}_1 + 0.02588\, \dfrac{gEI}{Wl^3}\, \eta_1 = \dfrac{P_0}{1.607\,(W/g)} \sin \Omega t \\[2ex] \ddot{\eta}_2 + 3.09908\, \dfrac{gEI}{Wl^3}\, \eta_2 = \dfrac{P_0}{-7.303\,(W/g)} \sin \Omega t \\[2ex] \ddot{\eta}_3 + 25.89415\, \dfrac{gEI}{Wl^3}\, \eta_3 = \dfrac{P_0}{68.255\,(W/g)} \sin \Omega t \end{array}\right\} \qquad (23\text{-}105)$$

The solution of the above differential equations was treated in Sec. 23-51. The steady-state forced vibration in the normal coordinates $\{\eta\}$ is described by (see Eq. 23-28):

$$\left.\begin{array}{l} \eta_1 = \dfrac{0.6223\, gP_0/W}{0.02588\, gEI/(Wl^3) - \Omega^2} \sin \Omega t \\[2ex] \eta_2 = -\dfrac{0.1369\, gP_0/W}{3.09908\, gEI/(Wl^3) - \Omega^2} \sin \Omega t \\[2ex] \eta_3 = \dfrac{0.0147\, gP_0/W}{25.89415\, gEI/(Wl^3) - \Omega^2} \sin \Omega t \end{array}\right\} \qquad (23\text{-}106)$$

Substitution of Eqs. 23-101 and 23-106 into Eq. 23-84 gives the displacements $\{D\}$.

23-10   RESPONSE OF STRUCTURES TO EARTHQUAKES

In an analysis of seismic effects, we determine the response of structures due to a given motion of the supports rather than due to the application of external forces. The support motion may be described by an acceleration-time curve detained from records of previous earthquakes.

Consider the response of a one-degree-of-freedom system (Fig. 23-9) subjected to support motion $u_s(t)$ specified in terms of the support acceleration $\ddot{u}_s(t)$. If at any time t the horizontal displacement of the mass m is $u(t)$, the relative displacement of the mass with respect to the ground is

$$D = u - u_s \qquad (23\text{-}107)$$

In the absence of an external disturbing force, the only forces acting on the mass are: a restoring force SD (where S is stiffness coefficient), a damping force $c\dot{D}$ assumed proportional to the relative velocity, and an inertia force $m\ddot{u}$. The equation of motion is

$$m\ddot{u} + c\dot{D} + SD = 0 \tag{23-108}$$

Differentiating Eq. 23-107 with respect to time and substituting into Eq. 23-108, we obtain

$$\ddot{D} + 2\beta\dot{D} + \omega^2 D = -\ddot{u}_s \tag{23-109}$$

where $\omega = \sqrt{S/m}$, $\beta = c/2m$ (see Eqs. 23-16 and 23-37).

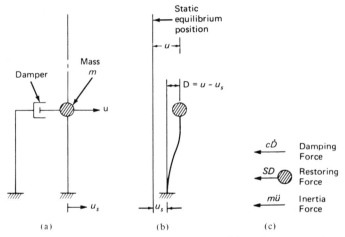

Fig. 23-9. A single-degree-of-freedom system subjected to support movements. (a) Positive directions of u and u_s. (b) Deformed shape at any time t. (c) Forces acting on mass at any time t.

A comparison of Eq. 23-109 with Eq. 23-36 shows that the effect of support motion is the same as that of a force $(-\ddot{u}_s m)$. The equations derived in Secs. 23-62 and 23-63 can be used to determine the response to support motion by substituting for $P(t)$ the quantity $(-\ddot{u}_s m)$.

With respect to a system with multidegrees of freedom subjected to support motion, we shall limit our discussion to the undamped case. Consider the multistorey frame of Fig. 23-2, for which the acceleration at the support $\ddot{u}_s(t)$ is given. In a way similar to that for a single-degree-of-freedom system, it can be shown that the equation of motion is

$$[m]\{\ddot{D}\} + [S]\{D\} = -\ddot{u}_s\{m^*\} \tag{23-110}$$

where $\{D\}$ and $\{m^*\}$ are defined as follows:

$$\{D\} = \begin{Bmatrix} u_1 - u_s \\ u_2 - u_s \\ \cdots \\ u_n - u_s \end{Bmatrix} \quad \text{and} \quad \{m^*\} = \sum_{j=1}^{n} \begin{Bmatrix} m_{1j} \\ m_{2j} \\ \cdots \\ m_{nj} \end{Bmatrix} = [m] \begin{Bmatrix} 1 \\ 1 \\ \cdots \\ 1 \end{Bmatrix} \quad (23\text{-}111)$$

$u_1, u_2, \ldots, u_n$ being the absolute displacements at floor levels. Equation 23-110 can be replaced by a system of uncoupled differential equations of the form of Eq. 23-99, but in this case we have

$$\mathscr{L}_r = -\ddot{u}_s \{\phi^{(r)}\}^T \{m^*\} \qquad (23\text{-}112)$$

A solution of the uncoupled equations will give the displacements $\{\eta\}$ and hence, by Eq. 23-84, the relative displacements $\{D\}$.

The above analysis is based on the elastic theory and is usually applied only for moderate ground motion. In practice, structures are designed to withstand such motion without damage, but some plastic deformations are tolerated in the case of strong earthquakes, provided that life is not endangered. Analysis for this case must allow for the inelastic behavior of the structure.

Example 23-4   Find the response of the system shown in Fig. 23-8 to a support motion described by $\ddot{u}_s = (g/5) \sin \Omega t$, where $\Omega = 4\pi(1/\text{sec})$, and g is the acceleration due to gravity.

Substitution of Eq. 23-71 into Eq. 23-111 gives

$$\{m^*\} = \frac{W}{g} \begin{Bmatrix} 4 \\ 1 \\ 1 \end{Bmatrix} \qquad (23\text{-}113)$$

If we substitute Eqs. 23-101 and 23-113 into Eq. 23-112, we obtain

$$\left. \begin{aligned} \mathscr{L}_1 &= -0.934\, W \sin \Omega t \\ \mathscr{L}_2 &= 1.381\, W \sin \Omega t \\ \mathscr{L}_3 &= -2.005\, W \sin \Omega t \end{aligned} \right\} \qquad (23\text{-}114)$$

Following the same procedure as in Example 23-3, but using the above values of $\mathscr{L}_r$, we obtain

$$\left. \begin{aligned} \eta_1 &= -\left[\frac{0.183g}{0.02588gEI/(Wl^3) - \Omega^2} \right] \sin \Omega t \\ \eta_2 &= \left[\frac{0.0212g}{3.09908gEI/(Wl^3) - \Omega^2} \right] \sin \Omega t \\ \eta_3 &= -\left[\frac{0.0037g}{25.89415gEI/(Wl^3) - \Omega^2} \right] \sin \Omega t \end{aligned} \right\} \qquad (23\text{-}115)$$

23-11 GENERAL

This chapter should be considered only as an introduction to the subject of structural dynamics. For more complete treatment reference can be made to specialized publications.[4]

PROBLEMS

Take the acceleration of gravity $g = 32.2$ ft/sec^2 or 386 in/sec^2 whenever it is needed in the solution of the following problems.

23-1   Compute the natural angular frequency of vibration in sidesway for the frame in the figure, and calculate the natural period of vibration. Idealize the frame as a one-degree-of-freedom system. Neglect the axial and shear deformations, and the weight of the columns. If initially the displacement is 1 in. and the velocity is 10 in./sec, what is the amplitude and what is the displacement at $t = 1$ sec?

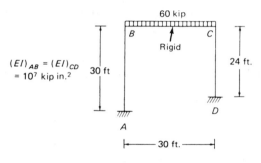

Prob. 23-1

23-2   Solve Prob. 23-1 assuming that BC has a flexural rigidity $(EI)_{BC} = 10^7$ kip in.^{2}

23-3   The prismatic cantilever AB is idealized by the two-degrees-of-freedom system shown in the figure. Using the consistent mass matrix, find the first natural angular frequency and the corresponding mode. The beam has a total mass m, length l, and flexural rigidity EI. Consider bending deformation only.

[4]See for example M. Paz (1985) *Structural Dynamics*, 2nd edn, Van Nostrand Reinhold, New York; R. W. Clough and J. Penzien, *Dynamics of Structures*, McGraw-Hill, New York, 1975.

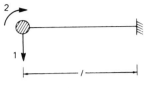

Prob. 23-3

23-4   Find the consistent mass matrix for the beam element in Fig. 23-3 assuming that the area of the cross section varies linearly between a_1 and a_2 at ends A and B respectively. Use the displacement functions of Eq. 23-11.

23-5   The frame in Prob. 23-1 is disturbed from rest by a horizontal force of 8 kip at C, suddenly applied at time $\tau = 0$ and removed at time $\tau = T/2$, where T is the natural period of vibration. What is the displacement and velocity at the removal of the force? What is the displacement at time $\tau = 11\ T/8$?

23-6   Solve Prob. 23-5 assuming that the disturbing force increases linearly from zero at $\tau = 0$ to 8 kip at $\tau = T/4$, then decreases linearly to zero at $\tau = T/2$, at which time the force is removed.

23-7   If the system of Prob. 23-1 has a damping coefficient $\xi = 0.1$, what are the damped natural circular frequency ω_d and the natural period of damped vibration T_d? What is the displacement at $t = 1$ sec, if $D_0 = 1$ in. and $\dot{D}_0 = 10$ in./sec?

23-8   If the amplitude of free vibration of a system with one degree of freedom decreases by 50 percent in 3 cycles, what is the damping coefficient?

23-9   Determine the maximum steady-state sidesway in the frame of Prob. 23-1 when it is subjected to a harmonic horizontal force at the level of BC of magnitude $4 \sin 14t$ (kip), and (a) no damping is present, (b) the damping coefficient $= 0.10$.

23-10   Assume that the frame in Prob. 23-1 has a damping coefficient $= 0.05$ and it is disturbed from rest by a horizontal force of 8 kip at C. The force is suddenly applied at time $\tau = 0$ and removed at time $\tau = T_d/2$, where T_d is the natural period of damped vibration. What are the displacements at the removal of the force and at time $\tau = 11\ T_d/8$? (Compare the answers with the undamped case, Prob. 23-5.)

23-11   Determine the natural circular frequencies and characteristic shapes for the two-degrees-of-freedom systems shown in the figure.

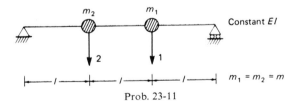

Prob. 23-11

23-12   Apply the requirement of Prob. 23-11 to the frame shown in the figure.

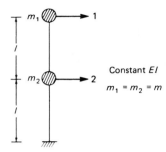

Prob. 23-12

23-13 Write the uncoupled equations of motion for the undamped system in Prob. 23-12, assuming that a force P_1 is suddenly applied at time $t = 0$ and continues to act after this. Find the time-displacement relations for D_1 and for D_2.

23-14 If in Prob. 23-13 the force $P_1 = P_0 \sin \Omega t$, what is the amplitude of the steady-state vibration of m_1?

23-15 The supports of the frame in Prob. 23-1 move horizontally with an acceleration indicated in the figure. What is the maximum displacement of *BC* relative to the support? Neglect damping.

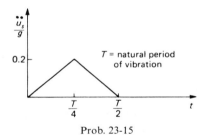

Prob. 23-15

23-16 If the support in Prob. 23-12 has a horizontal acceleration $\ddot{u}_s = (g/4) \sin \Omega t$, determine the maximum displacement of the mass m_1 relative to the support.

CHAPTER 24

Computer analysis of framed structures

24-1   INTRODUCTION

This chapter discusses the use of computers in the analysis of large structures by the displacement (stiffness) method.[1] Although the concepts discussed apply to all types of structure, the matrices involved in the analysis are given explicitly only for framed structures composed of prismatic straight members. Five types of framed structure are considered: plane and space trusses with pin joints, plane and space frames with rigid joints, and grids with rigid joints.

The nodes (joints) and the members are numbered sequentially 1 to n_j and 1 to n_m, respectively. The coordinates representing the degrees of freedom are also numbered sequentially, following at each node the order indicated in Fig. 24-1. For example, at the ith node of a plane truss, the translations in the global x and y directions are represented by coordinate numbers $(2i - 1)$ and $2i$, respectively. The origin and the directions of the orthogonal x, y (and z) axes are chosen arbitrarily.

The analysis of framed structures is commonly performed to determine the nodal displacements and the member end-forces. While the nodal displacements are determined in the global directions, the member end-forces are calculated in local directions of the individual members. These are defined in the following section.

Before proceeding with this chapter, some readers may find it beneficial to go over the summary given in Sec. 24-13.

24-2   MEMBER LOCAL COORDINATES

The stiffness matrix $[S^*]$ of a member of a framed structure is first generated with respect to local coordinates in the directions of the centroidal axis and the principal axes of the member cross section; $[S^*]$ is then

---

[1]A summary of the steps involved in the analysis by the displacement method is given in Sec. 3-6. It is proposed to prepare manuals and diskettes for the analysis of framed structures on IBM micro computers; enquiries should be made to Liliane Ghali, 3911 Vincent Drive N.W., Calgary, Alberta, T3A OG9, Canada.

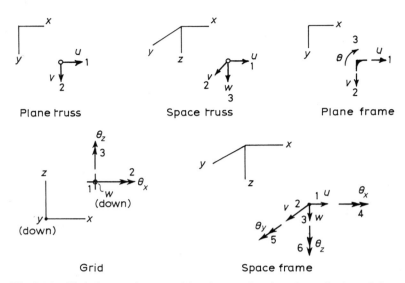

Fig. 24-1. Global axes, degrees of freedom and order of numbering of the coordinates at typical nodes of framed structures.

transformed to coordinates parallel to the global directions. This is necessary before the stiffness matrix of individual members can be assembled to obtain the structure stiffness matrix $[S]$. Also, the member end-actions are generally determined in local coordinates.

Typical nodal coordinates in global directions are shown in Fig. 24-1, while the local coordinates for typical members of framed structures are shown in their positive directions in Fig. 24-2.

The local coordinates are parallel to local axes x^*, y^* and z^*. The local x^* axis is directed along the centroidal axis from the first node to the second node of the member. The first node of a member is specified arbitrarily by the analyst by the order in which the two nodes at the ends of a member are listed in the member information in the input data (see Sec. 24-4).

Each of the sets of global and local axes is an orthogonal right-hand triad.

With the above system of numbering of coordinates and the choice of member local coordinates, the first step of the displacement method is completed when the numbering of nodes and elements is terminated. We recall that the first step involves the definition of the degrees of freedom and of the required actions and their positive directions.

24-3 BAND WIDTH

As mentioned in Sec. 4-3, any element S_{ij} of the stiffness matrix $[S]$ of a structure is nonzero only when the coordinates i and j are adjacent to each

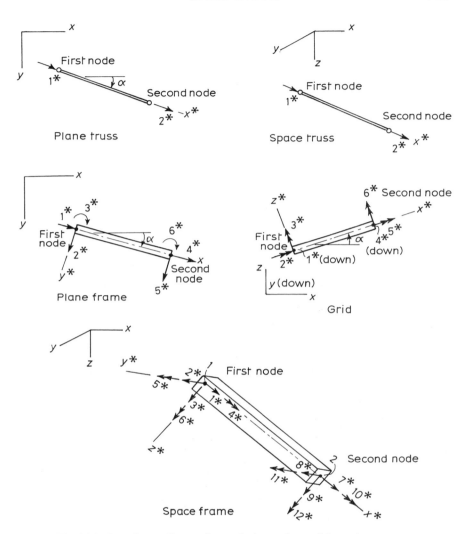

Fig. 24-2. Local coordinates for typical members of framed structures.

other. In a framed structure, $S_{ij} \neq 0$ only when i and j are at one node or when i and j are at the two ends of one member. Generally, the nonzero elements of $[S]$ are limited to a band adjacent to the diagonal (Fig. 24-3). This property of $[S]$, combined with the fact that the matrix is symmetrical, is used to conserve computer storage space and to reduce the number of computations. Only the diagonal elements of $[S]$ and the part of the band above the diagonal are generated and stored in a rectangular matrix of n rows and n_b columns (Fig. 24-3). Here n is the number of degrees of freedom; n_b is referred to as the

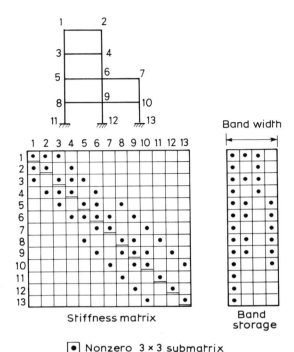

Fig. 24-3. Node numbering and band width for a plane frame.

half-band width or simply the band width and is given by

$$n_b = s(|k - j|_{\text{largest}} + 1) \tag{24-1}$$

where s is the number of degrees of freedom per node, namely 2, 3, 3, 3 and 6, respectively, for plane truss, plane frame, grid, space truss and space frame. The term $|k - j|$ is the absolute value of the difference between the node numbers at the ends of a member. The band width is determined by the member for which $|k - j|$ is a maximum.

A large portion of the computer time in the analysis of a structure is spent in the solution of the equilibrium equations

$$[S][D] = -[F]$$

where $[D]$ and $[F]$ are, respectively, the unknown displacements and the forces at the coordinates to prevent the displacements due to loading. The respective columns of $[D]$ and $[F]$ correspond to the same loading case. The number of arithmetic operations in the solution of the equilibrium equations increases linearly with n and with n_b^2. Thus we have an interest in reducing n_b.

For a given structure, a narrower band width is generally achieved by numbering the nodes sequentially across the side which has a smaller number of nodes. An example of a plane frame with suggested node numbering, the corresponding $[S]$, and the band to be generated and stored in computing is shown in Fig. 24-3. For the same structure, $n_b = 18$ (instead of 12) when the nodes are numbered sequentially down the columns. Note that $[S]$ generally has zeros within the band (as in Fig. 24-3).

There exist several techniques[2] aimed at efficiency and accuracy in equation solvers by automatic renumbering of the nodes to minimize n_b, by generating and operating only on the nonzero entries of $[S]$ and by estimating and controlling round-off errors.

The stiffness matrix discussed above (Fig. 24-3) is for a free (unsupported) structure. Elimination of rows and columns corresponding to the coordinates where the displacements are zero results in a condensed stiffness matrix of the supported structure (see Sec. 4-5). The condensed matrix is the one to be used in the equilibrium equations to be solved.

It is of course possible[3] to avoid generating the nonrequired rows and columns of $[S]$. However, for simplicity in computer programming, at the expense of more computing, it is possible to work with the stiffness matrix of the free structure by adjusting $[S]$. The adjustment, discussed in Sec. 24-10, causes the displacement to be equal to zero or equal to a prescribed value at any specified coordinates. This can be used when the analysis is for the effect of support movements.

24-4 INPUT DATA

The input data for computer analysis must be sufficient to define the material and geometric properties of the structure as well as the loading. Two basic units are used: unit of length and unit of force. These units must be employed consistently in the data.

The data for the geometric properties of a plane truss or a space truss are defined by the following: the modulus of elasticity E; the (x, y) or (x, y, z) coordinates of the nodes; the cross-sectional area of each member; two integers identifying the nodes at its ends; and the support conditions. The support conditions are defined by giving the number of the node and a restraint indicator for each of its two coordinates (u, v). The convention used here is to put 1 to indicate a free (unrestrained) displacement and 0 to indicate that the

[2]See Bathe, K. J. and E. L. Wilson, *Numerical Methods in Finite Element Analysis*, Prentice-Hall, Englewood Cliffs, NJ, 1976. See also Holzer, S. M., *Computer Analysis of Structures*, Elsevier, New York, 1985 (a list of references on this topic is given on pp. 325-327 of Holzer).

[3]A computer coding for this purpose is discussed in Section 2.9 of Cook, R. D., *Concepts and Application of the Finite Elements* (2nd edn), Wiley, New York, 1981.

displacement at this coordinate is prescribed. When the restraint indicator is zero the magnitude of the prescribed displacement must be given (zero for a support with no settlement). In Example 24-1, the input data for the plane truss in Fig. 24-4 are given in Table 24-1.

When preparing the data for support conditions, it is important to ensure that the structure cannot translate or rotate freely as a rigid body. This requirement for stability should be verified before the computer analysis of any structure (particularly spatial structures).

Ideally, a truss should be loaded only at the nodes so that there are only axial forces in the members. The input data for the load at a node are the node number and the components (F_x, F_y) or (F_x, F_y, F_z).

When a truss member is subjected to a temperature rise, the data are the member number and the two end-forces $\{A_r\}$ (along the local coordinates) which occur when the nodal displacements are prevented. This produces an axial compressive force of magnitude $\alpha T E a$, where α is the coefficient of thermal expansion, T is the temperature rise, E is the modulus of elasticity, and a is the cross-sectional area.

As mentioned in Sec. 24-3, the effect of a prescribed displacement at a support can be accounted for by adjusting the stiffness matrix, but this can be done only when the analysis is for a single loading case. To avoid this restriction, we calculate the member end-forces when the prescribed displacements occur while the other displacements are zero (see Sec. 3-2). These member end-forces are then included in the data in the same way as the data

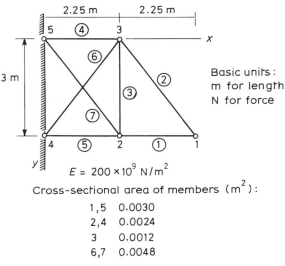

Fig. 24-4. Plane truss of Example 24-1.

for temperature. It is obvious that this needs to be done only for the members connected to nodes with prescribed displacements.

Example 24-1 Prepare the input data for the analysis of the plane truss in Fig. 24-4 with the following loading cases: (1) A single downward force of 100×10^3 N at node 1; (2) a temperature rise of 20 degrees of member 4, the coefficient of thermal expansion being 10^{-5} per degree; and (3) a downward movement of 0.002 m of node 4.

Table 24-1 gives the required input data. The member end-forces in the data for loading case (3) are calculated in a similar way to the procedure included in Example 3-1.

Table 24-1. Input Data for a Plane Truss (Fig. 24-4)

$E = 200 \times 10^9$

Nodal coordinates

Node	x	y
1	4.5	3.0
2	2.25	3.0
3	2.25	0.0
4	0.0	3.0
5	0.0	0.0

Element information

Element	First node	Second node	a
1	2	1	0.0030
2	3	1	0.0024
3	2	3	0.0012
4	5	3	0.0024
5	4	2	0.0030
6	4	3	0·0048
7	5	2	0.0048

Support conditions

Node	Restraint indicator		Prescribed displacement	
	u	v	u	v
4	0	0	0.0	0.0
5	0	0	0.0	0.0

Loading case 1
Forces applied at nodes

Node	F_x	F_y
1	0.0	100×10^3

Loading case 2
Member end-forces with displacements restrained

Member	A_{r1}	A_{r2}
4	96×10^3	-96×10^3

Loading case 3
Member end-forces with displacements restrained

Member	A_{r1}	A_{r2}
6	-410×10^3	410×10^3

For a plane frame, the input data differ from a plane truss in that the second moment of area of the cross sections must also be given; at a supported node, three displacements u, v and θ must be described; the forces at each member end have three components along the local coordinates (Fig. 24-2); and when a member is subjected to forces away from the nodes, the data should include the member end-forces $\{A_r\}$ with the nodal displacements restrained. Example 24-2 gives the data for a plane frame.

The input data for a grid are similar to those for a plane frame except that the cross-sectional torsion constant J (Appendix G) is required for the members instead of their area; the modulus of elasticity in shear G is also required. The displacement components at a typical node and the local coordinates of a typical member of a grid are shown in Figs. 24-1 and 24-2.

The data defining the geometric properties of a space frame are the (x, y, z) coordinates of the nodes and, for each member, two integers identifying the nodes at its ends, the area a of the cross section, its torsion constant J, and its second moments of area I_{y^*} and I_{z^*} about the centroidal principal axes parallel to the local y^* and z^* axes (Fig. 24-2). While the direction of the local x^* axis is defined by the (x, y, z) coordinates of the nodes, the y^* direction needs further definition. For this reason, we include in the data for each member the direction cosines of y^*: $\lambda_{y^*x}, \lambda_{y^*y}, \lambda_{y^*z}$. Here the λ values are the cosines of the angles between the local y^* axis and the global directions. The third direction z^* is orthogonal to the other two local axes and thus does not require additional data.

In most practical cases, the global axes are selected in horizontal and vertical directions and the member cross sections have one of their principal axes horizontal or vertical. The y^* axis can be considered in such a direction and its direction cosines will be easy to find (see Prob. 25-1).

The data for loading of a space frame are similar to those for a plane frame, noting that the forces at the nodes have six components in the global directions and each member end has six forces defined in Figs. 24-1 and 24-2.

Example 24-2 Prepare the input data for the analysis of the plane frame in Fig. 24-5 for the following loadings: (1) the loads shown; (2) displacements $u = 0.2b$ and $v = 0.5b$ of the support at node 1.

Table 24-2 gives the required input data. The member end-forces included in the data may be calculated using Appendices C and D. See also answer to Problem 25-2.

24-5 DIRECTION COSINES OF ELEMENT LOCAL AXES

The direction cosines λ of the member local x^*, y^*, z^* axes will be used in the transformation matrices. The coordinates (x, y) defining the position of the nodes give the λ values for the local x^* axis. For a member of a plane truss or a

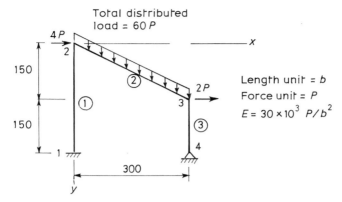

Properties of cross section of members

Member	a (b^2)	I (b^4)
1	30	3000
2	40	5000
3	30	3000

Fig. 24-5. Plane frame of Example 24-2.

plane frame,

$$\lambda_{x^*x} = \frac{x_2 - x_1}{l_{12}} \qquad \lambda_{x^*y} = \frac{y_2 - y_1}{l_{12}} \tag{24-2}$$

where

$$l_{12} = [(x_2 - x_1)^2 + (y_2 - y_1)^2]^{\frac{1}{2}} \tag{24-3}$$

The subscripts 1 and 2 refer to the first and second nodes of the member.

In a plane frame, the y^* axis for any member is perpendicular to x^* (Fig. 24-2), so that the λ values for the y^* axis are

$$\lambda_{y^*x} = -\lambda_{x^*y} \qquad \lambda_{y^*y} = \lambda_{x^*x} \tag{24-4}$$

For a member of a grid in the xz plane, the direction cosines required for the local x^* and z^* axes are λ_{x^*x}, λ_{x^*z}, λ_{z^*x} and λ_{z^*z}. These are given by Eqs. 24-2 to 24-4, replacing y by z and y^* by z^*.

The direction cosines of the local x^* axis of a member of a space truss or a space frame are

$$\lambda_{x^*x} = \frac{x_2 - x_1}{l_{12}} \qquad \lambda_{x^*y} = \frac{y_2 - y_1}{l_{12}} \qquad \lambda_{x^*z} = \frac{z_2 - z_1}{l_{12}} \tag{24-5}$$

where

$$l_{12} = [(x_2 - x_1)^2 + (y_2 - y_1)^2 + (z_2 - z_1)^2]^{1/2} \tag{24-6}$$

Table 24-2   Input Data for a Plane Frame (Fig. 24-5)

$E = 30000.0$

Nodal coordinates

Node	x	y
1	0.0	300.0
2	0.0	0.0
3	300.0	150.0
4	300.0	300.0

Element information

Element	First node	Second node	a	I
1	1	2	30.0	3000.0
2	2	3	40.0	5000.0
3	3	4	30.0	3000.0

Support conditions

Node	Restraint indicator			Prescribed displacement		
	u	v	θ	u	v	θ
1	0	0	0	0.0	0.0	0.0
4	0	0	1	0.0	0.0	—

Loading case 1
Forces applied at nodes

Node	F_x	F_y
2	4.0	0.0
3	2.0	0.0

Member end-forces with displacements restrained

Member	A_{r1}	A_{r2}	A_{r3}	A_{r4}	A_{r5}	A_{r6}
2	-13.42	-26.83	-1500.0	-13.42	-26.83	1500.0

Loading case 2
Member end-forces with displacements restrained

Member	A_{r1}	A_{r2}	A_{r3}	A_{r4}	A_{r5}	A_{r6}
1	-1500.0	8.0	1200.0	1500.0	-8.0	1200.0

As mentioned in the preceding section, the λ values for the local y^* axis of each member are included in the input data. The direction cosines of the local z^* axis are obtained from the cross product of two unit vectors in the x^* and y^* directions:

$$\lambda_{z^*x} = \lambda_{x^*y}\lambda_{y^*z} - \lambda_{x^*z}\lambda_{y^*y} \tag{24-7}$$

$$\lambda_{z^*y} = \lambda_{x^*z}\lambda_{y^*x} - \lambda_{x^*x}\lambda_{y^*z} \tag{24-8}$$

$$\lambda_{z^*z} = \lambda_{x^*x}\lambda_{y^*y} - \lambda_{x^*y}\lambda_{y^*x} \tag{24-9}$$

24-6   ELEMENT STIFFNESS MATRICES

The stiffness matrices of individual members of a framed structure with respect to local coordinates (Fig. 24-2) can be generated from Appendix D.

For a member of a *plane truss* or a *space truss* (Fig. 24-2), the stiffness matrix is

$$[S^*] = \frac{Ea}{l} \left[\begin{array}{c|c} 1 & -1 \\ \hline -1 & 1 \end{array} \right]$$

(24-10)

The stiffness matrix for a member of a *plane frame* (Fig. 24-2), given in Eq. 4-6 is

$$[S^*] = \left[\begin{array}{ccc|ccc} Ea/l & & & \text{symmetrical} \\ & 12EI/l^3 & & & \text{elements not shown} \\ & 6EI/l^2 & 4EI/l & & \text{are zero} \\ \hline -Ea/l & & & Ea/l \\ & -12EI/l^3 & -6EI/l^2 & & 12EI/l^3 \\ & 6EI/l^2 & 2EI/l & & -6EI/l^2 & 4EI/l \end{array} \right]$$

(24-11)

In the above equation, each member has its centroidal axis in the xy plane, and $I \equiv I_{z*}$, with z^* perpendicular to the same plane.

The stiffness matrix for a member of a *grid* is (Fig. 24-2)

$$[S^*] = \left[\begin{array}{ccc|ccc} 12EI/l^3 & & & \text{symmetrical} \\ 0 & GJ/l & & \text{elements not shown are zero} \\ 6EI/l^2 & 0 & 4EI/l \\ \hline -12EI/l^3 & 0 & -6EI/l^2 & 12EI/l^3 \\ 0 & -GJ/l & 0 & 0 & GJ/l \\ 6EI/l^2 & 0 & 2EI/l & -6EI/l^2 & 0 & 4EI/l \end{array} \right]$$

(24-12)

Here the centroidal axes of the members are in the xz plane, and $I \equiv I_{z*}$, with z^* situated in the same plane.

The stiffness matrix for a member of a *space frame* (Fig. 24-2) is given by Eq. 4-5.

In the above matrices the shear deformations are ignored. To take shear deformations into account, the elements of $[S^*]$ associated with coordinates representing shear or bending have to be adjusted as discussed in Sec. 16-2. The stiffness matrix of a typical member of a plane frame accounting for deformations due to axial force, bending and shear is given by Eq. 16-5.

24-7 TRANSFORMATION MATRICES

The displacements $\{D^*\}$ in local directions at the end of a member of a framed structure of any type (Fig. 24-2) are related to the displacements at the same node in global directions (Fig. 24-1) by the geometrical relation

$$\{D^*\} = [t]\{D\}$$

(24-13)

where $[t]$ is a transformation matrix which is generated in terms of the direction cosines of the local axes x^*, y^*, z^* with respect to the global directions. The $[t]$ matrix for the different types of framed structures is given below.

$$[t] = [\lambda_{x^*x} \quad \lambda_{x^*y}] \tag{24-14}$$

Plane truss
Space truss

$$[t] = [\lambda_{x^*x} \quad \lambda_{x^*y} \quad \lambda_{x^*z}] \tag{24-15}$$

Plane frame

$$[t] = \begin{bmatrix} \lambda_{x^*x} & \lambda_{x^*y} & 0 \\ \lambda_{y^*x} & \lambda_{y^*y} & 0 \\ 0 & 0 & 1 \end{bmatrix} \tag{24-16}$$

Grid

$$[t] = \begin{bmatrix} 1 & 0 & 0 \\ 0 & \lambda_{x^*x} & \lambda_{x^*z} \\ 0 & \lambda_{z^*x} & \lambda_{z^*z} \end{bmatrix} \tag{24-17}$$

Space frame

$$[t] = \begin{bmatrix} \lambda_{x^*x} & \lambda_{x^*y} & \lambda_{x^*z} \\ \lambda_{y^*x} & \lambda_{y^*y} & \lambda_{y^*z} \\ \lambda_{z^*x} & \lambda_{z^*y} & \lambda_{z^*z} \end{bmatrix} \tag{24-18}$$

The transformation matrix given in the last equation is used to relate the three translations $\{u^*, v^*, w^*\}$ to $\{u, v, w\}$. The same matrix can be used to relate $\{\theta_x^*, \theta_y^*, \theta_z^*\}$ to $\{\theta_x, \theta_y, \theta_z\}$.

A member stiffness matrix $[S^*]$, with respect to local coordinates, can be transformed into a stiffness matrix $[S_m]$ corresponding to coordinates in the global directions (see Eq. 8-17) by

$$[S_m] = [T]^T [S^*][T] \tag{24-19}$$

where

$$[T] = \begin{bmatrix} [t] & [0] \\ [0] & [t] \end{bmatrix} \tag{24-20}$$

The above equation may be used for members of a plane truss, a space truss, a plane frame or a grid, substituting for $[t]$ from one of the Eqs. 24-14 to 24-17. However, for a member of a space frame.

$$[T] = \begin{bmatrix} [t] & & & \\ & [t] & & \\ & & [t] & \\ & & & [t] \end{bmatrix} \quad \begin{array}{l} \text{submatrices not} \\ \text{shown are null} \end{array} \tag{24-21}$$

where $[t]$ is given by Eq. 24-18.

The $[T]$ matrices may be used to transform the actions in local directions at the ends of a member into equivalent forces along nodal coordinates in the global directions (see Eq. 8-16):

$$\{F\}_m = [T]_m^T \{A_r\}_m \tag{24-22}$$

where the subscript m refers to element number.

The elements of the matrices $[S_m]$ and $\{F\}_m$ are forces in the global directions; they represent contributions of one member. Summing the matrices for all members gives the stiffness matrix and load vector of the assembled structure, as shown in Sec. 24-9.

24-8 MEMBER STIFFNESS MATRICES WITH RESPECT TO GLOBAL COORDINATES

Equation 24-19 expresses the stiffness matrix $[S_m]$ for a member of a framed structure with respect to coordinates at its two ends in the direction of the global axes. Instead of the matrix product in Eq. 24-19, the stiffness matrix $[S_m]$ is given below in explicit form for a member of framed structures of various types.

Plane truss

$$[S_m] = \frac{Ea}{l}
\begin{bmatrix}
c^2 & cs & -c^2 & -cs \\
cs & s^2 & -cs & -s^2 \\
-c^2 & -cs & c^2 & cs \\
-cs & -s^2 & cs & s^2
\end{bmatrix} \tag{24-23}$$

where $c = \cos\alpha = \lambda_{x^*x}$; $s = \sin\alpha = \lambda_{x^*y}$; and α is the angle defined in Fig. 24-2 with its positive sign convention.

Space truss

$$[S_m] = \frac{Ea}{l}
\begin{bmatrix}
[S_{11}] & [S_{12}] \\
[S_{21}] & [S_{22}]
\end{bmatrix}_m \tag{24-24}$$

where

$$[S_{11}]_m = \frac{Ea}{l}
\begin{bmatrix}
\lambda_{x^*x}^2 & & \text{symmetrical} \\
\lambda_{x^*y}\lambda_{x^*x} & \lambda_{x^*y}^2 & \\
\lambda_{x^*z}\lambda_{x^*x} & \lambda_{x^*z}\lambda_{x^*y} & \lambda_{x^*z}^2
\end{bmatrix} \tag{24-25}$$

$$[S_{22}]_m = [S_{11}]_m \qquad [S_{21}]_m = -[S_{11}]_m \qquad [S_{12}]_m = [S_{21}]_m^T \tag{24-26}$$

Plane frame

$$
\begin{bmatrix}
\dfrac{Eac^2}{l}+\dfrac{12EIs^2}{l^3} & & & & & \text{symmetrical} \\[2.5ex]
\left(\dfrac{Ea}{l}-\dfrac{12EI}{l^3}\right)cs & \dfrac{Eas^2}{l}+\dfrac{12EIc^2}{l^3} & & & & \\[2.5ex]
-\dfrac{6EI}{l^2}s & \dfrac{6EI}{l^2}c & \dfrac{4EI}{l} & & & \\[2.5ex]
-\dfrac{Eac^2}{l}-\dfrac{12EIs^2}{l^3} & -\left(\dfrac{Ea}{l}-\dfrac{12EI}{l^3}\right)cs & \dfrac{6EI}{l^2}s & \dfrac{Eac^2}{l}+\dfrac{12EIs^2}{l^3} & & \\[2.5ex]
-\left(\dfrac{Ea}{l}-\dfrac{12EI}{l^3}\right)cs & -\dfrac{Eas^2}{l}-\dfrac{12EIc^2}{l^3} & -\dfrac{6EI}{l^2}c & \left(\dfrac{Ea}{l}-\dfrac{12EI}{l^3}\right)cs & \dfrac{Eas^2}{l}+\dfrac{12EIc^2}{l^3} & \\[2.5ex]
-\dfrac{6EI}{l^2}s & \dfrac{6EI}{l^2}c & \dfrac{2EI}{l} & \dfrac{6EI}{l^2}s & -\dfrac{6EI}{l^2}c & \dfrac{4EI}{l}
\end{bmatrix}
$$

(24-27)

where $c=\cos\alpha=\lambda_{x^*x}=\lambda_{y^*y}$; $s=\sin\alpha=\lambda_{x^*y}=-\lambda_{y^*x}$; and α is the angle defined in Fig. 24-2 with its positive sign convention.

Grid

$$
[S_m]=
\begin{bmatrix}
\dfrac{12EI}{l^3} & & & & & \text{symmetrical} \\[2.5ex]
-\dfrac{6EI}{l^2}s & \dfrac{GJc^2}{l}+\dfrac{4EIs^2}{l} & & & & \\[2.5ex]
\dfrac{6EI}{l^2}c & \left(\dfrac{GJ}{l}-\dfrac{4EI}{l}\right)cs & \dfrac{GJs^2}{l}+\dfrac{4EIc^2}{l} & & & \\[2.5ex]
-\dfrac{12EI}{l^3} & \dfrac{6EI}{l^2}s & -\dfrac{6EI}{l^2}c & \dfrac{12EI}{l^3} & & \\[2.5ex]
-\dfrac{6EI}{l^2}s & -\dfrac{GJc^2}{l}+\dfrac{2EIs^2}{l} & \left(\dfrac{GJ}{l}-\dfrac{2EI}{l}\right)cs & \dfrac{6EIs}{l^2} & \dfrac{GJc^2}{l}+\dfrac{4EIs^2}{l} & \\[2.5ex]
\dfrac{6EI}{l^2}c & \left(-\dfrac{GJ}{l}-\dfrac{2EI}{l}\right)cs & -\dfrac{GJs^2}{l}+\dfrac{2EIc^2}{l} & -\dfrac{6EI}{l^2}c & \left(\dfrac{GJ}{l}-\dfrac{4EI}{l}\right)cs & \dfrac{GJs^2}{l}+\dfrac{4EIc^2}{l}
\end{bmatrix}
$$

(24-28)

where $c=\cos\alpha=\lambda_{x^*x}=\lambda_{z^*z}$; $s=\sin\alpha=\lambda_{x^*z}=-\lambda_{z^*x}$; and α is the angle defined in Fig. 24-2 with its positive sign convention.

To save computing, Eqs. 24-23 to 24-28 may be used in lieu of the matrix product in Eq. 24-19.

The stiffness matrix $[S_m]$ for a member of a *space frame* with respect to global coordinates is too long to be included here.

24-9 ASSEMBLAGE OF STIFFNESS MATRICES AND LOAD VECTORS

Consider a typical member of a framed structure whose first and second nodes are numbered j and k, respectively. Let $[S_m]$ be the stiffness matrix of the member with respect to coordinates in the global directions at the two nodes. $[S_m]$ may be partitioned as follows:

$$[S_m] = \left[\begin{array}{c|c} [S_{11}] & [S_{12}] \\ \hline [S_{21}] & [S_{22}] \end{array} \right]_m \qquad (24\text{-}29)$$

The same partitioning is made by dashed lines in Eqs. 24-23, 24-24, 24-27 and 24-28. The elements above the horizontal dashed line represent the forces at the first node (the jth node), and the elements below the same line represent forces at the second node (the kth node). The elements to the left of the vertical dashed line are produced by unit displacements at node j, and the elements to the right of the same line correspond to unit displacements at node k. Each of the submatrices in Eq. 24-29 is of size $s \times s$, where s is the number of degrees of freedom per node.

Let the structure stiffness matrix $[S]$ be also partitioned into $n_j \times n_j$ submatrices, each of size $s \times s$, where n_j is the number of nodes. The submatrices of the element stiffness matrix in Eq. 24-29 may be arranged in a matrix of the same size and partitioned in the same way as $[S]$:

$$[\bar{S}_m] = \begin{array}{c} \\ j \\ \\ k \\ \\ \end{array} \begin{array}{cc} j & k \\ \left[\begin{array}{ccccc} & \cdots & & & \\ \cdots & [S_{11}] & \cdots & [S_{12}] & \cdots \\ & \cdots & & & \\ & [S_{12}] & \cdots & [S_{22}] & \cdots \\ & \cdots & & \cdots & \end{array} \right]_m \end{array} \qquad (24\text{-}30)$$

In this equation only the submatrices associated with nodes j and k are shown while all other submatrices are null. The above matrix represents the contribution of the mth member to the structure stiffness. The structure stiffness matrix can now be obtained by a summation:

$$[S] = \sum_{m=1}^{n_m} [\bar{S}_m] \qquad (24\text{-}31)$$

where n_m is the number of members.

We should note that each column of the stiffness matrix $[\bar{S}_m]$ for an element represents a set of forces in equilibrium, and an assemblage of such matrices gives a structure stiffness matrix $[S]$ which has the same property.

The matrix $[S]$ generated as described above represents the stiffness of a free unsupported structure. It is a singular matrix which cannot be used to solve for the unknown displacements before it has been adjusted to account for the support conditions (see Sec. 24-10).

If the structure is to be analyzed for one case of loading, we need to solve the equilibrium equations $[S]\{D\} = -\{F\}$. The vector $\{F\}$ of the forces necessary to prevent the nodal displacements may be considered as the sum of two vectors:

$$\{F\} = \{F_a\} + \{F_b\} \tag{24.32}$$

Here $\{F_a\}$ accounts for the external forces at the nodes, and $\{F_b\}$ for other forces (between the nodes), for temperature variation and for support movements. $\{F_a\}$ is generated simply by listing the nodal forces given in the input data with a reversed sign. The vector $\{F_b\}$ is generated from the restraining forces $\{A_r\}_m$ given in the input data for individual members.

First, Eq. 24-22 is used to transform $\{A_r\}_m$ into equivalent nodal forces $\{F\}_m$ in global directions at member ends. The vector $\{F\}_m$, which has $2s$ elements (s being the number of degrees of freedom per node), may be partitioned in two submatrices, $\{\{F_1\}\ \{F_2\}\}_m$, each having s elements representing forces at one end. The two submatrices may be rearranged into a vector of the same size as $\{F_b\}$ (or $\{F\}$):

$$\{\bar{F}\}_m = \begin{array}{c} \\ j \\ \\ k \\ \\ \end{array} \left\{ \begin{array}{c} \cdots \\ \{F_1\} \\ \cdots \\ \{F_2\} \\ \cdots \end{array} \right\}_m \tag{24-33}$$

This vector has n_j submatrices, all of which are null with the exception of the jth and the kth, where j and k are the node numbers at the two ends of the mth member. The vector $\{\bar{F}_m\}$ represents the contribution of the mth member to the load vector $\{F_b\}$; thus, summing for all members,

$$\{F_b\} = \sum_{m=1}^{n_m} \{\bar{F}\}_m \tag{24-34}$$

where n_m is the number of members.

Enlarging the element stiffness and the force matrices to the size of the structure stiffness matrix and of its load vector (Eqs. 24-30 and 24-33) requires formidable computer storage space. The actual computations are carried out using one matrix for stiffness and another for the load vector(s) of the assembled structure. The matrices are null at the start, and then, for each member, the elements of $[S_m]$ and $\{F\}_m$ are generated and added in the appropriate location by an algorithm. Furthermore, advantage is taken of the symmetry of the structure stiffness matrix and of its banded nature; thus, for $[S]$, only the elements within the band on and above the diagonal are generated, as discussed in Sect. 24-3.

Example 24-3 Generate the stiffness matrix and the load vectors for the three loading cases in Example 24-1 (Fig. 24-4). The input data are given in Table 24-1.

To save space, we give below the member stiffness matrices $[S_m]$ and $[\bar{S}_m]$ for two members only. The submatrices not shown are null.

We start with the structure stiffness matrix $[S] = [0]$.

For member 1, having its first node $j = 2$ and second node $k = 1$ (see Table 24-1), Eqs. 24-23 and 24-30 give

$$[S_1] = \frac{200 \times 10^9 (0.0030)}{2.25} \begin{bmatrix} 1 & 0 & -1 & 0 \\ 0 & 0 & 0 & 0 \\ \hline -1 & 0 & 1 & 0 \\ 0 & 0 & 0 & 0 \end{bmatrix}$$

and

$$[\bar{S}_1] = \begin{array}{c} \\ 1 \\ \\ 2 \\ \\ 3 \\ \\ 4 \\ \\ 5 \end{array} \begin{array}{cccccc} 1 & 2 & 3 & 4 & 5 \\ \left[\begin{array}{ll} 267\ 0 & -267\ 0 \\ 0\ 0 & 0\ 0 \end{array}\right. & & & & \\ -267\ 0 & 267\ 0 & & & \\ 0\ 0 & 0\ 0 & & & \\ & & & & \\ & & & & \\ & & & & \left.\begin{array}{l} \\ \end{array}\right] \end{array} \times 10^6$$

Adding $[\bar{S}_1]$ to the existing structure stiffness matrix $[S]$ gives the new matrix $[S]$, which in this case is the same as $[\bar{S}_1]$ and therefore not repeated here.

A repetition of the above calculations for member 2, which has $j = 3$ and $k = 1$, gives

$$[S_2] = \frac{200 \times 10^9 (0.0024)}{3.75} \begin{bmatrix} 0.36 & 0.48 & -0.36 & -0.48 \\ 0.48 & 0.64 & -0.48 & -0.64 \\ \hline -0.36 & -0.48 & 0.36 & 0.48 \\ -0.48 & -0.64 & 0.48 & 0.64 \end{bmatrix}$$

and

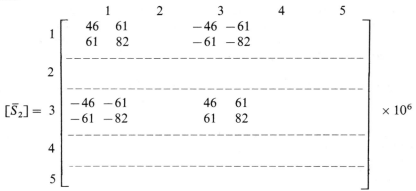

$$[\bar{S}_2] = \begin{array}{c} \\ 1 \\ \\ 2 \\ \\ 3 \\ \\ 4 \\ \\ 5 \end{array} \begin{array}{ccccc} 1 & 2 & 3 & 4 & 5 \\ \begin{array}{cc} 46 & 61 \\ 61 & 82 \end{array} & & \begin{array}{cc} -46 & -61 \\ -61 & -82 \end{array} & & \\ & & & & \\ \begin{array}{cc} -46 & -61 \\ -61 & -82 \end{array} & & \begin{array}{cc} 46 & 61 \\ 61 & 82 \end{array} & & \\ & & & & \\ & & & & \end{array} \times 10^6$$

Adding this matrix to the existing matrix $[S]$, we obtain the new matrix $[S]$:

$$[S] = \begin{array}{c} 1 \\ \\ 2 \\ \\ 3 \\ \\ 4 \\ \\ 5 \end{array} \begin{array}{ccccc} 1 & 2 & 3 & 4 & 5 \\ \begin{array}{cc} 313 & 61 \\ 61 & 82 \end{array} & \begin{array}{cc} -267 & 0 \\ 0 & 0 \end{array} & \begin{array}{cc} -46 & -61 \\ -61 & -82 \end{array} & & \\ \begin{array}{cc} -267 & 0 \\ 0 & 0 \end{array} & \begin{array}{cc} 267 & 0 \\ 0 & 0 \end{array} & & & \\ \begin{array}{cc} -46 & -61 \\ -61 & -82 \end{array} & & \begin{array}{cc} 46 & 61 \\ -61 & 82 \end{array} & & \\ & & & & \\ & & & & \end{array} \times 10^6$$

Repetition of the above computations for members 3 to 7 leads to the structure stiffness matrix:

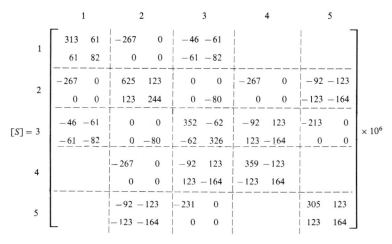

The load vector for case 1 is

$$\{F\} = \{0 \quad -100\,000.0 \mid 0 \ 0 \mid 0 \ 0 \mid 0 \ 0 \mid 0 \ 0\}$$

For cases 2 and 3, we need a transformation matrix for member 4, which has its first node 5 and the second node 3. Equations 24-2, 24-14 and 24-20 give

$$[t] = [1 \quad 0] \quad \text{and} \quad [T] = \begin{bmatrix} 1 & 0 & 0 & 0 \\ 0 & 0 & 1 & 0 \end{bmatrix}$$

For the same member, Eqs. 24-22 and 24-33 give for loading case 2

$$\{F\}_4 = \begin{bmatrix} 1 & 0 \\ 0 & 0 \\ 0 & 1 \\ 0 & 0 \end{bmatrix} \begin{Bmatrix} 96 \times 10^3 \\ -96 \times 10^3 \end{Bmatrix} = \begin{Bmatrix} 96 \\ 0 \\ -96 \\ 0 \end{Bmatrix} \times 10^3$$

and

$$\{\bar{F}\}_4 = \{0 \ 0 \quad 0 \ 0 \quad -96 \times 10^3 \ 0 \quad 0 \ 0 \quad 96 \times 10^3 \ 0\}$$

Because member 4 is the only member which has restraining forces listed in the input data (Table 24-1), this vector is the same as the vector $\{F\}$ for load case 2. Similar calculations are made for case 3, and the load vectors for the three cases are combined into one matrix

$$[F] = \begin{array}{c} 1 \\ \\ 2 \\ \\ 3 \\ \\ 4 \\ \\ 5 \end{array} \begin{bmatrix} 0 & & \\ -100 & & \\ \hline & & \\ \hline & -96 & 246 \\ & 0 & -328 \\ \hline & & -246 \\ & & 328 \\ \hline & 96 & \\ & 0 & \end{bmatrix} \times 10^3$$

24-10 DISPLACEMENT SUPPORT CONDITIONS AND SUPPORT REACTIONS

To find the unknown displacements we need to solve the equilibrium equation

$$[S]\{D\} = -\{F\} \tag{24-35}$$

where $[S]$ is the structure stiffness matrix, $\{D\}$ is a vector of nodal displacements, and $\{F\}$ is a vector of artificial restraining forces which prevent the nodal displacements.

In general, some elements of $\{D\}$ are known to be zero or have prescribed values at the supports, while the corresponding elements of $\{F\}$ are unknown. The total number of unknowns is n, which is equal to the number of equations.

The stiffness matrix $[S]$, generated as described in the preceding section, is the stiffness of a free (unsupported) structure. It is theoretically possible to rearrange the rows and columns in the matrices in Eq. 24-35 so that the known elements of $\{D\}$ are listed first. The equation can then be separated into two sets (see Sec. 4-5). One set, containing only the unknown displacements, can be solved and the result is then substituted in the second set to give the reactions. However, rearranging the equations increases the band width.

Without disturbing the arrangement, it is possible to adjust Eq. 24-35 so that its solution ensures that, at any coordinate k, the displacement $D_k = c_k$, where c_k can be zero in the case of a support without movement or can be a known value of support movement. In one method, the kth row and column of the matrices in Eq. 24-35 are modified as follows:

$$
\begin{array}{c}
\begin{array}{ccc} 1 & \quad k & \quad n \end{array} \\
\begin{array}{c} 1 \\ \\ k \\ \\ n \end{array}
\begin{bmatrix}
& 0 & \\
& \cdots & \\
0 & \cdots & 1 & \cdots & 0 \\
& \cdots & \\
& 0 &
\end{bmatrix}
\begin{Bmatrix} D_1 \\ \cdots \\ D_k \\ \cdots \\ D_n \end{Bmatrix}
=
\begin{bmatrix} -F_1 - S_{1k}c_k \\ \cdots \\ c_k \\ \cdots \\ -F_n - S_{nk}c_k \end{bmatrix}
\end{array}
\qquad (24\text{-}36)
$$

The kth equation is now replaced by $D_k = c_k$; the terms $S_{1k}c_k, S_{2k}c_k, \ldots$ represent the forces at nodes $1, 2, \ldots$ when $D_k = c_k$ and the other displacements are prevented. The modifications shown in Eq. 24-36 have to be repeated for all the prescribed displacements. The solution of the modified equations gives the unknown displacements, but not the unknown reactions. To determine the force F_k we substitute in the kth row of Eq. 24-35 the D values, which are now all known:

$$
F_k = \sum_{j=1}^{n} S_{kj}D_j
\qquad (24\text{-}37)
$$

The reaction at a supported node in the direction of the kth coordinate is given by

$$
R_k = F_k + F'_k
\qquad (24\text{-}38)
$$

where F'_k is the kth element of the original load vector, i.e. before Eq. 24-35 was modified. This element represents a reaction at coordinate k with $\{D\} = \{0\}$. The value F'_k will be nonzero only when an external force is applied at the supported node or over the members meeting at the node, or when the members meeting at the node are subjected to temperature change.

·As mentioned in Sec. 24-3, computer storage space is commonly saved by storing $[S]$ in a compact form which takes advantage of its symmetry and its banded nature. The same space is used to store the reduced matrix needed for the solution; thus the original $[S]$ is lost when $\{D\}$ is determined. To be able to use Eq. 24-37 at this stage, the necessary rows of $[S]$ must be stored (usually on a peripheral device). Also, the element F'_k of the original load vector must be retained for use in Eq. 24-38.

Another method of prescribing a displacement $D_k = c_k$ which requires fewer operations is to modify the kth row of the original equations as follows:

$$
\begin{bmatrix}
\cdots & & & & \\
S_{k1} & \cdots & S_{kk} \times 10^6 & \cdots & S_{kn} \\
\cdots & & & &
\end{bmatrix}
\begin{Bmatrix}
D_1 \\
\cdots \\
D_k \\
\cdots \\
D_n
\end{Bmatrix}
=
\begin{Bmatrix}
-F_1 \\
\cdots \\
S_{kk} \times 10^6 c_k \\
\cdots \\
-F_n
\end{Bmatrix}
\tag{24-39}
$$

where the 10^6 is an arbitrarily chosen large number. The change in Eq. 24-39 is equivalent to providing a fixed support connected to the structure by a very strong spring of stiffness $(10^6 - 1)S_{kk}$ at the kth coordinate and applying a large force of $S_{kk} \times 10^6 c_k$. For all practical purposes, Eq. 24-39 satisfies with sufficient accuracy the condition that D_k equals c_k. This is so because the division of both sides of the kth row in Eq. 24-39 by $10^6 S_{kk}$ reduces the term $S_{kk} \times 10^6$ to 1; the other S terms become negligible, and the kth row reduces to $D_k \simeq c_k$.

After modification of the original equations for each prescribed displacement, a solution will give a vector $\{D\}$ of the displacements at all nodes. The force F_k can now be determined by Eq. 24-37, which requires storage of the kth row of the original $[S]$ matrix. Alternatively, F_k can be obtained by combining Eq. 24-37 and the kth row of Eq. 24-39:

$$
F_k = S_{kk} \times 10^6 (c_k - D_k) + S_{kk} D_k
\tag{24-40}
$$

The difference $c_k - D_k$ needs to be determined with sufficient significant figures (three or more). After calculating F_k by the above equation, substitution in Eq. 24-38 gives the support reaction.

The condition that D_k equals c_k can also be satisfied by replacing S_{kk} on both sides of Eq. 24-39 by $(S_{kk} + 1)$. In this case, F_k can be calculated using Eq. 24-40 and replacing S_{kk} in the first term on the right-hand side by $(S_{kk} + 1)$. This option should be used when S_{kk} can be zero (e.g. in trusses).

When a structure is analyzed for a number of load cases, the vectors $\{D\}$ and $\{F\}$ in Eq. 24-35 are replaced by rectangular matrices $[D]$ and $[F]$, with the respective columns of the two matrices representing one load case; however, the same stiffness matrix $[S]$ is used for all cases. Modification of $[S]$, as suggested by Eq. 24-36 or Eq. 24-39, indicates that a loading case representing the effect of support movements cannot be solved simultaneously with other loading cases unless the prescribed displacements are valid for all cases. This will be the condition when all the prescribed displacements are zero, thus representing supports without movement.

In order to analyze for a nonzero prescribed displacement $D_k = c_k$ at the same time as for other loading cases, we may prepare the support conditions in the input data as if c_k were zero. However, the effect of $D_k = c_k$ must then be presented in the forces $\{A_r\}$ included in the data; $\{A_r\}$ represents the end-forces for the members connected to the coordinate k when $D_k = c_k$ when the displacements at the other coordinates are restrained (see Example 24-1).

The techniques presented in this section require the solution of a larger number of equations than the number of unknown displacements. The extra computing is justified by simpler coding and by the fact that the solution gives the unknown displacements as well as the unknown forces.

As mentioned in Sec. 24-3, it is possible to generate only the equilibrium equations corresponding to the unknown displacements. After the solution of these equations, the displacements, which become known at all coordinates, are used to determine the forces $\{A\}$ at member ends in the directions of local axes (see Sec. 24-12). The reactions at a supported node (with or without support movement) can then be obtained by summing the forces at the ends of the members meeting at the node:

$$\{R\} = \sum ([t]^T \{A\})_i - \{F_S\} \qquad (24\text{-}41)$$

Here $\{R\}$ represents s reaction components in global directions, s being the number of degrees of freedom per node; the $[t]$ matrix is included to transform the member end-forces from local to global directions (Eq. 24-22); the subscript i refers to a member number, and the summation is performed for the members meeting at the supported node; and $\{F_S\}$ represents forces applied direct at the supported node, producing no displacements and no member end-forces.

24-11   SOLUTION OF BANDED EQUATIONS

The solution of the equilibrium Eq. 24-35 was expressed in earlier chapters in the succinct form $\{D\} = -[S]^{-1}\{F\}$. However, a solution by matrix inversion involves more operations than the Gauss elimination or the Cholesky method, discussed later in this section.

Consider a system of n linear equations

$$[a]\{x\} = \{c\} \qquad (24\text{-}42)$$

The elements of $[a]$ and $\{c\}$ are known, and it is required to determine the vector of unknowns $\{x\}$.

The standard Gauss elimination (see Eq. A-44, Appendix A) reduces the original equations to

$$[u]\{x\} = \{d\} \qquad (24\text{-}43)$$

where $[u]$ is a unit upper triangular matrix in which all the diagonal elements are equal to 1, and the elements below the diagonal are zero. Back substitution, starting from the last row, gives successively the unknowns $x_n, x_{n-1}, \ldots, x_1$.

The Gauss elimination can be achieved by a computer, using compact storage space. In Crout's procedure, presented in Sec. A-9, the Gauss elimination is achieved by generating auxiliary matrices $[b]$ and $\{d\}$ of the same size as $[a]$ and $\{c\}$. The elements above the diagonal in $[b]$ are the same as the corresponding elements of $[u]$; thus

$$u_{ij} \equiv b_{ij} \qquad \text{with } i < j \tag{24-44}$$

For structural analysis, we are interested in the case when $[a]$ is symmetrical and banded. In such a case, $[b]$ is also banded and has the same band width. Also, any element b_{ij} above the diagonal is equal to b_{ji}/b_{ii}. Thus, when only the half-band of $[a]$ is stored in a rectangular matrix (Fig. 24-3), $[b]$ can be stored in a matrix of the same size with no need to store the elements below the diagonal.

Examination of the equations which give the elements of $[b]$ (see Eqs. A-42 and A-43 and the numerical example in Sec. A-9) will indicate that, starting from the top row, the elements of $[b]$ can be calculated one by one. Each element b_{ij} can be stored in the same space occupied by a_{ij} because the latter term is no longer needed. In other words, the $[b]$ matrix can be generated in a band form and can replace $[a]$ using the same storage space. The vector $\{d\}$ is generated in a similar way and is stored in the same space originally occupied by $\{c\}$.

The Cholesky method offers some advantages when $[a]$ is symmetrical and banded.[4] In this method, the symmetrical matrix $[a]$ is decomposed into the product of three matrices:

$$[a] = [u]^T [e][u] \tag{24-45}$$

Here again, $[u]$ is a unit upper triangular matrix and therefore its transpose $[u]^T$ is a unit lower triangular matrix; $[e]$ is a diagonal matrix.

Let

$$[e][u] = [h] \tag{24-46}$$

where $[h]$ is an upper triangular matrix. Also, let

$$[h]\{x\} = \{g\} \tag{24-47}$$

Substitution of Eqs. 24-45 to 24-47 into Eq. 24-42 gives

$$[u]^T\{g\} = \{c\} \tag{24-48}$$

Forward substitution in this equation gives $\{g\}$, and using the result with backward substitution in Eq. 24-47 gives the required vector $\{x\}$.

---

[4]See the references in footnote 2 of this chapter.

The operations involved in the matrix product in Eq. 24-45 and in the forward and backward substitutions are relatively small, compared with the operations involved in the decomposition to generate $[u]$ and $[e]$.

Performing the matrix product in Eq. 24-45 gives

$$a_{ij} = e_{ii}u_{ij} + \sum_{k=1}^{i-1} e_{kk}u_{kj}u_{ki} \qquad \text{with } i < j \qquad (24\text{-}49)$$

and

$$a_{jj} = e_{jj} + \sum_{k=1}^{j-1} e_{kk}u_{kj}^2 \qquad (24\text{-}50)$$

To avoid the triple products in the above summations and thereby to reduce the number of operations, let us introduce the symbol

$$\bar{u}_{ij} = e_{ii}u_{ij} \qquad (24\text{-}51)$$

Substitution of Eq. 24-51 into Eqs. 24-49 and 24-50 and rearrangement of terms give

$$\bar{u}_{ij} = a_{ij} - \sum_{k=1}^{i-1} \bar{u}_{kj}u_{ki} \qquad \text{with } i < j \qquad (24\text{-}52)$$

and

$$e_{jj} = a_{jj} - \sum_{k=1}^{j-1} \bar{u}_{kj}u_{kj} \qquad (24\text{-}53)$$

Here again, when $[a]$ is symmetrically banded, only the elements within the band, on and above the diagonal, are stored in the computer. The same storage space is reused; the elements e_{jj} replace a_{jj}, and u_{ij} replace a_{ij} (with $i < j$).

Working down the jth column, the nonzero elements $\bar{u}_{ij}$ are calculated by Eq. 24-52 and temporarily stored in place of a_{ij}. Then the first nonzero u_{ij} in the column is calculated from $\bar{u}_{ij}$ by Eq. 24-51; the product $\bar{u}_{ij}u_{ij}$ is deducted from a_{jj}; and u_{ij} replaces $\bar{u}_{ij}$. Repetition of this step for the remaining $\bar{u}_{ij}$ in the jth column results in the replacement of a_{ii} by e_{ii} (see Eq. 24-53).

Unlike the Gauss eliminations, with Cholesky's method only the matrix $[a]$ is reduced, while $\{c\}$ is used in the forward substitution without change (Eq. 24-46).

In the following example the same equations which are solved in Sec. A-9 are solved here by Cholesky's method.

Example 24-4   Use Cholesky's method to solve the equations $[a]\{x\} = \{c\}$. The vector $\{c\}$ is $\{1, -4, 11, -5\}$. The matrix $[a]$ is

$$\begin{bmatrix} 5 & -4 & 1 & 0 \\ & 6 & -4 & 1 \\ & & 6 & -4 \\ \text{symmetrical} & & & 7 \end{bmatrix}$$

Equations 24-49 to 24-53 are used to generate

$$\begin{bmatrix} e_{11} & u_{12} & u_{13} & 0 \\ - & e_{22} & u_{23} & u_{24} \\ - & - & e_{33} & u_{34} \\ - & - & - & e_{44} \end{bmatrix} = \begin{bmatrix} 5 & -4/5 & 1/5 & 0 \\ - & 14/5 & -8/7 & 5/14 \\ - & - & 15/7 & -4/3 \\ - & - & - & 17/6 \end{bmatrix}$$

Equation 24-48 and forward substitution give

$$\begin{bmatrix} 1 & 0 & 0 & 0 \\ -4/5 & 1 & 0 & 0 \\ 1/5 & -8/7 & 1 & 0 \\ 0 & 5/14 & -4/3 & 1 \end{bmatrix} \begin{Bmatrix} g_1 \\ g_2 \\ g_3 \\ g_4 \end{Bmatrix} = \begin{Bmatrix} 1 \\ -4 \\ 11 \\ -5 \end{Bmatrix}$$

whence

$$\{g\} = \{1, \ -16/5, \ 50/7, \ 17/3\}$$

Equations 24-46 and 24-47 and backward substitution give

$$\begin{bmatrix} 5 & -4 & 1 & 0 \\ & 14/5 & -16/5 & 1 \\ & & 15/7 & -20/7 \\ & & & 17/6 \end{bmatrix} \begin{Bmatrix} x_1 \\ x_2 \\ x_3 \\ x_4 \end{Bmatrix} = \begin{Bmatrix} 1 \\ -16/5 \\ 50/7 \\ 17/3 \end{Bmatrix}$$

whence

$$\{x\} = \{3, 5, 6, 2\}$$

24-12 MEMBER END-FORCES

The last step in the displacement method (see Sec. 3-6) is the superposition to determine the required actions:

$$\{A\} = \{A_r\} + [A_u]\{D\} \tag{24-54}$$

where $\{A_r\}$ are the values of the actions with $\{D\} = \{0\}$, and $[A_u]$ are the values of the actions due to unit displacements introduced separately at each coordinate.

In a framed structure, the required actions are commonly the forces at the member ends in the local directions for individual members. The superposition is done separately for individual members.

After solution of the equilibrium equations, we have a vector $\{D\}$ of the nodal displacements in the global directions at all the nodes. Let $\{D\}$ be partitioned into submatrices, each having s values of the displacements at a node, where s is the number of degrees of freedom per node. For the mth member, having its first and second node j and k, respectively, the displacements at the two ends in local coordinates are given by (Eq. 24-13)

$$\{D^*\}_m = \begin{Bmatrix} \{D^*\}_1 \\ \{D^*\}_2 \end{Bmatrix}_m = \begin{bmatrix} [t] & | & [0] \\ \hline [0] & | & [t] \end{bmatrix}_m \begin{Bmatrix} \{D\}_j \\ \{D\}_k \end{Bmatrix} \tag{24-55}$$

where $\{D\}_j$ and $\{D\}_k$ are submatrices of $\{D\}$, corresponding to nodes j and k, and $[t]_m$ is a transformation matrix for the mth member given by one of Eqs. 24-14 to 24-18.

To apply the superposition Eq. 24-54 in order to find member end-forces, we have $\{A_r\}$ given in the input data. The values of the member end-forces due to unit displacements introduced separately at each local coordinate (Fig. 24-2) are the elements of the member stiffness matrix $[S^*]$, given by one of Eqs. 24-10 to 24-12 or by Eq. 4-5. Thus, in this application, the product $[S^*]\{D^*\}$ stands for $[A_u]\{D\}$. The member end-forces for any member are therefore given by

$$\{A\} = \{A_r\} + [S^*]\{D^*\} \qquad (24\text{-}56)$$

The stiffness matrices of individual members need to be calculated again in this step, unless they have been stored.

Example 24-5   Find the reactions and the end-forces in member 6 in the truss of Fig. 24-4 for each of the three cases in Example 24-1.

The stiffness matrix and the load vectors for the three cases have been determined in Example 24-3. Zero displacement is prescribed in the x and y directions at nodes 4 and 5; therefore, multiply the diagonal coefficient in each of rows 7, 8, 9 and 10 by 10^6 and set zeros in the same rows of the load vectors (see Eq. 24-39). Because we accounted for the support movement in the member end-forces by $\{A_r\}$ used in the input data, we put here $c_k = 0$. The adjusted equilibrium equations are

$$10^6 \begin{bmatrix} 313 & 61 & -267 & 0 & -46 & -61 & & & & \\ & 82 & 0 & 0 & -61 & -82 & & & & \\ \hline & & 625 & 123 & 0 & 0 & -267 & 0 & -92 & -123 \\ & & & 244 & 0 & -80 & 0 & 0 & & -164 \\ \hline & & & & 352 & -62 & -92 & 123 & -213 & 0 \\ & & & & & 326 & 123 & -164 & 0 & 0 \\ \hline & & \text{symmetrical} & & & & 359 \times 10^6 & -123 & & \\ & & \text{submatrices not shown are null} & & & & & 164 \times 10^6 & & \\ & & & & & & & & 305 \times 10^6 & 123 \\ & & & & & & & & & 164 \times 10^6 \end{bmatrix} [D]$$

$$= \begin{bmatrix} 0 & 0 & 0 \\ 100 & 0 & 0 \\ 0 & 0 & 0 \\ 0 & 0 & 0 \\ \hline 0 & +96 & -246 \\ 0 & 0 & +328 \\ \hline 0 & 0 & 0 \\ 0 & 0 & 0 \\ 0 & 0 & 0 \\ 0 & 0 & 0 \end{bmatrix} \times 10^3$$

The solution is

$$[D] = \begin{bmatrix} -0.6510 \times 10^{-3} & -0.3223 \times 10^{-5} & -0.1912 \times 10^{-3} \\ 0.3016 \times 10^{-2} & 0.5688 \times 10^{-3} & 0.1372 \times 10^{-2} \\ \hline -0.3698 \times 10^{-3} & -0.3223 \times 10^{-5} & -0.1912 \times 10^{-3} \\ 0.4694 \times 10^{-3} & 0.9411 \times 10^{-4} & 0.5582 \times 10^{-3} \\ \hline 0.5925 \times 10^{-3} & 0.4097 \times 10^{-3} & -0.2390 \times 10^{-3} \\ 0.8627 \times 10^{-3} & 0.2374 \times 10^{-3} & 0.1408 \times 10^{-2} \\ \hline -0.4180 \times 10^{-9} & -0.1013 \times 10^{-15} & -0.6856 \times 10^{-9} \\ 0.4183 \times 10^{-9} & -0.6994 \times 10^{-10} & 0.1587 \times 10^{-8} \\ \hline 0.4910 \times 10^{-9} & 0.3142 \times 10^{-9} & 0.3782 \times 10^{-15} \\ 0.1921 \times 10^{-9} & 0.6994 \times 10^{-10} & 0.4149 \times 10^{-9} \end{bmatrix} \begin{matrix} 1 \\ \\ 2 \\ \\ 3 \\ \\ 4 \\ \\ 5 \end{matrix}$$

The forces at the supported nodes 4 and 5 (Eq. 24-40) are

$$\begin{Bmatrix} F_7 \\ F_8 \\ F_9 \\ F_{10} \end{Bmatrix} = 10^3 \times \begin{bmatrix} 150 & 0 & 246 \\ -69 & 11 & -260 \\ \hline -150 & -96 & 0 \\ -31 & -11 & -68 \end{bmatrix}$$

These forces are calculated by the multiplication of the nodal displacement with reversed sign by the diagonal elements of the adjusted stiffness matrix. (The last term in Eq. 24-40 is negligible and $c_k = 0$.) Addition of the above forces to the corresponding rows of the original load vector (generated in Example 24-3) gives the reactions (Eq. 24-38):

$$\begin{Bmatrix} R_7 \\ R_8 \\ R_9 \\ R_{10} \end{Bmatrix} = 10^3 \times \begin{bmatrix} 150 & 0 & 0 \\ -69 & 11 & 68 \\ \hline -150 & 0 & 0 \\ -31 & -11 & -68 \end{bmatrix}$$

The first and second nodes of member 6 are $j = 4$ and $k = 3$ (see input data, Table 24-1). The transformation matrix is (Eq. 24-14)

$$[t] = [0.6 - 0.8]$$

Let us transform the nodal displacements for the three cases of nodes 4 and 3 from global to local directions (Eq. 24-55);

$$[D^*] = \begin{bmatrix} 0.6 & -0.8 & 0 & 0 \\ 0 & 0 & 0.6 & -0.8 \end{bmatrix} \times$$

$$\begin{bmatrix} -0.4180 \times 10^{-9} & -0.1013 \times 10^{-15} & -0.6856 \times 10^{-9} \\ 0.4183 \times 10^{-9} & -0.6994 \times 10^{-10} & 0.1587 \times 10^{-8} \\ 0.5925 \times 10^{-3} & 0.4097 \times 10^{-3} & -0.2390 \times 10^{-3} \\ 0.8687 \times 10^{-3} & 0.2374 \times 10^{-3} & 0.1408 \times 10^{-2} \end{bmatrix}$$

whence

$$[D^*] = \begin{bmatrix} -0.5854 \times 10^{-9} & 0.5595 \times 10^{-10} & -0.1681 \times 10^{-8} \\ -0.3347 \times 10^{-3} & 0.5590 \times 10^{-4} & -0.1270 \times 10^{-2} \end{bmatrix}$$

The end-forces for member 6 in the three cases are (Eq. 24-54)

$$[A] = 10^3 \times \begin{bmatrix} 0 & 0 & -410 \\ 0 & 0 & 410 \end{bmatrix} + 256 \times 10^6 \begin{bmatrix} 1 & -1 \\ -1 & 1 \end{bmatrix} [D^*]$$

whence

$$[A] = 10^3 \times \begin{bmatrix} 86 & -14 & -85 \\ -86 & +14 & 85 \end{bmatrix}$$

24-13 GENERAL

The analysis of framed structures by computer to give the nodal displacements and the member end-forces is summarized below. The nodal displacements are in the directions of arbitrarily chosen global axes x, y (and z) (Fig. 24-1), while the member end-forces are in local coordinates pertaining to individual members (Fig. 24-2).

The input data are composed of: the material properties E (and G); the x, y (and z) coordinates of the nodes; the number of the two nodes at the ends of each member; the cross-sectional properties a, I, J etc. of each member; the support conditions; the external forces applied at the nodes; and the end-forces $\{A_r\}$ for the members subjected to loads away from the nodes. The elements of $\{A_r\}$ are forces in local coordinates at the member ends when the nodal displacements are restrained.

The five steps of the displacement method summarized in Sec. 3-6 are executed as follows:

Step 1 The nodes and the members are numbered sequentially by the analyst, from 1 to n_j and from 1 to n_m, respectively. This will automatically define the degrees of freedom and the required end-actions according to the systems specified in Figs. 24-1 and 24-2.

Step 2 The forces $\{A_r\}$ included in the input data are transformed into global directions and assembled to give a vector $\{F_b\}$ of restraining nodal forces (Eqs. 24-22 and 24-34). A vector $\{F_a\}$ is also generated, simply by listing, from the input data, the nodal forces with a reversed sign. The sum $\{F_a\}$ plus $\{F_b\}$ gives a vector of the restraining forces $\{F\}$ which are necessary to prevent the nodal displacements (Eq. 24-32).

Step 3 To generate the stiffness matrix of the structure, start with $[S] = [0]$ and partition this matrix into $n_j \times n_j$ submatrices each of size $s \times s$, where s is the number of degrees of freedom per node. The stiffness matrix of the first member is generated by one of Eqs. 24-23, 24-24, 24-27, 24-28 or 4-5 and partitioned into 2×2 submatrices, each of size $s \times s$. The four submatrices are added to the appropriate submatrices of $[S]$ according to Eqs. 24-30 and

24-31. By repeating this procedure for all the members, the stiffness matrix of a free unsupported structure is generated.

When one of Eqs. 24-23, 24-24, 24-27 or 24-28 is used, the stiffness matrix of a member is obtained with respect to coordinates in global directions. However, Eq. 4-5 gives the stiffness of a member of a space frame with respect to local coordinates. Transformation is necessary according to Eq. 24-19 before the assemblage of the stiffness matrices can be performed by Eq. 24-31.

The structure stiffness matrix $[S]$ is symmetrical and generally has the nonzero elements limited to a band adjacent to the diagonal. To save computer space, only the diagonal elements of $[S]$ and the elements above the diagonal within the band are stored in a rectangular matrix as shown in Fig. 24-3. The node numbering selected by the analyst in step 1 affects the band width and hence the width of the rectangular matrix used to store $[S]$. A narrower band width is generally obtained by numbering the nodes sequentially across that side of the frame which has a smaller number of nodes (as example, see Fig. 24-3).

Step 4   Before the equilibrium equations $[S]\{D\} = -\{F\}$ can be solved, $[S]$ and $\{F\}$ must be adjusted according to the displacements prescribed in the input data (Eqs. 24-36 or 24-39).

Two methods of solution of the equilibrium equations are discussed in Sec. 24-11. The solution gives the unknown displacements $\{D\}$ and these are used to calculate the reactions by Eq. 24-38 and one of Eqs. 24-37 or 24-40.

Step 5   The end-forces for each member are obtained by the superposition Eq. 24-56, which sums $\{A_r\}$ given in the input data and the product $([S^*]\{D^*\})$, with $[S^*]$ and $\{D^*\}$ being, respectively, the member stiffness matrix and the displacements at the member ends in local coordinates. For this reason, the displacements obtained in step 4 have to be transformed from global directions to local directions by Eq. 24-55 before the superposition can proceed.

The general approach in this chapter is followed in Chapter 25 by a discussion of special topics related to use of computers in structural analysis.

Problems for Chapter 24 are included at the end of Chapter 25.

CHAPTER 25

Implementation of computer analysis

25-1 INTRODUCTION

Analysis of structures by the displacement method requires the solution of the equilibrium equations $[S]\{D\} = \{-F\}$. The solution must satisfy the displacement boundary conditions. In Chapter 24 we assumed that the displacements $\{D\}$ and the forces $\{F\}$ are in directions of global axes. In Sec. 24-10, we discussed a method in which the stiffness matrix $[S]$ generated for a free, unsupported structure is adjusted, together with the vector $\{F\}$ to satisfy the condition that the displacement at a coordinate equals zero or a prescribed value. In Sec. 25-2, we shall consider the case when the prescribed displacement is in a direction inclined to global axes; this may be so at a roller support.

In the analysis of a large structure, it is often possible to consider only a part of the structure rather than the whole. This approach is useful to reduce the cost of preparing the data, of computing and of interpretation of the results. When an isolated part of a structure is analyzed, it is crucial that the displacement boundary conditions accurately represent the conditions in the actual structure. If this is not so, the results can be grossly erroneous.

Structures which are composed of symmetrical sectors can be analyzed by considering one sector with appropriate displacement constraints imposed at the boundaries with adjacent sectors. The displacement boundary conditions for different types of symmetry are discussed in Sec. 25-3 and 25-5.

A displacement component may be constrained so as to be equal to a prescribed value or to zero. For instance, the constraint may be specified by imposing the value of the displacement at one coordinate to be equal to the displacement at another coordinate; more generally, the displacement at one coordinate may be required to be a linear combination of the displacement at two other coordinates. Section 25-4 shows how the equilibrium equations can be adjusted to satisfy the displacement constraints, and gives examples of structures in which such constraints are required.

The analysis of a large structure can be reduced to a series of analyses of substructures which are easier to handle one at a time. This is discussed in Sec. 25-6.

Section 25-7 is concerned with the use of the computer for the plastic analysis of plane frames on the basis of the assumptions adopted in Chapter 21.

Section 25-8 gives a method for generating the stiffness matrix of a member with a variable section or with a curved axis. The stiffness matrix is derived from the flexibility matrix of the member treated as a cantilever.

25-2 DISPLACEMENT BOUNDARY CONDITIONS IN INCLINED COORDINATES

In the analysis of structures by computer, using the displacement method, the coordinates at the nodes typically represent displacement components in the directions of one set of global axes. It is, however, sometimes necessary to express the boundary conditions at a node as prescribed displacements in directions inclined to the global axes. An example of this situation is at the roller support at node i of the plane frame in Fig. 25-1a. The support allows a free translation $\bar{u}$ in the $\bar{x}$ direction, parallel to a plane inclined at an angle γ_i to

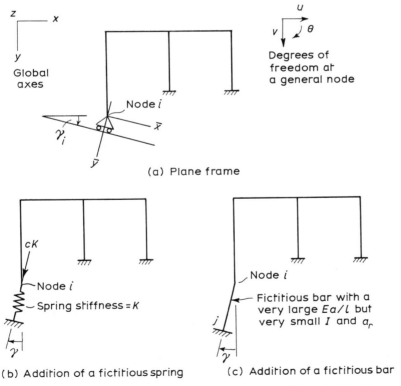

Fig. 25-1. Introduction of displacement boundary conditions in directions inclined to the global axes.

the global x axis, but prevents translation $\bar{v}$ in the $\bar{y}$ direction. Alternatively, $\bar{v}$ may take a prescribed value c.

For the analysis of the structure in Fig. 25-1a, we need a stiffness matrix $[\bar{S}]$ corresponding to degrees of freedom u, v, θ in the global x, y, z directions at each node except node i, where the degrees of freedom are $\bar{u}, \bar{v}$ and $\bar{\theta}$ in the $\bar{x}, \bar{y}, \bar{z}$ directions. One way to achieve this is first to generate the stiffness matrix $[S]$, in the usual way, corresponding to degrees of freedom in the global x, y, z directions, and subsequently to adjust $[S]$ to $[\bar{S}]$ as discussed below.

In general, a geometrical relation can be written:

$$\{D\}_i = [H]_i \{\bar{D}\}_i \tag{25-1}$$

where $\{D\}_i$ and $\{\bar{D}\}_i$ are nodal displacements at i in the global x, y, z directions and in the $\bar{x}, \bar{y}, \bar{z}$ directions, respectively. The matrix $[H]_i$ is a transformation matrix for node i. The forces at node i in the two coordinate systems can also be related by using the same transformation matrix (see Eq. 8-16):

$$\{\bar{F}\}_i = [H]_i^T \{F\}_i \tag{25-2}$$

In the example considered in Fig. 25-1a, the transformation matrix is

$$[H]_i = \begin{bmatrix} \cos\gamma & -\sin\gamma & 0 \\ \sin\gamma & \cos\gamma & 0 \\ 0 & 0 & 1 \end{bmatrix}_i \tag{25-3}$$

Using the stiffness matrix $[S]$, we can write the equilibrium equations:

$$[S]\{D\} = -\{F\} \tag{25-4}$$

Let us now partition $[S]$ into $n_j \times n_j$ submatrices each of size $s \times s$, where n_j is the number of nodes and s is the number of degrees of freedom per node. The partitioned equilibrium equation will appear as follows:

$$\begin{bmatrix} & & S_{1i} & \\ & & S_{2i} & \\ & & \cdots & \\ S_{i1} & S_{i2} & \cdots & S_{ii} & \cdots \\ & & \cdots & \end{bmatrix} \begin{Bmatrix} D_1 \\ D_2 \\ \cdots \\ D_i \\ \cdots \end{Bmatrix} = - \begin{Bmatrix} F_1 \\ F_2 \\ \cdots \\ F_i \\ \cdots \end{Bmatrix} \tag{25-5}$$

In this equation and in the following two, only the elements in the ith row and ith column of the square matrix on the left-hand side are written.

The term $\{D\}_i$ in Eq. 25-5 may be eliminated by the use of Eq. 25-1, giving

$$\begin{bmatrix} & & S_{1i}H_i & \\ & & S_{2i}H_i & \\ & & \cdots & \\ S_{i1} & S_{i2} & \cdots & S_{ii}H_i & \cdots \\ & & \cdots & \end{bmatrix} \begin{Bmatrix} D_1 \\ D_2 \\ \cdots \\ \bar{D}_i \\ \cdots \end{Bmatrix} = - \begin{Bmatrix} F_1 \\ F_2 \\ \cdots \\ F_i \\ \cdots \end{Bmatrix} \tag{25-6}$$

The submatrix $\{F\}_i$ in this equation may also be eliminated by multiplication of the ith row by $[H]_i^T$ and substitution of Eq. 25-2:

$$
\begin{bmatrix}
& & & S_{1i}H_i & & \\
& & & S_{2i}H_i & & \\
& & & \cdots & & \\
H_i^T S_{i1} & H_i^T S_{i2} & \cdots & H_i^T S_{ii}H_i & \cdots \\
& & & \cdots & &
\end{bmatrix}
\begin{Bmatrix}
D_1 \\
D_2 \\
\cdots \\
\bar{D}_i \\
\cdots
\end{Bmatrix}
= -
\begin{Bmatrix}
F_1 \\
F_2 \\
\cdots \\
\bar{F}_i \\
\cdots
\end{Bmatrix}
\tag{25-7}
$$

The square symmetrical matrix on the left-hand side of Eq. 25-7 is the stiffness matrix $[\bar{S}]$ in which the boundary conditions can be introduced as discussed in Sec. 24-10. In practice, the adjustment to obtain $[\bar{S}]$ can be conveniently made to the stiffness matrices of the elements connected to node i (see Example 25-1).

An alternative to the above procedure is to add a spring of stiffness K at node i in the $\bar{y}$ direction (Fig. 25-1b). The stiffness K is much larger than the diagonal elements of $[S]$, e.g. 10^5 times greater than their values. When a displacement c is prescribed in the $\bar{y}$ direction, a force cK has to be applied in the $\bar{y}$ direction at node i. Because the stiffness of the structure is negligible compared with that of the spring, the force cK produces the prescribed displacement c at i.

The stiffness matrix of the added spring with respect to the coordinates $\bar{u}$, $\bar{v}$ and $\bar{\theta}$ in Fig. 25-1b is

$$
[\bar{S}]_K =
\begin{bmatrix}
0 & 0 & 0 \\
0 & K & 0 \\
0 & 0 & 0
\end{bmatrix}
\tag{25-8}
$$

Before assemblage of $[S]$ (see Sec. 24-9), the above stiffness matrix for the spring must be transformed to correspond to the global directions:

$$
[S]_K = [H]_i [\bar{S}]_K [H]_i^T
\tag{25-9}
$$

where $[H]_i$ is the transformation matrix in Eq. 25-3. Equation 25-9 is derived from Eq. 8-17, noting that the inverse of $[H]_i$ is equal to its transpose.

The concept discussed above is general and it can be used for any structure in which a displacement is prescribed in an inclined direction.

For the frame in Fig. 25-1a, the conditions that, at node i, $\bar{v} = c$ while $\bar{u}$ is free, can be achieved by connecting to i a fictitious bar with end j fixed (Fig. 25-1c). The length l and cross-sectional area a of the fictitious bar must be chosen so that $Ea/l = K$, K being the same as for the spring considered earlier. Very small values should be assigned to the second moment of area I (and to the reduced area a_r when shear deformations are considered). Prescribed displacements in the global directions $u_j = -c \sin \gamma$ and $v_j = c \cos \gamma$, introduced at node j instead of node i, will produce the desired effect. This method has the advantage that the required boundary conditions are achieved by entering appropriate input data without changing the computer program which performs the analysis. It should be mentioned, however, that a numerical difficulty in solving the equilibrium equations may be encountered if the stiffness of the added member is excessive.

Example 25-1   Generate the stiffness matrix of a prismatic member of a plane frame with respect to the coordinates shown in Fig. 25-2a.

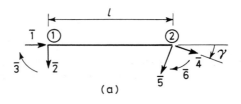

Fig. 25-2. Member of a plane frame. (a) Inclined coordinates. (b) Coordinates parallel to member local axes.

The transformation matrix relating $\{\bar{D}\}$ to $\{D^*\}$ at node 2 is (see Eqs. 25-1 and 25-2)

$$[H]_2 = \begin{bmatrix} c & -s & 0 \\ s & c & 0 \\ 0 & 0 & 1 \end{bmatrix}_2$$

where $c = \cos\gamma$ and $s = \sin\gamma$.

The stiffness matrix $[S^*]$ with respect to the coordinates parallel to the member local axes (Fig. 25-2b) is given by Eq. 24-11. Application of Eq. 25-7 gives

$$\left[\begin{array}{c|c} S_{11} & S_{12}H_2 \\ \hline H_2^T S_{21} & H_2^T S_{22}H_2 \end{array}\right]\left\{\begin{array}{c} \bar{D}_1 \\ \hline \bar{D}_2 \end{array}\right\} = \left\{\begin{array}{c} \bar{F}_1 \\ \hline \bar{F}_2 \end{array}\right\}$$

where S_{ij} represents the 3×3 submatrix given in Eq. 24-11; $\bar{D}_1 \equiv D_1^*$ and $\bar{F}_1 \equiv F_1^*$ are subvectors representing, respectively, the three displacements and forces at member end 1.

Substitution for S_{ij} and H_2 gives the stiffness matrix with respect to the coordinates in Fig. 25-2a:

$$[\bar{S}] = \begin{bmatrix} \dfrac{Ea}{l} & & & & & \\[3mm] 0 & \dfrac{12EI}{l^3} & & & & \\[3mm] 0 & \dfrac{6EI}{l^2} & \dfrac{4EI}{l} & & & \\[3mm] \hline -\dfrac{cEa}{l} & -\dfrac{12sEI}{l^3} & -\dfrac{6sEI}{l^2} & \dfrac{c^2Ea}{l}+\dfrac{12s^2EI}{l^3} & & \\[3mm] +\dfrac{sEa}{l} & -\dfrac{12cEI}{l^3} & -\dfrac{6cEI}{l^2} & -\dfrac{scEa}{l}+\dfrac{12scEI}{l^3} & \dfrac{s^2Ea}{l}+\dfrac{12c^2EI}{l^3} & \\[3mm] 0 & \dfrac{6EI}{l^2} & \dfrac{2EI}{l} & -\dfrac{6sEI}{l^2} & -\dfrac{6cEI}{l^2} & \dfrac{4EI}{l} \end{bmatrix}$$ (25-10)

25-3 STRUCTURAL SYMMETRY

Large structures are commonly analyzed by computer and, when possible, advantage is taken of symmetry so as to reduce the effort needed in data preparation and in the interpretation of the results. When a structure has one or more planes of symmetry it is possible to perform the analysis on one-half, one-quarter or an even smaller part of the structure, provided that appropriate boundary conditions are applied at the nodes on the plane(s) of symmetry. In the following sections, these boundary conditions are discussed for spatial structures in which the symmetry exists both for the structure configuration and for the loading, and also for the case when the structure is symmetrical but is subjected to nonsymmetrical loads. The discussion applies to structures composed of bars or other finite elements.

25-31 Symmetrical Structures Subjected to Symmetrical Loading

Figure 25-3a is a top view of a spatial structure with two vertical planes of symmetry: xz and yz. It follows that for all the elements and applied forces (not shown in the figure) on one side of a plane of symmetry, there exists a set of mirror-image elements and forces on the opposite side. The analysis of a structure of this type needs to be performed on one-quarter of the structure only (Fig. 25-3b).

Let us assume the degrees of freedom at a typical node to be three translations u, v, w in the global x, y, z directions and three rotations $\theta_x, \theta_y, \theta_z$ in the same directions. The global directions x, y, z are chosen parallel to the

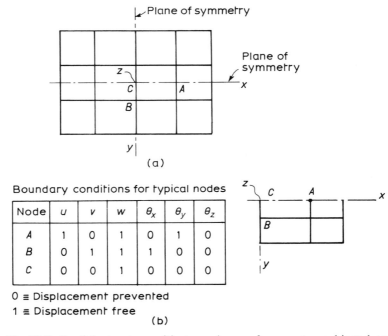

Boundary conditions for typical nodes

Node	u	v	w	θ_x	θ_y	θ_z
A	1	0	1	0	1	0
B	0	1	1	1	0	0
C	0	0	1	0	0	0

0 ≡ Displacement prevented
1 ≡ Displacement free

(b)

Fig. 25-3. Spatial structure with two planes of symmetry subjected to symmetrical loading. (a) Top view. (b) Quarter structure to be analyzed.

planes of symmetry. In the part of the structure to be analyzed, the elements situated on a plane of symmetry must have adjusted properties. For example, when the elements on a plane of symmetry are bars, their cross-sectional properties $(a, I, J$ and $a_r)$ to be used in the analysis must be assumed to be equal to one-half of the values in the actual structure. It is also possible, instead of changing the cross-sectional properties, to reduce correspondingly (i.e. by one-half) the material properties E and G. Also the forces applied at the nodes on a plane of symmetry must be assumed to be one-half of the actual values.

Because of symmetry, some of the displacement components at the nodes on the planes of symmetry are known to be zero. The table in Fig. 25-3b indicates 0 or 1 for each of the six degrees of freedom at typical nodes A, B and C on the planes of symmetry. The value 0 indicates zero displacement, and 1 indicates that the displacement is free to occur.

As mentioned in Sec. 24-4, the displacement boundary conditions must be sufficient to ensure that the structure analyzed cannot translate or rotate as a free rigid body. Now, the boundary conditions prescribed in Fig. 25-3b are not sufficient to prevent translation in the z direction; to prevent this translation, at least one node must have $w = 0$.

25-32 Symmetrical Structures Subjected to Nonsymmetrical Loading

Let us assume that the spatial structure shown in Fig. 25-3a is subjected to forces P_1 and P_2 (where $P_1 \neq P_2$) at the symmetrical nodes D, E, F and G, as shown in Fig. 25-4a. The system of forces in Fig. 25-4a may be considered to be equivalent to the sum of a symmetrical and an antisymmetrical system, shown in Fig. 25-4b. The magnitude of the symmetrical component P_s and that of the antisymmetrical component P_a must satisfy the equations

$$P_s + P_a = P_1 \qquad -P_s + P_a = P_2 \qquad (25\text{-}11)$$

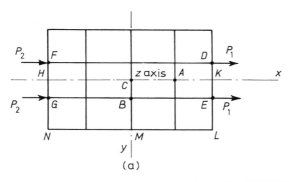

(a)

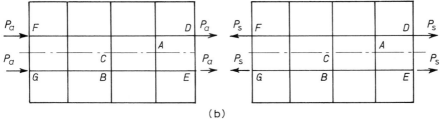

(b)

Boundary conditions for antisymmetrical case

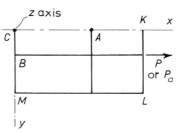

Node	u	v	w	θ_x	θ_y	θ_z
A	1	0	1	0	1	0
B	1	0	0	0	1	1
C	1	0	0	0	1	0

$0 \equiv$ displacement prevented
$1 \equiv$ displacement free

(c)

Fig. 25-4. Spatial structure with two planes of symmetry subjected to nonsymmetrical loading. (a) Top view. (b) Symmetrical and antisymmetrical load components. (c) Quarter structure to be analyzed.

whence

$$P_s = \frac{P_1 - P_2}{2} \qquad P_a = \frac{P_1 + P_2}{2} \tag{25-12}$$

The analysis for the symmetrical and for the antisymmetrical loadings can be performed separately for the quarter structure $CKLM$. For the symmetrical loading, the boundary conditions are as indicated in Fig. 25-3b. The boundary conditions for the antisymmetrical loading case are shown in Fig. 25-4c.

The results of the two analyses, when added, give the displacements and the internal forces for the quarter structure $CKLM$ due to the actual nonsymmetrical loading (Fig. 25-4a). By subtracting the results for the antisymmetrical loading from the results for the symmetrical loading, we obtain a mirror image of the displacements and the internal forces in the quarter structure $CHNM$ due to the actual nonsymmetrical loading. In other words, the results for the antisymmetrical loading subtracted from the results for the symmetrical loading give the displacements and the internal forces in the quarter structure $CKLM$ when the two forces at F and G (Fig. 25-4a) are interchanged with the two forces at D and E and the four forces are reversed in direction.

It should be noted that the procedure described above involves two separate analyses of quarter structures, each requiring the solution of a set of equations which differ both in the right-hand side and in the left-hand side. However, each set involves a smaller number of unknown displacements than if a half structure were analyzed. The results of the two analyses of the quarter structure must be stored and the two solutions are combined. The procedure is advantageous only in very large structures to save computing time for the solution of the equations or when the capacity of the computer would otherwise be exceeded.

25-4 DISPLACEMENT CONSTRAINTS

In Sec. 24-10 we discussed the methods to be used in order to satisfy the condition that the displacement at a coordinate is zero or has a prescribed value. In this section, we consider a method to impose relations between nodal displacements.

As an example we can consider the plane frame shown in Fig. 25-5a, which is composed of an infinite number of bays identical in geometry and in loading. The analysis needs to be performed for a single isolated bay $ABCB'$ with the following relation between the displacements at nodes B and B':

$$\{u, v, \theta\}_{B'} = \{u, v, \theta\}_B \tag{25-13}$$

Another example is shown in Fig. 25-5b, which represents a square plate in bending with symmetry in geometry and in loading about the vertical planes

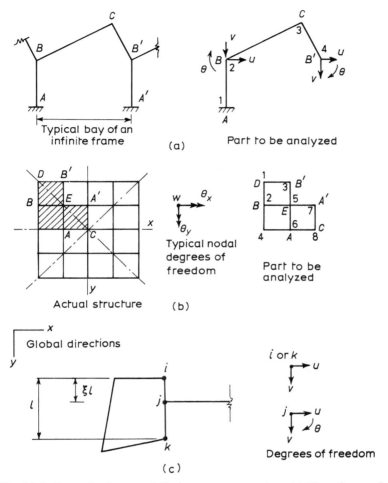

Fig. 25-5. Examples for use of displacement constraints. (a) Plane frame of
many bays. (b) Plate in bending. (c) Beam-to-wall connection.

through the x and y axes and about the two diagonals. A finite-element
analysis of this structure can be performed using the square elements for the
shaded part of the plate. The degrees of freedom at a typical node can be taken
as $\{w, \theta_x, \theta_y\}$, as shown in the figure. For symmetry of displacements, the
following constraints apply:

$$(\theta_x)_C = -(\theta_y)_C \qquad (\theta_x)_D = -(\theta_y)_D \qquad (\theta_x)_E = -(\theta_y)_E \qquad (25\text{-}14)$$

$$\begin{aligned}
\{w, \theta_x, \theta_y\}_{A'} &= \{w, -\theta_y, -\theta_x\}_A \\
\{w, \theta_x, \theta_y\}_{B'} &= \{w, -\theta_y, -\theta_x\}_B
\end{aligned} \qquad (25\text{-}15)$$

Another example where displacement constraints can be used in finite-element analysis is shown in Fig. 25-5c. A plane-stress quadrilateral element with nodes at its corners is connected to a beam element. The constraint equations in this case can be taken as

$$u_j = (1 - \xi)u_i + \xi u_k$$

$$v_j = (1 - \xi)v_i + \xi v_k$$

$$\theta_j = \left(\frac{1}{l}\right)u_i - \left(\frac{1}{l}\right)u_k \qquad (25\text{-}16)$$

Displacement constraints can also be used in structures with cyclic symmetry (see Sec. 25-5).

A typical constraint equation may be written in the form

$$D_j = \beta_i D_i \qquad (25\text{-}17)$$

where β_i is a constant. (A second term $\beta_k D_k$ is added to the right-hand side for the constraints of Eq. 25-16.)

Each constraint equation reduces the number of unknown displacements by one. Thus, in the equilibrium equations $[S]\{D\} = -\{F\}$, the constraint Eq. 25-17 may be used to eliminate D_j so that we can solve the equations for the remaining unknowns, but this requires renumbering of the unknowns. A method[1] of adjusting the equilibrium equations $[S]\{D\} = -\{F\}$ to satisfy any number of constraints of this type is discussed below. The adjustment does not require changes in the number of equations to be solved or in the arrangement of the unknowns.

For each constraint Eq. 25-17, the adjustments are limited to the ith and jth columns of $[S]$ and the ith and jth rows of $[S]$ and of $\{F\}$. The adjustments are performed in steps. First the elements in the jth column of $[S]$ are multiplied by β_i and added to the ith column. Then the elements in the jth row of $[S]$ and of $\{F\}$ are multiplied by β_i and added to the ith row. Element S_{jj} is replaced by 1 and the off-diagonal elements in the jth row and column of $[S]$ are replaced by zero. The adjusted columns and rows now become

$$
\begin{array}{c}
 & 1 & & i & & j \\
\begin{array}{c}1\\2\\ \\i\\ \\j\end{array}
&
\left[
\begin{array}{ccccc}
 & & (S_{1i} + \beta_i S_{1j}) & & 0 \\
 & & (S_{2i} + \beta_i S_{2j}) & & 0 \\
\cdots & & \cdots & & \cdots \\
(S_{i1} + \beta_i S_{j1}) & \cdots & (S_{ii} + 2\beta_i S_{ij} + \beta_i^2 S_{jj}) & \cdots & 0 & \cdots \\
 & & \cdots & & \cdots \\
0 & \cdots & 0 & \cdots & 1 \\
 & & \cdots & & \cdots
\end{array}
\right]
& \{D\}
\end{array}
$$

---

[1] Other methods are discussed in the reference in footnote 3 of Chapter 24.

$$= - \left\{ \begin{array}{c} F_1 \\ F_2 \\ \cdots \\ F_i + \beta_i F_j \\ \cdots \\ 0 \\ \cdots \end{array} \right\} \qquad (25\text{-}18)$$

The above adjustments consist of the elimination of D_j in all equations and multiplication of both sides of the ith equation by β_i to maintain symmetry; the jth equation is replaced by the dummy equation $D_j = 0$. Solution of the adjusted equations gives the unknown displacements, including a zero for D_j. This answer must be replaced by $D_j = \beta_i D_i$.

The adjustment of the equilibrium equations as described above can be repeated for each constraint equation.

As mentioned earlier, each constraint equation, for the example in Fig. 26-5c can be written in the form: $D_j = \beta_i D_i + \beta_k D_k$. The adjustment of the equilibrium equation to satisfy this equation has to be done as described above for the ith column of $[S]$ and the ith row of $[S]$ and $\{F\}$. Similar adjustments have to be made for the kth column and row of the same matrices before replacing S_{jj} by 1 and the off-diagonal elements in the jth column and row by zero. This procedure causes external forces at node j to be replaced by static equivalents added to the external forces at nodes i and k.

The adjusted stiffness matrix will continue to be symmetrical, but the band width can increase. Equation 24-1 can be used to calculate the new band width, noting that any element connected to node i is also "connected" to node j by the constraint Eq. 25-17. For the plane frame $ABCD$ in Fig. 25-5a, with the number of degrees of freedom per node $s = 3$ and a constraint equation relating displacements at D and B, the band width is governed by member AB. Thus element AB is considered to be connected to nodes 1, 2 and 4 and the band width is (Eq. 24-1)

$$n_b = 3[(4 - 1) + 1] = 12$$

We should note that, in the adjusted equations, F_i is added to $\beta_i F_j$. For this reason, external forces at B in the actual frame in Fig. 25-5a, which are the same as at D, must be applied in the analysis at either node 2 or node 4, but not at both nodes. Similarly, in the structure in Fig. 25-5b, the forces at A, which are the same as at A', should be applied at either node 6 or at node 7; and so on.

25-5 CYCLIC SYMMETRY

Each of the structures shown in top view in Fig. 25-6 is composed of r sectors which are identical in terms of geometry, material properties and

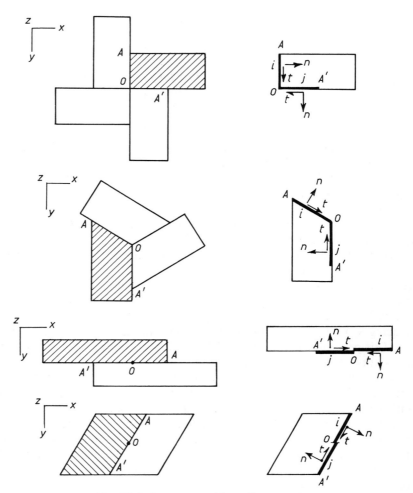

Fig. 25-6. Structures with cyclic symmetry.

loading.[2] The global structure can be generated by successive rotations of the hatched sector through an angle $2\pi/r$ about a vertical axis through O, where r is an integer. The structures can be two-dimensional or three-dimensional and they may be idealized by bar elements or other finite elements. The analysis

---

[2]Further discussion of symmetry can be found in Glockner, P. G., "Symmetry in Structural Mechanics," *Proceedings American Society of Civil Engineers*, 99 (ST1), 1973, pp. 71–89.

can be performed for the hatched sectors isolated as shown by imposing the following displacement constraints:

$$\{D_n, D_t, D_z\}_j = \{D_n, D_t, D_z\}_i \tag{25-19}$$

$$(D_n)_O = (D_t)_O = 0 \tag{25-20}$$

where i and j are nodes at equal distances from O on the common sector boundaries OA and OA' (indicated by heavy lines); D_n and D_t are displacement components (translations or rotations) in the directions of the normal n and the tangent t to the common boundaries; and D_z is a displacement component in a vertical downward direction. Equation 25-20 means that, at nodes on a vertical axis through O, the displacements can be nonzero only in the z direction.

The equilibrium equations $[S]\{D\} = \{-F\}$ have to be generated for all the degrees of freedom in the sector; the equations are then adjusted as discussed in Sec. 25-4.

The use of Eq. 25-19 requires that the degrees of freedom at the boundaries be in the n, t and z directions. The necessary transformation is discussed in Sec. 25-2.

Members or elements located on the common boundaries may be situated either on plane OA or on plane OA', but not on both. Similarly, forces at the nodes on the common boundaries must be applied on the nodes situated on one of the two vertical planes. At a node on the vertical through O the forces on the actual structure (in the z direction) must be divided by r when applied on the sector analyzed.

Example 25-2 Find the nodal displacements for the horizontal grid shown in top view in Fig. 25-7. The grid is subjected to a uniform downward load of q per unit length on DOF only. Assume that all members have the same cross section with $GJ/EI = 0.6$.

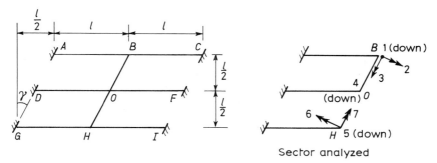

Fig. 25-7. Horizontal grid of Example 25-2.

The structure is composed of two identical sectors, one of which is shown in the figure. This sector will be analyzed using seven degrees of freedom: downward deflection and two rotations at each of the nodes B and H, and downward deflection only at O.The restraining forces are:

$$\{F\} = \{0, 0, 0, \ -0.5ql, 0, 0, 0\}$$

We should note that member OB is included in the sector analyzed but OH is omitted. For the cyclic symmetry to be valid, the two members must be identical, and OH could have been included in the sector analyzed instead of OB.

The stiffness matrix corresponding to the coordinate system shown in Fig. 25-7 is

$$[S] = \begin{bmatrix}
\dfrac{12+96c^3}{l^3} & & & & & & \text{symmetrical} \\[2ex]
\dfrac{6s+24c^2}{l^2} & \dfrac{4s^2+0.6c^2+8c}{l} & & & & & \\[2ex]
\dfrac{6c}{l^2} & \dfrac{4sc-0.6sc}{l} & \dfrac{1.2c+4c^2+0.60s^2}{l} & & & & \\[2ex]
-\dfrac{96c^3}{l^3} & -\dfrac{24c^2}{l^2} & 0 & \dfrac{12+96c^3}{l^3} & & & \\[2ex]
0 & 0 & 0 & 0 & \dfrac{12}{l^3} & & \\[2ex]
0 & 0 & 0 & 0 & -\dfrac{6s}{l^2} & \dfrac{0.6c^2+4s^2}{l} & \\[2ex]
0 & 0 & 0 & 0 & -\dfrac{6c}{l^2} & \dfrac{4cs-0.6cs}{l} & \dfrac{4c^2+0.6s^2}{l}
\end{bmatrix} EI$$

where $s = \sin\gamma = 1/\sqrt{5}$ and $c = \cos\gamma = 2/\sqrt{5}$.

The displacement constraint equations are

$$D_1 = D_5 \qquad D_2 = D_6 \qquad D_3 = D_7$$

Substitution for s and c by their values and adjustment of the equilibrium equations gives (Eq. 25-18)

$$EI \begin{bmatrix} \dfrac{92.692}{l^3} & & \text{symmetrical} & \\[2mm] \dfrac{19.2}{l^2} & \dfrac{9.715}{l} & & \\[2mm] 0 & \dfrac{2.720}{l} & \dfrac{7.713}{l} & \\[2mm] -\dfrac{68.692}{l^3} & -\dfrac{19.20}{l^2} & 0 & \dfrac{80.692}{l^3} \end{bmatrix} \begin{Bmatrix} D_1 \\[2mm] D_2 \\[2mm] D_3 \\[2mm] D_4 \end{Bmatrix} = ql \begin{Bmatrix} 0 \\[2mm] 0 \\[2mm] 0 \\[2mm] +0.5 \end{Bmatrix}$$

The last three equations, which become dummy, have been omitted here. The solution of the above equations yields

$$\{D\} = \{10.68l,\ 21.13,\ -7.45,\ 20.31l\}\ \frac{9l^3}{1000\,EI}$$

25-6 SUBSTRUCTURING

A large structure may be divided into substructures to make its analysis possible by a small computer. Such substrructuring breaks a large problem into smaller parts, thus replacing a long computer analysis by several shorter analyses. We shall explain the relevant procedure using the structure shown in top view in Fig. 25-8; this structure is partitioned into seven substructures, with six connection boundaries. The steps in the analysis are as follows:

Step 1 The equilibrium equations for a substructure are written in the form

$$\begin{bmatrix} S_{bb}^* & S_{bi}^* \\ \hline S_{ib}^* & S_{ii}^* \end{bmatrix} \begin{Bmatrix} D_b^* \\ \hline D_i^* \end{Bmatrix} = - \begin{Bmatrix} F_b^* \\ \hline F_i^* \end{Bmatrix} \tag{25-21}$$

In this equation, the displacement and force vectors are partitioned into connection-boundary coordinates and interior coordinates, referred to by subscripts b and i, respectively. Coordinates at real supports are treated as interior coordinates.

The displacements of the substructure are now prevented at the supported nodes and also at the connection-boundary nodes. With the connection-boundary nodes artificially restrained, $\{D_b^*\} = \{0\}$, the displacement at interior coordinates are given by

$$\{D_i^*\}_{\text{boundaries fixed}} = -[S_{ii}^*]^{-1}\{F_i^*\} \tag{25-22}$$

The restraining forces are

$$\{F_b^*\} = [S_{bi}^*][S_{ii}^*]^{-1}\{F_i^*\} \tag{25-23}$$

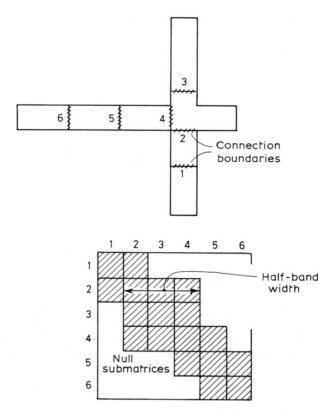

Fig. 25-8. Substructuring. (a) Top view of actual structure. (b) Band width of assembled stiffness matrix $[S_b]$.

Step 2 For each substructure, a condensed stiffness matrix is calculated (Eq. 4-17):

$$[\bar{S}_b^*] = [S_{bb}^*] - [S_{bi}^*][S_{ii}]^{-1}[S_{ib}^*] \tag{25-24}$$

A star is used here as a superscript to refer to a substructure.

Step 3 The condensed stiffness matrices are assembled into the stiffness matrix $[S_b]$ of the global structure. The connection boundary should be numbered in sequence so that the $[S_b]$ have small band width. An example of boundary numbering and of the resulting band width is shown in Figs. 25-8a and b.

Step 4 A vector is generated for the restraining forces at the boundaries:

$$F_{bj} = \sum F_{bj}^* + P_{bj} \tag{25-25}$$

where j refers to a coordinate, F_{bj}^* is determined in step 1 (Eq. 25-23), and P_{bj} is the external force at j. The summation is for the two substructures connected at the boundary considered.

Step 5   The displacements $\{D_b\}$ are determined by solving

$$[S_b]\{D_b\} = -\{F_b\} \tag{25-26}$$

Here, $\{D_b\}$ represents the displacements at the boundary nodes in the actual structure.

Step 6   Elimination of restraint produces new displacements at interior nodes. These displacements have to be added to the displacements determined by Eq. 25-22 so as to give the displacements in the actual structure:

$$\{D_i^*\} = \{D_i^*\}_{\text{boundaries fixed}} - [S_{ii}^*]^{-1}[S_{ib}^*]\{D_b^*\} \tag{25-27}$$

The last term in this equation is determined from the second row of Eq. 25-21 by setting $\{F_i^*\} = \{0\}$.

Step 7   By steps 5 and 6, the displacements at all nodes in the actual structure have been determined; these are used in the usual way to determine the internal forces or stresses in individual elements.

Substructuring requires more complicated coding and may not always be advantageous, particularly when the substructures involve no repetition.

25-7 PLASTIC ANALYSIS OF PLANE FRAMES

This section discusses the use of a computer for the plastic analysis of plane frames[3] on the basis of the assumptions made in Chapter 21, in particular the bilinear moment–curvature relationship of Fig. 21-3. The frames are assumed to have prismatic members with known cross-sectional properties a, I and M_p, where a and I are the area and the second moment of area, respectively, and M_p is the fully plastic moment (see Sec. 21-2). A set of forces is applied and their magnitudes are increased in stages, without changing their relative values, until a collapse mechanism is formed. A plastic hinge is assumed to occur when M_p is reached, ignoring the effects of axial and shear forces on the plastic moment capacity (see Secs. 21-8 and 21-9).

An elastic analysis is performed for each load increment and the member end-moments are recorded. When M_p is reached at a section, a hinge is inserted there and the structure stiffness matrix $[S]$ is changed accordingly. Collapse is reached when (a) $[S]$ becomes singular (determinant close to zero) or (b) a diagonal element S_{ii} becomes zero or (c) very large deflections are obtained.

[3]The method presented in this section and the example included are taken from Wang, C. K., "General Computer Program for Limit Analysis," *Proceedings American Society of Civil Engineers*, 89 (ST6) (December 1963), pp. 101–117.

For each load stage, the analysis is carried our for a unit increment in one of the loads ($P_i = 1$), with proportionate increments in the other loads. The corresponding member end-moments are used to determine a load multiplier which causes M_p to be reached at any one section, thus developing a new plastic hinge. The sum of the multipliers in all stages is the value of P_j at collapse.

For each loading stage, we generate a stiffness matrix for a structure with hinges located at the ends of members, as determined in earlier stages. The stiffness matrix for an individual member with a hinge at one or both ends is discussed below.

25-71 Stiffness Matrix of a Member with a Hinged End

The stiffness matrix of a prismatic member of a plane frame (Fig. 24-2) is given by Eq. 24-11, which ignores the effect of shear deformation and the beam-column effect (Chapter 15). After the development of a hinge at end 1 (left-hand end), the stiffness matrix of the member becomes

$$[S_{H1}^*] = \begin{bmatrix} Ea/l & & & & & \\ 0 & 3EI/l^3 & & \text{symmetrical} & & \\ 0 & 0 & 0 & & & \\ -Ea/l & 0 & 0 & Ea/l & & \\ 0 & -3EI/l^3 & 0 & 0 & 3EI/l^3 & \\ 0 & 3EI/l^2 & 0 & 0 & -3EI/l^2 & 3EI/l \end{bmatrix} \quad (25\text{-}28)$$

This matrix can be generated by condensation of $[S^*]$ in Eq. 24-11 (using Eq. 4-17) to obtain a 5 × 5 matrix corresponding to the coordinates in Fig. 24-2, with coordinate 3 omitted. The original size of the matrix and the coordinate numbering are maintained by the insertion of a column and a row of zeros, as shown in Eq. 25-28.

In a similar way, when a hinge develops at end 2 instead of end 1, the member stiffness will be

$$[S_{H2}^*] = \begin{bmatrix} Ea/l & & & & & \\ 0 & 3EI/l^3 & & \text{symmetrical} & & \\ 0 & 3EI/l^2 & 3EI/l & & & \\ -Ea/l & 0 & 0 & Ea/l & & \\ 0 & -3EI/l^3 & -3EI/l^2 & 0 & 3EI/l^3 & \\ 0 & 0 & 0 & 0 & 0 & 0 \end{bmatrix} \quad (25\text{-}29)$$

When hinges are developed at both ends, the member stiffness matrix will be

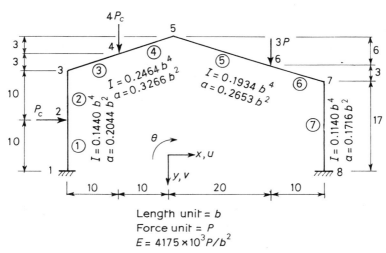

Length unit = b
Force unit = P
$E = 4175 \times 10^3 P/b^2$

Fig. 25-9. Plane frame of Example 25-3.

Table 25-1   Member End-Moments at the Termination of Stage 1 (in Terms of Pb)

Member	Node at member end	Plastic moment capacity M_p	Moment due to $P_c = 1$ $M_u^{(1)}$	Load multiplier	Number of the stage in which a hinge is formed	Moment at the end of stage 1 $M^{(1)}$
1	1	765.0	7.56			260.61
	2	765.0	0.56			19.36
2	2	765.0	−0.56			−19.36
	3	765.0	18.69			643.92
3	3	1275.0	−18.69			−643.92
	4	1275.0	−12.95			−446.23
4	4	1275.0	12.95			446.23
	5	1275.0	−4.59			−158.06
5	5	1015.0	4.59			158.06
	6	1015.0	−9.61			−331.18
6	6	1015.0	9.61			331.18
	7	1015.0	17.88			616.00
7	7	616.0	−17.88	34.458		−616.00
	8	616.0	−12.94			−445.75

the same as Eq. 25-28 or Eq. 25-29 with zero substituted for all the matrix elements which include I.

It should be pointed out that in this section we limit the discussion to frames subjected to concentrated loads applied at the nodes only. Thus plastic hinges can develop only at member ends.

Example 25-3   Determine the value of P_c to produce collapse for the load system shown in Fig. 25-9. For the seven members in the figure, the values of M_p, in terms of Pb, are $\{765, 765, 1275, 1275, 1015, 1015, 616\}$; the values of a and I are given in the same figure.

Load stage 1   With the value $P_c = 1$, the load system produces the end-moments $\{M_u^{(1)}\}$ listed in Table 25-1. From these values, it can be seen that a multiplier $P_c = 34.458$ will cause M_p to be reached so that the first plastic hinge is developed at node 7 in member 7. The corresponding nodal displacements, listed for nodes 1 to 8 in the order u, v, θ and given in terms of length unit b or radian, are

$$\{D^{(1)}\} = 34.458 \times 10^{-6} \mid \times$$
$$\{0, 0, 0 \mid -403.5, 43.44, -58.19 \mid -436.50, 86.92, 101.8 \mid$$
$$-27.80, 1524, -130.9 \mid 198.2, 2321, 41.94 \mid$$
$$399.4, 1531, -141.6 \mid 808.8, 78.11, -88.20 \mid 0, 0, 0\}$$

Table 25-2   Member End-Moments at Termination of Stage 2 (in Terms of Pb)

Member	Node at member end	$M_p - \lvert M^{(1)} \rvert$	Moment due to $P_c = 1$ $M_u^{(2)}$	Load multi-plier	Number of the stage in which a hinge is formed	$5.871\, M_u^{(2)}$	Moment at the end of stage 2 $M^{(2)}$
1	1	504.39	−1.42			−8.31	252.30
	2	745.64	6.02			35.34	54.71
2	2	745.64	−6.02			−35.34	−54.71
	3	121.08	20.62	5.871	2	121.08	756.00
3	3	631.08	−20.62			−121.08	−756.00
	4	828.77	−16.24			−95.38	−541.61
4	4	828.77	16.24			95.38	541.61
	5	1116.94	−13.11			−76.98	−235.04
5	5	856.94	13.11			76.98	235.04
	6	683.82	−24.37			−143.08	−474.26
6	6	683.82	24.37			143.08	474.26
	7	399.00	0.00			0.00	616.00
7	7	0	0		1	0	−616.00
	8	170.25	−24.83			−145.79	−591.54

Load stage 2   A new stiffness matrix is generated for the structure with a hinge introduced in member 7 at node 7. The member end-moments $\{M_u^{(2)}\}$, for $P_c = 1$, given in Table 25-2, show that a multiplier $P_c = 5.871$ will cause a second hinge to develop at node 3 in member 2. The member end-moments after application of this load increment are

$$\{M^{(2)}\} = \{M^{(1)}\} + 5.891\{M_u^{(2)}\}$$

The corresponding displacements are calculated in a way similar to that used for stage 1, giving

$$\{D^{(2)}\} = 5.871 \times 10^{-6} \times$$

$$\{0, 0, 0 \mid 245.2, 48.33, 61.79 \mid 1768, 96.67, 283.2 \mid$$

$$2726, 3358, 305.4 \mid 3400, 5643, 156.5 \mid$$

$$3744, 4406, -328.1 \mid 5024, 68.21, -485.6 \mid 0, 0, 0\}$$

We should note that, with a hinge existing at the connection of member 7 to node 7, the rotation at node 7, included in the above equation, represents the rotation at the node, and not at the top end of member 7.

Load stage 3   A second hinge is introduced and the pocess described in stage 2 is repeated, giving (Table 25-3)

$$\{M^{(3)}\} = \{M^{(2)}\} + 1.522\{M_u^{(3)}\}$$

Table 25-3   Member End-Moments at Termination of Stage 3 (in Terms of *Pb*)

Member	Node at member end	$M_p - \|M^{(2)}\|$	Moment due to $P_c = 1$ $M_u^{(3)}$	Load multi-plier	Number of the stage in which a hinge is formed	$1.522\,M_u^{(3)}$	Moment at the end of stage 3 $M^{(3)}$
1	1	512.70	8.91			13.56	265.86
	2	710.29	−9.45			−14.39	40.32
2	2	710.29	9.45			14.39	−40.32
	3	0	0		2	0	765.00
3	3	515.00	−0.00			−0.00	−765.00
	4	733.39	−34.60			−52.66	−594.27
4	4	733.39	34.60			52.66	594.27
	5	1039.96	−29.19			−44.44	−279.48
5	5	779.96	29.19			44.44	279.48
	6	540.74	−29.73			−45.25	−519.52
6	6	540.74	29.73			45.25	519.52
	7	399.00	0.00			−0.00	616.00
7	7	0	0		1	0	−616.00
	8	24.46	−16.07	1.522	3	−24.46	−616.00

The corresponding displacement after the third load stage is

$$\{D^{(3)}\} = 1.522 \times 10^{-6} \times$$
$$\{0, 0, 0 \mid -755.4, 43.86, -152.6 \mid -2805, 87.75, 709.3 \mid$$
$$-867.5, 6600, 533.8 \mid 288.9, 10275, 210.2 \mid$$
$$1228, 6880, -551.6 \mid 3251, 77.26, -743.9 \mid 0, 0, 0\}$$

Load stage 4   Introducing a third hinge at node 8 in member 7 and repeating the above procedure gives (Table 25-4)

$$\{M^{(4)}\} = \{M^{(3)}\} + 15.484\{M_u^{(4)}\}$$
$$\{D^{(4)}\} = 15.484 \times 10^{-6} \times$$
$$\{0, 0, 0 \mid 554.1, 44.53, 83.11 \mid 1385, 89.08, 812.6 \mid$$
$$3623, 7575, 619.8 \mid 4914, 11883, 245.0 \mid$$
$$6136, 7805, -634.7 \mid 8442, 75.92, -841.6 \mid 0, 0, 0\}$$

The fourth hinge changes the structure into a mechanism so that an additional loading stage would give extremely large displacements. The collapse load is

Table 25-4   Member End-Moments at Termination of Stage 4 (in Terms of *Pb*)

Member	Node at member end	$M_p - \lvert M^{(3)} \rvert$	Moment due to $P_c = 1$ $M_u^{(4)}$	Load multi-plier	Number of the stage in which a hinge is formed	$15.489 \, M_u^{(2)}$	Moment at the end of stage 4 $M^{(4)}$
1	1	499.14	−10.00			−154.84	111.01
	2	714.68	0.00			0	40.32
2	2	714.68	−0.00			−0.00	−40.32
	3	0	0		2	0	765.00
3	3	515.00	−0.00			−0.00	−765.00
	4	680.73	−38.00			−588.402	−1182.67
4	4	680.73	38.00			588.40	1182.67
	5	995.52	−36.00			−557.44	−836.92
5	5	735.52	36.00			557.44	836.92
	6	495.48	−32.00			−495.48	−1015.00
6	6	495.48	32.00	15.484	4	495.48	1015.00
	7	399.00	0.00			−0.00	616.00
7	7	0	0		1	0	−616.00
	8	0	0		3	0	−616.00

therefore the sum of the load increments in the four stages, namely

$$P_c = (34.458 + 5.871 + 1.522 + 15.484)P = 57.335P$$

25-8 STIFFNESS MATRIX OF MEMBER WITH VARIABLE SECTION OR WITH CURVED AXIS

A procedure for generating the stiffness matrix of a member of a framed structure of any type is presented below. The procedure can be used when the member has a variable section or a curved axis, and also in nonlinear analysis of reinforced concrete structures where cracking reduces the effective area[4] of the cross section.

Any of the stiffness matrices given in Chapter 24 for individual members of framed structures (Fig. 24-2) may be partitioned as follows:

$$[S^*] = \begin{bmatrix} [S_{11}^*] & [S_{12}^*] \\ [S_{21}^*] & [S_{22}^*] \end{bmatrix}$$
(25-30)

The submatrices in the first row contain forces at the first node of the member. Equilibrants of these forces at the second node form the elements in the submatrices in the second row. Because of this equilibrium relationship and the symmetry of $[S^*]$, Eq. 25-30 may be rewritten as

$$[S^*] = \begin{bmatrix} [S_{11}^*] & [S_{11}^*][R]^T \\ [R][S_{11}^*] & [R][S_{11}^*][R]^T \end{bmatrix}$$
(25-31)

where $[R]$ is a matrix generated by static equilibrium. As an example, considering a member of a plane frame (Fig. 24-2), we have

$$[R] = \begin{bmatrix} -1 & 0 & 0 \\ 0 & -1 & 0 \\ 0 & l & -1 \end{bmatrix}$$
(25-32)

Here, the elements in the first column of $[R]$ are values of forces at coordinates 4*, 5* and 6* in equilibrium with $F_1^* = 1$. Similarly, the second and third columns correspond to $F_2^* = 1$ and $F_3^* = 1$, respectively.

The submatrix $[S_{11}^*]$ can be determined by

$$[S_{11}^*] = [f]^{-1}$$
(25-33)

where $[f]$ is the flexibility matrix of the member when it is treated as a cantilever fixed at the second node. Any element f_{ij} of the flexibility matrix can be calculated by the unit-load theorem (Sec. 5-6):

$$f_{ij} = \int N_{ui}N_{uj}\frac{dl}{Ea} + \int M_{ui}M_{uj}\frac{dl}{EI} + \int V_{ui}V_{uj}\frac{dl}{Ga_r} + \int T_{ui}T_{uj}\frac{dl}{GJ}$$
(25-34)

[4]See the first reference mentioned in footnote 3 of Chapter 4.

where N, M, V and T are axial force, bending moment, shearing force and twisting moment, respectively; subscripts ui and uj refer to the effect of unit force $F_i = 1$ and $F_j = 1$, respectively, applied separately at coordinate i and j; E is the modulus of elasticity in tension or compression, and G is the modulus of elasticity in shear, respectively; a, I and J are the cross-sectional area, second moment of area and torsion constant; and a_r is the reduced area of the cross section (see Sec. 5-33).

When the cross-sectional properties vary arbitrarily, the integrals in this equation are evaluated numerically. Substitution in Eqs. 25-33 and 25-31 generates the stiffness matrix of the member.

Example 25-4   Generate the stiffness matrix for a horizontal grid member curved in the form of circular arc of radius r (Fig. 25-10a). Consider deformations due to bending and torsion only. The member has a contant cross section with $GJ/EI = 0.8$; $\theta = 1$ radian.

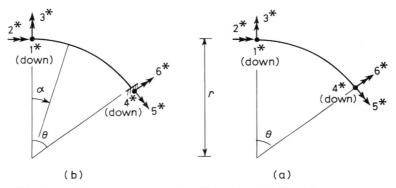

Fig. 25-10. Top view of a curved grid member. (a) Coordinate system.
(b) Member fixed at one end to generate the flexibility matrix.

A unit load applied at coordinate 1, 2 or 3 on the cantilever in Fig. 25-10b produces the following internal forces:

$$M_{u1} = -rs \qquad M_{u2} = s \qquad M_{u3} = c \qquad (25\text{-}35)$$
$$T_{u1} = r(1 - c) \qquad T_{u2} = c \qquad T_{u3} = -s \qquad (25\text{-}36)$$

where $s = \sin \alpha$ and $c = \cos \alpha$, with α defined in Fig. 25-10b. The elements of the flexibility matrix of the cantilever shown in Fig. 25-10b are given by (Eq. 25-34)

$$f_{ij} = \frac{r}{EI} \int_0^\theta M_{ui} M_{uj} \, d\alpha + \frac{r}{GJ} \int_0^\theta T_{ui} T_{uj} \, d\alpha \qquad (25\text{-}37)$$

The flexibility matrix can thus be considered as

$$[f] = [f_M] + [f_T] \qquad (25\text{-}38)$$

where $[f_M]$ and $[f_T]$ represent bending and torsion contributions to be calculated by the first and second term of Eq. 25-37, respectively.

Substitution of Eqs. 25-35 and 25-36 into Eq. 25-37 and evaluation of the integrals gives

$$[f_M] = \frac{r}{EI} \begin{bmatrix} 0.2727r^2 & \text{symmetrical} & \\ -0.2727r & 0.2727 & \\ -0.3540r & 0.3540 & 0.7273 \end{bmatrix}$$

$$[f_T] = \frac{r}{GJ} \begin{bmatrix} 0.0444r^2 & \text{symmetrical} & \\ 0.1141r & 0.7273 & \\ -0.1057r & -0.3540 & 0.2727 \end{bmatrix}$$

and

$$[f] = \frac{r}{EI} \begin{bmatrix} 0.3282r^2 & \text{symmetrical} & \\ -0.1294r & 1.1818 & \\ -0.4861r & -0.008\,85 & 1.0682 \end{bmatrix}$$

Considering the reactions at the fixed end of the cantilever shown in Fig. 25-10a when unit forces are applied separate at each of coordinates 1*, 2* and 3*, we write

$$[R] = \begin{bmatrix} -1 & 0 & 0 \\ -r(1-c) & -c & s \\ rs & -s & -c \end{bmatrix} \qquad (25\text{-}39)$$

Substitution in Eq. 25-31 gives

$$[S^*] = \frac{EI}{r} \left[\begin{array}{ccc|ccc} 12.1240/r^2 & & & & \text{symmetrical} & \\ 1.7515/r & 1.1045 & & & & \\ 5.6623/r & 0.8886 & 3.5865 & & & \\ \hline -12.124/r^2 & -1.7515/r & -5.6623/r & 12.1240/r^2 & & \\ -1.7549/r & -0.6542 & -0.0650 & 1.7549/r & 1.1055 & \\ 5.6691/r & 0.0643 & 2.0793 & -5.6691/r & -0.8911 & 3.5930 \end{array} \right]$$

25-9 GENERAL

The techniques presented in Chapters 24 and 25 can be used to analyze fairly large structures idealized as bars or other finite elements. There exist more sophisticated techniques to make optimum use of computer storage and to minimize computing time.

The problems given below are related to Chapters 24 and 25.

PROBLEMS

25-1   Change the space truss of Prob. 1-15 into a space frame with rigid joints. Assume that each member has a hollow rectangular cross section, arranged so that at least two opposite sides are in vertical planes. With this arrangement, local axis y^* (see figure here) for each member will be parallel to one of the global axes. Generate the transformation matrices $[t]$ for members AC and DG, which have local x^* axes along AC and DG. The global x, y, z axes are indicated in the figure for Prob. 1-15.

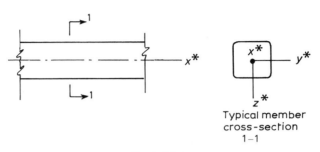

Typical member
cross-section
1-1

Prob. 25-1

25-2   The right-hand end of the plane frame member in Fig. 24-2 is subjected to prescribed displacements $\{D\} = \{u, v, \theta\}$ in global x, y, z directions, while the displacements are prevented at the left-hand end. The corresponding member end-forces, in local coordinates, can be expressed as $\{A_r\} = [G]\{D\}$. Generate matrix $[G]$.

25-3   Find the forces at the ends of member CD of the plane frame of Prob. 3-12 corresponding to a unit downward displacement of D while the displacements are prevented at C. Consider the end-forces in local directions as shown in Fig. 24-2, with the first three coordinates at C. Ignore deformations due to shear.

25-4   Use the answer of Prob. 25-2 to calculate the member end-forces $\{A_r\}$ of member 1 of the frame in Fig. 24-5 due to $u = 0.2b$ and $v = 0.5b$ at the bottom end while the displacements are prevented at the top end. The answers to this problem are given in the last row of Table 24-2 (see Example 24-2).

25-5   Use Eq. 24-41 to verify the reactions determined at node 4 in Example 24-5 for loading case 3, Fig. 24-4.

25-6   Use the nodal displacements determined in Example 24-5 to calculate the end-forces for member 4 in the truss of Fig. 24-4 due to a rise in temperature of the same member (loading case 2).

25-7   Analysis of the frame in Fig. 24-5 for the loads shown gives the following displacements: $\{D\} = 10^{-3}\{-68.14 \times 10^{-6}, 8.119 \times 10^{-6}, -0.1601 \times 10^{-6}, 39.83, 8.119, 0.7198, 38.25, 5.941, -0.4722, 27.28 \times 10^{-6}, 5.941 \times 10^{-6}, 0.6186\}$. The displacements are listed in the order u, v, θ for nodes 1, 2, 3 and 4. In obtaining the solution, the displacements at the supports were prescribed as indicated by Eq. 24-39. Find:
(a) The reactions at node 1 using Eq. 24-40 and check the results by Eq. 24-41.
(b) The forces at the ends of member 2.

25-8 Generate the stiffness matrix for a member of a plane frame corresponding to the coordinates in Fig. 24-2, assuming that the member has a rectangular cross section of width b and a depth which varies as shown in Prob. 10-12.

25-9 The stiffness matrix of a grid member shown in Fig. 24-2 is given by Eq. 24-12. Adjust the stiffness matrix so that it will correspond to coordinates 5* and 6* rotated through an angle γ in the clockwise direction, without a change in the remaining coordinates.

25-10 Derive the stiffness matrix for a member of a plane truss with the coordinates as shown. Verify that when $\alpha_1 = \alpha_2 = \alpha$, $[S]$ will be given by Eq. 24-23.

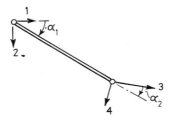

Prob. 25-10

25-11 The figure represents the top view of a typical panel of a horizontal pipeline composed of an infinite number of identical panels. Assuming that, in each panel, the pipe has supports at A, B, C and D, which can provide vertical reaction components only, find θ_x and θ_z at B. Obtain the bending-moment and twisting-moment diagrams for EBH due to uniform downward load of q per unit length, representing the self-weight of the pipe and its contents. Take $GJ/EI = 0.8$. (By symmetry, the unknown displacement components can be reduced to two.)

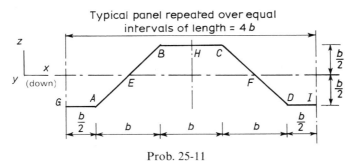

Prob. 25-11

25-12 Solve Prob. 25-11 for a uniform rise in temperature of T degrees. The coefficient of thermal expansion is α; the cross-sectional area of the pipe is $a = 100I/b^2$. Do not consider the self-weight of the pipe in this problem.

25-13 The figure is a top view of a horizontal beam EF subjected to a gravity load of q per unit length. AB and CD are rigid bars connecting E and F to supports at A, B, C and F which can provide vertical components only. Obtain the bending-moment and twisting-moment diagrams for EF, assuming $GJ/EI = 0.6$. Take advantage of symmetry to reduce the unknown displacement components to one. (This structure may be considered as a simplified idealization of a bridge girder over skew supports.)

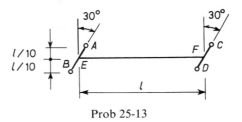

Prob 25-13

25-14   The figure is a top view of a horizontal beam EF subjected to a gravity load of q per unit length. AB, CD and GH are rigid bars connecting E, O and F to supports at A, B, C, D, G and H which can provide vertical components only. Obtain the bending-moment and twisting-moment diagrams for EF, assuming $GJ/EI = 0.6$. Also find the reactions at supports G and H. Take advantage of cyclic symmetry to reduce the unknown displacement components to one.

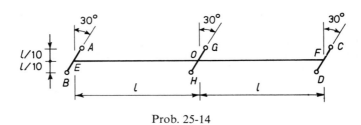

Prob. 25-14

25-15   The figure is the top view of a horizontal grid. Find the displacements at node B due to the self-weight of the members of q per unit length. All members have the same cross section, with $GJ = 0.6EI$. Take advantage of symmetry to reduce the unknown displacement components to three.

For all members, $GJ/EI = 0.6$ and EI = constant

Prob. 25-15

25-16   The grid of Prob. 25-15 is subjected to a downward force P at each of B and C. Replace the actual load by symmetrical and antisymmetrical components and perform analyses for a quarter structure to determine the three displacement components at B and F.

25-17   The figure is the top view of a horizontal grid with all members subjected to a uniform downward load of q per unit length. Find the displacement components at node E by analyzing part AEF, considering cyclic symmetry. Also, take advantage of symmetry about an appropriate plane to reduce the unknown displacement components to two. All members have same length l and the same cross section, with $GJ/EI = 0.5$.

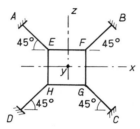

Prob. 27-17

25-18   The figure is the top view of a horizontal grid which has supports at A, B, C, D, E and F providing vertical reaction components only. Find the deflection at O and bending moments at member ends marked 1, 2 and 3 due to uniform load of q per unit length on all members. All members have the same length l and the same cross section. Considering cyclic symmetry, perform the analysis for FAO. In addition, take advantage of symmetry about an appropriate plane to reduce the number of unknown displacements to two.

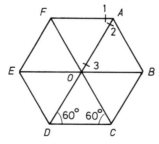

Prob. 25-18

Matrix algebra

A-1 INTRODUCTION

Matrix algebra is used in structural analysis for two reasons. First, it allows to represent a group of algebraic or numerical quantities by a single symbol; thus matrix notation may be regarded as a kind of shorthand writing. The second advantage is that the use of matrices permits a systematic compilation of the quantities required in the analysis so that a solution can be obtained by a series of matrix operations. This is particularly convenient for programming for digital computers, and standard computer programs are readily available for all the necessary matrix operations. We should not forget, though, that the organizational properties of matrices make their use advantageous also in the case of hand computation.

In this appendix only a few basic matrix operations, usually needed in structural analysis, are introduced.

A-2 DEFINITIONS

Consider the equations

$$
\left.\begin{aligned}
a_{11}x_1 + a_{12}x_2 + a_{13}x_3 = c_1 \\
a_{21}x_1 + a_{22}x_2 + a_{23}x_3 = c_2
\end{aligned}\right\}
\tag{A-1}
$$

The coefficients on the left-hand side of Eq. A-1 form a rectangular array or a matrix

$$
[a] = \begin{bmatrix} a_{11}\ a_{12}\ a_{13} \\ a_{21}\ a_{22}\ a_{23} \end{bmatrix}
$$

which has two rows and three columns. A matrix which has m rows and n columns is defined as matrix of order $m \times n$, and we can indicate the order by writing $[a]_{m \times n}$. A typical element a_{ij} is an element in the ith row and in the jth column.

If the matrix has one column only ($n = 1$), the matrix is called a *column matrix* or a *column vector*. If $m = 1$, that is the matrix has one row only, it is referred to as a *row matrix*.

A matrix with the same number of rows and columns is called a *square matrix*. Computation of determinants and inversion can be done for square matrices only, as discussed in Secs. A-8 and A-9.

Equation A-1 can be written in matrix notation as

$$\begin{bmatrix} a_{11} \, a_{12} \, a_{13} \\ a_{21} \, a_{22} \, a_{23} \end{bmatrix} \begin{Bmatrix} x_1 \\ x_2 \\ x_3 \end{Bmatrix} = \begin{Bmatrix} c_1 \\ c_2 \end{Bmatrix} \tag{A-2}$$

Braces are used for the column matrices to distinguish them from rectangular matrices, usually put in square brackets. There is no deep significance in this, and in some works square brackets are used for both types. To save space in printing, column matrices may be written horizontally, thus $\{x_1, x_2, x_3\}$.

We should distinguish the square brackets from two straight bars enclosing a determinant

$$\begin{vmatrix} a_{11} \, a_{12} \, a_{13} \\ a_{21} \, a_{22} \, a_{23} \\ a_{31} \, a_{32} \, a_{33} \end{vmatrix}$$

which is simply an arithmetical operator.

In shorthand matrix notation Eq. A-1 can be written as

$$[a]\{x\} = \{c\} \tag{A-3}$$

or

$$\mathbf{a}\,\mathbf{x} = \mathbf{c} \tag{A-4}$$

In this book, brackets are used to indicate matrices as in Eq. A-3.

Some further definitions may be useful. A *diagonal matrix* is a square matrix which has zero elements everywhere except on the *main diagonal*, that is

$$\begin{bmatrix} a_{11} & 0 & 0 \\ 0 & a_{22} & 0 \\ 0 & 0 & a_{33} \end{bmatrix}$$

In other words, in a diagonal matrix, $a_{ij} = 0$ when $i \neq j$ and not all elements a_{ii} are zero.

In order to save space, a diagonal matrix can be written thus:

$$\lceil a_{11}, a_{22}, ..., a_{nn} \rfloor$$

A *unit matrix* or *identity matrix* is a diagonal matrix in which each of the elements on the main diagonal equals one, that is $a_{ii} = 1$. Such a matrix is usually denoted by the symbol $[I]$, that is

$$[I] = \begin{bmatrix} 1 & 0 & 0 \\ 0 & 1 & 0 \\ 0 & 0 & 1 \end{bmatrix} \tag{A-5}$$

A *symmetrical matrix* is a square matrix in which the elements are symmetrical about the main diagonal; thus, $a_{ij} = a_{ji}$. Symmetrical matrices are often met with in structural analysis.

A matrix in which all elements are zero is called a *null matrix* or *zero matrix*.

A-3 MATRIX TRANSPOSITION

The *transpose of a matrix* $[a]$ is formed by interchanging the first row with the first column, the second row with the second column, and so on — that is, a_{ij} is interchanged with a_{ji}. The transpose of a matrix is indicated by a superscript T, thus

$$\begin{bmatrix} a_{11} \, a_{12} \, a_{13} \\ a_{21} \, a_{22} \, a_{23} \end{bmatrix}^T = \begin{bmatrix} a_{11} \, a_{21} \\ a_{12} \, a_{22} \\ a_{13} \, a_{23} \end{bmatrix} = [a]^T \qquad (A\text{-}6)$$

where $[a]$ represents the matrix being transposed.

We can see that if the order of $[a]$ is $m \times n$, then the order of $[a]^T$ is $n \times m$.

It is evident that if a matrix is symmetrical, its transpose is identical with the original matrix. Conversely, if a matrix $[a]$ and its transpose are equal, $[a]^T = [a]$, then $[a]$ is symmetrical.

A-4 MATRIX ADDITION AND SUBTRACTION

Addition and subtraction are possible only for matrices of the same order. We define the sum of matrix $[a]$ and matrix $[b]$ as a matrix $[c]$, in which any element

$$c_{ij} = a_{ij} + b_{ij} \qquad (A\text{-}7)$$

For example, if

$$[a] = \begin{bmatrix} 1 & 0 & 2 \\ 3 & 4 & 1 \end{bmatrix} \quad \text{and} \quad [b] = \begin{bmatrix} 2 & 3 & 1 \\ 5 & 2 & 0 \end{bmatrix}$$

then,

$$[c] = [a] + [b] = \begin{bmatrix} 3 & 3 & 3 \\ 8 & 6 & 1 \end{bmatrix}$$

Similarly, subtracting $[b]$ from $[a]$ results in a matrix $[d]$ in which

$$d_{ij} = a_{ij} - b_{ij}$$

For the above example,

$$[d] = [a] - [b] = \begin{bmatrix} -1 & -3 & 1 \\ -2 & 2 & 1 \end{bmatrix}$$

A-5 MATRIX MULTIPLICATION

Multiplication of a matrix $[a]$ by a constant α results in a matrix $[e]$, in which any element $e_{ij} = \alpha a_{ij}$.

Multiplication of two matrices $[a]$ and $[b]$ is possible only if the number of columns in $[a]$ is equal to the number of rows in $[b]$. The multiplication results in a matrix $[c]$ which has the same number of rows as $[a]$ and the same number of columns as $[b]$, that is

$$[a]_{m \times n} [b]_{n \times p} = [c]_{m \times p} \tag{A-8}$$

The product $[c]$ is defined as a matrix whose elements are given by

$$c_{ij} = \sum_{r=1}^{n} a_{ir} b_{rj} \tag{A-9}$$

with $i = 1, 2, ..., m$ and $j = 1, 2, ..., p$; that is, c_{ij} is the sum of the products of elements in the ith row of $[a]$ and of elements in jth column of $[b]$. It is evident that the product of a rectangular matrix by a column vector is a column vector.

The procedure of matrix multiplication can be illustrated by the following example. Given

$$[A] = \begin{bmatrix} 2 & 1 & 0 \\ 3 & -1 & 1 \end{bmatrix} \quad \text{and} \quad [B] = \begin{bmatrix} 1 & 0 \\ -1 & 0 \\ 2 & 2 \end{bmatrix}$$

we write

$$[A]_{2 \times 3} [B]_{3 \times 2} = [C]_{2 \times 2}$$

where

$$C_{11} = 2 \times 1 - 1 \times 1 + 2 \times 0 = 1$$
$$C_{12} = 2 \times 0 + 1 \times 0 + 0 \times 2 = 0$$
$$C_{21} = 3 \times 1 - 1 \times (-1) + 1 \times 2 = 6$$
$$C_{22} = 3 \times 0 - 1 \times 0 + 1 \times 2 = 2$$

Thus

$$[C] = \begin{bmatrix} 1 & 0 \\ 6 & 2 \end{bmatrix}$$

The product $[A]\,[B]$ is generally not equal to the product $[B]\,[A]$. For this reason we use the term $[A]$ *postmultiplied* by $[B]$ or $[B]$ *premultiplied* by $[A]$ to indicate the product $[A]\,[B]$. In the preceding example, the product $[B]\,[A]$ is a matrix of order 3×3.

Any number of matrices can be multiplied provided that they are of the proper order. For example,

$$[A]_{m \times n} [B]_{n \times p} [C]_{p \times r} = [D]_{m \times r} \tag{A-10}$$

To find $[D]$, we determine first the product $[A][B]$ and then multiply it by $[C]$. The same result will be obtained if $[A]$ is multiplied by the product $[B][C]$, that is

$$[A][B][C] = [AB][C] = [A][BC] \tag{A-11}$$

The distributive law also applies to matrices. For example,

$$[A]([B] + [C]) = [A][B] + [A][C] \tag{A-12}$$

The transpose of the product of two or more matrices is the product of the transposes of the matrix in a reversed order. For example, the transpose of both sides of Eq. A-10 is

$$[C]^T[B]^T[A]^T = [D]^T \tag{A-13}$$

The order of the matrices on the left-hand side of this equation is $r \times p$, $p \times n, n \times m$, and that of the product is $r \times m$.

Multiplication of a matrix by a unit matrix does not change the matrix. Thus

$$[A][I] = [I][A] = [A] \tag{A-14}$$

In structural analysis we often use a matrix product of the form $[A]^T[f][A]$, where $[A]$ is a rectangular matrix of the order $n \times m$, and $[f]$ is a square symmetrical matrix of the order $n \times n$. It can be easily proved following the rules discussed in this section that the product $[A]^T[f][A]$ results in a symmetrical matrix of the order $m \times m$.

A-6 PARTITIONED MATRICES

A matrix can be partitioned into smaller matrices by vertical and horizontal lines running between the rows or columns. The matrix is then called a *partitioned matrix*. This procedure is often used in structural analysis to refer to, or to operate on, a part of a matrix. The following is an example of a partitioned matrix:

$$[a] = \begin{bmatrix} a_{11} & a_{12} & a_{13} & a_{14} \\ a_{21} & a_{22} & a_{23} & a_{24} \\ a_{31} & a_{32} & a_{33} & a_{34} \end{bmatrix} = \begin{bmatrix} [A_{11}] & [A_{12}] \\ [A_{21}] & [A_{22}] \end{bmatrix} \tag{A-15}$$

where the *submatrices* are

$$\begin{aligned} [A_{11}] &= \begin{bmatrix} a_{11} & a_{12} & a_{13} \\ a_{21} & a_{22} & a_{23} \end{bmatrix} & [A_{12}] &= \begin{bmatrix} a_{14} \\ a_{24} \end{bmatrix} \\[2mm] [A_{21}] &= \begin{bmatrix} a_{31} & a_{32} & a_{33} \end{bmatrix} & [A_{22}] &= \begin{bmatrix} a_{34} \end{bmatrix} \end{aligned} \tag{A-16}$$

The submatrices can be treated as if they were ordinary matrix elements. For example, if $[a]$ and $[b]$ are partitioned matrices, such that

$$[a] = \begin{bmatrix} [A_{11}] & [A_{12}] \\ \hline [A_{21}] & [A_{22}] \end{bmatrix} \quad \text{and} \quad [b] = \begin{bmatrix} [B_1] \\ [B_2] \end{bmatrix}$$

then

$$[a][b] = \begin{bmatrix} [A_{11}][B_1] + [A_{12}][B_2] \\ [A_{21}][B_1] + [A_{22}][B_2] \end{bmatrix}$$

The transpose of a partitioned matrix is formed by interchanging rows of submatrices for the corresponding columns and transposing each submatrix. For example, the transpose of the matrix $[A]$ (Eq. A-15) is

$$[a]^T = \begin{bmatrix} [A_{11}]^T & [A_{21}]^T \\ [A_{12}]^T & [A_{22}]^T \end{bmatrix} \tag{A-17}$$

A-7 MATRIX INVERSION

The inverse of a square matrix $[S]$ is a matrix $[S]^{-1}$ of the same order as $[S]$ which satisfies the condition that

$$[S][S]^{-1} = [S]^{-1}[S] = [I] \tag{A-18}$$

where $[I]$ is a unit matrix.

Consider the equation

$$[S]_{n \times n} \{D\}_{n \times 1} = \{F\}_{n \times 1} \tag{A-19}$$

which represents in ordinary algebra a system of n simultaneous equations with n unknowns which are the elements of the column vector $\{D\}$. Premultiplying both sides of Eq. A-19 by $[S]^{-1}$, we obtain

$$[S]^{-1}[S]\{D\} = [S]^{-1}\{F\}$$

Hence, using Eq. A-18 and A-14,

$$\{D\} = [S]^{-1}\{F\} \tag{A-20}$$

The inverse of a matrix performs a function analogous to the reciprocal in ordinary algebra. This is illustrated by considering the algebraic equation $SD = F$. We can solve for D by dividing both sides by S or by multiplying by $(1/S)$, so that $D = \frac{1}{S}F$ the equation $S\frac{1}{S} = \frac{1}{S}S = 1$ is analogous to the matrix Eq. A-18.

Consider a matrix $[b]$ which is the inverse of a matrix $[a]$. Therefore,

$$[a][b] = [I] \tag{A-21}$$

or

$$
\begin{bmatrix} a_{11} & a_{12} & a_{13} \\ a_{21} & a_{22} & a_{23} \\ a_{31} & a_{32} & a_{33} \end{bmatrix}
\begin{bmatrix} b_{11} & b_{12} & b_{13} \\ b_{21} & b_{22} & b_{23} \\ b_{31} & b_{32} & b_{33} \end{bmatrix}
=
\begin{bmatrix} 1 & 0 & 0 \\ 0 & 1 & 0 \\ 0 & 0 & 1 \end{bmatrix}
$$

The first column of $[b]$ is the same as $\{x_1, x_2, x_3\}$ obtained by solving the equation

$$
[a]\{x\} = \begin{Bmatrix} 1 \\ 0 \\ 0 \end{Bmatrix} \tag{A-22}
$$

The solution of an equation similar to Eq. A-22 but with the right-hand side $\{0,1,0\}$, gives the second column of $[b]$. Similarly, when the right-hand side is $\{0,0,1\}$ the solution gives the third column of $[b]$. This then is a method of obtaining the inverse of a matrix; it involves the solution of a system of simultaneous equations several times. It is apparent, therefore, that the use of an inverse matrix in Eq. A-20 to solve for $\{D\}$ in Eq. A-19 is not the easiest way to solve a set of simultaneous equations. However, Eq. A-20 is often used when a solution of equations is required for several column vectors on the right-hand side of the equation. This occurs in the analysis of a structure for several loading cases.

The amount of computation required to invert a matrix increases rapidly with the order of the matrix, and it is therefore convenient sometimes to carry out the inversion by matrix operations on submatrices of the original large matrix.

Assume that we want to find a matrix $[b]$ which is the inverse of a given matrix $[a]$. Let $[a]$ and $[b]$ be partitioned into submatrices, such that each submatrix of $[a]$ is of the same order as the corresponding submatrix of $[b]$ and the diagonal submatrices are square. We can therefore write,

$$
\begin{bmatrix} [A_{11}] & [A_{12}] \\ [A_{21}] & [A_{22}] \end{bmatrix}
\begin{bmatrix} [B_{11}] & [B_{12}] \\ [B_{21}] & [B_{22}] \end{bmatrix}
=
\begin{bmatrix} [I] & [0] \\ [0] & [I] \end{bmatrix} \tag{A-23}
$$

Performing the multiplication gives

$$
\left.
\begin{aligned}
[A_{11}][B_{11}] + [A_{12}][B_{21}] &= [I] \\
[A_{21}][B_{11}] + [A_{22}][B_{21}] &= [0] \\
[A_{11}][B_{12}] + [A_{12}][B_{22}] &= [0] \\
[A_{21}][B_{12}] + [A_{22}][B_{22}] &= [I]
\end{aligned}
\right\} \tag{A-24}
$$

Premultiplying the second equation by $[A_{22}]^{-1}$, we obtain

$$
[B_{21}] = -[A_{22}]^{-1}[A_{21}][B_{11}] \tag{A-25}
$$

Substituting in the first equation gives

$$\left[[A_{11}] - [A_{12}] [A_{22}]^{-1} [A_{21}]\right] [B_{11}] = [I]$$

whence

$$[B_{11}] = \left[[A_{11}] - [A_{12}] [A_{22}]^{-1} [A_{21}]\right]^{-1} \qquad \text{(A-26)}$$

Similarly, we can solve for $[B_{22}]$ and $[B_{12}]$ from the last two equations of A-24. Thus, four matrices have to be inverted in order to obtain the inverse of a larger matrix $[A]$. These matrices are

$$[A_{22}], \quad \left[[A_{11}] - [A_{12}] [A_{22}]^{-1} [A_{21}]\right], \quad [A_{11}]$$

and

$$\left[[A_{22}] - [A_{21}] [A_{11}]^{-1} [A_{12}]\right]$$

We can further limit the inversion to the first two of the above matrices as follows.

From the definition of the inverse matrix (Eq. A-18), we can write (similarly to Eq. A-23)

$$\begin{bmatrix} [B_{11}] & [B_{12}] \\ [B_{21}] & [B_{22}] \end{bmatrix} \begin{bmatrix} [A_{11}] & [A_{12}] \\ [A_{21}] & [A_{22}] \end{bmatrix} = \begin{bmatrix} [I] & [0] \\ [0] & [I] \end{bmatrix} \qquad \text{(A-27)}$$

whence

$$[B_{11}] [A_{12}] + [B_{12}] [A_{22}] = [0]$$

and

$$[B_{21}] [A_{12}] + [B_{22}] [A_{22}] = [I]$$

$$\qquad \text{(A-28)}$$

Solving for $[B_{12}]$ and $[B_{22}]$,

$$[B_{12}] = - [B_{11}] [A_{12}] [A_{22}]^{-1} \qquad \text{(A-29)}$$

$$[B_{22}] = [A_{22}]^{-1} - [B_{21}] [A_{12}] [A_{22}]^{-1} \qquad \text{(A-30)}$$

Therefore, the four submatrices in $[b]$ can be determined from Eqs. A-25, A-26, A-29, and A-30.

A-8 DETERMINANTS

The determinant of a square matrix is denoted by $|S|$. This indicates specified arithmetical operations with the elements of the square matrix $[S]$, which result in a single number. A determinant $|S|$ is said to be of order n if $[S]$ is a square matrix of order $n \times n$.

A second-order determinant is defined as

$$|S| = \begin{vmatrix} S_{11} & S_{12} \\ S_{21} & S_{22} \end{vmatrix} = S_{11} S_{22} - S_{12} S_{21} \qquad \text{(A-31)}$$

The operations for a determinant of higher order will be defined later in this section.

The *minor* M_{ij} of element S_{ij} in determinant $|S|$ is defined as the determinant obtained by omission of the ith row and jth column of $|S|$. For example, in the third-order determinant

$$|S| = \begin{vmatrix} S_{11} & S_{12} & S_{13} \\ S_{21} & S_{22} & S_{23} \\ S_{31} & S_{32} & S_{33} \end{vmatrix}$$

the minor M_{32} is given by

$$M_{32} = \begin{vmatrix} S_{11} & S_{13} \\ S_{21} & S_{23} \end{vmatrix}$$

It is apparent that if $|S|$ is of order n, any minor is a determinant of order $n - 1$.

The *cofactor* Co_{ij} of element S_{ij} is obtained from the minor M_{ij} as follows:

$$Co_{ij} = (-1)^{i+j} M_{ij} \qquad \text{(A-32)}$$

Thus in the above example

$$Co_{32} = (-1)^{3+2} M_{32} = -M_{32}$$

A determinant of order n can be evaluated from the cofactors of any row i, thus:

$$|S| = \sum_{j=1}^{n} S_{ij} Co_{ij} \qquad \text{(A-33)}$$

Alternatively, the determinant can be evaluated from the cofactors of any column j, thus:

$$|S| = \sum_{i=1}^{n} S_{ij} Co_{ij} \qquad \text{(A-33a)}$$

Equations A-33 and A-33a are called *Laplace expansion equations*.

Properties of determinants. Some useful properties of determinants are listed below

1. If all elements in one row or one column are zero, the determinant is zero.
2. When any two rows or two columns are interchanged, the sign of the determinant is changed.
3. The determinant of a matrix is the same as the determinant of its transpose.
4. If the elements in a row (or column) are multiplied by a constant and the result added to the corresponding elements in another row (or column), the determinant is not changed.

5. If one row (or column) can be generated by linear combination of other row(s) (or columns), the determinant is zero. From this it follows that if two rows or columns are identical, the determinant is zero.

6. The sum of the product of the elements in one row i by the corresponding cofactors of another row m is zero. Thus for a matrix $[S]_{n \times n}$,

$$\sum_{j=1}^{n} S_{ij} Co_{mj} = 0 \qquad \text{(when } i \neq m) \qquad \text{(A-34)}$$

This is the general form of Laplace expansion; it can be compared with Eq. A-33, which gives the value of the determinant $|S|$ when $m = i$. Similarly, when the product is for the elements in a column j with the cofactors of another column m, we can write

$$\sum_{i=1}^{n} S_{ij} Co_{im} = 0 \qquad \text{(when } j \neq m) \qquad \text{(A-34a)}$$

The summation is equal to $|S|$ when $m = j$ (see Eq. A-33a).

This last property will be used to derive the inverse of a matrix.

Matrix inversion using cofactors: Consider a matrix $[S]_{n \times n}$ for which the inverse $[S]^{-1}$ is to be derived. We form a new matrix $[Co]$ consisting of cofactors Co_{ij} of the elements of $[S]$, and then form the product

$$[S]_{n \times n} [Co]_{n \times n}^{T} = [B]_{n \times n} \qquad \text{(A-35)}$$

From the definition of matrix multiplication (Eq. A-9), any element of $[B]$ is

$$B_{ij} = \sum_{r=1}^{n} S_{ir} Co_{jr}$$

From Eq. A-33 and A-34 we can see that the summation in the above equation is zero when $i \neq j$, and equals $|S|$ when $i = j$. Therefore $[B]$ is a diagonal matrix with all the diagonal elements equal to $|S|$. Thus we can write $[B] = |S| \, [I]$. Hence, by substituting in Eq. A-35 and dividing by $|S|$, we obtain

$$\frac{1}{|S|} [S] [Co]^{T} = [S] [S]^{-1} = [I]$$

whence

$$[S]^{-1} = \frac{[Co]^{T}}{|S|} \qquad \text{(A-36)}$$

From Eq. A-36 it is apparent that the division by $|S|$ is possible only when $|S| \neq 0$. It follows that only matrices which have a nonzero determinant have an inverse. When its determinant is zero, a matrix is said to be *singular* (and its inverse does not exist).

A-9　SOLUTION OF SIMULTANEOUS LINEAR EQUATIONS

Consider a system of n linear equations with n unknowns

$$[a]_{n \times n} \{x\}_{n \times 1} = \{c\}_{n \times 1} \tag{A-37}$$

If all the elements in $\{c\}$ are zero, the system is called *homogeneous*. This system occurs in *eigenvalue* or *characteristic-value* problems, such as stability and vibration problems in structural analysis. In this section, we shall deal only with *nonhomogeneous* equations for which not all the elements of $\{c\}$ are zero.

A unique solution of a system of nonhomogeneous equations exists when the determinant $|a|$ is nonzero. If $|a| = 0$, that is if matrix $[a]$ is singular, Eq. A-37 may have no solution or may have an infinite number of solutions. If the determinant $|a|$ is small compared with the cofactors of $[a]$, the system of equations is said to be *ill-conditioned*, in contrast to *well-conditioned* equations. The determinant $|a|$ of ill-conditioned equations is said to be *almost singular*.

For the purposes of this book, it is useful to mention two methods of solving a system of simultaneous equations. According to one of these, known as *Cramer's rule*, the solution of Eq. A-37 is

$$
\left.
\begin{aligned}
x_1 &= \frac{1}{|a|}
\begin{vmatrix}
c_1 & a_{12} & \dots & a_{1n} \\
c_2 & a_{22} & \dots & a_{2n} \\
\dots & \dots & \dots & \dots \\
c_n & a_{n2} & \dots & a_{nn}
\end{vmatrix} \\[2em]
x_2 &= \frac{1}{|a|}
\begin{vmatrix}
a_{11} & c_1 & \dots & a_{1n} \\
a_{21} & c_2 & \dots & a_{2n} \\
\dots & \dots & \dots & \dots \\
a_{n1} & c_n & \dots & a_{nn}
\end{vmatrix} \\[2em]
x_n &= \frac{1}{|a|}
\begin{vmatrix}
a_{11} & a_{12} & \dots & c_1 \\
a_{21} & a_{22} & \dots & c_2 \\
\dots & \dots & \dots & \dots \\
a_{n1} & a_{n2} & \dots & c_n
\end{vmatrix}
\end{aligned}
\right\} \tag{A-38}
$$

Each of the unknowns x_i is found by multiplying the reciprocal of the determinant $|a|$ by the determinant of a matrix which has the same elements as $|a|$ except that the ith column is replaced by the column vector $\{c\}$. The evaluation of the determinants involves a large number of operations, and Cramer's rule is perhaps the most suitable method when the number of equations is 2 or 3. When the number of equations is greater, other methods require fewer operations. However, Cramer's rule serves to explain some of the preceding statements related to the case when the determinant $|a|$ is small or zero.

As example of the use of Cramer's rule, let us consider the equations

$$\begin{bmatrix} S_{11} & S_{12} \\ S_{21} & S_{22} \end{bmatrix} \begin{Bmatrix} D_1 \\ D_2 \end{Bmatrix} = \begin{Bmatrix} F_1 \\ F_2 \end{Bmatrix} \tag{A-39}$$

Applying Eq. A-38,

$$\left. \begin{aligned} D_1 &= \frac{\begin{vmatrix} F_1 & S_{12} \\ F_2 & S_{22} \end{vmatrix}}{\begin{vmatrix} S_{11} & S_{12} \\ S_{21} & S_{22} \end{vmatrix}} = \frac{F_1 S_{22} - F_2 S_{12}}{S_{11} S_{22} - S_{12} S_{21}} \\[2em] \text{and} \qquad D_2 &= \frac{\begin{vmatrix} S_{11} & F_1 \\ S_{21} & F_2 \end{vmatrix}}{\begin{vmatrix} S_{11} & S_{12} \\ S_{21} & S_{22} \end{vmatrix}} = \frac{F_2 S_{11} - F_1 S_{21}}{S_{11} S_{22} - S_{12} S_{21}} \end{aligned} \right\} \tag{A-40}$$

The inverse of $[S]$ may be found as explained in Sec. A-7 by solving Eq. A-39 twice: once with $\{F_1, F_2\} = \{1,0\}$ and the other with $\{F_1 F_2\} = \{0,1\}$. Thus

$$\begin{bmatrix} S_{11} & S_{12} \\ S_{21} & S_{22} \end{bmatrix}^{-1} = \frac{1}{(S_{11} S_{22} - S_{12} S_{21})} \begin{bmatrix} S_{22} & -S_{12} \\ -S_{21} & S_{11} \end{bmatrix} \tag{A-41}$$

We can see then that the inverse of a matrix of order 2×2 is the product of the reciprocal of the determinant and of a matrix in which the two elements on the main diagonal of the original matrix are interchanged, and the other two elements change sign.

The method which is most commonly used to solve linear equations (and to invert matrices) is the method of elimination, introduced in elementary algebra. Several procedures of the elimination process have been developed,[1] and standard programs are now available for digital computers. A discussion of these procedures and of their advantages is beyond the scope of this book, and we shall limit ourselves here to one procedure suitable for the use of a small desk calculator.

In the procedure suggested by *P. D. Crout*,[2] the calculation is arranged in two tables. Let us consider it with reference to a system of four equations $[a]_{4 \times 4} \{x\}_{4 \times 1} = \{c\}_{4 \times 1}$. The tables are:

---

[1] See references on numerical procedures, for example, S. H. Crandall, *Engineering Analysis*, McGraw-Hill, New York, 1956.

[2] P. D. Crout, "A Short Method for Evaluating Determinants and Solving Systems of Linear Equations with Real or Complex Coefficients," *Trans. AIEE*, 60 (1941), pp. 1235–1240.

Given equations

a_{11}	a_{12}	a_{13}	a_{14}	c_1
a_{21}	a_{22}	a_{23}	a_{24}	c_2
a_{31}	a_{32}	a_{33}	a_{34}	c_3
a_{41}	a_{42}	a_{43}	a_{44}	c_4

Auxiliary quantities

b_{11}	b_{12}	b_{13}	b_{14}	d_1
b_{21}	b_{22}	b_{23}	b_{24}	d_2
b_{31}	b_{32}	b_{33}	b_{34}	d_3
b_{41}	b_{42}	b_{43}	b_{44}	d_4

Solution

x_1	x_2	x_3	x_4

The first table requires no computation, and is simply formed by the elements of $[a]$ and $\{c\}$.

The second table contains auxiliary quantities which are determined following a set pattern of computation. The elements in the first column of this table $\{b_{11}, b_{21}, b_{31}, b_{41}\}$ are the same as the column $\{a_{11}, a_{21}, a_{31}, a_{41}\}$, and are simply copied from the first table. The quantities in the first row are determined as follows:

$$[b_{12}\, b_{13}\, b_{14}\, d_1] = \frac{1}{b_{11}}[a_{12}\, a_{13}\, a_{14}\, c_1]$$

Then the second column (b_{22} to b_{42}) is completed from the following:

$$\begin{Bmatrix} b_{22} \\ b_{32} \\ b_{42} \end{Bmatrix} = \begin{Bmatrix} (a_{22} - b_{21}\, b_{12}) \\ (a_{32} - b_{31}\, b_{12}) \\ (a_{42} - b_{41}\, b_{12}) \end{Bmatrix}$$

Next, the second row (b_{23} to d_2) is completed from

$$[b_{23}\, b_{24}\, d_2] = \frac{1}{b_{22}}\left[(a_{23} - b_{21}\, b_{13}), (a_{24} - b_{21}b_{14}), (c_2 - b_{21}\, d_1)\right]$$

Proceeding diagonally downward, a column is completed starting from the diagonal element b_{ii}, followed by a row to the right of the diagonal. The elements required to complete the third column and the third row in the above example are calculated as follows:

$$\begin{Bmatrix} b_{33} \\ b_{43} \end{Bmatrix} = \begin{Bmatrix} (a_{33} - b_{31} b_{13} - b_{32} b_{23}) \\ (a_{43} - b_{41} b_{13} - b_{42} b_{23}) \end{Bmatrix}$$

$$[b_{34} \, d_3] = \frac{1}{b_{33}} [(a_{34} - b_{31} b_{14} - b_{32} b_{24}), (c_3 - b_{31} d_1 - b_{32} d_2)]$$

Finally, to complete the last row, the elements b_{44} and d_4 are determined:

$$b_{44} = a_{44} - b_{41} b_{14} - b_{42} b_{24} - b_{43} b_{34}$$

$$d_4 = \frac{1}{b_{44}} (c_4 - b_{41} d_1 - b_{42} d_2 - b_{43} d_3)$$

From the above we can see that the diagonal elements b_{11}, b_{22}, b_{33}, and b_{44} and the elements below the diagonal in Crout's second table are determined by one pattern of operations. The same pattern is followed for the remainder of the table, except that division by the diagonal element b_{ii} has also to be carried out.

All these operations can be summarized as follows. Any element b_{ij}, such that $j \leq i$, is calculated by

$$b_{ij} = a_{ij} - \sum_{r=1}^{j-1} b_{ir} b_{rj} \tag{A-42}$$

An element b_{ij}, such that $i < j$, is determined by

$$b_{ij} = \frac{1}{b_{ii}} \left[a_{ij} - \sum_{r=1}^{i-1} b_{ir} b_{rj} \right] \tag{A-43}$$

The elements $d_1, \ldots, d_4$ are determined by Eq. A-43, treating the columns $\{c\}$ and $\{d\}$ as if they were the last columns of matrices $[a]$ and $[b]$ respectively.

The above procedure is in fact a process of elimination in which the original equations are reduced to

$$\begin{bmatrix} 1 & b_{12} & b_{13} & b_{14} \\ 0 & 1 & b_{23} & b_{24} \\ 0 & 0 & 1 & b_{34} \\ 0 & 0 & 0 & 1 \end{bmatrix} \begin{Bmatrix} x_1 \\ x_2 \\ x_3 \\ x_4 \end{Bmatrix} = \begin{Bmatrix} d_1 \\ d_2 \\ d_3 \\ d_4 \end{Bmatrix} \tag{A-44}$$

The values of x_i can now be determined by a process of back substitution starting from the last equation, thus

$$x_4 = d_4$$
$$x_3 = d_3 - b_{34}\,x_4$$
$$x_2 = d_2 - b_{23}\,x_3 - b_{24}\,x_4$$
$$x_1 = d_1 - b_{12}\,x_2 - b_{13}\,x_3 - b_{14}\,x_4$$

$$(A\text{-}45)$$

In practice, Eqs. A-44 and A-45 need not be written down, and the answers are put directly in a row below the tables. From Eq. A-45 we can see that, in a general case, each of the unknowns x_j is determined by

$$x_j = d_j - \sum_{r=j+1}^{n} b_{jr}\,x_r \qquad (A\text{-}46)$$

where n is the number of the unknowns (or of equations).

If we want to invert $[a]$, the columns $\{c\}$ is replaced by n columns forming a unit matrix. These columns are treated in the same way as $\{c\}$ in the preceding discussion, and the column $\{d\}$ is replaced by a matrix $[d]_{n \times n}$. Back substitution in Eq. A-46 with the elements of d in each column of this matrix gives a column of the inverse matrix.

If matrix $[a]$ is symmetrical, which is usually the case in equations needed in structural analysis, the arithmetical operations are reduced considerably as each of the auxiliary quantities b_{ij} to the right of the diagonal is related to an element b_{ji} below the diagonal by

$$b_{ij} = \frac{b_{ji}}{b_{ii}} \quad i < j \qquad (A\text{-}47)$$

The following tables are an example of the application of Crout's procedure to the solution of four symmetrical equations.

Given equations

5	-4	1	0	1
-4	6	-4	1	-4
1	-4	6	-4	11
0	1	-4	7	-5

Auxiliary quantities

5	$-\frac{4}{5}$	$\frac{1}{5}$	0	$\frac{1}{5}$
-4	$\frac{14}{5}$	$-\frac{8}{7}$	$\frac{5}{14}$	$-\frac{8}{7}$
1	$-\frac{16}{5}$	$\frac{15}{7}$	$-\frac{4}{3}$	$\frac{10}{3}$
0	1	$-\frac{20}{7}$	$\frac{17}{6}$	2

Solution

3	5	6	2

A-10 EIGENVALUES

In the preceding section we mentioned the solution of the equation $[a]_{n \times n} \{x\}_{n \times 1} = \{0\}$. This of interest when $[a] = [b]_{n \times n} - \lambda [I]_{n \times n}$ and λ is unknown. In other words, we require the solution of

$$([b]_{n \times n} - \lambda [I]_{n \times n}) \{x\}_{n \times 1} = \{0\} \tag{A-48}$$

Equation A-48 can also be written

$$[b]_{n \times n} \{x\}_{n \times 1} = \lambda \{x\}_{n \times 1} \tag{A-49}$$

which means that premultiplying a column vector by a matrix results in a multiple of the column vector. Such a column vector is known as an *eigenvector* of the matrix and the multiplier is called an *eigenvalue* or characteristic value.

Equation A-48 represents a system of n homogeneous linear equations and its solution arising from $\{x\} = \{0\}$, is given by the determinant

$$|[b] - \lambda [I]| = 0 \tag{A-50}$$

For example, if we want to find the eigenvalues of the matrix

$$[b] = \begin{bmatrix} 3 & 7 & 9 \\ 7 & 11 & 7 \\ 9 & 7 & 9 \end{bmatrix}$$

we can write from Eq. A-50

$$\begin{vmatrix} 3 - \lambda & 7 & 9 \\ 7 & 11 - \lambda & 7 \\ 9 & 7 & 9 - \lambda \end{vmatrix} = 0$$

Expanding this determinant with the aid of Eq. A-33a,

$$|a| = \sum_{j=1}^{3} a_{ij} Co_{ij}$$

and choosing $i = 1$, we obtain

$$|a| = (3 - \lambda) [(11 - \lambda)(9 - \lambda) - 7 \times 7] - 7[7(9 - \lambda) - 7 \times 9]$$
$$+ 9[7 \times 7 - (11 - \lambda) \times 9] = 0$$

from which

$$\lambda^3 - 23\lambda^2 - 20\lambda + 300 = 0$$

Hence the three values of λ are 23.31, 3.44, and -3.75. As expected, the number of eigenvalues is equal to the order of the matrix. In many practical problems it is only the largest eigenvalue that is of interest.

In general, when $[b]$ is of order n, expansion of the determinant in Eq. A-50 gives a polynomial in λ of the form

$$\lambda^n + c_1 \lambda^{n-1} + c_2 \lambda^{n-2} + \cdots + c_n = 0 \qquad \text{(A-51)}$$

This equation is called the *characteristic equation* of the matrix $[b]$, and its solution gives the eigenvalues $\lambda_1, \lambda_2, \ldots, \lambda_n$ which satisfy the conditions

$$\sum_{i=1}^{n} \lambda_i = \sum_{i=1}^{n} b_{ii} \qquad \text{(A-52)}$$

and

$$\lambda_1 \times \lambda_2 \times \cdots \times \lambda_n = |b| \qquad \text{(A-53)}$$

The eigenvalues obtained in the above examples can be checked by these two equations.

The eigenvector corresponding to the λ-values can be found by substituting for λ in Eq. A-48 and solving. Since Eq. A-48 is homogeneous, only relative values of x can be determined for each eigenvalue. We can assume $x_1 = 1$, solve for the other values using $(n-1)$ equations and the last equation can be used to check the calculations.

Instead of evaluating the determinant, we may use an iteration method which leads to the eigenvalue with the *largest absolute value*. In this we assume the value of the eigenvector $\{x\}$ and substitute it in the left-hand side of Eq. A-49. Hence we find a value of λ and a new value of $\{x\}$. If this new value of $\{x\}$ is equal to the assumed value, then we have solved the problem: λ is the required eigenvalue, and $\{x\}$ the eigenvector. If however there is a discrepancy between the two values of $\{x\}$, we use the new value as an assumed value in the second cycle of evaluating $[b]\{x\}$. We repeat this procedure until Eq. A-49 is satisfied.

In many practical problems, a good estimate of $\{x\}$ corresponding to the largest λ can be made from the physical data available, and rapid convergence is obtained.

As an example, let us consider the matrix $[b]$ given earlier, and assume the vector $\{x^{(1)}\}$ as below. Then

$$
\begin{bmatrix} 3 & 7 & 9 \\ 7 & 11 & 7 \\ 9 & 7 & 9 \end{bmatrix}
\overset{\{x^{(1)}\}}{\begin{Bmatrix} 1.0 \\ 1.2 \\ 1.1 \end{Bmatrix}}
=
\begin{Bmatrix} 22.2 \\ 28.6 \\ 27.3 \end{Bmatrix}
= 22.2
\overset{\{x^{(2)}\}}{\begin{Bmatrix} 1.00 \\ 1.29 \\ 1.23 \end{Bmatrix}}
$$

Since $\{x^{(2)}\} \neq \{x^{(1)}\}$, we repeat the multiplication with $\{x^{(2)}\}$

$$
\begin{array}{c}
\{x^{(2)}\} \\
\downarrow \\
\begin{bmatrix} 3 & 7 & 9 \\ 7 & 11 & 7 \\ 9 & 7 & 9 \end{bmatrix}
\begin{Bmatrix} 1.00 \\ 1.29 \\ 1.23 \end{Bmatrix}
=
\begin{Bmatrix} 23.10 \\ 29.80 \\ 29.10 \end{Bmatrix}
= 23.10
\begin{Bmatrix} 1.00 \\ 1.29 \\ 1.26 \end{Bmatrix}
\begin{array}{c} \{x^{(3)}\} \\ \downarrow \\ \\ \\ \end{array}
\end{array}
$$

Since $\{x^{(2)}\} \cong \{x^{(3)}\}$, we have found the eigenvector and $\lambda \cong 23.10$ is the largest eigenvalue. A third repetition of the multiplication using $\{x^{(3)}\}$ gives $\lambda = 23.37$ which is very close to the exact solution.

PROBLEMS

A-1   Given the matrices

$$
[A] = \begin{bmatrix} 1 & 0 & 2 \\ 3 & 1 & 1 \end{bmatrix} \qquad [B] = \begin{bmatrix} 4 & 2 \\ 2 & 3 \end{bmatrix}
$$

execute the following matrix operations, if possible:

(a) $[A][B]$ (b) $[A]^T [B]$

(c) $[A]^T [B][A]$ (d) $[A][B]^T [A]^T$

(e) Prove that if $[B]$ is symmetrical the product $[A]^T [B][A]$ must also be symmetrical.

A-2   Given $[A]$ and $[B]$ as in Prob. A-1 and

$$
[C] = \begin{bmatrix} 3 & 2 & 0 \\ 1 & 0 & 2 \end{bmatrix}
$$

execute the following operations

(a) $[B][A + C]$ (b) $[B][A] + [B][C]$

Note that the distributive law $[B][A + C] = [B][A] + [B][C]$ applies to matrices.

A-3   Prove Eq. A-14.

A-4   Find the inverse of the following matrices.

(a) $\begin{bmatrix} 5 & -2 \\ -2 & 3 \end{bmatrix}$ (b) $\begin{bmatrix} 2 & -1 & 0 \\ -1 & 2 & -1 \\ 0 & -1 & 2 \end{bmatrix}$

(c) $\begin{bmatrix} 5 & -4 & 1 \\ -4 & 6 & -4 \\ 1 & -4 & 7 \end{bmatrix}$

A-5   The coordinates 1 and 2 in part (a) of the figure represent the positive directions of forces F_1 and F_2 and displacements D_1 and D_2 at the end A of cantilever AB. The displacements (deflection or rotation) due to unit values of the forces (vertical load or couple) are given in parts (b) and (c); these displacements are arranged in the flexibility matrix

$$[f] = \begin{bmatrix} f_{11} & f_{12} \\ f_{21} & f_{22} \end{bmatrix} = \begin{bmatrix} l/(EI) & -l^2/(2EI) \\ -l^2/(2EI) & l^3/(3EI) \end{bmatrix}$$

Assuming that the displacements $\{D_1, D_2\}$ due to forces $\{F_1, F_2\}$ can be obtained by the superposition equation

$$\begin{bmatrix} f_{11} & f_{12} \\ f_{21} & f_{22} \end{bmatrix} \begin{Bmatrix} F_1 \\ F_2 \end{Bmatrix} = \begin{Bmatrix} D_1 \\ D_2 \end{Bmatrix}$$

(a) Find the displacements due to $F_1 = \dfrac{Pl}{2}$ and $F_2 = P$ acting simultaneously.

(b) Find the forces S_{11} and S_{21} corresponding to the displacements $D_1 = 1$ and $D_2 = 0$. Sketch the deflected shape of the beam

(c) Find the forces S_{12} and S_{22} corresponding to the displacement $D_1 = 0$ and $D_2 = 1$. Sketch the deflected shape of the beam.

(d) Show that the stiffness matrix formed by the forces obtained in (b) and (c)

$$[S] = \begin{bmatrix} S_{11} & S_{12} \\ S_{21} & S_{22} \end{bmatrix}$$

is the inverse of the flexibility matrix $[f]$.

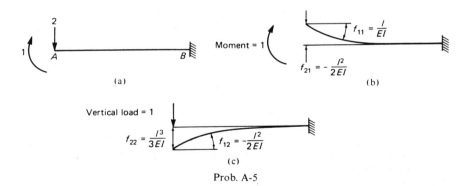

(a)   (b)

(c)

Prob. A-5

A-6   The stiffness matrix for the beam in the figure is

$$[S] = \frac{EI}{l^3} \begin{bmatrix} 9.6 & -8.4 \\ -8.4 & 9.6 \end{bmatrix}$$

The forces $\{F\} = \{F_1, F_2\}$ and the corresponding displacements $\{D\} = \{D_1, D_2\}$ along the coordinate 1 and 2 are related as follows $[S]\{D\} = \{F\}$.

(a) Sketch the deflected shape of the beam when the forces applied are $F_1 = S_{11}$ and $F_2 = S_{21}$.

(b) Find the inverse of the stiffness matrix $[S]$. What does the resulting matrix signify?

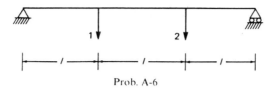

Prob. A-6

A-7   The stiffness matrix of the beam shown is part (a) of the figure

$$[S] = \frac{4EI_0}{l} \begin{bmatrix} 4 & 0 & 2 & \dfrac{12}{l} \\[2mm] 0 & 2 & 1 & -\dfrac{6}{l} \\[2mm] 2 & 1 & 6 & \dfrac{6}{l} \\[2mm] \dfrac{12}{l} & -\dfrac{6}{l} & \dfrac{6}{l} & \dfrac{72}{l^2} \end{bmatrix}$$ (a)

The forces P_1, P_2, P_3 and P_4 and the corresponding displacements D_1, D_2, D_3 and D_4 along the coordinates are related as follows:

$$\begin{bmatrix} [S_{11}] & [S_{12}] \\ [S_{21}] & [S_{22}] \end{bmatrix} \begin{Bmatrix} \{D_1\} \\ \{D_2\} \end{Bmatrix} = \begin{Bmatrix} \{F_1\} \\ \{F_2\} \end{Bmatrix}$$ (b)

where $[S_{11}]$, $[S_{12}]$ etc., refer to the partitioned matrices in $[S]$ and

$$\{D_1\} = \begin{Bmatrix} D_1 \\ D_2 \end{Bmatrix} \qquad \{D_2\} = \begin{Bmatrix} D_3 \\ D_4 \end{Bmatrix} \qquad \{F_1\} = \begin{Bmatrix} P_1 \\ P_2 \end{Bmatrix} \qquad \{F_2\} = \begin{Bmatrix} P_3 \\ P_4 \end{Bmatrix}$$

By putting $\{F_2\} = \{0\}$ in Eq. (b), find a matrix $[S^*]$ relating the forces to the displacements along the coordinates indicated in Prob. A-7(b), such that

$$[S^*] \{D_1\} = \{F_1\}$$ (c)

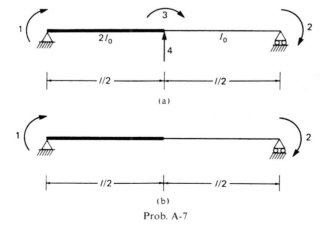

(a)

(b)

Prob. A-7

A-8   Solve the following simultaneous equations:

(a)

$$\begin{bmatrix} 8 & -2 \\ -3 & 11 \end{bmatrix} \qquad \begin{Bmatrix} x_1 \\ x_2 \end{Bmatrix} = \begin{Bmatrix} 18 \\ -17 \end{Bmatrix}$$

(b)

$$\begin{bmatrix} 4 & -2 & 1 \\ -3 & 8 & -2 \\ 2 & -5 & 10 \end{bmatrix} \begin{Bmatrix} x_1 \\ x_2 \\ x_3 \end{Bmatrix} = \begin{Bmatrix} 2 \\ 13 \\ 27 \end{Bmatrix}$$

(c)

$$\begin{bmatrix} 2 & 1 & 0 \\ 1 & 2 & 1 \\ 0 & 1 & 2 \end{bmatrix} \begin{Bmatrix} x_1 \\ x_2 \\ x_3 \end{Bmatrix} = \begin{Bmatrix} 1 \\ 1 \\ 1 \end{Bmatrix}$$

(d)

$$\frac{EI}{l} \begin{bmatrix} \dfrac{108}{l^2} & -\dfrac{6}{l} & -\dfrac{24}{l} \\ -\dfrac{6}{l} & 8 & 2 \\ -\dfrac{24}{l} & 2 & 12 \end{bmatrix} \begin{Bmatrix} D_1 \\ D_2 \\ D_3 \end{Bmatrix} = P \begin{Bmatrix} 0.5 \\ l \\ -\dfrac{l}{8} \end{Bmatrix}$$

(e)

$$\frac{l}{6EI} \begin{bmatrix} 4 & 1 & 0 \\ 1 & 4 & 1 \\ 0 & 1 & 4 \end{bmatrix} \begin{Bmatrix} F_1 \\ F_2 \\ F_3 \end{Bmatrix} = \begin{bmatrix} \dfrac{1}{l} \\ 0 \\ 0 \end{bmatrix}$$

A-9   Find the eigenvalues for the matrices in Prob. A-4 (a) and (c). Calculate the eigenvector corresponding to the highest eigenvalue.

A-10   Using iteration, evaluate the smallest eigenvalue of the matrix

$$[a] = \begin{bmatrix} 2 & -1 & 0.5 \\ -1 & 2 & 0 \\ -1 & -1 & 3.5 \end{bmatrix}$$

The eigenvalues of $[a]^{-1}$ are the reciprocals of those for $[a]$. To find by iteration the largest eigenvalue of $[a]^{-1}$, assume that the eigenvector represents the buckling mode of a strut hinged at one end and encastré at the other. The elements of the eigenvector are proportional to the deflections at equally-spaced points on the axis of the strut. (See Fig. 17-14.)

Displacements of prismatic members

The following table gives the displacements in beams of constant flexural rigidity EI and constant torsional rigidity GJ, subjected to the loading shown on each beam. The positive directions of the displacements are downward for translation, clockwise for rotation. The deformations due to shearing forces are neglected.

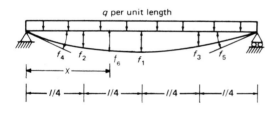

$$f_1 = \frac{5}{384}\frac{ql^4}{EI}$$

$$f_2 = f_3 = \frac{19}{2048}\frac{ql^4}{EI}$$

$$f_4 = -f_5 = \frac{ql^3}{24EI}$$

$$f_6 = \frac{qx}{24EI}(l^3 - 2lx^2 + x^3)$$

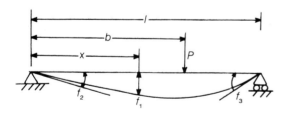

if force applied is upward or displ. upward then +ve.

$$f_1 = \frac{P(l-b)x}{6lEI}(2lb - b^2 - x^2) \qquad \text{when } x \leqslant b$$

$$f_1 = \frac{Pb(l-x)}{6lEI}(2lx - x^2 - b^2) \qquad \text{when } x \geqslant b$$

$$f_2 = \frac{Pb(l-b)}{6lEI}(2l-b) \qquad f_3 = -\frac{Pb}{6lEI}(l^2 - b^2)$$

When $b = l/2$, $f_2 = -f_3 = Pl^2/(16EI)$, and $f_1 = Pl^3/48EI$ at $x = l/2$.

797

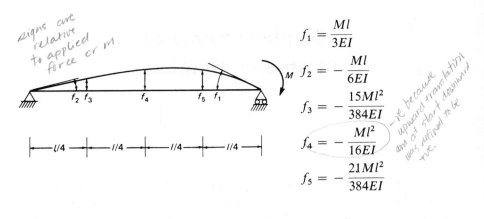

signs are relative to applied force or M.

$$f_1 = \frac{Ml}{3EI}$$

$$f_2 = -\frac{Ml}{6EI}$$

$$f_3 = -\frac{15Ml^2}{384EI}$$

$$f_4 = -\frac{Ml^2}{16EI}$$

$$f_5 = -\frac{21Ml^2}{384EI}$$

-ve because upward translation and at start downward was defined to be +ve.

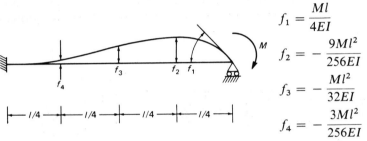

$$f_1 = \frac{Ml}{4EI}$$

$$f_2 = -\frac{9Ml^2}{256EI}$$

$$f_3 = -\frac{Ml^2}{32EI}$$

$$f_4 = -\frac{3Ml^2}{256EI}$$

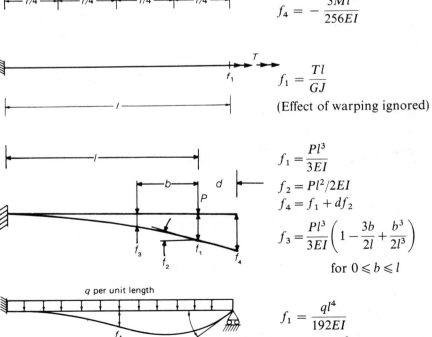

$$f_1 = \frac{Tl}{GJ}$$

(Effect of warping ignored)

$$f_1 = \frac{Pl^3}{3EI}$$

$$f_2 = Pl^2/2EI$$

$$f_4 = f_1 + df_2$$

$$f_3 = \frac{Pl^3}{3EI}\left(1 - \frac{3b}{2l} + \frac{b^3}{2l^3}\right)$$

for $0 \leqslant b \leqslant l$

q per unit length

$$f_1 = \frac{ql^4}{192EI}$$

$$f_2 = -\frac{ql^3}{48EI}$$

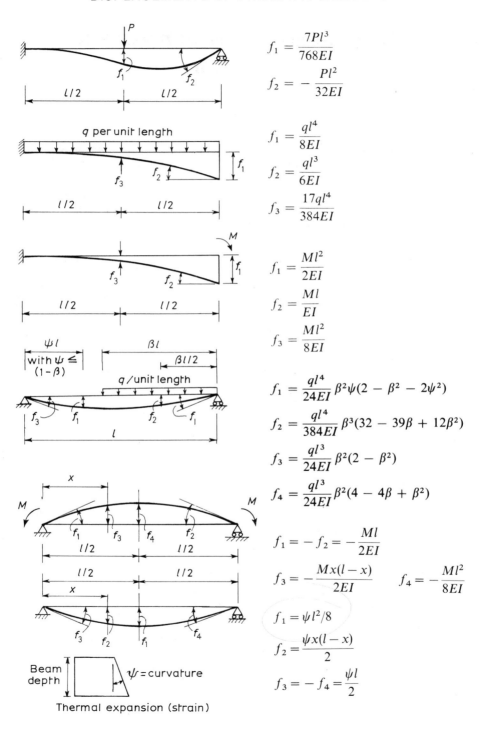

$$f_1 = \frac{7Pl^3}{768EI}$$

$$f_2 = -\frac{Pl^2}{32EI}$$

q per unit length

$$f_1 = \frac{ql^4}{8EI}$$

$$f_2 = \frac{ql^3}{6EI}$$

$$f_3 = \frac{17ql^4}{384EI}$$

$$f_1 = \frac{Ml^2}{2EI}$$

$$f_2 = \frac{Ml}{EI}$$

$$f_3 = \frac{Ml^2}{8EI}$$

ψl with $\psi \leq (1-\beta)$

βl / $\beta l/2$

q /unit length

$$f_1 = \frac{ql^4}{24EI}\, \beta^2\psi(2 - \beta^2 - 2\psi^2)$$

$$f_2 = \frac{ql^4}{384EI}\, \beta^3(32 - 39\beta + 12\beta^2)$$

$$f_3 = \frac{ql^3}{24EI}\, \beta^2(2 - \beta^2)$$

$$f_4 = \frac{ql^3}{24EI}\, \beta^2(4 - 4\beta + \beta^2)$$

$$f_1 = -f_2 = -\frac{Ml}{2EI}$$

$$f_3 = -\frac{Mx(l-x)}{2EI} \qquad f_4 = -\frac{Ml^2}{8EI}$$

$$f_1 = \psi l^2/8$$

$$f_2 = \frac{\psi x(l-x)}{2}$$

$$f_3 = -f_4 = \frac{\psi l}{2}$$

Beam depth ψ = curvature

Thermal expansion (strain)

Fixed-end forces of prismatic members

The following table gives the fixed-end forces in beams of constant flexural rigidity and constant torsional rigidity due to applied loads. The forces are considered positive if upwards or in the clockwise direction. A twisting couple is positive if it acts in the direction of rotation of a right-hand screw progressing to the right. When the end-forces are used in the displacement method, appropriate signs have to be assigned according to the chosen coordinate system.

Beam   Fixed-End Force

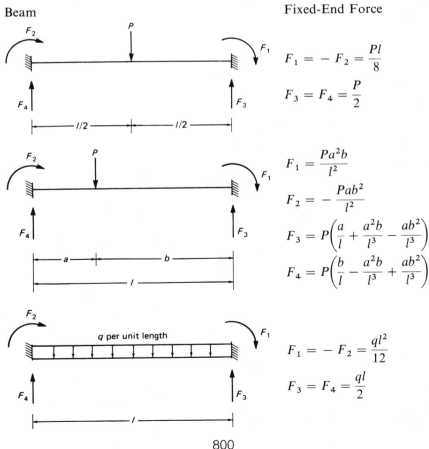

$$F_1 = -F_2 = \frac{Pl}{8}$$

$$F_3 = F_4 = \frac{P}{2}$$

$$F_1 = \frac{Pa^2b}{l^2}$$

$$F_2 = -\frac{Pab^2}{l^2}$$

$$F_3 = P\left(\frac{a}{l} + \frac{a^2b}{l^3} - \frac{ab^2}{l^3}\right)$$

$$F_4 = P\left(\frac{b}{l} - \frac{a^2b}{l^3} + \frac{ab^2}{l^3}\right)$$

$$F_1 = -F_2 = \frac{ql^2}{12}$$

$$F_3 = F_4 = \frac{ql}{2}$$

Fixed-End Force

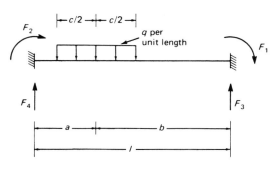

$$F_1 = \frac{qc}{12l^2} \left[12a^2b + c^2(l - 3a) \right]$$

$$F_2 = -\frac{qc}{12l^2} \left[12ab^2 + c^2(l - 3b) \right]$$

$$F_3 = \frac{qca}{l} + \frac{F_1 + F_2)}{l}$$

$$F_4 = \frac{qcb}{l} + \frac{F_2 + F_1)}{l}$$

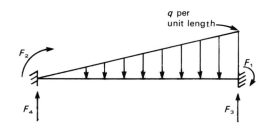

$$F_1 = \frac{Ma}{l} \left(2 - \frac{3a}{l} \right)$$

$$F_2 = \frac{Mb}{l} \left(2 - \frac{3b}{l} \right)$$

$$F_3 = -F_4 = \frac{6Mab}{l^3}$$

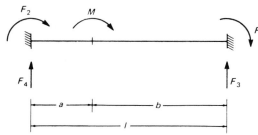

$$F_1 = \frac{ql^2}{20}$$

$$F_2 = -\frac{ql^2}{30}$$

$$F_3 = \frac{7}{20} ql$$

$$F_4 = \frac{3}{20} ql$$

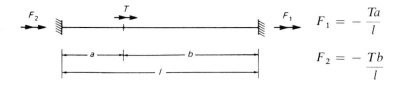

$$F_1 = -\frac{Ta}{l}$$

$$F_2 = -\frac{Tb}{l}$$

If the totally fixed support in any of the above cases, except the last, is changed to a hinge or a roller, the fixed-end moment at the other end can be calculated using the equations of this appendix and Eq. 11-23. Examples are as follows:

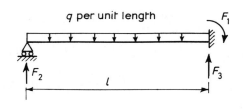

$$F_1 = \frac{ql^2}{8}$$

$$F_2 = \frac{3ql}{8}$$

$$F_3 = \frac{5ql}{8}$$

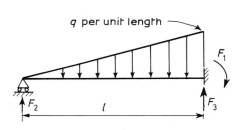

$$F_1 = \frac{Pab}{l^2}\left(a + \frac{b}{2}\right)$$

$$F_2 = P\left[\frac{b}{l} - \frac{ab}{l^3}\left(a + \frac{b}{2}\right)\right]$$

$$F_3 = P\left[\frac{a}{l} + \frac{ab}{l^3}\left(a + \frac{b}{2}\right)\right]$$

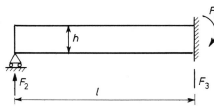

$$F_1 = \frac{ql^2}{15}$$

$$F_2 = \frac{ql}{10}$$

$$F_3 = \frac{2ql}{5}$$

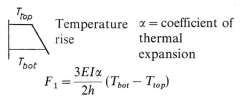

Temperature α = coefficient of
rise   thermal
              expansion

$$F_1 = \frac{3EI\alpha}{2h}(T_{bot} - T_{top})$$

For the case when both ends are
encastré, see Fig. 4-8g and Eq. 4-42

$$F_2 = -F_3 = -\frac{3EI}{2hl}(T_{bot} - T_{top})$$

End-forces caused by
end-displacements
of prismatic members

The following table gives the forces at the ends of beams due to a unit translation or unit rotation of one end. The positive directions for the forces are upward and clockwise. The effect of the deformation caused by the shearing forces is neglected; this topic is considered in Sec. 16-2. Moreover, the equations do not account for the bending moment due to axial forces; if a member is subjected to a large axial force, its effect may be included using Table 15-5 instead of this appendix. The beams have a constant flexural rigidity EI and a constant torsional rigidity GJ.

Beam   Force

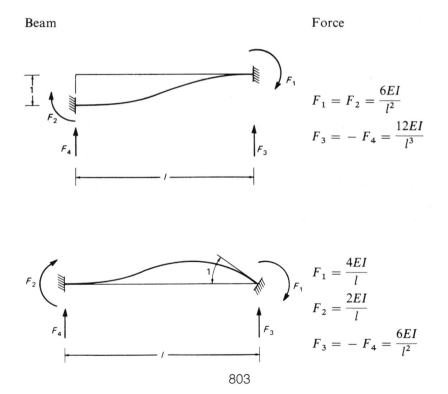

$$F_1 = F_2 = \frac{6EI}{l^2}$$

$$F_3 = -F_4 = \frac{12EI}{l^3}$$

$$F_1 = \frac{4EI}{l}$$

$$F_2 = \frac{2EI}{l}$$

$$F_3 = -F_4 = \frac{6EI}{l^2}$$

Beam   Force

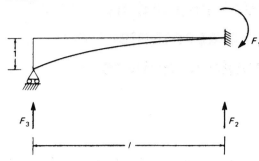

$$F_1 = \frac{3EI}{l^2}$$

$$F_2 = -F_3 = \frac{3EI}{l^3}$$

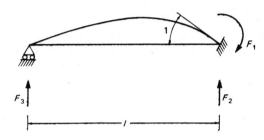

$$F_1 = \frac{3EI}{l}$$

$$F_2 = -F_3 = \frac{3EI}{l^2}$$

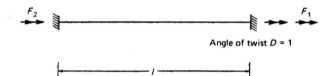

Angle of twist $D = 1$

$$F_1 = -F_2 = \frac{GJ}{l}$$

(Effect of warping ignored)

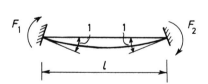

$$F_1 = -F_2 = \frac{2EI}{l}$$

APPENDIX E

Reactions and bending moments at supports of continuous beams due to unit displacement of supports

The following tables give the reactions and the bending moments at the supports of a continuous beam due to a unit downward translation of each of the supports separately. All spans are of equal length l and have a constant flexural rigidity EI. The number of spans is 2 (or 1) to 5. The end supports are hinged (Table E-1), fixed (Table E-2), and fixed at the left end and hinged at the right end (Table E-3). The bending moment at the hinged end is known to be zero and is not listed in the table.

The values given in each row are the bending moments or the reactions at consecutive supports starting from the left-hand end. The first row after the heading gives the effect of the settlement of the first support from the left, the second row gives the effect of the settlement of the second support from the left, and so on.

Figure E-1 shows an example of the use of the tables: the number of

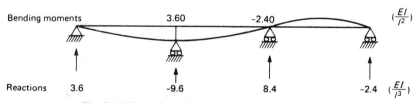

Fig. E-1. Illustration of the use of tables of Appendix E.

spans is 3, the second support from left settles a unit distance and the values are taken from the second row of the appropriate table.

In these tables, the reaction is considered positive when acting upward, and the bending moment is positive if it causes tension in the bottom fiber of the beam. When the reactions are used to generate a stiffness matrix, appropriate signs should be given according to the chosen coordinate system.

The effect of deformation caused by the shearing forces is neglected.

805

Table E-1.   Effect of a Unit Downward Displacement of One Support of Continuous Beams. End Supports Hinged. *EI* Constant. All Spans are of Equal Length *l*

Number of spans = 2
Support moments in terms of EI/l^2
   −1.50000
    3.00000
   −1.50000

Reactions in terms of EI/l^3

−1.50000	3.00000	−1.50000
3.00000	−6.00000	3.00000
−1.50000	3.00000	−1.50000

Number of spans = 3
Support moments in terms of EI/l^2

−1.60000	0.40000
3.60000	−2.40000
−2.40000	3.60000
0.40000	−1.60000

Reactions in terms of EI/l^3

−1.60000	3.60000	−2.40000	0.40000
3.60000	−9.60000	8.40000	−2.40000
−2.40000	8.40000	−9.60000	3.60000
0.40000	−2.40000	3.60000	−1.60000

matrix

Number of spans = 4
Support moments in terms of EI/l^2

−1.60714	0.42857	−0.10714
3.64286	−2.57143	0.64286
−2.57143	4.28571	−2.57143
0.64286	−2.57143	3.64286
−0.10714	0.42857	−1.60714

Reactions in terms of EI/l^3

−1.60714	3.64286	−2.57143	0.64286	−0.10714
3.64286	−9.85714	9.42857	−3.85714	0.64286
−2.57143	9.42857	−13.71428	9.42857	−2.57143
0.64286	−3.85714	9.42857	−9.85714	3.64286
−0.10714	0.64286	−2.57143	3.64286	−1.60714

Number of spans = 5
Support moments in terms of EI/l^2

−1.60765	0.43062	−0.11483	0.02871
3.64593	−2.58373	0.68900	−0.17225
−2.58373	4.33493	−2.75598	0.68900
0.68900	−2.75598	4.33493	−2.58373
−0.17225	0.68900	−2.58373	3.64593
0.02871	−0.11483	0.43062	−1.60765

Reactions in terms of EI/l^3

−1.60765	3.64593	−2.58373	0.68900	−0.17225	0.02871
3.64593	−9.87560	9.50239	−4.13397	1.03349	−0.17225
−2.58373	9.50239	−14.00956	10.53588	−4.13397	0.68900
0.68900	−4.13397	10.53588	−14.00957	9.50239	−2.58373
−0.17225	1.03349	−4.13397	9.50239	−9.87560	3.64593
0.02871	−0.17225	0.68900	−2.58373	3.64593	−1.60765

Table E-2. Effect of a Unit Downward Displacement of one Support of Continuous Beams. Two End Supports Fixed, EI Constant, all Spans of Equal Length l

Number of spans $= 1$
Support moments in terms of EI/l^2

6.00000	-6.00000
-6.00000	6.00000

Reactions in terms of EI/l^3

-12.00000	12.00000
12.00000	-12.00000

Number of spans $= 2$
Support moments in terms of EI/l^2

4.50000	-3.00000	1.50000
-6.00000	6.00000	-6.00000
1.50000	-3.00000	4.50000

Reactions in terms of EI/l^3

-7.50000	12.00000	-4.50000
12.00000	-23.99998	12.00000
-4.50000	12.00000	-7.50000

Number of spans $= 3$
Support moments in terms of EI/l^2

4.40000	-2.80000	0.80000	-0.40000
-5.60000	5.20000	-3.20000	1.60000
1.60000	-3.20000	5.20000	-5.60000
-0.40000	0.80000	-2.80000	4.40000

Reactions in terms of EI/l^3

-7.20000	10.80000	-4.80000	1.20000
10.80000	-19.20000	13.20000	-4.80000
-4.80000	13.20000	-19.20000	10.80000
1.20000	-4.80000	10.80000	-7.20000

Number of spans $= 4$
Support moments in terms of EI/l^2

4.39286	-2.78571	0.75000	-0.21429	0.10714
-5.57143	5.14286	-3.00000	0.85714	-0.42857
1.50000	-3.00000	4.50000	-3.00000	1.50000
-0.42857	0.85714	-3.00000	5.14286	-5.57143
0.10714	-0.21429	0.75000	-2.78571	4.39286

Reactions in terms of EI/l^3

-7.17857	10.71428	-4.50000	1.28571	-0.32143
10.71428	-18.85713	12.00000	-5.14285	1.28571
-4.50000	12.00000	-14.99999	12.00000	-4.50000
1.28571	-5.14285	12.00000	-18.85713	10.71428
-0.32143	1.28571	-4.50000	10.71428	-7.17857

Number of spans $= 5$
Support moments in terms of EI/l^2

4.39235	-2.78469	0.74641	-0.20096	0.05742	-0.02871
-5.56938	5.13875	-2.98564	0.80383	-0.22966	0.11483
1.49282	-2.98564	4.44976	-2.81340	0.80383	-0.40191
-0.40191	0.80383	-2.81339	4.44976	-2.98564	1.49282
0.11483	-0.22966	0.80383	-2.98564	5.13875	-5.56938
-0.02871	0.05742	-0.20096	0.74641	-2.78469	4.39235

Reactions in terms of EI/l^3

-7.17703	10.70813	-4.47847	1.20574	-0.34450	0.08612
10.70813	-18.83252	11.91387	-4.82296	1.37799	-0.34450
-4.47847	11.91387	-14.69856	10.88038	-4.82296	1.20574
1.20574	-4.82296	10.88038	-14.69856	11.91387	-4.47847
-0.34450	1.37799	-4.82296	11.91387	-18.83252	10.70813
0.08612	-0.34450	1.20574	-4.47847	10.70813	-7.17703

Table E-3.   Effect of a Unit Downward Displacement of One Support of Continuous Beams. Support at Left End Fixed, Hinged at Right End, *EI* Constant, all Spans of Equal Length *l*

Number of spans $= 1$
Support moments in terms of EI/l^2
   3.00000
  −3.00000

Reactions in terms of EI/l^3
  −3.00000   3.00000
   3.00000   −3.00000

Number of spans $= 2$
Support moments in terms of EI/l^2
   4.28571   −2.57143
  −5.14286   4.28571
   0.85714   −1.71428

Reactions in terms of EI/l^3
  −6.85714   9.42857   −2.57143
   9.42857   −13.71428   4.28571
  −2.57143   4.28571   −1.71428

Number of spans $= 3$
Support moments in terms of EI/l^2
   4.38462   −2.76923   0.69231
  −5.53846   5.07692   −2.76923
   1.38461   −2.76923   3.69231
  −0.23077   0.46154   −1.61538

Reactions in terms of EI/l^3
  −7.15385   10.61538   −4.15384   0.69231
  10.61539   −18.46153   10.61538   −2.76923
  −4.15384   10.61538   −10.15384   3.69231
   0.69231   −2.76923   3.69230   −1.61538

Number of spans $= 4$
Support moments in terms of EI/l^2
   4.39175   −2.78350   0.74227   −0.18557
  −5.56701   5.13402   −2.96907   0.74227
   1.48454   −2.96907   4.39175   −2.59794
  −0.37113   0.74227   −2.59794   3.64948
   0.06186   −0.12371   0.43299   −1.60825

Reactions in terms of EI/l^3
  −7.17526   10.70103   −4.45361   1.11340   −0.18557
  10.70103   −18.80411   11.81443   −4.45361   0.74227
  −4.45361   11.81443   −14.35051   9.58763   −2.59794
   1.11340   −4.45361   9.58763   −9.89690   3.64948
  −0.18557   0.74227   −2.59794   3.64948   −1.60825

Number of spans $= 5$
Support moments in terms of EI/l^2
   4.39227   −2.78453   0.74586   −0.19889   0.04972
  −5.56906   5.13812   −2.98342   0.79558   −0.19889
   1.49171   −2.98342   4.44199   −2.78453   0.69613
  −0.39779   0.79558   −2.78453   4.34254   −2.58563
   0.09945   −0.19889   0.69613   −2.58563   3.64641
  −0.01657   0.03315   −0.11602   0.43094   −1.60773

Reactions in terms of EI/l^3
  −7.17680   10.70718   −4.47513   1.19337   −0.29834   0.04972
  10.70718   −18.82872   11.90055   −4.77348   1.19337   −0.19889
  −4.47513   11.90055   −14.65193   10.70718   −4.17679   0.69613
   1.19337   −4.77348   10.70718   −14.05524   9.51381   −2.58563
  −0.29834   1.19337   −4.17679   9.51381   −9.87845   3.64641
   0.04972   −0.19889   0.69613   −2.58563   3.64641   −1.60773

Properties of geometrical figures

	Area	x and y coordinates of centroid

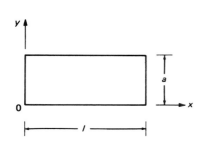

$a \times l$

$$\bar{x} = \frac{l}{2}$$

$$\bar{y} = \frac{a}{2}$$

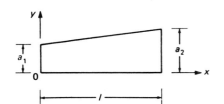

$$\frac{(a_1 + a_2)}{2} \times l$$

$$\bar{x} = \frac{l\,(a_1 + 2a_2)}{3\,(a_1 + a_2)}$$

$$\bar{y} = \frac{(a_1^2 + a_1 a_2 + a_2^2)}{3(a_1 + a_2)}$$

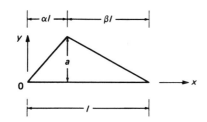

$$\frac{a \times l}{2}$$

$$\bar{x} = \frac{1}{3}\,(\alpha l + l)$$

$$\bar{y} = \frac{a}{3}$$

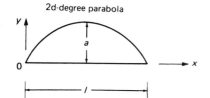

2d-degree parabola

$$\frac{2}{3}a \times l$$

$$\bar{x} = \frac{l}{2}$$

$$\bar{y} = \frac{2}{5}a$$

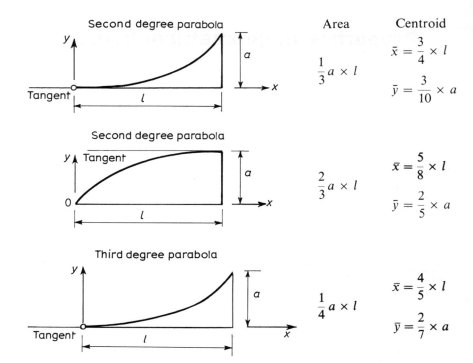

	Area	Centroid
Second degree parabola	$\frac{1}{3} a \times l$	$\bar{x} = \frac{3}{4} \times l$ $\bar{y} = \frac{3}{10} \times a$
Second degree parabola	$\frac{2}{3} a \times l$	$\bar{x} = \frac{5}{8} \times l$ $\bar{y} = \frac{2}{5} \times a$
Third degree parabola	$\frac{1}{4} a \times l$	$\bar{x} = \frac{4}{5} \times l$ $\bar{y} = \frac{2}{7} \times a$

Construction of tangents of a second-degree parabola:

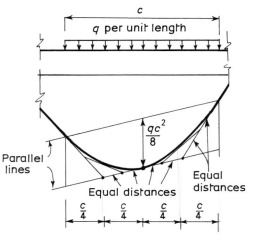

Bending-moment diagram for a part of a member subjected to a uniform load.

Torsional Constants *J*

If a circular bar of constant section and of length l is subjected to a constant torque T, the angle of twist between the two bar ends is

$$\theta = \frac{Tl}{GJ}$$

where G is the shear modulus and J the polar moment of inertia.

When the cross section of the bar is noncircular, plane cross sections do not remain plane after deformation and warping will occur caused by longitudinal displacements of points in the cross section. Nevertheless, the above equation can be used with good accuracy for noncircular cross sections, but J should be taken as the appropriate torsion constant. The torsion constants for several shapes of cross sections are listed below.

Section	Torsional Constant *J*
	$J = \dfrac{\pi r^4}{2}$
	$J = 0.1406b^4$
	$J = \dfrac{\pi(r_1^4 - r_2^4)}{2}$

Section	Torsional Constant J

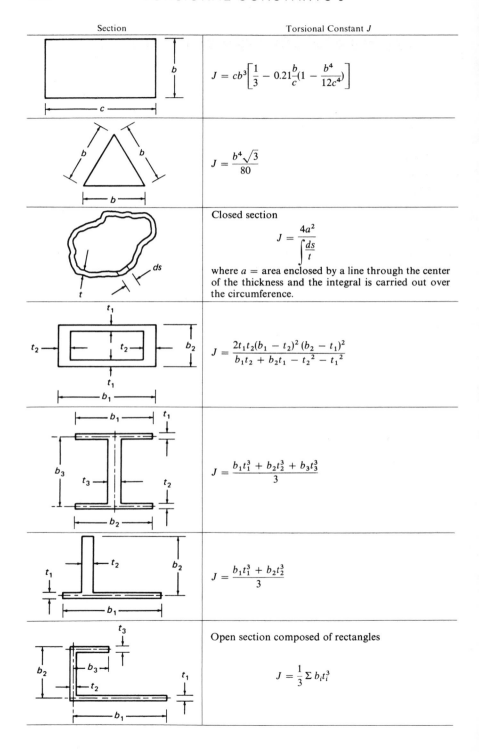

$$J = cb^3 \left[\frac{1}{3} - 0.21\frac{b}{c}(1 - \frac{b^4}{12c^4}) \right]$$

$$J = \frac{b^4\sqrt{3}}{80}$$

Closed section

$$J = \frac{4a^2}{\displaystyle\int \frac{ds}{t}}$$

where a = area enclosed by a line through the center of the thickness and the integral is carried out over the circumference.

$$J = \frac{2t_1t_2(b_1 - t_2)^2 (b_2 - t_1)^2}{b_1t_2 + b_2t_1 - t_2^2 - t_1^2}$$

$$J = \frac{b_1t_1^3 + b_2t_2^3 + b_3t_3^3}{3}$$

$$J = \frac{b_1t_1^3 + b_2t_2^3}{3}$$

Open section composed of rectangles

$$J = \frac{1}{3} \Sigma b_i t_i^3$$

Values of the integral $\int M_u M\, dl$

The following table gives the values of the integral $\int^l M_u M\, dl$, needed in the calculation of displacement of framed structures by virtual work (Eq. 7-2). The same tables can be used for the evaluation of the integrals $\int^l N_u N\, dl$, $\int^l V_u V\, dl$, $\int^l T_u T\, dl$, or for the integral over a length l of any two functions which vary in the manner indicated in the diagrams at the top and at the left-hand edge of the table.

M \ M_u	Rectangle, a	Trapezoid, a_1, a_2	Triangle, αl | βl, a	Parabola, a *	Parabola with tangent, a *	Parabola with tangent, a *
Triangle, αl | βl, b	$\tfrac{1}{2}abl$	$\dfrac{bl}{6}\left[(1+\beta)a_1 + (1+\alpha)a_2\right]$	$\dfrac{abl}{3}(1+\alpha\beta)$	$\dfrac{1}{3}abl$	$\dfrac{abl}{12}(1+\alpha+\alpha^2)$	$\dfrac{abl}{12}(5-\beta-\beta^2)$
Trapezoid, b_1, b_2	$\dfrac{al}{2}(b_1+b_2)$	$\dfrac{l}{6}\left(2a_1b_1 + a_1b_2 + a_2b_1 + 2a_2b_2\right)$	$\dfrac{al}{6}\left[(1+\beta)b_1 + (1+\alpha)b_2\right]$	$\dfrac{1}{3}al(b_1+b_2)$	$\dfrac{al}{12}(b_1+3b_2)$	$\dfrac{al}{12}(3b_1+5b_2)$
Triangle, b	$\tfrac{1}{2}abl$	$\dfrac{bl}{6}(2a_1+a_2)$	$\dfrac{abl}{6}(1+\beta)$	$\dfrac{1}{3}abl$	$\dfrac{1}{12}abl$	$\dfrac{1}{4}abl$
Triangle, b	$\tfrac{1}{2}abl$	$\dfrac{bl}{6}(a_1+2a_2)$	$\dfrac{abl}{6}(1+\alpha)$	$\dfrac{1}{3}abl$	$\dfrac{1}{4}abl$	$\dfrac{5}{12}abl$
Rectangle, b	abl	$\dfrac{bl}{2}(a_1+a_2)$	$\dfrac{1}{2}abl$	$\dfrac{2}{3}abl$	$\dfrac{1}{3}abl$	$\dfrac{2}{3}abl$

*Second-degree parabola

Deflection of a simple beam of constant *EI* subjected to unit end-moments

The values of the deflection given below are helpful in the calculation of ordinates of influence lines (see Sec. 13-5). The deflection y due to end-moments $M_{AB} = 1$ and $M_{BA} = 0$ (Fig. I-1a) is given from Eq. 13-5 by

$$y = \frac{l^2}{6EI}(2\varepsilon - 3\varepsilon^2 + \varepsilon^3) \tag{I-1}$$

where $\varepsilon = x/l$, x is the distance from the left-hand end, and l is the member

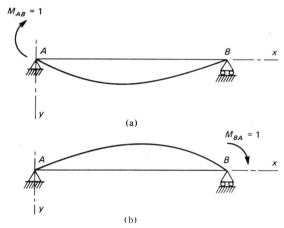

Fig. 1-1. Deflection of a simple beam due to unit end-moments.

length. The values of the deflections at different values of x/l are given in Table I-1.

Table 1-1. Deflection Due to Unit Clockwise Moment at Left-Hand End

$\varepsilon = \dfrac{x}{l}$	0	0.1	0.2	0.3	0.4	0.5	0.6	0.7	0.8	0.9	1.0	Multiplier
y	0	285	480	595	640	625	560	455	320	165	0	$10^{-4}\dfrac{l^2}{EI}$

The deflection y due to $M_{BA} = 1$ and $M_{AB} = 0$ (Fig. I-1b) is given by

$$y = -\frac{l^2}{6EI}(\varepsilon - \varepsilon^3) \tag{I-2}$$

The values of the deflection at different values of $\varepsilon = x/l$ are given in Table I-2.

Table I-2. Deflections Due to Unit Clockwise Moment at Right-Hand End

$\varepsilon = \dfrac{x}{l}$	0	0.1	0.2	0.3	0.4	0.5	0.6	0.7	0.8	0.9	1.0	*Multiplier*
y	0	-165	-320	-455	-560	-625	-640	-595	-480	-285	0	$10^{-4}\dfrac{l^2}{EI}$

Geometrical properties of some plane areas commonly used in the method of column analogy

Property	Figure Representing Analogous Column
Area $= a = \dfrac{l}{EI}$ $I_x = \dfrac{ac^2}{12}$, $I_y = \dfrac{ab^2}{12}$ $I_{xy} = \dfrac{abc}{12}$, positive if $0 < \theta < \dfrac{\pi}{2}$ and negative when $\dfrac{\pi}{2} < \theta < \pi$ $I_{x1} = \dfrac{ac^2}{3}$	 Straight member of constant EI
θ is measured in radians Area $= a = \dfrac{2\theta r}{EI}$ $\bar{y} = \dfrac{r \sin \theta}{\theta}$ $I_{x1} = \dfrac{r^3(\theta + \sin \theta \cos \theta)}{EI}$ $I_y = \dfrac{r^3(\theta - \sin \theta \cos \theta)}{EI}$ $r = \dfrac{(4b^2 + l^2)}{8b}$ $y = -b + r - \sqrt{r^2 - x^2}$	 Circular arch of constant EI
Area $= a = \dfrac{l}{EI_0}$ $I_x = \dfrac{4b^2 l}{45 EI_0}$ $I_y' = \dfrac{l^3}{12 EI_0}$	 Parabolic arch with EI assumed to vary as the secant of the inclination of the arch axis, EI_o is the flexural rigidity at the crown

Forces due to prestressing

The figures below show sets of forces in equilibrium representing the effect of prestressed tendons of various profiles on straight members. The profiles are composed of one or more straight lines or of second-degree parabolas. The symbol P represents the absolute value of the prestressing force which is assumed constant, and y represents the vertical distance between the tendon and the centroidal axis of the member. θ represents the angle between a tangent to the tendon and the horizontal; θ is assumed small so that $\theta \simeq \tan \theta \simeq \sin \theta$ and $\cos \theta \simeq 1$.

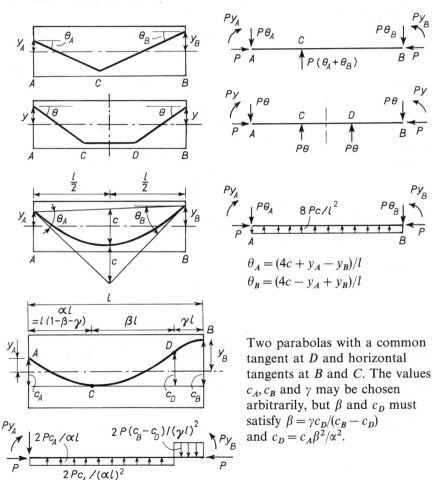

$$\theta_A = (4c + y_A - y_B)/l$$
$$\theta_B = (4c - y_A + y_B)/l$$

Two parabolas with a common tangent at D and horizontal tangents at B and C. The values c_A, c_B and γ may be chosen arbitrarily, but β and c_D must satisfy $\beta = \gamma c_D/(c_B - c_D)$ and $c_D = c_A \beta^2/\alpha^2$.

General references

1. Argyris, J. H. and Kelsey, S., *Energy Theorems and Structural Analysis*, Butterworth, London, 1960.
2. Au, T. and Christiano, P., *Structural Analysis*, Prentice-Hall, Englewood Cliffs, NJ, 1987.
3. Bathe, K. J., *Finite Element Procedures*, Prentice-Hall, Englewood Cliffs, NJ, 1982.
4. Beaufait, F. W., *Basic Concepts of Structural Analysis*, Prentice-Hall, Englewood Cliffs, NJ, 1977.
5. Cook, R. D., *Concepts and Applications of Finite Element Analysis*, Wiley, New York, 1981.
6. Gallagher, R., *Finite Element Analysis Fundamentals*, Prentice-Hall, Englewood Cliffs, NJ, 1975.
7. Gere, J. M., *Moment Distribution*, Van Nostrand, New York, 1963.
8. Ghali, A. and Favre, R., *Concrete Structures: Stresses and Deformations*, Chapman and Hall, London, New York, 1986.
9. Holzer, S. M., *Computer Analysis of Structures*, Elsevier, New York, 1985.
10. Irons, B. M. and Ahmad, S., *Techniques of Finite Elements*, Ellis Horwood, Chichester, England, 1980.
11. Irons, B. M. and Shrive, N. G., *Finite Element Primer*, Ellis Horwood, Chichester, England, 1983.
12. Irons, B. M. and Shrive, N. G., *Numerical Methods in Engineering and Applied Science*, Ellis Horwood, Chichester, England, 1987.
13. Jirousek, J., *Calcul des Structures par Ordinateur*, Swiss Federal Institute of Technology, Lausanne, Switzerland, 1982.
14. Lightfoot, E., *Moment Distribution*, Spon, London, 1961.
15. Livesley, R. K., *Matrix Methods of Structural Analysis*, Macmillan, New York, 1964.
16. Martin, H. C., *Introduction to Matrix Methods of Structural Analysis*, McGraw-Hill, New York, 1966.
17. Neal, B. G., *Structural Theorems and Their Applications*, Macmillan, New York, 1964.
18. Przemieniecki, J. S., *Theory of Matrix Structural Analysis*, McGraw-Hill, New York, 1968.

19. Rubenstein, M. F., *Matrix Computer Analysis of Structures*, Prentice-Hall, Englewood Cliffs, NJ, 1966.
20. Sack, R. L., *Structural Analysis*, McGraw-Hill, New York, 1984.
21. Shames, I. R. and Dym, C. L., *Energy and Finite Element Methods in Structural Mechanics*, McGraw-Hill, New York, 1985.
22. Wang, C. K., *Intermediate Structural Analysis*, McGraw-Hill, New York, 1983.
23. Wang, P. C., *Numerical and Matrix Methods in Structural Mechanics*, Wiley, New York, 1960.
24. Weaver, W. Jr and Gere, J. M., *Matrix Analysis of Framed Structures*, Van Nostrand, New York, 1980.
25. West, H. H., *Analysis of Structures*, Wiley, New York, 1980.
26. Willems, N. and Lucas, W. M. Jr, *Matrix Analysis for Engineers*, McGraw-Hill, New York, 1978.

Answers to problems

CHAPTER 1

1-1 $i = 1$.   Introduce a hinge in the beam at B.

1-2 $i = 1$.   Delete members CF or ED.

1-3 $i = 3$.   Make A a free end.

1-4 $i = 2$.   Introduce hinges in the beam at B and D.

1-5 $i = 4$.   Change support B to a roller and cut member DE at any section.

1-6 $i = 5$.   Remove support B and delete members AE, IF, FJ and GC.

1-7 $i = 3$.

1-8   (a) $i = 15$.   (b) $i = 7$.

1-9 $i = 2$.   Delete members IJ and EG.

1-10   5 degrees and 3 degrees when axial deformation is ignored.

1-11   14 degrees and 8 degrees when axial deformation is ignored.

1-12   9 degrees.

1-13   Cut member BC at any section. The member end-moments in the released structure are: $M_{AD} = -M_{AB} = Pb$, all other end-moments are zero. A positive end-moment acts in a clockwise direction on the member end.

1-14   (a) $R_A = 0.68ql$; $R_C = 1.42ql$; $M_B = 0.272ql^2$; $M_C = -0.1ql^2$.

       (b) $R_A = 1.207P$ down and $1.207P$ to the left; $R_D = 0.5P$ up and $0.5P$ to the left; $M_B = 1.207Pl$, tension side inside frame; $M_A = M_C = M_D = 0$.

       (c) $R_A = 0.57ql$; $R_B = 1.23ql$; $R_C = 1.08ql$; $R_D = 0.42ql$; $M_G = 0.16ql^2$; $M_B = M_C = -0.08ql^2$.

       (d) $R_A = 0.661ql$ up; $R_C = 0.489ql$ up. Member end-mements: $M_{BD} = 0.03125qb^2$; $M_{BA} = -0.06610qb^2$; $M_{BC} = 0.03485qb^2$.

       (e) $R_F = 0.75ql$ up and $0.0857ql$ to the right. Member end-moments: $M_{BF} = 0.0429ql^2$; $M_{BA} = 0.05ql^2$; $M_{BC} = -0.0929ql^r$.

1-15.   Cut six members: AC, BD, AH, DG, CF and BE. Forces in members of the released structure: $AD = -P$; $AE = P$; $DE = P\sqrt{2}$; $AF = -P\sqrt{2}$; $DH = -P$; forces in other members are zero.

1-16.   (a) Cut EF and CD; each results in three releases.

       (b) Member end-moments: $M_{AC} = M_{CA} = M_{BD} = M_{DB} = -Pl/2$; $M_{CE} = M_{EC} = M_{DF} = M_{FD} = -Pl/4$; $M_{EF} = M_{FE} = Pl/4$; $M_{CD} = M_{DC} = 3Pl/4$.

1-17.   (a) $M_{\max+} = 0.405Wl$, at $0.05l$ on either side of mid-span.

       (b) $M_{\max+} = 0.476Wl$, at $0.02l$ on either side of mid-span (assuming reversible loading).

       (c) $M_{\max+} = 0.2628Wl$, at $0.3625l$ from the left support (when one of the two loads is at this section and the other is at $0.9125l$ from the same support).

1-18 $M_{\max+} = 0.352Wl$, at $l/30$ on either side of mid-span. $V_{\max+} = -V_{\max-} = 1.56W$ at the supports.

1-20 The maximum bending-moment diagrams are symmetrical about the center line of the beam. The ordinates for the left half are as follows:
  (a) M_{max+} is zero for the overhang and is the same as in Fig. 1-22a for the remaining part. M_{max-} is 0, $-0.3Pl$, $-0.39Pl$ and $-0.195Pl$ at distances 0, $0.3l$, $0.35l$ and $0.85l$ from the left end; join the ordinates by straight lines.
  (b) M_{max+} is zero for the overhang; between the support and the centre, M_{max+} is $P(1.32\xi - 1.8\xi^2)$, where ξ is the distance from the support divided by l. M_{max-} is 0, $-0.35Pl$ and $-0.175Pl$ at distances 0, $0.35l$ and $0.85l$ from the left end; join ordinates by straight lines.
1-21 (a) $M_{n\,max+} = 0.2775Pl$; $M_{n\,max-} = -0.2925Pl$; $V_{n\,max+} = 1.11P$.
  (b) $M_{n\,max+} = 93.75 \times 10^{-3}ql^2$; $M_{n\,max-} = -61.25 \times 10^{-3}ql^2$; $V_{n\,max+} = 0.3425ql$.

CHAPTER 2

2-1 (a) $[f] = \dfrac{l}{6EI}\begin{bmatrix} 4 & 1 \\ 1 & 2 \end{bmatrix}$. (b) $[f] = \dfrac{l^3}{6EI}\begin{bmatrix} 16 & 5 \\ 5 & 2 \end{bmatrix}$.

2-2 (a) $\{F\} = -ql^2\left\{\dfrac{3}{28}, \dfrac{1}{14}\right\}$. (b) $\{F\} = ql\left\{\dfrac{11}{28}, \dfrac{8}{7}\right\}$.

2-3 $M_B = 0.0106ql^2$, and $M_C = -0.0385ql^2$.

2-4 $M_B = 0.00380\,\dfrac{EI}{l}$, and $M_C = -0.00245\,\dfrac{EI}{l}$.

2-5 $M_B = -0.0377ql^2 = M_C$.
2-6 Forces in the springs:

$$\{F_A, F_B, F_C\} = P\{0.209, 0.139, 0.039\}.$$

Reactions:

$$\{R_D, R_E, R_F, R_G\} = P\{0.51, 0.10, 0.24, 0.15\}.$$

2-7 Forces in the springs:

$$\{F_A, F_B\} = P\left\{-\dfrac{9}{400}, \dfrac{9}{400}\right\}.$$

2-8 Forces in the cables:

$$\{F_C, F_D\} = \{0,427, 4.658\} \text{ k.}$$
$$M_D = 1.28\,k\,ft.$$

2-9 $M_C = -0.075ql^2$.
2-10 $\{M_B, M_C\} = ql^2\{-0.158, -0.128\}$.
2-11 The answers are in terms of $\alpha EI(T_t - T_b)/h$. Bending moments: $M_B = 9/7$; $M_C = 6/7$. Reaction components: at A, $9/7l$ up; at B, $12/7l$ down; and at C, $3/7l$ up with an anticlockwise couple of $6/7$.
2-12 $M_A = -0.0556ql^2$; $M_B = -0.1032ql^2$; $M = -0.1151ql^2$. Upward reactions at A, B and C are $0.785ql$, $1.036ql$ and $0.512ql$.
2-13 At end of stage 1: $M_A = M_D = 0$; $M_B = -20 \times 10^{-3}ql^2$; $R_A = 0.480ql$; $R_B = 0.720ql$. At end of stage 2: $M_A = M_C = 0$; $M_B = -74.4 \times 10^{-3}ql^2$; $M_D = 20.5 \times 10^{-3}ql^2$. $R_A = 0.426ql$; $R_B = 1.148ql$; $R_C = 0.426ql$.
2-15 Maximum negative bending moment at each of the two interior supports is

$-0.2176ql^2$. Maximum positive bending moment at the center of each exterior span is $0.175ql^2$, and at the center of the interior span is $0.100ql^2$. Maximum reaction at an interior support is $2.300ql$. Absolute maximum shear is $1.2167ql$ at the section which is at a distance l from the ends, just before reaching the interior supports.

CHAPTER 3

3-1 Forces in members AB, AC, AD, and AE are:

$$\{A\} = P\{-0.33, 0.15, 0.58, 0.91\}.$$

3-2 $A_i = PH(x_i \sin \theta_i / \sum_{i=1}^{n} x_i^2 \sin^2 \theta_i).$

3-3 See answers to Prob. 2-7.

3-4 $[S] = \dfrac{Ea}{b} \begin{bmatrix} 1.707 & & & \\ -1.000 & 1.707 & \text{symmetrical} & \\ 0 & 0 & 2.707 & \\ 0 & 0 & 0 & 2.707 \end{bmatrix}$

3-5 $\{M_{BC}, M_{CF}\} = Pl\{-0.0393, 0.0071\}.$

3-6 $\{M_{BC}, M_{CF}\} = \dfrac{EI}{l^2} \Delta \left\{ \dfrac{186}{35}, -\dfrac{72}{35} \right\}.$

3-7 $[S] = \dfrac{EI}{l} \begin{bmatrix} 8 & 2 \\ 2 & 12 \end{bmatrix}$, $M_{BC} = -0.0507ql^2$, and $M_{CB} = 0.0725ql^2$.

Reaction at A: $0.478ql$ upward, $0.076ql$ to the right, and a couple of $0.0254ql^2$ clockwise.

3-8 $\{M_{BC}, M_{CB}\} = \{-0.704, -0.415\}$ k ft.

3-9 The first column of $[S]$ is

$$\left\{ \left(\dfrac{12EI_2}{l_2^3} + \dfrac{Ea_1}{l_1} \right), 0, -\dfrac{6EI_2}{l_2^2}, -\dfrac{Ea_1}{l_1}, 0, 0 \right\}.$$

3-10 $S_{11} = \dfrac{4EI_2}{l_2} + \dfrac{4EI_1}{l_1}$, $S_{12} = \dfrac{2EI_1}{l_1}$, $S_{13} = -\dfrac{6EI_2}{l_2^2}$.

$S_{22} = \dfrac{4EI_2}{l_2} + \dfrac{4EI_1}{l_1}$, $S_{32} = -\dfrac{6EI_2}{l_2^2}$, $S_{33} = \dfrac{24EI_2}{l_2^3}$.

3-11 The end-moments for the members are (in terms of Pl_1):

$M_{AB} = -0.010$, $M_{BA} = 0.135$, $M_{BC} = -0.135$, $M_{CB} = 0.366$,

$M_{CD} = -0.366$, $M_{DC} = -0.260$.

3-12 The end-moments for the members are (in terms of Pl):

$M_{AB} = -0.123$, $M_{BA} = 0.133$, $M_{BC} = -0.133$, $M_{CB} = 0.090$,

$M_{CD} = -0.090$, $M_{DC} = -0.051$.

3-13 $\{D\} = \dfrac{Pl^2}{EI} \{0.038l, 0.022, -0.048\}$. D_1 is a downward deflection, D_2 and D_3 are rotations represented by vectors in the positive x and z directions.

3-14 The first column of $[S]$ is

$$\left\{\frac{48EI}{l^3}, 0, 0, -\frac{12EI}{l^3}, 0, \frac{6EI}{l^2}, -\frac{12EI}{l^3}, \frac{6EI}{l^2}, 0, 0, 0, 0\right\}.$$

and the second column is

$$\left\{0, \left(\frac{8EI}{l} + \frac{2GJ}{l}\right), 0, 0, -\frac{GJ}{l}, 0, -\frac{6EI}{l^2}, \frac{2EI}{l}, 0, 0, 0, 0\right\}.$$

3-15 See answers to Prob. 2-5.

3-16 Girder BE:

$$\{M_B, M_I, M_J, M_E\} = Pl\{-0.184, 0.142, -0.032, -0.035\}.$$

Girder AF:

$$\{M_A, M_K, M_L, M_F\} = Pl\{-0.077, 0.051, 0.006, -0.038\}.$$

3-17

$$[S] = \sum_{l=1}^{n} \begin{bmatrix} 12\dfrac{EI_i}{l_i^3} & & & \text{symmetrical} \\ -6s\dfrac{EI_i}{l_i^2} & 4s^2\dfrac{EI_i}{l_i} + c^2\dfrac{GJ}{l_i} & & \\ 6c\dfrac{EI_i}{l_i^2} & -sc\left(\dfrac{4EI_i}{l_i} - \dfrac{GJ_i}{l_i}\right) & 4c^2\dfrac{EI_i}{l_i} + s^2\dfrac{GJ_i}{l_i} \end{bmatrix}$$

where $s = \sin \alpha_i$ and $c = \cos \alpha_i$.

3-19 $\{D\} = (Pl^2/EI)\{0.0498l, 0.1504, 0.0172\}$
$\{A\} = Pl\{0, 0.49, 0.51, 0.49, -0.49, 0\}$

3-20 $\{D\} = \dfrac{Pl^2}{EI}\{0.0057, -0.0399, 0.0730\}$

$$\{A\} = \frac{Pl}{1000}\{0, 23, -23, 783, -783, -884\}$$

CHAPTER 4

4-1 $\{D\} = \dfrac{Pl^3}{384EI}\{21, 21, 11, 11\}$. The values of the bending moments at four sections of DG are:

$$\{M_D, M_J, M_L, M_G\} = \frac{Pl}{384}\{-100, 74, -10, -28\}.$$

4-2 The bending-moment values at four sections of beam AD are:

$$M_A = M_D = 0, \quad M_B = M_C = 4\,Pl/13,$$

and for beam BF:

$$M_B = 0, \quad \text{and} \quad M_F = -18Pl/13.$$

4-3 (a)

$$[S] = \begin{bmatrix} \dfrac{24EI}{l^3} & & \text{symmetrical} & \\ 0 & \dfrac{8EI}{l} & & \\ -\dfrac{12EI}{l^3} & -\dfrac{6EI}{l^2} & \dfrac{24EI}{l^3} & \\ \dfrac{6EI}{l^2} & \dfrac{2EI}{l} & 0 & \dfrac{8EI}{l} \end{bmatrix}$$

(b)

$$[S] = \begin{bmatrix} \dfrac{24EI}{l^3} & \text{symmetrical} & \\ 0 & \dfrac{8EI}{l} & \\ \dfrac{6EI}{l^2} & \dfrac{2EI}{l} & \dfrac{8EI}{l} \end{bmatrix}.$$

(c) $[S] = \dfrac{EI}{l^3} \begin{bmatrix} 19.2 & -13.2 \\ -13.2 & 19.2 \end{bmatrix}$

4-4 (a)

$$[S] = \begin{bmatrix} \dfrac{24EI}{l^3} & & \text{symmetrical} & \\ 0 & \dfrac{8EI}{l} & & \\ -\dfrac{12EI}{l^3} & -\dfrac{6EI}{l^2} & \dfrac{12EI}{l^3} & \\ \dfrac{6EI}{l^2} & \dfrac{2EI}{l} & -\dfrac{6EI}{l^2} & \dfrac{4EI}{l} \end{bmatrix}.$$

(b)

$$[S] = \begin{bmatrix} \dfrac{24EI}{l^3} & \text{symmetrical} & \\ 0 & \dfrac{8EI}{l} & \\ \dfrac{6EI}{l^2} & \dfrac{2EI}{l} & \dfrac{4EI}{l} \end{bmatrix}.$$

(c) $[S] = \dfrac{EI}{l^3} \begin{bmatrix} 13.71 & -4.29 \\ -4.29 & 1.71 \end{bmatrix}.$

4-5

$$[S] = \dfrac{EI}{l^3} \begin{bmatrix} 2.5 & \text{symmetrical} & \\ -3.0 & 7.0 & \\ 1.5 & -3.0 & 2.5 \end{bmatrix}.$$

$$\{D\} = \dfrac{Pl^3}{EI}\{2.85, 1.30, -0.15\}.$$

4-7

$$[S] = \begin{bmatrix} \dfrac{24EI}{l^3} & & & \text{symmetrical} \\[2ex] 0 & \dfrac{12EI}{l} & & \\[2ex] -\dfrac{12EI}{l^3} & -\dfrac{6EI}{l^2} & \dfrac{12EI}{l^3} & \\[2ex] \dfrac{6EI}{l^2} & \dfrac{2EI}{l} & -\dfrac{6EI}{l^2} & \dfrac{8EI}{l} \end{bmatrix}.$$

$$[S^*] = \dfrac{EI}{l^3}\begin{bmatrix} 19.304 & -8.087 \\ -8.087 & 5.743 \end{bmatrix}.$$

4-8 $\quad S^*_{11} = S_{33} - \dfrac{S^2_{31} S_{22} - 2S_{31} S_{32} S_{21} + S^2_{32} S_{11}}{S_{11} S_{22} - S^2_{21}}.$

4-9 $\quad$ The value of P which makes the structure unstable is $P = \dfrac{EI}{2l^2}$.

4-10 $\quad$ The smallest value of P for the instability of the system is $P = \dfrac{Kl}{3}$.

4-11 $\quad$ Bending-moment ordinates: $M_A = Pl/24$; $M_B = -Pl/12$; $M_E = Pl/6$. Reactions: $R_A = P/8$ down; $R_B = 5P/8$ up.

4-13 $\quad$ Bending-moment ordinates: $M_A = M_B = -Pl/9$; $M_C = Pl/9$. Shearing-force ordinates: for AC, $V = 2P/3$; for CB, $V = -P/3$. Reactions: $R_A = R_B = P$.

4-14 $\quad$ Member end-moments: $M_{AB} = 0.0703qb^2$; $M_{BA} = 0.1406qb^2$; $M_{BC} = -0.1406qb^2$. Reactions at A: $0.75qb$ up, $0.211qb$ horizontal to the right, and a clockwise couple $0.070qb^2$.

4-15 $\quad$ Member end-moments: $M_{AB} = ql^2/48$; $M_{BA} = ql^2/24$; $M_{BD} = -ql^2/24$; $M_{DB} = 5ql^2/48$. Reactions at A: $7ql/16$ up, $ql/16$ to the right, and $ql^2/48$ clockwise. Reaction at $C = 9ql/8$ up.

4-16 $\quad$ Member end-moments: $M_{AB} = -M_{BA} = -ql^2/12$. $V_A = -V_B = 0.5ql\cos\theta$.

4-17 $\quad \{D\} = 10^{-3}\dfrac{Pl^3}{EI}\{92.01l, -145.8, -76.39, 149.3l, -208.3\}$

4-18 $\quad \{D\} = 10^{-3}\dfrac{ql^4}{EI}\{11.11, \dfrac{22.22}{l}, 19.44\}$

4-19 $\quad$ Stresses at top, mid-height and bottom, in terms of αET, are $-0.25, 0.25$ and -0.25, respectively. The variation between these values is linear. The change in length of centroidal axis is $0.25\alpha Tl$. The deflection at mid-span is $\alpha Tl^2/8d$ upward.

4-20 $\quad$ At the central support, $M = 1.5EI\alpha T/d$. The stresses at top, mid-height and bottom are, in terms of αET, $-1.00, 0.25$ and 0.50, respectively. Reactions at outer supports are $1.5EI\alpha T/dl$ up, and at interior support $3E\alpha T/dl$ down.

4-21 $\quad$ Member end-moments in terms of $EI\alpha T/d$: $M_{AB} = -0.2344$; $M_{BA} = -M_{BC} = -0.6938$. Reactions at A: horizontal outward component is $0.0928EI\alpha T/d^2$, and anticlockwise couple is $0.2344EI\alpha T/d$.

CHAPTER 6

6-1 $D_1 = 0.248$, $D_2 = 0.228$ (inch or cm).

6-2 $D_1 = 2.121\ PB$. Forces in members of the indeterminate truss are (in terms of P):
$CB = -0.53$, $AD = 0.88$, $AC = 0.38$, $CD = 0.38$, $DB = -0.62$, $BA = 0.38$.

6-3 $D = 3.73\,\dfrac{hP}{Ea}$.

6-4   Downward deflection at E in cases (a) and (b) $= b\left\{9.84\dfrac{P}{Ea},\, -0.22 \times 10^{-3}\right\}$.
Forces in members in case (c) are (in terms of P):

$AE = 1.78$, $EF = 1.46$, $FB = 1.56$, $AC = -2.22$, $CD = -1.88$,
$DB = -1.94$, $CE = 1.26$, $DF = 0.93$, $DE = 0.40$, and $CF = 0.12$.

6-5   Vertical deflection at C in cases (a) and (b) $= \{0.11,\ -0.34\}$ in. Forces in members in case (c) are (kip):

$AB = -18.9$, $BC = -15.0$, $DE = 26.1$, $EC = 25.0$, $EA = -18.5$,
$EB = -5.2$, and $BD = 6.5$.

6-6   Forces in members are (in terms of P):

$AB = -1.95$, $BC = -0.38$, $CD = -0.38$, $DE = 0.62$, $EF = 2.06$,
$BE = -0.33$, $CE = 0.54$, $DB = -0.88$, $BF = 1.34$, and $EA = -1.49$.

CHAPTER 7

7-1 $M_{AB} = -0.056\ ql^2$, $M_{BA} = 0.043\ ql^2$.

7-2 $M_{AB} = -0.057\ ql^2$, $M_{BA} = 0.130\ ql^2$.

7-3 $[f] = \dfrac{l}{24EI}\begin{bmatrix} 4.5 & -3 \\ -3 & 7.5 \end{bmatrix}$.

7-5   Force in tie $= 2.34P$. Bending-moment values (in terms of Pb):

$M_A = 0$, $M_B = M_D = 0.099$, and $M_C = 0.132$.

7-6   Force in tie $= 25.6$ kip. Horizontal reaction component at $A = 2.44$ k (inward).
Bending-moment values (kip ft):

$M_A = M_E = 0$, $M_B = M_D = -73.2$, $M_C = -43.8$.

7-7   Downward deflection at $A = 0.104\ l^3/EI$.

7-8   Downward deflection at $D = 0.015\ Pl^3/EI$, and rotation at $A = 0.047\ Pl^2/EI$ (clockwise).

7-9   Points E and C move away from each other a distance $= 0.0142\ Pl^3/EI$.

7-10 $[f] = \begin{bmatrix} 1.093 & 1.320 \\ 1.320 & 1.947 \end{bmatrix}$ in./kip.

CHAPTER 8

8-1
$$[B] = \begin{bmatrix} 0.8 & -0.6 & 0 \\ 0.6 & 0.8 & 0 \\ 0 & 0 & 1 \\ -0.8 & 0.6 & 0 \\ -0.6 & -0.8 & 0 \\ 0.6l & 0.8l & -1 \\ 0.8 & 0.6 & 0 \\ -0.6 & 0.8 & 0 \\ -0.6l & -0.8l & 1 \\ -0.8 & -0.6 & 0 \\ 0.6 & -0.8 & 0 \\ 0 & 1.6l & -1 \end{bmatrix}.$$

8-2
$$[f] = \frac{1}{EI} \begin{bmatrix} 0.24l^3 & \text{symmetrical} \\ 0.48l^3 & 1.707l^3 \\ -0.6l^2 & -1.6l^2 & 2l \end{bmatrix}.$$

8-3
$$[f] = \frac{l}{6EI} \begin{bmatrix} 2 & \text{symmetrical} \\ 1 & 4 \\ 0 & -1 & 2 \end{bmatrix}.$$

8-4
$$[f] = \frac{1}{24EI} \begin{bmatrix} 9l^3 & \text{symmetrical} \\ -15l^2 & 60l \\ -12l^3 & 36l^2 & 32l^3 \end{bmatrix}.$$

8-5
$$[C]^T = \begin{bmatrix} 0 & 1 & 0 & 1 & 0 & 0 & 0 & 0 \\ 0 & 0 & 0 & 0 & 0 & 1 & 0 & 1 \\ 1.155 & 0 & 1 & 0 & 1 & 0 & 1.155 & 0 \end{bmatrix}.$$

$$[S] = \frac{EI}{l} \begin{bmatrix} 12 & \text{symmetrical} \\ 4 & 12 \\ -6.93/l & 6.93/l & 32/l^2 \end{bmatrix}.$$

8-6 $\{F\} = q\{0.041l^2, -0.041l^2, -1.366l\}.$

8-8 For any member i

$$[S_M]_i = \begin{bmatrix} 12EI/l^3 & & & \text{symmetrical} \\ 0 & GJ/l \\ 6EI/l^2 & 0 & 4EI/l \\ -12EI/l^3 & 0 & -6EI/l^2 & 12EI/l^3 \\ 0 & -GJ/l & 0 & 0 & GJ/l \\ 6EI/l^2 & 0 & 2EI/l & -6EI/l^2 & 0 & 4EI/l \end{bmatrix}_i.$$

The displacements $\{D^*\}_i$ at the coordinates of the ith member are related to the displacements $\{D\}$ at the structure coordinates by the equation $\{D^*\}_i = [C]_i \{D\}$ where

$$[C]_i = \begin{bmatrix} [0] & [0] \\ \hline [t]_i & [0] \end{bmatrix} \quad (i = 1, 2), \quad [C]_3 = [I],$$

$$[C]_i = \begin{bmatrix} [0] & [t]_i \\ \hline [0] & [0] \end{bmatrix} \quad (i = 4, 5), \quad [t]_i = \begin{bmatrix} 1 & 0 & 0 \\ 0 & \cos\alpha_i & \sin\alpha_i \\ 0 & -\sin\alpha & \cos\alpha_i \end{bmatrix}.$$

8-9

$$[S] = \begin{bmatrix} \dfrac{Ea}{l} & \text{symmetrical} & \\ 0 & \dfrac{12EI}{l^3} & \\ 0 & \dfrac{6EI}{l^2} & \dfrac{4EI}{l} \end{bmatrix}.$$

$$[S^*] = \begin{bmatrix} \dfrac{12EI}{l^3} & \text{symmetrical} & \\ \dfrac{6EI}{bl^2} & \dfrac{Ea}{4l} + \dfrac{4EI}{b^2 l} & \\ \dfrac{6EI}{bl^2} & -\dfrac{Ea}{4l} + \dfrac{4EI}{b^2 l} & \dfrac{Ea}{4l} + \dfrac{4EI}{b^2 l} \end{bmatrix}.$$

8-10

$$[f] = \begin{bmatrix} \dfrac{l}{Ea} & \text{symmetrical} & \\ 0 & \dfrac{l^3}{3EI} & \\ 0 & -\dfrac{l^2}{2EI} & \dfrac{l}{EI} \end{bmatrix}.$$

$$[f^*] = \begin{bmatrix} \dfrac{l^3}{3EI} & \text{symmetrical} & \\ -\dfrac{bl^2}{4EI} & \dfrac{l}{aE} + \dfrac{b^2 l}{4EI} & \\ -\dfrac{bl^2}{4EI} & -\dfrac{l}{aE} + \dfrac{b^2 l}{4EI} & \dfrac{l}{aE} + \dfrac{b^2 l}{4EI} \end{bmatrix}.$$

8-11

$$[S] = EI \begin{bmatrix} \dfrac{4}{l} & \text{symmetrical} & & \\ 0 & \dfrac{4}{l} & & \\ \dfrac{2}{l} & \dfrac{2}{l} & \dfrac{8}{l} & \\ -\dfrac{6}{l^2} & \dfrac{6}{l^2} & 0 & \dfrac{24}{l^3} \end{bmatrix}.$$

$$[S^*] = EI \begin{bmatrix} \dfrac{16}{b^2 l} & & \text{symmetrical} & \\ -\dfrac{12}{b^2 l} & \dfrac{16}{b^2 l} & & \\ -\dfrac{16}{b^2 l} & \dfrac{8}{b^2 l} & \dfrac{32}{b^2 l} & \\ -\dfrac{12}{b l^2} & \dfrac{12}{b l^2} & 0 & \dfrac{24}{l^3} \end{bmatrix}.$$

8-12 $F_1 = \dfrac{\pi^4}{8} \dfrac{EI}{l^3} D_1.$

8-13 $F_1 = D_1 \left(\dfrac{\pi^4}{8} - \dfrac{\pi^2}{2} \right) \dfrac{EI}{l^3}.$

8-16 Forces in members AB, AC, AD, and AE are (in terms of P):
$\{ -0.3544, 0.1530, 0.6003, 0.8885 \}$

CHAPTER 9

9-1   Vertical displacement of joint $G = 1.298$ in. downward.
9-2   Vertical displacement of joint $G = 0.543$ in. downward.
9-3   Vertical deflection of joint $H = 0.248$ in. downward.
9-4   Vertical deflection of joint $H = 1.297$ in. downward.
9-6   Deflection at $C = 0.197$ in. downward. Rotation at $C = 0.00413$ rad clockwise.
9-7   Deflection at $D = 2.40 Pb^3/EI_0$. Deflection at $C = Pb^3/EI_0$ (both downward).
9-9   Vertical deflections are: at C or E, $1.212 \dfrac{Hl^3}{EI}$ and at D, $1.687 \dfrac{Hl^3}{EI}$, both upward.

9-10 $[S] = \dfrac{EI_0}{l} \begin{bmatrix} 6.59 & 2.92 \\ 2.92 & 4.35 \end{bmatrix}.$

9-11 $M = \dfrac{3}{16} \dfrac{EI}{l}.$ Deflection at $D = (13/32)\, l$ downward and at $E = (3/32)\, l$ upward.

9-12 (a) Deflection at $D = -0.0070 q b^4/EI.$
       (b) The settlement $= 0.0790 q b^4/EI$ and the rotation $= 0.1267 q b^3/EI.$

9-13 $\{y_1, y_2, y_3\} = \dfrac{l^3}{48 EI_0} \{0.147, 0.239, 0.182\}$ and the rotation at $A = \dfrac{0.147 l^2}{12 EI_0}.$

9-14 $y = \dfrac{4ql^4}{\pi^5 EI} \displaystyle\sum_{n=1,3,\dots}^{\infty} \dfrac{\sin(n\pi x/l)}{n^3 [n^2 - Pl^2/(\pi^2 EI)]}.$

9-15 $a_1 = 0.02511\, ql^4/EI$, $a_2 = 0$, and $a_3 = 0.0000569\, ql^4/EI.$

9-16 $y = \dfrac{4Pl^3}{\pi^4 EI} \displaystyle\sum_{n=1,2,\dots}^{\infty} \dfrac{[1 - \cos(2n\pi c/l)][1 - \cos(2n\pi x/l)]}{16n^4 + 3k\, l^4/(\pi^4 EI)}.$

CHAPTER 10

10-1 The member end-moments are (in terms of Pb):
$M_{AB} = -0.54$, $M_{BA} = -M_{BC} = -0.09$, $M_{CB} = -M_{CD} = -0.70$,
$M_{DC} = -M_{DE} = 1.11$, $M_{ED} = -1.27.$

Reaction components at A are $0.7P$ (upward), $0.21P$ (to the left) and $0.535Pb$ (anticlockwise). Reaction components at E are: $1.3P$ (upward), 0.79 (to the left) and $1.265Pb$ (anticlockwise).

10-2 The member end-moments are (in terms of Pb):

$$M_{AB} = -M_{AF} = -0.461, \quad M_{BA} = -M_{BC} = M_{CB} = -M_{CD} = -0.149.$$

Use symmetry to find the moments in the other half of the frame.

10-3 The member end-moments in terms of Pb are:

$$M_{BC} = 0.180, \quad M_{CB} = -M_{CD} = 0.070, \quad M_{DC} = 0.320.$$

Use symmetry about vertical and horizontal axes to find the remaining moments.

10-4 The member end-moments are (in terms of $qb^2/100$):

$$M_{AB} = -188.9, \quad M_{BA} = -M_{BC} = -45.7, \quad M_{CB} = 111.8,$$
$$M_{CE} = -40.0, \quad M_{CD} = -71.8.$$

The reaction components at A are: $0.79qb$ (downward), $2.28qb$ (to the left) and $-1.89qb^2$ (anticlockwise). The reaction components at D are $1.79qb$ (upward) and $0.72qb$ (to the left).

10-5 The member end-moments in terms of Pb:

$$M_{CA} = -0.500, \quad M_{CD} = 0.249, \quad M_{CG} = 0.251, \quad M_{DC} = -M_{DE} = 0.215,$$
and $M_{ED} = -0.535$.

Use symmetry to find the moments in the other half of the frame.

10-6 The required values of the bending moments are (kip ft):

(a) $M_A = M_B = -20.7$, and $M_C = M_D = -8.3$.
(b) $M_A = -M_B = 27.8$, and $M_C = -M_D = 27.8$.
(c) $M_A = -111.5$, $M_B = -251.2$, and $M_C = -2.7$, $M_D = -142.4$.

As usual, a positive bending moment causes tensile stress in the inner fiber of the frame.

10-7 The required values of the bending moments are (kip ft):

(a) $M_A = M_B = -32.8$, $M_C = -7.0$ and $M_D = 18.7$
(b) $M_A = -M_B = 34.0$, $M_C = 29.4$ and $M_D = 0$
(c) $M_A = -137.9$, $M_B = -308.6$, $M_C = 7.5$, and $M_D = -60.5$.

10-8

$$[S] = \frac{EI_0}{4lr^2} \begin{bmatrix} 45 & & \text{symmetrical} \\ 30r & 36r^2 & & \\ -45 & -30r & 45 & \\ -30r & -12r^2 & 30r & 36r^2 \end{bmatrix}.$$

10-9 (a)

$$[S] = \frac{EI}{b^3} \begin{bmatrix} 2.08 & & & & & \text{symmetrical} \\ 0 & 0.29 & & & & \\ 1.25b & -0.47b & 1.75b^2 & & & \\ -2.08b & 0 & -1.25b & 2.08 & & \\ 0 & -0.29 & 0.47b & 0 & 0.29 & \\ -1.25b & -0.47b & -0.25b^2 & 1.25b & 0.47b & 1.75b^2 \end{bmatrix}$$

(b) $$[S] = \frac{EI}{b^3} \begin{bmatrix} 1.19 & 1.07b \\ 1.07b & 1.71b^2 \end{bmatrix}$$

10-10   The forces at end A are: $1.6qb$ (upward), $1.07qb$ (to the right) and $0.213qb^2$ (anticlockwise). Use the symmetry of the frame to find the forces at end B.

10-11 $S_{AB} = 7.272EI_0/l$, $S_{BA} = 4.364EI_0/l$, $C_{AB} = 2/5$, and $C_{BA} = 2/3$.

10-12 $S_{AB} = 9.12EI_B/l$, $S_{BA} = 4.75EI_B/l$, $C_{AB} = 0.428$, and $C_{BA} = 0.824$, where EI_B is the flexural rigidity at B.

10-13 $M_{AB} = -0.0966wl^2$, and $M_{BA} = 0.0739wl^2$.

10-14 $M_{AB} = -0.2171Pl$, and $M_{BA} = 0.0866Pl$.

CHAPTER 12

12-1   Rotations at B and C are:

$$\{D_1, D_2\} = 10^{-4}\frac{ql^3}{EI}\{27.9, 111.1\}.$$

$M_{BC} = -0.0944ql^2$, and $M_{CB} = 0.0444ql^2$.

12-2   Downward deflection at B is $0.125ql^4/EI$. Rotation at B is $0.125ql^3/EI$, clockwise. The displacements at C are symmetrical with the displacements at B.

$$M_{BC} = -M_{CB} = 0.1667ql^2.$$

12-3   Translation of B is $0.0230Pl^3/EI$ to the left. Translation of C is $0.0306Pl^3/EI$ downward. Rotation at B is $0.00127Pl^2/EI$ anticlockwise. Because of symmetry, rotation at C is zero and the displacement at D and B are equal and opposite.

$$M_{BC} = -0.143Pl, \; M_{CB} = -0.145Pl.$$

12-4   Translation of B is $0.03384Pl^3/EI$ to the left. Translation of C is $0.01713Pl^3/EI$ to the left and $0.02399Pl^3/EI$ downward. Rotation at B is $0.0135Pl^2/EI$ anticlockwise and at C, $0.0030Pl^2/EI$ clockwise.

$$M_{BC} = -0.149Pl, \text{ and } M_{CB} = -0.1179Pl.$$

The structure is symmetrical about the column DF.

12-5   Considering sidesway to the right and clockwise rotation positive, we have (in terms of $Pb^3/10^4$ and $Pb^2/10^3$): sidesway of B or C is $2584/EI$, and of E or F, $14660/EI$. Rotation of B is $65/EI$, of C is $257/EI$, of E is $1341/EI$, and of F is $-713/EI$.

12-6   The end-moments on the members are (in terms of $ql^2/1000$):

$M_{AB} = 25.3$, $M_{BA} = -M_{BC} = 50.7$, $M_{CB} = 72.5$,
$M_{CD} = -36.3$, $M_{DC} = -18.1$, $M_{CE} = -36.3$.

Reaction components at A: $0.479ql$ upward, $0.077ql$ to the right and $0.025ql^2$ clockwise.

12-7   The end-moments on the members are (in terms of $ql^2/1000$):

$M_{AB} = 19.6$, $M_{BA} = -M_{BC} = 46.6$, $M_{CB} = 75.9$,
$M_{CD} = -41.7$, $M_{DC} = -24.5$, $M_{CE} = -34.3$.

Reaction components at A: $0.471ql$ upward, $0.067ql$ to the right, and $0.020ql^2$ elockwise.

12-8   The member end-moments are (in terms of $ql^2/1000$):

$M_{BA} = 93.7$, $M_{BE} = 6.1$, $M_{BC} = -99.8$, and $M_{EB} = 3.1$.

The end-moments in the right-hand half of the frame are equal and opposite in sign.

12-9   The member end-moments are (in terms of $Hl/100$):

$M_{BA} = -3.39$, $M_{BE} = 7.92$, $M_{BC} = -4.53$, and $M_{EB} = 8.73$.

Because of antisymmetry the end-moments in the right-hand half are equal.

12-10   The member end-moments in terms of EI_A/l^2 are:

$M_{BA} = 10.2$, $M_{BE} = 3.5$, $M_{BC} = -13.8$, $M_{EB} = 1.2$,
$M_{CB} = -17.1$, $M_{CF} = -2.7$, $M_{CD} = 19.8$, $M_{FC} = -2.0$.

12-11   The values of the bending moment at support B for the four cases of loading are: $ql^2\{-0.2422, -0.3672, -0.2422, -0.1016\}$.

12-12   The member end-moments are (in terms of $qb^2/100$):

$M_{AB} = 2.62$, $M_{BA} = -M_{BC} = 7.95$, $M_{CB} = 24.15$,
$M_{CD} = -6.15$, $M_{CE} = -18.00$, $M_{DC} = -4.42$.

12-13   (a)   The member end-moments are (k ft):

$M_{AB} = 122$, $M_{BA} = -M_{BC} = 126$.

           Because of symmetry, the end-moments in the right-hand half of the frame are equal and opposite.

      (b)   The reaction components at A are: 0.782 k to the right and 27.4 k ft clockwise, and the vertical component is zero.

12-14   (a)   The end-moment at the top or bottom of any of the three piers is $Pl/12$.

      (b)   Shear at F or H is $P/6$ and shear at G is $2P/3$.

12-15   The member end-moments in terms of $10^{-3}Pl$ are columns:

$M_{AB} = 575$, $M_{BA} = 421$, $M_{FC} = 618$, $M_{CF} = 506$, $M_{ED} = 537$,
$M_{DE} = 344$, $M_{CG} = 288$, $M_{GC} = 293$, $M_{DH} = 153$, $M_{HD} = 266$.

Beams:

$M_{BC} = -421$, $M_{CB} = -378$, $M_{CD} = -416$, $M_{GH} = -293$,
$M_{HG} = -266$.

12-16   The member end-moments in the left-hand half of the frame are (in terms of Pb):

$M_{AB} = -1.19$, $M_{BA} = -0.31$, $M_{BG} = 0.89$, $M_{BC} = -0.58$,
$M_{CB} = -0.68$, $M_{CF} = 0.79$, $M_{CD} = -0.11$, and $M_{DC} = -0.89$.

The corresponding end-moments in the right-hand half are equal.

12-17   The member end-moments in the left-hand half of the frame are (in terms of $Pb/10$):

$M_{AB} = M_{FG} = -5.71$, $M_{BA} = M_{GF} = -5.54$,
$M_{BC} = M_{GH} = -1.41$, $M_{CB} = M_{HG} = -2.34$,
$M_{AF} = M_{FA} = 5.71$, $M_{BG} = M_{GB} = 6.95$, $M_{CH} = M_{HC} = 0$.

The end-moments in the right-hand half are equal and opposite.

12-18   The member end-moments in the left-hand half of the frame (in terms of Pl) are:

$M_{AB} = -1.545$, $M_{BA} = -0.455$, $M_{BE} = 0.614$,
$M_{BC} = -0.159$, $M_{CB} = -M_{CD} = -0.341$.

The corresponding end-moments in the right-hand half are equal.

12-20   The member end-moments are (in terms of $Pb/100$):

$M_{AB} = -46.1$, $M_{BA} = -42.1$, $M_{BC} = 3.5$, $M_{CB} = -16.1$,
$M_{CD} = 28.0$, $M_{DC} = 16.9$, $M_{DE} = 24.2$, $M_{ED} = 31.8$, $M_{FG} = -50.6$,
$M_{GF} = -48.6$, $M_{GH} = 9.1$, $M_{HG} = -34.0$, $M_{HI} = 46.7$,
$M_{IH} = 20.9$, $M_{IJ} = 22.5$, $M_{JI} = 34.0$, $M_{AF} = 46.1$, $M_{FA} = 50.6$,
$M_{BG} = 38.7$, $M_{GB} = 39.5$, $M_{CH} = -11.9$, $M_{HC} = -12.7$,
$M_{DI} = -41.1$, $M_{ID} = -43.4$, $M_{EJ} = -31.8$, $M_{JE} = -34.0$.

12-21 The member end-moments are (in terms of $Pb/100$):

$M_{AB} = -22.1$, $M_{BA} = -13.9$, $M_{HG} = -66.4$, $M_{GH} = -41.6$,
$M_{LK} = -66.4$, $M_{KL} = -41.6$, $M_{PO} = -22.1$, $M_{OP} = -13.9$,
$M_{BC} = -10.9$, $M_{CB} = -13.1$, $M_{GF} = -32.7$, $M_{FG} = -39.3$,
$M_{KJ} = -32.7$, $M_{JK} = -39.3$, $M_{ON} = -10.9$, $M_{NO} = -13.1$,
$M_{CD} = -5.1$, $M_{DC} = -6.9$, $M_{FE} = -15.4$, $M_{EF} = -20.6$,
$M_{JI} = -15.4$, $M_{IJ} = -20.6$, $M_{NM} = -5.1$, $M_{MN} = -6.9$,
$M_{BG} = 24.8$, $M_{GB} = 24.8$, $M_{GK} = 49.6$, $M_{KG} = 49.6$, $M_{KO} = 24.8$,
$M_{OK} = 24.8$, $M_{CF} = 18.2$, $M_{FC} = 18.2$, $M_{FJ} = 36.4$, $M_{JF} = 36.4$,
$M_{JN} = 18.2$, $M_{NJ} = 18.2$, $M_{DE} = 6.9$, $M_{ED} = 6.9$, $M_{EI} = 13.8$,
$M_{IE} = 13.8$, $M_{IM} = 6.9$, $M_{MI} = 6.9$.

12-22 The member end-moments are (in terms of $Pb/100$):

$M_{AB} = -37.6$, $M_{BA} = -23.2$, $M_{HG} = -45.1$, $M_{GH} = -38.1$,
$M_{LK} = -45.1$, $M_{KL} = -38.1$, $M_{PO} = -37.6$,
$M_{OP} = -23.2$, $M_{BC} = -12.5$, $M_{CB} = -17.4$, $M_{GF} = -32.0$,
$M_{FG} = -34.1$, $M_{KJ} = -32.0$, $M_{JK} = -34.1$, $M_{ON} = -12.5$,
$M_{NO} = -17.4$, $M_{CD} = -6.7$, $M_{DC} = -9.3$, $M_{FE} = -15.1$,
$M_{EF} = -16.8$, $M_{JI} = -15.1$, $M_{IJ} = -16.8$, $M_{NM} = -6.7$,
$M_{MN} = -9.3$, $M_{BG} = 35.8$, $M_{GB} = 28.3$, $M_{GK} = 41.8$,
$M_{KG} = 41.8$, $M_{KO} = 28.3$, $M_{OK} = 35.8$, $M_{CF} = 24.1$,
$M_{FC} = 19.5$, $M_{FJ} = 29.7$, $M_{JF} = 29.7$, $M_{JN} = 19.5$, $M_{NJ} = 24.1$,
$M_{DE} = 9.3$, $M_{ED} = 7.1$, $M_{EI} = 9.7$, $M_{IE} = 9.7$, $M_{IM} = 7.1$,
$M_{MI} = 9.3$.

12-23 The member end-moments are (in terms of $Pb/100$):

$M_{AB} = -51.3$, $M_{BA} = -25.9$, $M_{JI} = -73.9$, $M_{IJ} = -45.6$,
$M_{ML} = -36.8$, $M_{LM} = -22.6$, $M_{BC} = -22.2$, $M_{CB} = -26.5$,
$M_{IH} = -42.3$, $M_{HI} = -44.3$, $M_{LK} = -20.2$, $M_{KL} = -20.5$,
$M_{CD} = -14.6$, $M_{DC} = -20.4$, $M_{HG} = -16.1$, $M_{GH} = -20.9$,
$M_{DE} = -4.5$, $M_{ED} = -7.6$, $M_{GF} = -4.2$, $M_{FG} = -7.6$,
$M_{BI} = 48.0$, $M_{IB} = 45.3$, $M_{IL} = 42.6$, $M_{LI} = 42.7$, $M_{CH} = 41.2$,
$M_{HC} = 40.3$, $M_{HK} = 20.1$, $M_{KH} = 20.5$, $M_{DG} = 24.9$,
$M_{GD} = 25.1$, $M_{EF} = 7.6$, $M_{FE} = 7.6$.

12-24 The member end-moments in terms of P are (in terms of $Pb/10$):

$M_{AB} = 4.5$, $M_{BA} = 3.3$, $M_{DE} = 5.1$, $M_{ED} = 4.5$, $M_{IJ} = 5.1$,
$M_{JI} = 4.5$, $M_{NO} = 4.6$, $M_{ON} = 3.4$, $M_{BC} = 2.5$, $M_{CB} = 3.1$,
$M_{EF} = 3.6$, $M_{FE} = 3.7$, $M_{JK} = 3.6$, $M_{KJ} = 3.7$, $M_{OP} = 2.2$,
$M_{PO} = 2.5$, $M_{FG} = 2.5$, $M_{GF} = 2.4$, $M_{KL} = 2.9$, $M_{LK} = 3.1$,
$M_{PQ} = 1.9$, $M_{QP} = 2.2$, $M_{GH} = 0.4$, $M_{HG} = 0.8$, $M_{LM} = 1.1$,
$M_{ML} = 1.3$, $M_{QR} = 0.5$, $M_{RQ} = 0.8$, $M_{BE} = -5.8$,
$M_{EB} = -4.6$, $M_{EJ} = -3.5$, $M_{JE} = -3.5$, $M_{JO} = -4.6$,
$M_{OJ} = -5.7$, $M_{CF} = -3.1$, $M_{FC} = -3.1$, $M_{FK} = -3.1$,

$M_{KF} = -3.0$, $M_{KP} = -3.6$, $M_{PK} = -4.4$, $M_{GL} = -2.8$,
$M_{LG} = -2.2$, $M_{LQ} = -2.1$, $M_{QL} = -2.7$, $M_{HM} = -0.8$,
$M_{MH} = -0.7$, $M_{MR} = -0.7$, $M_{RM} = -0.8$.

CHAPTER 14

14-1   The influence lines are composed of straight segments joining the following ordinates:

Reaction at B: $\eta_A = 0$, $\eta_D = 1.2$, and $\eta_C = 0$.

Bending moment at $E(l)$: $\eta_A = 0$, $\eta_E = 0.234$, $\eta_D = -0.075$, and $\eta_C = 0$.

Shear at E: $\eta_A = 0$, $\eta_{E\ left} = -\dfrac{3}{8}$, $\eta_{E\ right} = \dfrac{5}{8}$, $\eta_D = -\dfrac{1}{5}$, and $\eta_C = 0$.

14-2   The influence lines are composed of straight segments joining the following ordinates:

Force in member Z_1 : $\eta_A = 0$, $\eta_C = \dfrac{3}{4}$, and $\eta_B = 0$.

Force in member Z_2 : $\eta_A = 0$, $\eta_C = -0.354$, $\eta_D = 0.707$, and $\eta_B = 0$.

14-3   With the line AB as a datum, the influence line is composed of straight segments joining the following ordinates:

Horizontal component of the reaction at A (inward):

$\eta_A = \eta_B = 0$, and $\eta_C = \dfrac{l}{4h}$.

Bending moment at D:

$\eta_A = \eta_B = 0$, $\eta_D = \dfrac{3l}{32}$, and $\eta_C = -\dfrac{l}{16}$.

14-4   The three required ordinates in the spans are

Bending moment at n:
Span AB: $\{-89, -123, -117\}$ $l/1000$.
Span BC: $\{-62, -65, -47\}$ $l/1000$.

Bending moment at center of AB:
Span AB: $\{105, 189, 92\}$ $l/1000$.
Span BC: $\{-31, -33, -24\}$ $l/1000$.

Shearing force at n:
Span AB: $\{-0.389, -0.623, -0.817\}$.
Span BC: $\{-0.062, -0.065, -0.047\}$.

14-5   Influence ordinates of R_A:

$\{\eta_A, \eta_D, \eta_B, \eta_E, \eta_C\} = \{1.0, 0.902, 0.677, 0.404, 0.119\}$.

Influence ordinates of R_B:

$\{\eta_A, \eta_D, \eta_B, \eta_E, \eta_C\} = \{0, 0.046, 0.126, 0.175, 0.199\}$.

Influence ordinates of the end-moment at A:

$\{\eta_A, \eta_D, \eta_B, \eta_E, \eta_C\} = l\{0, -0.350, -0.480, -0.483, -0.437\}$.

14-6   The three required ordinates in the spans are

Span AB: $\{1.36, 1.88, 1.79\}$ $l/100$.
Span BC: $\{-1.79, -1.88, -1.36\}$ $l/100$.

14-7 $\eta_G = 0.132l$, $\eta_H = 0.236l$, and $\eta_I = 0.132l$.
14-8 $\eta_G = 0.125l$, $\eta_H = 0.250l$, $\eta_I = 0.125l$, and $\eta_K = 0.078l$.
14-11   The influence lines are composed of straight segments joining the following ordinates:

For X_A: $\{\eta_A, \eta_B, \eta_C, \eta_D, \eta_E\} = \{0, 0.535, 0.931, 0.535, 0\}$.

For Y_A: $\{\eta_A, \eta_B, \eta_C, \eta_D, \eta_E\} = \{1.0, 0.840, 0.500, 0.160, 0\}$.

For M_{AB}: $\{\eta_A, \eta_B, \eta_C, \eta_D, \eta_E\} = b\{0, -0.222, 0.084, 0.139, 0\}$.

14-12   The influence lines are composed of straight segments joining the following ordinates:

For X_A: $\{\eta_C, \eta_D, \eta_E, \eta_F, \eta_G\} = \{0.036, 0.427, 0.643, 0.427, 0.036\}$.

For the force in DE: $\{\eta_C, \eta_D, \eta_E, \eta_F, \eta_G\} = \{0.072, -0.146, -0.714, -0.146, 0.072\}$.

14-13   The influence lines are composed of straight segments joining the following ordinates:

For the force in Z_1: $\{\eta_A, \eta_D, \eta_E, \eta_F, \eta_B\} = \{0, -0.317, -0.634, -0.853, 0\}$, and the line is symmetrical about B.

For the force in Z_2, the ordinates in the spans are

Span AB: $\{\eta_A, \eta_D, \eta_E, \eta_F, \eta_B\} = \{0, -0.034, -0.067, -0.052, 0\}$.
Span BC: $\{\eta_B, \eta_G, \eta_H, \eta_I, \eta_C\} = \{0, 0.448, 0.183, -0.034, 0\}$.

14-15   The influence line for $\bar{M}_n$ is the same as a bending-moment diagram with the following member end-moments $(10^{-2}/(4b))$:

$M_{AB} = 0$, $M_{BA} = 33$, $M_{BE} = -41$, $M_{EB} = -21$, $M_{BC} = -22$, $M_{CB} = -8$, $M_{CF} = 13$, $M_{FC} = 7$, $M_{CD} = 5$ and $M_{DC} = 0$.

These values are calculated by moment distribution in Fig. 13-10c. To find the influence ordinates of the statically indeterminate reaction at A due to prestressing divide the ordinates of $\eta_{\bar{M}_n}$ by the length of span AB, 6.5b.

14-16   The required influence line is composed of straight segments joining the following ordinates:

$\{\eta_A, \eta_B, \eta_C, \eta_D\} = \{0, -0.204, 0.057, 0\}$ b^{-1}.

The total moment at $B = 0.523|P|h$.

14-17   The ordinates of the required influence line in the spans are:

Span AB: $\{\eta_A, \eta_H, \eta_I, \eta_J, \eta_B\}$
$= \{0, 1.01, 1.24, 0.95, 0.61\}$ l^{-1}.

Span BC: $\{\eta_B, \eta_K, \eta_J, \eta_L, \eta_C\}$
$= \{1.07, 2.79, 2.33, -0.41, -0.49\}$ l^{-1}.

Span CD: $\{\eta_C, \eta_M, \eta_N, \eta_O, \eta_D\}$
$= \{-0.27, -0.43, -0.56, -0.46, 0\}$ l^{-1}.

14-18   The required influence line is the same as a bending-moment diagram with the following member end-moments (in terms of $10^{-3}/l$):

$M_{FG} = -112, M_{GF} = 100, M_{GH} = -112, M_{HG} = -344,$
$M_{HI} = 292, M_{IH} = 110, M_{IJ} = -98, M_{JI} = 104.$

The secondary moment at $n = -0.063|P|h$.

CHAPTER 15

15-1 $M_{BC} = -M_{BA} = 271.9$ k ft, $M_{AB} = -135.9$ k ft.

15-2 $\{M_{AB}, M_{BA}, M_{BC}, M_{CB}\} = \dfrac{QI}{al} \{-4.7, -6.0, 6.0, 8.2\}.$

15-3 $\{M_{BA}, M_{BC}, M_{BD}, M_{DB}\} = \{24.1, -29.8, 5.7, -5.7\}$ k ft.

15-4   Member end-moments for the left-hand half of the truss are

$\{M_{AD}, M_{AC}, M_{DA}, M_{DC}, M_{CD}, M_{CA}\} = \dfrac{\sigma I}{l} \{3.8, -3.8, -4.5, 0, 0, -14.4\}.$

Ratios of the maximum bending stress to the primary stress at the member ends, listed in the same order as the moments, are

$\pm (d/l) \{1.9, 1.9, 2.3, 0, 0, 7.2\}.$

15-5   The end-moments in kip in. are

$M_{AC} = -4.0, M_{AB} = 4.0, M_{BA} = 3.9, M_{BC} = 0.2, M_{BD} = -4.0,$
$M_{CA} = -3.2, M_{CB} = 0.9, M_{CD} = 2.4, M_{DC} = 2.7, M_{DB} = -2.7.$

15-6   The end-moments for the members in the left-hand half of the truss in kip ft are:

$M_{AF} = 4.4, M_{AJ} = -1.5, M_{AB} = -2.9, M_{FA} = 3.8, M_{FJ} = 1.4, M_{FG} = -5.2,$
$M_{JF} = 0.6, M_{JA} = -2.8, M_{JB} = 2.4, M_{JC} = -0.1, M_{BJ} = 2.7,$
$M_{BA} = -3.5, M_{BC} = 0.9, M_{CB} = 0.4, M_{CJ} = -2.3, M_{CG} = 0,$
$M_{GF} = -12.5, M_{GC} = 0.$

15-7   The end moments in terms of $Pb/1000$ are:

$M_{AB} = 8.7, M_{AC} = -8.7, M_{BA} = 12.1$
$M_{BC} = -12.1, M_{CB} = -0.9, M_{CA} = 0.9$

15-8 $M_{BC} = -M_{BA} = 247.9$ k ft, $M_{AB} = -166.1$ k ft.
15-9 $\{M_{BA}, M_{BC}, M_{BD}, M_{DB}\} = \{28.0, -25.9, -2.2, -11.6\}$ k ft.
15-11 $\{M_{AB}, M_{BA}, M_{BC}, M_{CB}\} = ql^2 \{-0.658, -0.360, 0.360, 0\}.$
15-12   Buckling occurs when $Q = 1.82EI/b^2.$
15-13 $Q = 93.3EI_{BC}/l^2.$
15-14 $Q = 6\dfrac{EI}{l^2}.$
15-17 $Q = 20.8EI/l^2.$
15-18 $Q = 3.22EI/l^2.$

CHAPTER 16

16-2   The required submatrices are

$$[S_{11}]_r = \begin{bmatrix} \dfrac{2(S_1 + t_1)}{h_1^2} & & & & \text{symmetrical} \\[2mm] -\dfrac{2(S_1 + t_1)}{h_1^2} & \dfrac{2(S_1 + t_1)}{h_1^2} + \dfrac{2(S_2 + t_2)}{h_2^2} \\[2mm] & -\dfrac{2(S_2 + t_2)}{h_2^2} & \dfrac{2(S_2 + t_2)}{h_2^2} + \dfrac{2(S_3 + t_3)}{h_3^2} \\[2mm] \text{elements not shown} & & & \cdots \\ \text{are zero} & & & & \cdots \\ & & & & -\dfrac{2(S_{n-1} + t_{n-1})}{h_{n-1}^2} & [2(S_{n-1} + t_{n-1})/h_{n-1}^2 + \\ & & & & & 2(S_n + t_n)/h_n^2] \end{bmatrix}$$

$$[S_{21}]_r = [S_{12}]_r^T = \begin{bmatrix} -\dfrac{(S_1 + t_1)}{h_1} & \dfrac{(S_1 + t_1)}{h_1} \\[2mm] -\dfrac{(S_1 + t_1)}{h_1} & \dfrac{(S_1 + t_1)}{h_1} - \dfrac{(S_2 + t_2)}{h_2} & \dfrac{(S_2 + t_2)}{h_2} \\[2mm] & -\dfrac{(S_2 + t_2)}{h_2} & \dfrac{(S_2 + t_2)}{h_2} - \dfrac{(S_3 + t_3)}{h_3} & \dfrac{(S_3 + t_3)}{h_3} \\[2mm] \text{elements not shown} & & \cdots & \cdots \\ \text{are zero} & & \cdots & \cdots \\ & & & -\dfrac{(S_{n-1} + t_{n-1})}{h_{n-1}} & [(S_{n-1} + t_{n-1})/h_{n-1} \\ & & & & -(S_n + t_n)/h_n] \end{bmatrix}$$

$$[S_{22}]_r = \begin{bmatrix} S_1 + 3E(I/l)_{b1} & & & \text{symmetrical} \\[2mm] t_1 & S_1 + S_2 + 3E(I/l)_{b2} \\[2mm] & t_2 & S_2 + S_3 + 3E(I/l)_{b3} \\[2mm] \text{elements not shown} & & \cdots & \cdots \\ \text{are zero} & & \cdots & \cdots & t_{n-1} & [S_{n-1} + S_n + 3E(I/l)_{bn}] \end{bmatrix}$$

where

$$S_i = \frac{(4 + \alpha_i)}{(1 + \alpha_i)} \frac{EI_{ci}}{h_i}$$

$$t_i = \frac{(2 - \alpha_i)}{(1 + \alpha_i)} \frac{EI_{ci}}{h_i}$$

$$\alpha_i = \frac{12EI_{ci}}{h_i^2 Ga_{rci}}$$

The subscripts i of S, t, h, I_c, α and a_{rc} in the above equations refers to the storey number starting from the top storey. The subscript i in $(I/l)_{bi}$ refers to the floor number starting from the roof.

16-3   The sums of moments in walls are (refer to Fig. 16-7)

$$\{M_A, M_B, M_C, M_D, M_E\} = \frac{Ph}{10} \{0, 4.41, 18.80, 43.27, 77.82\}.$$

The sums of moments in columns are (in terms of $Ph/10$):

$M_{FG} = -0.318$, $M_{GF} = -0.271$, $M_{GH} = -0.287$, $M_{HG} = -0.316$.
$M_{HI} = -0.260$, $M_{IH} = -0.275$, $M_{IJ} = -0.196$, $M_{JI} = -0.262$.

16-4   **Bending moment at Sec. A–A is 6.07Ph. The moment on each end of the lower beam (at its intersection with the wall) is 0.295 Ph.**

16-6　Forces on the walls are:

$$\{F_1, F_2, F_3\}_A = P\{0.0003, -0.1050, -0.03272b\}$$
$$\{F_1, F_2, F_3\}_B = P\{-0.0003, -0.3150, -0.00027b\}$$
$$\{F_1, F_2, F_3\}_C = P\{0, -0.5800, -0.00027b\}$$

where F_1, F_2 are components in the x and y directions respectively, and F_3 is a clockwise couple.

16-7　Forces on the walls are:

$$\{F_1, F_2, F_3\}_A = P\{0.1183, 0.2739, 0.00072b\}$$
$$\{F_1, F_2, F_3\}_B = P\{0.0008, 0.4603, 0.00036b\}$$
$$\{F_1, F_2, F_3\}_C = P\{-0.1176, 0.0047, 0.00036b\}$$
$$\{F_1, F_2, F_3\}_D = P\{-0.0015, 0.2611, 0.00036b\}$$

where F_1, F_2 are components in the x and y directions, respectively, and F_3 is a clockwise couple.

16-8　The translation of point B in the x and y directions and the rotation of the deck in the clockwise direction are (kip, ft units):

$$10^{-6}P\{330.68, -2.51, -1.29\}.$$

The forces on the supporting elements are (kip, ft units):

$$\{F_1, F_2, F_3\}_A = P\{0.4571, 0.0748, -0.0002\}$$
$$\{F_1, F_2, F_3\}_B = P\{0.0475, -0.0056, -0.1448\}$$
$$\{F_1, F_2, F_3\}_C = P\{0.1207, 0.1194, -0.1448\}$$
$$\{F_1, F_2, F_3\}_D = P\{0.3747, -0.1886, -0.0002\}$$

where F_1, F_2 are components in the x and y directions, respectively, and F_3 is a clockwise couple.

CHAPTER 17

17-1　The first error term $= -\dfrac{\lambda^2}{8} y_{i+1/2}^{(5)}$.

17-2　The first error term $= -\dfrac{\lambda^3}{4} y_A^{(4)}$.

17-3　$\left(\dfrac{d^2 y}{dx^2}\right)_B = \dfrac{4}{3\lambda^2}(2y_A - 3y_B + y_C)$.

The first error term $= -\dfrac{\lambda}{6} y_B^{(3)}$.

17-4　$\{y_1, y_2, y_3\} = \dfrac{ql^4}{384EI_A}\{2.1, 2.7, 1.8\}$.

17-5　$\{y_1, y_2, y_3\} = (ql^4/1000EI)\{0.698, 0.843, 0.566\}$
$\{M_1, M_2, M_3\} = (ql^2/100)\{0.885, 0.675, 0.462\}$
$M_B = -1.811$, $R_A = 0.160ql$ and $R_B = 0.432ql$

17-6　Rotation at A is $l/(3.882EI)$.

17-7　
$$[f] = \begin{bmatrix} \dfrac{l^3}{2.842EI} & -\dfrac{l^2}{2EI} \\[3mm] -\dfrac{l^2}{2EI} & \dfrac{l}{EI} \end{bmatrix}$$

17-8   End-rotational stiffness $S = 2.102EI/l$. Carryover moment $t = 2.008EI/l$. From Table 15-4, we have:

$S = 2.624EI/l$,   and $t = 2.411EI/l$

17-9 $M_B = -0.324Ql$.

17-10 $\{y_1, y_2, y_3\} = 10^{-3} \dfrac{\gamma l^5}{EI} \{0.571, 0.480, 0.469\}$

17-11   Radial outward deflections at nodes 1, 2, ..., 5 are:

$\dfrac{\gamma H^2}{10E} \{15.853, 43.974, 72.552, 83.022, 54.410\}$.

The hoop forces per unit height at the nodes are:

$10^{-2} \times \gamma H^2 \{6.605, 21.987, 42.322, 55.348, 40.807\}$.

The bending moments in the vertical direction at the nodes are:

$10^{-3} \times \gamma H^3 \{0, -0.01, 0.77, 2.48, 2.32\}$.

The radial reaction at the bottom $= 0.172\gamma H^2$ (inward). The fixing moment at the base $= 0.013\gamma H^3$ (producing tensile stress at inner face). With $\lambda = H/20$, the solution gives the following values at the nodes at fifth points of the height.

Radial outward deflections at nodes 1, 2, ..., 5 are:

$\dfrac{\gamma H^2}{10^3 E} \{4.284, 43.109, 74.776, 81.176, 43.749\}$.

The hoop forces per unit height at the nodes are:

$10^{-2} \times \gamma H^2 \{1.785, 21,554, 43.619, 54.117, 32.811\}$.

The bending moments in the vertical direction at the nodes are:

$10^{-3} \times \gamma H^3 \{0, 0.17, 1.05, 3.03, 1.62\}$.

The radial reaction at the bottom $= 0.190\gamma H^2$ and the fixing moment $= 0.019\gamma H^3$.

17-12 $P_{cr} = 11.9EI/l^2$.

17-13 $P_{cr} = 20.7EI/l^2$.

CHAPTER 18

18-1   Using the ϕ values at the boundaries as shown in Fig. 18-6d, the values of ϕ at the interior nodes are:

$\{\phi\} = \dfrac{Pl}{h} \{0.1112, 0.0913, 0.0628, 0.0762, 0.0665, 0.0517\}$.

The stresses at nodes G, 1, 2, 3, and H are:

$\{\sigma_x\} = \dfrac{P}{lh} \{-0.440, -0.275, -0.218, 0.084, 1.263\}$.

18-2   Stresses at nodes G, 1, 2, 3, and H are:

$\{\sigma_x\} = \dfrac{P}{lh} \{-0.440, -0.276, -0.219, 0.084, 1.263\}$.

18-3 $(\sigma_x)_4 = -0.1530\dfrac{P}{bh}$, $(\sigma_y)_4 = -0.8106\dfrac{P}{bh}$, $(\tau_{xy})_A = 0.3986\dfrac{P}{bh}$.

18-4 $\{w\} = \dfrac{q\lambda^4}{N}\{1.2950, 0.8456, 0.6918, 0.4586\}$.

$(M_y)_1 = 0.0241qb^2$, and $(M_y)_2 = 0.0155qb^2$.

18-5 $\{w\} = \dfrac{qb^4}{N}\{0.00305, 0.00235, 0.00155, 0.00122\}$.

$(M_y)_1 = 0.0375qb^2$, and $(M_y)_2 = 0.0283qb^2$

18-6 $\{w\} = \dfrac{M\lambda^2}{N}\{0.4494, 0.3962, 0.2332, 0.2254\}$.

$(M_y)_1 = 0.2162M$. and $(M_y)_2 = 0.1708M$.

18-7 $\{w\} = \dfrac{q\lambda^4}{N}\{1.0614, 0.6764, 0.5894, 0.3869\}$.

$(M_y)_1 = 0.0189qb^2$, $(M_y)_2 = 0.0116qb^2$.
Bending moment in beam EF:
$M_1 = 0.00616qb^3$.

18-8 $\{w\} = \dfrac{P\lambda^2}{N}\{0.2981, 0.2506, 0.2006, 0.8871, 0.7423, 0.6281\}$.

M_y at A, B, and C are:

$\{M_y\} = P\{-0.5962, -0.5012, -0.4012\}$.

18-9 $\{w\} = \dfrac{P\lambda^2}{N}\{0.2740, 0.2525, 0.2211, 0.8159, 0.7531, 0.6780\}$.

M_y at A, B, and C are:

$\{M_y\} = -P\{0.5480, 0.5050, 0.4422\}$.

18-10 $\{w\} = \dfrac{qb^4}{N}\{0.0143, 0.0095, 0.0047\}$.

Reaction $R_A = 0.1039qb^2$.

18-11 $\sigma = 1.72E\,(h/b)^2$.

CHAPTER 19

19-1 $[S^*] = \dfrac{EI_0}{l^3}\begin{bmatrix} 18 & & & \text{symmetrical} \\ 8l & 5l^2 & & \\ -18 & -8l & 18 & \\ 10l & 3l^2 & -10l & 7l^2 \end{bmatrix}$

$[S^*] = \dfrac{EI_0}{l^3}\begin{bmatrix} 17.45 & & & \text{symmetrical} \\ 7.72l & 4.86l & & \\ -17.45 & -7.72l & 17.45 & \\ 9.72l & 2.86l^2 & -9.72l & 6.86l^2 \end{bmatrix}$

19-2 Deflection by finite-element method is $0.1923Pl^3/EI_0$. Exact answer is $0.1928Pl^3/EI_0$.

19-3
$$[S] = \frac{Ea}{3l}\begin{bmatrix} 7 & 1 & -8 \\ 1 & 7 & -8 \\ -8 & -8 & 16 \end{bmatrix} \qquad [S^*] = \frac{Ea}{l}\begin{bmatrix} 1 & -1 \\ -1 & 1 \end{bmatrix}$$

19-4 $\quad D_2 = 0.375\alpha T_0 l \qquad D_3 = 0.0625\alpha T_0 l \qquad \sigma = -0.125E\alpha T_0$

19-5 $\quad S_{11}^* = Eh\left(0.347\dfrac{c}{b} + 0.139\dfrac{b}{c}\right) \qquad S_{21}^* = 0.156Eh$

$$S_{31}^* = Eh\left(-0.347\frac{c}{b} + 0.069\frac{b}{c}\right) \qquad S_{41}^* = -0.052Eh$$

$$S_{51}^* = Eh\left(-0.174\frac{c}{b} - 0.069\frac{b}{c}\right)$$

19-6 $\quad \{F_b^*\} = 0.625\alpha T Eh\{c, b, -c, b, -c, -b, c, -b\}$

19-7
$$[S^*] = \frac{Eh}{1000}\begin{bmatrix} 486 & & & & & & & \text{symmetrical} \\ 156 & 486 \\ -278 & 52 & 486 \\ -52 & 35 & -156 & 486 \\ -243 & -156 & 35 & 52 & 486 \\ -156 & -243 & -52 & -278 & 156 & 486 \\ 35 & -52 & -243 & 156 & -278 & 52 & 486 \\ 52 & -278 & 156 & -243 & -52 & 35 & -156 & 486 \end{bmatrix}$$

19-8 Deflection by finite-element method is $1.202Pb/Ehc$; deflection by beam theory is $(P/Eh)[0.5(b/c)^3 + 1.44(b/c)]$.

19-9 $\quad \{\sigma\} = \{0, 0, -0.5P/hc\}$

19-10 $\quad \{F_b^*\} = \dfrac{\alpha EhT}{1-v}\left\{\dfrac{c}{2}, \dfrac{b}{2}, \dfrac{b^2}{12}, -\dfrac{c}{2}, \dfrac{b}{2}, -\dfrac{b^2}{12}, -\dfrac{c}{2}, -\dfrac{b}{2}, \dfrac{b^2}{12}, \dfrac{c}{2}, -\dfrac{b}{2}, -\dfrac{b^2}{12}\right\}$

19-11 $\quad S_{11}^* = \dfrac{Eh^3}{(1-v^2)cb}\left[\dfrac{1}{3}\left(\dfrac{b^2}{c^2} + \dfrac{c^2}{b^2}\right) - \dfrac{v}{15} + \dfrac{7}{30}\right]$

Deflection at center of rectangular clamped plate is $D_1^* = (S_{11}^*)^{-1}(P/4)$. M_x at node 2 is $-[Eh^3/2(1-v^2)b^2]D_1^*$.

19-12 $\quad [S^*] = \dfrac{Eh}{4}\begin{bmatrix} 3 & & & & \text{symmetrical} \\ -1 & 3 \\ -1 & 1 & 1 \\ 0 & -2 & 0 & 2 \\ -2 & 0 & 0 & 0 & 2 \\ 1 & -1 & -1 & 0 & 0 & 1 \end{bmatrix}$
$\qquad [T] = \begin{bmatrix} [t] & [0] & [0] \\ [0] & [t] & [0] \\ [0] & [0] & [t] \end{bmatrix}$

$[t] = \begin{bmatrix} c & s \\ -s & c \end{bmatrix} \qquad$ where $c = \cos\alpha$ and $s = \sin\alpha$

For the system in (b), $[S] = [T]^T[S^*][T]$. For the system in (c), substitute $\alpha = 180°$; thus $[t] = -[I]$ and $[\bar{S}] = [S^*]$.

19-13 The nonzero nodal displacements are

$$\{v_1, v_2, v_3, u_4, v_4, u_5, v_5, u_6, v_6\}$$

$$= \frac{P}{Eh}\{2.68, 2.21, 2.57, -0.44, 1.16, 0.03, 1.19, 0.39, 1.28\}$$

Stresses in element 2–6–3 are

$$\{\sigma_x, \sigma_y, \tau_{xy}\} = \frac{P}{bh}\{0.392, 0.358, -0.642\}$$

19-14

$$[\{B\}_1\{B\}_2\{B\}_3] = \begin{bmatrix} -(1-\eta)/b & 0 & 0 \\ 0 & -(1-3\xi^2+2\xi^3)/c & -\xi(\xi-1)^2 b/c \\ -(1-\xi)/c & (-6\xi+6\xi^2)(1-\eta)/b & (3\xi^2-4\xi+1)(1-\eta) \end{bmatrix}$$

19-15

$$[S^*] = (Eh/1000) \times \begin{bmatrix} 486 & & & & & & & & & & & \\ 156 & 552 & & & & \text{symmetrical} & & & & & & \\ -8.7b & 68b & 28b^2 & & & & & & & & & \\ -278 & 52 & 8.7b & 486 & & & & & & & & \\ -52 & -31 & 19b & -156 & 552 & & & & & & & \\ 8.68b & -19b & -12b^2 & -8.7b & -68b & 28b^2 & & & & & & \\ -243 & -156 & -26b & 35 & 52 & 26b & 486 & & & & & \\ -156 & -219 & -40b & -52 & -302 & 47b & 156 & 552 & & & & \\ 26b & 40b & 5.5b^2 & -26b & 47b & -0.3b^2 & 8.7b & -68b & 28b^2 & & & \\ 35 & -52 & 26b & -243 & 156 & -26b & -278 & 52 & -8.7b & 486 & & \\ 52 & -302 & -47b & 156 & -219 & 40b & -52 & -31 & -19b & 156 & 552 & \\ -26b & -47b & -0.3b^2 & 26b & -40b & 5.5b^2 & -8.7b & 19b & -12b^2 & 8.7b & 68b & 28b^2 \end{bmatrix}$$

CHAPTER 20

20-1
$$[S] = \frac{Ea_0}{l} \begin{bmatrix} 3.775 & \text{symmetrical} & \\ 0.442 & 2.443 & \\ -4.221 & -2.889 & 7.110 \end{bmatrix}$$

The function integrated is quartic; three sampling points are sufficient for exact integration.

20-2 When both q and the displacement are quadratic, the consistent restraining forces are: $F_{b1} = -(b/30)(4q_1 + 2q_2 - q_3)$; $F_{b2} = -(b/30)(2q_1 + 16q_2 + 2q_3)$; $F_{b3} = -(b/30)(-q_1 + 2q_2 + 4q_3)$. When q is constant and the displacement is quadratic, substitute q for q_1, q_2 and q_3, giving $\{F_b\} = -qb\{1/6, 2/3, 1/6\}$.

20-4

$$L_1 = \frac{1}{32}\eta(1-\eta)(3\xi+1)(3\xi-1)(\xi-1) \qquad L_2 = -\frac{9}{32}\eta(1-\eta)(\xi+1)(3\xi-1)(\xi-1);$$

$$L_{10} = -\frac{1}{16}(1-\eta^2)(3\xi+1)(3\xi-1)(\xi-1) \qquad L_{11} = \frac{9}{16}(1-\eta^2)(\xi+1)(3\xi-1)(\xi-1)$$

Consistent restraining forces in the y direction at the nodes 1 to 12 are

$$-q_y bch \left\{ \frac{1}{48}, \frac{1}{16}, \frac{1}{16}, \frac{1}{48}, \frac{1}{12}, \frac{1}{48}, \frac{1}{16}, \frac{1}{16}, \frac{1}{48}, \frac{1}{12}, \frac{1}{4}, \frac{1}{4} \right\}$$

20-5
$$\begin{Bmatrix} \{\varepsilon\}_1 \\ \{\varepsilon\}_2 \\ \{\varepsilon\}_3 \end{Bmatrix} = [G] \begin{Bmatrix} D_1^* \\ D_2^* \end{Bmatrix}$$

$$[G]^T = \frac{1}{2\Delta} \left[\begin{array}{ccc|ccc|ccc} 3b_1 & 0 & 3c_1 & -b_1 & 0 & -c_1 & -b_1 & 0 & -c_1 \\ 0 & 3c_1 & 3b_1 & 0 & -c_1 & -b_1 & 0 & -c_1 & -b_1 \end{array} \right]$$

20-6
$$[\{B\}_1 \{B\}_2] = \frac{1}{2\Delta} \begin{bmatrix} 3b_1\alpha_1 - b_1\alpha_2 - b_1\alpha_3 & 0 \\ 0 & 3c_1\alpha_1 - c_1\alpha_2 - c_1\alpha_3 \\ 3c_1\alpha_1 - c_1\alpha_2 - c_1\alpha_3 & 3b_1\alpha_1 - b_1\alpha_2 - b_1\alpha_3 \end{bmatrix}$$

20-7
$$S_{11}^* = \frac{h}{4\Delta}(d_{11}b_1^2 + d_{33}c_1^2) \qquad S_{12}^* = (c_1 b_1 h/4\Delta)(d_{12} + d_{33})$$

20-8
$$[B] = \frac{1}{l} \left[\begin{array}{c|c|c|c|c|c} -1/l & 2/l & -1/l & -1.732 & -1.732 & 0 \\ 1/l & -2/l & 1/l & -0.577 & -0.577 & 2.309 \\ -3.464/l & 0 & 3.464/l & -2 & -2 & 0 \end{array} \right]$$

$$[S^*] = \frac{\Delta E h^3}{1000l^4} \begin{bmatrix} 513 & & & & symmetrical & \\ -256 & 513 & & & & \\ -256 & -256 & 513 & & & \\ -148l & -148l & 295l & 488l^2 & & \\ 296l & -148l & -148l & 232l^2 & 488l^2 & \\ 148l & -296l & 148l & -232l^2 & -232l^2 & 488l^2 \end{bmatrix}$$

20-9 Deflection at center $= 1.154 \, (Pl^2/Eh^3)$; $\{M_x, M_y, M_{xy}\}_A = (P/1000) \{192, 0, 0\}$.

20-10 No deflection. $\{M_x, M_y, M_{xy}\}_A = \alpha E h^2 T(0.119, 0.119, 0)$.

20-11 Deflection at $A = 0.381qb^4/Eh^3$; $M_{xB} = -0.200qb^2$.

20-12 Deflection at $A = 0.1536pb^3/Eh^3$; $M_{xB} = -0.074pb$.

20-13 Because of symmetry, one strip only needs to be considered. The nodal displacement parameters are: $\theta_{\text{at support}} = -0.1515ql^3/Eh^3$; $w_{\text{at center}} = 0.0452ql^4/Eh^3$.

20-14 Deflection along center line $= 10^{-3}q_0(c^4/Eh^3) [58 \sin(\pi y/l) - 11 \sin(2\pi y/l) + 2.8 \sin(3\pi y/l)]$.

20-15 Beam flexural stiffness $= (EIl/2)(k\pi/l)^4$; beam torsional stiffness $= (GJl/2)(k\pi/l)^2$. Deflection at center $= 0.204Pl^2/Eh^3$; bending moment in an edge beam at mid-span $= 0.045Pl$.

20-16
$$S_{11}^* = \frac{Eh}{1-v^2} \left[\frac{c}{12b}(4 - v^2) + \frac{b}{8c}(1-v) \right] \qquad S_{21}^* = \frac{Eh}{8(1-v)}$$

$$S_{31}^* = \frac{Eh}{1-v^2} \left[-\frac{c}{12b}(4 - v^2) + \frac{b}{8c}(1-v) \right] \qquad S_{41}^* = \frac{Eh(3v-1)}{8(1-v^2)}$$

$$S_{51}^* = -\frac{Eh}{(1-v^2)} \left[\frac{c}{12b}(2 + v^2) + \frac{b}{8c}(1-v) \right]$$

20-17

$$[S^*] = \frac{Eh}{1000} \begin{bmatrix} 448 & & & & & & & \\ 156 & 448 & & & \text{symmetrical} & & & \\ -240 & 52 & 448 & & & & & \\ -52 & 73 & -156 & 448 & & & & \\ -281 & -156 & 73 & 52 & 448 & & & \\ -156 & -281 & -52 & -240 & 156 & 448 & & \\ 73 & -52 & -281 & 156 & -240 & 52 & 448 & \\ 52 & -240 & 156 & -281 & -52 & 73 & -156 & 488 \end{bmatrix}$$

CHAPTER 21

21-1 (a) $M_p = 0.1458ql^2$.   (b) For AC, $M_p = 0.0996ql^2$.
                                      For CD, $M_p = 0.1656ql^2$.

21-2 (a) $M_p = 0.29Pb$.   (b) $M_p = 0.1875Pl$.
      (c) $M_p = 1.2Pb$ (d) $M_p = 1.59qb^2$.
      (e) $M_p = 0.11\ Wl$.

21-3 $M_p = 0.1884Pl$.

21-4 $M_p = 0.1461ql^2$.

CHAPTER 22

22-1 (a) $m = 0.090\ W$.   (b) $m = W/18$.
      (c) $m = W/6$.   (d) $m = W/14.14$.
      (e) $m = 0.0637ql^2$.   (f) $m = 0.0262ql^2$.

22-2 (a) $m = 0.00961ql^2$.   (b) $m = 0.299ql^2$.
      (c) $m = 0.113ql^2$.   (d) $m = 25P/72$.
      (e) $m = 0.1902qb^2$.

CHAPTER 23

23-1 $\omega = 7.0$ rad/sec, $T = 0.898$ sec. The amplitude $= 1.745$ in., and $D_{t=1} = 1.692$ in.

23-2 $\omega = 5.714$ rad/sec, $T = 1.09$ sec. The amplitude $= 2.02$ in., and $D_{t=1} = -0.101$ in.

23-3 $\omega_1 = \dfrac{3.464}{l^2} \sqrt{\dfrac{EIl}{m}}$, $\{D^{(1)}\} = \left\{ 1.00, \ -\dfrac{1.375}{l} \right\}$.

23-4 The consistent mass matrix is

$$m = \frac{\gamma l}{840} \begin{bmatrix} (240a_1 + 72a_2) & & & \text{symmetrical} \\ (30a_1 + 14a_2)l & (5a_1 + 3a_2)l^2 & & \\ (54a_1 + 54a_2) & (-93a_1 + 119a_2)l & (72a_1 + 240a_2) & \\ -(14a_1 + 12a_2)l & -(3a_1 + 3a_2)l^2 & -(14a_1 + 30a_2)l & (3a_1 + 5a_2)l^2 \end{bmatrix}$$

23-5 $D_{\tau=T/2} = 2.107$ in., $\dot{D}_{\tau=T/2} = 0$, and $D_{\tau=11T/8} = 1.488$ in.

23-6 $D_{\tau=T/2} = 1.342$ in., $\dot{D}_{\tau=T/2} = 0$, and $D_{\tau=11T/8} = 0.95$ in.

23-7 $\omega_d = 6.964$ rad/sec, $T_d = 0.902$ sec, $D_{t=1} = 0.866$ in.

23-8 $\xi = 0.0368$.

23-9 (a) $D_{max} = 0.176$ in.   (b) $D_{max} = 0.174$ in.

23-10 $D_{\tau=T_d/2} = 1.953$ in., and $D_{\tau=11T_d/8} = 0.995$ in.

23-11 $\omega_1^2 = 1.2\dfrac{EI}{ml^3}$, $\{D^{(1)}\} = \{1, 1\}$.

$\omega_2^2 = 18.0\dfrac{EI}{ml^3}$, and $\{D^{(2)}\} = \{1, -1\}$.

23-12 $\omega_1^2 = 0.341\dfrac{EI}{ml^3}$, $\{D^{(1)}\} = \{1, 0.32\}$.

$\omega_2^2 = 15.088\dfrac{EI}{ml^3}$, and $\{D^{(2)}\} = \{1, -3.12\}$.

23-13 Equations of motion in the $\{\eta\}$ coordinates:

$$\ddot{\eta} + 0.34\,\frac{EI}{ml^3}\,\eta = 0.906\,P_1/m.$$

$$\ddot{\eta} + 15.09\,\frac{EI}{ml^3}\,\eta = 0.103\,P_1/m.$$

$$\eta_1 = \frac{P_1}{0.3758}\frac{l^3}{EI}\left[1 - \cos\left(\tau\sqrt{0.341\frac{EI}{ml^3}}\right)\right].$$

$$\eta_2 = \frac{P_1}{131.78}\frac{l^3}{EI}\left[1 - \cos\left(\tau\sqrt{15.09\frac{EI}{ml^3}}\right)\right].$$

$$D_1 = \eta_1 + \eta_2, \quad D_2 = 0.32\,\eta_1 - 3.12\,\eta_2.$$

23-14 Amplitude of $m_1 = \dfrac{P_0}{0.3758\,\dfrac{EI}{l^3} - 1.1024\,m\Omega^2}$.

23-15 Maximum displacement of BC relative to the support $= 2.01$ in.

23-16 Maximum displacement of mass m_1 relative to the support

$$= \frac{(g/4)}{1.2EI/(ml^3) - \Omega^2}$$

CHAPTER 25

25-1 $[t]_{AC} = \begin{bmatrix} 0.707 & 0.707 & 0 \\ 0 & 0 & 1 \\ 0.707 & -0.707 & 0 \end{bmatrix}$ $[t]_{DG} = \begin{bmatrix} 0.707 & 0 & 0.707 \\ 0 & 1 & 0 \\ -0.707 & 0 & 0.707 \end{bmatrix}$

25-2 $[G]^T = E\begin{bmatrix} -ac/l & 12Is/l^3 & 6Is/l^2 & ac/l & -12Is/l^3 & 6Is/l^2 \\ -as/l & -12Ic/l^3 & -6Ic/l^2 & as/l & 12Ic/l^3 & -6Ic/l^2 \\ 0 & 6I/l^2 & 2I/l & 0 & -6I/l^2 & 4I/l \end{bmatrix}$

where $c = \cos\alpha$ and $s = \sin\alpha$.

25-3 $\{A_r\} = E\left\{-0.894\dfrac{a}{l}, -5.367\dfrac{I}{l^3}, -2.683\dfrac{I}{l^2}, 0.894\dfrac{a}{l}, 5.367\dfrac{I}{l^3}, -2.683\dfrac{I}{l^2}\right\}_{CD}$

25-6   Matrix $[G]$ given in the answer to Prob. 25-2 may be used in this problem. Member end-forces at nodes 5 and 3 are $\{A\} = \{8597, -8597\}$ N.

25-7   Reactions at node 1 are: $R_1 = 2.73P$; $R_2 = -24.4P$; $R_3 = 193Pb$. Forces at the ends of member 2 are $\{A\} = \{-4.89, -24.8, -625, -22.0, -28.9, 1310\}$.

25-8

$$[S^*] = \frac{Ebh}{l} \begin{bmatrix} 1.12 & & & & & \text{symmetrical} \\ 0 & 1.81h^2/l^2 & & & & \\ 0 & 1.09h^2/l & 0.762h^2 & & & \\ -1.12 & 0 & 0 & 1.120 & & \\ 0 & -1.81h^2/l & -1.09h^2/l & 0 & 1.81h^2/l^2 & \\ 0 & 0.722h^2/l & 0.326h^2 & 0 & -0.722h^2/l & 0.396h^2 \end{bmatrix}$$

25-9   The stiffness matrix elements on and below the diagonal are: six elements above the horizontal dashed line in Eq. 24-12 unchanged; $S_{41} = -12EI/l^3$; $S_{42} = 0$; $S_{43} = -6EI/l^2$; $S_{44} = 12EI/l^3$; $S_{51} = -6EIs/l^2$; $S_{52} = -GJc/l$; $S_{53} = -2EIs/l$; $S_{54} = 6EIs/l^2$; $S_{55} = GJc^2/l + 4EIs^2/l$; $S_{61} = 6EIc/l^2$; $S_{62} = -GJs/l$; $S_{63} = 2EIc/l$; $S_{64} = -6EIc/l^2$; $S_{65} = GJcs/l - 4EIcs/l$; $S_{66} = 4EIc^2/l + GJs^2/l$; where $s = \sin\gamma$, $c = \cos\gamma$.

25-10   The elements of $[S]$ on and below the diagonal are (in terms of Ea/l): $S_{11} = c_1^2$; $S_{21} = c_1 s_1$; $S_{31} = -c_1 c_2$; $S_{41} = -c_1 s_2$; $S_{22} = s_1^2$; $S_{32} = -c_2 s_1$; $S_{42} = -s_1 s_2$; $S_{33} = c_2^2$; $S_{43} = c_2 s_2$; $S_{44} = s_2^2$; where $c_1 = \cos\alpha_1$, $s_1 = \sin\alpha_1$, $c_2 = \cos\alpha_2$, $s_2 = \sin\alpha_2$.

25-11   Rotations at B: $\theta_z = -0.00658qb^3/EI$; $\theta_x = 0.09186qb^3/EI$. Bending and torsional moments in terms of $10^{-3}qb^2$: in member EB, $M_E = 181.8$, $M_B = -68.2$ and $T_E = T_B = 68.2$; in member BC, $M_B = -96.5$, $M_H = 28.5$ and $T_B = T_H = 0$. Positive bending moment produces tension in bottom fiber; twisting moment is given as an absolute value.

25-12   Member end-moments in terms of $EI\alpha T/b$: $M_{EB} = 0$; $M_{BE} = -M_{BH} = M_{HB} = 2.59$. No torsion.

25-13   Bending and twisting moments are given in terms of ql^2 and reactions in terms of ql. Positive bending moment produces tension in bottom fiber; twisting moment is given as an absolute value; upward reaction is positive. $M_E = M_F = -13.9 \times 10^{-3}$; $T_E = T_F = 24.1 \times 10^{-3}$; $R_A = R_D = 0.370$; $R_B = R_C = 0.130$.

25-14   Bending and twisting moments are given in terms of $10^{-3}ql^2$ and reactions in terms of ql. $M_E = M_F = -4.0$; $M_O = -123.0$; $T = 6.9 =$ constant over whole length; $R_A = R_D = 0.225$; $R_B = R_C = 0.156$; $R_G = R_H = 0.619$. Positive bending moment produces tension in bottom fiber; twisting moment is given as an absolute value.

25-15   Displacement components at B: $\{D\} = (qb^3/EI)\{0.187b, 0.0235, 0.209\}$.

25-16   Due to $P/2$ at each of B, C, G and F, the displacements at B are $\{D\} = (Pb^2/EI)\{0.0544b, 0.00608, 0.0635\}$. Due to $P/2$ at each of B and C and $-P/2$ at F and G, the displacements at B are $\{D\} = (Pb^2/EI)\{0.0172b, -0.0493, 0.0206\}$. Due to actual loading, the displacements are $\{D\}_B = (Pb^2/EI)\{0.0716b, -0.0432, 0.0841\}$, $\{D\}_F = (Pb^2/EI)\{0.0372b, -0.0554, 0.0429\}$.

25-17   Displacements at E: downward deflection is $0.256ql^4/EI$; rotation vector perpendicular to AE is $0.262ql^3/EI$.

25-18   Deflection at O is $7ql^4/48EI$. Bending moments in terms of ql^2: $M_1 = 0.125$; $M_2 = -0.125$; $M_3 = 0.375$. Positive bending moment produces tension in bottom fiber.

APPENDIX A

A-1   (a) $[A][B]$: not possible

     (b)
$$[A]^T[B] = \begin{bmatrix} 10 & 11 \\ 2 & 3 \\ 10 & 7 \end{bmatrix}$$

     (c)
$$[A]^T[B][A] = \begin{bmatrix} 43 & 11 & 31 \\ 11 & 3 & 7 \\ 31 & 7 & 27 \end{bmatrix}.$$

     (d) $[A][B]^T[A]^T$: not possible.

A-2   (a) $[B][A+C] = \begin{bmatrix} 24 & 10 & 14 \\ 20 & 7 & 13 \end{bmatrix}.$

     (b) $[B][A]+[B][C] = \begin{bmatrix} 24 & 10 & 14 \\ 20 & 7 & 13 \end{bmatrix}.$

A-4   (a) $\dfrac{1}{11}\begin{bmatrix} 3 & 2 \\ 2 & 5 \end{bmatrix}.$

     (b)
$$\begin{bmatrix} 0.750 & \text{symmetrical} & \\ 0.500 & 1.000 & \\ 0.250 & 0.500 & 0.750 \end{bmatrix}.$$

     (c)
$$\begin{bmatrix} 0.591 & \text{symmetrical} & \\ 0.545 & 0.773 & \\ 0.227 & 0.364 & 0.318 \end{bmatrix}.$$

A-5   (a) $\{D\} = \left\{ 0, \dfrac{Pl^3}{12EI} \right\}.$

     (b) $S_{11} = \dfrac{4EI}{l}$, and $S_{21} = \dfrac{6EI}{l^2}.$

     (c) $S_{12} = \dfrac{6EI}{l^2}, S_{22} = \dfrac{12EI}{l^3}.$

Additional problems using SI units

A certain number of the problems given at the end of the various chapters of this book involve the use of British units. For the convenience of readers who prefer the SI (International System) units, the following problems are presented in these units. Each problem is identified by two numbers followed by an asterisk, the first number being that of the chapter to which the problem belongs. Problems identified by the same numbers here and at the end of a chapter are similar and differ only in the units used. The answers to the additional problems are given in the following section.

2-8*   A steel beam AB is supported by two steel cables at C and D. Using the force method, find the tension in the cables and the bending moment at D due to a load $P = 25\text{kN}$ and a drop of temperature of 20 degrees Celsius in the two cables. For the beam $I = 16 \times 10^6 \text{ mm}^4$, for the cables $a = 100 \text{ mm}^2$; the modulus of elasticity for both is $E = 200 \text{ GN/m}^2$, and the coefficient of thermal expansion for steel is 1×10^{-5} per degree Celsius.

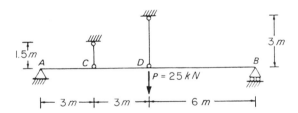

3-8*   Find the end-moments M_{BC} and M_{CB} in the frame of Prob. 3-7 due to a rise in temperature of 20 degrees Celsius in part BD only. Assume the coefficient of thermal expansion to be $\alpha = 1 \times 10^{-5}$ per degree Celsius, take $EI = 4000 \text{ kN m}^2$ and $l = 6$ m. The members are assumed to have infinite axial rigidity. The load q is not acting in this case.

6-5*   For the plane truss shown in the figure, find: (a) the deflection at C due to the load P, (b) the vertical deflection at C in the unloaded truss if members DE and EC are each shortened by 3 mm, and (c) the forces in all members if a member of cross-sectional area of 2400 mm^2 is added between B and D and the truss is subjected to the load P. Assume $E = 200 \text{ GN/m}^2$ and the cross-sectional area of the members as indicated in the figure.

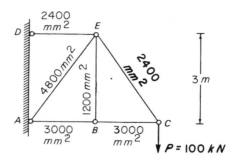

|—2.25 m —+—2.25m—|

7-6* Using the force method, find the tension in the tie and the horizontal reaction component at A for the frame shown in the figure and draw the bending-moment diagram. Consider only the bending deformation in the frame $ABCDE$ and only the axial deformation in the tie BD. $E = 200 \text{ GN/m}^2$, I for $ABCDE = 3.1 \times 10^9 \text{ mm}^4$, area of tie = 5000 mm^2.

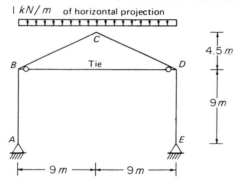

7-10* Considering the tension F_1 and F_2 in the cables at C and D in Prob. 2-8* as redundants, find the flexibility matrix of the released structure using Eq. 7-23. Consider only the bending deformation in AB and only the axial deformation in the cables.

9-1* Draw a Williot diagram for the truss shown in the figure. Use this diagram to find the vertical displacement of joint G, and check this value by the

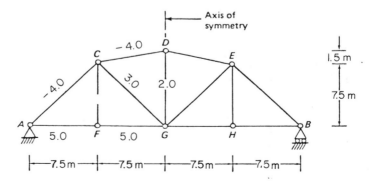

method of virtual work. The changes in length (mm) of the members due to loading and temperature variation are indicated in the figure.

9-2*   Apply the requirements of Prob. 9-1* to the truss shown in the figure.

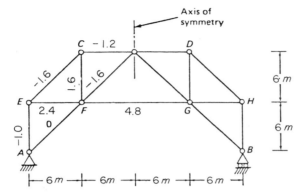

9-3*   Draw a Williot-Mohr diagram for the truss shown in the figure. The changes in lengths (mm) of the members are indicated in the figure. From the Williot-Mohr diagram find the vertical deflection of joint H.

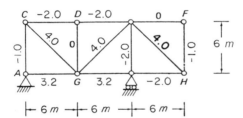

9-4*   Apply the requirements of Prob. 9-3* to the truss shown in the figure.

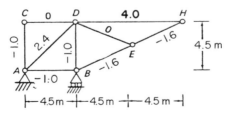

9-6*   Determine the deflection and the angular rotation at C for the prismatic beam shown in the figure. Assume $I = 330 \times 10^6$ mm^4 and $E = 200$ GN/m^2.

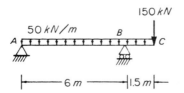

10-6*   Find the bending moment at points *A*, *B*, *C* and *D* of the frame shown in
the figure due to:
(a) a uniform shrinkage strain = 0.0002
(b) vertical settlement of 10 mm at support *A*
(c) rotation of *B* in the clockwise direction by 0.20°
Assume $I = 30 \times 10^9$ mm⁴ and $E = 15$ GN/m²

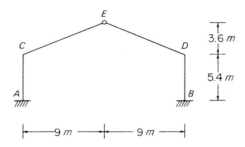

10-7*   Apply the requirements of Prob. 10-6 to the structure shown in the figure.

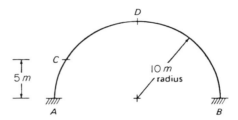

12-13* For the frame shown in the figure:
(a) draw the bending-moment diagram due to a distributed vertical load
1 kN/m of horizontal projection.
(b) find the reactions due to a rise of temperature of 25 degrees Celsius
with no loading on the frame.
Take $EI = 800$ MN m².
Coefficient of thermal expansion $\alpha = 1 \times 10^{-5}$ per degree Celsius.

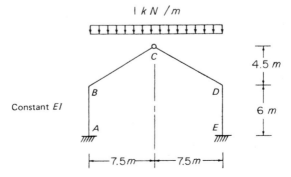

15-1*   Determine the end-moments in the frame shown in the figure, taking into
account the change in length of members.

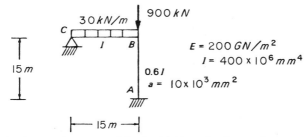

15-3*   Apply the requirements of Prob. 15-1* to the frame shown.

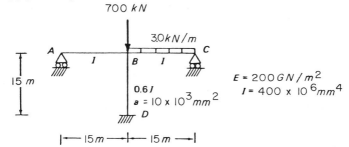

15-5*   Find the member end-moments for the rigid-jointed truss shown in the figure. Assume for all members $I = 800 \times 10^3$ mm^4 and take the primary stress caused by the axial tensile or compressive forces as 100 N/mm^2.

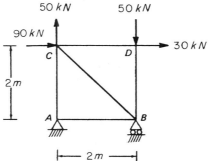

15-6*   Determine the end-moments in the members in the left-hand half of the rigid-jointed truss shown in the figure. $E = 200$ GN/m^2.

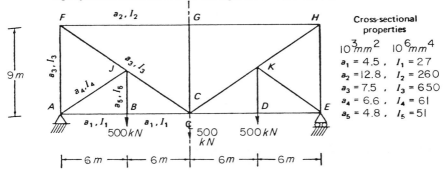

15-8*   Solve Prob. 15-1* taking into account the beam-column effect.

15-9*   Solve Prob. 15-3* taking into account both the change in length of members and the beam-column effect.

16-8*   The figure shows a plan view of a curved slab bridge deck, supported on piers at B and C and on bearing pads above rigid abutments at A and D. The piers are assumed to be pin-connected to an infinitely rigid bridge deck. Each pier has a cross-section 0.6 m × 2.4 m and is 12 m high and encastré at the base. The bearing pads at A and D have an area of 0.36 m², thickness 32 mm and shear modulus of elasticity = 2.1 N/mm². Find the three displacement components of the deck and the forces on each of the supporting elements A, B, C and D due to a force P at the deck level, as shown. Assume that for the pier material $E = 2.3G = 28$ GN/m², and that the bearing pads at A and D are strips of length 2.4 m, width 0.15 m and height 32 mm.

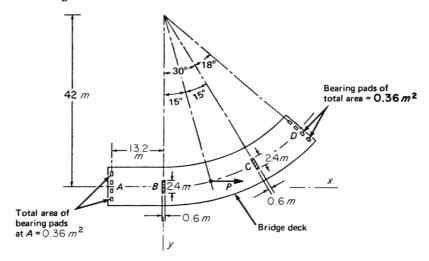

Take the acceleration of gravity to be $g = 9.81$ m/sec² whenever it is needed in the solution of the following problems:

23-1*   Compute the natural angular frequency of vibration in sidesway for the frame in the figure and calculate the natural period of vibration. Idealize

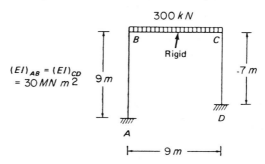

the frame as one-degree-of-freedom system. Neglect the axial and shear deformations and the weight of the columns. If initially the displacement is 25 mm and the velocity is 0.25 m/sec, what is the amplitude and what is the displacement at $t = 1$ sec?

23-2* Solve Prob. 23-1* assuming that BC has a flexural rigidity $(EI)_{BC} = 30$ MN m^2.

23-5* The frame in Prob. 23-1* is disturbed from rest by a horizontal force of 40 kN at C, suddenly applied at time $\tau = 0$ and removed at time $\tau = T/2$, where T is the natural period of vibration. What is the displacement and velocity at the removal of the force? What is the displacement at time $\tau = 11T/8$?

23-6* Solve Prob. 23-5* assuming that the disturbing force increases linearly from zero at $\tau = 0$ to 40 kN at $\tau = T/4$, then decreases linearly to zero at $\tau = T/2$, at which time the force is removed.

23-7* If the system of Prob. 23-1* has a damping coefficient $\xi = 0.1$, what are the damped natural circular frequency ω_d and the natural period of damped vibration T_d? What is the displacement at $t = 1$ sec, if $D_0 = 25$ mm and $\dot{D}_0 = 0.25$ m/sec.?

23-9* Determine the maximum steady-state sidesway in the frame of Prob. 23-1* when it is subjected to a harmonic horizontal force at the level of BC of magnitude $20 \sin 14t$ (kN) and (a) no damping is present, (b) the damping coefficient $= 0.10$.

23-10* Assume that the frame in Prob. 23-1* has a damping coefficient $= 0.05$ and it is disturbed from rest by a horizontal force of 40 kN at C. The force is suddenly applied at time $\tau = 0$ and removed at time $\tau = T_d/2$, where T_d is the natural period of damped vibration. What are the displacements at the removal of the force and at time $\tau = 11T_d/8$? (Compare the answers with the undamped case, Prob. 23-5.)

23-15* The supports of the frame of Prob. 23-1* move horizontally with an acceleration indicated in the figure for Prob. 23-15, p. 716. What is the maximum displacement of BC relative to the support? Neglect damping.

Answers to additional problems using SI units

2-8*   Forces in the cables:

$$\{F_C, F_D\} = \{2.12, 23.29\} \text{ kN}$$

$$M_D = 1.96 \text{ kN m}$$

3-8* $\{M_{BC}, M_{CB}\} = \{-0.722, -0.426\}$ kN m

6-5*   Vertical deflection at C in cases (a) and (b) = $\{3.3, -8.3\}$ mm. Forces in members in case (c) are (kN):

$AB = 94.5, BC = -75.0, DE = 130.5, EC = 125.0, EA = -92.5,$
$EB = -26.0$ and $BD = 32.5$

7-6*   Force in tie = 7.68 kN. Horizontal reaction component at $A = 0.73$ kN (inward). Bending moment values (kN m):

$$M_A = M_E = 0, \; M_B = M_D = -6.59, \; M_C = -3.94$$

7-10* $[f] = 10^{-3}\begin{bmatrix} 6.403 & 7.734 \\ 7.734 & 11.400 \end{bmatrix}$ m/kN

9-1*   Vertical displacement of joint $G = 25.95$ mm downward
9-2*   Vertical displacement of joint $G = 10.86$ mm downward
9-3*   Vertical deflection of joint $H = 4.96$ mm downward
9-4*   Vertical deflection of joint $H = 25.94$ mm downward
9-6*   Deflection at $C = 5.59$ mm downward. Rotation at $C = 0.00469$ radian clockwise.

10-6*   The required values of the bending moments are (kN m):

(a) $M_A = M_B = -27.0$, and $M_C = M_D = -10.8$
(b) $M_A = -M_B = 28.5$, and $M_C = -M_D = 28.5$
(c) $M_A = -145.2, M_B = -327.1, M_C = -3.5$ and $M_D = -185.4$

As usual, a positive bending moment causes tensile stress in the inner fibre of the frame.

10-7*   The required values of the bending moments are (kN m):

(a) $M_A = M_B = -38.4, M_C = -8.2$ and $M_D = 21.9$
(b) $M_A = -M_B = 31.4, M_C = 27.1$ and $M_D = 0$
(c) $M_A = -161.6, M_B = -361.6, M_C = 8.8$ and $M_D = -70.9$

12-13*   (a) The member end-moments are (kN m):

$$M_{AB} = 10.94, \; M_{BA} = -M_{BC} = 11.38$$

(b) The reaction components at A are: 3.62 kN to the right and 38.0 kN m clockwise, and the vertical component is zero.

15-1* $M_{BC} = -M_{BA} = 370$ kN m, $M_{AB} = -185$ kN m

15-3* $\{M_{BA}, M_{BC}, M_{BD}, M_{DB}\} = \{35.2, -43.4, 8.2, -8.2\}$ kN m

15-5*   The end-moments in kN m are:

$$M_{AC} = -0.43, \ M_{AB} = 0.43, \ M_{BA} = 0.41, \ M_{BC} = 0.01, \ M_{BD} = -0.42$$
$$M_{CA} = -0.34, \ M_{CB} = 0.09, \ M_{CD} = 0.25, \ M_{DC} = 0.29, \ M_{DB} = -0.29$$

15-6*   The end-moments for the members in the left-hand half of the truss in kN m are:

$$M_{AF} = 6.8, \ M_{AJ} = -2.3, \ M_{AB} = -4.5, \ M_{FA} = 5.7, \ M_{FJ} = 2.0, \ M_{FG} = -7.6$$
$$M_{JF} = 0.9, \ M_{JA} = -4.5, \ M_{JB} = 3.6, \ M_{JC} = -0.0, \ M_{BJ} = 4.1, \ M_{BA} = -5.5$$
$$M_{BC} = 1.4, \ M_{CB} = -0.7, \ M_{CJ} = -3.7, \ M_{CG} = 0, \ M_{GF} = -19.5, \ M_{GC} = 0$$

15-8* $M_{BC} = -M_{BA} = 327$ kN m, $M_{AB} = -224$ kN m

15-9* $\{M_{BA}, M_{BC}, M_{BD}, M_{DB}\} = \{41.7, -36.8, -4.9, -18.2\}$ kN m

16-8*   The translation of point B in the x and y directions and the rotation of the deck in the clockwise direction are (N, m units):

$$10^{-6}P\{0.02315, -0.00019, -0.000302\}$$

The forces on the supporting elements are (N, m units):

$$\{F_1, F_2, F_3\}_A = P\{0.4557, 0.0747, 0.0000\}$$
$$\{F_1, F_2, F_3\}_B = P\{0.0485, -0.0062, 0.0446\}$$
$$\{F_1, F_2, F_3\}_C = P\{0.1226, 0.1207, -0.0446\}$$
$$\{F_1, F_2, F_3\}_D = P\{0.3732, -0.1892, 0.0000\}$$

23-1* $\omega = 7.103$ rad/sec, $T = 0.885$ sec. The amplitude $= 43.2$ mm, and $D_{t=1} = 42.8$ mm.

23-2* $\omega = 5.777$ rad/sec, $T = 1.088$ sec. The amplitude $= 50.0$ mm, and $D_{t=1} = 0.9$ mm.

23-5* $D_{t=T/2} = 51.9$ mm, $\dot{D}_{t=T/2} = 0$ and $D_{t=11T/8} = 36.7$ mm

23-6* $D_{t=T/2} = 33.0$ mm, $\dot{D}_{t=T/2} = 0$ and $D_{t=11T/8} = 23.3$ mm

23-7* $\omega_d = 7.067$ rad/sec, $T_d = 0.889$ sec, $D_{t=1} = 21.9$ mm

23-9*   (a) $D_{max} = 4.49$ mm   (b) $D_{max} = 4.45$ mm

23-10* $D_{t=T_d/2} = 48.1$ mm and $D_{t=11T_d/8} = 24.5$ mm

23-15*   Maximum displacement of BC relative to the support $= 49.5$ mm

Author index

Subject index